EDITION 10

Statistics for Behavioral Sciences

© Deborah Batt

FREDERICK J GRAVETTER

The College at Brockport, State University of New York

LARRY B. WALLNAU

The College at Brockport, State University of New York

CENGAGE
Learning®

Australia • Brazil • Mexico • Singapore • United Kingdom • United States

Statistics for the Behavioral Sciences,
Tenth Edition
Frederick J Gravetter and Larry B. Wallnau

Product Director: Jon-David Hague

Product Manager: Timothy Matray

Content Developer: Stefanie Chase

Product Assistant: Kimiya Hojjat

Marketing Manager: Melissa Larmon

Content Project Manager: Michelle Clark

Art Director: Vernon Boes

Manufacturing Planner: Karen Hunt

Production Service: Lynn Lustberg,
MPS Limited

Text and Photo Researcher:
Lumina Datamatics

Copy Editor: Sara Kreisman

Illustrator: MPS Limited

Text and Cover Designer: Lisa Henry

Cover Image: © Deborah Batt
"Community 2-1"

Compositor: MPS Limited

For product information and technology assistance, contact us at
Cengage Learning Customer & Sales Support, 1-800-354-9706

For permission to use material from this text or product,
submit all requests online at **www.cengage.com/permissions**
Further permissions questions can be e-mailed to
permissionrequest@cengage.com

Library of Congress Control Number: 2015940372

Student Edition:
ISBN: 978-1-305-50491-2

Loose-leaf Edition:
ISBN: 978-1-305-86280-7

Cengage Learning
20 Channel Center Street
Boston, MA 02210
USA

Cengage Learning is a leading provider of customized learning solutions with employees residing in nearly 40 different countries and sales in more than 125 countries around the world. Find your local representative at **www.cengage.com**

Cengage Learning products are represented in Canada by Nelson Education, Ltd.

To learn more about Cengage Learning Solutions,
visit www.cengage.com

Purchase any of our products at your local college store or at our preferred online store **www.cengagebrain.com**

Printed in the U.S.A.
Print Number: 01   Print Year: 2015

BRIEF CONTENTS

CONTENTS

CHAPTER **6** **Probability**   **159**

CHAPTER **7** **Probability and Samples: The Distribution
of Sample Means**   **193**

CHAPTER **8**   # Introduction to Hypothesis Testing   223

CHAPTER **9**   # Introduction to the *t* Statistic   267

CHAPTER **13**   **Repeated-Measures Analysis of Variance**   413

CHAPTER **14**   **Two-Factor Analysis of Variance (Independent Measures)**   447

CHAPTER **18**  **The Binomial Test**   603

APPENDIXES

Many students in the behavioral sciences view the required statistics course as an intimidating obstacle that has been placed in the middle of an otherwise interesting curriculum. They want to learn about human behavior—not about math and science. As a result, the statistics course is seen as irrelevant to their education and career goals. However, as long as the behavioral sciences are founded in science, knowledge of statistics will be necessary. Statistical procedures provide researchers with objective and systematic methods for describing and interpreting their research results. Scientific research is the system that we use to gather information, and statistics are the tools that we use to distill the information into sensible and justified conclusions. The goal of this book is not only to teach the methods of statistics, but also to convey the basic principles of objectivity and logic that are essential for science and valuable for decision making in everyday life.

Those of you who are familiar with previous editions of *Statistics for the Behavioral Sciences* will notice that some changes have been made. These changes are summarized in the section entitled "To the Instructor." In revising this text, our students have been foremost in our minds. Over the years, they have provided honest and useful feedback. Their hard work and perseverance has made our writing and teaching most rewarding. We sincerely thank them. Students who are using this edition should please read the section of the preface entitled "To the Student."

The book chapters are organized in the sequence that we use for our own statistics courses. We begin with descriptive statistics, and then examine a variety of statistical procedures focused on sample means and variance before moving on to correlational methods and nonparametric statistics. Information about modifying this sequence is presented in the *To The Instructor* section for individuals who prefer a different organization. Each chapter contains numerous examples, many based on actual research studies, learning checks, a summary and list of key terms, and a set of 20–30 problems.

Ancillaries

Ancillaries for this edition include the following.

- **MindTap® Psychology:** *MindTap® Psychology for Gravetter/Wallnau's Statistics for The Behavioral Sciences, 10th Edition* is the digital learning solution that helps instructors engage and transform today's students into critical thinkers. Through paths of dynamic assignments and applications that you can personalize, real-time course analytics, and an accessible reader, MindTap helps you turn cookie cutter into cutting edge, apathy into engagement, and memorizers into higher-level thinkers.

  As an instructor using MindTap you have at your fingertips the right content and unique set of tools curated specifically for your course, such as video tutorials that walk students through various concepts and interactive problem tutorials that provide students opportunities to practice what they have learned, all in an interface designed to improve workflow and save time when planning lessons and course structure. The control to build and personalize your course is all yours, focusing on the most relevant material while also lowering costs for your students. Stay connected and informed in your course through real time student tracking that provides the opportunity to adjust the course as needed based on analytics of interactivity in the course.

- **Online Instructor's Manual:**   The manual includes learning objectives, key terms, a detailed chapter outline, a chapter summary, lesson plans, discussion topics, student activities, "What If" scenarios, media tools, a sample syllabus and an expanded test bank. The learning objectives are correlated with the discussion topics, student activities, and media tools.

- **Online PowerPoints:**   Helping you make your lectures more engaging while effectively reaching your visually oriented students, these handy Microsoft PowerPoint® slides outline the chapters of the main text in a classroom-ready presentation. The PowerPoint® slides are updated to reflect the content and organization of the new edition of the text.

- **Cengage Learning Testing, powered by Cognero®:**   Cengage Learning Testing, Powered by Cognero®, is a flexible online system that allows you to author, edit, and manage test bank content. You can create multiple test versions in an instant and deliver tests from your LMS in your classroom.

Acknowledgments

It takes a lot of good, hard-working people to produce a book. Our friends at Cengage have made enormous contributions to this textbook. We thank: Jon-David Hague, Product Director; Timothy Matray, Product Team Director; Jasmin Tokatlian, Content Development Manager; Kimiya Hojjat, Product Assistant; and Vernon Boes, Art Director. Special thanks go to Stefanie Chase, our Content Developer and to Lynn Lustberg who led us through production at MPS.

Reviewers play a very important role in the development of a manuscript. Accordingly, we offer our appreciation to the following colleagues for their assistance: Patricia Case, University of Toledo; Kevin David, Northeastern State University; Adia Garrett, University of Maryland, Baltimore County; Carrie E. Hall, Miami University; Deletha Hardin, University of Tampa; Angela Heads, Prairie View A&M University; Roberto Heredia, Texas A&M International University; Alisha Janowski, University of Central Florida; Matthew Mulvaney, The College at Brockport (SUNY); Nicholas Von Glahn, California State Polytechnic University, Pomona; and Ronald Yockey, Fresno State University.

To the Instructor

Those of you familiar with the previous edition of *Statistics for the Behavioral Sciences* will notice a number of changes in the 10th edition. Throughout this book, research examples have been updated, real world examples have been added, and the end-of-chapter problems have been extensively revised. Major revisions for this edition include the following:

1. Each section of every chapter begins with a list of Learning Objectives for that specific section.

2. Each section ends with a Learning Check consisting of multiple-choice questions with at least one question for each Learning Objective.

3. The former Chapter 19, Choosing the Right Statistics, has been eliminated and an abridged version is now an Appendix replacing the Statistics Organizer, which appeared in earlier editions.

Other examples of specific and noteworthy revisions include the following.

Chapter 1 The section on data structures and research methods parallels the new Appendix, Choosing the Right Statistics.

Chapter 2 The chapter opens with a new Preview to introduce the concept and purpose of frequency distributions.

Chapter 3 Minor editing clarifies and simplifies the discussion the median.

Chapter 4 The chapter opens with a new Preview to introduce the topic of Central Tendency. The sections on standard deviation and variance have been edited to increase emphasis on concepts rather than calculations.

Chapter 5 The section discussion relationships between z, X, μ, and σ has been expanded and includes a new demonstration example.

Chapter 6 The chapter opens with a new Preview to introduce the topic of Probability. The section, Looking Ahead to Inferential Statistics, has been substantially shortened and simplified.

Chapter 7 The former Box explaining difference between standard deviation and standard error was deleted and the content incorporated into Section 7.4 with editing to emphasize that the standard error is the primary new element introduced in the chapter. The final section, Looking Ahead to Inferential Statistics, was simplified and shortened to be consistent with the changes in Chapter 6.

Chapter 8 A redundant example was deleted which shortened and streamlined the remaining material so that most of the chapter is focused on the same research example.

Chapter 9 The chapter opens with a new Preview to introduce the t statistic and explain why a new test statistic is needed. The section introducing Confidence Intervals was edited to clarify the origin of the confidence interval equation and to emphasize that the interval is constructed at the sample mean.

Chapter 10 The chapter opens with a new Preview introducing the independent-measures t statistic. The section presenting the estimated standard error of $(M_1 - M_2)$ has been simplified and shortened.

Chapter 11 The chapter opens with a new Preview introducing the repeated-measures t statistic. The section discussing hypothesis testing has been separated from the section on effect size and confidence intervals to be consistent with the other two chapters on t tests. The section comparing independent- and repeated-measures designs has been expanded.

Chapter 12 The chapter opens with a new Preview introducing ANOVA and explaining why a new hypothesis testing procedure is necessary. Sections in the chapter have been reorganized to allow flow directly from hypothesis tests and effect size to post tests.

Chapter 13   Substantially expanded the section discussing factors that influence the outcome of a repeated-measures hypothesis test and associated measures of effect size.

Chapter 14   The chapter opens with a new Preview presenting a two-factor research example and introducing the associated ANOVA. Sections have been reorganized so that simple main effects and the idea of using a second factor to reduce variance from individual differences are now presented as extra material related to the two-factor ANOVA.

Chapter 15   The chapter opens with a new Preview presenting a correlational research study and the concept of a correlation. A new section introduces the t statistic for evaluating the significance of a correlation and the section on partial correlations has been simplified and shortened.

Chapter 16   The chapter opens with a new Preview introducing the concept of regression and its purpose. A new section demonstrates the equivalence of testing the significance of a correlation and testing the significance of a regression equation with one predictor variable. The section on residuals for the multiple-regression equation has been edited to simplify and shorten.

Chapter 17   A new chapter Preview presents an experimental study with data consisting of frequencies, which are not compatible with computing means and variances. Chi-square tests are introduced as a solution to this problem. A new section introduces Cohen's w as a means of measuring effect size for both chi-square tests.

Chapter 18   Substantial editing clarifies the section explaining how the real limits for each score can influence the conclusion from a binomial test.

The former Chapter 19 covering the task of matching statistical methods to specific types of data has been substantially shortened and converted into an Appendix.

■ Matching the Text to Your Syllabus

The book chapters are organized in the sequence that we use for our own statistics courses. However, different instructors may prefer different organizations and probably will choose to omit or deemphasize specific topics. We have tried to make separate chapters, and even sections of chapters, completely self-contained, so they can be deleted or reorganized to fit the syllabus for nearly any instructor. Some common examples are as follows.

- It is common for instructors to choose between emphasizing analysis of variance (Chapters 12, 13, and 14) or emphasizing correlation/regression (Chapters 15 and 16). It is rare for a one-semester course to complete coverage of both topics.

- Although we choose to complete all the hypothesis tests for means and mean differences before introducing correlation (Chapter 15), many instructors prefer to place correlation much earlier in the sequence of course topics. To accommodate this, Sections 15.1, 15.2, and 15.3 present the calculation and interpretation of the Pearson correlation and can be introduced immediately following Chapter 4 (variability). Other sections of Chapter 15 refer to hypothesis testing and should be delayed until the process of hypothesis testing (Chapter 8) has been introduced.

- It is also possible for instructors to present the chi-square tests (Chapter 17) much earlier in the sequence of course topics. Chapter 17, which presents hypothesis tests for proportions, can be presented immediately after Chapter 8, which introduces the process of hypothesis testing. If this is done, we also recommend that the Pearson correlation (Sections 15.1, 15.2, and 15.3) be presented early to provide a foundation for the chi-square test for independence.

To the Student

A primary goal of this book is to make the task of learning statistics as easy and painless as possible. Among other things, you will notice that the book provides you with a number of opportunities to practice the techniques you will be learning in the form of Learning Checks, Examples, Demonstrations, and end-of-chapter problems. We encourage you to take advantage of these opportunities. Read the text rather than just memorizing the formulas. We have taken care to present each statistical procedure in a conceptual context that explains why the procedure was developed and when it should be used. If you read this material and gain an understanding of the basic concepts underlying a statistical formula, you will find that learning the formula and how to use it will be much easier. In the "Study Hints," that follow, we provide advice that we give our own students. Ask your instructor for advice as well; we are sure that other instructors will have ideas of their own.

Over the years, the students in our classes and other students using our book have given us valuable feedback. If you have any suggestions or comments about this book, you can write to either Professor Emeritus Frederick Gravetter or Professor Emeritus Larry Wallnau at the Department of Psychology, SUNY College at Brockport, 350 New Campus Drive, Brockport, New York 14420. You can also contact Professor Emeritus Gravetter directly at fgravett@brockport.edu.

■ Study Hints

You may find some of these tips helpful, as our own students have reported.

- The key to success in a statistics course is to keep up with the material. Each new topic builds on previous topics. If you have learned the previous material, then the new topic is just one small step forward. Without the proper background, however, the new topic can be a complete mystery. If you find that you are falling behind, get help immediately.

- You will learn (and remember) much more if you study for short periods several times per week rather than try to condense all of your studying into one long session. For example, it is far more effective to study half an hour every night than to have a single 3½-hour study session once a week. We cannot even work on *writing* this book without frequent rest breaks.

- Do some work before class. Keep a little ahead of the instructor by reading the appropriate sections before they are presented in class. Although you may not fully understand what you read, you will have a general idea of the topic, which will make the lecture easier to follow. Also, you can identify material that is particularly confusing and then be sure the topic is clarified in class.

- Pay attention and think during class. Although this advice seems obvious, often it is not practiced. Many students spend so much time trying to write down every example presented or every word spoken by the instructor that they do not actually understand and process what is being said. Check with your instructor—there may not be a need to copy every example presented in class, especially if there are many examples like it in the text. Sometimes, we tell our students to put their pens and pencils down for a moment and just listen.

- Test yourself regularly. Do not wait until the end of the chapter or the end of the week to check your knowledge. After each lecture, work some of the end-of-chapter problems and do the Learning Checks. Review the Demonstration Problems, and be sure you can define the Key Terms. If you are having trouble, get your questions answered *immediately*—reread the section, go to your instructor, or ask questions in class. By doing so, you will be able to move ahead to new material.

- Do not kid yourself! Avoid denial. Many students observe their instructor solve problems in class and think to themselves, "This looks easy, I understand it." Do you really understand it? Can you really do the problem on your own without having to leaf through the pages of a chapter? Although there is nothing wrong with using examples in the text as models for solving problems, you should try working a problem with your book closed to test your level of mastery.

- We realize that many students are embarrassed to ask for help. It is our biggest challenge as instructors. You must find a way to overcome this aversion. Perhaps contacting the instructor directly would be a good starting point, if asking questions in class is too anxiety-provoking. You could be pleasantly surprised to find that your instructor does not yell, scold, or bite! Also, your instructor might know of another student who can offer assistance. Peer tutoring can be very helpful.

Frederick J Gravetter
Larry B. Wallnau

ABOUT THE AUTHORS

Frederick Gravetter

FREDERICK J GRAVETTER is Professor Emeritus of Psychology at the State University of New York College at Brockport. While teaching at Brockport, Dr. Gravetter specialized in statistics, experimental design, and cognitive psychology. He received his bachelor's degree in mathematics from M.I.T. and his Ph.D in psychology from Duke University. In addition to publishing this textbook and several research articles, Dr. Gravetter co-authored *Research Methods for the Behavioral Science* and *Essentials of Statistics for the Behavioral Sciences.*

Larry B. Wallnau

LARRY B. WALLNAU is Professor Emeritus of Psychology at the State University of New York College at Brockport. While teaching at Brockport, he published numerous research articles in biopsychology. With Dr. Gravetter, he co-authored *Essentials of Statistics for the Behavioral Sciences.* Dr. Wallnau also has provided editorial consulting for numerous publishers and journals. He has taken up running and has competed in 5K races in New York and Connecticut. He takes great pleasure in adopting neglected and rescued dogs.

Introduction to Statistics

© Deborah Batt

Before we begin our discussion of statistics, we ask you to read the following paragraph taken from the philosophy of Wrong Shui (Candappa, 2000).

The Journey to Enlightenment
In Wrong Shui, life is seen as a cosmic journey, a struggle to overcome unseen and unexpected obstacles at the end of which the traveler will find illumination and enlightenment. Replicate this quest in your home by moving light switches away from doors and over to the far side of each room.*

Why did we begin a statistics book with a bit of twisted philosophy? In part, we simply wanted to lighten the mood with a bit of humor—starting a statistics course is typically not viewed as one of life's joyous moments. In addition, the paragraph is an excellent counterexample for the purpose of this book. Specifically, our goal is to do everything possible to prevent you from stumbling around in the dark by providing lots of help and illumination as you journey through the world of statistics. To accomplish this, we begin each section of the book with clearly stated learning objectives and end each section with a brief quiz to test your mastery of the new material. We also introduce each new statistical procedure by explaining the purpose it is intended to serve. If you understand why a new procedure is needed, you will find it much easier to learn.

The objectives for this first chapter are to provide an introduction to the topic of statistics and to give you some background for the rest of the book. We discuss the role of statistics within the general field of scientific inquiry, and we introduce some of the vocabulary and notation that are necessary for the statistical methods that follow.

As you read through the following chapters, keep in mind that the general topic of statistics follows a well-organized, logically developed progression that leads from basic concepts and definitions to increasingly sophisticated techniques. Thus, each new topic serves as a foundation for the material that follows. The content of the first nine chapters, for example, provides an essential background and context for the statistical methods presented in Chapter 10. If you turn directly to Chapter 10 without reading the first nine chapters, you will find the material confusing and incomprehensible. However, if you learn and use the background material, you will have a good frame of reference for understanding and incorporating new concepts as they are presented.

_____

*Candappa, R. (2000). *The little book of wrong shui.* Kansas City: Andrews McMeel Publishing. Reprinted by permission.

1.1 | Statistics, Science, and Observations

LEARNING OBJECTIVES

1. Define the terms population, sample, parameter, and statistic, and describe the relationships between them.

2. Define descriptive and inferential statistics and describe how these two general categories of statistics are used in a typical research study.

3. Describe the concept of sampling error and explain how this concept creates the fundamental problem that inferential statistics must address.

■ Definitions of Statistics

By one definition, *statistics* consist of facts and figures such as the average annual snowfall in Denver or Derrick Jeter's lifetime batting average. These statistics are usually informative and time-saving because they condense large quantities of information into a few simple figures. Later in this chapter we return to the notion of calculating statistics (facts and figures) but, for now, we concentrate on a much broader definition of statistics. Specifically, we use the term statistics to refer to a general field of mathematics. In this case, we are using the term *statistics* as a shortened version of *statistical procedures*. For example, you are probably using this book for a statistics course in which you will learn about the statistical techniques that are used to summarize and evaluate research results in the behavioral sciences.

Research in the behavioral sciences (and other fields) involves gathering information. To determine, for example, whether college students learn better by reading material on printed pages or on a computer screen, you would need to gather information about students' study habits and their academic performance. When researchers finish the task of gathering information, they typically find themselves with pages and pages of measurements such as preferences, personality scores, opinions, and so on. In this book, we present the statistics that researchers use to analyze and interpret the information that they gather. Specifically, statistics serve two general purposes:

1. Statistics are used to organize and summarize the information so that the researcher can see what happened in the research study and can communicate the results to others.

2. Statistics help the researcher to answer the questions that initiated the research by determining exactly what general conclusions are justified based on the specific results that were obtained.

DEFINITION

> The term **statistics** refers to a set of mathematical procedures for organizing, summarizing, and interpreting information.

Statistical procedures help ensure that the information or observations are presented and interpreted in an accurate and informative way. In somewhat grandiose terms, statistics help researchers bring order out of chaos. In addition, statistics provide researchers with a set of standardized techniques that are recognized and understood throughout the scientific community. Thus, the statistical methods used by one researcher will be familiar to other researchers, who can accurately interpret the statistical analyses with a full understanding of how the analysis was done and what the results signify.

■ Populations and Samples

Research in the behavioral sciences typically begins with a general question about a specific group (or groups) of individuals. For example, a researcher may want to know what factors are associated with academic dishonesty among college students. Or a researcher may want to examine the amount of time spent in the bathroom for men compared to women. In the first example, the researcher is interested in the group of *college students*. In the second example, the researcher wants to compare the group of *men* with the group of *women*. In statistical terminology, the entire group that a researcher wishes to study is called a *population*.

DEFINITION

> A **population** is the set of all the individuals of interest in a particular study.

As you can well imagine, a population can be quite large—for example, the entire set of women on the planet Earth. A researcher might be more specific, limiting the population for study to women who are registered voters in the United States. Perhaps the investigator would like to study the population consisting of women who are heads of state. Populations can obviously vary in size from extremely large to very small, depending on how the investigator defines the population. The population being studied should always be identified by the researcher. In addition, the population need not consist of people—it could be a population of rats, corporations, parts produced in a factory, or anything else an investigator wants to study. In practice, populations are typically very large, such as the population of college sophomores in the United States or the population of small businesses.

Because populations tend to be very large, it usually is impossible for a researcher to examine every individual in the population of interest. Therefore, researchers typically select a smaller, more manageable group from the population and limit their studies to the individuals in the selected group. In statistical terms, a set of individuals selected from a population is called a *sample*. A sample is intended to be representative of its population, and a sample should always be identified in terms of the population from which it was selected.

DEFINITION

> A **sample** is a set of individuals selected from a population, usually intended to represent the population in a research study.

Just as we saw with populations, samples can vary in size. For example, one study might examine a sample of only 10 students in a graduate program and another study might use a sample of more than 10,000 people who take a specific cholesterol medication.

So far we have talked about a sample being selected from a population. However, this is actually only half of the full relationship between a sample and its population. Specifically, when a researcher finishes examining the sample, the goal is to generalize the results back to the entire population. Remember that the research started with a general question about the population. To answer the question, a researcher studies a sample and then generalizes the results from the sample to the population. The full relationship between a sample and a population is shown in Figure 1.1.

■ Variables and Data

Typically, researchers are interested in specific characteristics of the individuals in the population (or in the sample), or they are interested in outside factors that may influence the individuals. For example, a researcher may be interested in the influence of the weather on people's moods. As the weather changes, do people's moods also change? Something that can change or have different values is called a *variable*.

DEFINITION

> A **variable** is a characteristic or condition that changes or has different values for different individuals.

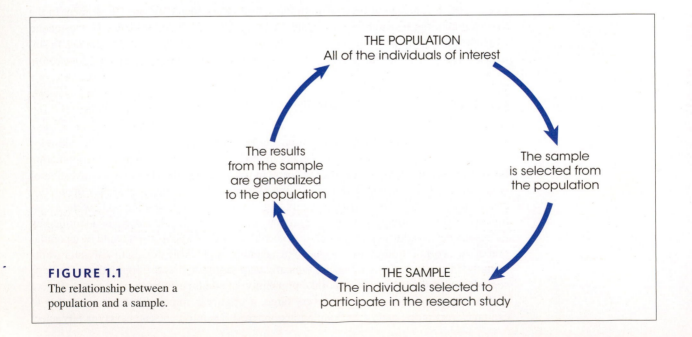

FIGURE 1.1

The relationship between a population and a sample.

Once again, variables can be characteristics that differ from one individual to another, such as height, weight, gender, or personality. Also, variables can be environmental conditions that change such as temperature, time of day, or the size of the room in which the research is being conducted.

To demonstrate changes in variables, it is necessary to make measurements of the variables being examined. The measurement obtained for each individual is called a *datum*, or more commonly, a *score* or *raw score*. The complete set of scores is called the *data set* or simply the *data*.

> **DEFINITION**
>
> **Data** (plural) are measurements or observations. A **data set** is a collection of measurements or observations. A **datum** (singular) is a single measurement or observation and is commonly called a **score** or **raw score**.

Before we move on, we should make one more point about samples, populations, and data. Earlier, we defined populations and samples in terms of *individuals*. For example, we discussed a population of graduate students and a sample of cholesterol patients. Be forewarned, however, that we will also refer to populations or samples of *scores*. Because research typically involves measuring each individual to obtain a score, every sample (or population) of individuals produces a corresponding sample (or population) of scores.

■ Parameters and Statistics

When describing data it is necessary to distinguish whether the data come from a population or a sample. A characteristic that describes a population—for example, the average score for the population—is called a *parameter*. A characteristic that describes a sample is called a *statistic*. Thus, the average score for a sample is an example of a statistic. Typically, the research process begins with a question about a population parameter. However, the actual data come from a sample and are used to compute sample statistics.

> **DEFINITION**
>
> A **parameter** is a value, usually a numerical value, that describes a population. A parameter is usually derived from measurements of the individuals in the population.
>
> A **statistic** is a value, usually a numerical value, that describes a sample. A statistic is usually derived from measurements of the individuals in the sample.

Every population parameter has a corresponding sample statistic, and most research studies involve using statistics from samples as the basis for answering questions about population parameters. As a result, much of this book is concerned with the relationship between sample statistics and the corresponding population parameters. In Chapter 7, for example, we examine the relationship between the mean obtained for a sample and the mean for the population from which the sample was obtained.

■ Descriptive and Inferential Statistical Methods

Although researchers have developed a variety of different statistical procedures to organize and interpret data, these different procedures can be classified into two general categories. The first category, *descriptive statistics*, consists of statistical procedures that are used to simplify and summarize data.

> **DEFINITION**
>
> **Descriptive statistics** are statistical procedures used to summarize, organize, and simplify data.

Descriptive statistics are techniques that take raw scores and organize or summarize them in a form that is more manageable. Often the scores are organized in a table or a graph so that it is possible to see the entire set of scores. Another common technique is to summarize a set of scores by computing an average. Note that even if the data set has hundreds of scores, the average provides a single descriptive value for the entire set.

The second general category of statistical techniques is called *inferential statistics*. Inferential statistics are methods that use sample data to make general statements about a population.

DEFINITION   **Inferential statistics** consist of techniques that allow us to study samples and then make generalizations about the populations from which they were selected.

Because populations are typically very large, it usually is not possible to measure everyone in the population. Therefore, a sample is selected to represent the population. By analyzing the results from the sample, we hope to make general statements about the population. Typically, researchers use sample statistics as the basis for drawing conclusions about population parameters. One problem with using samples, however, is that a sample provides only limited information about the population. Although samples are generally *representative* of their populations, a sample is not expected to give a perfectly accurate picture of the whole population. There usually is some discrepancy between a sample statistic and the corresponding population parameter. This discrepancy is called *sampling error*, and it creates the fundamental problem inferential statistics must always address.

DEFINITION   **Sampling error** is the naturally occurring discrepancy, or error, that exists between a sample statistic and the corresponding population parameter.

The concept of sampling error is illustrated in Figure 1.2. The figure shows a population of 1,000 college students and 2 samples, each with 5 students who were selected from the population. Notice that each sample contains different individuals who have different characteristics. Because the characteristics of each sample depend on the specific people in the sample, statistics will vary from one sample to another. For example, the five students in sample 1 have an average age of 19.8 years and the students in sample 2 have an average age of 20.4 years.

It is also very unlikely that the statistics obtained for a sample will be identical to the parameters for the entire population. In Figure 1.2, for example, neither sample has statistics that are exactly the same as the population parameters. You should also realize that Figure 1.2 shows only two of the hundreds of possible samples. Each sample would contain different individuals and would produce different statistics. This is the basic concept of sampling error: sample statistics vary from one sample to another and typically are different from the corresponding population parameters.

One common example of sampling error is the error associated with a sample proportion. For example, in newspaper articles reporting results from political polls, you frequently find statements such as this:

Candidate Brown leads the poll with 51% of the vote. Candidate Jones has 42% approval, and the remaining 7% are undecided. This poll was taken from a sample of registered voters and has a margin of error of plus-or-minus 4 percentage points.

The "margin of error" is the sampling error. In this case, the percentages that are reported were obtained from a sample and are being generalized to the whole population. As always, you do not expect the statistics from a sample to be perfect. There always will be some "margin of error" when sample statistics are used to represent population parameters.

FIGURE 1.2

A demonstration of sampling error. Two samples are selected from the same population. Notice that the sample statistics are different from one sample to another and all the sample statistics are different from the corresponding population parameters. The natural differences that exist, by chance, between a sample statistic and population parameter are called sampling error.

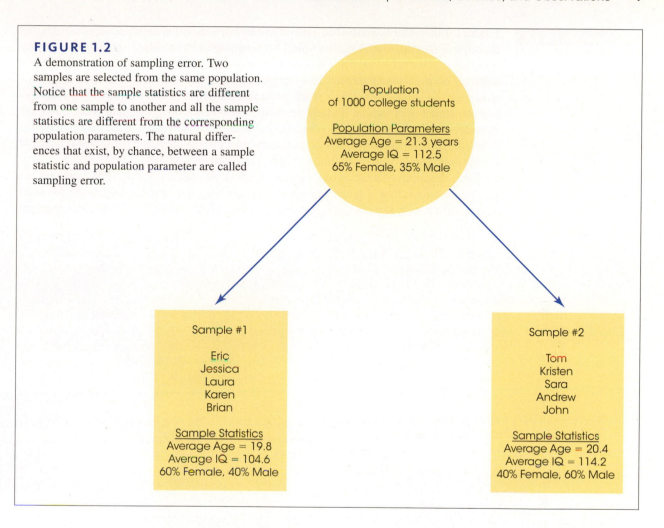

Population
of 1000 college students

Population Parameters
Average Age = 21.3 years
Average IQ = 112.5
65% Female, 35% Male

Sample #1

Eric
Jessica
Laura
Karen
Brian

Sample Statistics
Average Age = 19.8
Average IQ = 104.6
60% Female, 40% Male

Sample #2

Tom
Kristen
Sara
Andrew
John

Sample Statistics
Average Age = 20.4
Average IQ = 114.2
40% Female, 60% Male

As a further demonstration of sampling error, imagine that your statistics class is separated into two groups by drawing a line from front to back through the middle of the room. Now imagine that you compute the average age (or height, or IQ) for each group. Will the two groups have exactly the same average? Almost certainly they will not. No matter what you chose to measure, you will probably find some difference between the two groups. However, the difference you obtain does not necessarily mean that there is a systematic difference between the two groups. For example, if the average age for students on the right-hand side of the room is higher than the average for students on the left, it is unlikely that some mysterious force has caused the older people to gravitate to the right side of the room. Instead, the difference is probably the result of random factors such as chance. The unpredictable, unsystematic differences that exist from one sample to another are an example of sampling error.

■ Statistics in the Context of Research

The following example shows the general stages of a research study and demonstrates how descriptive statistics and inferential statistics are used to organize and interpret the data. At the end of the example, note how sampling error can affect the interpretation of experimental results, and consider why inferential statistical methods are needed to deal with this problem.

EXAMPLE 1.1   Figure 1.3 shows an overview of a general research situation and demonstrates the roles that descriptive and inferential statistics play. The purpose of the research study is to address a question that we posed earlier: Do college students learn better by studying text on printed pages or on a computer screen? Two samples are selected from the population of college students. The students in sample A are given printed pages of text to study for 30 minutes and the students in sample B study the same text on a computer screen. Next, all of the students are given a multiple-choice test to evaluate their knowledge of the material. At this point, the researcher has two sets of data: the scores for sample A and the scores for sample B (see the figure). Now is the time to begin using statistics.

First, descriptive statistics are used to simplify the pages of data. For example, the researcher could draw a graph showing the scores for each sample or compute the average score for each sample. Note that descriptive methods provide a simplified, organized

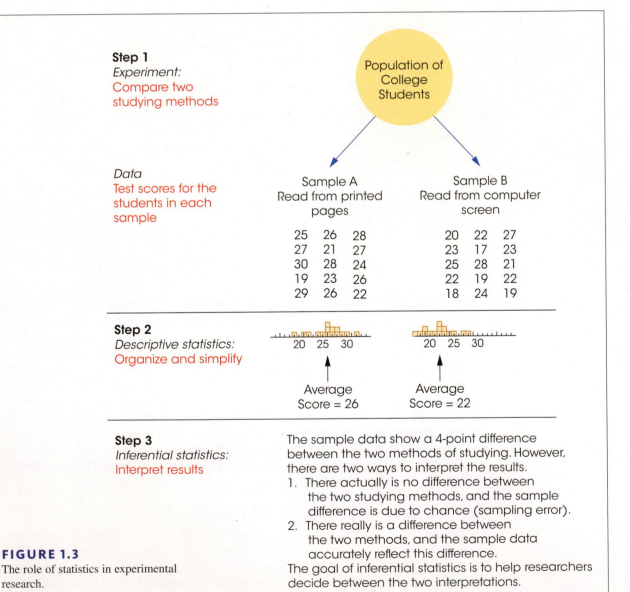

FIGURE 1.3

The role of statistics in experimental research.

description of the scores. In this example, the students who studied printed pages had an average score of 26 on the test, and the students who studied text on the computer averaged 22.

Once the researcher has described the results, the next step is to interpret the outcome. This is the role of inferential statistics. In this example, the researcher has found a difference of 4 points between the two samples (sample A averaged 26 and sample B averaged 22). The problem for inferential statistics is to differentiate between the following two interpretations:

1. There is no real difference between the printed page and a computer screen, and the 4-point difference between the samples is just an example of sampling error (like the samples in Figure 1.2).

2. There really is a difference between the printed page and a computer screen, and the 4-point difference between the samples was caused by the different methods of studying.

In simple English, does the 4-point difference between samples provide convincing evidence of a difference between the two studying methods, or is the 4-point difference just chance? The purpose of inferential statistics is to answer this question.   ■

LEARNING CHECK

1. A researcher is interested in the sleeping habits of American college students. A group of 50 students is interviewed and the researcher finds that these students sleep an average of 6.7 hours per day. For this study, the average of 6.7 hours is an example of a(n) _____.
   a. parameter
   b. statistic
   c. population
   d. sample

2. A researcher is curious about the average IQ of registered voters in the state of Florida. The entire group of registered voters in the state is an example of a _____.
   a. sample
   b. statistic
   c. population
   d. parameter

3. Statistical techniques that summarize, organize, and simplify data are classified as _____.
   a. population statistics
   b. sample statistics
   c. descriptive statistics
   d. inferential statistics

4. In general, _____ statistical techniques are used to summarize the data from a research study and _____ statistical techniques are used to determine what conclusions are justified by the results.
   a. inferential, descriptive
   b. descriptive, inferential
   c. sample, population
   d. population, sample

5. IQ tests are standardized so that the average score is 100 for the entire group of people who take the test each year. However, if you selected a group of 20 people who took the test and computed their average IQ score you probably would not get 100. What statistical concept explains the difference between your mean and the mean for the entire group?

  a. statistical error

  b. inferential error

  c. descriptive error

  d. sampling error

ANSWERS   **1.** B, **2.** C, **3.** C, **4.** B, **5.** D

1.2 | Data Structures, Research Methods, and Statistics

LEARNING OBJECTIVES   **4.** Differentiate correlational, experimental, and nonexperimental research and describe the data structures associated with each.

5. Define independent, dependent, and quasi-independent variables and recognize examples of each.

■ Individual Variables: Descriptive Research

Some research studies are conducted simply to describe individual variables as they exist naturally. For example, a college official may conduct a survey to describe the eating, sleeping, and study habits of a group of college students. When the results consist of numerical scores, such as the number of hours spent studying each day, they are typically described by the statistical techniques that are presented in Chapters 3 and 4. Non-numerical scores are typically described by computing the proportion or percentage in each category. For example, a recent newspaper article reported that 34.9% of Americans are obese, which is roughly 35 pounds over a healthy weight.

■ Relationships Between Variables

Most research, however, is intended to examine relationships between two or more variables. For example, is there a relationship between the amount of violence in the video games played by children and the amount of aggressive behavior they display? Is there a relationship between the quality of breakfast and academic performance for elementary school children? Is there a relationship between the number of hours of sleep and grade point average for college students? To establish the existence of a relationship, researchers must make observations—that is, measurements of the two variables. The resulting measurements can be classified into two distinct data structures that also help to classify different research methods and different statistical techniques. In the following section we identify and discuss these two data structures.

I. One Group with Two Variables Measured for Each Individual: The Correlational Method   One method for examining the relationship between variables is to observe the two variables as they exist naturally for a set of individuals. That is, simply measure the two variables for each individual. For example, research has demonstrated a relationship between sleep habits, especially wake-up time, and academic performance for college students (Trockel, Barnes, and Egget, 2000). The researchers used a survey to measure wake-up time and school records to measure academic performance for each student. Figure 1.4 shows an example of the kind of data obtained in the study. The researchers then look for consistent patterns in the data to provide evidence for a relationship between variables. For example, as wake-up time changes from one student to another, is there also a tendency for academic performance to change?

Consistent patterns in the data are often easier to see if the scores are presented in a graph. Figure 1.4 also shows the scores for the eight students in a graph called a scatter plot. In the scatter plot, each individual is represented by a point so that the horizontal position corresponds to the student's wake-up time and the vertical position corresponds to the student's academic performance score. The scatter plot shows a clear relationship between wake-up time and academic performance: as wake-up time increases, academic performance decreases.

A research study that simply measures two different variables for each individual and produces the kind of data shown in Figure 1.4 is an example of the *correlational method*, or the *correlational research strategy*.

DEFINITION

> In the **correlational method**, two different variables are observed to determine whether there is a relationship between them.

■ Statistics for the Correlational Method

When the data from a correlational study consist of numerical scores, the relationship between the two variables is usually measured and described using a statistic called a correlation. Correlations and the correlational method are discussed in detail in Chapters 15 and 16. Occasionally, the measurement process used for a correlational study simply classifies individuals into categories that do not correspond to numerical values. For example, a researcher could classify a group of college students by gender (male

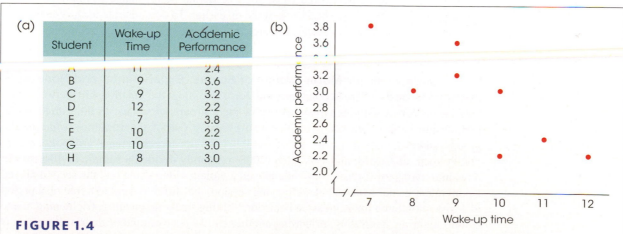

FIGURE 1.4

One of two data structures for evaluating the relationship between variables. Note that there are two separate measurements for each individual (wake-up time and academic performance). The same scores are shown in a table (a) and in a graph (b).

or female) and by cell-phone preference (talk or text). Note that the researcher has two scores for each individual but neither of the scores is a numerical value. This type of data is typically summarized in a table showing how many individuals are classified into each of the possible categories. Table 1.1 shows an example of this kind of summary table. The table shows for example, that 30 of the males in the sample preferred texting to talking. This type of data can be coded with numbers (for example, male = 0 and female = 1) so that it is possible to compute a correlation. However, the relationship between variables for non-numerical data, such as the data in Table 1.1, is usually evaluated using a statistical technique known as a *chi-square test*. Chi-square tests are presented in Chapter 17.

TABLE 1.1   Correlational data consisting of non-numerical scores. Note that there are two measurements for each individual: gender and cell phone preference. The numbers indicate how many people are in each category. For example, out of the 50 males, 30 prefer text over talk.

	Cell Phone Preference		
	Text	Talk	
Males	30	20	50
Females	25	25	50

■ Limitations of the Correlational Method

The results from a correlational study can demonstrate the existence of a relationship between two variables, but they do not provide an explanation for the relationship. In particular, a correlational study cannot demonstrate a cause-and-effect relationship. For example, the data in Figure 1.4 show a systematic relationship between wake-up time and academic performance for a group of college students; those who sleep late tend to have lower performance scores than those who wake early. However, there are many possible explanations for the relationship and we do not know exactly what factor (or factors) is responsible for late sleepers having lower grades. In particular, we cannot conclude that waking students up earlier would cause their academic performance to improve, or that studying more would cause students to wake up earlier. To demonstrate a cause-and-effect relationship between two variables, researchers must use the experimental method, which is discussed next.

II. Comparing Two (or More) Groups of Scores: Experimental and Nonexperimental Methods   The second method for examining the relationship between two variables involves the comparison of two or more groups of scores. In this situation, the relationship between variables is examined by using one of the variables to define the groups, and then measuring the second variable to obtain scores for each group. For example, Polman, de Castro, and van Aken (2008) randomly divided a sample of 10-year-old boys into two groups. One group then played a violent video game and the second played a nonviolent game. After the game-playing session, the children went to a free play period and were monitored for aggressive behaviors (hitting, kicking, pushing, frightening, name calling, fighting, quarreling, or teasing another child). An example of the resulting data is shown in Figure 1.5. The researchers then compare the scores for the violent-video group with the scores for the nonviolent-video group. A systematic difference between the two groups provides evidence for a relationship between playing violent video games and aggressive behavior for 10-year-old boys.

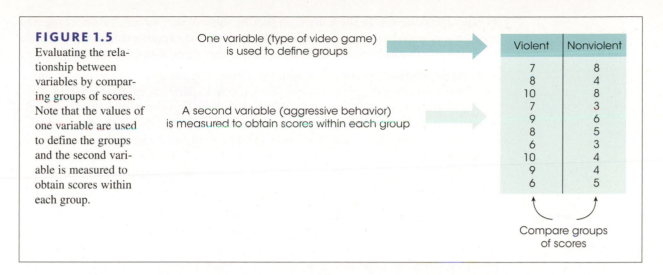

FIGURE 1.5

Evaluating the relationship between variables by comparing groups of scores. Note that the values of one variable are used to define the groups and the second variable is measured to obtain scores within each group.

One variable (type of video game) is used to define groups

A second variable (aggressive behavior) is measured to obtain scores within each group

Violent	Nonviolent
7	8
8	4
10	8
7	3
9	6
8	5
6	3
10	4
9	4
6	5

Compare groups of scores

■ Statistics for Comparing Two (or More) Groups of Scores

Most of the statistical procedures presented in this book are designed for research studies that compare groups of scores like the study in Figure 1.5. Specifically, we examine descriptive statistics that summarize and describe the scores in each group and we use inferential statistics to determine whether the differences between the groups can be generalized to the entire population.

When the measurement procedure produces numerical scores, the statistical evaluation typically involves computing the average score for each group and then comparing the averages. The process of computing averages is presented in Chapter 3, and a variety of statistical techniques for comparing averages are presented in Chapters 8–14. If the measurement process simply classifies individuals into non-numerical categories, the statistical evaluation usually consists of computing proportions for each group and then comparing proportions. Previously, in Table 1.1, we presented an example of non-numerical data examining the relationship between gender and cell-phone preference. The same data can be used to compare the proportions for males with the proportions for females. For example, using text is preferred by 60% of the males compared to 50% of the females. As before, these data are evaluated using a chi-square test, which is presented in Chapter 17.

■ Experimental and Nonexperimental Methods

There are two distinct research methods that both produce groups of scores to be compared: the experimental and the nonexperimental strategies. These two research methods use exactly the same statistics and they both demonstrate a relationship between two variables. The distinction between the two research strategies is how the relationship is interpreted. The results from an experiment allow a cause-and-effect explanation. For example, we can conclude that changes in one variable are responsible for causing differences in a second variable. A nonexperimental study does not permit a cause-and effect explanation. We can say that changes in one variable are accompanied by changes in a second variable, but we cannot say why. Each of the two research methods is discussed in the following sections.

■ The Experimental Method

One specific research method that involves comparing groups of scores is known as the *experimental method* or the *experimental research strategy*. The goal of an experimental study is to demonstrate a cause-and-effect relationship between two variables. Specifically, an experiment attempts to show that changing the value of one variable causes changes to occur in the second variable. To accomplish this goal, the experimental method has two characteristics that differentiate experiments from other types of research studies:

1. **Manipulation** The researcher manipulates one variable by changing its value from one level to another. In the Polman et al. (2008) experiment examining the effect of violence in video games (Figure 1.5), the researchers manipulate the amount of violence by giving one group of boys a violent game to play and giving the other group a nonviolent game. A second variable is observed (measured) to determine whether the manipulation causes changes to occur.

2. **Control** The researcher must exercise control over the research situation to ensure that other, extraneous variables do not influence the relationship being examined.

> In more complex experiments, a researcher may systematically manipulate more than one variable and may observe more than one variable. Here we are considering the simplest case, in which only one variable is manipulated and only one variable is observed.

To demonstrate these two characteristics, consider the Polman et al. (2008) study examining the effect of violence in video games (see Figure 1.5). To be able to say that the difference in aggressive behavior is caused by the amount of violence in the game, the researcher must rule out any other possible explanation for the difference. That is, any other variables that might affect aggressive behavior must be controlled. There are two general categories of variables that researchers must consider:

1. **Participant Variables** These are characteristics such as age, gender, and intelligence that vary from one individual to another. Whenever an experiment compares different groups of participants (one group in treatment A and a different group in treatment B), researchers must ensure that participant variables do not differ from one group to another. For the experiment shown in Figure 1.5, for example, the researchers would like to conclude that the violence in the video game causes a change in the participants' aggressive behavior. In the study, the participants in both conditions were 10-year-old boys. Suppose, however, that the participants in the nonviolent condition were primarily female and those in the violent condition were primarily male. In this case, there is an alternative explanation for the difference in aggression that exists between the two groups. Specifically, the difference between groups may have been caused by the amount of violence in the game, but it also is possible that the difference was caused by the participants' gender (females are less aggressive than males). Whenever a research study allows more than one explanation for the results, the study is said to be *confounded* because it is impossible to reach an unambiguous conclusion.

2. **Environmental Variables** These are characteristics of the environment such as lighting, time of day, and weather conditions. A researcher must ensure that the individuals in treatment A are tested in the same environment as the individuals in treatment B. Using the video game violence experiment (see Figure 1.5) as an example, suppose that the individuals in the nonviolent condition were all tested in the morning and the individuals in the violent condition were all tested in the evening. Again, this would produce a confounded experiment because the researcher could not determine whether the differences in aggressive behavior were caused by the amount of violence or caused by the time of day.

Researchers typically use three basic techniques to control other variables. First, the researcher could use *random assignment*, which means that each participant has an equal chance of being assigned to each of the treatment conditions. The goal of random assignment is to distribute the participant characteristics evenly between the two groups so that neither group is noticeably smarter (or older, or faster) than the other. Random assignment can also be used to control environmental variables. For example, participants could be assigned randomly for testing either in the morning or in the afternoon. A second technique for controlling variables is to use *matching* to ensure equivalent groups or equivalent environments. For example, the researcher could match groups by ensuring that every group has exactly 60% females and 40% males. Finally, the researcher can control variables by *holding them constant.* For example, in the video game violence study discussed earlier (Polman et al., 2008), the researchers used only 10-year-old boys as participants (holding age and gender constant). In this case the researchers can be certain that one group is not noticeably older or has a larger proportion of females than the other.

DEFINITION

> In the **experimental method**, one variable is manipulated while another variable is observed and measured. To establish a cause-and-effect relationship between the two variables, an experiment attempts to control all other variables to prevent them from influencing the results.

■ Terminology in the Experimental Method

Specific names are used for the two variables that are studied by the experimental method. The variable that is manipulated by the experimenter is called the *independent variable*. It can be identified as the treatment conditions to which participants are assigned. For the example in Figure 1.5, the amount of violence in the video game is the independent variable. The variable that is observed and measured to obtain scores within each condition is the *dependent variable*. For the example in Figure 1.5, the level of aggressive behavior is the dependent variable.

DEFINITION

> The **independent variable** is the variable that is manipulated by the researcher. In behavioral research, the independent variable usually consists of the two (or more) treatment conditions to which subjects are exposed. The independent variable consists of the *antecedent* conditions that were manipulated *prior* to observing the dependent variable.
>
> The **dependent variable** is the one that is observed to assess the effect of the treatment.

Control Conditions in an Experiment   An experimental study evaluates the relationship between two variables by manipulating one variable (the independent variable) and measuring one variable (the dependent variable). Note that in an experiment only one variable is actually measured. You should realize that this is different from a correlational study, in which both variables are measured and the data consist of two separate scores for each individual.

Often an experiment will include a condition in which the participants do not receive any treatment. The scores from these individuals are then compared with scores from participants who do receive the treatment. The goal of this type of study is to demonstrate that the treatment has an effect by showing that the scores in the treatment condition are substantially different from the scores in the no-treatment condition. In this kind of research, the no-treatment condition is called the *control condition*, and the treatment condition is called the *experimental condition*.

DEFINITION

> Individuals in a **control condition** do not receive the experimental treatment. Instead, they either receive no treatment or they receive a neutral, placebo treatment. The purpose of a control condition is to provide a baseline for comparison with the experimental condition.
>
> Individuals in the **experimental condition** do receive the experimental treatment.

Note that the independent variable always consists of at least two values. (Something must have at least two different values before you can say that it is "variable.") For the video game violence experiment (see Figure 1.5), the independent variable is the amount of violence in the video game. For an experiment with an experimental group and a control group, the independent variable is treatment versus no treatment.

■ Nonexperimental Methods: Nonequivalent Groups and Pre-Post Studies

In informal conversation, there is a tendency for people to use the term *experiment* to refer to any kind of research study. You should realize, however, that the term only applies to studies that satisfy the specific requirements outlined earlier. In particular, a real experiment must include manipulation of an independent variable and rigorous control of other, extraneous variables. As a result, there are a number of other research designs that are not true experiments but still examine the relationship between variables by comparing groups of scores. Two examples are shown in Figure 1.6 and are discussed in the following paragraphs. This type of research study is classified as nonexperimental.

The top part of Figure 1.6 shows an example of a *nonequivalent groups* study comparing boys and girls. Notice that this study involves comparing two groups of scores (like an experiment). However, the researcher has no ability to control which participants go into

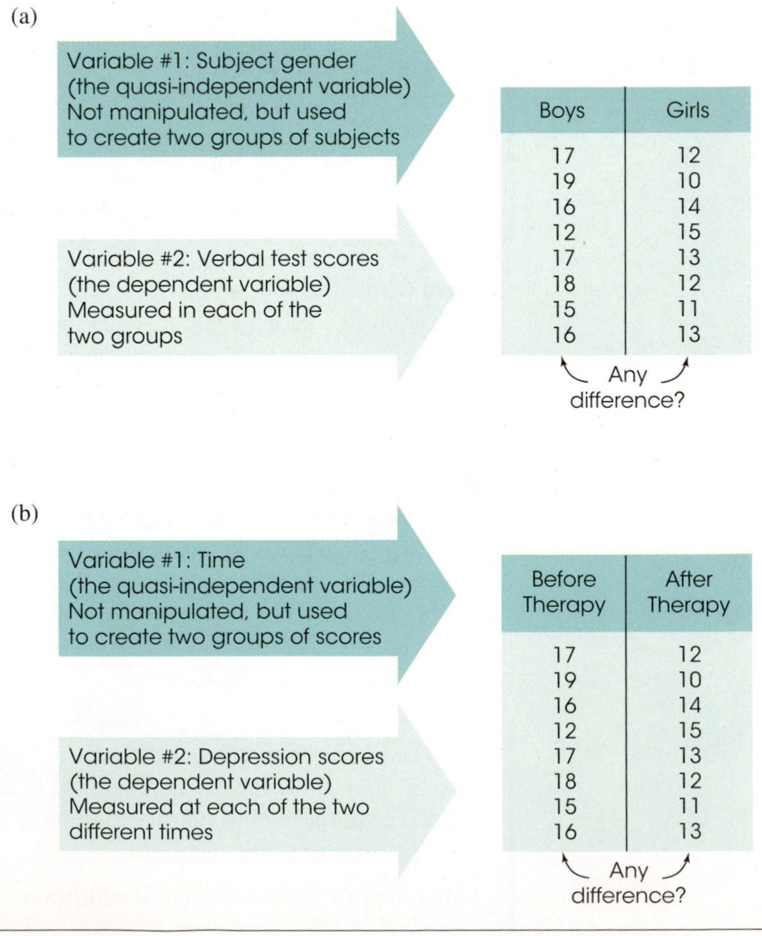

FIGURE 1.6

Two examples of nonexperimental studies that involve comparing two groups of scores. In (a) the study uses two preexisting groups (boys/girls) and measures a dependent variable (verbal scores) in each group. In (b) the study uses time (before/after) to define the two groups and measures a dependent variable (depression) in each group.

(a)

Variable #1: Subject gender
(the quasi-independent variable)
Not manipulated, but used
to create two groups of subjects

Variable #2: Verbal test scores
(the dependent variable)
Measured in each of the
two groups

Boys	Girls
17	12
19	10
16	14
12	15
17	13
18	12
15	11
16	13

Any difference?

(b)

Variable #1: Time
(the quasi-independent variable)
Not manipulated, but used
to create two groups of scores

Variable #2: Depression scores
(the dependent variable)
Measured at each of the two
different times

Before Therapy	After Therapy
17	12
19	10
16	14
12	15
17	13
18	12
15	11
16	13

Any difference?

which group—all the males must be in the boy group and all the females must be in the girl group. Because this type of research compares preexisting groups, the researcher cannot control the assignment of participants to groups and cannot ensure equivalent groups. Other examples of nonequivalent group studies include comparing 8-year-old children and 10-year-old children, people with an eating disorder and those with no disorder, and comparing children from a single-parent home and those from a two-parent home. Because it is impossible to use techniques like random assignment to control participant variables and ensure equivalent groups, this type of research is not a true experiment.

Correlational studies are also examples of nonexperimental research. In this section, however, we are discussing nonexperimental studies that compare two or more groups of scores.

The bottom part of Figure 1.6 shows an example of a *pre–post* study comparing depression scores before therapy and after therapy. The two groups of scores are obtained by measuring the same variable (depression) twice for each participant; once before therapy and again after therapy. In a pre-post study, however, the researcher has no control over the passage of time. The "before" scores are always measured earlier than the "after" scores. Although a difference between the two groups of scores may be caused by the treatment, it is always possible that the scores simply change as time goes by. For example, the depression scores may decrease over time in the same way that the symptoms of a cold disappear over time. In a pre–post study the researcher also has no control over other variables that change with time. For example, the weather could change from dark and gloomy before therapy to bright and sunny after therapy. In this case, the depression scores could improve because of the weather and not because of the therapy. Because the researcher cannot control the passage of time or other variables related to time, this study is not a true experiment.

Terminology in Nonexperimental Research   Although the two research studies shown in Figure 1.6 are not true experiments, you should notice that they produce the same kind of data that are found in an experiment (see Figure 1.5). In each case, one variable is used to create groups, and a second variable is measured to obtain scores within each group. In an experiment, the groups are created by manipulation of the independent variable, and the participants' scores are the dependent variable. The same terminology is often used to identify the two variables in nonexperimental studies. That is, the variable that is used to create groups is the independent variable and the scores are the dependent variable. For example, the top part of Figure 1.6, gender (boy/girl), is the independent variable and the verbal test scores are the dependent variable. However, you should realize that gender (boy/girl) is not a true independent variable because it is not manipulated. For this reason, the "independent variable" in a nonexperimental study is often called a *quasi-independent variable.*

DEFINITION

> In a nonexperimental study, the "independent variable" that is used to create the different groups of scores is often called the **quasi-independent variable**.

LEARNING CHECK

1. In a correlational study, how many variables are measured for each individual and how many groups of scores are obtained?
   a. 1 variable and 1 group
   b. 1 variable and 2 groups
   c. 2 variables and 1 group
   d. 2 variables and 2 groups

2. A research study comparing alcohol use for college students in the United States and Canada reports that more Canadian students drink but American students drink more (Kuo, Adlaf, Lee, Gliksman, Demers, and Wechsler, 2002). What research design did this study use?

a. correlational

b. experimental

c. nonexperimental

d. noncorrelational

3. Stephens, Atkins, and Kingston (2009) found that participants were able to tolerate more pain when they shouted their favorite swear words over and over than when they shouted neutral words. For this study, what is the independent variable?

a. the amount of pain tolerated

b. the participants who shouted swear words

c. the participants who shouted neutral words

d. the kind of word shouted by the participants

ANSWERS   **1.** C, **2.** C, **3.** D

1.3 | Variables and Measurement

LEARNING OBJECTIVES

6. Explain why operational definitions are developed for constructs and identify the two components of an operational definition.

7. Describe discrete and continuous variables and identify examples of each.

8. Differentiate nominal, ordinal, interval, and ratio scales of measurement.

■ Constructs and Operational Definitions

The scores that make up the data from a research study are the result of observing and measuring variables. For example, a researcher may finish a study with a set of IQ scores, personality scores, or reaction-time scores. In this section, we take a closer look at the variables that are being measured and the process of measurement.

Some variables, such as height, weight, and eye color are well-defined, concrete entities that can be observed and measured directly. On the other hand, many variables studied by behavioral scientists are internal characteristics that people use to help describe and explain behavior. For example, we say that a student does well in school because he or she is *intelligent*. Or we say that someone is *anxious* in social situations, or that someone seems to be *hungry*. Variables like intelligence, anxiety, and hunger are called *constructs*, and because they are intangible and cannot be directly observed, they are often called hypothetical constructs.

Although constructs such as intelligence are internal characteristics that cannot be directly observed, it is possible to observe and measure behaviors that are representative of the construct. For example, we cannot "see" intelligence but we can see examples of intelligent behavior. The external behaviors can then be used to create an operational definition for the construct. An *operational definition* defines a construct in terms of external behaviors that can be observed and measured. For example, your intelligence is measured and defined by your performance on an IQ test, or hunger can be measured and defined by the number of hours since last eating.

> **DEFINITION**
>
> **Constructs** are internal attributes or characteristics that cannot be directly observed but are useful for describing and explaining behavior.
>
> An **operational definition** identifies a measurement procedure (a set of operations) for measuring an external behavior and uses the resulting measurements as a definition and a measurement of a hypothetical construct. Note that an operational definition has two components. First, it describes a set of operations for measuring a construct. Second, it defines the construct in terms of the resulting measurements.

■ Discrete and Continuous Variables

The variables in a study can be characterized by the type of values that can be assigned to them. A *discrete variable* consists of separate, indivisible categories. For this type of variable, there are no intermediate values between two adjacent categories. Consider the values displayed when dice are rolled. Between neighboring values—for example, seven dots and eight dots—no other values can ever be observed.

> **DEFINITION**
>
> A **discrete variable** consists of separate, indivisible categories. No values can exist between two neighboring categories.

Discrete variables are commonly restricted to whole, countable numbers—for example, the number of children in a family or the number of students attending class. If you observe class attendance from day to day, you may count 18 students one day and 19 students the next day. However, it is impossible ever to observe a value between 18 and 19. A discrete variable may also consist of observations that differ qualitatively. For example, people can be classified by gender (male or female), by occupation (nurse, teacher, lawyer, etc.), and college students can by classified by academic major (art, biology, chemistry, etc.). In each case, the variable is discrete because it consists of separate, indivisible categories.

On the other hand, many variables are not discrete. Variables such as time, height, and weight are not limited to a fixed set of separate, indivisible categories. You can measure time, for example, in hours, minutes, seconds, or fractions of seconds. These variables are called *continuous* because they can be divided into an infinite number of fractional parts.

> **DEFINITION**
>
> For a **continuous variable**, there are an infinite number of possible values that fall between any two observed values. A continuous variable is divisible into an infinite number of fractional parts.

Suppose, for example, that a researcher is measuring weights for a group of individuals participating in a diet study. Because weight is a continuous variable, it can be pictured as a continuous line (Figure 1.7). Note that there are an infinite number of possible points on

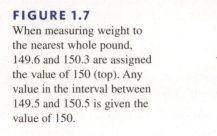
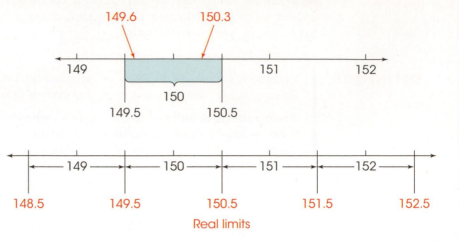

FIGURE 1.7
When measuring weight to the nearest whole pound, 149.6 and 150.3 are assigned the value of 150 (top). Any value in the interval between 149.5 and 150.5 is given the value of 150.

the line without any gaps or separations between neighboring points. For any two different points on the line, it is always possible to find a third value that is between the two points. Two other factors apply to continuous variables:

1. When measuring a continuous variable, it should be very rare to obtain identical measurements for two different individuals. Because a continuous variable has an infinite number of possible values, it should be almost impossible for two people to have exactly the same score. If the data show a substantial number of tied scores, then you should suspect that the measurement procedure is very crude or that the variable is not really continuous.

2. When measuring a continuous variable, each measurement category is actually an *interval* that must be defined by boundaries. For example, two people who both claim to weigh 150 pounds are probably not *exactly* the same weight. However, they are both around 150 pounds. One person may actually weigh 149.6 and the other 150.3. Thus, a score of 150 is not a specific point on the scale but instead is an interval (see Figure 1.7). To differentiate a score of 150 from a score of 149 or 151, we must set up boundaries on the scale of measurement. These boundaries are called *real limits* and are positioned exactly halfway between adjacent scores. Thus, a score of $X = 150$ pounds is actually an interval bounded by a *lower real limit* of 149.5 at the bottom and an *upper real limit* of 150.5 at the top. Any individual whose weight falls between these real limits will be assigned a score of $X = 150$.

DEFINITION

> **Real limits** are the boundaries of intervals for scores that are represented on a continuous number line. The real limit separating two adjacent scores is located exactly halfway between the scores. Each score has two real limits. The **upper real limit** is at the top of the interval, and the **lower real limit** is at the bottom.

The concept of real limits applies to any measurement of a continuous variable, even when the score categories are not whole numbers. For example, if you were measuring time to the nearest tenth of a second, the measurement categories would be 31.0, 31.1, 31.2, and so on. Each of these categories represents an interval on the scale that is bounded by real limits. For example, a score of $X = 31.1$ seconds indicates that the actual measurement is in an interval bounded by a lower real limit of 31.05 and an upper real limit of 31.15. Remember that the real limits are always halfway between adjacent categories.

Students often ask whether a value of exactly 150.5 should be assigned to the $X = 150$ interval or the $X = 151$ interval. The answer is that 150.5 is the *boundary* between the two intervals and is not necessarily in one or the other. Instead, the placement of 150.5 depends on the rule that you are using for rounding numbers. If you are rounding up, then 150.5 goes in the higher interval ($X = 151$) but if you are rounding down, then it goes in the lower interval ($X = 150$).

Later in this book, real limits are used for constructing graphs and for various calculations with continuous scales. For now, however, you should realize that real limits are a necessity whenever you make measurements of a continuous variable.

Finally, we should warn you that the terms *continuous* and *discrete* apply to the variables that are being measured and not to the scores that are obtained from the measurement. For example, measuring people's heights to the nearest inch produces scores of 60, 61, 62, and so on. Although the scores may appear to be discrete numbers, the underlying variable is continuous. One key to determining whether a variable is continuous or discrete is that a continuous variable can be divided into any number of fractional parts. Height can be measured to the nearest inch, the nearest 0.5 inch, or the nearest 0.1 inch. Similarly, a professor evaluating students' knowledge could use a pass/fail system that classifies students into two broad categories. However, the professor could choose to use a 10-point quiz that divides student knowledge into 11 categories corresponding to quiz scores from 0 to 10. Or the professor could use a 100-point exam that potentially divides student knowledge into 101 categories from 0 to 100. Whenever you are free to choose the degree of precision or the number of categories for measuring a variable, the variable must be continuous.

■ Scales of Measurement

It should be obvious by now that data collection requires that we make measurements of our observations. Measurement involves assigning individuals or events to categories. The categories can simply be names such as male/female or employed/unemployed, or they can be numerical values such as 68 inches or 175 pounds. The categories used to measure a variable make up a *scale of measurement*, and the relationships between the categories determine different types of scales. The distinctions among the scales are important because they identify the limitations of certain types of measurements and because certain statistical procedures are appropriate for scores that have been measured on some scales but not on others. If you were interested in people's heights, for example, you could measure a group of individuals by simply classifying them into three categories: tall, medium, and short. However, this simple classification would not tell you much about the actual heights of the individuals, and these measurements would not give you enough information to calculate an average height for the group. Although the simple classification would be adequate for some purposes, you would need more sophisticated measurements before you could answer more detailed questions. In this section, we examine four different scales of measurement, beginning with the simplest and moving to the most sophisticated.

■ The Nominal Scale

The word *nominal* means "having to do with names." Measurement on a nominal scale involves classifying individuals into categories that have different names but are not related to each other in any systematic way. For example, if you were measuring the academic majors for a group of college students, the categories would be art, biology, business, chemistry, and so on. Each student would be classified in one category according to his or her major. The measurements from a nominal scale allow us to determine whether two individuals are different, but they do not identify either the direction or the size of the difference. If one student is an art major and another is a biology major we can say that they are different, but we cannot say that art is "more than" or "less than" biology and we cannot specify how much difference there is between art and biology. Other examples of nominal scales include classifying people by race, gender, or occupation.

DEFINITION

A **nominal scale** consists of a set of categories that have different names. Measurements on a nominal scale label and categorize observations, but do not make any quantitative distinctions between observations.

Although the categories on a nominal scale are not quantitative values, they are occasionally represented by numbers. For example, the rooms or offices in a building may be identified by numbers. You should realize that the room numbers are simply names and do not reflect any quantitative information. Room 109 is not necessarily bigger than Room 100 and certainly not 9 points bigger. It also is fairly common to use numerical values as a code for nominal categories when data are entered into computer programs. For example, the data from a survey may code males with a 0 and females with a 1. Again, the numerical values are simply names and do not represent any quantitative difference. The scales that follow do reflect an attempt to make quantitative distinctions.

■ The Ordinal Scale

The categories that make up an *ordinal scale* not only have different names (as in a nominal scale) but also are organized in a fixed order corresponding to differences of magnitude.

DEFINITION

> An **ordinal scale** consists of a set of categories that are organized in an ordered sequence. Measurements on an ordinal scale rank observations in terms of size or magnitude.

Often, an ordinal scale consists of a series of ranks (first, second, third, and so on) like the order of finish in a horse race. Occasionally, the categories are identified by verbal labels like small, medium, and large drink sizes at a fast-food restaurant. In either case, the fact that the categories form an ordered sequence means that there is a directional relationship between categories. With measurements from an ordinal scale, you can determine whether two individuals are different and you can determine the direction of difference. However, ordinal measurements do not allow you to determine the size of the difference between two individuals. In a NASCAR race, for example, the first-place car finished faster than the second-place car, but the ranks don't tell you how much faster. Other examples of ordinal scales include socioeconomic class (upper, middle, lower) and T-shirt sizes (small, medium, large). In addition, ordinal scales are often used to measure variables for which it is difficult to assign numerical scores. For example, people can rank their food preferences but might have trouble explaining "how much" they prefer chocolate ice cream to steak.

■ The Interval and Ratio Scales

Both an *interval scale* and a *ratio scale* consist of a series of ordered categories (like an ordinal scale) with the additional requirement that the categories form a series of intervals that are all exactly the same size. Thus, the scale of measurement consists of a series of equal intervals, such as inches on a ruler. Other examples of interval and ratio scales are the measurement of time in seconds, weight in pounds, and temperature in degrees Fahrenheit. Note that, in each case, one interval (1 inch, 1 second, 1 pound, 1 degree) is the same size, no matter where it is located on the scale. The fact that the intervals are all the same size makes it possible to determine both the size and the direction of the difference between two measurements. For example, you know that a measurement of 80° Fahrenheit is higher than a measure of 60°, and you know that it is exactly 20° higher.

The factor that differentiates an interval scale from a ratio scale is the nature of the zero point. An interval scale has an arbitrary zero point. That is, the value 0 is assigned to a particular location on the scale simply as a matter of convenience or reference. In particular, a value of zero does not indicate a total absence of the variable being measured. For example a temperature of 0° Fahrenheit does not mean that there is no temperature, and it does not prohibit the temperature from going even lower. Interval scales with an arbitrary zero point are relatively rare. The two most common examples are the Fahrenheit and Celsius temperature scales. Other examples include golf scores (above and below par) and relative measures such as above and below average rainfall.

A ratio scale is anchored by a zero point that is not arbitrary but rather is a meaningful value representing none (a complete absence) of the variable being measured. The existence of an absolute, non-arbitrary zero point means that we can measure the absolute amount of the variable; that is, we can measure the distance from 0. This makes it possible to compare measurements in terms of ratios. For example, a gas tank with 10 gallons (10 more than 0) has twice as much gas as a tank with only 5 gallons (5 more than 0). Also note that a completely empty tank has 0 gallons. To recap, with a ratio scale, we can measure the direction and the size of the difference between two measurements and we can describe the difference in terms of a ratio. Ratio scales are quite common and include physical measures such as height and weight, as well as variables such as reaction time or the number of errors on a test. The distinction between an interval scale and a ratio scale is demonstrated in Example 1.2.

DEFINITION

> An **interval scale** consists of ordered categories that are all intervals of exactly the same size. Equal differences between numbers on scale reflect equal differences in magnitude. However, the zero point on an interval scale is arbitrary and does not indicate a zero amount of the variable being measured.
>
> A **ratio scale** is an interval scale with the additional feature of an absolute zero point. With a ratio scale, ratios of numbers do reflect ratios of magnitude.

EXAMPLE 1.2

A researcher obtains measurements of height for a group of 8-year-old boys. Initially, the researcher simply records each child's height in inches, obtaining values such as 44, 51, 49, and so on. These initial measurements constitute a ratio scale. A value of zero represents no height (absolute zero). Also, it is possible to use these measurements to form ratios. For example, a child who is 60 inches tall is one and a half times taller than a child who is 40 inches tall.

Now suppose that the researcher converts the initial measurement into a new scale by calculating the difference between each child's actual height and the average height for this age group. A child who is 1 inch taller than average now gets a score of $+1$; a child 4 inches taller than average gets a score of $+4$. Similarly, a child who is 2 inches shorter than average gets a score of -2. On this scale, a score of zero corresponds to average height. Because zero no longer indicates a complete absence of height, the new scores constitute an interval scale of measurement.

Notice that original scores and the converted scores both involve measurement in inches, and you can compute differences, or distances, on either scale. For example, there is a 6-inch difference in height between two boys who measure 57 and 51 inches tall on the first scale. Likewise, there is a 6-inch difference between two boys who measure $+9$ and $+3$ on the second scale. However, you should also notice that ratio comparisons are not possible on the second scale. For example, a boy who measures $+9$ is not three times taller than a boy who measures $+3$. ∎

Statistics and Scales of Measurement   For our purposes, scales of measurement are important because they help determine the statistics that are used to evaluate the data. Specifically, there are certain statistical procedures that are used with numerical scores from interval or ratio scales and other statistical procedures that are used with non-numerical scores from nominal or ordinal scales. The distinction is based on the fact that numerical scores are compatible with basic arithmetic operations (adding, multiplying, and so on) but non-numerical scores are not. For example, if you measure IQ scores for a group of students, it is possible to add the scores together to find a total and then calculate the average score for the group. On the other hand, if you measure the academic major for each student, you cannot add the scores to obtain a total. (What is the total for three psychology majors plus an English major plus two chemistry majors?) The vast majority of the statistical techniques presented in this book are designed for numerical scores from interval or ratio scales. For most statistical applications, the distinction between an interval scale and a ratio scale is not important because both scales produce numerical values that permit us to compute differences between scores, to add scores, and to calculate mean scores. On the other hand, measurements from nominal or ordinal scales are typically not numerical values, do not measure distance, and are not compatible with many basic arithmetic operations. Therefore, alternative statistical techniques are necessary for data from nominal or ordinal scales of measurement (for example, the median and the mode in Chapter 3, the Spearman correlation in Chapter 15, and the chi-square tests in Chapter 17). Additional statistical methods for measurements from ordinal scales are presented in Appendix E.

LEARNING CHECK

1. An operational definition is used to _____ a hypothetical construct.
   a. define
   b. measure
   c. measure and define
   d. None of the other choices is correct.

2. A researcher studies the factors that determine the number of children that couples decide to have. The variable, number of children, is an example of a _____ variable.
   a. discrete
   b. continuous
   c. nominal
   d. ordinal

3. When measuring height to the nearest half inch, what are the real limits for a score of 68.0 inches?
   a. 67 and 69
   b. 67.5 and 68.5
   c. 67.75 and 68.75
   d. 67.75 and 68.25

4. The teacher in a communications class asks students to identify their favorite reality television show. The different television shows make up a _____ scale of measurement.
   a. nominal
   b. ordinal
   c. interval
   d. ratio

ANSWERS   1. C, 2. A, 3. D, 4. A

1.4 | Statistical Notation

LEARNING OBJECTIVES

9. Identify what is represented by each of the following symbols: X, Y, N, n, and Σ.

10. Perform calculations using summation notation and other mathematical operations following the correct order of operations.

The measurements obtained in research studies provide the data for statistical analysis. Most of the statistical analyses use the same general mathematical operations, notation, and basic arithmetic that you have learned during previous years of school. In case you are unsure of your mathematical skills, there is a mathematics review section in Appendix A at the back of this book. The appendix also includes a skills assessment exam (p. 626) to help you determine whether you need the basic mathematics review. In this section, we introduce some of the specialized notation that is used for statistical calculations. In later chapters, additional statistical notation is introduced as it is needed.

■ Scores

Measuring a variable in a research study yields a value or a score for each individual. Raw scores are the original, unchanged scores obtained in the study. Scores for a particular variable are typically represented by the letter X. For example, if performance in your statistics course is measured by tests and you obtain a 35 on the first test, then we could state that $X = 35$. A set of scores can be presented in a column that is headed by X. For example, a list of quiz scores from your class might be presented as shown below (the single column on the left).

Quiz Scores	Height	Weight
X	X	Y
37	72	165
35	68	151
35	67	160
30	67	160
25	68	146
17	70	160
16	66	133

When observations are made for two variables, there will be two scores for each individual. The data can be presented as two lists labeled X and Y for the two variables. For example, observations for people's height in inches (variable X) and weight in pounds (variable Y) can be presented as shown in the double column in the margin. Each pair X, Y represents the observations made of a single participant.

The letter N is used to specify how many scores are in a set. An uppercase letter N identifies the number of scores in a population and a lowercase letter n identifies the number of scores in a sample. Throughout the remainder of the book you will notice that we often use notational differences to distinguish between samples and populations. For the height and weight data in the preceding table, $n = 7$ for both variables. Note that by using a lowercase letter n, we are implying that these data are a sample.

■ Summation Notation

Many of the computations required in statistics involve adding a set of scores. Because this procedure is used so frequently, a special notation is used to refer to the sum of a set of scores. The Greek letter *sigma*, or Σ, is used to stand for summation. The expression ΣX

means to add all the scores for variable X. The summation sign Σ can be read as "the sum of." Thus, ΣX is read "the sum of the scores." For the following set of quiz scores,

$$10, 6, 7, 4, \quad \Sigma X = 27 \text{ and } N = 4.$$

To use summation notation correctly, keep in mind the following two points:

1. The summation sign, Σ, is always followed by a symbol or mathematical expression. The symbol or expression identifies exactly which values are to be added. To compute ΣX, for example, the symbol following the summation sign is X, and the task is to find the sum of the X values. On the other hand, to compute $\Sigma(X - 1)^2$, the summation sign is followed by a relatively complex mathematical expression, so your first task is to calculate all of the $(X - 1)^2$ values and then add the results.

2. The summation process is often included with several other mathematical operations, such as multiplication or squaring. To obtain the correct answer, it is essential that the different operations be done in the correct sequence. Following is a list showing the correct *order of operations* for performing mathematical operations. Most of this list should be familiar, but you should note that we have inserted the summation process as the fourth operation in the list.

Order of Mathematical Operations

More information on the order of operations for mathematics is available in the Math Review Appendix, page 625.

1. Any calculation contained within parentheses is done first.
2. Squaring (or raising to other exponents) is done second.
3. Multiplying and/or dividing is done third. A series of multiplication and/or division operations should be done in order from left to right.
4. Summation using the Σ notation is done next.
5. Finally, any other addition and/or subtraction is done.

The following examples demonstrate how summation notation is used in most of the calculations and formulas we present in this book. Notice that whenever a calculation requires multiple steps, we use a computational table to help demonstrate the process. The table simply lists the original scores in the first column and then adds columns to show the results of each successive step. Notice that the first three operations in the order-of-operations list all create a new column in the computational table. When you get to summation (number 4 in the list), you simply add the values in the last column of your table to obtain the sum.

EXAMPLE 1.3

A set of four scores consists of values 3, 1, 7, and 4. We will compute ΣX, ΣX^2, and $(\Sigma X)^2$ for these scores. To help demonstrate the calculations, we will use a computational table showing the original scores (the X values) in the first column. Additional columns can then be added to show additional steps in the series of operations. You should notice that the first three operations in the list (parentheses, squaring, and multiplying) all create a new column of values. The last two operations, however, produce a single value corresponding to the sum.

The table to the left shows the original scores (the X values) and the squared scores (the X^2 values) that are needed to compute ΣX^2.

The first calculation, ΣX, does not include any parentheses, squaring, or multiplication, so we go directly to the summation operation. The X values are listed in the first column of the table, and we simply add the values in this column:

X	X^2
3	9
1	1
7	49
4	16

$$\Sigma X = 3 + 1 + 7 + 4 = 15$$

To compute ΣX^2, the correct order of operations is to square each score and then find the sum of the squared values. The computational table shows the original scores and the results obtained from squaring (the first step in the calculation). The second step is to find the sum of the squared values, so we simply add the numbers in the X^2 column:

$$\Sigma X^2 = 9 + 1 + 49 + 16 = 75$$

The final calculation, $(\Sigma X)^2$, includes parentheses, so the first step is to perform the calculation inside the parentheses. Thus, we first find ΣX and then square this sum. Earlier, we computed $\Sigma X = 15$, so

$$(\Sigma X)^2 = (15)^2 = 225$$ ∎

EXAMPLE 1.4   Use the same set of four scores from Example 1.3 and compute $\Sigma(X - 1)$ and $\Sigma(X - 1)^2$. The following computational table will help demonstrate the calculations.

X	(X − 1)	(X − 1)²	
3	2	4	The first column lists the
1	0	0	original scores. A second
7	6	36	column lists the $(X - 1)$
4	3	9	values, and a third column
			shows the $(X - 1)^2$ values.

To compute $\Sigma(X - 1)$, the first step is to perform the operation inside the parentheses. Thus, we begin by subtracting one point from each of the X values. The resulting values are listed in the middle column of the table. The next step is to add the $(X - 1)$ values, so we simply add the values in the middle column.

$$\Sigma(X - 1) = 2 + 0 + 6 + 3 + = 11$$

The calculation of $\Sigma(X + 1)^2$ requires three steps. The first step (inside parentheses) is to subtract 1 point from each X value. The results from this step are shown in the middle column of the computational table. The second step is to square each of the $(X - 1)$ values. The results from this step are shown in the third column of the table. The final step is to add the $(X - 1)^2$ values, so we add the values in the third column to obtain

$$\Sigma(X - 1)^2 = 4 + 0 + 36 + 9 = 49$$

Notice that this calculation requires squaring before adding. A common mistake is to add the $(X - 1)$ values and then square the total. Be careful! ∎

EXAMPLE 1.5   In both of the preceding examples, and in many other situations, the summation operation is the last step in the calculation. According to the order of operations, parentheses, exponents, and multiplication all come before summation. However, there are situations in which extra addition and subtraction are completed after the summation. For this example, use the same scores that appeared in the previous two examples, and compute $\Sigma X - 1$.

With no parentheses, exponents, or multiplication, the first step is the summation. Thus, we begin by computing ΣX. Earlier we found $\Sigma X = 15$. The next step is to subtract one point from the total. For these data,

$$\Sigma X - 1 = 15 - 1 = 14$$ ∎

EXAMPLE 1.6   For this example, each individual has two scores. The first score is identified as X, and the second score is Y. With the help of the following computational table, compute ΣX, ΣY, and ΣXY.

Person	X	Y	XY
A	3	5	15
B	1	3	3
C	7	4	28
D	4	2	8

To find ΣX, simply add the values in the X column.

$$\Sigma X = 3 + 1 + 7 + 4 = 15$$

Similarly, ΣY is the sum of the Y values in the middle column.

$$\Sigma Y = 5 + 3 + 4 + 2 = 14$$

To compute ΣXY, the first step is to multiply X times Y for each individual. The resulting products (XY values) are listed in the third column of the table. Finally, we add the products to obtain

$$\Sigma XY = 15 + 3 + 28 + 8 = 54$$ ■

The following example is an opportunity for you to test your understanding of summation notation.

EXAMPLE 1.7   Calculate each value requested for the following scores: 5, 2, 4, 2

a. ΣX^2 **b.** $\Sigma(X + 1)$ **c.** $\Sigma(X + 1)^2$

You should obtain answers of 49, 17, and 79 for a, b, and c, respectively. Good luck.   ■

LEARNING CHECK   **1.** What value is represented by the lowercase letter n?

    a. the number of scores in a population

    b. the number of scores in a sample

    c. the number of values to be added in a summation problem

    d. the number of steps in a summation problem

2. What is the value of $\Sigma(X - 2)$ for the following scores: 6, 2, 4, 2?

    a. 12

    b. 10

    c. 8

    d. 6

3. What is the first step in the calculation of $(\Sigma X)^2$?

    a. Square each score.

    b. Add the scores.

    c. Subtract 2 points from each score.

    d. Add the $X - 2$ values.

ANSWERS   **1.** B, **2.** D, **3.** B

SUMMARY

1. The term *statistics* is used to refer to methods for organizing, summarizing, and interpreting data.

2. Scientific questions usually concern a population, which is the entire set of individuals one wishes to study. Usually, populations are so large that it is impossible to examine every individual, so most research is conducted with samples. A sample is a group selected from a population, usually for purposes of a research study.

3. A characteristic that describes a sample is called a statistic, and a characteristic that describes a population is called a parameter. Although sample statistics are usually representative of corresponding population parameters, there is typically some discrepancy between a statistic and a parameter. The naturally occurring difference between a statistic and a parameter is called sampling error.

4. Statistical methods can be classified into two broad categories: descriptive statistics, which organize and summarize data, and inferential statistics, which use sample data to draw inferences about populations.

5. The correlational method examines relationships between variables by measuring two different variables for each individual. This method allows researchers to measure and describe relationships, but cannot produce a cause-and-effect explanation for the relationship.

6. The experimental method examines relationships between variables by manipulating an independent variable to create different treatment conditions and then measuring a dependent variable to obtain a group of scores in each condition. The groups of scores are then compared. A systematic difference between groups provides evidence that changing the independent variable from one condition to another also caused a change in the dependent variable. All other variables are controlled to prevent them from influencing the relationship. The intent of the experimental method is to demonstrate a cause-and-effect relationship between variables.

7. Nonexperimental studies also examine relationships between variables by comparing groups of scores, but they do not have the rigor of true experiments and cannot produce cause-and-effect explanations. Instead of manipulating a variable to create different groups, a nonexperimental study uses a preexisting participant characteristic (such as male/female) or the passage of time (before/after) to create the groups being compared.

8. A measurement scale consists of a set of categories that are used to classify individuals. A nominal scale consists of categories that differ only in name and are not differentiated in terms of magnitude or direction. In an ordinal scale, the categories are differentiated in terms of direction, forming an ordered series. An interval scale consists of an ordered series of categories that are all equal-sized intervals. With an interval scale, it is possible to differentiate direction and magnitude (or distance) between categories. Finally, a ratio scale is an interval scale for which the zero point indicates none of the variable being measured. With a ratio scale, ratios of measurements reflect ratios of magnitude.

9. A discrete variable consists of indivisible categories, often whole numbers that vary in countable steps. A continuous variable consists of categories that are infinitely divisible and each score corresponds to an interval on the scale. The boundaries that separate intervals are called real limits and are located exactly halfway between adjacent scores.

10. The letter X is used to represent scores for a variable. If a second variable is used, Y represents its scores. The letter N is used as the symbol for the number of scores in a population; n is the symbol for a number of scores in a sample.

11. The Greek letter sigma (Σ) is used to stand for summation. Therefore, the expression ΣX is read "the sum of the scores." Summation is a mathematical operation (like addition or multiplication) and must be performed in its proper place in the order of operations; summation occurs after parentheses, exponents, and multiplying/dividing have been completed.

KEY TERMS

statistics (3)	datum (5)	sampling error (6)
population (3)	raw score (5)	correlational method (11)
sample (4)	parameter (5)	experimental method (15)
variable (4)	statistic (5)	independent variable (15)
data (5)	descriptive statistics (5)	dependent variable (15)
data set (5)	inferential statistics (6)	control condition (15)

experimental condition (15)   operational definition (19)   lower real limit (20)
nonequivalent groups study (16) discrete variable (19)   nominal scale (21)
pre–post study (17)   continuous variable (19)   ordinal scale (22)
quasi-independent variable (17) real limits (20)   interval scale (23)
construct (19)   upper real limit (20)   ratio scale (23)

SPSS®

The Statistical Package for the Social Sciences, known as SPSS, is a computer program that performs most of the statistical calculations that are presented in this book, and is commonly available on college and university computer systems. Appendix D contains a general introduction to SPSS. In the Resource section at the end of each chapter for which SPSS is applicable, there are step-by-step instructions for using SPSS to perform the statistical operations presented in the chapter.

FOCUS ON PROBLEM SOLVING

It may help to simplify summation notation if you observe that the summation sign is always followed by a symbol or symbolic expression—for example, ΣX or $\Sigma(X + 3)$. This symbol specifies which values you are to add. If you use the symbol as a column heading and list all the appropriate values in the column, your task is simply to add up the numbers in the column. To find $\Sigma(X + 3)$ for example, start a column headed with $(X + 3)$ next to the column of Xs. List all the $(X + 3)$ values; then find the total for the column.

Often, summation notation is part of a relatively complex mathematical expression that requires several steps of calculation. The series of steps must be performed according to the order of mathematical operations (see page 26). The best procedure is to use a computational table that begins with the original X values listed in the first column. Except for summation, each step in the calculation creates a new column of values. For example, computing $\Sigma(X + 1)^2$ involves three steps and produces a computational table with three columns. The final step is to add the values in the third column (see Example 1.4).

DEMONSTRATION 1.1

SUMMATION NOTATION

A set of scores consists of the following values:

$$7 \quad 3 \quad 9 \quad 5 \quad 4$$

For these scores, compute each of the following:

$$\Sigma X$$
$$(\Sigma X)^2$$
$$\Sigma X^2$$
$$\Sigma X + 5$$
$$\Sigma(X - 2)$$

X	X^2
7	49
3	9
9	81
5	25
4	16

Compute ΣX To compute ΣX, we simply add all of the scores in the group.

$$\Sigma X = 7 + 3 + 9 + 5 + 4 = 28$$

Compute $(\Sigma X)^2$ The first step, inside the parentheses, is to compute ΣX. The second step is to square the value for ΣX.

$$\Sigma X = 28 \text{ and } (\Sigma X)^2 = (28)^2 = 784$$

Compute ΣX^2   The first step is to square each score. The second step is to add the squared scores. The computational table shows the scores and squared scores. To compute ΣX^2 we add the values in the X^2 column.

$$\Sigma X^2 = 49 + 9 + 81 + 25 + 16 = 180$$

Compute $\Sigma X + 5$ The first step is to compute ΣX. The second step is to add 5 points to the total.

$$\Sigma X = 28 \text{ and } \Sigma X + 5 = 28 + 5 = 33$$

Compute $\Sigma(X - 2)$ The first step, inside parentheses, is to subtract 2 points from each score. The second step is to add the resulting values. The computational table shows the scores and the $(X - 2)$ values. To compute $\Sigma(X - 2)$, add the values in the $(X - 2)$ column

$$\Sigma(X - 2) = 5 + 1 + 7 + 3 + 2 = 18$$

X	X − 2
7	5
3	1
9	7
5	3
4	2

PROBLEMS

*1. A researcher is interested in the texting habits of high school students in the United States. The researcher selects a group of 100 students, measures the number of text messages that each individual sends each day, and calculates the average number for the group.
  a. Identify the population for this study.
  b. Identify the sample for this study.
  c. The average number that the researcher calculated is an example of a _____.

2. Define the terms population, sample, parameter, and statistic.

3. Statistical methods are classified into two major categories: descriptive and inferential. Describe the general purpose for the statistical methods in each category.

4. Define the concept of sampling error and explain why this phenomenon creates a problem to be addressed by inferential statistics.

5. Describe the data for a correlational research study. Explain how these data are different from the data obtained in experimental and nonexperimental studies, which also evaluate relationships between two variables.

6. Describe how the goal of an experimental research study is different from the goal for nonexperimental or correlational research. Identify the two elements that are necessary for an experiment to achieve its goal.

7. Stephens, Atkins, and Kingston (2009) conducted an experiment in which participants were able to tolerate more pain when they were shouting their favorite swear words than when they were shouting neutral words. Identify the independent and dependent variables for this study.

8. The results of a recent study showed that children who routinely drank reduced fat milk (1% or skim) were more likely to be overweight or obese at age

2 and age 4 compared to children who drank whole or 2% milk (Scharf, Demmer, and DeBoer, 2013). Is this an example of an experimental or a nonexperimental study?

9. Gentile, Lynch, Linder, and Walsh (2004) surveyed over 600 8th- and 9th-grade students asking about their gaming habits and other behaviors. Their results showed that the adolescents who experienced more video game violence were also more hostile and had more frequent arguments with teachers. Is this an experimental or a nonexperimental study? Explain your answer.

10. Weinstein, McDermott, and Roediger (2010) conducted an experiment to evaluate the effectiveness of different study strategies. One part of the study asked students to prepare for a test by reading a passage. In one condition, students generated and answered questions after reading the passage. In a second condition, students simply read the passage a second time. All students were then given a test on the passage material and the researchers recorded the number of correct answers.
  a. Identify the dependent variable for this study.
  b. Is the dependent variable discrete or continuous?
  c. What scale of measurement (nominal, ordinal, interval, or ratio) is used to measure the dependent variable?

11. A research study reports that alcohol consumption is significantly higher for students at a state university than for students at a religious college (Wells, 2010). Is this study an example of an experiment? Explain why or why not.

12. In an experiment examining the effects Tai Chi on arthritis pain, Callahan (2010) selected a large sample of individuals with doctor-diagnosed arthritis. Half of the participants immediately began a Tai Chi course and the other half (the control group) waited 8 weeks before beginning. At the end of 8 weeks, the individuals

*Solutions for odd-numbered problems are provided in Appendix C.

who had experienced Tai Chi had less arthritis pain that those who had not participated in the course.

a. Identify the independent variable for this study.
b. What scale of measurement is used for the independent variable?
c. Identify the dependent variable for this study.
d. What scale of measurement is used for the dependent variable?

13. A tax form asks people to identify their annual income, number of dependents, and social security number. For each of these three variables, identify the scale of measurement that probably is used and identify whether the variable is continuous or discrete.

14. Four scales of measurement were introduced in this chapter: nominal, ordinal, interval, and ratio.
a. What additional information is obtained from measurements on an ordinal scale compared to measurements on a nominal scale?
b. What additional information is obtained from measurements on an interval scale compared to measurements on an ordinal scale?
c. What additional information is obtained from measurements on a ratio scale compared to measurements on an interval scale?

15. Knight and Haslam (2010) found that office workers who had some input into the design of their office space were more productive and had higher well-being compared to workers for whom the office design was completely controlled by an office manager. For this study, identify the independent variable and the dependent variable.

16. Explain why *honesty* is a hypothetical construct instead of a concrete variable. Describe how shyness might be measured and defined using an operational definition.

17. Ford and Torok (2008) found that motivational signs were effective in increasing physical activity on a college campus. Signs such as "Step up to a healthier lifestyle" and "An average person burns 10 calories a minute walking up the stairs" were posted by the elevators and stairs in a college building. Students and faculty increased their use of the stairs during times that the signs were posted compared to times when there were no signs.
a. Identify the independent and dependent variables for this study.
b. What scale of measurement is used for the independent variable?

18. For the following scores, find the value of each expression:
a. ΣX
b. ΣX^2
c. $\Sigma X + 1$
d. $\Sigma(X + 1)$

X
3
5
0
2

19. For the following set of scores, find the value of each expression:
a. ΣX^2
b. $(\Sigma X)^2$
c. $\Sigma(X - 1)$
d. $\Sigma(X - 1)^2$

X
3
2
5
1
3

20. For the following set of scores, find the value of each expression:
a. ΣX
b. ΣX^2
c. $\Sigma(X + 3)$.

X
6
-2
0
-3
-1

21. Two scores, X and Y, are recorded for each of $n = 4$ subjects. For these scores, find the value of each expression.
a. ΣX
b. ΣY
c. ΣXY

Subject	X	Y
A	3	4
B	0	7
C	-1	5
D	2	2

22. Use summation notation to express each of the following calculations:
a. Add the scores and then add then square the sum.
b. Square each score and then add the squared values.
c. Subtract 2 points from each score and then add the resulting values.
d. Subtract 1 point from each score and square the resulting values. Then add the squared values.

23. For the following set of scores, find the value of each expression:
a. ΣX^2
b. $(\Sigma X)^2$
c. $\Sigma(X - 3)$
d. $\Sigma(X - 3)^2$

X
1
6
2
3

Frequency Distributions

Tools You Will Need

The following items are considered essential background material for this chapter. If you doubt your knowledge of any of these items, you should review the appropriate chapter or section before proceeding.

- Proportions (Appendix A)
  - Fractions
  - Decimals
  - Percentages
- Scales of measurement (Chapter 1): Nominal, ordinal, interval, and ratio
- Continuous and discrete variables (Chapter 1)
- Real limits (Chapter 1)

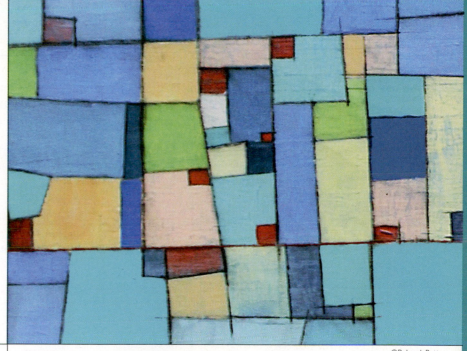

©Deborah Batt

There is some evidence that people with a visible tattoo are viewed more negatively than are people without a visible tattoo (Resenhoeft, Villa, and Wiseman, 2008). In the study, one group of community college students was shown a color photograph of a 24-year-old woman with a tattoo of a dragon on her arm. A second group of students was shown the same photograph but with the tattoo removed. Each participant was asked to rate the participant on several characteristics, including attractiveness, using a 5-point scale (5 = most positive). Data similar to the attractiveness ratings obtained in the study are shown in Table 2.1.

TABLE 2.1

Attractiveness ratings of a woman shown in a color photograph for two samples of college students. For the first group of students the woman in the photograph had a visible tattoo. The students in the second group saw the same photograph with the tattoo removed.

Visible Tattoo				No Visible Tattoo			
1	2	4	3	2	4	4	3
2	2	1	3	5	4	2	4
2	5	4	3	4	5	3	3

You probably find it difficult to see any clear pattern simply by looking at the list of numbers. Can you tell whether the ratings for one group are generally higher than those for the other group? One solution to this problem is to organize each group of scores into a frequency distribution, which provides a clearer view of the entire group.

For example, the same attractiveness ratings that are shown in Table 2.1 have been organized in a frequency distribution graph in Figure 2.1. In the figure, each individual is represented by a block that is placed above that individual's score. The resulting pile of blocks shows a picture of how the individual scores are distributed. For this example, it is now easy to see that the attractiveness scores for the woman without a tattoo are generally higher than the scores for the woman with a tattoo; on average, the rated attractiveness of the woman was around 2 with a tattoo and around 4 without a tattoo.

In this chapter we present techniques for organizing data into tables and graphs so that an entire set of scores can be presented in a relatively simple display or illustration.

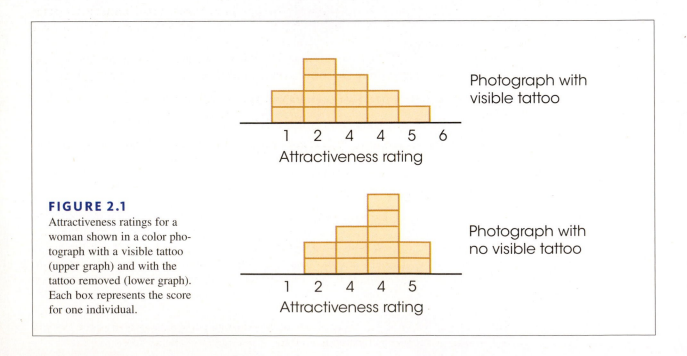

FIGURE 2.1

Attractiveness ratings for a woman shown in a color photograph with a visible tattoo (upper graph) and with the tattoo removed (lower graph). Each box represents the score for one individual.

2.1 | Frequency Distributions and Frequency Distribution Tables

LEARNING OBJECTIVES

1. Describe the basic elements of a frequency distribution table and explain how they are related to the original set of scores.

2. Calculate the following from a frequency table: ΣX, ΣX^2, and the proportion and percentage of the group associated with each score.

The results from a research study usually consist of pages of numbers corresponding to the measurements or scores collected during the study. The immediate problem for the researcher is to organize the scores into some comprehensible form so that any patterns in the data can be seen easily and communicated to others. This is the job of descriptive statistics: to simplify the organization and presentation of data. One of the most common procedures for organizing a set of data is to place the scores in a *frequency distribution*.

DEFINITION

> A **frequency distribution** is an organized tabulation of the number of individuals located in each category on the scale of measurement.

A frequency distribution takes a disorganized set of scores and places them in order from highest to lowest, grouping together individuals who all have the same score. If the highest score is $X = 10$, for example, the frequency distribution groups together all the 10s, then all the 9s, then the 8s, and so on. Thus, a frequency distribution allows the researcher to see "at a glance" the entire set of scores. It shows whether the scores are generally high or low, whether they are concentrated in one area or spread out across the entire scale, and generally provides an organized picture of the data. In addition to providing a picture of the entire set of scores, a frequency distribution allows you to see the location of any individual score relative to all of the other scores in the set.

A frequency distribution can be structured either as a table or as a graph, but in either case, the distribution presents the same two elements:

1. The set of categories that make up the original measurement scale.

2. A record of the frequency, or number of individuals in each category.

Thus, a frequency distribution presents a picture of how the individual scores are distributed on the measurement scale—hence the name *frequency distribution*.

■ Frequency Distribution Tables

It is customary to list categories from highest to lowest, but this is an arbitrary arrangement. Many computer programs list categories from lowest to highest.

The simplest frequency distribution table presents the measurement scale by listing the different measurement categories (X values) in a column from highest to lowest. Beside each X value, we indicate the frequency, or the number of times that particular measurement occurred in the data. It is customary to use an X as the column heading for the scores and an f as the column heading for the frequencies. An example of a frequency distribution table follows.

EXAMPLE 2.1

The following set of $N = 20$ scores was obtained from a 10-point statistics quiz. We will organize these scores by constructing a frequency distribution table. Scores:

$$8, \quad 9, \quad 8, \quad 7, \quad 10, \quad 9, \quad 6, \quad 4, \quad 9, \quad 8,$$
$$7, \quad 8, \quad 10, \quad 9, \quad 8, \quad 6, \quad 9, \quad 7, \quad 8, \quad 8$$

X	f
10	2
9	5
8	7
7	3
6	2
5	0
4	1

1. The highest score is $X = 10$, and the lowest score is $X = 4$. Therefore, the first column of the table lists the categories that make up the scale of measurement (X values) from 10 down to 4. Notice that all of the possible values are listed in the table. For example, no one had a score of $X = 5$, but this value is included. With an ordinal, interval, or ratio scale, the categories are listed in order (usually highest to lowest). For a nominal scale, the categories can be listed in any order.

2. The frequency associated with each score is recorded in the second column. For example, two people had scores of $X = 10$, so there is a 2 in the f column beside $X = 10$.

Because the table organizes the scores, it is possible to see very quickly the general quiz results. For example, there were only two perfect scores, but most of the class had high grades (8s and 9s). With one exception (the score of $X = 4$), it appears that the class has learned the material fairly well.

Notice that the X values in a frequency distribution table represent the scale of measurement, *not* the actual set of scores. For example, the X column lists the value 10 only one time, but the frequency column indicates that there are actually two values of $X = 10$. Also, the X column lists a value of $X = 5$, but the frequency column indicates that no one actually had a score of $X = 5$.

You also should notice that the frequencies can be used to find the total number of scores in the distribution. By adding up the frequencies, you obtain the total number of individuals:

$$\Sigma f = N \qquad \blacksquare$$

Obtaining ΣX from a Frequency Distribution Table   There may be times when you need to compute the sum of the scores, ΣX, or perform other computations for a set of scores that has been organized into a frequency distribution table. To complete these calculations correctly, you must use all the information presented in the table. That is, it is essential to use the information in the f column as well as the X column to obtain the full set of scores.

When it is necessary to perform calculations for scores that have been organized into a frequency distribution table, the safest procedure is to use the information in the table to recover the complete list of individual scores before you begin any computations. This process is demonstrated in the following example.

EXAMPLE 2.2

X	F
5	1
4	2
3	3
2	3
1	1

Consider the frequency distribution table shown in the margin. The table shows that the distribution has one 5, two 4s, three 3s, three 2s, and one 1, for a total of 10 scores. If you simply list all 10 scores, you can safely proceed with calculations such as finding ΣX or ΣX^2. For example, to compute ΣX you must add all 10 scores:

$$\Sigma X = 5 + 4 + 4 + 3 + 3 + 3 + 2 + 2 + 2 + 1$$

For the distribution in this table, you should obtain $\Sigma X = 29$. Try it yourself.

Similarly, to compute ΣX^2 you square each of the 10 scores and then add the squared values.

$$\Sigma X^2 = 5^2 + 4^2 + 4^2 + 3^2 + 3^2 + 3^2 + 2^2 + 2^2 + 2^2 + 1^2$$

This time you should obtain $\Sigma X^2 = 97$. $\blacksquare$

Caution: **Doing calculations within the table works well for ΣX but can lead to errors for more complex formulas.**

An alternative way to get ΣX from a frequency distribution table is to multiply each X value by its frequency and then add these products. This sum may be

No matter which method you use to find ΣX, the important point is that you must use the information given in the frequency column in addition to the information in the X column.

expressed in symbols as ΣfX. The computation is summarized as follows for the data in Example 2.2:

X	f	fX	
5	1	5	(the one 5 totals 5)
4	2	8	(the two 4s total 8)
3	3	9	(the three 3s total 9)
2	3	6	(the three 2s total 6)
1	1	1	(the one 1 totals 1)
		$\Sigma X = 29$	

The following example is an opportunity for you to test your understanding by computing ΣX and ΣX^2 for scores in a frequency distribution table.

EXAMPLE 2.3

Calculate ΣX and ΣX^2 for scores shown in the frequency distribution table in Example 2.1 (p. 37). You should obtain $\Sigma X = 158$ and $\Sigma X^2 = 1288$, Good luck. ∎

■ Proportions and Percentages

In addition to the two basic columns of a frequency distribution, there are other measures that describe the distribution of scores and can be incorporated into the table. The two most common are proportion and percentage.

Proportion measures the fraction of the total group that is associated with each score. In Example 2.2, there were two individuals with $X = 4$. Thus, 2 out of 10 people had $X = 4$, so the proportion would be $\frac{2}{10} = 0.20$. In general, the proportion associated with each score is

$$\text{proportion} = p = \frac{f}{N}$$

Because proportions describe the frequency (f) in relation to the total number (N), they often are called *relative frequencies*. Although proportions can be expressed as fractions (for example, $\frac{2}{10}$), they more commonly appear as decimals. A column of proportions, headed with a p, can be added to the basic frequency distribution table (see Example 2.4).

In addition to using frequencies (f) and proportions (p), researchers often describe a distribution of scores with percentages. For example, an instructor might describe the results of an exam by saying that 15% of the class earned As, 23% Bs, and so on. To compute the percentage associated with each score, you first find the proportion (p) and then multiply by 100:

$$\text{percentage} = p(100) = \frac{f}{N}(100)$$

Percentages can be included in a frequency distribution table by adding a column headed with %. Example 2.4 demonstrates the process of adding proportions and percentages to a frequency distribution table.

EXAMPLE 2.4

The frequency distribution table from Example 2.2 is repeated here. This time we have added columns showing the proportion (p) and the percentage (%) associated with each score.

X	f	$p = \frac{f}{N}$	$\% = p(100)$
5	1	$\frac{1}{10} = 0.10$	10%
4	2	$\frac{2}{10} = 0.20$	20%
3	3	$\frac{3}{10} = 0.30$	30%
2	3	$\frac{3}{10} = 0.30$	30%
1	1	$\frac{1}{10} = 0.10$	10%

∎

LEARNING CHECK

1. For the following frequency distribution, how many individuals had a score of $X = 2$?

   a. 1
   b. 2
   c. 3
   d. 4

X	f
5	1
4	2
3	4
2	3
1	2

2. The following is a distribution of quiz scores. If a score of $X = 2$ or lower is failing, then how many individuals failed the quiz?

   a. 2
   b. 3
   c. 5
   d. 9

X	f
5	1
4	2
3	4
2	3
1	2

3. For the following frequency distribution, what is ΣX^2?

   a. 30
   b. 45
   c. 77
   d. $(17)^2 = 289$

X	f
4	1
3	2
2	2
1	3
0	1

ANSWERS   **1.** C, **2.** C, **3.** B

2.2 | Grouped Frequency Distribution Tables

LEARNING OBJECTIVE

3. Identify when it is useful to set up a grouped frequency distribution table, and explain how to construct this type of table for a set of scores.

When a set of data covers a wide range of values, it is unreasonable to list all the individual scores in a frequency distribution table. Consider, for example, a set of exam scores that range from a low of $X = 41$ to a high of $X = 96$. These scores cover a *range* of more than 50 points.

If we were to list all the individual scores from $X = 96$ down to $X = 41$, it would take 56 rows to complete the frequency distribution table. Although this would organize the data, the table would be long and cumbersome. Remember: The purpose for constructing a table is to obtain a relatively simple, organized picture of the data. This can be accomplished by grouping the scores into intervals and then listing the intervals in the table instead of listing each individual score. For example, we could construct a table showing the number of students who had scores in the 90s, the number with scores in the 80s, and

> When the scores are whole numbers, the total number of rows for a regular table can be obtained by finding the difference between the highest and the lowest scores and adding 1:
>
> rows = highest
>   − lowest + 1

so on. The result is called a *grouped frequency distribution table* because we are present-ing groups of scores rather than individual values. The groups, or intervals, are called *class intervals*.

There are several guidelines that help guide you in the construction of a grouped frequency distribution table. Note that these are simply guidelines, rather than absolute requirements, but they do help produce a simple, well-organized, and easily understood table.

GUIDELINE 1   The grouped frequency distribution table should have about 10 class intervals. If a table has many more than 10 intervals, it becomes cumbersome and defeats the purpose of a frequency distribution table. On the other hand, if you have too few intervals, you begin to lose information about the distribution of the scores. At the extreme, with only one interval, the table would not tell you anything about how the scores are distributed. Remember that the purpose of a frequency distribution is to help a researcher see the data. With too few or too many intervals, the table will not provide a clear picture. You should note that 10 inter-vals is a general guide. If you are constructing a table on a blackboard, for example, you probably want only 5 or 6 intervals. If the table is to be printed in a scientific report, you may want 12 or 15 intervals. In each case, your goal is to present a table that is relatively easy to see and understand.

GUIDELINE 2   The width of each interval should be a relatively simple number. For example, 2, 5, 10, or 20 would be a good choice for the interval width. Notice that it is easy to count by 5s or 10s. These numbers are easy to understand and make it possible for someone to see quickly how you have divided the range of scores.

GUIDELINE 3   The bottom score in each class interval should be a multiple of the width. If you are using a width of 10 points, for example, the intervals should start with 10, 20, 30, 40, and so on. Again, this makes it easier for someone to understand how the table has been constructed.

GUIDELINE 4   All intervals should be the same width. They should cover the range of scores completely with no gaps and no overlaps, so that any particular score belongs in exactly one interval.

The application of these rules is demonstrated in Example 2.5.

EXAMPLE 2.5   An instructor has obtained the set of $N = 25$ exam scores shown here. To help organize these scores, we will place them in a frequency distribution table. The scores are

82,   75,   88,   93,   53,   84,   87,   58,   72,   94,   69,   84,   61,
91,   64,   87,   84,   70,   76,   89,   75,   80,   73,   78,   60

Remember, when the scores are whole numbers, the number of rows is determined by

highest − lowest + 1

The first step is to determine the range of scores. For these data, the smallest score is $X = 53$ and the largest score is $X = 94$, so a total of 42 rows would be needed for a table that lists each individual score. Because 42 rows would not provide a simple table, we have to group the scores into class intervals.

The best method for finding a good interval width is a systematic trial-and-error approach that uses guidelines 1 and 2 simultaneously. Specifically, we want about 10 inter-vals and we want the interval width to be a simple number. For this example, the scores cover a range of 42 points, so we will try several different interval widths to see how many intervals are needed to cover this range. For example, if each interval were 2 points wide, it would take 21 intervals to cover a range of 42 points. This is too many, so we move on to an interval width of 5 or 10 points. The following table shows how many intervals would be needed for these possible widths:

Width	Number of Intervals Needed to Cover a Range of 42 Points	
2	21	(too many)
5	9	(OK)
10	5	(too few)

Notice that an interval width of 5 will result in about 10 intervals, which is exactly what we want.

The next step is to actually identify the intervals. The lowest score for these data is $X = 53$, so the lowest interval should contain this value. Because the interval should have a multiple of 5 as its bottom score, the interval should begin at 50. The interval has a width of 5, so it should contain 5 values: 50, 51, 52, 53, and 54. Thus, the bottom interval is 50–54. The next interval would start at 55 and go to 59. Note that this interval also has a bottom score that is a multiple of 5, and contains exactly 5 scores (55, 56, 57, 58, and 59). The complete frequency distribution table showing all of the class intervals is presented in Table 2.2.

TABLE 2.2

This grouped frequency distribution table shows the data from Example 2.4. The original scores range from a high of $X = 94$ to a low of $X = 53$. This range has been divided into 9 intervals with each interval exactly 5 points wide. The frequency column (f) lists the number of individuals with scores in each of the class intervals.

X	f
90–94	3
85–89	4
80–84	5
75–79	4
70–74	3
65–69	1
60–64	3
55–59	1
50–54	1

Once the class intervals are listed, you complete the table by adding a column of frequencies. The values in the frequency column indicate the number of individuals who have scores located in that class interval. For this example, there were three students with scores in the 60–64 interval, so the frequency for this class interval is $f = 3$ (see Table 2.2). The basic table can be extended by adding columns showing the proportion and percentage associated with each class interval.

Finally, you should note that after the scores have been placed in a grouped table, you lose information about the specific value for any individual score. For example, Table 2.2 shows that one person had a score between 65 and 69, but the table does not identify the exact value for the score. In general, the wider the class intervals are, the more information is lost. In Table 2.2 the interval width is 5 points, and the table shows that there are three people with scores in the lower 60s and one person with a score in the upper 60s. This information would be lost if the interval width were increased to 10 points. With an interval width of 10, all of the 60s would be grouped together into one interval labeled 60–69. The table would show a frequency of four people in the 60–69 interval, but it would not tell whether the scores were in the upper 60s or the lower 60s. ■

■ Real Limits and Frequency Distributions

Recall from Chapter 1 that a continuous variable has an infinite number of possible values and can be represented by a number line that is continuous and contains an infinite number of points. However, when a continuous variable is measured, the resulting measurements correspond to *intervals* on the number line rather than single points. If you are measuring time in seconds, for example, a score of $X = 8$ seconds actually represents an interval bounded by the real limits 7.5 seconds and 8.5 seconds. Thus, a frequency distribution table showing a frequency of $f = 3$ individuals all assigned a score of $X = 8$ does not mean that all three individuals had exactly the same measurement. Instead, you should realize that the three measurements are simply located in the same interval between 7.5 and 8.5.

The concept of real limits also applies to the class intervals of a grouped frequency distribution table. For example, a class interval of 40–49 contains scores from $X = 40$ to $X = 49$. These values are called the *apparent limits* of the interval because it appears that they form the upper and lower boundaries for the class interval. If you are measuring a continuous variable, however, a score of $X = 40$ is actually an interval from 39.5 to 40.5. Similarly, $X = 49$ is an interval from 48.5–49.5. Therefore, the real limits of the interval are 39.5 (the lower real limit) and 49.5 (the upper real limit). Notice that the next higher class interval is 50–59, which has a lower real limit of 49.5. Thus, the two intervals meet at the real limit 49.5, so there are no gaps in the scale. You also should notice that the width of each class interval becomes easier to understand when you consider the real limits of an interval. For example, the interval 50–59 has real limits of 49.5 and 59.5. The distance between these two real limits (10 points) is the width of the interval.

LEARNING CHECK

1. For this distribution, how many individuals had scores lower than $X = 20$?
   a. 2
   b. 3
   c. 4
   d. cannot be determined

X	f
24–25	2
22–23	4
20–21	6
18–19	3
16–17	1

2. In a grouped frequency distribution one interval is listed as 20–24. Assuming that the scores are measuring a continuous variable, what is the width of this interval?
   a. 3 points
   b. 4 points
   c. 5 points
   d. 54 points

3. A set of scores ranges from a high of $X = 48$ to a low of $X = 13$. If these scores are placed in a grouped frequency distribution table with an interval width of 5 points, the bottom interval in the table would be _____.
   a. 13–18
   b. 13–19
   c. 10–14
   d. 10–15

ANSWERS 1. C, 2. C, 3. C

2.3 | Frequency Distribution Graphs

LEARNING OBJECTIVES

4. Describe how the three types of frequency distribution graphs—histograms, polygons, and bar graphs—are constructed and identify when each is used.

5. Describe the basic elements of a frequency distribution graph and explain how they are related to the original set of scores.

6. Explain how frequency distribution graphs for populations differ from the graphs used for samples.

7. Identify the shape—symmetrical, and positively or negatively skewed—of a distribution in a frequency distribution graph.

A frequency distribution graph is basically a picture of the information available in a frequency distribution table. We will consider several different types of graphs, but all start with two perpendicular lines called *axes*. The horizontal line is the *X*-axis, or the abscissa (ab-SIS-uh). The vertical line is the *Y*-axis, or the ordinate. The measurement scale (set of *X* values) is listed along the *X*-axis with values increasing from left to right. The frequencies are listed on the *Y*-axis with values increasing from bottom to top. As a general rule, the point where the two axes intersect should have a value of zero for both the scores and the frequencies. A final general rule is that the graph should be constructed so that its height (*Y*-axis) is approximately two-thirds to three-quarters of its length (*X*-axis). Violating these guidelines can result in graphs that give a misleading picture of the data (see Box 2.1).

■ Graphs for Interval or Ratio Data

When the data consist of numerical scores that have been measured on an interval or ratio scale, there are two options for constructing a frequency distribution graph. The two types of graphs are called *histograms* and *polygons*.

Histograms   To construct a histogram, you first list the numerical scores (the categories of measurement) along the *X*-axis. Then you draw a bar above each *X* value so that

a. The height of the bar corresponds to the frequency for that category.

b. For continuous variables, the width of the bar extends to the real limits of the category. For discrete variables, each bar extends exactly half the distance to the adjacent category on each side.

For both continuous and discrete variables, each bar in a histogram extends to the midpoint between adjacent categories. As a result, adjacent bars touch and there are no spaces or gaps between bars. An example of a histogram is shown in Figure 2.2.

When data have been grouped into class intervals, you can construct a frequency distribution histogram by drawing a bar above each interval so that the width of the bar extends exactly half the distance to the adjacent category on each side. This process is demonstrated in Figure 2.3.

For the two histograms shown in Figures 2.2 and 2.3, notice that the values on both the vertical and horizontal axes are clearly marked and that both axes are labeled. Also note that, whenever possible, the units of measurement are specified; for example, Figure 2.3 shows a distribution of heights measured in inches. Finally, notice that the horizontal axis in Figure 2.3 does not list all of the possible heights starting from zero and going up to 48 inches. Instead, the graph clearly shows a break between zero and 30, indicating that some scores have been omitted.

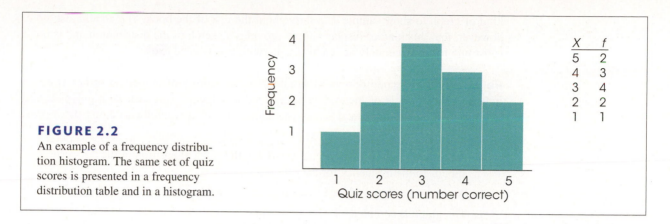

FIGURE 2.2

An example of a frequency distribution histogram. The same set of quiz scores is presented in a frequency distribution table and in a histogram.

X	f
5	2
4	3
3	4
2	2
1	1

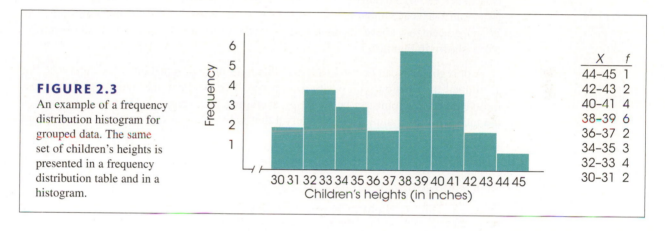

FIGURE 2.3

An example of a frequency distribution histogram for grouped data. The same set of children's heights is presented in a frequency distribution table and in a histogram.

X	f
44–45	1
42–43	2
40–41	4
38–39	6
36–37	2
34–35	3
32–33	4
30–31	2

A Modified Histogram   A slight modification to the traditional histogram produces a very easy to draw and simple to understand sketch of a frequency distribution. Instead of drawing a bar above each score, the modification consists of drawing a stack of blocks. Each block represents one individual, so the number of blocks above each score corresponds to the frequency for that score. An example is shown in Figure 2.4.

Note that the number of blocks in each stack makes it very easy to see the absolute frequency for each category. In addition, it is easy to see the exact difference in frequency from one category to another. In Figure 2.4, for example, there are exactly two more people with scores of $X = 2$ than with scores of $X = 1$. Because the frequencies are clearly displayed by the number of blocks, this type of display eliminates the need for a vertical line (the Y-axis) showing frequencies. In general, this kind of graph provides a simple and concrete picture of the distribution for a sample of scores. Note that we often will use this

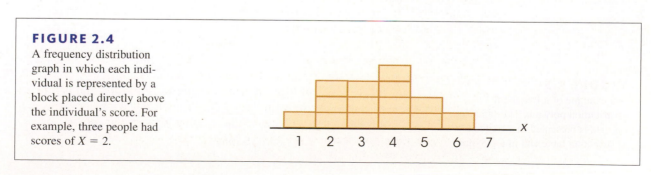

FIGURE 2.4

A frequency distribution graph in which each individual is represented by a block placed directly above the individual's score. For example, three people had scores of $X = 2$.

kind of graph to show sample data throughout the rest of the book. You should also note, however, that this kind of display simply provides a sketch of the distribution and is not a substitute for an accurately drawn histogram with two labeled axes.

Polygons   The second option for graphing a distribution of numerical scores from an interval or ratio scale of measurement is called a polygon. To construct a polygon, you begin by listing the numerical scores (the categories of measurement) along the *X*-axis. Then,

    a. A dot is centered above each score so that the vertical position of the dot corresponds to the frequency for the category.

    b. A continuous line is drawn from dot to dot to connect the series of dots.

    c. The graph is completed by drawing a line down to the *X*-axis (zero frequency) at each end of the range of scores. The final lines are usually drawn so that they reach the *X*-axis at a point that is one category below the lowest score on the left side and one category above the highest score on the right side. An example of a polygon is shown in Figure 2.5.

A polygon also can be used with data that have been grouped into class intervals. For a grouped distribution, you position each dot directly above the midpoint of the class interval. The midpoint can be found by averaging the highest and the lowest scores in the interval. For example, a class interval that is listed as 20–29 would have a midpoint of 24.5.

$$\text{midpoint} = \frac{20 + 29}{2} = \frac{49}{2} = 24.5$$

An example of a frequency distribution polygon with grouped data is shown in Figure 2.6.

■ Graphs for Nominal or Ordinal Data

When the scores are measured on a nominal or ordinal scale (usually non-numerical values), the frequency distribution can be displayed in a *bar graph*.

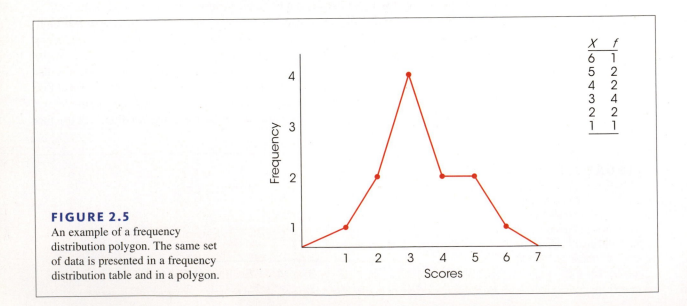

FIGURE 2.5

An example of a frequency distribution polygon. The same set of data is presented in a frequency distribution table and in a polygon.

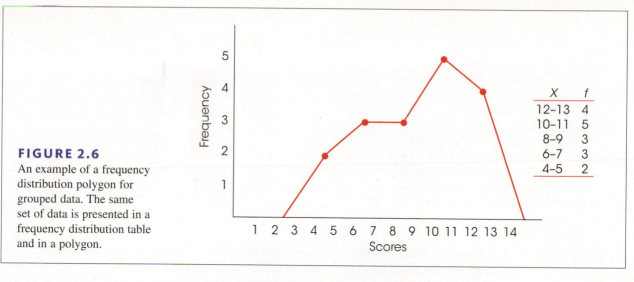

X	f
12–13	4
10–11	5
8–9	3
6–7	3
4–5	2

FIGURE 2.6
An example of a frequency distribution polygon for grouped data. The same set of data is presented in a frequency distribution table and in a polygon.

Bar Graphs   A bar graph is essentially the same as a histogram, except that spaces are left between adjacent bars. For a nominal scale, the space between bars emphasizes that the scale consists of separate, distinct categories. For ordinal scales, separate bars are used because you cannot assume that the categories are all the same size.

To construct a bar graph, list the categories of measurement along the X-axis and then draw a bar above each category so that the height of the bar corresponds to the frequency for the category. An example of a bar graph is shown in Figure 2.7.

■ Graphs for Population Distributions

When you can obtain an exact frequency for each score in a population, you can construct frequency distribution graphs that are exactly the same as the histograms, polygons, and bar graphs that are typically used for samples. For example, if a population is defined as a specific group of $N = 50$ people, we could easily determine how many have IQs of $X = 110$. However, if we were interested in the entire population of adults in the United States, it would be impossible to obtain an exact count of the number of people with an IQ of 110. Although it is still possible to construct graphs showing frequency distributions for extremely large populations, the graphs usually involve two special features: relative frequencies and smooth curves.

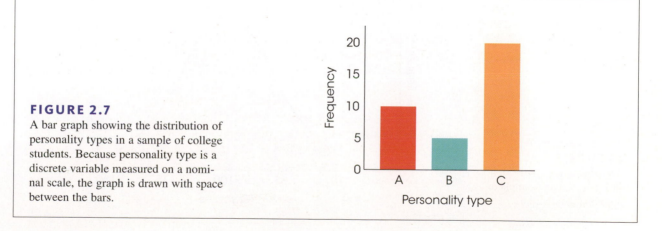

FIGURE 2.7
A bar graph showing the distribution of personality types in a sample of college students. Because personality type is a discrete variable measured on a nominal scale, the graph is drawn with space between the bars.

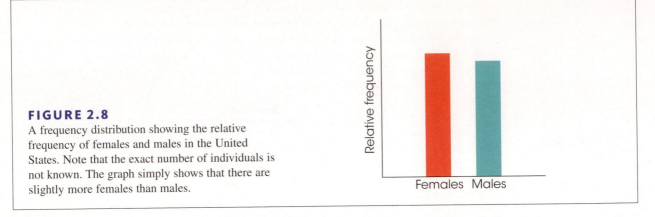

FIGURE 2.8

A frequency distribution showing the relative frequency of females and males in the United States. Note that the exact number of individuals is not known. The graph simply shows that there are slightly more females than males.

Relative Frequencies   Although you usually cannot find the absolute frequency for each score in a population, you very often can obtain *relative frequencies*. For example, no one knows the exact number of male and female human beings living in the United States because the exact numbers keep changing. However, based on past census data and general trends, we can estimate that the two numbers are very close, with women slightly outnumbering men. You can represent these relative frequencies in a bar graph by making the bar above female slightly taller than the bar above male (Figure 2.8). Notice that the graph does not show the absolute number of people. Instead, it shows the relative number of females and males.

Smooth Curves   When a population consists of numerical scores from an interval or a ratio scale, it is customary to draw the distribution with a smooth curve instead of the jagged, step-wise shapes that occur with histograms and polygons. The smooth curve indicates that you are not connecting a series of dots (real frequencies) but instead are showing the relative changes that occur from one score to the next. One commonly occurring population distribution is the normal curve. The word *normal* refers to a specific shape that can be precisely defined by an equation. Less precisely, we can describe a normal distribution as being symmetrical, with the greatest frequency in the middle and relatively smaller frequencies as you move toward either extreme. A good example of a normal distribution is the population distribution for IQ scores shown in Figure 2.9. Because normal-shaped

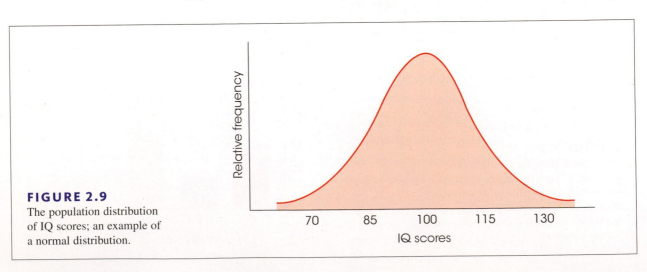

FIGURE 2.9

The population distribution of IQ scores; an example of a normal distribution.

distributions occur commonly and because this shape is mathematically guaranteed in certain situations, we give it extensive attention throughout this book.

In the future, we will be referring to *distributions of scores*. Whenever the term *distribution* appears, you should conjure up an image of a frequency distribution graph. The graph provides a picture showing exactly where the individual scores are located. To make this concept more concrete, you might find it useful to think of the graph as showing a pile of individuals just like we showed a pile of blocks in Figure 2.3. For the population of IQ scores shown in Figure 2.9, the pile is highest at an IQ score around 100 because most people have average IQs. There are only a few individuals piled up at an IQ of 130; it must be lonely at the top.

BOX 2.1 The Use and Misuse Of Graphs

Although graphs are intended to provide an accurate picture of a set of data, they can be used to exaggerate or misrepresent a set of scores. These misrepresentations generally result from failing to follow the basic rules for graph construction. The following example demonstrates how the same set of data can be presented in two entirely different ways by manipulating the structure of a graph.

For the past several years, the city has kept records of the number of homicides. The data are summarized as follows:

Year	Number of Homicides
2011	42
2012	44
2013	47
2014	49

These data are shown in two different graphs in Figure 2.10. In the first graph, we have exaggerated the height and started numbering the Y-axis at 40 rather than at zero. As a result, the graph seems to indicate a rapid rise in the number of homicides over the 4-year period. In the second graph, we have stretched out the X-axis and used zero as the starting point for the Y-axis. The result is a graph that shows little change in the homicide rate over the 4-year period.

Which graph is correct? The answer is that neither one is very good. Remember that the purpose of a graph is to provide an accurate display of the data. The first graph in Figure 2.10 exaggerates the differences between years, and the second graph conceals the differences. Some compromise is needed. Also note that in some cases a graph may not be the best way to display information. For these data, for example, showing the numbers in a table would be better than either graph.

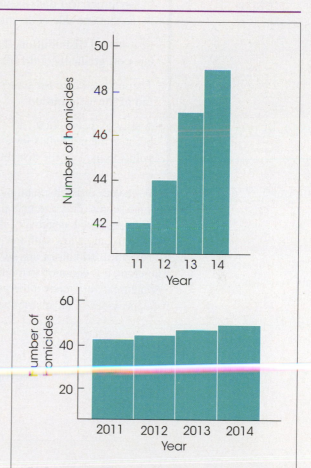

FIGURE 2.10

Two graphs showing the number of homicides in a city over a 4-year period. Both graphs show exactly the same data. However, the first graph gives the appearance that the homicide rate is high and rising rapidly.

The second graph gives the impression that the homicide rate is low and has not changed over the 4-year period.

■ The Shape of a Frequency Distribution

Rather than drawing a complete frequency distribution graph, researchers often simply describe a distribution by listing its characteristics. There are three characteristics that completely describe any distribution: shape, central tendency, and variability. In simple terms, central tendency measures where the center of the distribution is located and variability measures the degree to which the scores are spread over a wide range or are clustered together. Central tendency and variability are covered in detail in Chapters 3 and 4. Technically, the shape of a distribution is defined by an equation that prescribes the exact relationship between each X and Y value on the graph. However, we will rely on a few less-precise terms that serve to describe the shape of most distributions.

Nearly all distributions can be classified as being either symmetrical or skewed.

DEFINITIONS

In a **symmetrical distribution**, it is possible to draw a vertical line through the middle so that one side of the distribution is a mirror image of the other (Figure 2.11).

In a **skewed distribution**, the scores tend to pile up toward one end of the scale and taper off gradually at the other end (see Figure 2.11).

The section where the scores taper off toward one end of a distribution is called the **tail of the distribution**.

A skewed distribution with the tail on the right-hand side is **positively skewed** because the tail points toward the positive (above-zero) end of the X-axis. If the tail points to the left, the distribution is **negatively skewed** (see Figure 2.11).

For a very difficult exam, most scores tend to be low, with only a few individuals earning high scores. This produces a positively skewed distribution. Similarly, a very easy exam tends to produce a negatively skewed distribution, with most of the students earning high scores and only a few with low values.

Not all distributions are perfectly symmetrical or obviously skewed in one direction. Therefore, it is common to modify these descriptions of shape with phrases likely "roughly symmetrical" or "tends to be positively skewed." The goal is to provide a general idea of the appearance of the distribution.

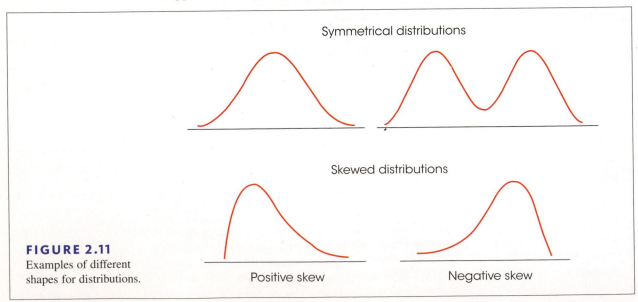

FIGURE 2.11
Examples of different shapes for distributions.

LEARNING CHECK

1. The seminar rooms in the library are identified by letters (A, B, C, and so on). A professor records the number of classes held in each room during the fall semester. If these values are presented in a frequency distribution graph, what kind of graph would be appropriate?

   a. a histogram

   b. a polygon

   c. a histogram or a polygon

   d. a bar graph

2. A group of quiz scores ranging from 4–9 are shown in a histogram. If the bars in the histogram gradually increase in height from left to right, what can you conclude about the set of quiz scores?

   a. There are more high scores than there are low scores.

   b. There are more low scores than there are high scores.

   c. The height of the bars always increases as the scores increase.

   d. None of the above

3. If a frequency distribution graph is drawn as a smooth curve, it is probably showing a _____ distribution.

   a. sample

   b. population

   c. skewed

   d. symmetrical

4. A set of scores is presented in a frequency distribution histogram. If the histogram shows a series of bars that tend to decrease in height from left to right, then what is the shape of the distribution?

   a. symmetrical

   b. positively skewed

   c. negatively skewed

   d. normal

ANSWERS   1. D, 2. A, 3. B, 4. B

2.4 | Percentiles, Percentile Ranks, and Interpolation

LEARNING OBJECTIVES

8. Define percentiles and percentile ranks.

9. Determine percentiles and percentile ranks for values corresponding to real limits in a frequency distribution table.

10. Estimate percentiles and percentile ranks using interpolation for values that do not correspond to real limits in a frequency distribution table.

Although the primary purpose of a frequency distribution is to provide a description of an entire set of scores, it also can be used to describe the position of an individual within the set. Individual scores, or *X* values, are called raw scores. By themselves, raw scores do not provide much information. For example, if you are told that your score on an exam is $X = 43$, you cannot tell how well you did relative to other students in the class. To evaluate your score, you need more information, such as the average score or the number of people who had scores above and below you. With this additional information, you would be able to determine your relative position in the class. Because raw scores do not provide much information, it is desirable to transform them into a more meaningful form. One transformation that we will consider changes raw scores into percentiles.

DEFINITIONS

> The **rank** or **percentile rank** of a particular score is defined as the percentage of individuals in the distribution with scores at or below the particular value.
>
> When a score is identified by its percentile rank, the score is called a **percentile**.

Suppose, for example, that you have a score of $X = 43$ on an exam and that you know that exactly 60% of the class had scores of 43 or lower. Then your score $X = 43$ has a percentile rank of 60%, and your score would be called the 60th percentile. Notice that *percentile rank* refers to a percentage and that *percentile* refers to a score. Also notice that your rank or percentile describes your exact position within the distribution.

■ Cumulative Frequency and Cumulative Percentage

To determine percentiles or percentile ranks, the first step is to find the number of individuals who are located at or below each point in the distribution. This can be done most easily with a frequency distribution table by simply counting the number of scores that are in or below each category on the scale. The resulting values are called *cumulative frequencies* because they represent the accumulation of individuals as you move up the scale.

EXAMPLE 2.6   In the following frequency distribution table, we have included a cumulative frequency column headed by *cf*. For each row, the cumulative frequency value is obtained by adding up the frequencies in and below that category. For example, the score $X = 3$ has a cumulative frequency of 14 because exactly 14 individuals had scores of $X = 3$ or less.

X	F	cf
5	1	20
4	5	19
3	8	14
2	4	6
1	2	2

The cumulative frequencies show the number of individuals located at or below each score. To find percentiles, we must convert these frequencies into percentages. The resulting values are called *cumulative percentages* because they show the percentage of individuals who are accumulated as you move up the scale.

EXAMPLE 2.7   This time we have added a cumulative percentage column (c%) to the frequency distribution table from Example 2.6. The values in this column represent the percentage of individuals who are located in and below each category. For example, 70% of the individuals (14 out of 20) had scores of $X = 3$ or lower. Cumulative percentages can be computed by

$$c\% = \frac{cf}{N}(100\%)$$

X	f	Cf	c%
5	1	20	100%
4	5	19	95%
3	8	14	70%
2	4	6	30%
1	2	2	10%

The cumulative percentages in a frequency distribution table give the percentage of individuals with scores at or below each X value. However, you must remember that the X values in the table are usually measurements of a continuous variable and, therefore, represent intervals on the scale of measurement (see page 20). A score of $X = 2$, for example, means that the measurement was somewhere between the real limits of 1.5 and 2.5. Thus, when a table shows that a score of $X = 2$ has a cumulative percentage of 30%, you should interpret this as meaning that 30% of the individuals have been accumulated by the time you reach the top of the interval for $X = 2$. Notice that each cumulative percentage value is associated with the upper real limit of its interval. This point is demonstrated in Figure 2.12, which shows the same data that were used in Example 2.7. Figure 2.12 shows that two people, or 10%, had scores of $X = 1$; that is, two people had scores between 0.5 and 1.5. You cannot be sure that both individuals have been accumulated until you reach 1.5, the upper real limit of the interval. Similarly, a cumulative percentage of 30% is reached at 2.5 on the scale, a percentage of 70% is reached at 3.5, and so on.

■ Interpolation

It is possible to determine some percentiles and percentile ranks directly from a frequency distribution table, provided the percentiles are upper real limits and the ranks are

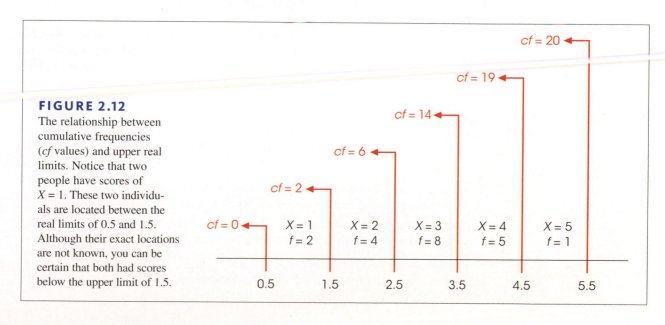

FIGURE 2.12

The relationship between cumulative frequencies (*cf* values) and upper real limits. Notice that two people have scores of $X = 1$. These two individuals are located between the real limits of 0.5 and 1.5. Although their exact locations are not known, you can be certain that both had scores below the upper limit of 1.5.

percentages that appear in the table. Using the table in Example 2.7, for example, you should be able to answer the following questions:

1. What is the 95th percentile? (Answer: $X = 4.5$.)
2. What is the percentile rank for $X = 3.5$? (Answer: 70%.)

However, there are many values that do not appear directly in the table, and it is impossible to determine these values precisely. Referring to the table in Example 2.7 again,

1. What is the 50th percentile?
2. What is the percentile rank for $X = 4$?

Because these values are not specifically reported in the table, you cannot answer the questions. However, it is possible to estimate these intermediate values by using a procedure known as *interpolation*.

Before we apply the process of interpolation to percentiles and percentile ranks, we will use a simple, commonsense example to introduce this method. Suppose that your friend offers you $60 to work for 8 hours on Saturday helping with spring cleaning in the house and yard. On Saturday morning, however, you realize that you have an appointment in the afternoon and will have to quit working after only 4 hours. What is a fair amount for your friend to pay you for 6 hours of work? Because this is a working-for-pay example, your automatic response probably is to calculate the hourly rate of pay —how many dollars per hour you are getting. However, the process of interpolation offers an alternative method for finding the answer. We begin by noting that the original total amount of time was 8 hours. You worked 4 hours, which corresponds to $\frac{1}{2}$ of the total time. Therefore, a fair payment would be $\frac{1}{2}$ of the original total amount: $\frac{1}{2}$ of $60 is $30.

The process of interpolation is pictured in Figure 2.13. In the figure, the line on the left shows the time for your agreed work, from 0 up to 8 hours, and the line on the right shows the agreed pay, from 0 to $60. We also have marked different fractions along the way. Using the figure, try answering the following questions about time and pay.

1. How long should you work to earn $45?
2. How much should you be paid after working 2 hours?

If you got answers of 6 hours and $15, you have mastered the process of interpolation.

Notice that interpolation provides a method for finding intermediate values—that is, values that are located between two specified numbers. This is exactly the problem we faced with percentiles and percentile ranks. Some values are given in the table, but others are not. Also notice that interpolation only *estimates* the intermediate values. The basic assumption underlying interpolation is that there is a constant rate of change from one end of the interval to the other. In the working-for-pay example, we assume a constant rate of pay for the entire job. Because interpolation is based on this assumption, the values we calculate are only estimates. The general process of interpolation can be summarized as follows:

1. A single interval is measured on two separate scales (for example, time and dollars). The endpoints of the interval are known for each scale.
2. You are given an intermediate value on one of the scales. The problem is to find the corresponding intermediate value on the other scale.
3. The interpolation process requires four steps:
   a. Find the width of the interval on both scales.

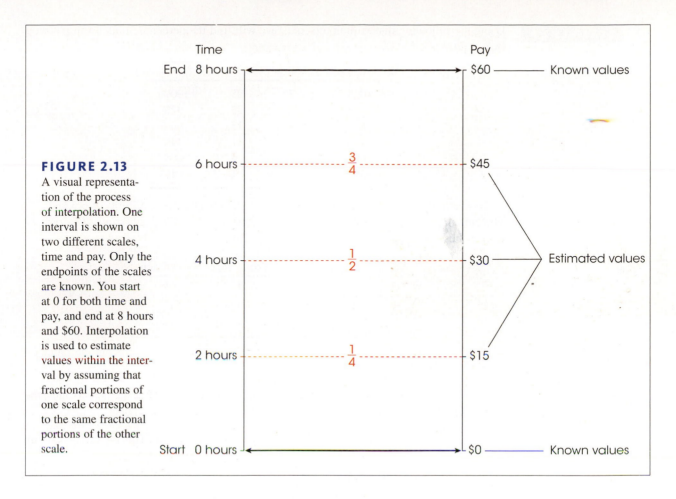

FIGURE 2.13
A visual representation of the process of interpolation. One interval is shown on two different scales, time and pay. Only the endpoints of the scales are known. You start at 0 for both time and pay, and end at 8 hours and $60. Interpolation is used to estimate values within the interval by assuming that fractional portions of one scale correspond to the same fractional portions of the other scale.

b. Locate the position of the intermediate value in the interval. This position corresponds to a fraction of the whole interval:

$$\text{fraction} = \frac{\text{distance from the top of the interval}}{\text{interval width}}$$

You may notice that in each of these problems we use interpolation working from the *top* of the interval. However, this choice is arbitrary, and you should realize that interpolation can be done just as easily working from the bottom of the interval.

c. Use the same fraction to determine the corresponding position on the other scale. First, use the fraction to determine the distance from the top of the interval:

$$\text{distance} = (\text{fraction}) \times (\text{width})$$

d. Use the distance from the top to determine the position on the other scale.

The following examples demonstrate the process of interpolation as it is applied to percentiles and percentile ranks. The key to success in solving these problems is that each cumulative percentage in the table is associated with the upper real limit of its score interval.

EXAMPLE 2.8   Using the following distribution of scores, we will find the percentile rank corresponding to $X = 7.0$:

X	f	cf	c%
10	2	25	100%
9	8	23	92%
8	4	15	60%
7	6	11	44%
6	4	5	20%
5	1	1	4%

Notice that $X = 7.0$ is located in the interval bounded by the real limits of 6.5 and 7.5. The cumulative percentages corresponding to these real limits are 20% and 44%, respectively. These values are shown in the following table:

	Scores (X)	Percentages
Top	7.5	44%
Intermediate value →	7.0	?
Bottom	6.5	20%

For interpolation problems, it is always helpful to create a table showing the range on both scales.

STEP 1   For the scores, the width of the interval is 1 point (from 6.5–7.5). For the percentages, the width is 24 points (from 20–44%).

STEP 2   Our particular score is located 0.5 point from the top of the interval. This is exactly halfway down in the interval.

STEP 3   On the percentage scale, halfway down is

$$\frac{1}{2} \,(24 \text{ points}) = 12 \text{ points}$$

STEP 4   For the percentages, the top of the interval is 44%, so 12 points down would be
$$44\% - 12\% = 32\%$$
This is the answer. A score of $X = 7.0$ corresponds to a percentile rank of 32%   ■

This same interpolation procedure can be used with data that have been grouped into class intervals. Once again, you must remember that the cumulative percentage values are associated with the upper real limits of each interval. The following example demonstrates the calculation of percentiles and percentile ranks using data in a grouped frequency distribution.

EXAMPLE 2.9   Using the following distribution of scores, we will use interpolation to find the 40th percentile:

X	F	cf	c%
20–24	2	20	100%
15–19	3	18	90%
10–14	3	15	75%
5–9	10	12	60%
0–4	2	2	10%

A percentage value of 40% is not given in the table; however, it is located between 10% and 60%, which are given. These two percentage values are associated with the upper real limits of 4.5 and 9.5, respectively. These values are shown in the following table:

	Scores (X)	Percentages	
Top	9.5	60%	
	?	40%	← Intermediate value
Bottom	4.5	10%	

STEP 1   For the scores, the width of the interval is 5 points. For the percentages, the width is 50 points.

STEP 2   The value of 40% is located 20 points from the top of the percentage interval. As a fraction of the whole interval, this is 20 out of 50, or $\frac{2}{5}$ of the total interval.

STEP 3   Using this same fraction for the scores, we obtain a distance of

$$\frac{2}{5} \ (5 \text{ points}) = 2 \text{ points}$$

The location we want is 2 points down from the top of the score interval.

STEP 4   Because the top of the interval is 9.5, the position we want is

$$9.5 - 2 = 7.5$$

This is the answer. The 40th percentile is $X = 7.5$.   ■

The following example is an opportunity for you to test your understanding by doing interpolation yourself.

EXAMPLE 2.10   Using the frequency distribution table in Example 2.9, use interpolation to find the percentile rank for $X = 9.0$. You should obtain 55%. Good luck.   ■

LEARNING CHECK

1. In a distribution of exam scores, which of the following would be the highest score?
   a. the 20th percentile
   b. the 80th percentile
   c. a score with a percentile rank of 15%
   d. a score with a percentile rank of 75%

2. Following are three rows from a frequency distribution table. For this distribution, what is the 90th percentile?
   a. $X = 24.5$
   b. $X = 25$
   c. $X = 29$
   d. $X = 29.5$

X	c%
30–34	100%
25–29	90%
20–24	60%

3. Following are three rows from a frequency distribution table. Using interpolation, what is the percentile rank for $X = 18$?

a. 52.5%

b. 30%

c. 29%

d. 25%

X	c%
20–24	60%
15–19	35%
10–14	15%

ANSWERS   1. B, **2.** D, **3.** C

2.5 | Stem and Leaf Displays

11. Describe the basic elements of a stem and leaf display and explain how the display shows the entire distribution of scores.

In 1977, J.W. Tukey presented a technique for organizing data that provides a simple alternative to a grouped frequency distribution table or graph (Tukey, 1977). This technique, called a *stem and leaf display*, requires that each score be separated into two parts: The first digit (or digits) is called the *stem*, and the last digit is called the *leaf*. For example, $X = 85$ would be separated into a stem of 8 and a leaf of 5. Similarly, $X = 42$ would have a stem of 4 and a leaf of 2. To construct a stem and leaf display for a set of data, the first step is to list all the stems in a column. For the data in Table 2.3, for example, the lowest scores are in the 30s and the highest scores are in the 90s, so the list of stems would be

Stems
3
4
5
6
7
8
9

The next step is to go through the data, one score at a time, and write the leaf for each score beside its stem. For the data in Table 2.3, the first score is $X = 83$, so you would write 3 (the leaf) beside the 8 in the column of stems. This process is continued for the entire set of scores. The complete stem and leaf display is shown with the original data in Table 2.3.

■ Comparing Stem and Leaf Displays with Frequency Distributions

Notice that the stem and leaf display is very similar to a grouped frequency distribution. Each of the stem values corresponds to a class interval. For example, the stem 3 represents

TABLE 2.3

A set of $N = 24$ scores presented as raw data and organized in a stem and leaf display.

Data			Stem and Leaf Display	
83	82	63	3	23
62	93	78	4	26
71	68	33	5	6279
76	52	97	6	283
85	42	46	7	1643846
32	57	59	8	3521
56	73	74	9	37
74	81	76		

all scores in the 30s—that is, all scores in the interval 30–39. The number of leaves in the display shows the frequency associated with each stem. It also should be clear that the stem and leaf display has one important advantage over a traditional grouped frequency distribution. Specifically, the stem and leaf display allows you to identify every individual score in the data. In the display shown in Table 2.3, for example, you know that there were three scores in the 60s and that the specific values were 62, 68, and 63. A frequency distribution would tell you only the frequency, not the specific values. This advantage can be very valuable, especially if you need to do any calculations with the original scores. For example, if you need to add all the scores, you can recover the actual values from the stem and leaf display and compute the total. With a grouped frequency distribution, however, the individual scores are not available.

LEARNING CHECK

1. For the scores shown in the following stem and leaf display, what is the lowest score in the distribution?   9 | 374

    a. 7   8 | 945

    b. 15   7 | 7042

    c. 50   6 | 68

    d. 51   5 | 14

2. For the scores shown in the following stem and leaf display, how many people had scores in the 70s?   9 | 374

    a. 1   8 | 945

    b. 2   7 | 7042

    c. 3   6 | 68

    d. 4   5 | 14

ANSWERS   **1.** D, **2.** D

SUMMARY

1. The goal of descriptive statistics is to simplify the organization and presentation of data. One descriptive technique is to place the data in a frequency distribution table or graph that shows exactly how many individuals (or scores) are located in each category on the scale of measurement.

2. A frequency distribution table lists the categories that make up the scale of measurement (the X values) in one column. Beside each X value, in a second column, is the frequency or number of individuals in that category. The table may include a proportion column showing the relative frequency for each category:

$$\text{proportion} = p = \frac{f}{n}$$

   The table may include a percentage column showing the percentage associated with each X value:

$$\text{percentage} = p(100) = \frac{f}{n}(100)$$

3. It is recommended that a frequency distribution table have a maximum of 10–15 rows to keep it simple. If the scores cover a range that is wider than this suggested maximum, it is customary to divide the range into sections called class intervals. These intervals are then listed in the frequency distribution table along with the frequency or number of individuals with scores in each interval. The result is called a grouped frequency distribution. The guidelines for constructing a grouped frequency distribution table are as follows:

   a. There should be about 10 intervals.

   b. The width of each interval should be a simple number (e.g., 2, 5, or 10).

   c. The bottom score in each interval should be a multiple of the width.

   d. All intervals should be the same width, and they should cover the range of scores with no gaps.

4. A frequency distribution graph lists scores on the horizontal axis and frequencies on the vertical axis. The type of graph used to display a distribution depends on the scale of measurement used. For interval or ratio scales, you should use a histogram or a polygon. For a histogram, a bar is drawn above each score so that the height of the bar corresponds to the frequency. Each bar extends to the real limits of the score, so that adjacent bars touch. For a polygon, a dot is placed above the midpoint of each score or class interval so that the height of the dot corresponds to the frequency; then lines are drawn to connect the dots. Bar graphs are used with nominal or ordinal scales. Bar graphs are similar to histograms except that gaps are left between adjacent bars.

5. Shape is one of the basic characteristics used to describe a distribution of scores. Most distributions can be classified as either symmetrical or skewed. A skewed distribution that tails off to the right is positively skewed. If it tails off to the left, it is negatively skewed.

6. The cumulative percentage is the percentage of individuals with scores at or below a particular point in the distribution. The cumulative percentage values are associated with the upper real limits of the corresponding scores or intervals.

7. Percentiles and percentile ranks are used to describe the position of individual scores within a distribution. Percentile rank gives the cumulative percentage associated with a particular score. A score that is identified by its rank is called a percentile.

8. When a desired percentile or percentile rank is located between two known values, it is possible to estimate the desired value using the process of interpolation. Interpolation assumes a regular linear change between the two known values.

9. A stem and leaf display is an alternative procedure for organizing data. Each score is separated into a stem (the first digit or digits) and a leaf (the last digit). The display consists of the stems listed in a column with the leaf for each score written beside its stem. A stem and leaf display is similar to a grouped frequency distribution table, however the stem and leaf display identifies the exact value of each score and the grouped frequency distribution does not.

KEY TERMS

frequency distribution (35)	bar graph (45)	percentile rank (50)
range (38)	relative frequency (46)	cumulative frequency (cf) (61)
grouped frequency distribution (39)	symmetrical distribution (48)	cumulative percentage (c%) (50)
class interval (39)	tail(s) of a distribution (48)	interpolation (51)
apparent limits (41)	positively skewed distribution (48)	stem and leaf display (56)
histogram (42)	negatively skewed distribution (48)	
polygon (44)	percentile (50)	

General instructions for using SPSS are presented in Appendix D. Following are detailed instructions for using SPSS to produce **Frequency Distribution Tables or Graphs**.

FREQUENCY DISTRIBUTION TABLES

Data Entry

1. Enter all the scores in one column of the data editor, probably VAR00001.

Data Analysis

1. Click **Analyze** on the tool bar, select **Descriptive Statistics**, and click on **Frequencies**.
2. Highlight the column label for the set of scores (VAR00001) in the left box and click the arrow to move it into the **Variable** box.
3. Be sure that the option to **Display Frequency Table** is selected.
4. Click **OK**.

SPSS Output

The frequency distribution table will list the score values in a column from smallest to largest, with the percentage and cumulative percentage also listed for each score. Score values that do not occur (zero frequencies) are not included in the table, and the program does not group scores into class intervals (all values are listed).

FREQUENCY DISTRIBUTION HISTOGRAMS OR BAR GRAPHS

Data Entry

1. Enter all the scores in one column of the data editor, probably VAR00001.

Data Analysis

1. Click **Analyze** on the tool bar, select **Descriptive Statistics**, and click on **Frequencies**.
2. Highlight the column label for the set of scores (VAR00001) in the left box and click the arrow to move it into the **Variable** box.
3. Click **Charts**.
4. Select either **Bar Graphs** or **Histogram**.
5. Click **Continue**.
6. Click **OK**.

SPSS Output

SPSS will display a frequency distribution table and a graph. Note that SPSS often produces a histogram that groups the scores in unpredictable intervals. A bar graph usually produces a clearer picture of the actual frequency associated with each score.

FOCUS ON PROBLEM SOLVING

1. When constructing or working with a grouped frequency distribution table, a common mistake is to calculate the interval width by using the highest and lowest values that define each interval. For example, some students are tricked into thinking that an

interval identified as 20–24 is only 4 points wide. To determine the correct interval width, you can

a. Count the individual scores in the interval. For this example, the scores are 20, 21, 22, 23, and 24 for a total of 5 values. Thus, the interval width is 5 points.

b. Use the real limits to determine the real width of the interval. For example, an interval identified as 20–24 has a lower real limit of 19.5 and an upper real limit of 24.5 (halfway to the next score). Using the real limits, the interval width is

$$24.5 - 19.5 = 5 \text{ points}$$

2. Percentiles and percentile ranks are intended to identify specific locations within a distribution of scores. When solving percentile problems, especially with interpolation, it is helpful to sketch a frequency distribution graph. Use the graph to make a preliminary estimate of the answer before you begin any calculations. For example, to find the 60th percentile, you would want to draw a vertical line through the graph so that slightly more than half (60%) of the distribution is on the left-hand side of the line. Locating this position in your sketch will give you a rough estimate of what the final answer should be. When doing interpolation problems, you should keep several points in mind:

a. Remember that the cumulative percentage values correspond to the upper real limits of each score or interval.

b. You should always identify the interval with which you are working. The easiest way to do this is to create a table showing the endpoints on both scales (scores and cumulative percentages). This is illustrated in Example 2.8 on page 54.

DEMONSTRATION 2.1

A GROUPED FREQUENCY DISTRIBUTION TABLE

For the following set of $N = 20$ scores, construct a grouped frequency distribution table using an interval width of 5 points. The scores are:

14, 8, 27, 16, 10, 22, 9, 13, 16, 12,
10, 9, 15, 17, 6, 14, 11, 18, 14, 11

STEP 1 Set up the class intervals.

The largest score in this distribution is $X = 27$, and the lowest is $X = 6$. Therefore, a frequency distribution table for these data would have 22 rows and would be too large. A grouped frequency distribution table would be better. We have asked specifically for an interval width of five points, and the resulting table has five rows.

X
25–29
20–24
15–19
10–14
5–9

Remember that the interval width is determined by the real limits of the interval. For example, the class interval 25–29 has an upper real limit of 29.5 and a lower real limit of 24.5. The difference between these two values is the width of the interval—namely, 5.

STEP 2   Determine the frequencies for each interval.

Examine the scores, and count how many fall into the class interval of 25–29. Cross out each score that you have already counted. Record the frequency for this class interval. Now repeat this process for the remaining intervals. The result is the following table:

X	f	
25–29	1	(the score X = 27)
20–24	1	(X = 22)
15–19	5	(the scores X = 16, 16, 15, 17, and 18)
10–14	9	(X = 14, 10, 13, 12, 10, 14, 11, 14, and 11)
5–9	4	(X = 8, 9, 9, and 6)

DEMONSTRATION 2.2

USING INTERPOLATION TO FIND PERCENTILES AND PERCENTILE RANKS

Find the 50th percentile for the set of scores in the grouped frequency distribution table that was constructed in Demonstration 2.1.

STEP 1   Find the cumulative frequency (cf) and cumulative percentage values, and add these values to the basic frequency distribution table.

Cumulative frequencies indicate the number of individuals located in or below each category (class interval). To find these frequencies, begin with the bottom interval, and then accumulate the frequencies as you move up the scale. For this example, there are 4 individuals who are in or below the 5–9 interval ($cf = 4$). Moving up the scale, the 10–14 interval contains an additional 9 people, so the cumulative value for this interval is $9 + 4 = 13$ (simply add the 9 individuals in the interval to the 4 individuals below). Continue moving up the scale, cumulating frequencies for each interval.

Cumulative percentages are determined from the cumulative frequencies by the relationship

$$c\% = \left(\frac{cf}{N}\right)100\%$$

For example, the cf column shows that 4 individuals (out of the total set of $N = 20$) have scores in or below the 5–9 interval. The corresponding cumulative percentage is

$$c\% = \left(\frac{4}{20}\right)100\% = \left(\frac{1}{5}\right)100\% = 20\%$$

The complete set of cumulative frequencies and cumulative percentages is shown in the following table:

X	f	cf	c%
25–29	1	20	100%
20–24	1	19	95%
15–19	5	18	90%
10–14	9	13	65%
5–9	4	4	20%

STEP 2 Locate the interval that contains the value that you want to calculate.

We are looking for the 50th percentile, which is located between the values of 20% and 65% in the table. The scores (upper real limits) corresponding to these two percentages are 9.5 and 14.5, respectively. The interval, measured in terms of scores and percentages, is shown in the following table:

X	$c\%$
14.5	65%
??	50%
9.5	20%

STEP 3 Locate the intermediate value as a fraction of the total interval.

Our intermediate value is 50%, which is located in the interval between 65% and 20%. The total width of the interval is 45 points ($65 - 20 = 45$), and the value of 50% is located 15 points down from the top of the interval. As a fraction, the 50th percentile is located $\frac{15}{45} = \frac{1}{3}$ down from the top of the interval.

STEP 4 Use the fraction to determine the corresponding location on the other scale.

Our intermediate value, 50%, is located $\frac{1}{3}$ of the way down from the top of the interval. Our goal is to find the score, the X value, that also is located $\frac{1}{3}$ of the way down from the top of the interval.

On the score (X) side of the interval, the top value is 14.5, and the bottom value is 9.5, so the total interval width is 5 points ($14.5 - 9.5 = 5$). The position we are seeking is $\frac{1}{3}$ of the way from the top of the interval. One-third of the total interval is

$$\left(\frac{1}{3}\right)5 = \frac{5}{3} = 1.67 \text{ points}$$

To find this location, begin at the top of the interval, and come down 1.67 points:

$$14.5 - 1.67 = 12.83$$

This is our answer. The 50th percentile is $X = 12.83$.

PROBLEMS

1. Place the following set of $n = 20$ scores in a frequency distribution table.

6, 2, 2, 1, 3, 2, 4, 7, 1, 2
5, 3, 1, 6, 2, 6, 3, 3, 7, 2

2. Construct a frequency distribution table for the following set of scores. Include columns for proportion and percentage in your table.

Scores: 2, 7, 5, 3, 2, 9, 6, 1, 1, 2
3, 3, 2, 4, 5, 2, 5, 4, 6, 5

3. Find each value requested for the distribution of scores in the following table.

a. n
b. ΣX
c. ΣX^2

X	f
5	1
4	3
3	4
2	5
1	2

4. Find each value requested for the distribution of scores in the following table.

  a. n

  b. ΣX

  c. ΣX^2

X	f
6	1
5	2
4	2
3	4
2	3
1	2

5. For the following scores, the smallest value is $X = 13$ and the largest value is $X = 52$. Place the scores in a grouped frequency distribution table,

  a. Using an interval width of 5 points.

  b. Using an interval width of 10 points.

  44, 19, 23, 17, 25, 47, 32, 26

  25, 30, 18, 24, 49, 51, 24, 13

  43, 27, 34, 16, 52, 18, 36, 25

6. The following scores are the ages for a random sample of $n = 30$ drivers who were issued speeding tickets in New York during 2008. Determine the best interval width and place the scores in a grouped frequency distribution table. From looking at your table, does it appear that tickets are issued equally across age groups?

  17, 30, 45, 20, 39, 53, 28, 19,

  24, 21, 34, 38, 22, 29, 64,

  22, 44, 36, 16, 56, 20, 23, 58,

  32, 25, 28, 22, 51, 26, 43

7. For each of the following samples, determine the interval width that is most appropriate for a grouped frequency distribution and identify the approximate number of intervals needed to cover the range of scores.

  a. Sample scores range from $X = 8$ to $X = 41$

  b. Sample scores range from $X = 16$ to $X = 33$

  c. Sample scores range from $X = 26$ to $X = 98$

8. What information is available about the scores in a regular frequency distribution table that you cannot obtain for the scores in a grouped table?

9. Describe the difference in appearance between a bar graph and a histogram and describe the circumstances in which each type of graph is used.

10. For the following set of scores:

  8,   5,   9,   6,   8,   7,   4,   10,   6,   7

  9,   7,   9,   9,   5,   8,   8,   6,   7,   10

  a. Construct a frequency distribution table to organize the scores.

  b. Draw a frequency distribution histogram for these data.

11. A survey given to a sample of college students contained questions about the following variables. For each variable, identify the kind of graph that should be used to display the distribution of scores (histogram, polygon, or bar graph).

  a. Number of brothers and sisters

  b. Birth-order position among siblings (oldest = 1st)

  c. Gender (male/female)

  d. Favorite television show during the previous year

12. Gaucher, Friesen, and Kay (2010) found that masculine-themed words (such as competitive, independent, analyze, strong) are commonly used in job recruitment materials, especially for job advertisements in male-dominated areas. In a similar study, a researcher counted the number of masculine-themed words in job advertisements for job areas, and obtained the following data.

Area	Number of Masculine Words
Plumber	14
Electrician	12
Security guard	17
Bookkeeper	9
Nurse	6
Early-childhood educator	7

Determine what kind of graph would be appropriate for showing this distribution and sketch the frequency distribution graph.

13. Find each of the following values for the distribution shown in the following polygon.
    a. n
    b. ΣX
    c. ΣX^2

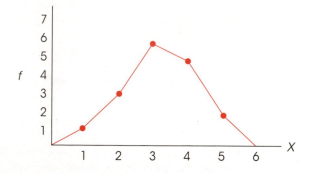

14. Place the following scores in a frequency distribution table. Based on the frequencies, what is the shape of the distribution?

    13, 14, 12, 15, 15, 14, 15, 11, 13, 14
    11, 13, 15, 12, 14, 14, 10, 14, 13, 15

15. For the following set of scores:

    8, 6, 7, 5, 4, 10, 8, 9, 5, 7, 2, 9
    9, 10, 7, 8, 8, 7, 4, 6, 3, 8, 9, 6

    a. Construct a frequency distribution table.
    b. Sketch a histogram showing the distribution.
    c. Describe the distribution using the following characteristics:
        (1) What is the shape of the distribution?
        (2) What score best identifies the center (average) for the distribution?
        (3) Are the scores clustered together, or are they spread out across the scale?

16. Recent research suggests that the amount of time that parents spend talking about numbers can have a big impact on the mathematical development of their children (Levine, Suriyakham, Rowe, Huttenlocher, & Gunderson, 2010). In the study, the researchers visited the children's homes between the ages of 14 and 30 months and recorded the amount of "number talk" they heard from the children's parents. The researchers then tested the children's knowledge of the meaning of numbers at 46 months. The following data are similar to the results obtained in the study.

Children's Knowledge-of-Numbers Scores for Two Groups of Parents	
Low Number-Talk Parents	High Number-Talk Parents
2, 1, 2, 3, 4	3, 4, 5, 4, 5
3, 3, 2, 2, 1	4, 2, 3, 5, 4
5, 3, 4, 1, 2	5, 3, 4, 5, 4

Sketch a polygon showing the frequency distribution for children with low number-talk parents. In the same graph, sketch a polygon showing the scores for the children with high number-talk parents. (Use two different colors or use a solid line for one polygon and a dashed line for the other.) Does it appear that there is a difference between the two groups?

17. Complete the final two columns in the following frequency distribution table and then find the percentiles and percentile ranks requested:

X	f	cf	c%
5	2		
4	5		
3	6		
2	4		
1	3		

    a. What is the percentile rank for $X = 2.5$?
    b. What is the percentile rank for $X = 4.5$?
    c. What is the 15th percentile?
    d. What is the 65th percentile?

18. Complete the final two columns in the following frequency distribution table and then find the percentiles and percentile ranks requested:

X	f	cf	c%
25–29	1		
20–24	4		
15–19	8		
10–14	7		
5–9	3		
0–4	2		

    a. What is the percentile rank for $X = 9.5$?
    b. What is the percentile rank for $X = 19.5$?
    c. What is the 48th percentile?
    d. What is the 96th percentile?

19. The following table shows four rows from a frequency distribution table for a sample of $n = 25$ scores. Use interpolation to find the percentiles and percentile ranks requested:

X	f	cf	c%
8	3	18	72
7	6	15	60
6	5	9	36
5	2	4	16

   a. What is the percentile rank for $X = 6$?
   b. What is the percentile rank for $X = 7$?
   c. What is the 20th percentile?
   d. What is the 66th percentile?

20. The following table shows four rows from a frequency distribution table for a sample of $n = 50$ scores. Use interpolation to find the percentiles and percentile ranks requested:

X	f	cf	c%
15	5	32	64
14	8	27	54
13	6	19	38
12	4	13	26

   a. What is the percentile rank for $X = 13$?
   b. What is the percentile rank for $X = 15$?
   c. What is the 50th percentile?
   d. What is the 60th percentile?

21. The following table shows four rows from a frequency distribution table for a sample of $n = 50$ scores. Use interpolation to find the percentiles and percentile ranks requested:

X	f	cf	c%
15–19	3	50	100
10–14	6	47	94
5–9	8	41	82
0–4	18	33	66

   a. What is the percentile rank for $X = 17$?
   b. What is the percentile rank for $X = 6$?
   c. What is the 70th percentile?
   d. What is the 90th percentile?

22. The following table shows four rows from a frequency distribution table for a sample of $n = 20$ scores. Use interpolation to find the percentiles and percentile ranks requested:

X	f	Cf	C%
40–49	4	20	100
30–39	7	16	80
20–29	4	9	45
10–19	3	5	25

   a. Find the 30th percentile.
   b. Find the 52nd percentile.
   c. What is the percentile rank for $X = 46$?
   d. What is the percentile rank for $X = 21$?

23. Construct a stem and leaf display for the data in problem 5 using one stem for the scores in the 50s, one for scores in the 40s, and so on.

24. A set of scores has been organized into the following stem and leaf display. For this set of scores:
   a. How many scores are in the 70s?
   b. Identify the individual scores in the 70s.
   c. How many scores are in the 40s?
   d. Identify the individual scores in the 40s.

4	8
5	421
6	3824
7	592374
8	24316
9	275

25. Use a stem and leaf display to organize the following distribution of scores. Use six stems with each stem corresponding to a 10-point interval.
   Scores:

36,	47,	14,	19,	65
52,	47,	42,	11,	25
28,	39,	32,	34,	58
57,	22,	49,	22,	16
33,	37,	23,	55,	44

Central Tendency

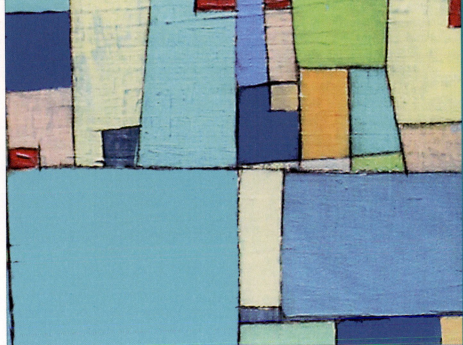

© Deborah Batt

Tools You Will Need

The following items are considered essential background material for this chapter. If you doubt your knowledge of any of these items, you should review the appropriate chapter or section before proceeding.

- Summation notation (Chapter 1)
- Frequency distributions (Chapter 2)

Research has now confirmed what you already suspected to be true—alcohol consumption increases the attractiveness of opposite-sex individuals (Jones, Jones, Thomas, and Piper, 2003). In the study, college-age participants were recruited from bars and restaurants near campus and asked to participate in a "market research" study. During the introductory conversation, they were asked to report their alcohol consumption for the day and were told that moderate consumption would not prevent them from taking part in the study. Participants were then shown a series of photographs of male and female faces and asked to rate the attractiveness of each face on a 1–7 scale. Figure 3.1 shows the general pattern of results obtained in the study. The two polygons in the figure show the distributions of attractiveness ratings for one male photograph obtained from two groups of females: those who had no alcohol and those with moderate alcohol consumption. Note that the attractiveness ratings from the alcohol group are noticeably higher than the ratings from the no-alcohol group. Incidentally, the same pattern of results was obtained for the male's ratings of female photographs.

Although it seems obvious that the alcohol-based ratings are noticeably higher than the no-alcohol ratings, this conclusion is based on a general impression, or a subjective interpretation, of the figure. In fact, this conclusion is not always true. For example, there is overlap between the two groups so that some of the no-alcohol females actually rate the photograph as more attractive than some of the alcohol-consuming females. What we need is a method to summarize each group as a whole so that we can objectively describe how much difference exists between the two groups.

The solution to this problem is to identify the typical or average rating as the representative for each group.

Then the research results can be described by saying that the typical rating for the alcohol group is higher than the typical rating for the no-alcohol group. On average, the male in the photograph really is seen as more attractive by the alcohol-consuming females.

In this chapter we introduce the statistical techniques used to identify the typical or average score for a distribution. Although there are several reasons for defining the average score, the primary advantage of an average is that it provides a single number that describes an entire distribution and can be used for comparison with other distributions.

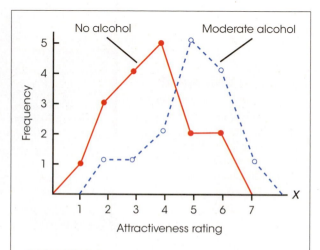

FIGURE 3.1

Frequency distributions for ratings of attractiveness of a male face shown in a photograph for two groups of female participants: those who had consumed no alcohol and those who had consumed moderate amounts of alcohol.

3.1 | Overview

The general purpose of descriptive statistical methods is to organize and summarize a set of scores. Perhaps the most common method for summarizing and describing a distribution is to find a single value that defines the average score and can serve as a typical example to represent the entire distribution. In statistics, the concept of an average or representative score is called *central tendency*. The goal in measuring central tendency is to describe a distribution of scores by determining a single value that identifies the center of the distribution. Ideally, this central value will be the score that is the best representative value for all of the individuals in the distribution.

| DEFINITION | **Central tendency** is a statistical measure to determine a single score that defines the center of a distribution. The goal of central tendency is to find the single score that is most typical or most representative of the entire group. |

In everyday language, central tendency attempts to identify the "average" or "typical" individual. This average value can then be used to provide a simple description of an entire population or a sample. In addition to describing an entire distribution, measures of central tendency are also useful for making comparisons between groups of individuals or between sets of data. For example, weather data indicate that for Seattle, Washington, the average yearly temperature is 53° and the average annual precipitation is 34 inches. By comparison, the average temperature in Phoenix, Arizona, is 71° and the average precipitation is 7.4 inches. The point of these examples is to demonstrate the great advantage of being able to describe a large set of data with a single, representative number. Central tendency characterizes what is typical for a large population and in doing so makes large amounts of data more digestible. Statisticians sometimes use the expression "number crunching" to illustrate this aspect of data description. That is, we take a distribution consisting of many scores and "crunch" them down to a single value that describes them all.

Unfortunately, there is no single, standard procedure for determining central tendency. The problem is that no single measure produces a central, representative value in every situation. The three distributions shown in Figure 3.2 should help demonstrate this fact. Before we discuss the three distributions, take a moment to look at the figure and try to identify the "center" or the "most representative score" for each distribution.

1. The first distribution (Figure 3.2(a)) is symmetrical, with the scores forming a distinct pile centered around $X = 5$. For this type of distribution, it is easy to identify the "center," and most people would agree that the value $X = 5$ is an appropriate measure of central tendency.

2. In the second distribution (Figure 3.2(b)), however, problems begin to appear. Now the scores form a negatively skewed distribution, piling up at the high end of the scale around $X = 8$, but tapering off to the left all the way down to $X = 1$. Where is the "center" in this case? Some people might select $X = 8$ as the center because more individuals had this score than any other single value. However, $X = 8$ is clearly not in the middle of the distribution. In fact, the majority of the scores (10 out of 16) have values less than 8, so it seems reasonable that the "center" should be defined by a value that is less than 8.

3. Now consider the third distribution (Figure 3.2(c)). Again, the distribution is symmetrical, but now there are two distinct piles of scores. Because the distribution is symmetrical with $X = 5$ as the midpoint, you may choose $X = 5$ as the "center." However, none of the scores is located at $X = 5$ (or even close), so this value is not particularly good as a representative score. On the other hand, because there are two separate piles of scores with one group centered at $X = 2$ and the other centered at $X = 8$, it is tempting to say that this distribution has two centers. But can one distribution have two centers?

Clearly, there can be problems defining the "center" of a distribution. Occasionally, you will find a nice, neat distribution like the one shown in Figure 3.2(a), for which everyone will agree on the center. But you should realize that other distributions are possible and that there may be different opinions concerning the definition of the center. To deal with these problems, statisticians have developed three different methods for measuring central tendency: the mean, the median, and the mode. They are computed differently and have

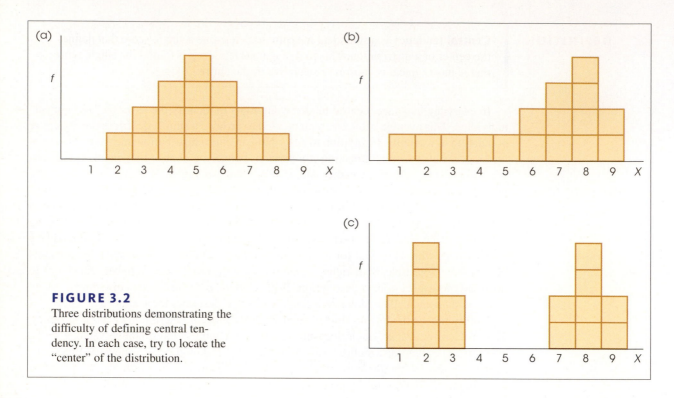

FIGURE 3.2

Three distributions demonstrating the difficulty of defining central tendency. In each case, try to locate the "center" of the distribution.

different characteristics. To decide which of the three measures is best for any particular distribution, you should keep in mind that the general purpose of central tendency is to find the single most representative score. Each of the three measures we present has been developed to work best in a specific situation. We examine this issue in more detail after we introduce the three measures.

3.2 | The Mean

LEARNING OBJECTIVES

1. Define the mean and calculate both the population mean and the sample mean.

2. Explain the alternative definitions of the mean as the amount each individual receives when the total is divided equally and as a balancing point.

3. Calculate a weighted mean.

4. Find n, ΣX, and M using scores in a frequency distribution table.

5. Describe how the mean is affected by each of the following: changing a score, adding or removing a score, adding or subtracting a constant from each score, and multiplying or dividing each score by a constant.

The *mean*, also known as the arithmetic average, is computed by adding all the scores in the distribution and dividing by the number of scores. The mean for a population is identified by the Greek letter mu, μ (pronounced "mew"), and the mean for a sample is identified by M or $\overline{X}$ (read "x-bar").

The convention in many statistics textbooks is to use $\overline{X}$ to represent the mean for a sample. However, in manuscripts and in published research reports the letter M is the standard notation for a sample mean. Because you will encounter the letter M when reading research reports and because you should use the letter *M* when writing research reports, we have decided to use the same notation in this text. Keep in mind that the $\overline{X}$ notation is still appropriate for identifying a sample mean, and you may find it used on occasion, especially in textbooks.

DEFINITION

> The **mean** for a distribution is the sum of the scores divided by the number of scores.

The formula for the *population mean* is

$$\mu = \frac{\Sigma X}{N} \tag{3.1}$$

First, add all the scores in the population, and then divide by *N*. For a sample, the computation exactly the same, but the formula for the *sample mean* uses symbols that signify sample values:

$$\text{sample mean} = M = \frac{\Sigma X}{n} \tag{3.2}$$

In general, we use Greek letters to identify characteristics of a population (parameters) and letters of our own alphabet to stand for sample values (statistics). If a mean is identified with the symbol *M*, you should realize that we are dealing with a sample. Also note that the equation for the sample mean uses a lowercase *n* as the symbol for the number of scores in the sample.

EXAMPLE 3.1

For the following population of *N* = 4 scores,

$$3, \quad 7, \quad 4, \quad 6$$

the mean is

$$\mu = \frac{\Sigma X}{N} = \frac{20}{4} = 5 \qquad \blacksquare$$

■ Alternative Definitions for the Mean

Although the procedure of adding the scores and dividing by the number of scores provides a useful definition of the mean, there are two alternative definitions that may give you a better understanding of this important measure of central tendency.

Dividing the Total Equally   The first alternative is to think of the mean as the amount each individual receives when the total (ΣX) is divided equally among all the individuals (*N*) in the distribution. Consider the following example.

EXAMPLE 3.2

A group of *n* = 6 boys buys a box of baseball cards at a garage sale and discovers that the box contains a total of 180 cards. If the boys divide the cards equally among themselves, how many cards will each boy get? You should recognize that this problem represents the standard procedure for computing the mean. Specifically, the total (ΣX) is divided by the number (*n*) to produce the mean, $\frac{180}{6} = 30$ cards for each boy. $\qquad \blacksquare$

The previous example demonstrates that it is possible to define the mean as the amount that each individual gets when the total is distributed equally. This somewhat socialistic technique is particularly useful in problems for which you know the mean and must find the total. Consider the following example.

EXAMPLE 3.3

Now suppose that the 6 boys from Example 3.2 decide to sell their baseball cards on eBay. If they make an average of $M = \$5$ per boy, what is the total amount of money for the whole group? Although you do not know exactly how much money each boy has, the new definition of the mean tells you that if they pool their money together and then distribute the total equally, each boy will get $5. For each of $n = 6$ boys to get $5, the total must be $6(\$5) = \$30$. To check this answer, use the formula for the mean:

$$M = \frac{\Sigma X}{n} = \frac{\$30}{6} = \$5$$ ∎

The Mean as a Balance Point   The second alternative definition of the mean describes the mean as a balance point for the distribution. Consider a population consisting of $N = 5$ scores (1, 2, 6, 6, 10). For this population, $\Sigma X = 25$ and $\mu = \frac{25}{5} = 5$. Figure 3.3 shows this population drawn as a histogram, with each score represented as a box that is sitting on a seesaw. If the seesaw is positioned so that it pivots at a point equal to the mean, then it will be balanced and will rest level.

The reason the seesaw is balanced over the mean becomes clear when we measures the distance of each box (score) from the mean:

Score	Distance from the Mean
$X = 1$	4 points below the mean
$X = 2$	3 points below the mean
$X = 6$	1 point above the mean
$X = 6$	1 point above the mean
$X = 10$	5 points above the mean

Notice that the mean balances the distances. That is, the total distance below the mean is the same as the total distance above the mean:

below the mean: $4 + 3 = 7$ points
above the mean: $1 + 1 + 5 = 7$ points

Because the mean serves as a balance point, the value of the mean will always be located somewhere between the highest score and the lowest score; that is, the mean can never be outside the range of scores. If the lowest score in a distribution is $X = 8$ and the highest is $X = 15$, then the mean *must* be between 8 and 15. If you calculate a value that is outside this range, then you have made an error.

The image of a seesaw with the mean at the balance point is also useful for determining how a distribution is affected if a new score is added or if an existing score is removed. For

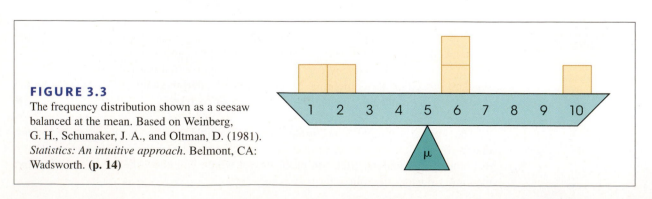

FIGURE 3.3
The frequency distribution shown as a seesaw balanced at the mean. Based on Weinberg, G. H., Schumaker, J. A., and Oltman, D. (1981). *Statistics: An intuitive approach.* Belmont, CA: Wadsworth. **(p. 14)**

the distribution in Figure 3.3, for example, what would happen to the mean (balance point) if a new score were added at $X = 10$?

■ The Weighted Mean

Often it is necessary to combine two sets of scores and then find the overall mean for the combined group. Suppose, for example, that we begin with two separate samples. The first sample has $n = 12$ scores and a mean of $M = 6$. The second sample has $n = 8$ and $M = 7$. If the two samples are combined, what is the mean for the total group?

To calculate the overall mean, we need two values:

1. the overall sum of the scores for the combined group (ΣX), and

2. the total number of scores in the combined group (n).

The total number of scores in the combined group can be found easily by adding the number of scores in the first sample (n_1) and the number in the second sample (n_2). In this case, there are 12 scores in the first sample and 8 in the second, for a total of $12 + 8 = 20$ scores in the combined group. Similarly, the overall sum for the combined group can be found by adding the sum for the first sample (ΣX_1) and the sum for the second sample (ΣX_2). With these two values, we can compute the mean using the basic equation

$$\text{overall mean} = M = \frac{\Sigma X \text{ (overall sum for the combined group)}}{n \text{ (total number in the combined group)}}$$

$$= \frac{\Sigma X_1 + \Sigma X_2}{n_1 + n_2}$$

To find the sum of the scores for each sample, remember that the mean can be defined as the amount each person receives when the total (ΣX) is distributed equally. The first sample has $n = 12$ and $M = 6$. (Expressed in dollars instead of scores, this sample has $n = 12$ people and each person gets $6 when the total is divided equally.) For each of 12 people to get $M = 6$, the total must be $\Sigma X = 12 \times 6 = 72$. In the same way, the second sample has $n = 8$ and $M = 7$ so the total must be $\Sigma X = 8 \times 7 = 56$. Using these values, we obtain an overall mean of

$$\text{overall mean} = M = \frac{\Sigma X_1 + \Sigma X_2}{n_1 + n_2} = \frac{72 + 56}{12 + 8} = \frac{128}{20} = 6.4$$

The following table summarizes the calculations.

First Sample	Second Sample	Combined Sample
$n = 12$	$n = 8$	$n = 20 \, (12 + 8)$
$\Sigma X = 72$	$\Sigma X = 56$	$\Sigma X = 128 \, (72 + 56)$
$M = 6$	$M = 7$	$M = 6.4$

Note that the overall mean is not halfway between the original two sample means. Because the samples are not the same size, one makes a larger contribution to the total group and therefore carries more weight in determining the overall mean. For this reason, the overall mean we have calculated is called the *weighted mean*. In this example, the overall mean of $M = 6.4$ is closer to the value of $M = 6$ (the larger sample) than it is to $M = 7$ (the smaller sample). An alternative method for finding the weighted mean is presented in Box 3.1.

BOX 3.1 An Alternative Procedure for Finding the Weighted Mean

In the text, the weighted mean was obtained by first determining the total number of scores (n) for the two combined samples and then determining the overall sum (ΣX) for the two combined samples. The following example demonstrates how the same result can be obtained using a slightly different conceptual approach.

We begin with the same two samples that were used in the text: One sample has $M = 6$ for $n = 12$ students, and the second sample has $M = 7$ for $n = 8$ students. The goal is to determine the mean for the overall group when the two samples are combined.

Logically, when these two samples are combined, the larger sample (with $n = 12$ scores) will make a greater contribution to the combined group than the smaller sample (with $n = 8$ scores). Thus, the larger sample will carry more weight in determining the mean for the combined group. We will accommodate this fact by assigning a weight to each sample mean so that the weight is determined by the size of the sample. To determine how much weight should be

assigned to each sample mean, you simply consider the sample's contribution to the combined group. When the two samples are combined, the resulting group has a total of 20 scores ($n = 12$ from the first sample and $n = 8$ from the second). The first sample contributes 12 out of 20 scores and, therefore, is assigned a weight of $\frac{12}{20}$. The second sample contributes 8 out of 20 scores, and its weight is $\frac{8}{20}$. Each sample mean is then multiplied by its weight, and the results are added to find the weighted mean for the combined sample. For this example,

$$
\begin{aligned}
\text{weighted mean} &= \left(\frac{12}{20}\right)(6) + \left(\frac{8}{20}\right)(7) \\
&= \frac{72}{20} + \frac{56}{20} \\
&= 3.6 + 2.8 \\
&= 6.4
\end{aligned}
$$

Note that this is the same result obtained using the method described in the text.

The following example is an opportunity for you to test your understanding by computing a weighted mean yourself.

EXAMPLE 3.4   One sample has $n = 4$ scores with a mean of $M = 8$ and a second sample has $n = 8$ scores with a mean of $M = 5$. If the two samples are combined, what is the mean for the combined group? For this example, you should obtain a mean of $M = 6$. Good luck and remember that you can use the example in the text as a model.   ■

■ Computing the Mean from a Frequency Distribution Table

When a set of scores has been organized in a frequency distribution table, the calculation of the mean is usually easier if you first remove the individual scores from the table. Table 3.1 shows a distribution of scores organized in a frequency distribution table. To compute the mean for this distribution you must be careful to use both the X values in the first column and the frequencies in the second column. The values in the table show that the distribution consists of one 10, two 9s, four 8s, and one 6, for a total of $n = 8$ scores. Remember that you can determine the number of scores by adding the frequencies, $n = \Sigma f$. To find the sum of the scores, you must add all eight scores:

$$\Sigma X = 10 + 9 + 9 + 8 + 8 + 8 + 8 + 6 = 66$$

Note that you can also find the sum of the scores by computing $\Sigma f X$ as we demonstrated in Chapter 2 (p. 36). Once you have found ΣX and n, you compute the mean as usual. For these data,

$$M = \frac{\Sigma X}{n} = \frac{66}{8} = 8.25$$

TABLE 3.1
Statistics quiz scores for a
section of $n = 8$ students.

Quiz Score (X)	f	fX
10	1	10
9	2	18
8	4	32
7	0	0
6	1	6

■ Characteristics of the Mean

The mean has many characteristics that will be important in future discussions. In general, these characteristics result from the fact that every score in the distribution contributes to the value of the mean. Specifically, every score adds to the total (ΣX) and every score contributes one point to the number of scores (n). These two values (ΣX and n) determine the value of the mean. We now discuss four of the more important characteristics of the mean.

Changing a Score   Changing the value of any score will change the mean. For example, a sample of quiz scores for a psychology lab section consists of 9, 8, 7, 5, and 1. Note that the sample consists of $n = 5$ scores with $\Sigma X = 30$. The mean for this sample is

$$M = \frac{\Sigma X}{n} = \frac{30}{5} = 6.00$$

Now suppose that the score of $X = 1$ is changed to $X = 8$. Note that we have added 7 points to this individual's score, which will also add 7 points to the total (ΣX). After changing the score, the new distribution consists of

$$9, 8, 7, 5, 8$$

There are still $n = 5$ scores, but now the total is $\Sigma X = 37$. Thus, the new mean is

$$M = \frac{\Sigma X}{n} = \frac{37}{5} = 7.40$$

Notice that changing a single score in the sample has produced a new mean. You should recognize that changing any score also changes the value of ΣX (the sum of the scores), and thus always changes the value of the mean.

Introducing a New Score or Removing a Score   Adding a new score to a distribution, or removing an existing score, will usually change the mean. The exception is when the new score (or the removed score) is exactly equal to the mean. It is easy to visualize the effect of adding or removing a score if you remember that the mean is defined as the balance point for the distribution. Figure 3.4 shows a distribution of scores represented as boxes on a seesaw that is balanced at the mean, $\mu = 7$. Imagine what would happen if we added a new score (a new box) at $X = 10$. Clearly, the seesaw would tip to the right and we would need to move the pivot point (the mean) to the right to restore balance.

Now imagine what would happen if we removed the score (the box) at $X = 9$. This time the seesaw would tip to the left and, once again, we would need to change the mean to restore balance.

Finally, consider what would happen if we added a new score of $X = 7$, exactly equal to the mean. It should be clear that the seesaw would not tilt in either direction, so the mean would stay in exactly the same place. Also note that if we remove the new score at $X = 7$, the seesaw will remain balanced and the mean will not change. In general, adding a new

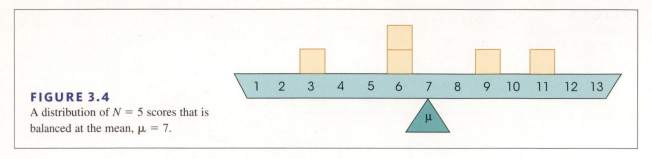

FIGURE 3.4

A distribution of $N = 5$ scores that is balanced at the mean, $\mu = 7$.

score or removing an existing score will cause the mean to change unless the new score (or existing score) is located exactly at the mean.

The following example demonstrates exactly how the new mean is computed when a new score is added to an existing sample.

EXAMPLE 3.5

Adding a score (or removing a score) has the same effect on the mean whether the original set of scores is a sample or a population. To demonstrate the calculation of the new mean, we will use the set of scores that is shown in Figure 3.4. This time, however, we will treat the scores as a sample with $n = 5$ and $M = 7$. Note that this sample must have $\Sigma X = 35$. What will happen to the mean if a new score of $X = 13$ is added to the sample?

To find the new sample mean, we must determine how the values for n and ΣX will be changed by a new score. We begin with the original sample and then consider the effect of adding the new score. The original sample had $n = 5$ scores, so adding one new score will produce $n = 6$. Similarly, the original sample had $\Sigma X = 35$. Adding a score of $X = 13$ will increase the sum by 13 points, producing a new sum of $\Sigma X = 35 + 13 = 48$. Finally, the new mean is computed using the new values for n and ΣX.

$$M = \frac{\Sigma X}{n} = \frac{48}{6} = 8$$

The entire process can be summarized as follows:

Original Sample	New Sample, Adding $X = 13$
$n = 5$	$n = 6$
$\Sigma X = 35$	$\Sigma X = 48$
$M = \frac{35}{5} = 7$	$M = \frac{48}{6} = 8$

The following example is an opportunity for you to test your understanding by determining how the mean is changed by removing a score from a distribution.

EXAMPLE 3.6

We begin with a sample of $n = 5$ scores with $\Sigma X = 35$ and $M = 7$. If one score with a value of $X = 11$ is removed from the sample, what is the mean for the remaining scores? You should obtain a mean of $M = 6$. Good luck and remember that you can use Example 3.5 as a model. ■

Adding or Subtracting a Constant from Each Score If a constant value is added to every score in a distribution, the same constant will be added to the mean. Similarly, if you subtract a constant from every score, the same constant will be subtracted from the mean.

In the Preview for this Chapter (page 68), we described a study that examined the effect of alcohol consumption on the perceived attractiveness of opposite-sex individuals shown in photographs (Jones et al., 2003). The results showed that alcohol significantly increased ratings of attractiveness.

Table 3.2 shows results for a sample of $n = 6$ female participants who are rating a specific male face. The first column shows the ratings when the participants have had no alcohol. Note that the total for this column is $\Sigma X = 17$ for a sample of $n = 6$ participants, so the mean is $M = \frac{17}{6} = 2.83$. Now suppose that the effect of alcohol is to add a constant amount (1 point) to each individual's rating score. The resulting scores, after moderate alcohol consumption, are shown in the second column of the table. For these scores, the total is $\Sigma X = 23$, so the mean is $M = \frac{23}{6} = 3.83$. Adding 1 point to each rating score has also added 1 point to the mean, from $M = 2.83$ to $M = 3.83$. (It is important to note that treatment effects are usually not as simple as adding or subtracting a constant amount. Nonetheless, the concept of adding a constant to every score is important and will be addressed in later chapters when we are using statistics to evaluate mean differences.)

TABLE 3.2

Attractiveness ratings of a male face for a sample of $n = 6$ females.

Participant	No Alcohol	Moderate Alcohol
A	4	5
B	2	3
C	3	4
D	3	4
E	2	3
F	3	4
	$\Sigma X = 17$	$\Sigma X = 23$
	$M = 2.83$	$M = 3.83$

Multiplying or Dividing Each Score by a Constant   If every score in a distribution is multiplied by (or divided by) a constant value, the mean will change in the same way.

Multiplying (or dividing) each score by a constant value is a common method for changing the unit of measurement. To change a set of measurements from minutes to seconds, for example, you multiply by 60; to change from inches to feet, you divide by 12. One common task for researchers is converting measurements into metric units to conform to international standards. For example, publication guidelines of the American Psychological Association call for metric equivalents to be reported in parentheses when most nonmetric units are used. Table 3.3 shows how a sample of $n = 5$ scores measured in inches would be transformed to a set of scores measured in centimeters. (Note that 1 inch equals 2.54 centimeters.) The first column shows the original scores that total $\Sigma X = 50$ with $M = 10$ inches. In the second column, each of the original scores has been multiplied by 2.54 (to convert from inches to centimeters) and the resulting values total $\Sigma X = 127$, with $M = 25.4$. Multiplying each score by 2.54 has also caused the mean to be multiplied by 2.54. You should realize, however, that although the numerical values for the individual scores and the sample mean have changed, the actual measurements are not changed.

TABLE 3.3

Measurements transformed from inches to centimeters.

Original Measurement in Inches	Conversion to Centimeters (Multiply by 2.54)
10	25.40
9	22.86
12	30.48
8	20.32
11	27.94
$\Sigma X = 50$	$\Sigma X = 127.00$
$M = 10$	$M = 25.40$

1. What is the mean for the following sample of scores?
   Scores: 1, 2, 5, 4
   a. 12
   b. 6
   c. 4
   d. 3

2. A sample has a mean of $M = 45$. If one person with a score of $X = 53$ is removed from the sample, what effect will it have on the sample mean?
   a. The sample mean will increase.
   b. The sample mean will decrease.
   c. The sample mean will remain the same.
   d. Cannot be determined from the information given

3. One sample has $n = 8$ scores and $M = 2$. A second sample has $n = 4$ scores and $M = 8$. If the two samples are combined, then what is the mean for the combined sample?
   a. 3
   b. 4
   c. 5
   d. 6

4. What is the mean for the population of scores shown in the frequency distribution table?
   a. $\frac{15}{10} = 1.5$
   b. $\frac{15}{5} = 3.0$
   c. $\frac{29}{10} = 2.9$
   d. $\frac{29}{5} = 5.8$

X	f
5	1
4	2
3	3
2	3
1	1

5. After 5 points are added to every score in a distribution, the mean is calculated and found to be $\mu = 30$. What was the value of the mean for the original distribution?
   a. 25
   b. 30
   c. 35
   d. Cannot be determined from the information given

1. D, **2.** B, **3.** B, **4.** C, **5.** A

3.3 | The Median

LEARNING OBJECTIVE   **6.** Define and calculate the median, and find the precise median for a continuous variable.

The second measure of central tendency we will consider is called the *median*. The goal of the median is to locate the midpoint of the distribution. Unlike the mean, there are no specific symbols or notation to identify the median. Instead, the median is simply identified by the word *median*. In addition, the definition and computations for the median are identical for a sample and for a population.

DEFINITION

> If the scores in a distribution are listed in order from smallest to largest, the **median** is the midpoint of the list. More specifically, the median is the point on the measurement scale below which 50% of the scores in the distribution are located.

■ Finding the Median for Most Distributions

Defining the median as the *midpoint* of a distribution means that that the scores are being divided into two equal-sized groups. We are not locating the midpoint between the highest and lowest *X* values. To find the median, list the scores in order from smallest to largest. Begin with the smallest score and count the scores as you move up the list. The median is the first point you reach that is greater than 50% of the scores in the distribution. The median can be equal to a score in the list or it can be a point between two scores. Notice that the median is not algebraically defined (there is no equation for computing the median), which means that there is a degree of subjectivity in determining the exact value. However, the following two examples demonstrate the process of finding the median for most distributions.

EXAMPLE 3.7   This example demonstrates the calculation of the median when *n* is an odd number. With an odd number of scores, you list the scores in order (lowest to highest), and the median is the middle score in the list. Consider the following set of *N* = 5 scores, which have been listed in order:

$$3, 5, 8, 10, 11$$

The middle score is $X = 8$, so the median is equal to 8. Using the counting method, with $N = 5$ scores, the 50% point would be $2\frac{1}{2}$ scores. Starting with the smallest scores, we must count the 3, the 5, and the 8 before we reach the target of at least 50%. Again, for this distribution, the median is the middle score, $X = 8$. ■

EXAMPLE 3.8   This example demonstrates the calculation of the median when *n* is an even number. With an even number of scores in the distribution, you list the scores in order (lowest to highest) and then locate the median by finding the average of the middle two scores. Consider the following population:

$$1, 1, 4, 5, 7, 8$$

Now we select the middle pair of scores (4 and 5), add them together, and divide by 2:

$$\text{median} = \frac{4+5}{2} = \frac{9}{2} = 4.5$$

Using the counting procedure, with $N = 6$ scores, the 50% point is 3 scores. Starting with the smallest scores, we must count the first 1, the second 1, and the 4 before we reach the target of 50%. Again, the median for this distribution is 4.5, which is the first point on the scale beyond $X = 4$. For this distribution, exactly 3 scores (50%) are located below 4.5. Note: If there is a gap between the middle two scores, the convention is to define the median as the midpoint between the two scores. For example, if the middle two scores are $X = 4$ and $X = 6$, the median would be defined as 5. ∎

The simple technique of listing and counting scores is sufficient to determine the median for most distributions and is always appropriate for discrete variables. Notice that this technique will always produce a median that is either a whole number or is halfway between two whole numbers. With a continuous variable, however, it is possible to divide a distribution precisely in half so that *exactly* 50% of the distribution is located below (and above) a specific point. The procedure for locating the precise median is discussed in the following section.

■ Finding the Precise Median for a Continuous Variable

Recall from Chapter 1 that a continuous variable consists of categories that can be split into an infinite number of fractional parts. For example, time can be measured in seconds, tenths of a second, hundredths of a second, and so on. When the scores in a distribution are measurements of a continuous variable, it is possible to split one of the categories into fractional parts and find the median by locating the precise point that separates the bottom 50% of the distribution from the top 50%. The following example demonstrates this process.

EXAMPLE 3.9   For this example, we will find the precise median for the following sample of $n = 8$ scores:

$$1, 2, 3, 4, 4, 4, 4, 6$$

The frequency distribution for this sample is shown in Figure 3.5(a). With an even number of scores, you normally would compute the average of the middle two scores to find the median. This process produces a median of $X = 4$. For a discrete variable, $X = 4$ is the correct value for the median. Recall from Chapter 1 that a discrete variable consists of indivisible categories such as the number of children in a family. Some families have 4 children and some have 5, but none have 4.31 children. For a discrete variable, the category $X = 4$ cannot be divided and the whole number 4 is the median.

However, if you look at the distribution histogram, the value $X = 4$ does not appear to be the exact midpoint. The problem comes from the tendency to interpret a score of $X = 4$ as meaning exactly 4.00. However, if the scores are measurements of a continuous variable, then the score $X = 4$ actually corresponds to an interval from 3.5–4.5, and the median corresponds to a point within this interval.

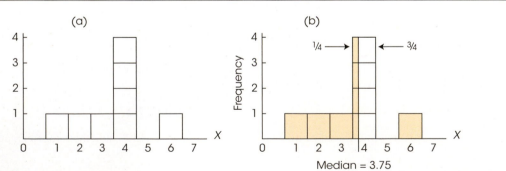

FIGURE 3.5

A distribution with several scores clustered at the median. The median for this distribution is positioned so that each of the four boxes at $X = 4$ is divided into two sections with $\frac{1}{4}$ of each box below the median (to the left) and $\frac{3}{4}$ of each box above the median (to the right). As a result, there are exactly 4 boxes, 50% of the distribution, on each side of the median.

To find the precise median, we first observe that the distribution contains $n = 8$ scores represented by 8 boxes in the graph. The median is the point that has exactly 4 boxes (50%) on each side. Starting at the left-hand side and moving up the scale of measurement, we accumulate a total of 3 boxes when we reach a value of 3.5 on the X-axis (see Figure 3.5(a)). What is needed is 1 more box to reach the goal of 4 boxes (50%). The problem is that the next interval contains four boxes. The solution is to take a fraction of each box so that the fractions combine to give you one box. For this example, if we take $\frac{1}{4}$ of each box, the four quarters will combine to make one whole box. This solution is shown in Figure 3.5(b). The fraction is determined by the number of boxes needed to reach 50% and the number that exist in the interval

$$\text{fraction} = \frac{\text{number needed to reach 50\%}}{\text{number in the interval}}$$

For this example, we needed 1 out of the 4 boxes in the interval, so the fraction is $\frac{1}{4}$. To obtain $\frac{1}{4}$ of each box, the median is the point that is located exactly $\frac{1}{4}$ of the way into the interval. The interval for $X = 4$ extends from 3.5–4.5. The interval width is 1 point, so $\frac{1}{4}$ of the interval corresponds to 0.25 points. Starting at the bottom of the interval and moving up 0.25 points produces a value of $3.50 + 0.25 = 3.75$. This is the median, with exactly 50% of the distribution (4 boxes) on each side. ∎

You may recognize that the process used to find the precise median in Example 3.7 is equivalent to the process of interpolation that was introduced in Chapter 2 (pp. 51-55). Specifically, the precise median is identical to the 50th percentile for a distribution, and interpolation can be used to locate the 50th percentile. The process of using interpolation is demonstrated in Box 3.2 using the same scores that were used in Example 3.9.

BOX 3.2 Using Interpolation to Locate the 50th Percentile (The Median)

The precise median and the 50th percentile are both defined as the point that separates the top 50% of a distribution from the bottom 50%. In Chapter 2, we introduced interpolation as a technique for finding specific percentiles. We now use that same process to find the 50th percentile for the scores in Example 3.9.

Looking at the distribution of scores shown in Figure 3.5, exactly 3 of the $n = 8$ scores, or 37.5%, are located below the real limit of 3.5. Also, 7 of the $n = 8$ scores (87.5%) are located below the real limit of 4.5. This interval of scores and percentages is shown in the following table. Note that the median, the 50th percentile, is located within this interval.

	Scores (X)	Percentages	
Top	4.5	87.5%	
	?	50%	← Intermediate value
Bottom	3.5	37.5%	

We will find the 50th percentile (the median) using the 4-step interpolation process that was introduced in Chapter 2.

1. For the scores, the width of the interval is 1 point. For the percentages, the width is 50 points.
2. The value of 50% is located 37.5 points down from the top of the percentage interval. As a fraction of the whole interval, this is 37.5 out of 50, or 0.75 of the total interval. Thus, the 50% point is located 0.75 or $\frac{3}{4}$ down from the top of the interval.
3. For the scores, the interval width is 1 point and 0.75 down from the top of the interval corresponds to a distance of $0.75(1) = 0.75$ points.
4. Because the top of the interval is 4.5, the position we want is

$$4.5 - 0.75 = 3.75$$

For this distribution, the 50% point (the 50th percentile) corresponds to a score of $X = 3.75$. Note that this is exactly the same value that we obtained for the median in Example 3.9.

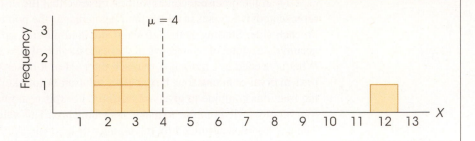

FIGURE 3.6
A population of N = 6 scores with a mean of μ = 4. Notice that the mean does not necessarily divide the scores into two equal groups. In this example, 5 out of the 6 scores have values less than the mean.

Remember, finding the precise midpoint by dividing scores into fractional parts is sensible for a continuous variable, however, it is not appropriate for a discrete variable. For example, a median time of 3.75 seconds is reasonable, but a median family size of 3.75 children is not.

■ The Median, the Mean, and the Middle

Earlier, we defined the mean as the "balance point" for a distribution because the distances above the mean must have the same total as the distances below the mean. One consequence of this definition is that the mean is always located inside the group of scores, somewhere between the smallest score and the largest score. You should notice, however, that the concept of a balance point focuses on distances rather than scores. In particular, it is possible to have a distribution in which the vast majority of the scores are located on one side of the mean. Figure 3.6 shows a distribution of $N = 6$ scores in which 5 out of 6 scores have values less than the mean. In this figure, the total of the distances above the mean is 8 points and the total of the distances below the mean is 8 points. Thus, the mean is located in the middle of the distribution if you use the concept of distance to define the "middle." However, you should realize that the mean is not necessarily located at the exact center of the group of scores.

The median, on the other hand, defines the middle of the distribution in terms of scores. In particular, the median is located so that half of the scores are on one side and half are on the other side. For the distribution in Figure 3.6, for example, the median is located at $X = 2.5$, with exactly 3 scores above this value and exactly 3 scores below. Thus, it is possible to claim that the median is located in the middle of the distribution, provided that the term "middle" is defined by the number of scores.

In summary, the mean and the median are both methods for defining and measuring central tendency. Although they both define the middle of the distribution, they use different definitions of the term "middle."

LEARNING CHECK   **1.** What is the median for the following set of scores:

Scores: 3, 4, 6, 8, 9, 10, 11

a. 7
b. 7.5
c. 8
d. 8.5

2. What is the median for the following set of scores:

$$\text{Scores: } 8, 10, 11, 12, 14, 15$$

    a. 11

    b. 11.5

    c. 12

    d. $\frac{70}{6} = 11.67$

3. For the sample shown in the frequency distribution table, what is the median?

    a. 3

    b. 3.5

    c. 4

    d. $\frac{15}{2} = 7.5$

X	f
5	1
4	5
3	4
2	1
1	1

ANSWERS   **1.** C, **2.** B, **3.** B

3.4 | The Mode

LEARNING OBJECTIVE   **7.** Define and identify the mode(s) for a distribution, including the major and minor modes for a binomial distribution.

The final measure of central tendency that we will consider is called the *mode*. In its common usage, the word *mode* means "the customary fashion" or "a popular style." The statistical definition is similar in that the mode is the most common observation among a group of scores.

DEFINITION | In a frequency distribution, the **mode** is the score or category that has the greatest frequency.

As with the median, there are no symbols or special notation used to identify the mode or to differentiate between a sample mode and a population mode. In addition, the definition of the mode is the same for a population and for a sample distribution.

The mode is a useful measure of central tendency because it can be used to determine the typical or most frequent value for any scale of measurement, including a nominal scale (see Chapter 1). Consider, for example, the data shown in Table 3.4. These data were obtained by asking a sample of 100 students to name their favorite restaurants in town. The result is a sample of $n = 100$ scores with each score corresponding to the restaurant that the student named.

For these data, the mode is Luigi's, the restaurant (score) that was named most frequently as a favorite place. Although we can identify a modal response for these data, you should notice that it would be impossible to compute a mean or a median. Specifically,

TABLE 3.4 Favorite restaurants named by a sample of $n = 100$ students.
Caution: The mode is a score or category, not a frequency. For this example, the mode is Luigi's, not $f = 42$.

Restaurant	f
College Grill	5
George & Harry's	16
Luigi's	42
Oasis Diner	18
Roxbury Inn	7
Sutter's Mill	12

you cannot add restaurants to obtain ΣX and you cannot list the scores (named restaurants) in order.

The mode also can be useful because it is the only measure of central tendency that corresponds to an actual score in the data; by definition, the mode is the most frequently occurring score. The mean and the median, on the other hand, are both calculated values and often produce an answer that does not equal any score in the distribution. For example, in Figure 3.6 (page 84) we presented a distribution with a mean of 4 and a median of 2.5. Note that none of the scores is equal to 4 and none of the scores is equal to 2.5. However, the mode for this distribution is $X = 2$ and there are three individuals who actually have scores of $X = 2$.

In a frequency distribution graph, the greatest frequency will appear as the tallest part of the figure. To find the mode, you simply identify the score located directly beneath the highest point in the distribution.

Although a distribution will have only one mean and only one median, it is possible to have more than one mode. Specifically, it is possible to have two or more scores that have the same highest frequency. In a frequency distribution graph, the different modes will correspond to distinct, equally high peaks. A distribution with two modes is said to be *bimodal,* and a distribution with more than two modes is called *multimodal.* Occasionally, a distribution with several equally high points is said to have no mode.

Incidentally, a bimodal distribution is often an indication that two separate and distinct groups of individuals exist within the same population (or sample). For example, if you measured height for each person in a set of 100 college students, the resulting distribution would probably have two modes, one corresponding primarily to the males in the group and one corresponding primarily to the females.

Technically, the mode is the score with the absolute highest frequency. However, the term *mode* is often used more casually to refer to scores with relatively high frequencies—that is, scores that correspond to peaks in a distribution even though the peaks are not the absolute highest points. For example, Athos et al. (2007) asked people to identify the pitch for both pure tones and piano tones. Participants were presented with a series of tones and had to name the note corresponding to each tone. Nearly half the participants (44%) had extraordinary pitch-naming ability (absolute pitch), and were able to identify most of the tones correctly. Most of the other participants performed around chance level, apparently guessing the pitch names randomly. Figure 3.7 shows a distribution of scores that is consistent with the results of the study. There are two distinct peaks in the distribution, one located at $X = 2$ (chance performance) and the other located at $X = 10$ (perfect performance). Each of these values is a mode in the distribution. Note, however, that the two modes do not have identical frequencies. Eight people scored at $X = 2$ and only seven had scores of $X = 10$. Nonetheless, both of these points are called modes. When two modes have unequal frequencies, researchers occasionally differentiate the two values by calling the taller peak the *major mode,* and the shorter one the *minor mode.*

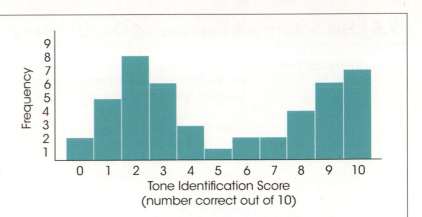

FIGURE 3.7

A frequency distribution for tone identifi-
cation scores. An example of a binomial
distribution.

1. During the month of October, an instructor recorded the number of absences
for each student in a class of $n = 20$ and obtained the following distribution.
What are the values for the mean, the median, and the mode for this
distribution?

    a. Mean = 2.35, Median = 2.5, Mode = 3

    b. Mean = 2.5, Median = 3, Mode = 3

    c. Mean = 2.35, Median = 3, Mode = 7

    d. Mean = 2.5, Median = 2.5, Mode = 7

Number of Absences	f
5	1
4	2
3	7
2	5
1	3
0	2

2. What is the mode for the following set of $n = 8$ scores? Scores: 0, 1, 1, 2, 2,
2, 2, 3

    a. 2

    b. 2.5

    c.

    d. 13

3. For the sample shown in the frequency distribution table, what is the mode?

    a. 3

    b. 3.5

    c. 4

    d. 5

X	f
5	1
4	3
3	4
2	2
1	1

1. A, **2.** A, **3.** A

3.5 | Selecting a Measure of Central Tendency

LEARNING OBJECTIVE **8.** Explain when each of the three measures of central tendency—mean, median, and mode—should be used, identify the advantages and disadvantages of each, and describe how each is presented in a report of research results.

You usually can compute two or even three measures of central tendency for the same set of data. Although the three measures often produce similar results, there are situations in which they are predictably different (see Section 3.6). Deciding which measure of central tendency is best to use depends on several factors. Before we discuss these factors, however, note that whenever the scores are numerical values (interval or ratio scale) the mean is usually the preferred measure of central tendency. Because the mean uses every score in the distribution, it typically produces a good representative value. Remember that the goal of central tendency is to find the single value that best represents the entire distribution. Besides being a good representative, the mean has the added advantage of being closely related to variance and standard deviation, the most common measures of variability (Chapter 4). This relationship makes the mean a valuable measure for purposes of inferential statistics. For these reasons, and others, the mean generally is considered to be the best of the three measures of central tendency. However, there are specific situations in which it is impossible to compute a mean or in which the mean is not particularly representative. It is in these situations that the mode and the median are used.

■ When to Use the Median

We will consider four situations in which the median serves as a valuable alternative to the mean. In the first three cases, the data consist of numerical values (interval or ratio scales) for which you would normally compute the mean. However, each case also involves a special problem so that it is either impossible to compute the mean, or the calculation of the mean produces a value that is not central or not representative of the distribution. The fourth situation involves measuring central tendency for ordinal data.

Extreme Scores or Skewed Distributions When a distribution has a few extreme scores, scores that are very different in value from most of the others, then the mean may not be a good representative of the majority of the distribution. The problem comes from the fact that one or two extreme values can have a large influence and cause the mean to be displaced. In this situation, the fact that the mean uses all of the scores equally can be a disadvantage. Consider, for example, the distribution of $n = 10$ scores in Figure 3.8. For this sample, the mean is

$$M = \frac{\Sigma X}{n} = \frac{203}{10} = 20.3$$

Notice that the mean is not very representative of any score in this distribution. Although most of the scores are clustered between 10 and 13, the extreme score of $X = 100$ inflates the value of ΣX and distorts the mean.

The median, on the other hand, is not easily affected by extreme scores. For this sample, $n = 10$, so there should be five scores on either side of the median. The median is 11.50. Notice that this is a very representative value. Also note that the median would be unchanged even if the extreme score were 1000 instead of only 100. Because it is relatively unaffected by extreme scores, the median commonly is used when reporting the average value for a skewed distribution. For example, the distribution of personal incomes is very

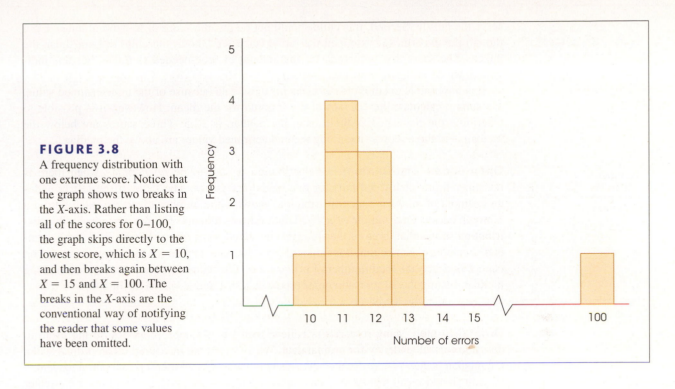

FIGURE 3.8
A frequency distribution with one extreme score. Notice that the graph shows two breaks in the X-axis. Rather than listing all of the scores for 0–100, the graph skips directly to the lowest score, which is $X = 10$, and then breaks again between $X = 15$ and $X = 100$. The breaks in the X-axis are the conventional way of notifying the reader that some values have been omitted.

skewed, with a small segment of the population earning incomes that are astronomical. These extreme values distort the mean, so that it is not very representative of the salaries that most of us earn. As in the previous example, the median is the preferred measure of central tendency when extreme scores exist.

Undetermined Values   Occasionally, you will encounter a situation in which an individual has an unknown or undetermined score. This often occurs when you are measuring the number of errors (or amount of time) required for an individual to complete a task. For example, suppose that preschool children are asked to assemble a wooden puzzle as quickly as possible. The experimenter records how long (in minutes) it takes each child to arrange all the pieces to complete the puzzle. Table 3.5 presents results for a sample of $n = 6$ children.

TABLE 3.5
Number of minutes needed to assemble a wooden puzzle.

Person	Time (Min.)
1	8
2	11
3	12
4	13
5	17
6	Never finished

Notice that one child never completed the puzzle. After an hour, this child still showed no sign of solving the puzzle, so the experimenter stopped him or her. This participant has an undetermined score. (There are two important points to be noted. First, the experimenter should not throw out this individual's score. The whole purpose for using a sample is to gain a picture of the population, and this child tells us that part of the population cannot solve the puzzle. Second, this child should not be given a score of $X = 60$ minutes. Even though the experimenter stopped the individual after 1 hour, the child did not finish the puzzle. The score that is recorded is the amount of time needed to finish. For this individual, we do not know how long this is.)

It is impossible to compute the mean for these data because of the undetermined value. We cannot calculate the ΣX part of the formula for the mean. However, it is possible to determine the median. For these data, the median is 12.5. Three scores are below the median, and three scores (including the undetermined value) are above the median.

Number of Pizzas (X)	f
5 or more	3
4	2
3	2
2	3
1	6
0	4

Open-ended Distributions   A distribution is said to be *open-ended* when there is no upper limit (or lower limit) for one of the categories. The table at the left provides an example of an open-ended distribution, showing the number of pizzas eaten during a 1 month period for a sample of $n = 20$ high school students. The top category in this distribution shows that three of the students consumed "5 or more" pizzas. This is an open-ended category. Notice that it is impossible to compute a mean for these data because you cannot find ΣX (the total number of pizzas for all 20 students). However, you can find the median. Listing the 20 scores in order produces $X = 1$ and $X = 2$ as the middle two scores. For these data, the median is 1.5.

Ordinal Scale   Many researchers believe that it is not appropriate to use the mean to describe central tendency for ordinal data. When scores are measured on an ordinal scale, the median is always appropriate and is usually the preferred measure of central tendency.

You should recall that ordinal measurements allow you to determine direction (greater than or less than) but do not allow you to determine distance. The median is compatible with this type of measurement because it is defined by direction: half of the scores are above the median and half are below the median. The mean, on the other hand, defines central tendency in terms of distance. Remember that the mean is the balance point for the distribution, so that the distances above the mean are exactly balanced by the distances below the mean. Because the mean is defined in terms of distances, and because ordinal scales do not measure distance, it is not appropriate to compute a mean for scores from an ordinal scale.

■ When to Use the Mode

We will consider three situations in which the mode is commonly used as an alternative to the mean, or is used in conjunction with the mean to describe central tendency.

Nominal Scales   The primary advantage of the mode is that it can be used to measure and describe central tendency for data that are measured on a nominal scale. Recall that the categories that make up a nominal scale are differentiated only by name, such as classifying people by occupation or college major. Because nominal scales do not measure quantity (distance or direction), it is impossible to compute a mean or a median for data from a nominal scale. Therefore, the mode is the only option for describing central tendency for nominal data. When the scores are numerical values from an interval or ratio scale, the mode is usually not the preferred measure of central tendency.

Discrete Variables   Recall that discrete variables are those that exist only in whole, indivisible categories. Often, discrete variables are numerical values, such as the number of children in a family or the number of rooms in a house. When these variables produce numerical scores, it is possible to calculate means. However, the calculated means are usually fractional values that cannot actually exist. For example, computing means will generate results such as "the average family has 2.4 children and a house with 5.33 rooms." The mode, on the other hand, always identifies an actual score (the most typical case) and, therefore, it produces more sensible measures of central tendency. Using the mode, our conclusion would be "the typical, or modal, family has 2 children and a house with 5 rooms." In many situations, especially with discrete variables, people are more comfortable using the realistic, whole-number values produced by the mode.

Describing Shape   Because the mode requires little or no calculation, it is often included as a supplementary measure along with the mean or median as a no-cost extra. The value of the mode (or modes) in this situation is that it gives an indication of the shape of the distribution as well as a measure of central tendency. Remember that the mode identifies the location of the peak (or peaks) in the frequency distribution graph. For example, if you are told that a set of exam scores has a mean of 72 and a mode of 80, you should have a better picture of the distribution than would be available from the mean alone (see Section 3.6).

IN THE LITERATURE

Reporting Measures of Central Tendency

Measures of central tendency are commonly used in the behavioral sciences to summarize and describe the results of a research study. For example, a researcher may report the sample means from two different treatments or the median score for a large sample. These values may be reported in text describing the results, or presented in tables or in graphs.

In reporting results, many behavioral science journals use guidelines adopted by the American Psychological Association (APA), as outlined in the *Publication Manual of the American Psychological Association* (2010). We will refer to the APA manual from time to time in describing how data and research results are reported in the scientific literature. The APA style uses the letter M as the symbol for the sample mean. Thus, a study might state:

> The treatment group showed fewer errors ($M = 2.56$) on the task than the control group ($M = 11.76$).

When there are many means to report, tables with headings provide an organized and more easily understood presentation. Table 3.6 illustrates this point.

TABLE 3.6

The mean number of errors made on the task for treatment and control groups according to gender.

	Treatment	Control
Females	1.45	8.36
Males	3.83	14.77

The median can be reported using the abbreviation *Mdn,* as in "Mdn = 8.5 errors," or it can simply be reported in narrative text, as follows:

> The median number of errors for the treatment group was 8.5, compared to a median of 13 for the control group.

There is no special symbol or convention for reporting the mode. If mentioned at all, the mode is usually just reported in narrative text.

■ Presenting Means and Medians in Graphs

Graphs also can be used to report and compare measures of central tendency. Usually, graphs are used to display values obtained for sample means, but occasionally you will see sample medians reported in graphs (modes are rarely, if ever, shown in a graph). The value of a graph is that it allows several means (or medians) to be shown simultaneously so it is possible to make quick comparisons between groups or treatment conditions. When preparing a graph, it is customary to list the different groups or treatment conditions on the horizontal axis. Typically, these are the different values that make up the independent variable or the quasi-independent variable. Values for the dependent variable (the scores) are listed on the vertical axis. The means (or medians) are then displayed using a *line graph, histogram,* or *bar graph,* depending on the scale of measurement used for the independent variable.

Figure 3.9 shows an example of a *line graph* displaying the relationship between drug dose (the independent variable) and food consumption (the dependent variable). In this study, there were five different drug doses (treatment conditions) and they are listed along the horizontal axis. The five means appear as points in the graph. To construct this graph, a point was placed above each treatment condition so that the vertical position of the point corresponds to the mean score for the treatment condition. The points are then connected with straight lines. A line graph is used when the values on the horizontal axis are measured on an interval or a ratio scale. An alternative to the line graph is a *histogram.* For this example, the histogram would show a bar above each drug dose so that the height of each bar corresponds to the mean food consumption for that group, with no space between adjacent bars.

Figure 3.10 shows a bar graph displaying the median selling price for single-family homes in different regions of the United States. Bar graphs are used to present means (or medians) when the groups or treatments shown on the horizontal axis are measured on a nominal or an ordinal scale. To construct a bar graph, you simply draw a bar directly above each group or treatment so that the height of the bar corresponds to the mean (or median) for that group or treatment. For a bar graph, a space is left between adjacent bars to indicate that the scale of measurement is nominal or ordinal.

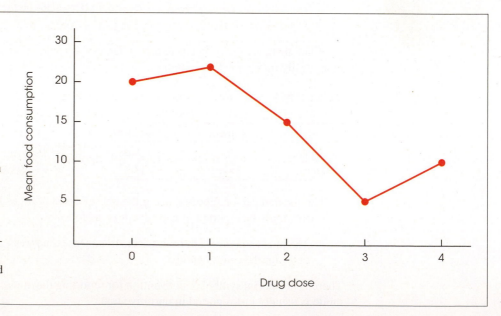

FIGURE 3.9

The relationship between an independent variable (drug dose) and a dependent variable (food consumption). Because drug dose is a continuous variable measured on a ratio scale, a line graph is used to show the relationship.

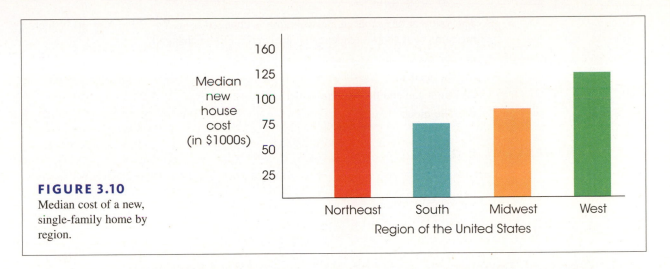

FIGURE 3.10

Median cost of a new, single-family home by region.

When constructing graphs of any type, you should recall the basic rules we introduced in Chapter 2:

1. The height of a graph should be approximately two-thirds to three-quarters of its length.

2. Normally, you start numbering both the X-axis and the Y-axis with zero at the point where the two axes intersect. However, when a value of zero is part of the data, it is common to move the zero point away from the intersection so that the graph does not overlap the axes (see Figure 3.9).

Following these rules will help produce a graph that provides an accurate presentation of the information in a set of data. Although it is possible to construct graphs that distort the results of a study (see Box 2.1), researchers have an ethical responsibility to present an honest and accurate report of their research results.

LEARNING CHECK

1. A teacher gave a reading test to a class of 5th-grade students and computed the mean, median, and mode for the test scores. Which of the following statements *cannot* be an accurate description of the scores?

   a. The majority of the students had scores above the mean.

   b. The majority of the students had scores above the median.

   c. The majority of the students had scores above the mode.

   d. All of the other options are false statements.

2. One item on a questionnaire asks, "How many siblings (brothers and sisters) did you have when you were a child?" A researcher computes the mean, the median, and the mode for a set of $n = 50$ responses to this question. Which of the following statements accurately describes the measures of central tendency?

   a. Because the scores are all whole numbers, the mean will be a whole number.

   b. Because the scores are all whole numbers, the median will be a whole number.

   c. Because the scores are all whole numbers, the mode will be a whole number.

   d. All of the other options are correct descriptions.

3. The value of one score in a distribution is changed from $X = 20$ to $X = 30$. Which measure(s) of central tendency is (are) certain to be changed?
   a. The mean
   b. The median
   c. The mean and the median
   d. The mode

ANSWERS   1. B, 2. C, 3. A

3.6 | Central Tendency and the Shape of the Distribution

LEARNING OBJECTIVE   9. Explain how the three measures of central tendency—mean, median, and mode—are related to each other for symmetrical and skewed distributions.

We have identified three different measures of central tendency, and often a researcher calculates all three for a single set of data. Because the mean, median, and mode are all trying to measure the same thing, it is reasonable to expect that these three values should be related. In fact, there are some consistent and predictable relationships among the three measures of central tendency. Specifically, there are situations in which all three measures will have exactly the same value. On the other hand, there are situations in which the three measures are guaranteed to be different. In part, the relationships among the mean, median, and mode are determined by the shape of the distribution. We will consider two general types of distributions.

■ Symmetrical Distributions

For a *symmetrical distribution,* the right-hand side of the graph is a mirror image of the left-hand side. If a distribution is perfectly symmetrical, the median is exactly at the center because exactly half of the area in the graph will be on either side of the center. The mean also is exactly at the center of a perfectly symmetrical distribution because each score on the left side of the distribution is balanced by a corresponding score (the mirror image) on the right side. As a result, the mean (the balance point) is located at the center of the distribution. Thus, for a perfectly symmetrical distribution, the mean and the median are the same (Figure 3.11). If a distribution is roughly symmetrical, but not perfect, the mean and median will be close together in the center of the distribution.

If a symmetrical distribution has only one mode, it will also be in the center of the distribution. Thus, for a perfectly symmetrical distribution with one mode, all three measures of central tendency—the mean, median, and mode—have the same value. For a roughly symmetrical distribution, the three measures are clustered together in the center of the distribution. On the other hand, a bimodal distribution that is symmetrical (see Figure 3.11(b)) will have the mean and median together in the center with the modes on each side. A rectangular distribution (see Figure 3.11(c)) has no mode because all X values occur with the same frequency. Still, the mean and the median are in the center of the distribution.

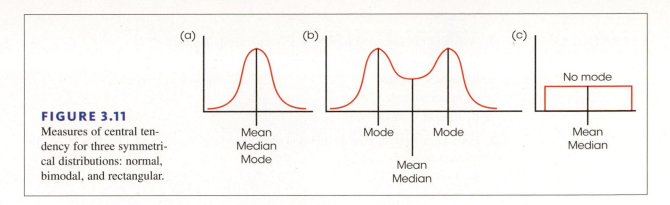

FIGURE 3.11
Measures of central tendency for three symmetrical distributions: normal, bimodal, and rectangular.

The positions of the mean, median, and mode are not as consistently predictable in distributions of discrete variables (see Von Hippel, 2005).

■ Skewed Distributions

In *skewed distributions,* especially distributions for continuous variables, there is a strong tendency for the mean, median, and mode to be located in predictably different positions. Figure 3.12(a), for example, shows a positively skewed distribution with the peak (highest frequency) on the left-hand side. This is the position of the mode. However, it should be clear that the vertical line drawn at the mode does not divide the distribution into two equal parts. To have exactly 50% of the distribution on each side, the median must be located to the right of the mode. Finally, the mean is typically located to the right of the median because it is influenced most by the extreme scores in the tail and is displaced farthest to the right toward the tail of the distribution. Therefore, in a positively skewed distribution, the most likely order of the three measures of central tendency from smallest to largest (left to right) is the mode, median, and mean.

Negatively skewed distributions are lopsided in the opposite direction, with the scores piling up on the right-hand side and the tail tapering off to the left. The grades on an easy exam, for example, tend to form a negatively skewed distribution (see Figure 3.12(b)). For a distribution with negative skew, the mode is on the right-hand side (with the peak), while the mean is displaced toward the left by the extreme scores in the tail. As before, the median is usually located between the mean and the mode. Therefore, in a negatively skewed distribution, the most probable order for the three measures of central tendency from smallest value to largest value (left to right), is the mean, median, and mode.

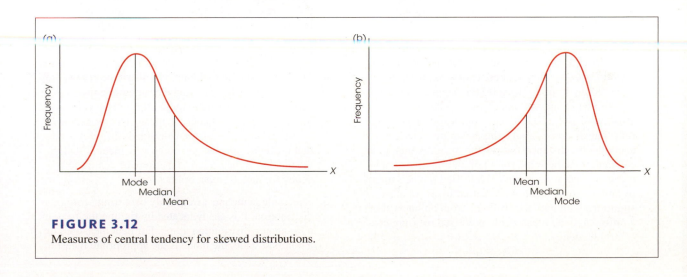

FIGURE 3.12
Measures of central tendency for skewed distributions.

LEARNING CHECK

1. Which of the following is true for a symmetrical distribution?
   a. the mean, median, and mode are all equal
   b. mean = median
   c. mean = mode
   d. median = mode

2. For a negatively skewed distribution with a mode of $X = 25$ and a median of 20, the mean is probably _____.
   a. greater than 25
   b. less than 20
   c. between 20 and 25
   d. cannot be determined from the information given

3. A distribution is positively skewed. Which is the most probable order for the three measures of central tendency?
   a. mean = 40, median = 50, mode = 60
   b. mean = 60, median = 50, mode = 40
   c. mean = 40, median = 60, mode = 50
   d. mean = 50, median = 50, mode = 50

ANSWERS   **1.** B, **2.** B, **3.** B

SUMMARY

1. The purpose of central tendency is to determine the single value that identifies the center of the distribution and best represents the entire set of scores. The three standard measures of central tendency are the mode, the median, and the mean.

2. The mean is the arithmetic average. It is computed by adding all the scores and then dividing by the number of scores. Conceptually, the mean is obtained by dividing the total (ΣX) equally among the number of individuals (N or n). The mean can also be defined as the balance point for the distribution. The distances above the mean are exactly balanced by the distances below the mean. Although the calculation is the same for a population or a sample mean, a population mean is identified by the symbol μ, and a sample mean is identified by M. In most situations with numerical scores from an interval or a ratio scale, the mean is the preferred measure of central tendency.

3. Changing any score in the distribution causes the mean to be changed. When a constant value is added to (or subtracted from) every score in a distribution, the same constant value is added to (or subtracted from) the mean. If every score is multiplied by a constant, the mean is multiplied by the same constant.

4. The median is the midpoint of a distribution of scores. The median is the preferred measure of central tendency when a distribution has a few extreme scores that displace the value of the mean. The median also is used when there are undetermined (infinite) scores that make it impossible to compute a mean. Finally, the median is the preferred measure of central tendency for data from an ordinal scale.

5. The mode is the most frequently occurring score in a distribution. It is easily located by finding the peak in a frequency distribution graph. For data measured on a nominal scale, the mode is the appropriate measure

of central tendency. It is possible for a distribution to have more than one mode.

6. For symmetrical distributions, the mean will equal the median. If there is only one mode, then it will have the same value, too.

7. For skewed distributions, the mode is located toward the side where the scores pile up, and the mean is pulled toward the extreme scores in the tail. The median is usually located between these two values.

KEY TERMS

central tendency (69)	mode (83)	line graph (90)
population mean (μ) (71)	bimodal (84)	symmetrical distribution (92)
sample mean (M) (71)	multimodal (84)	skewed distribution (93)
weighted mean (73)	major mode (84)	positive skew (93)
median (79)	minor mode (84)	negative skew (93)

SPSS®

General instructions for using SPSS are presented in Appendix D. Following are detailed instructions for using SPSS to compute the **Mean and ΣX** for a set of scores.

Data Entry

1. Enter all of the scores in one column of the data editor, probably VAR00001.

Data Analysis

1. Click **Analyze** on the tool bar, select **Descriptive Statistics**, and click on **Descriptives**.
2. Highlight the column label for the set of scores (VAR00001) in the left box and click the arrow to move it into the **Variable** box.
3. If you want ΣX as well as the mean, click on the **Options** box, select **Sum**, then click **Continue**.
4. Click **OK**.

SPSS Output

SPSS will produce a summary table listing the number of scores (N), the maximum and minimum scores, the sum of the scores (if you selected this option), the mean, and the standard deviation. Note: The standard deviation is a measure of variability that is presented in Chapter 4.

FOCUS ON PROBLEM SOLVING

1. Although the three measures of central tendency appear to be very simple to calculate, there is always a chance for errors. The most common sources of error are listed next.

X	f
4	1
3	4
2	3
1	2

a. Many students find it very difficult to compute the mean for data presented in a frequency distribution table. They tend to ignore the frequencies in the table and simply average the score values listed in the X column. You must use the frequencies *and* the scores! Remember that the number of scores is found by $N = \Sigma f$, and the sum of all N scores is found by ΣfX. For the distribution shown in the margin, the mean is $\frac{24}{10} = 2.40$.

b. The median is the midpoint of the distribution of scores, not the midpoint of the scale of measurement. For a 100-point test, for example, many students incorrectly assume that the median must be $X = 50$. To find the median, you must have the *complete set* of individual scores. The median separates the individuals into two equal-sized groups.

c. The most common error with the mode is for students to report the highest frequency in a distribution rather than the score with the highest frequency. Remember that the purpose of central tendency is to find the most representative score. For the distribution in the margin, the mode is $X = 3$, not $f = 4$.

DEMONSTRATION 3.1

COMPUTING MEASURES OF CENTRAL TENDENCY

For the following sample, find the mean, median, and mode. The scores are:

$$5, 6, 9, 11, 5, 11, 8, 14, 2, 11$$

Compute the Mean   The calculation of the mean requires two pieces of information; the sum of the scores, ΣX, and the number of scores, n. For this sample, $n = 10$ and

$$\Sigma X = 5 + 6 + 9 + 11 + 5 + 11 + 8 + 14 + 2 + 11 = 82$$

Therefore, the sample mean is

$$M = \frac{\Sigma X}{n} = \frac{82}{10} = 8.2$$

Find the Median   To find the median, first list the scores in order from smallest to largest. With an even number of scores, the median is the average of the middle two scores in the list. Listed in order, the scores are:

$$2, 5, 5, 6, 8, 9, 11, 11, 11, 14$$

The middle two scores are 8 and 9, and the median is 8.5.

Find the mode   For this sample, $X = 11$ is the score that occurs most frequently. The mode is $X = 11$.

PROBLEMS

1. Find the mean, median, and mode for the following sample of scores:

    $$5, 4, 5, 2, 7, 1, 3, 5$$

2. Find the mean, median, and mode for the following sample of scores:

    $$3, 6, 7, 3, 9, 8, 3, 7, 5$$

3. Find the mean, median, and mode for the scores in the following frequency distribution table:

X	F
6	1
5	2
4	2
3	2
2	2
1	5

4. Find the mean, median, and mode for the scores in the following frequency distribution table:

X	f
8	1
7	1
6	2
5	5
4	2
3	2

5. For the following sample of $n = 10$ scores:
    a. Assume that the scores are measurements of a discrete variable and find the median.
    b. Assume that the scores are measurements of a continuous variable and find the median by locating the precise midpoint of the distribution.

    Scores: 2, 3, 4, 4, 5, 5, 5, 6, 6, 7

6. A population of $N = 15$ scores has $\Sigma X = 120$. What is the population mean?

7. A sample of $n = 8$ scores has a mean of $M = 12$. What is the value of ΣX for this sample?

8. A population with a mean of $\mu = 8$ has $\Sigma X = 40$. How many scores are in the population?

9. One sample of $n = 12$ scores has a mean of $M = 7$ and a second sample of $n = 8$ scores has a mean of $M = 12$. If the two samples are combined, what is the mean for the combined sample?

10. One sample has a mean of $M = 8$ and a second sample has a mean of $M = 16$. The two samples are combined into a single set of scores.
  a. What is the mean for the combined set if both of the original samples have $n = 4$ scores?
  b. What is the mean for the combined set if the first sample has $n = 3$ and the second sample has $n = 5$?
  c. What is the mean for the combined set if the first sample has $n = 5$ and the second sample has $n = 3$?

11. One sample has a mean of $M = 5$ and a second sample has a mean of $M = 10$. The two samples are combined into a single set of scores.
  a. What is the mean for the combined set if both of the original samples have $n = 5$ scores?
  b. What is the mean for the combined set if the first sample has $n = 4$ scores and the second sample has $n = 6$?
  c. What is the mean for the combined set if the first sample has $n = 6$ scores and the second sample has $n = 4$?

12. A population of $N = 15$ scores has a mean of $\mu = 8$. One score in the population is changed from $X = 20$ to $X = 5$. What is the value for the new population mean?

13. A sample of $n = 7$ scores has a mean of $M = 16$. One score in the sample is changed from $X = 6$ to $X = 20$. What is the value for the new sample mean?

14. A sample of $n = 9$ scores has a mean of $M = 20$. One of the scores is changed and the new mean is found to be $M = 22$. If the changed score was originally $X = 7$, what is its new value?

15. A sample of $n = 7$ scores has a mean of $M = 9$. If one new person with a score of $X = 1$ is added to the sample, what is the value for the new mean?

16. A sample of $n = 6$ scores has a mean of $M = 13$. If one person with a score of $X = 3$ is removed from the sample, what is the value for the new mean?

17. A sample of $n = 15$ scores has a mean of $M = 6$. One person with a score of $X = 22$ is added to the sample. What is the value for the new sample mean?

18. A sample of $n = 10$ scores has a mean of $M = 9$. One person with a score of $X = 0$ is removed from the sample. What is the value for the new sample mean?

19. A sample of $n = 7$ scores has a mean of $M = 5$. After one new score is added to the sample, the new mean is found to be $M = 6$. What is the value of the new score? (Hint: Compare the values for ΣX before and after the score was added.)

20. A population of $N = 8$ scores has a mean of $\mu = 16$. After one score is removed from the population, the new mean is found to be $\mu = 15$. What is the value of the score that was removed? (Hint: Compare the values for ΣX before and after the score was removed.)

21. A sample of $n = 9$ scores has a mean of $M = 13$. After one score is added to the sample the mean is found to be $M = 12$. What is the value of the score that was added?

22. Explain why the median is often preferred to the mean as a measure of central tendency for a skewed distribution.

23. A researcher conducts a study comparing two different treatments with a sample of $n = 16$ participants in each treatment. The study produced the following data:

Treatment 1: 6, 7, 11, 4, 19, 17, 2, 5, 9, 13, 6, 23, 11, 4, 6, 1

Treatment 2: 10, 9, 6, 6, 1, 11, 8, 6, 3, 2, 11, 1, 12, 7, 10, 9

  a. Calculate the mean for each treatment. Based on the two means, which treatment produces the higher scores?
  b. Calculate the median for each treatment. Based on the two medians, which treatment produces the higher scores?
  c. Calculate the mode for each treatment. Based on the two modes, which treatment produces the higher scores?

24. Schmidt (1994) conducted a series of experiments examining the effects of humor on memory. In one study, participants were shown a list of sentences, of which half were humorous and half were nonhumorous. A humorous example is, "If at first you don't succeed, you are probably not related to the boss." Other participants would see a nonhumorous version of this sentence, such as "People who are related to the boss often succeed the very first time."

Schmidt then measured the number of each type of sentence recalled by each participant. The following scores are similar to the results obtained in the study.

Number of Sentences Recalled							
Humorous Sentences				Nonhumorous Sentences			
4	5	2	4	5	2	4	2
6	7	6	6	2	3	1	6
2	5	4	3	3	2	3	3
1	3	5	5	4	1	5	3

Calculate the mean number of sentences recalled for each of the two conditions. Do the data suggest that humor helps memory?

25. Stephens, Atkins, and Kingston (2009) conducted a research study demonstrating that swearing can help reduce pain. In the study, each participant was asked to plunge a hand into icy water and keep it there as long as the pain would allow. In one condition, the participants repeatedly yelled their favorite curse words while their hands were in the water. In the other condition the participants yelled a neutral word. Data similar to the results obtained in the study are shown in the following table. Calculate the mean number of seconds that the participants could tolerate the pain for each of the two treatment conditions. Does it appear that swearing helped with pain tolerance?

| Participant | Amount of Time (in seconds) | |
	Swear words	Neutral words
1	94	59
2	70	61
3	52	47
4	83	60
5	46	35
6	117	92
7	69	53
8	39	30
9	51	56
10	73	61

Variability

Tools You Will Need

The following items are considered essential background material for this chapter. If you doubt your knowledge of any of these items, you should review the appropriate chapter or section before proceeding.

- Summation notation (Chapter 1)
- Central tendency (Chapter 3)
    - Mean
    - Median

© Deborah Batt

In a controversial study, Bem (2011) appears to show that participants are able to anticipate and benefit from randomly selected events that will occur in the future. That's right. The individuals in the study were able to improve their performance by using information that was not made available until after their performance was tested. One of the many experiments reported was a demonstration of how practice and rehearsal can improve memory performance, which is not surprising except that the practice occurred after the memory test. In the study, participants were shown a list of 48 words (common nouns), one at a time for 3 seconds each. The words corresponded to 12 foods, 12 animals, 12 occupations, and 12 items of clothing. As each word was presented, the participants were asked to form a mental image of the object represented by the word. At the end of the list, they were given a surprise memory test and asked to type as many words as they could remember in any order. After the memory test, 24 of the words were randomly selected with the restriction that the 24 words consisted of 6 words from each of the 4 categories (6 foods, 6 animals, and so on). The participants then practiced these 24 words using the following procedure:

1. The complete list of 24 words was shown on a computer screen in a randomized order. The participant was asked to identify the 6 food words by clicking on them, which highlighted them. The participant then typed the 6 selected words into 6 empty boxes on the screen.

2. This process was then repeated for the other 3 categories until each participant had identified and typed all of the 24 words.

The researchers then returned to the results from the original memory test for each participant and computed a memory score for the 24 words that had been practiced and a separate memory score for the 24 no-practice words. The results showed that memory for the practiced words was significantly better than memory for the no-practice words, even though the practice occurred after the memory test.

Is this a scientific demonstration of psychic power? You may be somewhat skeptical, especially when you learn that the participants were regular Cornell undergraduate students who were not selected because they had unique ESP abilities. Because this backward cause-and-effect relationship is completely counterintuitive, many behavioral scientists were immediately skeptical. Several groups of researchers attempted to replicate Bem's results but repeatedly failed (Yong, 2012).

If Bem's study is not an example of ESP, then is there another possible explanation for the result? The answer is "yes" and the result might be explained by variability. Variability is an everyday concept as well as a statistic concept. In both cases, it simply means that things aren't always the same; they vary. If you are selecting individuals from the class list for an Introductory Psychology course, you will get a different person on each selection. They will have different ages, heights, personalities, IQs, and so on. In the same way, if you conduct a research study over and over, you will get different results every time. Usually, however, the results will be similar. The students you select will usually be 18- or 19-years-old, above average intelligence, and fairly typical college freshmen and sophomores. The results you obtain from repeating a research study will also tend to be similar from one time to the next. In some cases, however, an extreme outcome occurs. Just as it is possible to toss a coin and get 10 heads in a row, you may select a student who happens to be a 48-year-old lawyer who is going back to school to start a new career as a social worker. It also is possible to obtain really extreme results for a research study.

Fortunately, these extreme outcomes are very rare and researchers use this fact as a demonstration that their research studies are very unlikely to be extreme outcomes that occurred just by chance. If you give a new cholesterol medication to 25 patients and all 25 show a drop in cholesterol, then you can be reasonably confident that the medication works. If only 14 out of 25 had lower levels, then the results could be explained as just chance, but 25 out of 25 is very unlikely to have occurred by chance alone. For Bem's experiment the obtained outcome was found to be "very unlikely" to have occurred by chance. Therefore, chance is eliminated as a plausible explanation and we are left with the conclusion that there is some force at work that causes better performance for the practice words.

The problem comes from the fact that "very unlikely" does not mean the same thing as "impossible." Because outcomes are variable, you will occasionally obtain an extreme, very unlikely outcome simply by chance. Perhaps it is just chance, rather than ESP, that explains Bem's results.

In this chapter we introduce the statistical concept of variability. We will describe the methods that are used to measure and objectively describe the differences that exist from one score to another within a distribution. In addition to describing distributions of scores, variability also helps us determine which outcomes are likely and which are very unlikely to be obtained. This aspect of variability will play an important role in inferential statistics.

4.1 | Introduction to Variability

LEARNING OBJECTIVES

1. Define variability and explain its use and importance as a statistical measure.

2. Define and calculate the range as a simple measure of variability and explain its limitations

The term *variability* has much the same meaning in statistics as it has in everyday language; to say that things are variable means that they are not all the same. In statistics, our goal is to measure the amount of variability for a particular set of scores, a distribution. In simple terms, if the scores in a distribution are all the same, then there is no variability. If there are small differences between scores, then the variability is small, and if there are large differences between scores, then the variability is large.

DEFINITION

> **Variability** provides a quantitative measure of the differences between scores in a distribution and describes the degree to which the scores are spread out or clustered together.

Figure 4.1 shows two distributions of familiar values for the population of adult males: part (a) shows the distribution of men's heights (in inches), and part (b) shows the distribution of men's weights (in pounds). Notice that the two distributions differ in terms of central tendency. The mean height is 70 inches (5 feet, 10 inches) and the mean weight is 170 pounds. In addition, notice that the distributions differ in terms of variability. For example, most heights are clustered close together, within 5 or 6 inches of the mean. On the other hand, weights are spread over a much wider range. In the weight distribution it is not unusual to find individuals who are located more than 30 pounds away from the mean, and it would not be surprising to find two individuals whose weights differ by more than 30 or 40 pounds. The purpose for measuring variability is to obtain an objective measure of how the scores are spread out in a distribution. In general, a good measure of variability serves two purposes:

1. Variability describes the distribution. Specifically, it tells whether the scores are clustered close together or are spread out over a large distance. Usually, variability is defined in terms of *distance*. It tells how much distance to expect between one

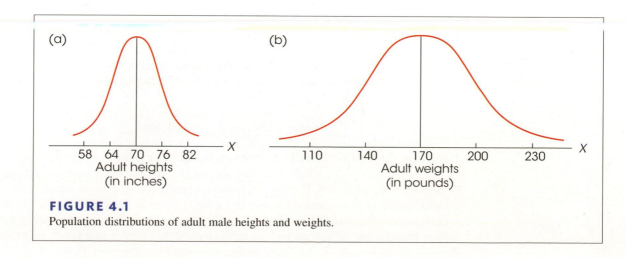

FIGURE 4.1

Population distributions of adult male heights and weights.

score and another, or how much distance to expect between an individual score and the mean. For example, we know that the heights for most adult males are clustered close together, within 5 or 6 inches of the average. Although more extreme heights exist, they are relatively rare.

2. Variability measures how well an individual score (or group of scores) represents the entire distribution. This aspect of variability is very important for inferential statistics, in which relatively small samples are used to answer questions about populations. For example, suppose that you selected a sample of one person to represent the entire population. Because most adult males have heights that are within a few inches of the population average (the distances are small), there is a very good chance that you would select someone whose height is within 6 inches of the population mean. On the other hand, the scores are much more spread out (greater distances) in the distribution of weights. In this case, you probably would *not* obtain someone whose weight was within 6 pounds of the population mean. Thus, variability provides information about how much error to expect if you are using a sample to represent a population.

In this chapter, we consider three different measures of variability: the range, standard deviation, and variance. Of these three, the standard deviation and the related measure of variance are by far the most important.

■ The Range

The obvious first step toward defining and measuring variability is the **range**, which is the distance covered by the scores in a distribution, from the smallest score to the largest score. Although the concept of the range is fairly straightforward, there are several distinct methods for computing the numerical value. One commonly used definition of the range simply measures the difference between the largest score (X_{max}) and the smallest score (X_{min}).

$$\text{range} = X_{max} - X_{max}$$

By this definition, scores having values from 1 to 5 cover a range of 4 points. Many computer programs, such as SPSS, use this definition. This definition works well for variables with precisely defined upper and lower boundaries. For example, if you are measuring proportions of an object, like pieces of a pizza, you can obtain values such as $\frac{1}{8}$, $\frac{1}{4}$, $\frac{1}{2}$, $\frac{3}{4}$, and so on. Expressed as decimal values, the proportions range from 0 to 1. You can never have a value less than 0 (none of the pizza) and you can never have a value greater than 1 (all of the pizza). Thus, the complete set of proportions is bounded by 0 at one end and by 1 at the other. As a result, the proportions cover a range of 1 point.

Continuous and discrete variables were discussed in Chapter 1 on pages 19–21.

When the scores are measurements of a continuous variable, the range can be defined as the difference between the upper real limit (URL) for the largest score (X_{max}) and the lower real limit (LRL) for the smallest score (X_{min}).

$$\text{range} = \text{URL for } X_{max} - \text{LRL for } X_{min}$$

If the scores have values from 1−5, for example, the range is 5.5 − 0.5 = 5 points.

When the scores are whole numbers, the range can also be defined as the number of measurement categories. If every individual is classified as 1, 2, or 3 then there are three measurement categories and the range is 3 points. Defining the range as the number of measurement categories also works for discrete variables that are measured with numerical scores. For example, if you are measuring the number of children in a family and the data produce values from 0−4, then there are five measurement categories (0, 1, 2, 3, and 4) and the range is 5 points. By this definition, when the scores are all whole numbers, the range can be obtained by

$$X_{max} - X_{min} + 1$$

Using either definition, the range is probably the most obvious way to describe how spread out the scores are—simply find the distance between the maximum and the minimum scores. The problem with using the range as a measure of variability is that it is completely determined by the two extreme values and ignores the other scores in the distribution. Thus, a distribution with one unusually large (or small) score will have a large range even if the other scores are all clustered close together.

Because the range does not consider all the scores in the distribution, it often does not give an accurate description of the variability for the entire distribution. For this reason, the range is considered to be a crude and unreliable measure of variability. Therefore, in most situations, it does not matter which definition you use to determine the range.

LEARNING CHECK

1. Which of the following sets of scores has the greatest variability?
   a. 2, 3, 7, 12
   b. 13, 15, 16, 17
   c. 24, 25, 26, 27
   d. 42, 44, 45, 46

2. What is the range for the following set of scores? 3, 7, 9, 10, 12
   a. 3 points
   b. 4 or 5 points
   c. 9 or 10 points
   d. 12 points

3. How many scores in the distribution are used to compute the range?
   a. only 1
   b. 2
   c. 50% of them
   d. all of the scores

ANSWERS   1. A   2. C   3. B

4.2 | Defining Standard Deviation and Variance

LEARNING OBJECTIVES

3. Define variance and standard deviation and describe what is measured by each.

4. Calculate variance and standard deviation for a simple set of scores.

5. Estimate the standard deviation for a set of scores based on a visual examination of a frequency distribution graph of the distribution.

The standard deviation is the most commonly used and the most important measure of variability. Standard deviation uses the mean of the distribution as a reference point and measures variability by considering the distance between each score and the mean.

In simple terms, the standard deviation provides a measure of the standard, or average, distance from the mean, and describes whether the scores are clustered closely around the mean or are widely scattered.

Although the concept of standard deviation is straightforward, the actual equations tend to be more complex. Therefore, we begin by looking at the logic that leads to these equations. If you remember that our goal is to measure the standard, or typical, distance from the mean, then this logic and the equations that follow should be easier to remember.

STEP 1   The first step in finding the standard distance from the mean is to determine the *deviation*, or distance from the mean, for each individual score. By definition, the deviation for each score is the difference between the score and the mean.

DEFINITION

> **Deviation** is distance from the mean:
>
> $$\text{deviation score} = X - \mu$$

A deviation score is often represented by a lowercase letter *x*.

For a distribution of scores with $\mu = 50$, if your score is $X = 53$, then your *deviation score* is

$$X - \mu = 53 - 50 = 3$$

If your score is $X = 45$, then your deviation score is

$$X - \mu = 45 - 50 = -5$$

Notice that there are two parts to a deviation score: the sign ($+$ or $-$) and the number. The sign tells the direction from the mean—that is, whether the score is located above ($+$) or below ($-$) the mean. The number gives the actual distance from the mean. For example, a deviation score of -6 corresponds to a score that is below the mean by a distance of 6 points.

STEP 2   Because our goal is to compute a measure of the standard distance from the mean, the obvious next step is to calculate the mean of the deviation scores. To compute this mean, you first add up the deviation scores and then divide by N. This process is demonstrated in the following example.

EXAMPLE 4.1   We start with the following set of $N = 4$ scores. These scores add up to $\Sigma X = 12$, so the mean is $\mu = \frac{12}{4} = 3$. For each score, we have computed the deviation.

X	$X - \mu$
8	$+5$
1	-2
3	0
0	-3
	$0 = \Sigma(X - \mu)$

■

Note that the deviation scores add up to zero. This should not be surprising if you remember that the mean serves as a balance point for the distribution. The total of the distances above the mean is exactly equal to the total of the distances below the mean (see page 72). Thus, the total for the positive deviations is exactly equal to the total for the negative deviations, and the complete set of deviations always adds up to zero.

Because the sum of the deviations is *always* zero, the mean of the deviations is also zero and is of no value as a measure of variability. Specifically, the mean of the deviations is zero if the scores are closely clustered and it is zero if the scores are widely scattered. (You should note, however, that the constant value of zero is useful in other ways. Whenever you are working with deviation scores, you can check your calculations by making sure that the deviation scores add up to zero.)

STEP 3 The average of the deviation scores will not work as a measure of variability because it is always zero. Clearly, this problem results from the positive and negative values canceling each other out. The solution is to get rid of the signs ($+$ and $-$). The standard procedure for accomplishing this is to square each deviation score. Using the squared values, you then compute the *mean squared deviation,* which is called *variance.*

DEFINITION

> **Variance** equals the mean of the squared deviations. Variance is the average squared distance from the mean.

Note that the process of squaring deviation scores does more than simply get rid of plus and minus signs. It results in a measure of variability based on *squared* distances. Although variance is valuable for some of the *inferential* statistical methods covered later, the concept of squared distance is not an intuitive or easy to understand *descriptive* measure. For example, it is not particularly useful to know that the squared distance from New York City to Boston is 26,244 miles squared. The squared value becomes meaningful, however, if you take the square root. Therefore, we continue the process one more step.

STEP 4 Remember that our goal is to compute a measure of the standard distance from the mean. Variance, which measures the average squared distance from the mean, is not exactly what we want. The final step simply takes the square root of the variance to obtain the *standard deviation,* which measures the standard distance from the mean.

DEFINITION

> **Standard deviation** is the square root of the variance and provides a measure of the standard, or average distance from the mean.
>
>
> $$\text{Standard deviation} = \sqrt{\text{variance}}$$

Figure 4.2 shows the overall process of computing variance and standard deviation. Remember that our goal is to measure variability by finding the standard distance from the mean. However, we cannot simply calculate the average of the distances because this value will always be zero. Therefore, we begin by squaring each distance, then we find the average of the squared distances, and finally we take the square root to obtain a measure of the standard distance. Technically, the standard deviation is the square root of the average squared deviation. Conceptually, however, the standard deviation provides a measure of the average distance from the mean.

Although we still have not presented any formulas for variance or standard deviation, you should be able to compute these two statistical values from their definitions. The following example demonstrates this process.

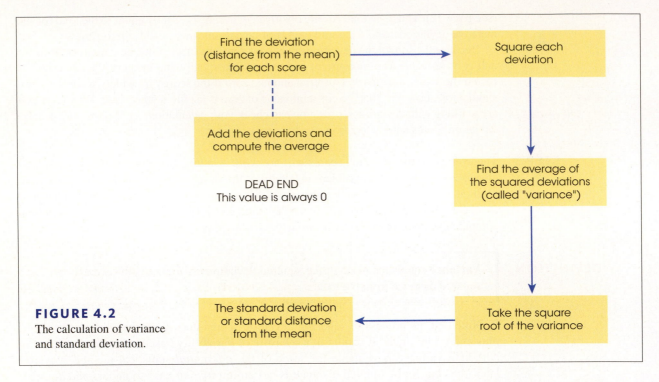

FIGURE 4.2

The calculation of variance and standard deviation.

EXAMPLE 4.2   We will calculate the variance and standard deviation for the following population of $N = 5$ scores:

$$1, \quad 9, \quad 5, \quad 8, \quad 7$$

Remember that the purpose of standard deviation is to measure the standard distance from the mean, so we begin by computing the population mean. These five scores add up to $\Sigma X = 30$ so the mean is $\mu = \frac{30}{5} = 6$. Next, we find the deviation (distance from the mean) for each score and then square the deviations. Using the population mean $\mu = 6$, these calculations are shown in the following table.

Score X	Deviation $X - \mu$	Squared Deviation $(X - \mu)^2$
1	-5	25
9	3	9
5	-1	1
8	2	4
7	1	1
		$40 =$ the sum of the squared deviations

For this set of $N = 5$ scores, the squared deviations add up to 40. The mean of the squared deviations, the variance, is $\frac{40}{5} = 8$, and the standard deviation is $\sqrt{8} = 2.83$. ∎

You should note that a standard deviation of 2.83 is a sensible answer for this distribution. The five scores in the population are shown in a histogram in Figure 4.3 so that you can see the distances more clearly. Note that the scores closest to the mean are only 1 point away. Also, the score farthest from the mean is 5 points away. For this distribution, the largest distance from the mean is 5 points and the smallest distance is 1 point. Thus, the standard

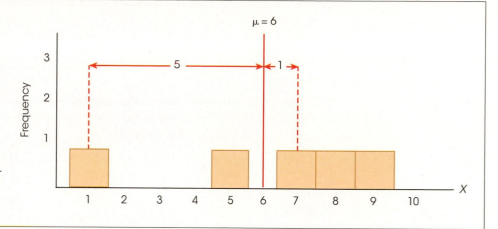

FIGURE 4.3

A frequency distribution histogram for a population of $N = 5$ scores. The mean for this population is $\mu = 6$. The smallest distance from the mean is 1 point and the largest distance is 5 points. The standard distance (or standard deviation) should be between 1 and 5 points.

distance should be somewhere between 1 and 5. By looking at a distribution in this way, you should be able to make a rough estimate of the standard deviation. In this case, the standard deviation should be between 1 and 5, probably around 3 points. The value we calculated for the standard deviation is in excellent agreement with this estimate.

Making a quick estimate of the standard deviation can help you avoid errors in calculation. For example, if you calculated the standard deviation for the scores in Figure 4.3 and obtained a value of 12, you should realize immediately that you have made an error. (If the biggest deviation is only 5 points, then it is impossible for the standard deviation to be 12.)

The following example is an opportunity for you to test your understanding by computing variance and standard deviation yourself.

EXAMPLE 4.3   Compute the variance and standard deviation for the following set of $N = 6$ scores: 12, 0, 1, 7, 4, and 6. You should obtain a variance of 16 and a standard deviation of 4. Good luck.   ∎

Because the standard deviation and variance are defined in terms of distance from the mean, these measures of variability are used only with numerical scores that are obtained from measurements on an interval or a ratio scale. Recall from Chapter 1 (page 22) that these two scales are the only ones that provide information about distance; nominal and ordinal scales do not. Also, recall from Chapter 3 (page 86–88) that it is inappropriate to compute a mean for ordinal data and it is impossible to compute a mean for nominal data. Because the mean is a critical component in the calculation of standard deviation and variance, the same restrictions that apply to the mean also apply to these two measures of variability. Specifically, the mean, standard deviation, and variance should be used only with numerical scores from interval or ratio scales of measurement.

LEARNING CHECK   **1.** Standard deviation is probably the most commonly used value to describe and measure variability. Which of the following accurately describes the concept of standard deviation?

  a. the average distance between one score and another

  b. the average distance between a score and the mean

  c. the total distance from the smallest score to the largest score

  d. one half of the total distance from the smallest score to the largest score

2. What is the variance for the following set of scores? 2, 2, 2, 2, 2

   a. 0
   b. 2
   c. 4
   d. 5

3. Which of the following values is the most reasonable estimate of the standard deviation for the set of scores in the following distribution? (It may help to imagine or sketch a histogram of the distribution.)

   a. 0
   b. 1
   c. 3
   d. 5

X	f
5	1
4	2
3	4
2	2
1	1

ANSWERS   **1.** B, **2.** A, **3.** B

4.3 | Measuring Variance and Standard Deviation for a Population

LEARNING OBJECTIVES

6. Calculate SS, the sum of the squared deviations, for a population using either the definitional or the computational formula and describe the circumstances in which each formula is appropriate.

7. Calculate the variance and the standard deviation for a population.

The concepts of standard deviation and variance are the same for both samples and populations. However, the details of the calculations differ slightly, depending on whether you have data from a sample or from a complete population. We first consider the formulas for populations and then look at samples in Section 4.4.

■ The Sum of Squared Deviations (SS)

Recall that variance is defined as the mean of the squared deviations. This mean is computed exactly the same way you compute any mean: first find the sum, and then divide by the number of scores.

$$\text{Variance} = \text{mean squared deviation} = \frac{\text{sum of squared deviations}}{\text{number of scores}}$$

The value in the numerator of this equation, the sum of the squared deviations, is a basic component of variability, and we will focus on it. To simplify things, it is identified by the notation *SS* (for sum of squared deviations), and it generally is referred to as the *sum of squares*.

DEFINITION   *SS*, or **sum of squares**, is the sum of the squared deviation scores.

You need to know two formulas to compute *SS*. These formulas are algebraically equivalent (they always produce the same answer), but they look different and are used in different situations.

The first of these formulas is called the definitional formula because the symbols in the formula literally define the process of adding up the squared deviations:

$$\text{Definitional formula: } SS = \Sigma(X - \mu)^2 \tag{4.1}$$

To find the sum of the squared deviations, the formula instructs you to perform the following sequence of calculations:

1. Find each deviation score $(X - \mu)$.

2. Square each deviation score, $(X - \mu)^2$.

3. Add the squared deviations.

The result is *SS*, the sum of the squared deviations. The following example demonstrates using this formula.

EXAMPLE 4.4

We will compute *SS* for the following set of $N = 4$ scores. These scores have a sum of $\Sigma X = 8$, so the mean is $\mu = \frac{8}{4} = 2$. The following table shows the deviation and the squared deviation for each score. The sum of the squared deviation is $SS = 22$.

Score X	Deviation $X - \mu$	Squared Deviation $(X - \mu)^2$	
1	-1	1	$\Sigma X = 8$
0	-2	4	$\mu = 2$
6	$+4$	16	
1	-1	$\dfrac{1}{22}$	$\Sigma(X - \mu)^2 = 22$

Although the definitional formula is the most direct method for computing *SS*, it can be awkward to use. In particular, when the mean is not a whole number, the deviations all contain decimals or fractions, and the calculations become difficult. In addition, calculations with decimal values introduce the opportunity for rounding error, which can make the result less accurate. For these reasons, an alternative formula has been developed for computing *SS*. The alternative, known as the computational formula, performs calculations with the scores (not the deviations) and therefore minimizes the complications of decimals and fractions.

$$\text{Computational formula: } SS = \Sigma X^2 - \frac{(\Sigma X)^2}{N} \tag{4.2}$$

The first part of this formula directs you to square each score and then add the squared values, ΣX^2. In the second part of the formula, you find the sum of the scores, ΣX, then square this total and divide the result by N. Finally, subtract the second part from the first. The use of this formula is shown in Example 4.5 with the same scores that we used to demonstrate the definitional formula.

EXAMPLE 4.5

The computational formula is used to calculate *SS* for the same set of $N = 4$ scores we used in Example 4.4. Note that the formula requires the calculation of two sums: first, compute ΣX, and then square each score and compute ΣX^2. These calculations are shown in the following table. The two sums are used in the formula to compute *SS*.

X	X^2
1	1
0	0
6	36
1	1
$\Sigma X = 8$	$\Sigma X^2 = 38$

$$SS = \Sigma X^2 - \frac{(\Sigma X)^2}{N}$$

$$= 38 - \frac{(8)^2}{4}$$

$$= 38 - \frac{64}{4}$$

$$= 38 - 16$$

$$= 22$$

Note that the two formulas produce exactly the same value for *SS*. Although the formulas look different, they are in fact equivalent. The definitional formula provides the most direct representation of the concept of *SS*; however, this formula can be awkward to use, especially if the mean includes a fraction or decimal value. If you have a small group of scores and the mean is a whole number, then the definitional formula is fine; otherwise, the computational formula is usually easier to use.

■ Final Formulas and Notation

In the same way that sum of squares, or *SS*, is used to refer to the sum of squared deviations, the term *mean square*, or *MS*, is often used to refer to variance, which is the mean squared deviation.

With the definition and calculation of *SS* behind you, the equations for variance and standard deviation become relatively simple. Remember that variance is defined as the mean squared deviation. The mean is the sum of the squared deviations divided by *N*, so the equation for the *population variance* is

$$\text{variance} = \frac{SS}{N}$$

Standard deviation is the square root of variance, so the equation for the *population standard deviation* is

$$\text{standard deviation} = \sqrt{\frac{SS}{N}}$$

There is one final bit of notation before we work completely through an example computing *SS*, variance, and standard deviation. Like the mean (μ), variance and standard deviation are parameters of a population and are identified by Greek letters. To identify the standard deviation, we use the Greek letter sigma (the Greek letter *s*, standing for standard deviation). The capital letter sigma (Σ) has been used already, so we now use the lowercase sigma, σ, as the symbol for the population standard deviation. To emphasize the relationship between standard deviation and variance, we use σ^2 as the symbol for population variance (standard deviation is the square root of the variance). Thus,

$$\text{population standard deviation} = \sigma = \sqrt{\sigma^2} = \sqrt{\frac{SS}{N}} \qquad (4.3)$$

$$\text{population variance} = \sigma^2 = \frac{SS}{N} \qquad (4.4)$$

DEFINITIONS

Population variance is represented by the symbol σ^2 and equals the mean squared distance from the mean. Population variance is obtained by dividing the sum of squares by *N*.

Population standard deviation is represented by the symbol σ and equals the square root of the population variance.

Earlier, in Examples 4.4 and 4.5, we computed the sum of squared deviations for a simple population of $N = 4$ scores $(1, 0, 6, 1)$ and obtained $SS = 22$. For this population, the variance is

$$\sigma^2 = \frac{SS}{N} = \frac{22}{4} = 5.50$$

and the standard deviation is $\sigma = \sqrt{5.50} = 2.345$

LEARNING CHECK

1. What is the value of SS, the sum of the squared deviations, for the following population of $N = 4$ scores? Scores: 1, 4, 6, 1
   a. 0
   b. 18
   c. 54
   d. $12^2 = 144$

2. Each of the following is the sum of the scores for a population of $N = 4$ scores. For which population would the definitional formula be a better choice than the computational formula for calculating SS.
   a. $\Sigma X = 9$
   b. $\Sigma X = 12$
   c. $\Sigma X = 15$
   d. $\Sigma X = 19$

3. What is the standard deviation for the following population of scores? Scores: 1, 3, 7, 4, 5
   a. 20
   b. 5
   c. 4
   d. 2

ANSWERS   1. B, 2. B, 3. D

4.4 | Measuring Standard Deviation and Variance for a Sample

LEARNING OBJECTIVES

8. Explain why it is necessary to make a correction to the formulas for variance and standard deviation when computing these statistics for a sample.

9. Calculate SS, the sum of the squared deviations, for a sample using either the definitional or the computational formula and describe the circumstances in which each formula is appropriate.

10. Calculate the variance and the standard deviation for a sample.

■ The Problem with Sample Variability

The goal of inferential statistics is to use the limited information from samples to draw general conclusions about populations. The basic assumption of this process is that samples should be

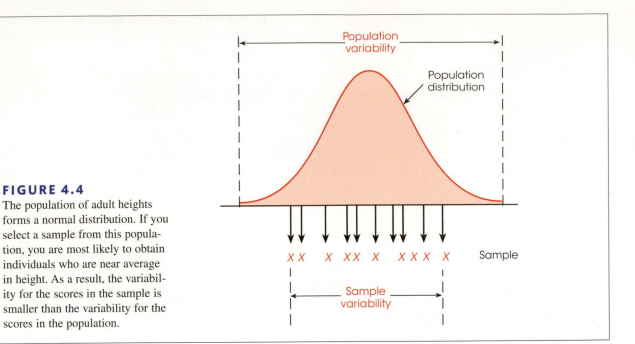

FIGURE 4.4

The population of adult heights forms a normal distribution. If you select a sample from this population, you are most likely to obtain individuals who are near average in height. As a result, the variability for the scores in the sample is smaller than the variability for the scores in the population.

A sample statistic is said to be *biased* if, on the average, it consistently overestimates or underestimates the corresponding population parameter.

representative of the populations from which they come. This assumption poses a special problem for variability because samples consistently tend to be less variable than their populations. The mathematical explanation for this fact is beyond the scope of this book but a simple demonstration of this general tendency is shown in Figure 4.4. Notice that a few extreme scores in the population tend to make the population variability relatively large. However, these extreme values are unlikely to be obtained when you are selecting a sample, which means that the sample variability is relatively small. The fact that a sample tends to be less variable than its population means that sample variability gives a *biased* estimate of population variability. This bias is in the direction of underestimating the population value rather than being right on the mark. (The concept of a biased statistic is discussed in more detail in Section 4.5.)

Fortunately, the bias in sample variability is consistent and predictable, which means it can be corrected. For example, if the speedometer in your car consistently shows speeds that are 5 mph slower than you are actually going, it does not mean that the speedometer is useless. It simply means that you must make an adjustment to the speedometer reading to get an accurate speed. In the same way, we will make an adjustment in the calculation of sample variance. The purpose of the adjustment is to make the resulting value for sample variance an accurate and unbiased representative of the population variance.

■ Formulas for Sample Variance and Standard Deviation

The calculations of variance and standard deviation for a sample follow the same steps that were used to find population variance and standard deviation. First, calculate the sum of squared deviations (SS). Second, calculate the variance. Third, find the square root of the variance, which is the standard deviation.

Except for minor changes in notation, calculating the sum of squared deviations, *SS*, is exactly the same for a sample as they were for a population. The changes in notation involve using *M* for the sample mean instead of μ, and using *n* (instead of *N*) for the number of scores. Thus, the definitional formula for *SS* for a sample is

Definitional formula: $SS = \Sigma(X - M)^2$ **(4.5)**

Note that the sample formula has exactly the same structure as the population formula (Equation 4.1 on p. 109) and instructs you to find the sum of the squared deviations using the following three steps:

1. Find the deviation from the mean for each score: deviation $= X - M$
2. Square each deviation: squared deviation $= (X - M)^2$
3. Add the squared deviations: $SS = \Sigma(X - M)^2$

The value of SS also can be obtained using a computational formula. Except for one minor difference in notation (using n in place of N), the computational formula for SS is the same for a sample as it was for a population (see Equation 4.2). Using sample notation, this formula is:

$$\text{Computational formula: } SS = \Sigma X^2 - \frac{(\Sigma X)^2}{n} \qquad (4.6)$$

Again, calculating SS for a sample is exactly the same as for a population, except for minor changes in notation. After you compute SS, however, it becomes critical to differentiate between samples and populations. To correct for the bias in sample variability, it is necessary to make an adjustment in the formulas for sample variance and standard deviation. With this in mind, *sample variance* (identified by the symbol s^2) is defined as

$$\text{sample variance} = s^2 = \frac{SS}{n - 1} \qquad (4.7)$$

Sample standard deviation (identified by the symbol s) is simply the square root of the variance.

$$\text{sample standard deviation} = s = \sqrt{s^2} = \sqrt{\frac{SS}{n - 1}} \qquad (4.8)$$

DEFINITIONS

> **Sample variance** is represented by the symbol s^2 and equals the mean squared distance from the mean. Sample variance is obtained by dividing the sum of squares by $n - 1$.
>
> **Sample standard deviation** is represented by the symbol s and equal the square root of the sample variance.

Remember, sample variability tends to underestimate population variability unless some correction is made.

Notice that the sample formulas divide by $n - 1$ unlike the population formulas, which divide by N (see Equations 4.3 and 4.4). This is the adjustment that is necessary to correct for the bias in sample variability. The effect of the adjustment is to increase the value you will obtain. Dividing by a smaller number ($n - 1$ instead of n) produces a larger result and makes sample variance an accurate and unbiased estimator of population variance. The following example demonstrates the calculation of variance and standard deviation for a sample.

EXAMPLE 4.6

We have selected a sample of $n = 8$ scores from a population. The scores are 4, 6, 5, 11, 7, 9, 7, 3. The frequency distribution histogram for this sample is shown in Figure 4.5. Before we begin any calculations, you should be able to look at the sample distribution and make a preliminary estimate of the outcome. Remember that standard deviation measures the standard distance from the mean. For this sample the mean is $M = \frac{52}{8} = 6.5$. The scores closest to the mean are $X = 6$ and $X = 7$, both of which are exactly 0.50 points away. The score farthest from the mean is $X = 11$, which is 4.50 points away. With the smallest distance from the mean equal to 0.50 and the largest distance equal to 4.50, we should obtain a standard distance somewhere between 0.50 and 4.50, probably around 2.5.

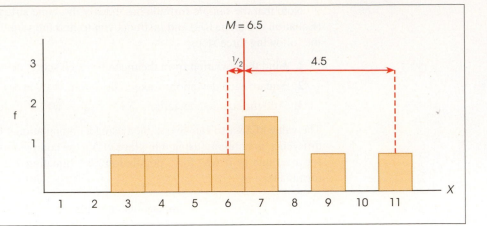

FIGURE 4.5

The frequency distribution histogram for a sample of $n = 8$ scores. The sample mean is $M = 6.5$. The smallest distance from the mean is 0.5 points, and the largest distance from the mean is 4.5 points. The standard distance (standard deviation) should be between 0.5 and 4.5 points, or about 2.5.

We begin the calculations by finding the value of SS for this sample. Because the mean is not a whole number ($M = 6.5$), the computational formula is easier to use. The scores, and the squared scores, needed for this formula are shown in the following table.

Scores X	Squared Scores X^2
4	16
6	36
5	25
11	121
7	49
9	81
7	49
3	9
$\Sigma X = 52$	$\Sigma X^2 = 386$

Using the two sums,

$$SS = \Sigma X^2 - \frac{(\Sigma X)^2}{n} = 386 - \frac{(52)^2}{8}$$

$$= 386 - 338$$

$$= 48$$

The sum of squared deviations for this sample is $SS = 48$. Continuing the calculations,

$$\text{sample variance} = s^2 = \frac{SS}{n - 1} = \frac{48}{8 - 1} = 6.86$$

Finally, the standard deviation is

$$s = \sqrt{s^2} = \sqrt{6.86} = 2.62$$

Note that the value we obtained is in excellent agreement with our preliminary prediction (see Figure 4.5). ■

The following example is an opportunity for you to test your understanding by computing sample variance and standard deviation yourself.

EXAMPLE 4.7

For the following sample of $n = 5$ scores, compute the variance and standard deviation: 1, 5, 5. 1, and 8. You should obtain $s^2 = 9$ and $s = 3$. Good luck. ∎

Remember that the formulas for sample variance and standard deviation were constructed so that the sample variability would provide a good estimate of population variability. For this reason, the sample variance is often called *estimated population variance,* and the sample standard deviation is called *estimated population standard deviation.* When you have only a sample to work with, the variance and standard deviation for the sample provide the best possible estimates of the population variability.

■ Sample Variability and Degrees of Freedom

Although the concept of a deviation score and the calculation *SS* are almost exactly the same for samples and populations, the minor differences in notation are really very important. Specifically, with a population, you find the deviation for each score by measuring its distance from the population mean. With a sample, on the other hand, the value of μ is unknown and you must measure distances from the sample mean. Because the value of the sample mean varies from one sample to another, you must first compute the sample mean before you can begin to compute deviations. However, calculating the value of M places a restriction on the variability of the scores in the sample. This restriction is demonstrated in the following example.

EXAMPLE 4.8

Suppose we select a sample of $n = 3$ scores and compute a mean of $M = 5$. The first two scores in the sample have no restrictions; they are independent of each other and they can have any values. For this demonstration, we will assume that we obtained $X = 2$ for the first score and $X = 9$ for the second. At this point, however, the third score in the sample is restricted.

X	A sample of $n = 3$ scores with a mean of $M = 5$
2	
9	
—	← What is the third score?

For this example, the third score must be $X = 4$. The reason that the third score is restricted to $X = 4$ is that the entire sample of $n = 3$ scores has a mean of $M = 5$. For 3 scores to have a mean of 5, the scores must have a total of $\Sigma X = 15$. Because the first two scores add up to 11 $(9 + 2)$, the third score must be $X = 4$. ∎

In Example 4.8, the first two out of three scores were free to have any values, but the final score was dependent on the values chosen for the first two. In general, with a sample of n scores, the first $n - 1$ scores are free to vary, but the final score is restricted. As a result, the sample is said to have $n - 1$ *degrees of freedom.*

DEFINITION

For a sample of n scores, the **degrees of freedom**, or *df*, for the sample variance are defined as $df = n - 1$. The degrees of freedom determine the number of scores in the sample that are independent and free to vary.

The $n - 1$ degrees of freedom for a sample is the same $n - 1$ that is used in the formulas for sample variance and standard deviation. Remember that variance is defined as the mean squared deviation. As always, this mean is computed by finding the sum and dividing by the number of scores:

$$\text{mean} = \frac{\text{sum}}{\text{number}}$$

To calculate sample variance (mean squared deviation), we find the sum of the squared deviations (*SS*) and divide by the number of scores that are free to vary. This number is $n - 1 = df$. Thus, the formula for sample variance is

$$s^2 = \frac{\text{sum of squared deviations}}{\text{number of scores free to vary}} = \frac{SS}{df} = \frac{SS}{n - 1}$$

Later in this book, we use the concept of degrees of freedom in other situations. For now, remember that knowing the sample mean places a restriction on sample variability. Only $n - 1$ of the scores are free to vary; $df = n - 1$.

LEARNING CHECK

1. If sample variance is computed by dividing by n, instead of $n - 1$, how will the obtained values be related to the corresponding population variance.
   a. They will consistently underestimate the population variance.
   b. They will consistently overestimate the population variance.
   c. The average value will be exactly equal to the population variance.
   d. The average value will be close to, but not exactly equal to, the population variance.

2. What is the value of *SS*, the sum of the squared deviations, for the following sample? Scores: 1, 4, 0, 1
   a. 36
   b. 18
   c. 9
   d. 3

3. What is the variance for the following sample of $n = 4$ scores? Scores; 2, 5, 1, 2
   a. 34/3 = 11.33
   b. 9/4 = 2.25
   c. 9/3 = 3
   d. $\sqrt{3}$ = 173

ANSWERS   1. A, 2. C, 3. C

4.5 | Sample Variance as an Unbiased Statistic

LEARNING OBJECTIVES

11. Define biased and unbiased statistics.

12. Explain why the sample mean and the sample variance (dividing by $n - 1$) are unbiased statistics.

■ Biased and Unbiased Statistics

Earlier we noted that sample variability tends to underestimate the variability in the corresponding population. To correct for this problem we adjusted the formula for sample variance by dividing by $n - 1$ instead of dividing by n. The result of the adjustment is that sample variance provides a much more accurate representation of the population variance. Specifically, dividing by $n - 1$ produces a sample variance that provides an *unbiased* estimate of the corresponding population variance. This does not mean that each individual sample variance will be exactly equal to its population variance. In fact, some sample variances will overestimate the population value and some will underestimate it. However, the average of all the sample variances will produce an accurate estimate of the population variance. This is the idea behind the concept of an unbiased statistic.

DEFINITIONS

> A sample statistic is **unbiased** if the average value of the statistic is equal to the population parameter. (The average value of the statistic is obtained from all the possible samples for a specific sample size, n.)
>
> A sample statistic is **biased** if the average value of the statistic either underestimates or overestimates the corresponding population parameter.

The following example demonstrates the concept of biased and unbiased statistics.

EXAMPLE 4.9

We begin with a population that consists of exactly $N = 6$ scores: 0, 0, 3, 3, 9, 9. With a few calculations you should be able to verify that this population has a mean of $\mu = 4$ and a variance of $\sigma^2 = 14$.

Next, we select samples of $n = 2$ scores from this population. In fact, we obtain every single possible sample with $n = 2$. The complete set of samples is listed in Table 4.1. Notice

TABLE 4.1

The set of all the possible samples for $n = 2$ selected from the population described in Example 4.9. The mean is computed for each sample, and the variance is computed two different ways: (1) dividing by n, which is incorrect and produces a biased statistic; and (2) dividing by $n - 1$, which is correct and produces an unbiased statistic.

				Sample Statistics	
Sample	First Score	Second Score	Mean M	Biased Variance (Using n)	Unbiased Variance (Using $n - 1$)
1	0	0	0.00	0.00	0.00
2	0	3	1.50	2.25	4.50
3	0	9	4.50	20.25	40.50
4	3	0	1.50	2.25	4.50
5	3	3	3.00	0.00	0.00
6	3	9	6.00	9.00	18.00
7	9	0	4.50	20.25	40.50
8	9	3	6.00	9.00	18.00
9	9	9	9.00	0.00	0.00
		Totals	36.00	63.00	126.00

We have structured this example to mimic "sampling with replacement," which is covered in Chapter 6.

that the samples are listed systematically to ensure that every possible sample is included. We begin by listing all the samples that have $X = 0$ as the first score, then all the samples with $X = 3$ as the first score, and so on. Notice that the table shows a total of 9 samples.

Finally, we have computed the mean and the variance for each sample. Note that the sample variance has been computed two different ways. First, we make no correction for bias and compute each sample variance as the average of the squared deviations by simply dividing SS by n. Second, we compute the correct sample variances for which SS is divided by $n - 1$ to produce an unbiased measure of variance. You should verify our calculations by computing one or two of the values for yourself. The complete set of sample means and sample variances is presented in Table 4.1. ■

First, consider the column of biased sample variances, which were calculated dividing by n. These 9 sample variances add up to a total of 63, which produces an average value of $\frac{63}{9} = 7$. The original population variance, however, is $\sigma^2 = 14$. Note that the average of the sample variances is *not* equal to the population variance. If the sample variance is computed by dividing by n, the resulting values will not produce an accurate estimate of the population variance. On average, these sample variances underestimate the population variance and, therefore, are biased statistics.

Next, consider the column of sample variances that are computed using $n - 1$. Although the population has a variance of $\sigma^2 = 14$, you should notice that none of the samples has a variance exactly equal to 14. However, if you consider the complete set of sample variances, you will find that the 9 values add up to a total of 126, which produces an average value of $\frac{126}{9} = 14.00$. Thus, the average of the sample variances is exactly equal to the original population variance. On average, the sample variance (computed using $n - 1$) produces an accurate, unbiased estimate of the population variance.

Finally, direct your attention to the column of sample means. For this example, the original population has a mean of $\mu = 4$. Although none of the samples has a mean exactly equal to 4, if you consider the complete set of sample means, you will find that the 9 sample means add up to a total of 36, so the average of the sample means is $\frac{36}{9} = 4$. Note that the average of the sample means is exactly equal to the population mean. Again, this is what is meant by the concept of an unbiased statistic. On average, the sample values provide an accurate representation of the population. In this example, the average of the 9 sample means is exactly equal to the population mean.

In summary, both the sample mean and the sample variance (using $n - 1$) are examples of unbiased statistics. This fact makes the sample mean and sample variance extremely valuable for use as inferential statistics. Although no individual sample is likely to have a mean and variance exactly equal to the population values, both the sample mean and the sample variance, on average, do provide accurate estimates of the corresponding population values.

LEARNING CHECK

1. What is meant by a biased statistic.

   a. The average value for the statistic overestimates the corresponding population parameter.

   b. The average value for the statistic underestimates the corresponding population parameter.

   c. The average value for the statistic either overestimates or underestimates the corresponding population parameter.

   d. The average value for the statistic is exactly equal to the corresponding population parameter.

2. A researcher selects all the possible samples with $n = 3$ scores from a population and computes the sample variance, dividing by $n - 1$, for each sample. If the population variance is $\sigma^2 = 6$, then what is the average value for all of the sample variances?

a. 6

b. greater than 6

c. less than 6

d. impossible to determine

3. Which of the following is an example of an unbiased statistic.

a. the sample mean

b. the sample variance (dividing by $n - 1$)

c. both the sample mean and the sample variance (dividing by $n - 1$)

d. neither the sample mean nor the sample variance (dividing by $n - 1$)

ANSWERS   **1.** C,   **2.** A,   **3.** C

4.6 | More about Variance and Standard Deviation

LEARNING OBJECTIVES

13. Describe how the mean and standard deviation are represented in a frequency distribution graph of a population or sample distribution.

14. Explain how the mean and standard deviation are affected when a constant is added to every score or when every score is multiplied by a constant.

15. Describe how the mean and standard deviation are reported in research journals.

16. Describe the appearance of a distribution based on the values for the mean and standard deviation.

17. Explain how patterns in sample data are affected by sample variance.

■ Presenting the Mean and Standard Deviation in a Frequency Distribution Graph

In frequency distribution graphs, we identify the position of the mean by drawing a vertical line and labeling it with μ or M. Because the standard deviation measures distance from the mean, it is represented by a line or an arrow drawn from the mean outward for a distance equal to the standard deviation and labeled with a σ or an s. Figure 4.6(a) shows an example of a population distribution with a mean of $\mu = 80$ and a standard deviation of $\sigma = 8$, and Figure 4.6(b) shows the frequency distribution for a sample with a mean of $M = 16$ and a standard deviation of $s = 2$. For rough sketches, you can identify the mean with a vertical line in the middle of the distribution. The standard deviation line should extend approximately halfway from the mean to the most extreme score. [*Note:* In Figure 4.6(a) we show the standard deviation as a line to the right of the mean. You should realize that we could have drawn the line pointing

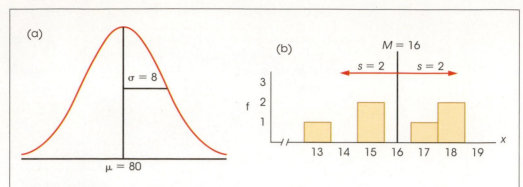

FIGURE 4.6

Showing means and standard deviations in frequency distribution graphs. (a) A population distribution with a mean of $\mu = 80$ and a standard deviation of $\sigma = 8$. (b) A sample with a mean of $M = 16$ and a standard deviation of $s = 2$.

to the left, or we could have drawn two lines (or arrows), with one pointing to the right and one pointing to the left, as in Figure 4.6(b). In each case, the goal is to show the standard distance from the mean.]

■ Transformations of Scale

Occasionally a set of scores is transformed by adding a constant to each score or by multiplying each score by a constant value. This happens, for example, when exposure to a treatment adds a fixed amount to each participant's score or when you want to change the unit of measurement (to convert from minutes to seconds, multiply each score by 60). What happens to the standard deviation when the scores are transformed in this manner?

The easiest way to determine the effect of a transformation is to remember that the standard deviation is a measure of distance. If you select any two scores and see what happens to the distance between them, you also will find out what happens to the standard deviation.

1. **Adding a constant to each score does not change the standard deviation.** If you begin with a distribution that has $\mu = 40$ and $\sigma = 10$, what happens to the standard deviation if you add 5 points to every score? Consider any two scores in this distribution: Suppose, for example, that these are exam scores and that you had a score of $X = 41$ and your friend had $X = 43$. The distance between these two scores is $43 - 41 = 2$ points. After adding the constant, 5 points, to each score, your score would be $X = 46$, and your friend would have $X = 48$. The distance between scores is still 2 points. Adding a constant to every score does not affect any of the distances and, therefore, does not change the standard deviation. This fact can be seen clearly if you imagine a frequency distribution graph. If, for example, you add 10 points to each score, then every score in the graph is moved 10 points to the right. The result is that the entire distribution is shifted to a new position 10 points up the scale. Note that the mean moves along with the scores and is increased by 10 points. However, the variability does not change because each of the deviation scores $(X - \mu)$ does not change.

2. **Multiplying each score by a constant causes the standard deviation to be multiplied by the same constant.** Consider the same distribution of exam scores we looked at earlier. If $\mu = 40$ and $\sigma = 10$, what would happen to the standard deviation if each score were multiplied by 2? Again, we will look at two scores, $X = 41$ and $X = 43$, with a distance between them equal to 2 points. After all the

scores have been multiplied by 2, these scores become $X = 82$ and $X = 86$. Now the distance between scores is 4 points, twice the original distance. Multiplying each score causes each distance to be multiplied, so the standard deviation also is multiplied by the same amount.

IN THE LITERATURE

Reporting the Standard Deviation

In reporting the results of a study, the researcher often provides descriptive information for both central tendency and variability. The dependent variables in psychology research are often numerical values obtained from measurements on interval or ratio scales. With numerical scores the most common descriptive statistics are the mean (central tendency) and the standard deviation (variability), which are usually reported together. In many journals, especially those following APA style, the symbol SD is used for the sample standard deviation. For example, the results might state:

Children who viewed the violent cartoon displayed more aggressive responses ($M = 12.45$, $SD = 3.7$) than those who viewed the control cartoon ($M = 4.22$, $SD = 1.04$).

When reporting the descriptive measures for several groups, the findings may be summarized in a table. Table 4.2 illustrates the results of hypothetical data.

TABLE 4.2
The number of aggressive behaviors for male and female adolescents after playing a violent or nonviolent video game.

	Type of Video Game	
	Violent	Control
Males	$M = 7.72$	$M = 4.34$
	$SD = 2.43$	$SD = 2.16$
Females	$M = 2.47$	$M = 1.61$
	$SD = 0.92$	$SD = 0.68$

Sometimes the table also indicates the sample size, n, for each group. You should remember that the purpose of the table is to present the data in an organized, concise, and accurate manner.

■ Standard Deviation and Descriptive Statistics

Because standard deviation requires extensive calculations, there is a tendency to get lost in the arithmetic and forget what standard deviation is and why it is important. Standard deviation is primarily a descriptive measure; it describes how variable, or how spread out, the scores are in a distribution. Behavioral scientists must deal with the variability that comes from studying people and animals. People are not all the same; they have different attitudes, opinions, talents, IQs, and personalities. Although we can calculate the average value for any of these variables, it is equally important to describe the variability. Standard deviation describes variability by measuring *distance from the mean*. In any distribution, some individuals will be close to the mean, and others will be relatively far from the mean. Standard deviation provides a measure of the typical, or standard, distance from the mean.

Describing an Entire Distribution   Rather than listing all of the individual scores in a distribution, research reports typically summarize the data by reporting only the mean

FIGURE 4.7

A sample of $n = 20$ scores with a mean of $M = 36$ and a standard deviation of $s = 4$.

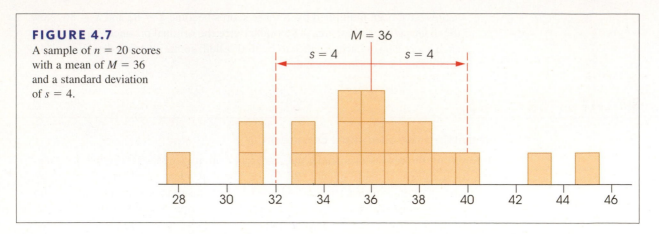

and the standard deviation. When you are given these two descriptive statistics, however, you should be able to visualize the entire set of data. For example, consider a sample with a mean of $M = 36$ and a standard deviation of $s = 4$. Although there are several different ways to picture the data, one simple technique is to imagine (or sketch) a histogram in which each score is represented by a box in the graph. For this sample, the data can be pictured as a pile of boxes (scores) with the center of the pile located at a value of $M = 36$. The individual scores or boxes are scattered on both sides of the mean with some of the boxes relatively close to the mean and some farther away. As a rule of thumb, roughly 70% of the scores in a distribution are located within a distance of one standard deviation from the mean, and almost all of the scores (roughly 95%) are within two standard deviations of the mean. In this example, the standard distance from the mean is $s = 4$ points so your image should have most of the boxes within 4 points of the mean, and nearly all of the boxes within 8 points. One possibility for the resulting image is shown in Figure 4.7.

Describing the Location of Individual Scores   Notice that Figure 4.7 not only shows the mean and the standard deviation, but also uses these two values to reconstruct the underlying scale of measurement (the X values along the horizontal line). The scale of measurement helps complete the picture of the entire distribution and helps to relate each individual score to the rest of the group. In this example, you should realize that a score of $X = 34$ is located near the center of the distribution, only slightly below the mean. On the other hand, a score of $X = 45$ is an extremely high score, located far out in the right-hand tail of the distribution.

   Notice that the relative position of a score depends in part on the size of the standard deviation. Earlier, in Figure 4.6 (p. 120), for example, we show a population distribution with a mean of $\mu = 80$ and a standard deviation of $\sigma = 8$, and a sample distribution with a mean of $M = 16$ and a standard deviation of $s = 2$. In the population distribution, a score that is 4 points above the mean is slightly above average but is certainly not an extreme value. In the sample distribution, however, a score that is 4 points above the mean is an extremely high score. In each case, the relative position of the score depends on the size of the standard deviation. For the population, a deviation of 4 points from the mean is relatively small, corresponding to only $\frac{1}{2}$ of the standard deviation. For the sample, on the other hand, a 4-point deviation is very large, equaling twice the size of the standard deviation.

   The general point of this discussion is that the mean and standard deviation are not simply abstract concepts or mathematical equations. Instead, these two values should be concrete and meaningful, especially in the context of a set of scores. The mean and standard deviation are central concepts for most of the statistics that are presented in the following chapters. A good understanding of these two statistics will help you with the more complex procedures that follow. (Box 4.1.)

BOX 4.1 An Analogy for the Mean and the Standard Deviation

Although the basic concepts of the mean and the standard deviation are not overly complex, the following analogy often helps students gain a more complete understanding of these two statistical measures.

In our local community, the site for a new high school was selected because it provides a central location. An alternative site on the western edge of the community was considered, but this site was rejected because it would require extensive busing for students living on the east side. In this example, the location of the high school is analogous to the concept of the mean; just as the high school is located in the center of the community, the mean is located in the center of the distribution of scores.

For each student in the community, it is possible to measure the distance between home and the new high school. Some students live only a few blocks from the new school and others live as much as 3 miles away. The average distance that a student must travel to school was calculated to be 0.80 miles. The average distance from the school is analogous to the concept of the standard deviation; that is, the standard deviation measures the standard distance from an individual score to the mean.

■ Variance and Inferential Statistics

In very general terms, the goal of inferential statistics is to detect meaningful and significant patterns in research results. The basic question is whether the patterns observed in the sample data reflect corresponding patterns that exist in the population, or are simply random fluctuations that occur by chance. Variability plays an important role in the inferential process because the variability in the data influences how easy it is to see patterns. In general, low variability means that existing patterns can be seen clearly, whereas high variability tends to obscure any patterns that might exist. The following example provides a simple demonstration of how variance can influence the perception of patterns.

EXAMPLE 4.10   In most research studies the goal is to compare means for two (or more) sets of data. For example:

- Is the mean level of depression lower after therapy than it was before therapy?
- Is the mean attitude score for men different from the mean score for women?
- Is the mean reading achievement score higher for students in a special program than for students in regular classrooms?

In each of these situations, the goal is to find a clear difference between two means that would demonstrate a significant, meaningful pattern in the results. Variability plays an important role in determining whether a clear pattern exists. Consider the following data representing hypothetical results from two experiments, each comparing two treatment conditions. For both experiments, your task is to determine whether there appears to be any consistent difference between the scores in treatment 1 and the scores in treatment 2.

Experiment A		Experiment B	
Treatment 1	Treatment 2	Treatment 1	Treatment 2
35	39	31	46
34	40	15	21
36	41	57	61
35	40	37	32

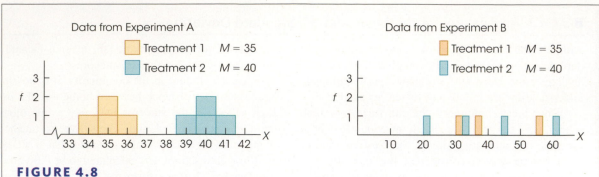

FIGURE 4.8

Graphs showing the results from two experiments. In experiment A, the variability is small and it is easy to see the 5-point mean difference between the two treatments. In experiment B, however, the 5-point mean difference between treatments is obscured by the large variability.

For each experiment, the data have been constructed so that there is a 5-point mean difference between the two treatments: On average, the scores in treatment 2 are 5 points higher than the scores in treatment 1. The 5-point difference is relatively easy to see in Experiment A, where the variability is low, but the same 5-point difference is difficult to see in Experiment B, where the variability is large. Again, high variability tends to obscure any patterns in the data. This general fact is perhaps even more convincing when the data are presented in a graph. Figure 4.8 shows the two sets of data from Experiments A and B. Notice that the results from Experiment A clearly show the 5-point difference between treatments. One group of scores piles up around 35 and the second group piles up around 40. On the other hand, the scores from Experiment B (Figure 4.8) seem to be mixed together randomly with no clear difference between the two treatments.

In the context of inferential statistics, the variance that exists in a set of sample data is often classified as *error variance*. This term is used to indicate that the sample variance represents unexplained and uncontrolled differences between scores. As the error variance increases, it becomes more difficult to see any systematic differences or patterns that might exist in the data. An analogy is to think of variance as the static that appears on a radio station or a cell phone when you enter an area of poor reception. In general, variance makes it difficult to get a clear signal from the data. High variance can make it difficult or impossible to see a mean difference between two sets of scores, or to see any other meaningful patterns in the results from a research study.

LEARNING CHECK

1. How is the standard deviation represented in a frequency distribution graph?
   a. By a vertical line located at a distance of one standard deviation above the mean.
   b. By two vertical lines located one standard deviation above the mean and one standard deviation below the mean.
   c. By a horizontal line or arrow extending from a location one standard deviation above the mean to a location one standard deviation below the mean.
   d. By a horizontal line or arrow extending from the mean for a distance equal to one standard deviation.

2. A population has a mean of $\mu = 35$ and a standard deviation of $\sigma = 5$. After 3 points are added to every score in the population, what are the new values for the mean and standard deviation?

   a. $\mu = 35$ and $\sigma = 5$

   b. $\mu = 35$ and $\sigma = 8$

   c. $\mu = 38$ and $\sigma = 5$

   d. $\mu = 38$ and $\sigma = 8$

3. What symbols are used for the mean and standard deviation for a sample in a research report?

   a. The mean is identified by the letter M and the standard deviation is represented by a lowercase s.

   b. The mean is identified by the letter M and the standard deviation is represented by SD.

   c. The mean is identified by a lowercase letter m and the standard deviation is represented by a lowercase s.

   d. The mean is identified by a lowercase letter m and the standard deviation is represented by SD.

4. Under what circumstances would a score that is above the mean by 5 points appear to be very close to the mean?

   a. When the mean is much greater than 5

   b. When the mean is much less than 5

   c. When the standard deviation is much greater than 5

   d. When the standard deviation is much less than 5

5. For which of the following pairs of distributions would the mean difference be easiest to see?

   a. M = 45 with s = 5 compared to M = 50 with s = 5.

   b. M = 45 with s = 5 compared to M = 55 with s = 5.

   c. M = 45 with s = 10 compared to M = 50 with s = 10.

   d. M = 45 with s = 10 compared to M = 55 with s = 10.

ANSWERS   1. D, 2. C, 3. B, 4. C, 5. B

SUMMARY

1. The purpose of variability is to measure and describe the degree to which the scores in a distribution are spread out or clustered together. There are three basic measures of variability: the range, variance, and standard deviation.

   The range is the distance covered by the set of scores, from the smallest score to the largest score. The range is completely determined by the two extreme scores and is considered to be a relatively crude measure of variability.

   Standard deviation and variance are the most commonly used measures of variability. Both of these measures are based on the idea that each score can be described in terms of its deviation or distance from the mean. The variance is the mean of the squared deviations. The standard deviation is the square root of the variance and provides a measure of the standard distance from the mean.

2. To calculate variance or standard deviation, you first need to find the sum of the squared deviations, SS.

Except for minor changes in notation, the calculation of SS is identical for samples and populations. There are two methods for calculating SS:

I. By definition, you can find SS using the following steps:
    a. Find the deviation $(X - \mu)$ for each score.
    b. Square each deviation.
    c. Add the squared deviations.
    This process can be summarized in a formula as follows:

    Definitional formula: $SS = \Sigma(X - \mu)^2$

II. The sum of the squared deviations can also be found using a computational formula, which is especially useful when the mean is not a whole number:

    Computational formula: $SS = \Sigma X^2 - \dfrac{(\Sigma X)^2}{N}$

3. Variance is the mean squared deviation and is obtained by finding the sum of the squared deviations and then dividing by the number of scores. For a population, variance is

    $$\sigma^2 = \frac{SS}{N}$$

    For a sample, only $n - 1$ of the scores are free to vary (degrees of freedom or $df = n - 1$), so sample variance is

    $$s^2 = \frac{SS}{n - 1} = \frac{SS}{df}$$

Using $n - 1$ in the sample formula makes the sample variance an accurate and unbiased estimate of the population variance.

4. Standard deviation is the square root of the variance. For a population, this is

    $$\sigma = \sqrt{\frac{SS}{N}}$$

    Sample standard deviation is

    $$s = \sqrt{\frac{SS}{n - 1}} = \sqrt{\frac{SS}{df}}$$

5. Adding a constant value to every score in a distribution does not change the standard deviation. Multiplying every score by a constant, however, causes the standard deviation to be multiplied by the same constant.

6. Because the mean identifies the center of a distribution and the standard deviation describes the average distance from the mean, these two values should allow you to create a reasonably accurate image of the entire distribution. Knowing the mean and standard deviation should also allow you to describe the relative location of any individual score within the distribution.

7. Large variance can obscure patterns in the data and, therefore, can create a problem for inferential statistics.

KEY TERMS

variability (101)

range (102)

deviation score (104)

population variance (σ^2) (110)

population standard deviation (σ) (110)

sum of squares (SS) (108)

sample variance (s^2) (113)

sample standard deviation (s) (113)

degrees of freedom (df) (115)

biased statistic (117)

unbiased statistic (117)

SPSS®

General instructions for using SPSS are presented in Appendix D. Following are detailed instructions for using SPSS to compute the **Range, Standard Deviation, and Variance** for a sample of scores.

Data Entry

1. Enter all of the scores in one column of the data editor, probably VAR00001.

Data Analysis

1. Click **Analyze** on the tool bar, select **Descriptive Statistics**, and click on **Descriptives**.

	N	Range	Minimum	Maximum	Mean	Std. Deviation	Variance
VAR00001	8	8.00	3.00	11.00	6.5000	2.61861	6.85714
Valid N (listwise)	8						

FIGURE 4.9

The SPSS summary table showing descriptive statistics for the sample of $n = 8$ scores in Example 4.6.

2. Highlight the column label for the set of scores (VAR00001) in the left box and click the arrow to move it into the **Variable** box.

3. If you want the variance and/or the range reported along with the standard deviation, click on the **Options** box, select **Variance** and/or **Range**, then click **Continue**.

4. Click **OK**.

SPSS Output

We used SPSS to compute the range, standard deviation, and variance for the sample data in Example 4.6 and the output is shown in Figure 4.9. The summary table lists the number of scores, maximum and minimum scores, mean, range, standard deviation, and variance. Note that the range and variance are included because these values were selected using the Options box during data analysis. Caution: SPSS computes the *sample* standard deviation and *sample* variance using $n - 1$. If your scores are intended to be a population, you can multiply the sample standard deviation by the square root of $(n - 1)/n$ to obtain the population standard deviation.

Note: You can also obtain the mean and standard deviation for a sample if you use SPSS to display the scores in a frequency distribution histogram (see the SPSS section at the end of Chapter 2). The mean and standard deviation are displayed beside the graph.

FOCUS ON PROBLEM SOLVING

1. The purpose of variability is to provide a measure of how spread out the scores are in a distribution. Usually this is described by the standard deviation. Because the calculations are relatively complicated, it is wise to make a preliminary estimate of the standard deviation before you begin. Remember that standard deviation provides a measure of the typical, or standard, distance from the mean. Therefore, the standard deviation must have a value somewhere between the largest and the smallest deviation scores. As a rule of thumb, the standard deviation should be about one-fourth of the range.

2. Rather than trying to memorize all the formulas for *SS,* variance, and standard deviation, you should focus on the definitions of these values and the logic that relates them to each other:

    SS is the sum of squared deviations.

    Variance is the mean squared deviation.

    Standard deviation is the square root of variance.

    The only formula you should need to memorize is the computational formula for *SS*.

3. A common error is to use $n - 1$ in the computational formula for *SS* when you have scores from a sample. Remember that the *SS* formula always uses n (or N). After you compute *SS* for a sample, you must correct for the sample bias by using $n - 1$ in the formulas for variance and standard deviation.

DEMONSTRATION 4.1

COMPUTING MEASURES OF VARIABILITY

For the following sample data, compute the variance and standard deviation. The scores are:

$$10, \quad 7, \quad 6, \quad 10, \quad 6, \quad 15$$

STEP 1 **Compute SS, the sum of squared deviations**　We will use the computational formula. For this sample, $n = 6$ and

$$\Sigma X = 10 + 7 + 6 + 10 + 6 + 15 = 54$$

$$\Sigma X^2 = 10^2 + 7^2 + 6^2 + 10^2 + 6^2 + 15^2 = 546$$

$$SS = \Sigma X^2 - \frac{(\Sigma X)^2}{N} = 546 - \frac{(54)^2}{6}$$

$$= 546 - 486$$

$$= 60$$

STEP 2 **Compute the sample variance**　For sample variance, SS is divided by the degrees of freedom, $df = n - 1$.

$$s^2 = \frac{SS}{n - 1} = \frac{60}{5} = 12$$

STEP 3 **Compute the sample standard deviation**　Standard deviation is simply the square root of the variance.

$$s = \sqrt{12} = 3.46$$

PROBLEMS

1. In words, explain what is measured by SS, variance, and standard deviation.

2. Is it possible to obtain a negative value for the variance or the standard deviation?

3. Describe the scores in a sample that has a standard deviation of zero.

4. There are two different formulas or methods that can be used to calculate SS.
   a. Under what circumstances is the definitional formula easy to use?
   b. Under what circumstances is the computational formula preferred?

5. Calculate the mean and SS (sum of squared deviations) for each of the following samples. Based on the value for the mean, you should be able to decide which SS formula is better to use.

Sample A:	1,	3,	4,	0
Sample B:	2,	5,	0,	3

6. Calculate SS, variance, and standard deviation for the following population of $N = 5$ scores: 2, 13, 4, 10, 6. (*Note:* The definitional formula works well with these scores.)

7. Calculate SS, variance, and standard deviation for the following population of $N = 7$ scores: 8, 1, 4, 3, 5, 3, 4. (*Note:* The definitional formula works well with these scores.)

8. For the following population of $N = 6$ scores:

   $$3, \quad 1, \quad 4, \quad 3, \quad 3, \quad 4$$

   a. Sketch a histogram showing the population distribution.
   b. Locate the value of the population mean in your sketch, and make an estimate of the standard deviation (as done in Example 4.2).
   c. Compute SS, variance, and standard deviation for the population. (How well does your estimate compare with the actual value of σ?)

9. For the following set of scores: 1, 4, 3, 5, 7
   a. If the scores are a population, what are the variance and standard deviation?
   b. If the scores are a sample, what are the variance and standard deviation?

10. Why is the formula for sample variance different from the formula for population variance?

11. For the following sample of $n = 6$ scores: 0, 11, 5, 10, 5, 5
    a. Sketch a histogram showing the sample distribution.
    b. Locate the value of the sample mean in your sketch, and make an estimate of the standard deviation (as done in Example 4.6).
    c. Compute SS, variance, and standard deviation for the sample. (How well does your estimate compare with the actual value of s?)

12. Calculate SS, variance, and standard deviation for the following sample of $n = 8$ scores: 0, 4, 1, 3, 2. 1, 1, 0.

13. Calculate SS, variance, and standard deviation for the following sample of $n = 5$ scores: 2, 9, 5, 5, 9.

14. A population has a mean of $\mu = 50$ and a standard deviation of $\sigma = 10$.
    a. If 3 points were added to every score in the population, what would be the new values for the mean and standard deviation?
    b. If every score in the population were multiplied by 2, then what would be the new values for the mean and standard deviation?

15. a. After 6 points have been added to every score in a sample, the mean is found to be $M = 70$ and the standard deviation is $s = 13$. What were the values for the mean and standard deviation for the original sample?
    b. After every score in a sample is multiplied by 3, the mean is found to be $M = 48$ and the standard deviation is $s = 18$. What were the values for the mean and standard deviation for the original sample?

16. Compute the mean and standard deviation for the following sample of $n = 4$ scores: 82, 88, 82, and 86. Hint: To simplify the arithmetic, you can subtracted 80 points from each score to obtain a new sample consisting of 2, 8, 2, and 6. Then, compute the mean and standard deviation for the new sample. Use the values you obtain to find the mean and standard deviation for the original sample.

17. For the following sample of $n = 8$ scores: 0, 1, $\frac{1}{2}$, 0, 3, $\frac{1}{2}$, 0, and 1:
    a. Simplify the arithmetic by first multiplying each score by 2 to obtain a new sample of 0, 2, 1, 0, 6, 1, 0, and 2. Then, compute the mean and standard deviation for the new sample.
    b. Using the values you obtained in part a, what are the values for the mean and standard deviation for the original sample?

18. For the following population of $N = 6$ scores:

    2, 9, 6, 8, 9, 8

    a. Calculate the range and the standard deviation. (Use either definition for the range.)
    b. Add 2 points to each score and compute the range and standard deviation again. Describe how adding a constant to each score influences measures of variability.

19. The range is completely determined by the two extreme scores in a distribution. The standard deviation, on the other hand, uses every score.
    a. Compute the range (choose either definition) and the standard deviation for the following sample of $n = 5$ scores. Note that there are three scores clustered around the mean in the center of the distribution, and two extreme values.

       Scores: 0, 6, 7, 8, 14.

    b. Now we will break up the cluster in the center of the distribution by moving two of the central scores out to the extremes. Once again compute the range and the standard deviation.

       New scores: 0, 0, 7, 14, 14.

    c. According to the range, how do the two distributions compare in variability? How do they compare according to the standard deviation?

20. For the data in the following sample:

    10, 6, 8, 6, 5

    a. Find the mean and the standard deviation.
    b. Now change the score of $X = 10$ to $X = 0$, and find the new mean and standard deviation.
    c. Describe how one extreme score influences the mean and standard deviation.

21. Within a population, the differences that exist from one person to another are often called diversity.

Researchers comparing cognitive skills for younger adults and older adults typically find greater differences (greater diversity) in the older population (Morse, 1993). Following are typical data showing problem-solving scores for two groups of participants.

Older Adults (average age 72)	Younger Adults (average age 31)
9 4 7 3 8	7 9 6 7 8
6 2 8 4 5	6 7 6 6 8
7 5 2 6 6	9 7 8 6 9

a. Compute the mean, variance, and standard deviation for each group.
b. Is one group of scores noticeably more variable (more diverse) than the other?

22. Wegesin and Stern (2004) found greater consistency (less variability) in the memory performance scores for younger women than for older women. The following data represent memory scores obtained for two women, one older and one younger, over a series of memory trials.
a. Calculate the variance of the scores for each woman.
b. Are the scores for the younger woman more consistent (less variable)?

Younger	Older
8	7
6	5
6	8
7	5
8	7
7	6
8	8
8	5

23. A population has a mean of $\mu = 50$ and a standard deviation of $\sigma = 20$.
a. Would a score of $X = 70$ be considered an extreme value (out in the tail) in this sample?
b. If the standard deviation were $\sigma = 5$, would a score of $X = 70$ be considered an extreme value?

24. On an exam with a mean of $M = 78$, you obtain a score of $X = 84$.
a. Would you prefer a standard deviation of $s = 2$ or $s = 10$? (*Hint:* Sketch each distribution and find the location of your score.)
b. If your score were $X = 72$, would you prefer $s = 2$ or $s = 10$? Explain your answer.

z-Scores: Location of Scores and Standardized Distributions

© Deborah Batt

Tools You Will Need

The following items are considered essential background material for this chapter. If you doubt your knowledge of any of these items, you should review the appropriate chapter and section before proceeding.

- The mean (Chapter 3)
- The standard deviation (Chapter 4)
- Basic algebra (Appendix A)

A common test of cognitive ability requires participants to search through a visual display and respond to specific targets as quickly as possible. This kind of test is called a perceptual-speed test. Measures of perceptual speed are commonly used for predicting performance on jobs that demand a high level of speed and accuracy. Although many different tests are used, a typical example is shown in Figure 5.1. This task requires the participant to search through the display of digit pairs as quickly as possible and circle each pair that adds up 10. Your score is determined by the amount of time required to complete the task with a correction for the number of errors that you make. One complaint about this kind of paper-and-pencil test is that it is tedious and time consuming to score because a researcher must also search through the entire display to identify errors to determine the participant's level of accuracy. An alternative, proposed by Ackerman and Beier (2007), is a computerized version of the task. The computer version presents a series of digit pairs and participants respond on a touch-sensitive monitor. The computerized test is very reliable and the scores are equivalent to the paper-and-pencil tests in terms of assessing cognitive skill. The advantage of the computerized test is that the computer produces a test score immediately when a participant finishes the test.

Suppose that you took Ackerman and Beier's test and your combined time and errors produced a score of 92. How did you do? Are you faster than average, fairly normal in perceptual speed, or does your score indicate a serious deficit in cognitive skill? The answer is that you have no idea how your score of 92 compares with scores for others who took the same test. Now suppose that you are also told that the distribution of perceptual speed scores has a mean of $\mu = 86.75$ and a standard deviation of $\sigma = 10.50$. With this additional information, you should realize that your score ($X = 92$) is somewhat higher than average but not an exceptionally high score.

In this chapter we introduce a statistical procedure for converting individual scores into *z-scores*, so that each *z*-score is a meaningful value that identifies exactly where the original score is located in the distribution. As you will see, *z*-scores use the mean as a reference point to determine whether the score is above or below average. A *z*-score also uses the standard deviation as a yardstick for describing how much an individual score differs from average. For example, a *z*-score will tell you if your score is above the mean by a distance equal to two standard deviations, or below the mean by one-half of a standard deviation. A good understanding of the mean and the standard deviation will be a valuable background for learning about *z*-scores.

64	23	19	31	19	46	31	91	83	82	82	46	19	87
11	42	94	87	64	44	19	55	82	46	57	98	39	46
78	73	72	66	63	71	67	42	62	73	45	22	62	99
73	91	52	37	55	97	91	51	44	23	46	64	97	62
97	31	21	49	93	91	89	46	73	82	55	98	12	56
73	82	37	55	89	83	73	27	83	82	73	46	97	62
57	96	46	55	46	19	13	67	73	26	58	64	32	73
23	94	66	55	91	73	67	73	82	55	64	62	46	39
87	11	99	73	56	73	63	73	91	82	63	33	16	88
19	42	62	91	12	82	32	92	73	46	68	19	11	64
93	91	32	82	63	91	46	46	36	55	19	92	62	71

FIGURE 5.1

An example of a perceptual speed task. The participant is asked to search through the display as quickly as possible and circle each pair of digits that add up to 10.

5.1 | Introduction to *z*-Scores

LEARNING OBJECTIVE **1.** Describe the two general purposes for transforming *X* values into *z*-scores.

In the previous two chapters, we introduced the concepts of the mean and standard deviation as methods for describing an entire distribution of scores. Now we shift attention to the individual scores within a distribution. In this chapter, we introduce a statistical technique that uses the mean and the standard deviation to transform each score (*X* value) into a *z-score* or a *standard score*. The purpose of *z*-scores, or standard scores, is to identify and describe the exact location of each score in a distribution.

The following example demonstrates why *z*-scores are useful and introduces the general concept of transforming *X* values into *z*-scores.

EXAMPLE 5.1

Suppose you received a score of $X = 76$ on a statistics exam. How did you do? It should be clear that you need more information to predict your grade. Your score of $X = 76$ could be one of the best scores in the class, or it might be the lowest score in the distribution. To find the location of your score, you must have information about the other scores in the distribution. It would be useful, for example, to know the mean for the class. If the mean were $\mu = 70$, you would be in a much better position than if the mean were $\mu = 85$. Obviously, your position relative to the rest of the class depends on the mean. However, the mean by itself is not sufficient to tell you the exact location of your score. Suppose you know that the mean for the statistics exam is $\mu = 70$ and your score is $X = 76$. At this point, you know that your score is 6 points above the mean, but you still do not know exactly where it is located. Six points may be a relatively big distance and you may have one of the highest scores in the class, or 6 points may be a relatively small distance and you are only slightly above the average. Figure 5.2 shows two possible distributions. Both distributions have a mean of $\mu = 70$, but for one distribution, the standard deviation is $\sigma = 3$, and for the other, $\sigma = 12$. The location of $X = 76$ is highlighted in each of the two distributions. When the standard deviation is $\sigma = 3$, your score of $X = 76$ is in the extreme right-hand tail, one of the highest scores in the distribution. However, in the other distribution, where $\sigma = 12$, your score is only slightly above average. Thus, the relative location of your score within the distribution depends on the standard deviation as well as the mean. ∎

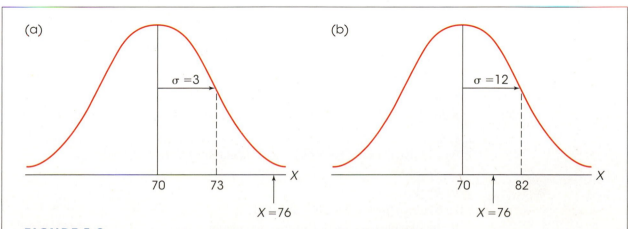

FIGURE 5.2
Two distributions of exam scores. For both distributions, $\mu = 70$, but for one distribution, $\sigma = 3$, and for the other, $\sigma = 12$. The relative position of $X = 76$ is very different for the two distributions.

The purpose of the preceding example is to demonstrate that a score *by itself* does not necessarily provide much information about its position within a distribution. These original, unchanged scores that are the direct result of measurement are called *raw scores*. To make raw scores more meaningful, they are often transformed into new values that contain more information. This transformation is one purpose for z-scores. In particular, we transform X values into z-scores so that the resulting z-scores tell exactly where the original scores are located.

A second purpose for z-scores is to *standardize* an entire distribution. A common example of a standardized distribution is the distribution of IQ scores. Although there are several different tests for measuring IQ, the tests usually are standardized so that they have a mean of 100 and a standard deviation of 15. Because most of the different tests are standardized, it is possible to understand and compare IQ scores even though they come from different tests. For example, we all understand that an IQ score of 95 is a little below average, *no matter which IQ test was used*. Similarly, an IQ of 145 is extremely high, *no matter which IQ test was used*. In general terms, the process of standardizing takes different distributions and makes them equivalent. The advantage of this process is that it is possible to compare distributions even though they may have been quite different before standardization.

In summary, the process of transforming X values into z-scores serves two useful purposes:

1. Each z-score tells the exact location of the original X value within the distribution.

2. The z-scores form a standardized distribution that can be directly compared to other distributions that also have been transformed into z-scores.

Each of these purposes is discussed in the following sections.

LEARNING CHECK

1. If your exam score is $X = 60$, which set of parameters would give you the best grade?
    a. $\mu = 65$ and $\sigma = 5$
    b. $\mu = 65$ and $\sigma = 2$
    c. $\mu = 70$ and $\sigma = 5$
    d. $\mu = 70$ and $\sigma = 2$

2. For a distribution of exam scores with $\mu = 70$, which value for the standard deviation would give the highest grade to a score of $X = 75$?
    a. $\sigma = 1$
    b. $\sigma = 2$
    c. $\sigma = 5$
    d. $\sigma = 10$

3. Last week Sarah had a score of $X = 43$ on a Spanish exam and a score of $X = 75$ on an English exam. For which exam should Sarah expect the better grade?
    a. Spanish
    b. English
    c. The two grades should be identical.
    d. Impossible to determine without more information

ANSWERS   **1.** A, **2.** A, **3.** D

5.2 | z-Scores and Locations in a Distribution

LEARNING OBJECTIVES

2. Explain how a z-score identifies a precise location in a distribution.

3. Using either the z-score definition or the z-score formula, transform X values into z-scores and transform z-scores into X values.

One of the primary purposes of a z-score is to describe the exact location of a score within a distribution. The z-score accomplishes this goal by transforming each X value into a signed number (+ or −) so that

1. the *sign* tells whether the score is located above (+) or below (−) the mean, and

2. the *number* tells the distance between the score and the mean in terms of the number of standard deviations.

Thus, in a distribution of IQ scores with $\mu = 100$ and $\sigma = 15$, a score of $X = 130$ would be transformed into $z = +2.00$. The z-score value indicates that the score is located above the mean (+) by a distance of 2 standard deviations (30 points).

DEFINITION

A **z-score** specifies the precise location of each X value within a distribution. The sign of the z-score (+ or −) signifies whether the score is above the mean (positive) or below the mean (negative). The numerical value of the z-score specifies the distance from the mean by counting the number of standard deviations between X and μ.

Whenever you are working with z-scores, you should imagine or draw a picture similar to Figure 5.3. Although you should realize that not all distributions are normal, we will use the normal shape as an example when showing z-scores for populations.

Notice that a z-score always consists of two parts: a sign (+ or −) and a magnitude. Both parts are necessary to describe completely where a raw score is located within a distribution.

Figure 5.3 shows a population distribution with various positions identified by their z-score values. Notice that all z-scores above the mean are positive and all z-scores below the mean are negative. The sign of a z-score tells you immediately whether the score is located above or below the mean. Also, note that a z-score of $z = +1.00$ corresponds to a position exactly 1 standard deviation above the mean. A z-score of $z = +2.00$ is always located exactly 2 standard deviations above the mean. The numerical value of the z-score tells you the number of standard deviations from the mean. Finally, you should notice that

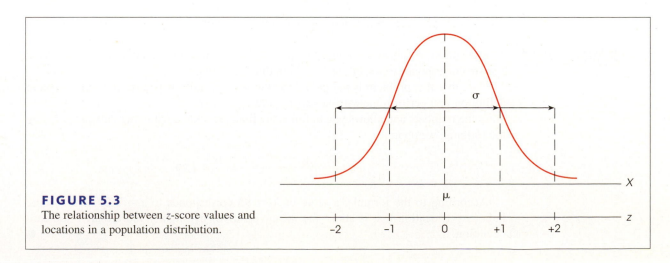

FIGURE 5.3

The relationship between z-score values and locations in a population distribution.

Figure 5.3 does not give any specific values for the population mean or the standard deviation. The locations identified by z-scores are the same for *all distributions,* no matter what mean or standard deviation the distributions may have.

Now we can return to the two distributions shown in Figure 5.2 and use a z-score to describe the position of $X = 76$ within each distribution as follows:

> In Figure 5.2(a), with a standard deviation of $\sigma = 3$, the score $X = 76$ corresponds to a z-score of $z = +2.00$. That is, the score is located above the mean by exactly 2 standard deviations.

> In Figure 5.2(b), with $\sigma = 12$, the score $X = 76$ corresponds to a z-score of $z = +0.50$. In this distribution, the score is located above the mean by exactly $\frac{1}{2}$ standard deviation.

■ The z-Score Formula

The z-score definition is adequate for transforming back and forth from X values to z-scores as long as the arithmetic is easy to do in your head. For more complicated values, it is best to have an equation to help structure the calculations. Fortunately, the relationship between X values and z-scores is easily expressed in a formula. The formula for transforming scores into z-scores is

$$z = \frac{X - \mu}{\sigma} \qquad \text{(5.1)}$$

The numerator of the equation, $X - \mu$, is a *deviation score* (Chapter 4, page 104); it measures the distance in points between X and μ and indicates whether X is located above or below the mean. The deviation score is then divided by σ because we want the z-score to measure distance in terms of standard deviation units. The formula performs exactly the same arithmetic that is used with the z-score definition, and it provides a structured equation to organize the calculations when the numbers are more difficult. The following examples demonstrate the use of the z-score formula.

EXAMPLE 5.2

A distribution of scores has a mean of $\mu = 100$ and a standard deviation of $\sigma = 10$. What z-score corresponds to a score of $X = 130$ in this distribution?

According to the definition, the z-score will have a value of $+3$ because the score is located above the mean by exactly 3 standard deviations. Using the z-score formula, we obtain

$$z = \frac{X - \mu}{\sigma} = \frac{130 - 100}{10} = \frac{30}{10} = 3.00$$

The formula produces exactly the same result that is obtained using the z-score definition. ■

EXAMPLE 5.3

A distribution of scores has a mean of $\mu = 86$ and a standard deviation of $\sigma = 7$. What z-score corresponds to a score of $X = 95$ in this distribution?

Note that this problem is not particularly easy, especially if you try to use the z-score definition and perform the calculations in your head. However, the z-score formula organizes the numbers and allows you to finish the final arithmetic with your calculator. Using the formula, we obtain

$$z = \frac{X - \mu}{\sigma} = \frac{95 - 86}{7} = \frac{9}{7} = 1.29$$

According to the formula, a score of $X = 95$ corresponds to $z = 1.29$. The z-score indicates a location that is above the mean (positive) by slightly more than 1 standard deviation. ■

When you use the z-score formula, it can be useful to pay attention to the definition of a z-score as well. For example, we used the formula in Example 5.3 to calculate the z-score corresponding to $X = 95$, and obtained $z = 1.29$. Using the z-score definition, we note that $X = 95$ is located above the mean by 9 points, which is slightly more than one standard deviation ($\sigma = 7$). Therefore, the z-score should be positive and have a value slightly greater than 1.00. In this case, the answer predicted by the definition is in perfect agreement with the calculation. However, if the calculations produce a different value, for example $z = 0.78$, you should realize that this answer is not consistent with the definition of a z-score. In this case, an error has been made and you should check the calculations.

■ Determining a Raw Score (*X*) from a *z*-Score

Although the z-score equation (Formula 5.1) works well for transforming X values into z-scores, it can be awkward when you are trying to work in the opposite direction and change z-scores back into X values. In general, it is easier to use the definition of a z-score, rather than a formula, when you are changing z-scores into X values. Remember, the z-score describes exactly where the score is located by identifying the direction and distance from the mean. It is possible, however, to express this definition as a formula, and we will use a sample problem to demonstrate how the formula can be created.

> For a distribution with a mean of $\mu = 60$ and $\sigma = 8$, what X value corresponds to a z-score of $z = -1.50$?

To solve this problem, we will use the z-score definition and carefully monitor the step-by-step process. The value of the z-score indicates that X is located below the mean by a distance equal to 1.5 standard deviations. Thus, the first step in the calculation is to determine the distance corresponding to 1.5 standard deviations. For this problem, the standard deviation is $\sigma = 8$ points, so 1.5 standard deviations is $1.5(8) = 12$ points. The next step is to find the value of X that is located below the mean by 12 points. With a mean of $\mu = 60$, the score is

$$X = \mu - 12 = 60 - 12 = 48$$

The two steps can be combined to form a single formula:

$$X = \mu + z\sigma \tag{5.2}$$

In the formula, the value of $z\sigma$ is the *deviation* of X and determines both the direction and the size of the distance from the mean. In this problem, $z\sigma = (-1.5)(8) = -12$, or 12 points below the mean. Formula 5.2 simply combines the mean and the deviation from the mean to determine the exact value of X.

Finally, you should realize that Formula 5.1 and Formula 5.2 are actually two different versions of the same equation. If you begin with either formula and use algebra to shuffle the terms around, you will soon end up with the other formula. We will leave this as an exercise for those who want to try it.

LEARNING CHECK

1. Of the following z-score values, which one represents the most extreme location on the left-hand side of the distribution?
   a. $z = +1.00$
   b. $z = +2.00$
   c. $z = -1.00$
   d. $z = -2.00$

2. Of the following z-score values, which one represents the location closest to the mean?

   a. $z = +0.50$
   b. $z = +1.00$
   c. $z = -1.00$
   d. $z = -2.00$

3. For a population with $\mu = 100$ and $\sigma = 20$, what is the z-score corresponding to $X = 105$?

   a. $+0.25$
   b. $+0.50$
   c. $+4.00$
   d. $+5.00$

ANSWERS   **1.** D, **2.** A, **3.** A

5.3 | Other Relationships Between z, X, μ, and σ

LEARNING OBJECTIVE   **4.** Explain how z-scores establish a relationship among X, μ, σ, and the value of z, and use that relationship to find an unknown mean when given a z-score, a score, and the standard deviation, or find an unknown standard deviation when given a z-score, a score, and the mean.

In most cases, we simply transform scores (X values) into z-scores, or change z-scores back into X values. However, you should realize that a z-score establishes a relationship between the score, mean, and standard deviation. This relationship can be used to answer a variety of different questions about scores and the distributions in which they are located. The following two examples demonstrate some possibilities.

EXAMPLE 5.4   In a population with a mean of $\mu = 65$, a score of $X = 59$ corresponds to $z = -2.00$. What is the standard deviation for the population?

To answer the question, we begin with the z-score value. A z-score of -2.00 indicates that the corresponding score is located below the mean by a distance of 2 standard deviations. You also can determine that the score ($X = 59$) is located below the mean ($\mu = 65$) by a distance of 6 points. Thus, 2 standard deviations correspond to a distance of 6 points, which means that 1 standard deviation must be $\sigma = 3$ points.   ■

EXAMPLE 5.5   In a population with a standard deviation of $\sigma = 6$, a score of $X = 33$ corresponds to $z = +1.50$. What is the mean for the population?

Again, we begin with the z-score value. In this case, a z-score of $+1.50$ indicates that the score is located above the mean by a distance corresponding to 1.50 standard deviations. With a standard deviation of $\sigma = 6$, this distance is $(1.50)(6) = 9$ points. Thus, the score is located 9 points above the mean. The score is $X = 33$, so the mean must be $\mu = 24$.   ■

Many students find problems like those in Examples 5.4 and 5.5 easier to understand if they draw a picture showing all of the information presented in the problem. For the

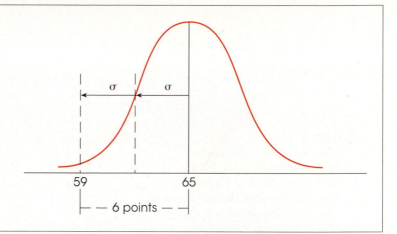

FIGURE 5.4

A visual representation of the question in Example 5.4. If 2 standard deviations correspond to a 6-point distance, then 1 standard deviation must equal 3 points.

problem in Example 5.4, the picture would begin with a distribution that has a mean of $\mu = 65$ (we use a normal distribution that is shown in Figure 5.4). The value of the standard deviation is unknown, but you can add arrows to the sketch pointing outward from the mean for a distance corresponding to 1 standard deviation. Finally, use standard deviation arrows to identify the location of $z = -2.00$ (2 standard deviations below the mean) and add $X = 59$ at that location. All of these factors are shown in Figure 5.4. In the figure, it is easy to see that $X = 59$ is located 6 points below the mean, and that the 6-point distance corresponds to exactly 2 standard deviations. Again, if 2 standard deviations equal 6 points, then 1 standard deviation must be $\sigma = 3$ points.

A slight variation on Examples 5.4 and 5.5 is demonstrated in the following example. This time you must use the z-score information to find both the population mean and the standard deviation.

EXAMPLE 5.6   In a population distribution, a score of $X = 54$ corresponds to $z = +2.00$ and a score of $X = 42$ corresponds to $z = -1.00$. What are the values for the mean and the standard deviation for the population? Once again, many students find this kind of problem easier to understand if they can see it in a picture, so we have sketched this example in Figure 5.5.

The key to solving this kind of problem is to focus on the distance between the two scores. Notice that the distance can be measured in points and in standard deviations. In points, the distance from $X = 42$ to $X = 54$ is 12 points. According to the two z-scores, $X = 42$ is located 1 standard deviation below the mean and $X = 54$ is located 2 standard deviations above the mean (see Figure 5.5). Thus, the total distance between the two scores is equal to 3 standard deviations. We have determined that the distance between the two scores is 12 points, which is equal to 3 standard deviations. As an equation

$$3\sigma = 12 \text{ points}$$

Dividing both sides by 3, we obtain

$$\sigma = 4 \text{ points}$$

Finally, note that $X = 42$ corresponds to $z = -1.00$, which means that $X = 42$ is located one standard deviation below the mean. With a standard deviation of $\sigma = 4$, the mean must be $\mu = 46$. Thus, the population has a mean of $\mu = 46$ and a standard deviation of $\sigma = 4$. ■

The following example is an opportunity for you to test your understanding by solving a problem similar to the demonstration in Example 5.6.

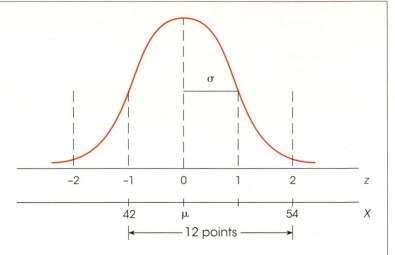

FIGURE 5.5

A visual presentation of the question in Example 5.6. The 12-point distance from 42–54 corresponds to 3 standard deviations. Therefore, the standard deviation must be $\sigma = 4$. Also, the score $X = 42$ is below the mean by one standard deviation, so the mean must be $\mu = 46$.

EXAMPLE 5.7   In a population distribution, a score of $X = 64$ has a *z*-score of $z = 0.50$ and a score of $X = 72$ has a *z*-score of $z = 1.50$. What are the values for the population mean and standard deviation? You should obtain $\mu = 60$ and $\sigma = 8$. Good luck   ∎

LEARNING CHECK

1. A population of scores has $\mu = 44$. In this population, an X value of 40 corresponds to $z = -0.50$. What is the population standard deviation?

   a. 2

   b. 4

   c. 6

   d. 8

2. A population of scores has $\sigma = 4$. In this population, an X value of 58 corresponds to $z = 2.00$. What is the population mean?

   a. 54

   b. 50

   c. 62

   d. 66

3. A population of scores has $\sigma = 10$. In this population, a score of $X = 60$ corresponds to $z = -1.50$. What is the population mean?

   a. -30

   b. 45

   c. 75

   d. 90

ANSWERS   **1.** D, **2.** B, **3.** C

5.4 | Using *z*-Scores to Standardize a Distribution

5. Describe the effects of standardizing a distribution by transforming the entire set of scores into z-scores and explain the advantages of this transformation.

It is possible to transform every X value in a distribution into a corresponding z-score. The result of this process is that the entire distribution of X values is transformed into a distribution of z-scores (Figure 5.6). The new distribution of z-scores has characteristics that make the *z-score transformation* a very useful tool. Specifically, if every X value is transformed into a z-score, then the distribution of z-scores will have the following properties:

1. **Shape**   The distribution of z-scores will have exactly the same shape as the original distribution of scores. If the original distribution is negatively skewed, for example, then the z-score distribution will also be negatively skewed. If the original distribution is normal, the distribution of z-scores will also be normal. Transforming raw scores into z-scores does not change anyone's position in the distribution. For example, any raw score that is above the mean by 1 standard deviation will be transformed to a z-score of $+1.00$, which is still above the mean by 1 standard deviation. Transforming a distribution from X values to z values does not move scores from one position to another; the procedure simply re-labels each score (see Figure 5.6). Because each individual score stays in its same position within the distribution, the overall shape of the distribution does not change.

2. **The Mean**   The z-score distribution will *always* have a mean of zero. In Figure 5.6, the original distribution of X values has a mean of $\mu = 100$. When this value, $X = 100$, is transformed into a z-score, the result is

$$z = \frac{X - \mu}{\sigma} = \frac{100 - 100}{10} = 0$$

Thus, the original population mean is transformed into a value of zero in the z-score distribution. The fact that the z-score distribution has a mean of zero makes the mean a convenient reference point. Recall from the definition of z-scores that all positive z-scores are above the mean and all negative z-scores are below the mean. In other words, for z-scores, $\mu = 0$.

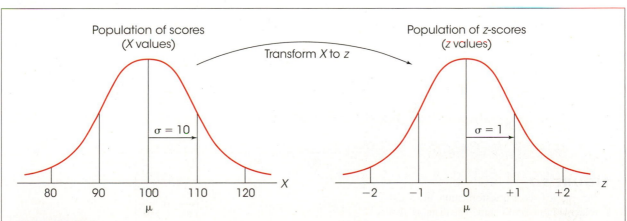

FIGURE 5.6

An entire population of scores is transformed into z-scores. The transformation does not change the shape of the distribution but the mean is transformed into a value of 0 and the standard deviation is transformed to a value of 1.

3. **The Standard Deviation**   The distribution of z-scores will *always* have a standard deviation of 1. In Figure 5.6, the original distribution of X values has $\mu = 100$ and $\sigma = 10$. In this distribution, a value of $X = 110$ is above the mean by exactly 10 points or 1 standard deviation. When $X = 110$ is transformed, it becomes $z = +1.00$, which is above the mean by exactly 1 point in the z-score distribution. Thus, the standard deviation corresponds to a 10-point distance in the X distribution and is transformed into a 1-point distance in the z-score distribution. The advantage of having a standard deviation of 1 is that the numerical value of a z-score is exactly the same as the number of standard deviations from the mean. For example, a z-score of $z = 1.50$ is exactly 1.50 standard deviations from the mean.

In Figure 5.6, we showed the z-score transformation as a process that changed a distribution of X values into a new distribution of z-scores. In fact, there is no need to create a whole new distribution. Instead, you can think of the z-score transformation as simply *relabeling* the values along the X-axis. That is, after a z-score transformation, you still have the same distribution, but now each individual is labeled with a z-score instead of an X value. Figure 5.7 demonstrates this concept with a single distribution that has two sets of labels: the X values along one line and the corresponding z-scores along another line. Notice that the mean for the distribution of z-scores is zero and the standard deviation is 1.

When *any* distribution (with any mean or standard deviation) is transformed into z-scores, the resulting distribution will always have a mean of $\mu = 0$ and a standard deviation of $\sigma = 1$. Because all z-score distributions have the same mean and the same standard deviation, the z-score distribution is called a *standardized distribution*.

DEFINITION   A **standardized distribution** is composed of scores that have been transformed to create predetermined values for μ and σ. Standardized distributions are used to make dissimilar distributions comparable.

A z-score distribution is an example of a standardized distribution with $\mu = 0$ and $\sigma = 1$. That is, when any distribution (with any mean or standard deviation) is transformed into z-scores, the transformed distribution will always have $\mu = 0$ and $\sigma = 1$.

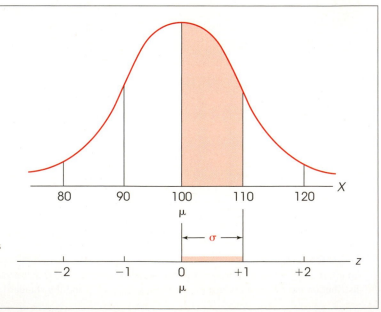

FIGURE 5.7
Following a z-score transformation, the X-axis is relabeled in z-score units. The distance that is equivalent to 1 standard deviation on the X-axis ($\sigma = 10$ points in this example) corresponds to 1 point on the z-score scale.

■ Demonstration of a z-Score Transformation

Although the basic characteristics of a z-score distribution have been explained logically, the following example provides a concrete demonstration that a z-score transformation creates a new distribution with a mean of zero, a standard deviation of 1, and the same shape as the original population.

EXAMPLE 5.8

We begin with a population of $N = 6$ scores consisting of the following values: 0, 6, 5, 2, 3, 2. This population has a mean of $\mu = \frac{18}{6} = 3$ and a standard deviation of $\sigma = 2$ (check the calculations for yourself).

Each of the X values in the original population is then transformed into a z-score as summarized in the following table.

$X = 0$	Below the mean by $1\frac{1}{2}$ standard deviations	$z = -1.50$
$X = 6$	Above the mean by $1\frac{1}{2}$ standard deviations	$z = +1.50$
$X = 5$	Above the mean by 1 standard deviation	$z = +1.00$
$X = 2$	Below the mean by $\frac{1}{2}$ standard deviation	$z = -0.50$
$X = 3$	Exactly equal to the mean—zero deviation	$z = 0$
$X = 2$	Below the mean by $\frac{1}{2}$ standard deviation	$z = -0.50$

The frequency distribution for the original population of X values is shown in Figure 5.8(a) and the corresponding distribution for the z-scores is shown in Figure 5.8(b). A simple comparison of the two distributions demonstrates the results of a z-score transformation.

1. The two distributions have exactly the same shape. Each individual has exactly the same relative position in the X distribution and in the z-score distribution.

2. After the transformation to z-scores, the mean of the distribution becomes $\mu = 0$. For these z-scores values, $N = 6$ and $\Sigma z = -1.50 + 1.50 + 1.00 + -0.50 + 0 + -0.50 = 0$. Thus, the mean for the z-scores is $\mu = \Sigma z/N = 0/6 = 0$.

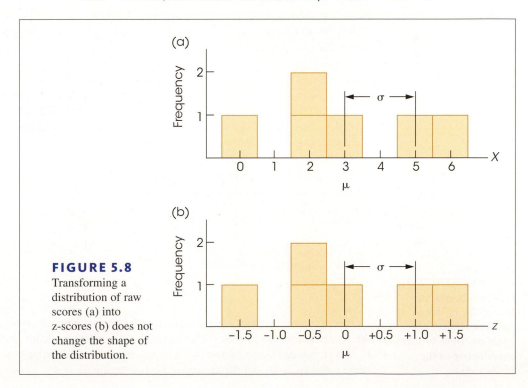

FIGURE 5.8

Transforming a distribution of raw scores (a) into z-scores (b) does not change the shape of the distribution.

Note that the individual with a score of $X = 3$ is located exactly at the mean in the X distribution and this individual is transformed into $z = 0$, exactly at the mean in the z-distribution.

3. After the transformation, the standard deviation becomes $\sigma = 1$. For these z-scores, $\Sigma z = 0$ and

$$\Sigma z^2 = (-1.50)^2 + (1.50)^2 + (1.00)^2 + (-0.50)^2 + (0)^2 + (-0.50)^2$$
$$= 2.25 + 2.25 + 1.00 + 0.25 + 0 + 0.25$$
$$= 6.00$$

Using the computational formula for SS, substituting z in place of X, we obtain

$$SS = \Sigma z^2 - \frac{(\Sigma z)^2}{N} = 6 - \frac{(0)^2}{6} = 6.00$$

For these z-scores, the variance is $\sigma^2 = \dfrac{SS}{N} = \dfrac{6}{6} = 1.00$ and the standard deviation is

$$\sigma = \sqrt{1.00} = 1.00$$

Note that the individual with $X = 5$ is located above the mean by 2 points, which is exactly one standard deviation in the X distribution. After transformation, this individual has a z-score that is located above the mean by 1 point, which is exactly one standard deviation in the z-score distribution. ■

■ Using z-Scores for Making Comparisons

Be sure to use the μ and σ values for the distribution to which X belongs.

One advantage of standardizing distributions is that it makes it possible to compare different scores or different individuals even though they come from completely different distributions. Normally, if two scores come from different distributions, it is impossible to make any direct comparison between them. Suppose, for example, Dave received a score of $X = 60$ on a psychology exam and a score of $X = 56$ on a biology test. For which course should Dave expect the better grade?

Because the scores come from two different distributions, you cannot make any direct comparison. Without additional information, it is even impossible to determine whether Dave is above or below the mean in either distribution. Before you can begin to make comparisons, you must know the values for the mean and standard deviation for each distribution. Suppose the biology scores had $\mu = 48$ and $\sigma = 4$, and the psychology scores had $\mu = 50$ and $\sigma = 10$. With this new information, you could sketch the two distributions, locate Dave's score in each distribution, and compare the two locations.

Note that Dave's z-score for biology is +2.0, which means that his test score is 2 standard deviations above the class mean. On the other hand, his z-score is +1.0 for psychology, or 1 standard deviation above the mean. In terms of relative class standing, Dave is doing much better in the biology class.

Instead of drawing the two distributions to determine where Dave's two scores are located, we simply can compute the two z-scores to find the two locations. For psychology, Dave's z-score is

$$z = \frac{X - \mu}{\sigma} = \frac{60 - 50}{10} = \frac{10}{10} = +1.0$$

For biology, Dave's z-score is

$$z = \frac{56 - 48}{4} = \frac{8}{4} = +2.0$$

Notice that we cannot compare Dave's two exam scores ($X = 60$ and $X = 56$) because the scores come from different distributions with different means and standard deviations. However, we can compare the two z-scores because all distributions of z-scores have the same mean ($\mu = 0$) and the same standard deviation ($\sigma = 1$).

LEARNING CHECK

1. A population with $\mu = 85$ and $\sigma = 12$ is transformed into z-scores. After the transformation, the population of z-scores will have a standard deviation of _____
   a. $\sigma = 12$
   b. $\sigma = 1.00$
   c. $\sigma = 0$
   d. cannot be determined from the information given

2. A population has $\mu = 50$ and $\sigma = 10$. If these scores are transformed into z-scores, the population of z-scores will have a mean of _____ and a standard deviation of _____.
   a. 50 and 10
   b. 50 and 1
   c. 0 and 10
   d. 0 and 1

3. Which of the following is an advantage of transforming X values into z-scores?
   a. All negative numbers are eliminated.
   b. The distribution is transformed to a normal shape.
   c. All scores are moved closer to the mean.
   d. None of the other options is an advantage.

ANSWERS   1. B, 2. D, 3. D

5.5 | Other Standardized Distributions Based on z-Scores

LEARNING OBJECTIVE   6. Use z-scores to transform any distribution into a standardized distribution with a predetermined mean and a predetermined standard deviation.

■ Transforming z-Scores to a Distribution with a Predetermined μ and σ

Although z-score distributions have distinct advantages, many people find them cumbersome because they contain negative values and decimals. For this reason, it is common to standardize a distribution by transforming the scores into a new distribution with a predetermined mean and standard deviation that are whole round numbers. The goal is to create a new (standardized) distribution that has "simple" values for the mean and standard deviation but does not change any individual's location within the distribution. *Standardized scores* of this type are frequently used in psychological or educational testing. For example, raw scores of the Scholastic Aptitude Test (SAT) are transformed to a standardized distribution that has $\mu = 500$ and $\sigma = 100$. For intelligence tests, raw scores are frequently converted to standard scores that have a mean of 100 and a standard deviation of 15. Because most IQ tests are standardized so that they have the same mean and standard deviation, it is possible to compare IQ scores even though they may come from different tests.

The procedure for standardizing a distribution to create new values for μ and σ is a two-step process:

1. The original raw scores are transformed into z-scores.

2. The z-scores are then transformed into new X values so that the specific μ and σ are attained.

This process ensures that each individual has exactly the same z-score location in the new distribution as in the original distribution. The following example demonstrates the standardization procedure.

EXAMPLE 5.9   An instructor gives an exam to a psychology class. For this exam, the distribution of raw scores has a mean of $\mu = 57$ with $\sigma = 14$. The instructor would like to simplify the distribution by transforming all scores into a new, standardized distribution with $\mu = 50$ and $\sigma = 10$. To demonstrate this process, we will consider what happens to two specific students: Maria, who has a raw score of $X = 64$ in the original distribution; and Joe, whose original raw score is $X = 43$.

STEP 1   Transform each of the original raw scores into z-scores. For Maria, $X = 64$, so her z-score is

$$z = \frac{X - \mu}{\sigma} = \frac{64 - 57}{14} = +0.5$$

For Joe, $X = 43$, and his z-score is

$$z = \frac{X - \mu}{\sigma} = \frac{43 - 57}{14} = -1.0$$

Remember: The values of μ and σ are for the distribution from which X was taken.

STEP 2   Change each z-score into an X value in the new standardized distribution that has a mean of $\mu = 50$ and a standard deviation of $\sigma = 10$.

Maria's z-score, $z = +0.50$, indicates that she is located above the mean by $\frac{1}{2}$ standard deviation. In the new, standardized distribution, this location corresponds to $X = 55$ (above the mean by 5 points).

Joe's z-score, $z = -1.00$, indicates that he is located below the mean by exactly 1 standard deviation. In the new distribution, this location corresponds to $X = 40$ (below the mean by 10 points).

The results of this two-step transformation process are summarized in Table 5.1. Note that Joe, for example, has exactly the same z-score ($z = -1.00$) in both the original distribution and the new standardized distribution. This means that Joe's position relative to the other students in the class has not changed.   ■

TABLE 5.1
A demonstration of how two individual scores are changed when a distribution is standardized. See Example 5.9.

	Original Scores $\mu = 57$ and $\sigma = 14$		z-Score Location		Standardized Scores $\mu = 50$ and $\sigma = 10$
Maria	$X = 64$	→	$z = +0.50$	→	$X = 55$
Joe	$X = 43$	→	$z = -1.00$	→	$X = 40$

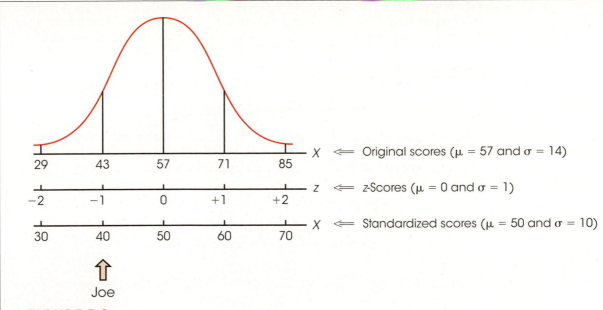

FIGURE 5.9

The distribution of exam scores from Example 5.9. The original distribution was standardized to produce a distribution with $\mu = 50$ and $\sigma = 10$. Note that each individual is identified by an original score, a z-score, and a new, standardized score. For example, Joe has an original score of 43, a z-score of -1.00, and a standardized score of 40.

Figure 5.9 provides another demonstration of the concept that standardizing a distribution does not change the individual positions within the distribution. The figure shows the original exam scores from Example 5.9, with a mean of $\mu = 57$ and a standard deviation of $\sigma = 14$. In the original distribution, Joe is located at a score of $X = 43$. In addition to the original scores, we have included a second scale showing the z-score value for each location in the distribution. In terms of z-scores, Joe is located at a value of $z = -1.00$. Finally, we have added a third scale showing the *standardized scores* where the mean is $\mu = 50$ and the standard deviation is $\sigma = 10$. For the standardized scores, Joe is located at $X = 40$. Note that Joe is always in the same place in the distribution. The only thing that changes is the number that is assigned to Joe: For the original scores, Joe is at 43; for the z-scores, Joe is at -1.00; and for the standardized scores, Joe is at 40.

LEARNING CHECK

1. A distribution with $\mu = 47$ and $\sigma = 6$ is being standardized so that the new mean and standard deviation will be $\mu = 100$ and $\sigma = 20$. What is the standardized score for a person with $X = 56$ in the original distribution?
   a. 110
   b. 115
   c. 120
   d. 130

2. A distribution with $\mu = 35$ and $\sigma = 8$ is being standardized so that the new mean and standard deviation will be $\mu = 50$ and $\sigma = 10$. In the new, standardized distribution your score is $X = 60$. What was your score in the original distribution?

   a. $X = 45$

   b. $X = 43$

   c. $X = 1.00$

   d. impossible to determine without more information

3. Using *z*-scores, a population with $\mu = 37$ and $\sigma = 6$ is standardized so that the new mean is $\mu = 50$ and $\sigma = 10$. How does an individual's *z*-score in the new distribution compare with his/her *z*-score in the original population?

   a. new $z = $ old $z + 13$

   b. new $z = (10/6)($old $z)$

   c. new $z = $ old z

   d. cannot be determined with the information given

ANSWERS   **1.** D, **2.** B, **3.** C

5.6 | Computing *z*-Scores for Samples

LEARNING OBJECTIVES

7. Transform X values into *z*-scores and transform *z*-scores into X values for a sample.

8. Describe the effects of transforming an entire sample into *z*-scores and explain the advantages of this transformation.

Although *z*-scores are most commonly used in the context of a population, the same principles can be used to identify individual locations within a sample. The definition of a *z*-score is the same for a sample as for a population, provided that you use the sample mean and the sample standard deviation to specify each *z*-score location. Thus, for a sample, each X value is transformed into a *z*-score so that

1. The sign of the *z*-score indicates whether the X value is above ($+$) or below ($-$) the sample mean, and

2. The numerical value of the *z*-score identifies the distance from the sample mean by measuring the number of sample standard deviations between the score (X) and the sample mean (M).

Expressed as a formula, each X value in a sample can be transformed into a *z*-score as follows:

$$z = \frac{X - M}{s} \qquad (5.3)$$

Similarly, each *z*-score can be transformed back into an X value, as follows:

$$X = M + zs \qquad (5.4)$$

You should recognize that these two equations are identical to the population equations (5.1 and 5.2) on pages 136 and 137, except that we are now using sample statistics, M and

s, in place of the population parameters μ and σ. The following example demonstrates the transformation of Xs and z-scores for a sample.

EXAMPLE 5.10

In a sample with a mean of $M = 40$ and a standard deviation of $s = 10$, what is the z-score corresponding to $X = 35$ and what is the X value corresponding to $z = +2.00$?

The score, $X = 35$, is located below the mean by 5 points, which is exactly half of the standard deviation. Therefore, the corresponding z-score is $z = -0.50$. The z-score, $z = +2.00$, corresponds to a location above the mean by 2 standard deviations. With a standard deviation of $s = 10$, this is a distance of 20 points. The score that is located 20 points above the mean is $X = 60$. Note that it is possible to find these answers using either the z-score definition or one of the equations (5.3 or 5.4). ∎

■ Standardizing a Sample Distribution

If all the scores in a sample are transformed into z-scores, the result is a sample of z-scores. The transformed distribution of z-scores will have the same properties that exist when a population of X values is transformed into z-scores. Specifically,

1. The sample of z-scores will have the same shape as the original sample of scores.
2. The sample of z-scores will have a mean of $M_z = 0$.
3. The sample of z-scores will have a standard deviation of $s_z = 1$.

Note that the set of z-scores is still considered to be a sample (just like the set of X values) and the sample formulas must be used to compute variance and standard deviation. The following example demonstrates the process of transforming the scores from a sample into z-scores.

EXAMPLE 5.11

We begin with a sample of $n = 5$ scores: 0, 2, 4, 4, 5. With a few simple calculations, you should be able to verify that the sample mean is $M = 3$, the sample variance is $s^2 = 4$, and the sample standard deviation is $s = 2$. Using the sample mean and sample standard deviation, we can convert each X value into a z-score. For example, $X = 5$ is located above the mean by 2 points. Thus, $X = 5$ is above the mean by exactly 1 standard deviation and has a z-score of $z = +1.00$. The z-scores for the entire sample are shown in the following table.

X	z
0	−1.50
2	−0.50
4	+0.50
4	+0.50
5	+1.00

Again, a few simple calculations demonstrate that the sum of the z-score values is $\Sigma z = 0$, so the mean is $M_z = 0$.

Because the mean is zero, each z-score value is its own deviation from the mean. Therefore, the sum of the squared deviations is simply the sum of the squared z-scores. For this sample of z-scores,

$$SS = \Sigma z^2 = (-1.50)^2 + (-0.50)^2 + (+0.50)^2 + (0.50)^2 + (+1.00)^2$$

$$= 2.25 + 0.25 + 0.25 + 0.25 + 1.00$$

$$= 4.00$$

Notice that the set of z-scores is considered to be a sample and the variance is computed using the sample formula with $df = n - 1$.

The variance for the sample of z-scores is

$$s_z^2 = \frac{SS}{n-1} = \frac{4}{4} = 1.00$$

Finally, the standard deviation for the sample of z-scores is $s_z^2 = \sqrt{1.00} = 1.00$. As always, the distribution of z-scores has a mean of 0 and a standard deviation of 1. ∎

LEARNING CHECK

1. For a sample with $M = 60$ and $s = 8$, what is the z-score corresponding to $X = 62$?
   a. 2
   b. 4
   c. 0.25
   d. 0.50

2. For a sample with a standard deviation of $s = 5$, what is the z-score corresponding to a score that is located 10 points below the mean?
   a. −10
   b. +2
   c. −2
   d. cannot answer without knowing the mean

3. If a sample with $M = 60$ and $s = 8$ is transformed into z-scores, then the resulting distribution of z-scores will have a mean of _____ and a standard deviation of _____.
   a. 0 and 1
   b. 60 and 1
   c. 0 and 8
   d. 60 and 8 (unchanged)

ANSWERS   1. C, 2. C, 3. A

5.7 | Looking Ahead to Inferential Statistics

LEARNING OBJECTIVE

9. Explain how z-scores can help researchers use the data from a sample to draw inferences about populations.

Recall that inferential statistics are techniques that use the information from samples to answer questions about populations. In later chapters, we will use inferential statistics to help interpret the results from research studies. A typical research study begins with a question about how a treatment will affect the individuals in a population. Because it is usually impossible to study an entire population, the researcher selects a sample and administers the treatment to the individuals in the sample. This general research situation is shown in Figure 5.10. To evaluate the effect of the treatment, the researcher simply compares the

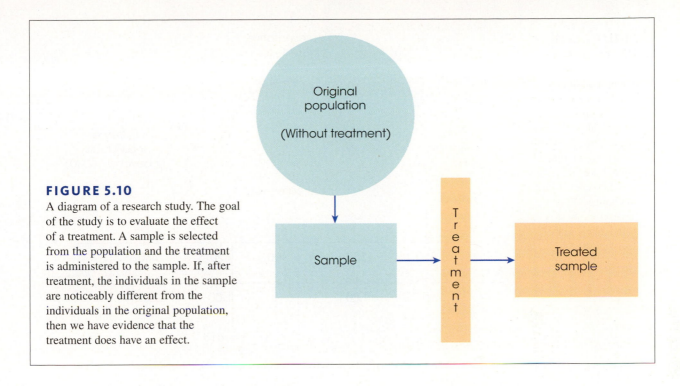

FIGURE 5.10

A diagram of a research study. The goal of the study is to evaluate the effect of a treatment. A sample is selected from the population and the treatment is administered to the sample. If, after treatment, the individuals in the sample are noticeably different from the individuals in the original population, then we have evidence that the treatment does have an effect.

treated sample with the original population. If the individuals in the sample are noticeably different from the individuals in the original population, the researcher has evidence that the treatment has had an effect. On the other hand, if the sample is not noticeably different from the original population, it would appear that the treatment has no effect.

Notice that the interpretation of the research results depends on whether the sample is *noticeably different* from the population. One technique for deciding whether a sample is noticeably different is to use z-scores. For example, an individual with a z-score near 0 is located in the center of the population and would be considered to be a fairly typical or representative individual. However, an individual with an extreme z-score, beyond $+2.00$ or -2.00 for example, would be considered "noticeably different" from most of the individuals in the population. Thus, we can use z-scores to help decide whether the treatment has caused a change. Specifically, if the individuals who receive the treatment finish the research study with extreme z-scores, we can conclude that the treatment does appear to have an effect. The following example demonstrates this process.

EXAMPLE 5.12   A researcher is evaluating the effect of a new growth hormone. It is known that regular adult rats weigh an average of $\mu = 400$ g. The weights vary from rat to rat, and the distribution of weights is normal with a standard deviation of $\sigma = 20$ g. The population distribution is shown in Figure 5.11. The researcher selects one newborn rat and injects the rat with the growth hormone. When the rat reaches maturity, it is weighed to determine whether there is any evidence that the hormone has an effect.

First, assume that the hormone-injected rat weighs $X = 418$ g. Although this is more than the average nontreated rat ($\mu = 400$ g), is it convincing evidence that the hormone has an effect? If you look at the distribution in Figure 5.11, you should realize that a rat weighing 418 g is not noticeably different from the regular rats that did not receive any

FIGURE 5.11

The distribution of weights for the population of adult rats. Note that individuals with z-scores near 0 are typical or representative. However, individuals with z-scores beyond +2.00 or −2.00 are extreme and noticeably different from most of the others in the distribution.

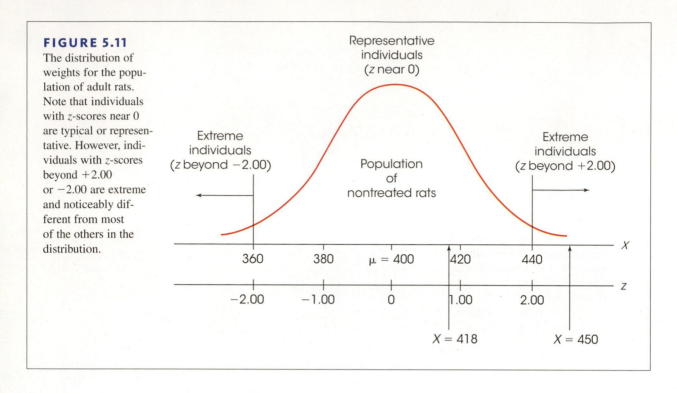

hormone injection. Specifically, our injected rat would be located near the center of the distribution for regular rats with a z-score of

$$z = \frac{X - \mu}{\sigma} = \frac{418 - 400}{20} = \frac{18}{20} = 0.90$$

Because the injected rat still looks the same as a regular, nontreated rat, the conclusion is that the hormone does not appear to have an effect.

Now, assume that our injected rat weighs $X = 450$ g. In the distribution of regular rats (see Figure 5.11), this animal would have a z-score of

$$z = \frac{X - \mu}{\sigma} = \frac{450 - 400}{20} = \frac{50}{20} = 2.50$$

In this case, the hormone-injected rat is substantially bigger than most ordinary rats, and it would be reasonable to conclude that the hormone does have an effect on weight.   ◼

In the preceding example, we used z-scores to help interpret the results obtained from a sample. Specifically, if the individuals who receive the treatment in a research study have extreme z-scores compared to those who do not receive the treatment, we can conclude that the treatment does appear to have an effect. The example, however, used an arbitrary definition to determine which z-score values are noticeably different. Although it is reasonable to describe individuals with z-scores near 0 as "highly representative" of the population, and individuals with z-scores beyond ±2.00 as "extreme," you should realize that these z-score boundaries were not determined by any mathematical rule. In the following chapter we introduce *probability*, which gives us a rationale for deciding exactly where to set the boundaries.

LEARNING CHECK

1. In $N = 25$ games last season, the college basketball team averaged $\mu = 74$ points with a standard deviation of $\sigma = 6$. In their final game of the season, the team scored 90 points. Based on this information, the number of points scored in the final game was _____.
   a. a little above average
   b. far above average
   c. above average, but it is impossible to describe how much above average
   d. There is not enough information to compare last year with the average.

2. Under what circumstances would a score that is 15 points above the mean be considered to be near the center of the distribution?
   a. when the population mean is much larger than 15
   b. when the population standard deviation is much larger than 15
   c. when the population mean is much smaller than 15
   d. when the population standard deviation is much smaller than 15

3. Under what circumstances would a score that is 20 points above the mean be considered to be an extreme, unrepresentative value?
   a. when the population mean is much larger than 20
   b. when the population standard deviation is much larger than 20
   c. when the population mean is much smaller than 20
   d. when the population standard deviation is much smaller than 20

ANSWERS   **1.** B, **2.** B, **3.** D

SUMMARY

1. Each X value can be transformed into a z-score that specifies the exact location of X within the distribution. The sign of the z-score indicates whether the location is above (positive) or below (negative) the mean. The numerical value of the z-score specifies the number of standard deviations between X and μ.

2. The z-score formula is used to transform X values into z-scores. For a population:

$$z = \frac{X - \mu}{\sigma}$$

For a sample:

$$z = \frac{X - M}{s}$$

3. To transform z-scores back into X values, it usually is easier to use the z-score definition rather than a formula. However, the z-score formula can be transformed into a new equation. For a population:

$$X = \mu + z\sigma$$

For a sample: $X = M + zs$

4. When an entire distribution of X values is transformed into z-scores, the result is a distribution of z-scores. The z-score distribution will have the same shape as the distribution of raw scores, and it always will have a mean of 0 and a standard deviation of 1.

5. When comparing raw scores from different distributions, it is necessary to standardize the distributions with a z-score transformation. The distributions will then be comparable because they will have the same parameters ($\mu = 0$, $\sigma = 1$). In practice, it is necessary to transform only those raw scores that are being compared.

6. In certain situations, such as psychological testing, a distribution may be standardized by converting the original X values into z-scores and then converting the z-scores into a new distribution of scores with predetermined values for the mean and the standard deviation.

7. In inferential statistics, z-scores provide an objective method for determining how well a specific score represents its population. A z-score near 0 indicates that the score is close to the population mean and therefore is representative. A z-score beyond $+2.00$ (or -2.00) indicates that the score is extreme and is noticeably different from the other scores in the distribution.

KEY TERMS

raw score (134)

z-score (135)

deviation score (136)

z-score transformation (141)

standardized distribution (142)

standardized score (145)

SPSS®

General instructions for using SPSS are presented in Appendix D. Following are detailed instructions for using SPSS to **Transform X Values into z-Scores for a Sample.**

Data Entry

1. Enter all of the scores in one column of the data editor, probably VAR00001.

Data Analysis

1. Click **Analyze** on the tool bar, select **Descriptive Statistics**, and click on **Descriptives**.

2. Highlight the column label for the set of scores (VAR0001) in the left box and click the arrow to move it into the **Variable** box.

3. Click the box to **Save standardized values as variables** at the bottom of the **Descriptives** screen.

4. Click **OK.**

SPSS Output

The program will produce the usual output display listing the number of scores (N), the maximum and minimum scores, the mean, and the standard deviation. However, if you go back to the Data Editor (use the tool bar at the bottom of the screen), SPSS will have produced a new column showing the z-score corresponding to each of the original X values.

Caution: The SPSS program computes the z-scores using the sample standard deviation instead of the population standard deviation. If your set of scores is intended to be a population, SPSS will not produce the correct z-score values. You can convert the SPSS values into population z-scores by multiplying each z-score value by the square root of $n/(n-1)$.

FOCUS ON PROBLEM SOLVING

1. When you are converting an X value to a z-score (or vice versa), do not rely entirely on the formula. You can avoid careless mistakes if you use the definition of a z-score (sign and numerical value) to make a preliminary estimate of the answer before you begin computations. For example, a z-score of $z = -0.85$ identifies a score located *below* the mean by almost 1 standard deviation. When computing the X value for this z-score, be

sure that your answer is smaller than the mean, and check that the distance between X and μ is slightly less than the standard deviation.

2. When comparing scores from distributions that have different standard deviations, it is important to be sure that you use the correct value for σ in the z-score formula. Use the σ value for the distribution from which the raw score in question was taken.

3. Remember that a z-score specifies a relative position within the context of a specific distribution. A z-score is a relative value, not an absolute value. For example, a z-score of $z = -2.0$ does not necessarily suggest a very low raw score—it simply means that the raw score is among the lowest within that specific group.

DEMONSTRATION 5.1

TRANSFORMING X VALUES INTO Z-SCORES

A distribution of scores has a mean of $\mu = 60$ with $\sigma = 12$. Find the z-score for $X = 75$.

STEP 1 **Determine the sign of the z-score.**   First, determine whether X is above or below the mean. This will determine the sign of the z-score. For this demonstration, X is larger than (above) μ, so the z-score will be positive.

STEP 2 **Convert the distance between X and μ into standard deviation units.**   For $X = 75$ and $\mu = 60$, the distance between X and μ is 15 points. With $\sigma = 12$ points, this distance corresponds to $\frac{15}{12} = 1.25$ standard deviations.

STEP 3 **Combine the sign from step 1 with the numerical value from Step 2.**   The score is above the mean $(+)$ by a distance of 1.25 standard deviations. Thus, $z = +1.25$.

STEP 4 **Confirm the answer using the z-score formula.**   For this example, $X = 75$, $\mu = 60$, and $\sigma = 12$.

$$z = \frac{X - \mu}{\sigma} = \frac{75 - 60}{12} = \frac{+15}{12} = +1.25$$

DEMONSTRATION 5.2

CONVERTING Z-SCORES TO X VALUES

For a population with $\mu = 60$ and $\sigma = 12$, what is the X value corresponding to $z = -0.50$? Notice that in this situation we know the z-score and must find X.

STEP 1 **Locate X in relation to the mean.**   A z-score of -0.50 indicates a location below the mean by half of a standard deviation.

STEP 2 **Convert the distance from standard deviation units to points.**   With $\sigma = 12$, half of a standard deviation is 6 points.

STEP 3 **Identify the X value.**   The value we want is located below the mean by 6 points. The mean is $\mu = 60$, so the score must be $X = 54$.

PROBLEMS

1. What information is provided by the sign ($+/-$) of a z-score? What information is provided by the numerical value of the z-score?

2. A distribution has a standard deviation of $\sigma = 10$. Find the z-score for each of the following locations in the distribution.
   a. Above the mean by 5 points.
   b. Above the mean by 2 points.
   c. Below the mean by 20 points.
   d. Below the mean by 15 points.

3. For a distribution with a standard deviation of $\sigma = 20$, describe the location of each of the following z-scores in terms of its position relative to the mean. For example, $z = +1.00$ is a location that is 20 points above the mean.
   a. $z = +2.00$
   b. $z = +0.50$
   c. $z = -1.00$
   d. $z = -0.25$

4. For a population with $\mu = 60$ and $\sigma = 12$,
   a. Find the z-score for each of the following X values. (*Note:* You should be able to find these values using the definition of a z-score. You should not need to use a formula or do any serious calculations.)

$X = 75$	$X = 48$	$X = 84$
$X = 54$	$X = 78$	$X = 51$

   b. Find the score (X value) that corresponds to each of the following z-scores. (Again, you should not need a formula or any serious calculations.)

$z = 1.00$	$z = 0.25$	$z = 1.50$
$z = -0.50$	$z = -1.25$	$z = -2.50$

5. For a population with $\mu = 40$ and $\sigma = 11$, find the z-score for each of the following X values. (*Note:* You probably will need to use a formula and a calculator to find these values.)

$X = 45$	$X = 52$	$X = 41$
$X = 30$	$X = 25$	$X = 38$

6. For a population with a mean of $\mu = 100$ and a standard deviation of $\sigma = 20$,
   a. Find the z-score for each of the following X values.

$X = 108$	$X = 115$	$X = 130$
$X = 90$	$X = 88$	$X = 95$

   b. Find the score (X value) that corresponds to each of the following z-scores.

$z = -0.40$	$z = -0.50$	$z = 1.80$
$z = 0.75$	$z = 1.50$	$z = -1.25$

7. A population has a mean of $\mu = 60$ and a standard deviation of $\sigma = 12$.
   a. For this population, find the z-score for each of the following X values.

$X = 69$	$X = 84$	$X = 63$
$X = 54$	$X = 48$	$X = 45$

   b. For the same population, find the score (X value) that corresponds to each of the following z-scores.

$z = 0.50$	$z = 1.50$	$z = -2.50$
$z = -0.25$	$z = -0.50$	$z = 1.25$

8. Find the z-score corresponding to a score of $X = 45$ for each of the following distributions.
   a. $\mu = 40$ and $\sigma = 20$
   b. $\mu = 40$ and $\sigma = 10$
   c. $\mu = 40$ and $\sigma = 5$
   d. $\mu = 40$ and $\sigma = 2$

9. Find the X value corresponding to $z = 0.25$ for each of the following distributions.
   a. $\mu = 40$ and $\sigma = 4$
   b. $\mu = 40$ and $\sigma = 8$
   c. $\mu = 40$ and $\sigma = 16$
   d. $\mu = 40$ and $\sigma = 32$

10. A score that is 6 points below the mean corresponds to a z-score of $z = -2.00$. What is the population standard deviation?

11. A score that is 9 points above the mean corresponds to a z-score of $z = 1.50$. What is the population standard deviation?

12. For a population with a standard deviation of $\sigma = 12$, a score of $X = 44$ corresponds to $z = -0.50$. What is the population mean?

13. For a population with a mean of $\mu = 70$, a score of $X = 64$ corresponds to $z = -1.50$. What is the population standard deviation?

14. In a population distribution, a score of $X = 28$ corresponds to $z = -1.00$ and a score of $X = 34$ corresponds to $z = -0.50$. Find the mean and standard deviation for the population. (*Hint:* Sketch the distribution and locate the two scores on your sketch.)

15. For each of the following populations, would a score of $X = 85$ be considered a central score (near the middle of the distribution) or an extreme score (far out in the tail of the distribution)?
    a. $\mu = 75$ and $\sigma = 15$
    b. $\mu = 80$ and $\sigma = 2$
    c. $\mu = 90$ and $\sigma = 20$
    d. $\mu = 93$ and $\sigma = 3$

16. A distribution of exam scores has a mean of $\mu = 78$.
   a. If your score is $X = 70$, which standard deviation would give you a better grade: $\sigma = 4$ or $\sigma = 8$?
   b. If your score is $X = 80$, which standard deviation would give you a better grade: $\sigma = 4$ or $\sigma = 8$?

17. For each of the following, identify the exam score that should lead to the better grade. In each case, explain your answer.
   a. A score of $X = 70$, on an exam with $M = 82$ and $\sigma = 8$; or a score of $X = 60$ on an exam with $\mu = 72$ and $\sigma = 12$.
   b. A score of $X = 58$, on an exam with $\mu = 49$ and $\sigma = 6$; or a score of $X = 85$ on an exam with $\mu = 70$ and $\sigma = 10$.
   c. A score of $X = 32$, on an exam with $\mu = 24$ and $\sigma = 4$; or a score of $X = 26$ on an exam with $\mu = 20$ and $\sigma = 2$.

18. A distribution with a mean of $\mu = 38$ and a standard deviation of $\sigma = 5$ is transformed into a standardized distribution with $\mu = 50$ and $\sigma = 10$. Find the new, standardized score for each of the following values from the original population.
   a. $X = 39$
   b. $X = 43$
   c. $X = 35$
   d. $X = 28$

19. A distribution with a mean of $\mu = 76$ and a standard deviation of $\sigma = 12$ is transformed into a standardized distribution with $\mu = 100$ and $\sigma = 20$. Find the new, standardized score for each of the following values from the original population.
   a. $X = 61$
   b. $X = 70$
   c. $X = 85$
   d. $X = 94$

20. A population consists of the following $N = 5$ scores: 0, 6, 4, 3, and 12.
   a. Compute μ and σ for the population.
   b. Find the z-score for each score in the population.
   c. Transform the original population into a new population of $N = 5$ scores with a mean of $\mu = 100$ and a standard deviation of $\sigma = 20$.

21. A sample has a mean of $M = 30$ and a standard deviation of $s = 8$. Find the z-score for each of the following X values from this sample.

$X = 32$	$X = 34$	$X = 36$
$X = 28$	$X = 20$	$X = 18$

22. A sample has a mean of $M = 25$ and a standard deviation of $s = 5$. For this sample, find the X value corresponding to each of the following z-scores.

$z = 0.40$	$z = 1.20$	$z = 2.00$
$z = -0.80$	$z = -0.60$	$z = -1.40$

23. For a sample with a standard deviation of $s = 8$, a score of $X = 65$ corresponds to $z = 1.50$. What is the sample mean?

24. For a sample with a mean of $M = 51$, a score of $X = 59$ corresponds to $z = 2.00$. What is the sample standard deviation?

25. In a sample distribution, $X = 56$ corresponds to $z = 1.00$, and $X = 47$ corresponds to $z = -0.50$. Find the mean and standard deviation for the sample.

26. A sample consists of the following $n = 7$ scores: 5, 0, 4, 5, 1, 2, and 4.
   a. Compute the mean and standard deviation for the sample.
   b. Find the z-score for each score in the sample.
   c. Transform the original sample into a new sample with a mean of $M = 50$ and $s = 10$.

Probability

Tools You Will Need

The following items are considered essential background material for this chapter. If you doubt your knowledge of any of these items, you should review the appropriate chapter or section before proceeding.

- Proportions (Appendix A)
  - Fractions
  - Decimals
  - Percentages
- Basic algebra (Appendix A)
- z-Scores (Chapter 5)

© Deborah Batt

How likely is it that you will marry someone whose last name begins with the same letter as your last name? This may sound like a silly question but it was the subject of a research study examining factors that contribute to interpersonal attraction (Jones, Pelham, Carvallo, & Mirenberg, 2004). In general, research has demonstrated that people tend to be attracted to others who are similar to themselves and it appears that one dimension of similarity is initials. The Jones et al. study found that individuals are disproportionately more likely to marry those with surnames that begin with the same letter as their own. The researchers began by looking at more than 14,000 marriage records from Georgia and Florida, and recording the surname for each groom and the maiden name of each bride. From these records it is possible to calculate the probability of randomly matching a bride and a groom whose

last names begin with the same letter. This value was found to be around 6.5%. Next, the researchers simply counted number of couples in the sample who shared the same last initial at the time they were married and found it to be around 7.7%. Although the difference between 6.5% and 7.7% may not seem large, for statisticians it is huge—certainly enough to conclude that it was not caused by random chance but is clear evidence of a systematic force that leads people to select partners who share the same initial. Jones et al. attribute the difference to egoism—people are attracted to people who are similar to them.

Although we used the term only once, the central theme of this discussion is *probability*. In this chapter we introduce the concept of probability, examine how it is applied in several different situations, and discuss its general role in the field of statistics.

6.1 | Introduction to Probability

LEARNING OBJECTIVE
1. Define probability and calculate (from information provided or from frequency distribution graph) the probability of a specific outcome as a proportion, decimal, and percentage.

In Chapter 1, we introduced the idea that research studies begin with a general question about an entire population, but the actual research is conducted using a sample. In this situation, the role of inferential statistics is to use the sample data as the basis for answering questions about the population. To accomplish this goal, inferential procedures are typically built around the concept of probability. Specifically, the relationships between samples and populations are usually defined in terms of probability.

Suppose, for example, you are selecting a single marble from a jar that contains 50 black and 50 white marbles. (In this example, the jar of marbles is the *population* and the single marble to be selected is the *sample*.) Although you cannot guarantee the exact outcome of your sample, it is possible to talk about the potential outcomes in terms of probabilities. In this case, you have a 50–50 chance of getting either color. Now consider another jar (population) that has 90 black and only 10 white marbles. Again, you cannot specify the exact outcome of a sample, but now you know that the sample probably will be a black marble. By knowing the makeup of a population, we can determine the probability of obtaining specific samples. In this way, probability gives us a connection between populations and samples, and this connection is the foundation for the inferential statistics to be presented in the chapters that follow.

You may have noticed that the preceding examples begin with a population and then use probability to describe the samples that could be obtained. This is exactly backward from what we want to do with inferential statistics. Remember that the goal of inferential statistics is to begin with a sample and then answer a general question about the population. We reach this goal in a two-stage process. In the first stage, we develop probability as

FIGURE 6.1

The role of probability in inferential statistics. Probability is used to predict the type of samples that are likely to be obtained from a population. Thus, probability establishes a connection between samples and populations. Inferential statistics rely on this connection when they use sample data as the basis for making conclusions about populations.

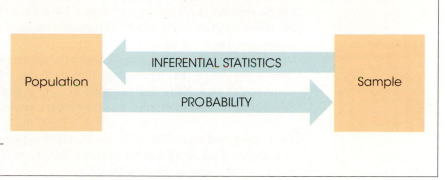

a bridge from populations to samples. This stage involves identifying the types of samples that probably would be obtained from a specific population. Once this bridge is established, we simply reverse the probability rules to allow us to move from samples to populations (Figure 6.1). The process of reversing the probability relationship can be demonstrated by considering again the two jars of marbles we looked at earlier. (Jar 1 has 50 black and 50 white marbles; jar 2 has 90 black and only 10 white marbles.) This time, suppose you are blindfolded when the sample is selected, so you do not know which jar is being used. Your task is to look at the sample that you obtain and then decide which jar is most likely. If you select a sample of $n = 4$ marbles and all are black, which jar would you choose? It should be clear that it would be relatively unlikely (low probability) to obtain this sample from jar 1; in four draws, you almost certainly would get at least 1 white marble. On the other hand, this sample would have a high probability of coming from jar 2, where nearly all the marbles are black. Your decision therefore is that the sample probably came from jar 2. Note that you now are using the sample to make an inference about the population.

■ Defining Probability

Probability is a huge topic that extends far beyond the limits of introductory statistics, and we will not attempt to examine it all here. Instead, we concentrate on the few concepts and definitions that are needed for an introduction to inferential statistics. We begin with a relatively simple definition of probability.

DEFINITION

For a situation in which several different outcomes are possible, the **probability** for any specific outcome is defined as a fraction or a proportion of all the possible outcomes. If the possible outcomes are identified as A, B, C, D, and so on, then

$$\text{probability of } A = \frac{\text{number of outcomes classified as } A}{\text{total number of possible outcomes}}$$

For example, if you are selecting a card from a complete deck, there are 52 possible outcomes. The probability of selecting the king of hearts is $p = \frac{1}{52}$. The probability of selecting an ace is $p = \frac{4}{52}$ because there are 4 aces in the deck.

To simplify the discussion of probability, we use a notation system that eliminates a lot of the words. The probability of a specific outcome is expressed with a p (for probability) followed by the specific outcome in parentheses. For example, the probability of selecting a king from a deck of cards is written as p(king). The probability of obtaining heads for a coin toss is written as p(heads).

Note that probability is defined as a proportion, or a part of the whole. This definition makes it possible to restate any probability problem as a proportion problem. For example, the probability problem "What is the probability of selecting a king from a deck of cards?" can be restated as "What proportion of the whole deck consists of kings?" In each case, the answer is $\frac{4}{52}$, or "4 out of 52." This translation from probability to proportion may seem trivial now, but it will be a great aid when the probability problems become more complex. In most situations, we are concerned with the probability of obtaining a particular sample from a population. The terminology of *sample* and *population* will not change the basic definition of probability. For example, the whole deck of cards can be considered as a population, and the single card we select is the sample.

Probability Values   The definition we are using identifies probability as a fraction or a proportion. If you work directly from this definition, the probability values you obtain are expressed as fractions. For example, if you are selecting a card at random,

$$p(\text{spade}) = \frac{13}{52} = \frac{1}{4}$$

Or, if you are tossing a coin,

$$p(\text{heads}) = \frac{1}{2}$$

If you are unsure how to convert from fractions to decimals or percentages, you should review the section on proportions in the math review, Appendix A.

You should be aware that these fractions can be expressed equally well as either decimals or percentages:

$$p = \frac{1}{4} = 0.25 = 25\%$$

$$p = \frac{1}{2} = 0.50 = 50\%$$

By convention, probability values most often are expressed as decimal values. But you should realize that any of these three forms is acceptable.

You also should note that all the possible probability values are contained in a limited range. At one extreme, when an event never occurs, the probability is zero, or 0% (Box 6.1). At the other extreme, when an event always occurs, the probability is 1, or 100%. Thus, all probability values are contained in a range from 0–1. For example, suppose that you have a jar containing 10 white marbles. The probability of randomly selecting a black marble is

$$p(\text{black}) = \frac{0}{10} = 0$$

The probability of selecting a white marble is

$$p(\text{white}) = \frac{10}{10} = 1$$

■ Random Sampling

For the preceding definition of probability to be accurate, it is necessary that the outcomes be obtained by a process called *random sampling*.

DEFINITION

> A **random sample** requires that each individual in the population has an *equal chance* of being selected. A sample obtained by this process is called a **simple random sample**.

A second requirement, necessary for many statistical formulas, states that if more than one individual is being selected, the probabilities must *stay constant* from one selection to the next. Adding this second requirement produces what is called *independent random sampling*. The term *independent* refers to the fact that the probability of selecting any particular individual is independent of the individuals already selected for the sample. For example, the probability that you will be selected is constant and does not change even when other individuals are selected before you are.

DEFINITION

> An **independent random sample** requires that each individual has an equal chance of being selected and that the probability of being selected stays constant from one selection to the next if more than one individual is selected.

Because independent random sample is usually a required component for most statistical applications, we will always assume that this is the sampling method being used. To simplify discussion, we will typically omit the word "independent" and simply refer to this sampling technique as *random sampling*. However, you should always assume that both requirements (equal chance and constant probability) are part of the process.

Each of the two requirements for random sampling has some interesting consequences. The first assures that there is no bias in the selection process. For a population with N individuals, each individual must have the same probability, $p = \frac{1}{N}$, of being selected. This means, for example, that you would not get a random sample of people in your city by selecting names from a yacht club membership list. Similarly, you would not get a random sample of college students by selecting individuals from your psychology classes. You also should note that the first requirement of random sampling prohibits you from applying the definition of probability to situations in which the possible outcomes are not equally likely. Consider, for example, the question of whether you will win a million dollars in the lottery tomorrow. There are only two possible alternatives.

1. You will win.

2. You will not win.

According to our simple definition, the probability of winning would be one out of two, or $p = \frac{1}{2}$. However, the two alternatives are not equally likely, so the simple definition of probability does not apply.

The second requirement also is more interesting than may be apparent at first glance. Consider, for example, the selection of $n = 2$ cards from a complete deck. For the first draw, the probability of obtaining the jack of diamonds is

$$p(\text{jack of diamonds}) = \frac{1}{52}$$

After selecting one card for the sample, you are ready to draw the second card. What is the probability of obtaining the jack of diamonds this time? Assuming that you still are holding the first card, there are two possibilities:

$$p(\text{jack of diamonds}) = \frac{1}{51} \text{ if the first card was not the jack of diamonds}$$

or

$$p(\text{jack of diamonds}) = 0 \text{ if the first card was the jack of diamonds}$$

In either case, the probability is different from its value for the first draw. This contradicts the requirement for random sampling, which says that the probability must stay constant. To keep the probabilities from changing from one selection to the next, it is necessary to return each individual to the population before you make the next selection. This process is called *sampling with replacement*. The second requirement for random samples (constant probability) demands that you sample with replacement.

(*Note*: We are using a definition of random sampling that requires equal chance of selection and constant probabilities. This kind of sampling is also known as independent random sampling, and often is called *random sampling with replacement*. Many of the statistics we will encounter later are founded on this kind of sampling. However, you should realize that other definitions exist for the concept of random sampling. In particular, it is very common to define random sampling without the requirement of constant probabilities—that is, *random sampling without replacement*. In addition, there are many different sampling techniques that are used when researchers are selecting individuals to participate in research studies.)

■ Probability and Frequency Distributions

The situations in which we are concerned with probability usually involve a population of scores that can be displayed in a frequency distribution graph. If you think of the graph as representing the entire population, then different portions of the graph represent different portions of the population. Because probabilities and proportions are equivalent, a particular portion of the graph corresponds to a particular probability in the population. Thus, whenever a population is presented in a frequency distribution graph, it will be possible to represent probabilities as proportions of the graph. The relationship between graphs and probabilities is demonstrated in the following example.

EXAMPLE 6.1   We will use a very simple population that contains only $N = 10$ scores with values 1, 1, 2, 3, 3, 4, 4, 4, 5, 6. This population is shown in the frequency distribution graph in Figure 6.2. If you are taking a random sample of $n = 1$ score from this population, what is the probability of obtaining an individual with a score greater than 4? In probability notation,

$$p(X > 4) = ?$$

Using the definition of probability, there are 2 scores that meet this criterion out of the total group of $N = 10$ scores, so the answer would be $p = \frac{2}{10}$. This answer can be obtained directly from the frequency distribution graph if you recall that probability and proportion measure the same thing. Looking at the graph (see Figure 6.2), what proportion of the population consists of scores greater than 4? The answer is the shaded part of the distribution—that is, 2 squares out of the total of 10 squares in the distribution. Notice that we now are defining probability as a proportion of area in the frequency distribution graph. This provides a very concrete and graphic way of representing probability.

Using the same population once again, what is the probability of selecting an individual with a score less than 5? In symbols,

$$p(X < 5) = ?$$

Going directly to the distribution in Figure 6.2, we now want to know what part of the graph is not shaded. The unshaded portion consists of 8 out of the 10 blocks ($\frac{8}{10}$ of the area of the graph), so the answer is $p = \frac{8}{10}$.   ■

FIGURE 6.2

A frequency distribution histogram for a population of $N = 10$ scores. The shaded part of the figure indicates the portion of the whole population that corresponds to scores greater than $X = 4$. The shaded portion is two-tenths $\left(\frac{2}{10}\right)$ of the whole distribution.

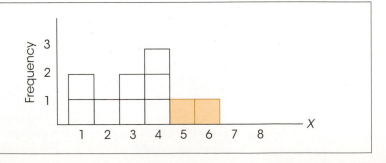

LEARNING CHECK

1. A survey of the students in a psychology class revealed that there were 19 females and 8 males. Of the 19 females, only 4 had no brothers or sisters, and 3 of the males were also the only child in the household. If a student is randomly selected from this class, then what is the probability of selecting a female who has no siblings?

   a. $\frac{4}{19}$

   b. $\frac{4}{27}$

   c. $\frac{19}{27}$

   d. $\frac{7}{27}$

2. A colony of laboratory rats contains 18 white rats and 7 spotted rats. What is the probability of randomly selecting a white rat from this colony?

   a. $\frac{1}{18}$

   b. $\frac{1}{25}$

   c. $\frac{17}{25}$

   d. $\frac{18}{25}$

3. A jar contains 20 red marbles and 10 blue marbles. If you take a *random sample* of $n = 3$ marbles from this jar, and the first two marbles are both red, what is the probability that the third marble also will be red?

   a. $\frac{20}{30}$

   b. $\frac{18}{28}$

   c. $\frac{1}{30}$

   d. $\frac{1}{28}$

ANSWERS   1. B, 2. D, 3. A

6.2 | Probability and the Normal Distribution

LEARNING OBJECTIVE   2. Use the unit normal table to find the following: (1) proportions/probabilities for specific z-score values and (2) z-score locations that correspond to specific proportions.

The normal distribution was first introduced in Chapter 2 as an example of a commonly occurring shape for population distributions. An example of a normal distribution is shown in Figure 6.3.

Note that the normal distribution is symmetrical, with the highest frequency in the middle and frequencies tapering off as you move toward either extreme. Although the exact shape for the normal distribution is defined by an equation (see Figure 6.3), the normal shape can also be described by the proportions of area contained in each section of the distribution. Statisticians often identify sections of a normal distribution by using z-scores. Figure 6.4 shows a normal distribution with several sections marked in z-score units. You should recall that z-scores measure positions in a distribution in terms of standard deviations from the mean. (Thus, $z = +1$ is 1 standard deviation above the mean, $z = +2$ is 2 standard deviations above the mean, and so on.) The graph shows the percentage of scores

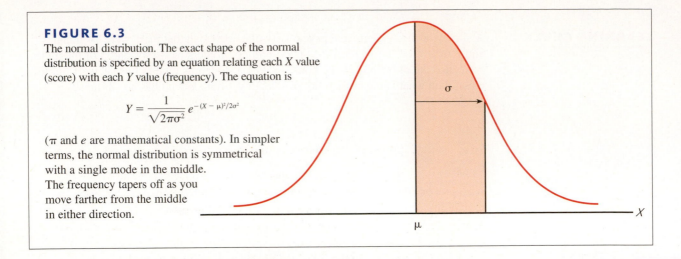

FIGURE 6.3

The normal distribution. The exact shape of the normal distribution is specified by an equation relating each X value (score) with each Y value (frequency). The equation is

$$Y = \frac{1}{\sqrt{2\pi\sigma^2}} \, e^{-(X - \mu)^2/2\sigma^2}$$

(π and e are mathematical constants). In simpler terms, the normal distribution is symmetrical with a single mode in the middle. The frequency tapers off as you move farther from the middle in either direction.

that fall in each of these sections. For example, the section between the mean ($z = 0$) and the point that is 1 standard deviation above the mean ($z = 1$) contains 34.13% of the scores. Similarly, 13.59% of the scores are located in the section between 1 and 2 standard deviations above the mean. In this way it is possible to define a normal distribution in terms of its proportions; that is, a distribution is normal if and only if it has all the right proportions.

There are two additional points to be made about the distribution shown in Figure 6.4. First, you should realize that the sections on the left side of the distribution have exactly the same areas as the corresponding sections on the right side because the normal distribution is symmetrical. Second, because the locations in the distribution are identified by z-scores, the percentages shown in the figure apply to *any normal distribution* regardless of the values for the mean and the standard deviation. Remember: When any distribution is transformed into z-scores, the mean becomes zero and the standard deviation becomes one.

Because the normal distribution is a good model for many naturally occurring distributions and because this shape is guaranteed in some circumstances (as you will see in Chapter 7), we devote considerable attention to this particular distribution. The process of answering probability questions about a normal distribution is introduced in the following example.

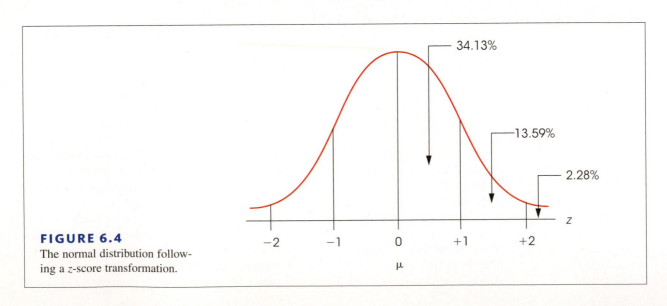

FIGURE 6.4

The normal distribution following a z-score transformation.

EXAMPLE 6.2   The population distribution of SAT scores is normal with a mean of $\mu = 500$ and a standard deviation of $\sigma = 100$. Given this information about the population and the known proportions for a normal distribution (see Figure 6.4), we can determine the probabilities associated with specific samples. For example, what is the probability of randomly selecting an individual from this population who has an SAT score greater than 700?

Restating this question in probability notation, we get

$$p(X > 700) = ?$$

We will follow a step-by-step process to find the answer to this question.

1. First, the probability question is translated into a proportion question: Out of all possible SAT scores, what proportion is greater than 700?

2. The set of "all possible SAT scores" is simply the population distribution. This population is shown in Figure 6.5. The mean is $\mu = 500$, so the score $X = 700$ is to the right of the mean. Because we are interested in all scores greater than 700, we shade in the area to the right of 700. This area represents the proportion we are trying to determine.

3. Identify the exact position of $X = 700$ by computing a z-score. For this example,

$$z = \frac{X - \mu}{\sigma} = \frac{700 - 500}{100} = \frac{200}{100} = 2.00$$

That is, an SAT score $X = 700$ inches is exactly 2 standard deviations above the mean and corresponds to a z-score of $z = +2.00$. We have also located this z-score in Figure 6.5.

4. The proportion we are trying to determine may now be expressed in terms of its z-score:

$$p(z > 2.00) = ?$$

According to the proportions shown in Figure 6.4, all normal distributions, regardless of the values for μ and σ, will have 2.28% of the scores in the tail beyond $z = +2.00$. Thus, for the population of SAT scores,

$$p(X > 700) = p(z > + 2.00) = 2.28\% \qquad ■$$

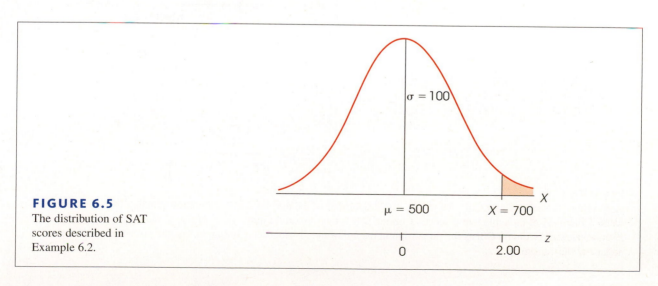

FIGURE 6.5
The distribution of SAT scores described in Example 6.2.

■ The Unit Normal Table

Before we attempt any more probability questions, we must introduce a more useful tool than the graph of the normal distribution shown in Figure 6.4. The graph shows proportions for only a few selected z-score values. A more complete listing of z-scores and proportions is provided in the *unit normal table*. This table lists proportions of the normal distribution for a full range of possible z-score values.

The complete unit normal table is provided in Appendix B Table B.1, and part of the table is reproduced in Figure 6.6. Notice that the table is structured in a four-column format. The first column (A) lists z-score values corresponding to different positions in a normal distribution. If you imagine a vertical line drawn through a normal distribution, then the exact location of the line can be described by one of the z-score values listed in column A. You should also realize that a vertical line separates the distribution into two sections: a larger section called the *body* and a smaller section called the *tail*. Columns B and C in the table identify the proportion of the distribution in each of the two sections. Column B presents the proportion in the body (the larger portion), and column C presents the proportion in the tail. Finally, we have added a fourth column, column D, that identifies the proportion of the distribution that is located *between* the mean and the z-score.

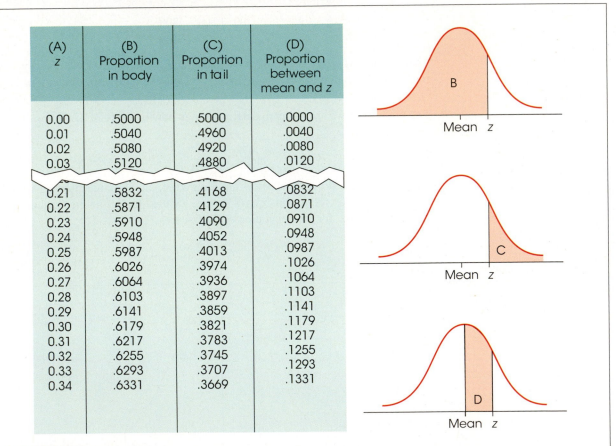

(A) z	(B) Proportion in body	(C) Proportion in tail	(D) Proportion between mean and z
0.00	.5000	.5000	.0000
0.01	.5040	.4960	.0040
0.02	.5080	.4920	.0080
0.03	.5120	.4880	.0120
0.21	.5832	.4168	.0832
0.22	.5871	.4129	.0871
0.23	.5910	.4090	.0910
0.24	.5948	.4052	.0948
0.25	.5987	.4013	.0987
0.26	.6026	.3974	.1026
0.27	.6064	.3936	.1064
0.28	.6103	.3897	.1103
0.29	.6141	.3859	.1141
0.30	.6179	.3821	.1179
0.31	.6217	.3783	.1217
0.32	.6255	.3745	.1255
0.33	.6293	.3707	.1293
0.34	.6331	.3669	.1331

FIGURE 6.6

A portion of the unit normal table. This table lists proportions of the normal distribution corresponding to each z-score value. Column A of the table lists z-scores. Column B lists the proportion in the body of the normal distribution for that z-score value and Column C lists the proportion in the tail of the distribution. Column D lists the proportion between the mean and the z-score.

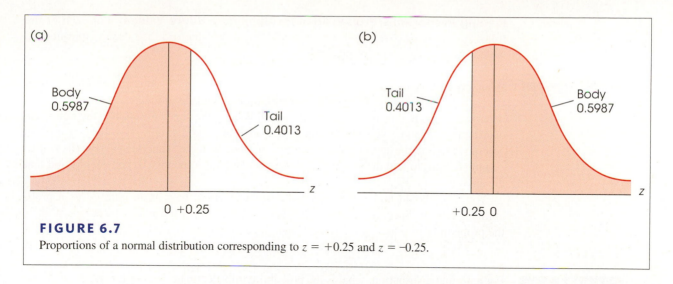

FIGURE 6.7

Proportions of a normal distribution corresponding to $z = +0.25$ and $z = -0.25$.

We will use the distribution in Figure 6.7(a) to help introduce the unit normal table. The figure shows a normal distribution with a vertical line drawn at $z = +0.25$. Using the portion of the table shown in Figure 6.6, find the row in the table that contains $z = 0.25$ in column A. Reading across the row, you should find that the line drawn $z = +0.25$ separates the distribution into two sections with the larger section (the body) containing 0.5987 or 59.87% of the distribution and the smaller section (the tail) containing 0.4013 or 40.13% of the distribution. Also, there is exactly 0.0987 or 9.87% of the distribution between the mean and $z = +0.25$.

To make full use of the unit normal table, there are a few facts to keep in mind:

1. The *body* always corresponds to the larger part of the distribution whether it is on the right-hand side or the left-hand side. Similarly, the *tail* is always the smaller section whether it is on the right or the left.

2. Because the normal distribution is symmetrical, the proportions on the right-hand side are exactly the same as the corresponding proportions on the left-hand side. Earlier, for example, we used the unit normal table to obtain proportions for $z = +0.25$. Figure 6.7(b) shows the same proportions for $z = -0.25$. For a negative z-score, however, notice that the tail of the distribution is on the left side and the body is on the right. For a positive z-score (Figure 6.7(a)), the positions are reversed. However, the proportions in each section are exactly the same, with 0.5987 in the body and 0.4013 in the tail. Once again, the table does not list negative z-score values. To find proportions for negative z-scores, you must look up the corresponding proportions for the positive value of z.

3. Although the z-score values change signs (+ and −) from one side to the other, the proportions are always positive. Thus, column C in the table always lists the proportion in the tail whether it is the right-hand tail or the left-hand tail.

■ Probabilities, Proportions, and z-Scores

The unit normal table lists relationships between z-score locations and proportions in a normal distribution. For any z-score location, you can use the table to look up the corresponding proportions. Similarly, if you know the proportions, you can use the table to find the specific z-score location. Because we have defined probability as equivalent to proportion, you can also use the unit normal table to look up probabilities for normal distributions. The following examples demonstrate a variety of different ways that the unit normal table can be used.

Finding Proportions/Probabilities for Specific *z*-Score Values   For each of the following examples, we begin with a specific *z*-score value and then use the unit normal table to find probabilities or proportions associated with the *z*-score.

EXAMPLE 6.3A   What proportion of the normal distribution corresponds to *z*-score values greater than $z = 1.00$? First, you should sketch the distribution and shade in the area you are trying to determine. This is shown in Figure 6.8(a). In this case, the shaded portion is the tail of the distribution beyond $z = 1.00$. To find this shaded area, you simply look for $z = 1.00$ in column A to find the appropriate row in the unit normal table. Then scan across the row to column C (tail) to find the proportion. Using the table in Appendix B, you should find that the answer is 0.1587.

You also should notice that this same problem could have been phrased as a probability question. Specifically, we could have asked, "For a normal distribution, what is the probability of selecting a *z*-score value greater than $z = +1.00$?" Again, the answer is $p(z > 1.00) = 0.1587$ (or 15.87%).   ∎

EXAMPLE 6.3B   For a normal distribution, what is the probability of selecting a *z*-score less than $z = 1.50$? In symbols, $p(z < 1.50) = ?$ Our goal is to determine what proportion of the normal distribution corresponds to *z*-scores less than 1.50. A normal distribution is shown in Figure 6.8(b) and $z = 1.50$ is located in the distribution. Note that we have shaded all the values to the left of (less than) $z = 1.50$. This is the portion we are trying to find. Clearly, the shaded portion is more than 50% so it corresponds to the body of the distribution. Therefore, find $z = 1.50$ in column A of the unit normal table and read across the row to obtain the proportion from column B. The answer is $p(z < 1.50) = 0.9332$ (or 93.32%).   ∎

EXAMPLE 6.3C   Many problems require that you find proportions for negative *z*-scores. For example, what proportion of the normal distribution is contained in the tail beyond $z = -0.50$? That is, $p(z < -0.50)$. This portion has been shaded in Figure 6.8(c). To answer questions with negative *z*-scores, simply remember that the normal distribution is symmetrical with a *z*-score of zero at the mean, positive values to the right, and negative values to the left. The proportion in the left tail beyond $z = -0.50$ is identical to the proportion in the right tail beyond $z = +0.50$. To find this proportion, look up $z = 0.50$ in column A, and read across the row to find the proportion in column C (tail). You should get an answer of 0.3085 (30.85%).   ∎

Moving to the left on the *X*-axis results in smaller *X* values and smaller *z*-scores. Thus, a *z*-score of −3.00 reflects a smaller value than a *z*-score of −1.

The following example is an opportunity for you to test your understanding by finding proportions in a normal distribution yourself.

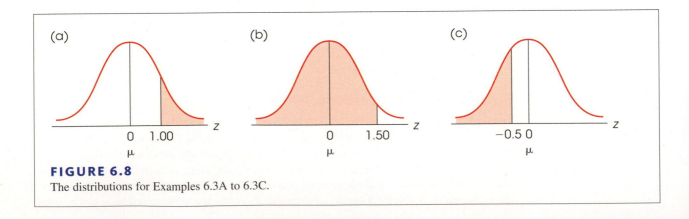

FIGURE 6.8
The distributions for Examples 6.3A to 6.3C.

EXAMPLE 6.4

Find proportion of a normal distribution corresponding to each of the following sections: a. $z > 0.80$ b. $z > -0.75$.

You should obtain answers of 0.2119 and 0.7734 for a and b, respectively. ■

Finding the z-Score Location that Corresponds to Specific Proportions The preceding examples all involved using a z-score value in column A to look up proportions in column B or C. You should realize, however, that the table also allows you to begin with a known proportion and then look up the corresponding z-score. The following examples demonstrate this process.

EXAMPLE 6.5A

For a normal distribution, what z-score separates the top 10% from the remainder of the distribution? To answer this question, we have sketched a normal distribution (Figure 6.9(a)) and drawn a vertical line that separates the highest 10% (approximately) from the rest. The problem is to locate the exact position of this line. For this distribution, we know that the tail contains 0.1000 (10%) and the body contains 0.9000 (90%). To find the z-score value, you simply locate the row in the unit normal table that has 0.1000 in column C or 0.9000 in column B. For example, you can scan down the values in column C (tail) until you find a proportion of 0.1000. Note that you probably will not find the exact proportion, but you can use the closest value listed in the table. For this example, a proportion of 0.1000 is not listed in column C but you can use 0.1003, which is listed. Once you have found the correct proportion in the table, simply read across the row to find the corresponding z-score value in column A.

For this example, the z-score that separates the extreme 10% in the tail is $z = 1.28$. At this point you must be careful because the table does not differentiate between the right-hand tail and the left-hand tail of the distribution. Specifically, the final answer could be either $z = +1.28$, which separates 10% in the right-hand tail, or $z = -1.28$, which separates 10% in the left-hand tail. For this problem we want the right-hand tail (the highest 10%), so the z-score value is $z = +1.28$. ■

EXAMPLE 6.5B

For a normal distribution, what z-score values form the boundaries that separate the middle 60% of the distribution from the rest of the scores?

Again, we sketched a normal distribution (Figure 6.9(b)) and drawn vertical lines so that roughly 60% of the distribution in the central section, with the remainder split equally between the two tails. The problem is to find the z-score values that define the exact locations for the lines. To find the z-score values, we begin with the known proportions: 0.6000 in the center and 0.4000 divided equally between the two tails. Although these proportions can be used in several different ways, this example provides an opportunity to demonstrate how column D in the table can be used to solve problems. For this problem, the 0.6000 in the center can be divided in half with exactly 0.3000 to the right of the mean

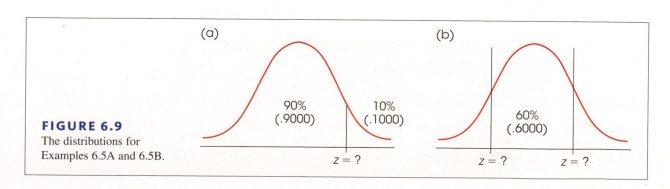

FIGURE 6.9
The distributions for Examples 6.5A and 6.5B.

and exactly 0.3000 to the left. Each of these sections corresponds to the proportion listed in column D. Begin by scanning down column D, looking for a value of 0.3000. Again, this exact proportion is not in the table, but the closest value is 0.2995. Reading across the row to column A, you should find a z-score value of $z = 0.84$. Looking again at the sketch (Figure 6.9(b)), the right-hand line is located at $z = +0.84$ and the left-hand line is located at $z = -0.84$. ∎

You may have noticed that we have sketched distributions for each of the preceding problems. As a general rule, you should always sketch a distribution, locate the mean with a vertical line, and shade in the portion you are trying to determine. Look at your sketch. It will help you determine which columns to use in the unit normal table. If you make a habit of drawing sketches, you will avoid careless errors when using the table.

1. What proportion of a normal distribution is located in the tail beyond a z-score of $z = -1.50$?
   a. −0.0668
   b. −0.9332
   c. 0.0668
   d. 0.9332

2. A vertical line is drawn through a normal distribution at $z = -1.00$. How much of the distribution is located between the line and the mean?
   a. 15.87%
   b. 34.13%
   c. 84.13%
   d. −15.87%

3. What z-score separates the lowest 10% of the distribution from the rest?
   a. $z = 0.90$
   b. $z = -0.90$
   c. $z = 1.28$
   d. $z = -1.28$

ANSWERS   1. C, 2. B, 3. D

6.3 | Probabilities and Proportions for Scores from a Normal Distribution

3. Calculate the probability for a specific X-value.

4. Calculate the score (X-value) corresponding to a specific proportion in a distribution.

In the preceding section, we used the unit normal table to find probabilities and proportions corresponding to specific z-score values. In most situations, however, it is necessary to find probabilities for specific *X* values. Consider the following example:

> It is known that IQ scores form a normal distribution with $\mu = 100$ and $\sigma = 15$. Given this information, what is the probability of randomly selecting an individual with an IQ score less than 120?

This problem is asking for a specific probability or proportion of a normal distribution. However, before we can look up the answer in the unit normal table, we must first transform the IQ scores (*X* values) into *z*-scores. Thus, to solve this new kind of probability problem, we must add one new step to the process. Specifically, to answer probability questions about scores (*X* values) from a normal distribution, you must use the following two-step procedure:

1. Transform the *X* values into *z*-scores.

2. Use the unit normal table to look up the proportions corresponding to the *z*-score values.

Caution: The unit normal table can be used only with normal-shaped distributions. If a distribution is not normal, transforming to *z*-scores will not make it normal.

This process is demonstrated in the following examples. Once again, we suggest that you sketch the distribution and shade the portion you are trying to find in order to avoid careless mistakes.

EXAMPLE 6.6

We will now answer the probability question about IQ scores that we presented earlier. Specifically, what is the probability of randomly selecting an individual with an IQ score less than 120? Restated in terms of proportions, we want to find the proportion of the IQ distribution that corresponds to scores less than 120. The distribution is drawn in Figure 6.10, and the portion we want has been shaded.

The first step is to change the *X* values into *z*-scores. In particular, the score of $X = 120$ is changed to

$$z = \frac{X - \mu}{\sigma} = \frac{120 - 100}{15} = \frac{20}{15} = 1.33$$

Thus, an IQ score of $X = 120$ corresponds to a *z*-score of $z = 1.33$, and IQ scores less than 120 correspond to *z*-scores less than 1.33.

Next, look up the *z*-score value in the unit normal table. Because we want the proportion of the distribution in the body to the left of $X = 120$ (see Figure 6.10), the answer will be found in column B. Consulting the table, we see that a *z*-score of 1.33 corresponds to a proportion of 0.9082. The probability of randomly selecting an individual with an IQ less than 120 is $p = 0.9082$. In symbols,

$$p(X < 120) = p(z < 1.33) = 0.9082 \text{ (or 90.82\%)}$$

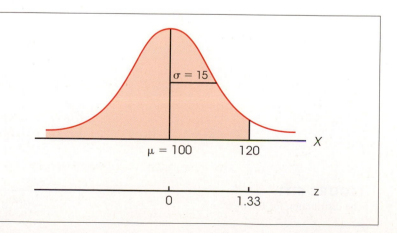

FIGURE 6.10

The distribution of IQ scores. The problem is to find the probability or proportion of the distribution corresponding to scores less than 120.

Finally, notice that we phrased this question in terms of a *probability*. Specifically, we asked, "What is the probability of selecting an individual with an IQ less than 120?" However, the same question can be phrased in terms of a *proportion*: "What proportion of all the individuals in the population have IQ scores less than 120?" Both versions ask exactly the same question and produce exactly the same answer. A third alternative for presenting the same question is introduced in Box 6.1.   ■

Finding Proportions/Probabilities Located between Two Scores   The next example demonstrates the process of finding the probability of selecting a score that is located *between* two specific values. Although these problems can be solved using the proportions of columns B and C (body and tail), they are often easier to solve with the proportions listed in column D.

EXAMPLE 6.7   The highway department conducted a study measuring driving speeds on a local section of interstate highway. They found an average speed of $\mu = 58$ miles per hour with a standard deviation of $\sigma = 10$. The distribution was approximately normal. Given this information, what proportion of the cars are traveling between 55 and 65 miles per hour? Using probability notation, we can express the problem as

$$p(55 < X < 65) = ?$$

The distribution of driving speeds is shown in Figure 6.11 with the appropriate area shaded. The first step is to determine the z-score corresponding to the X value at each end of the interval.

$$\text{For } X = 55: z = \frac{X - \mu}{\sigma} = \frac{55 - 58}{10} = \frac{-3}{10} = -0.30$$

$$\text{For } X = 65: z = \frac{X - \mu}{\sigma} = \frac{65 - 58}{10} = \frac{7}{10} = 0.70$$

Looking again at Figure 6.11, we see that the proportion we are seeking can be divided into two sections: (1) the area left of the mean and (2) the area right of the mean. The first area is the proportion between the mean and $z = -0.30$ and the second is the proportion between the mean and $z = +0.70$. Using column D of the unit normal table, these two proportions are 0.1179 and 0.2580. The total proportion is obtained by adding these two sections:

$$p(55 < X < 65) = p(-0.30 < z < +0.70) = 0.1179 + 0.2580 = 0.3759 \quad ■$$

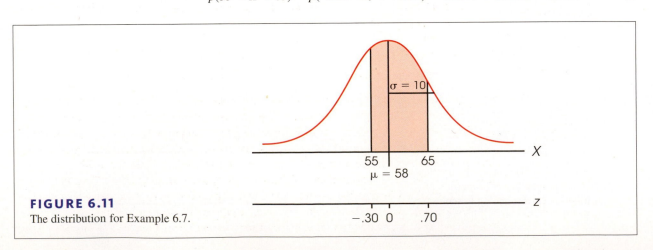

FIGURE 6.11
The distribution for Example 6.7.

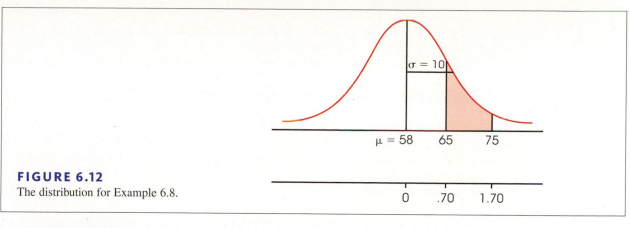

FIGURE 6.12
The distribution for Example 6.8.

EXAMPLE 6.8   Using the same distribution of driving speeds from the previous example, what proportion of cars is traveling between 65 and 75 miles per hour?

$$p(65 < X < 75) = ?$$

The distribution is shown in Figure 6.12 with the appropriate area shaded. Again, we start by determining the z-score corresponding to each end of the interval.

$$\text{For } X = 65: z = \frac{X - \mu}{\sigma} = \frac{65 - 58}{10} = \frac{7}{10} = 0.70$$

$$\text{For } X = 75: z = \frac{X - \mu}{\sigma} = \frac{75 - 58}{10} = \frac{17}{10} = 1.70$$

There are several different ways to use the unit normal table to find the proportion between these two z-scores. For this example, we will use the proportions in the tail of the distribution (column C). According to column C in the unit normal table, the proportion in the tail beyond $z = 0.70$ is $p = 0.2420$. Note that this proportion includes the section that we want, but it also includes an extra, unwanted section located in the tail beyond $z = 1.70$. Locating $z = 1.70$ in the table, and reading across the row to column C, we see that the unwanted section is $p = 0.0446$. To obtain the correct answer, we subtract the unwanted portion from the total proportion between the mean and $z = 0.70$.

$$p(65 < X, 75) = p(0.70 < z, 1.70) = 0.2420 - 0.0446 = 0.1974 \quad \blacksquare$$

The following example is an opportunity for you to test your understanding by finding probabilities for scores in a normal distribution yourself.

EXAMPLE 6.9   For a normal distribution with $\mu = 60$ and a standard deviation of $\sigma = 12$, find each probability requested.
a. $p(X > 66)$
b. $p(48 < X < 72)$

You should obtain answers of 0.3085 and 0.6826 for a and b, respectively. $\blacksquare$

Finding Scores Corresponding to Specific Proportions or Probabilities   In the previous three examples, the problem was to find the proportion or probability corresponding to specific X values. The two-step process for finding these proportions is shown in Figure 6.13. Thus far, we have only considered examples that move in a clockwise direction around the triangle shown in the figure; that is, we start with an X value that is transformed into a z-score, and then we use the unit normal table to look up the appropriate proportion. You should realize, however, that it is possible to reverse this two-step process so that we move backward, or counterclockwise, around the triangle. This reverse process allows us to find the score (X value) corresponding to a specific proportion in the distribution. Following the lines in Figure 6.13, we

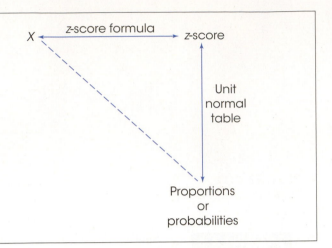

FIGURE 6.13

Determining probabilities or proportions for a normal distribution is shown as a two-step process with z-scores as an intermediate stop along the way. Note that you cannot move directly along the dashed line between X values and probabilities and proportions. Instead, you must follow the solid lines around the corner.

begin with a specific proportion, use the unit normal table to look up the corresponding z-score, and then transform the z-score into an X value. The following example demonstrates this process.

EXAMPLE 6.10

The U.S. Census Bureau (2005) reports that Americans spend an average of $\mu = 24.3$ minutes commuting to work each day. Assuming that the distribution of commuting times is normal with a standard deviation of $\sigma = 10$ minutes, how much time do you have to spend commuting each day to be in the highest 10% nationwide? (An alternative form of the same question is presented in Box 6.1.). The distribution is shown in Figure 6.14 with a portion representing approximately 10% shaded in the right-hand tail.

In this problem, we begin with a proportion (10% or 0.10), and we are looking for a score. According to the map in Figure 6.13, we can move from p (proportion) to X (score) via z-scores. The first step is to use the unit normal table to find the z-score that corresponds to a proportion of 0.10 in the tail. First, scan the values in column C to locate the row that has a proportion of 0.10 in the tail of the distribution. Note that you will not find 0.1000 exactly, but locate the closest value possible. In this case, the closest value is 0.1003. Reading across the row, we find $z = 1.28$ in column A.

The next step is to determine whether the z-score is positive or negative. Remember that the table does not specify the sign of the z-score. Looking at the distribution in Figure 6.14, you should realize that the score we want is above the mean, so the z-score is positive, $z = +1.28$.

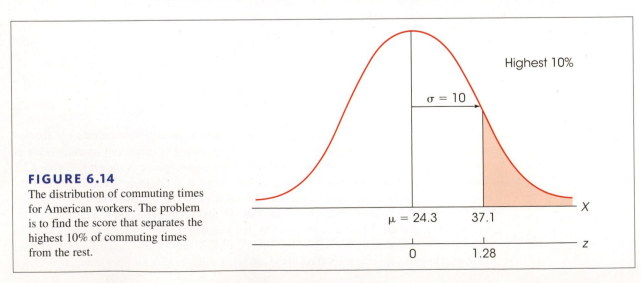

FIGURE 6.14

The distribution of commuting times for American workers. The problem is to find the score that separates the highest 10% of commuting times from the rest.

The final step is to transform the z-score into an X value. By definition, a z-score of $+1.28$ corresponds to a score that is located above the mean by 1.28 standard deviations. One standard deviation is equal to 10 points ($\sigma = 10$), so 1.28 standard deviations is

$$1.28\ \sigma = 1.28(10) = 12.8 \text{ points}$$

Thus, our score is located above the mean ($\mu = 24.3$) by a distance of 12.8 points. Therefore,

$$X = 24.3 + 12.8 = 37.1$$

The answer for our original question is that you must commute at least 37.1 minutes a day to be in the top 10% of American commuters.   ■

EXAMPLE 6.11

Again, the distribution of commuting times for American workers is normal with a mean of $\mu = 24.3$ minutes and a standard deviation of $\sigma = 10$ minutes. For this example, we will find the range of values that defines the middle 90% of the distribution. The entire distribution is shown in Figure 6.15 with the middle portion shaded.

The 90% (0.9000) in the middle of the distribution can be split in half with 45% (0.4500) on each side of the mean. Looking up 0.4500, in column D of the unit normal table, you will find that the exact proportion is not listed. However, you will find 0.4495 and 0.4505, which are equally close. Technically, either value is acceptable, but we will use 0.4505 so that the total area in the middle is at least 90%. Reading across the row, you should find a z-score of $z = 1.65$ in column A. Thus, the z-score at the right boundary is $z = +1.65$ and the z-score at the left boundary is $z = -1.65$. In either case, a z-score of 1.65 indicates a location that is 1.65 standard deviations away from the mean. For the distribution of commuting times, one standard deviation is $\sigma = 10$, so 1.65 standard deviations is a distance of

$$1.65\ \sigma = 1.65(10) = 16.5 \text{ points}$$

Therefore, the score at the right-hand boundary is located above the mean by 16.5 points and corresponds to $X = 24.3 + 16.5 = 40.8$. Similarly, the score at the left-hand boundary is below the mean by 16.5 points and corresponds to $X = 24.3 - 16.5 = 7.8$. The middle 90% of the distribution corresponds to values between 7.8 and 40.8. Thus, 90% of American commuters spend between 7.8 and 40.8 minutes commuting to work each day. Only 10% of commuters spend either more time or less time.   ■

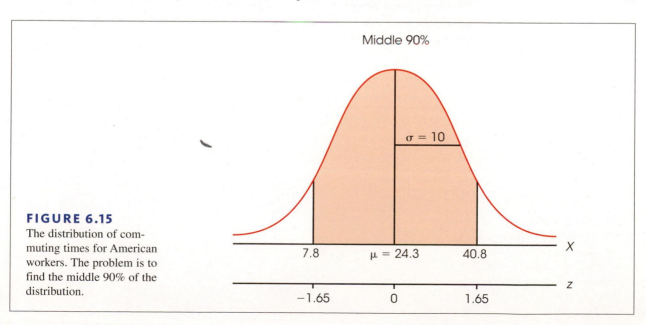

FIGURE 6.15

The distribution of commuting times for American workers. The problem is to find the middle 90% of the distribution.

BOX 6.1 Probabilities, Proportions, and Percentile Ranks

Thus far we have discussed parts of distributions in terms of proportions and probabilities. However, there is another set of terminology that deals with many of the same concepts. Specifically, in Chapter 2 we defined the *percentile rank* for a specific score as the percentage of the individuals in the distribution who have scores that are less than or equal to the specific score. For example, if 70% of the individuals have scores of $X = 45$ or lower, then $X = 45$ has a percentile rank of 70%. When a score is referred to by its percentile rank, the score is called a *percentile*. For example, a score with a percentile rank of 70% is called the 70th percentile.

Using this terminology, it is possible to rephrase some of the probability problems that we have

been working. In Example 6.6, the problem was presented as "What is the probability of randomly selecting an individual with an IQ of less than 120?" Exactly the same question could be phrased as "What is the percentile rank for an IQ score of 120?" In each case, we are looking for the proportion of the distribution corresponding to scores equal to or less than 120. Similarly, Example 6.10 asked "How much time do you have to spend commuting each day to be in the highest 10% nationwide?" Because this score separates the top 10% from the bottom 90%, the same question could be rephrased as "What is the 90th percentile for the distribution of commuting times?"

LEARNING CHECK

1. For a normal distribution with a mean of $\mu = 500$ and $\sigma = 100$, what is the probability of selecting an individual with a score less than 400?
   a. 0.1587
   b. 0.8413
   c. 0.34.13
   d. −0.15.87

2. For a normal distribution with a mean of $\mu = 40$ and $\sigma = 4$, what is the probability of selecting an individual with a score greater than 46?
   a. 0.0668
   b. 0.4452
   c. 0.9332
   d. 0.0548

3. SAT scores for a normal distribution with mean of $\mu = 500$ with $\sigma = 100$. What SAT score separates the top 10% of the distribution from the rest?
   a. 128
   b. 628
   c. 501.28
   d. 372

ANSWERS   **1.** A, **2.** A, **3.** B

6.4 | Probability and the Binomial Distribution

LEARNING OBJECTIVES

5. Describe the circumstances in which the normal distribution serves as a good approximation to the binomial distribution.

6. Calculate the probability associated with a specific *X* value in a binomial distribution and find the score associated with a specific proportion of a binomial distribution.

When a variable is measured on a scale consisting of exactly two categories, the resulting data are called binomial. The term *binomial* can be loosely translated as "two names," referring to the two categories on the measurement scale.

Binomial data can occur when a variable naturally exists with only two categories. For example, people can be classified as male or female, and a coin toss results in either heads or tails. It also is common for a researcher to simplify data by collapsing the scores into two categories. For example, a psychologist may use personality scores to classify people as either high or low in aggression.

In binomial situations, the researcher often knows the probabilities associated with each of the two categories. With a balanced coin, for example, $p(\text{heads}) = p(\text{tails}) = \frac{1}{2}$. The question of interest is the number of times each category occurs in a series of trials or in a sample of individuals. For example:

What is the probability of obtaining 15 heads in 20 tosses of a balanced coin?

What is the probability of obtaining more than 40 introverts in a sampling of 50 college freshmen?

As we shall see, the normal distribution serves as an excellent model for computing probabilities with binomial data.

■ The Binomial Distribution

To answer probability questions about binomial data, we need to examine the binomial distribution. To define and describe this distribution, we first introduce some notation.

1. The two categories are identified as *A* and *B*.

2. The probabilities (or proportions) associated with each category are identified as

$$p = p(A) = \text{the probability of } A$$
$$q = p(B) = \text{the probability of } B$$

Notice that $p + q = 1.00$ because *A* and *B* are the only two possible outcomes.

3. The number of individuals or observations in the sample is identified by *n*.

4. The variable *X* refers to the number of times category *A* occurs in the sample.

Notice that *X* can have any value from 0 (none of the sample is in category *A*) to *n* (all the sample is in category *A*).

DEFINITION

> Using the notation presented here, the **binomial distribution** shows the probability associated with each value of *X* from $X = 0$ to $X = n$.

A simple example of a binomial distribution is presented next.

EXAMPLE 6.12

Figure 6.16 shows the binomial distribution for the number of heads obtained in 2 tosses of a balanced coin. This distribution shows that it is possible to obtain as many as 2 heads or as few as 0 heads in 2 tosses. The most likely outcome (highest probability) is to obtain exactly 1 head in 2 tosses. The construction of this binomial distribution is discussed in detail next.

For this example, the event we are considering is a coin toss. There are two possible outcomes: heads and tails. We assume the coin is balanced, so

$$p = p(\text{heads}) = \frac{1}{2}$$

$$q = p(\text{tails}) = \frac{1}{2}$$

We are looking at a sample of $n = 2$ tosses, and the variable of interest is

$$X = \text{the number of heads}$$

To construct the binomial distribution, we will look at all the possible outcomes from tossing a coin 2 times. The complete set of 4 outcomes is listed below.

1st Toss	2nd Toss	
Heads	Heads	(both heads)
Heads	Tails	(each sequence has exactly 1 head)
Tails	Heads	
Tails	Tails	(no heads)

Notice that there are 4 possible outcomes when you toss a coin 2 times. Only 1 of the 4 outcomes has 2 heads, so the probability of obtaining 2 heads is $p = \frac{1}{4}$. Similarly, 2 of the 4 outcomes have exactly 1 head, so the probability of 1 head is $p = \frac{2}{4} = \frac{1}{2}$. Finally, the probability of no heads ($X = 0$) is $p = \frac{1}{4}$. These are the probabilities shown in Figure 6.16.

Note that this binomial distribution can be used to answer probability questions. For example, what is the probability of obtaining at least 1 head in 2 tosses? According to the distribution shown in Figure 6.16, the answer is $\frac{3}{4}$. ■

Similar binomial distributions have been constructed for the number of heads in 4 tosses of a balanced coin and in 6 tosses of a coin (Figure 6.17). It should be obvious from the binomial distributions shown in Figures 6.16 and 6.17 that the binomial distribution tends toward a normal shape, especially when the sample size (n) is relatively large.

It should not be surprising that the binomial distribution tends to be normal. With $n = 10$ coin tosses, for example, the most likely outcome would be to obtain around $X = 5$ heads. On the other hand, values far from 5 would be very unlikely—you would not expect

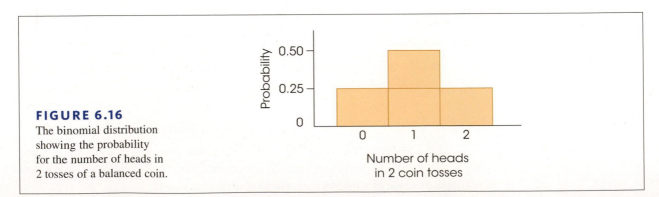

FIGURE 6.16
The binomial distribution showing the probability for the number of heads in 2 tosses of a balanced coin.

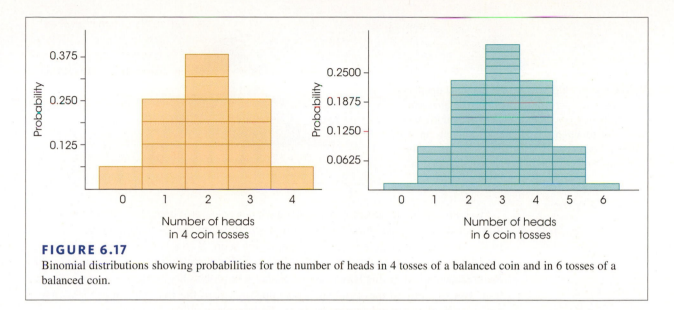

FIGURE 6.17

Binomial distributions showing probabilities for the number of heads in 4 tosses of a balanced coin and in 6 tosses of a balanced coin.

to get all 10 heads or all 10 tails (0 heads) in 10 tosses. Notice that we have described a normal-shaped distribution: The probabilities are highest in the middle (around $X = 5$), and they taper off as you move toward either extreme.

■ The Normal Approximation to the Binomial Distribution

The value of 10 for *pn* or *qn* is a general guide, not an absolute cutoff. Values slightly less than 10 still provide a good approximation. However, with smaller values the normal approximation becomes less accurate as a substitute for the binomial distribution.

We have stated that the binomial distribution tends to approximate a normal distribution, particularly when *n* is large. To be more specific, the binomial distribution will be a nearly perfect normal distribution when *pn* and *qn* are both equal to or greater than 10. Under these circumstances, the binomial distribution will approximate a normal distribution with the following parameters:

$$\text{Mean: } \mu = pn \tag{6.1}$$

$$\text{standard deviation: } \sigma = \sqrt{npq} \tag{6.2}$$

Within this normal distribution, each value of *X* has a corresponding *z*-score,

$$z = \frac{X - \mu}{\sigma} = \frac{X - pn}{\sqrt{npq}} \tag{6.3}$$

The fact that the binomial distribution tends to be normal in shape means that we can compute probability values directly from *z*-scores and the unit normal table.

It is important to remember that the normal distribution is only an approximation of a true binomial distribution. Binomial values, such as the number of heads in a series of coin tosses, are *discrete* (see Chapter 1, p. 19). For example, if you are recording the number of heads in four tosses of a coin, you may observe 2 heads or 3 heads but there are no values between 2 and 3. The normal distribution, on the other hand, is *continuous*. However, the *normal approximation* provides an extremely accurate model for computing binomial probabilities in many situations. Figure 6.18 shows the difference between the true binomial distribution, the discrete histogram, and the normal curve that approximates the binomial distribution. Although the two distributions are slightly different, the area under the distributions is nearly equivalent. *Remember, it is the area under the distribution that is used to find probabilities.*

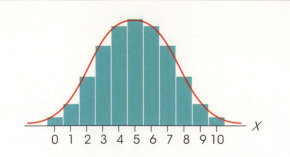

FIGURE 6.18

The relationship between the binomial distribution and the normal distribution. The binomial distribution is a discrete histogram and the normal distribution is a continuous, smooth curve. Each X value is represented by a bar in the histogram or a section of the normal distribution.

To gain maximum accuracy when using the normal approximation, you must remember that each X value in the binomial distribution actually corresponds to a bar in the histogram. In the histogram in Figure 6.18, for example, the score $X = 6$ is represented by a bar that is bounded by real limits of 5.5 and 6.5. The actual probability of $X = 6$ is determined by the area contained in this bar. To approximate this probability using the normal distribution, you should find the area that is contained between the two real limits. Similarly, if you are using the normal approximation to find the probability of obtaining a score greater than $X = 6$, you should use the area beyond the real limit boundary of 6.5. The following example demonstrates how the normal approximation to the binomial distribution is used to compute probability values.

EXAMPLE 6.13

Suppose that you plan to test for ESP (extra-sensory perception) by asking people to predict the suit of a card that is randomly selected from a complete deck. Before you begin your test, however, you need to know what kind of performance is expected from people who do not have ESP and are simply guessing. For these people, there are two possible outcomes, correct or incorrect, on each trial. Because there are four different suits, the probability of a correct prediction (assuming that there is no ESP) is $p = \frac{1}{4}$ and the probability of an incorrect prediction is $q = \frac{3}{4}$. With a series of $n = 48$ trials, this situation meets the criteria for the normal approximation to the binomial distribution:

$$pn = \frac{1}{4}(48) = 12 \qquad qn = \frac{3}{4}(48) = 36 \quad \text{Both are greater than 10.}$$

Caution: If the question had asked for the probability of *15 or more* correct predictions, we would find the area beyond $X = 14.5$. Read the question carefully.

Thus, the distribution of correct predictions will form a normal shaped distribution with a mean of $\mu = pn = 12$ and a standard deviation of $\sigma = \sqrt{npq} = \sqrt{9} = 3$. We can use this distribution to determine probabilities for different levels of performance. For example, we can calculate the probability that a person without ESP would guess correctly more than 15 times in a series of 48 trials.

Figure 6.19 shows the binomial distribution that we are considering. Because we want the probability of obtaining *more than* 15 correct predictions, we must find the shaded area in the tail of the distribution beyond $X = 15.5$. (Remember that a score of 15 corresponds to an interval from 14.5–15.5. We want scores beyond this interval.) The first step is to find the z-score corresponding to $X = 15.5$.

$$z = \frac{X - \mu}{\sigma} = \frac{15.5 - 12}{3} = 1.17$$

Next, look up the probability in the unit normal table. In this case, we want the proportion in the tail beyond $z = 1.17$. The value from the table is $p = 0.1210$. This is the answer we want. The probability of correctly predicting the suit of a card more than 15 times in a series of 48 trials is only $p = 0.1210$ or 12.10%. Thus, it is very unlikely for an individual without ESP to guess correctly more than 15 out of 48 trials. ∎

FIGURE 6.19

The normal approximation of the binomial distribution discussed in Example 6.13.

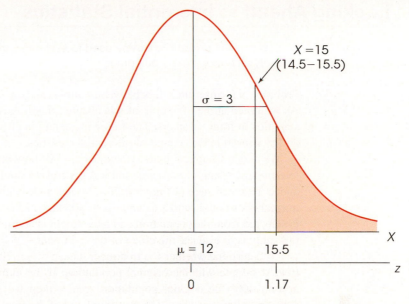

1. What is the standard deviation for a binomial distribution with $p = q = \frac{1}{2}$ and $n = 36$?

   a. 3

   b. 9

   c. 12

   d. 18

2. For a binomial distribution with $p = \frac{1}{4}$, what is the probability that A will occur more than 18 times in a series of 48 trials?

   a. 0.2514

   b. 0.0336

   c. 0.0228

   d. 0.0150

3. If you are simply guessing on a true/false test with 36 questions, what is the probability that you will get 20 or more correct?

   a. 0.1587

   b. 0.2033

   c. 0.2514

   d. 0.3085

ANSWERS   1. A, 2. D, 3. D

6.5 | Looking Ahead to Inferential Statistics

7. Explain how probability can be used to evaluate a treatment effect by identifying likely and very unlikely outcomes.

Probability forms a direct link between samples and the populations from which they come. As we noted at the beginning of this chapter, this link is the foundation for the inferential statistics in future chapters. The following example provides a brief preview of how probability is used in the context of inferential statistics.

We ended Chapter 5 with a demonstration of how inferential statistics are used to help interpret the results of a research study. A general research situation was shown in Figure 5.10 and is repeated here in Figure 6.20. The research begins with a population that forms a normal distribution with a mean of $\mu = 400$ and a standard deviation of $\sigma = 20$. A sample is selected from the population and a treatment is administered to the sample. The goal for the study is to evaluate the effect of the treatment.

To determine whether the treatment has an effect, the researcher simply compares the treated sample with the original population. If the individuals in the sample have scores around 400 (the original population mean), then we must conclude that the treatment appears to have no effect. On the other hand, if the treated individuals have scores that are noticeably different from 400, then the researcher has evidence that the treatment does have an effect. Notice that the study is using a sample to help answer a question about a population; this is the essence of inferential statistics.

The problem for the researcher is determining exactly what is meant by "noticeably different" from 400. If a treated individual has a score of $X = 415$, is that enough to say that the treatment has an effect? What about $X = 420$ or $X = 450$? In Chapter 5, we suggested that z-scores provide one method for solving this problem. Specifically, we suggested that a z-score value beyond $z = 2.00$ (or -2.00) was an extreme value and therefore noticeably different. However, the choice of $z = \pm2.00$ was purely arbitrary. Now we have another tool, *probability,* to help us decide exactly where to set the boundaries.

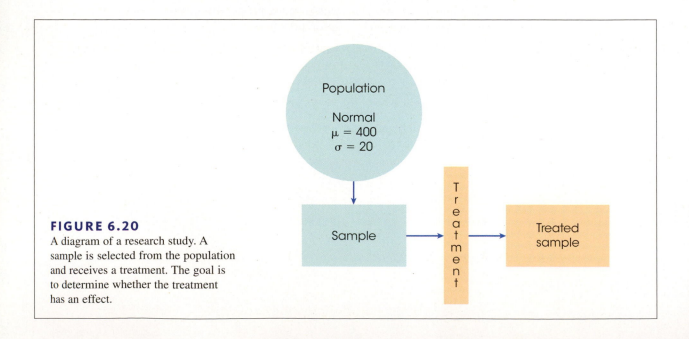

FIGURE 6.20

A diagram of a research study. A sample is selected from the population and receives a treatment. The goal is to determine whether the treatment has an effect.

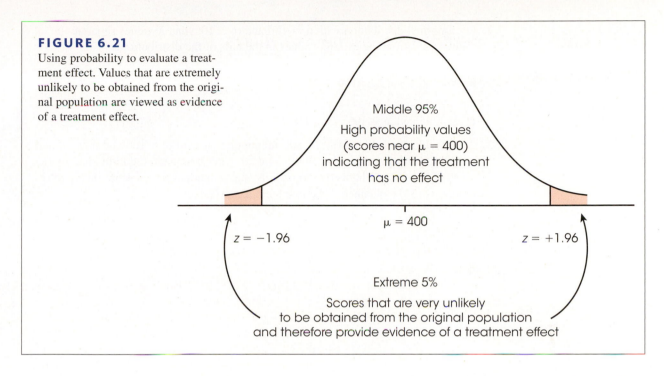

FIGURE 6.21

Using probability to evaluate a treatment effect. Values that are extremely unlikely to be obtained from the original population are viewed as evidence of a treatment effect.

Figure 6.21 shows the original population from our hypothetical research study. Note that most of the scores are located close to $\mu = 400$. Also note that we have added boundaries separating the middle 95% of the distribution from the extreme 5% or 0.0500 in the two tails. Dividing the 0.0500 in half produces a proportion of 0.0250 in the right-hand tail and 0.0250 in the left-hand tail. Using column C of the unit normal table, the z-score boundaries for the right and left tails are $z = +1.96$ and $z = -1.96$, respectively. If we are selecting an individual from the original untreated population, then it is very unlikely that we would obtain a score beyond the $z = \pm 1.96$ boundaries.

The boundaries set at $z = \pm 1.96$ provide objective criteria for deciding whether our sample provides evidence that the treatment has an effect. Specifically, if our sample is located in the tail beyond one of the ± 1.96 boundaries, then we can conclude:

1. The sample is an extreme value, nearly 2 standard deviations away from average, and therefore is noticeably different from most individuals in the original population.

2. The sample is a very unlikely value with a very low probability if the treatment has no effect.

Therefore, the sample provides clear evidence that the treatment has had an effect.

1. For a normal distribution with $\mu = 100$ and $\sigma = 10$, which of the following accurately describes the location of $X = 80$?

   a. It is an extreme, very unlikely score.

   b. It is a low score, but not extreme or unlikely.

   c. It is a fairly representative score.

   d. None of the other options is an accurate description.

2. For a normal distribution with $\mu = 100$ and $\sigma = 10$, what X values separate the middle 95% from the extreme values in the tails of the distribution?

a. $X = 83.5$ and $X = 116.5$

b. $X = 80.4$ and $X = 119.6$

c. $X = 87.2$ and $X = 112.8$

d. $X = 95$ and $X = 105$

3. An individual is selected from a normal population with $\mu = 100$ and $\sigma = 10$ and a treatment is administered to the individual. After treatment, the individual has a score of $X = 125$. What is the probability that individual with a score this high or higher would be obtained if the treatment has no effect?

a. 0.0668

b. 0.0228

c. 0.0062

d. 0.9938

ANSWERS   **1.** A, **2.** B, **3.** C

SUMMARY

1. The probability of a particular event A is defined as a fraction or proportion:

$$p(A) = \frac{\text{number of outcomes classified as } A}{\text{total number of possible outcomes}}$$

2. Our definition of probability is accurate only for random samples. There are two requirements that must be satisfied for a random sample:

a. Every individual in the population has an equal chance of being selected.

b. When more than one individual is being selected, the probabilities must stay constant. This means there must be sampling with replacement.

3. All probability problems can be restated as proportion problems. The "probability of selecting a king from a deck of cards" is equivalent to the "proportion of the deck that consists of kings." For frequency distributions, probability questions can be answered by determining proportions of area. The "probability of selecting an individual with an IQ greater than 108" is equivalent to the "proportion of the whole population that consists of IQs greater than 108."

4. For normal distributions, probabilities (proportions) can be found in the unit normal table. The table provides a listing of the proportions of a normal distribution that correspond to each z-score value. With the table, it is possible to move between X values and probabilities using a two-step procedure:

a. The z-score formula (Chapter 5) allows you to transform X to z or to change z back to X.

b. The unit normal table allows you to look up the probability (proportion) corresponding to each z-score or the z-score corresponding to each probability.

5. Percentiles and percentile ranks measure the relative standing of a score within a distribution (see Box 6.1). Percentile rank is the percentage of individuals with scores at or below a particular X value. A percentile is an X value that is identified by its rank. The percentile rank always corresponds to the proportion to the left of the score in question.

6. The binomial distribution is used whenever the measurement procedure simply classifies individuals into exactly two categories. The two categories are identified as A and B, with probabilities of

$$p(A) = p \quad \text{and} \quad p(B) = q$$

7. The binomial distribution gives the probability for each value of X, where X equals the number of occurrences of category A in a series of n events. For example, X equals the number of heads in $n = 10$ tosses of a coin.

8. When pn and qn are both at least 10, the binomial distribution is closely approximated by a normal distribution with

$$\mu = pn$$
$$\sigma = \sqrt{npq}$$

9. In the normal approximation to the binomial distribution, each value of X has a corresponding z-score:

$$z = \frac{X - \mu}{\sigma} = \frac{X - pn}{\sqrt{npq}}$$

With the z-score and the unit normal table, you can find probability values associated with any value of X. For maximum accuracy, you should use the appropriate real limits for X when computing z-scores and probabilities.

KEY TERMS

probability (161)

random sample (162)

sampling with replacement (163)

unit normal table (168)

percentile rank (178)

percentile (178)

binomial distribution (179)

normal approximation (binomial) (181)

SPSS®

The statistics computer package SPSS is not structured to compute probabilities. However, the program does report probability values as part of the inferential statistics that we will examine later in this book. In the context of inferential statistics, the probabilities are called *significance levels*, and they warn researchers about the probability of misinterpreting their research results.

FOCUS ON PROBLEM SOLVING

1. We have defined probability as being equivalent to a proportion, which means that you can restate every probability problem as a proportion problem. This definition is particularly useful when you are working with frequency distribution graphs in which the population is represented by the whole graph and probabilities (proportions) are represented by portions of the graph. When working problems with the normal distribution, you always should start with a sketch of the distribution. You should shade the portion of the graph that reflects the proportion you are looking for.

2. Remember that the unit normal table shows only positive z-scores in column A. However, since the normal distribution is symmetrical, the proportions in the table apply to both positive and negative z-score values.

3. A common error for students is to use negative values for proportions on the left-hand side of the normal distribution. Proportions (or probabilities) are always positive: 10% is 10% whether it is in the left or right tail of the distribution.

4. The proportions in the unit normal table are accurate only for normal distributions. If a distribution is not normal, you cannot use the table.

5. For maximum accuracy when using the normal approximation to the binomial distribution, you must remember that each X value is an interval bounded by real limits. For example, to find the probability of obtaining an X value greater than 10, you should use the real limit 10.5 in the z-score formula. Similarly, to find the probability of obtaining an X value less than 10, you should use the real limit 9.5.

DEMONSTRATION 6.1

FINDING PROBABILITY FROM THE UNIT NORMAL TABLE

A population is normally distributed with a mean of $\mu = 45$ and a standard deviation of $\sigma = 4$. What is the probability of randomly selecting a score that is greater than 43? In other words, what proportion of the distribution consists of scores greater than 43?

STEP 1 **Sketch the distribution.**   For this demonstration, the distribution is normal with $\mu = 45$ and $\sigma = 4$. The score of $X = 43$ is lower than the mean and therefore is placed to the left of the mean. The question asks for the proportion corresponding to scores greater than 43, so shade in the area to the right of this score. Figure 6.22 shows the sketch.

STEP 2 **Transform the X value to a z-score.**

$$z = \frac{X - \mu}{\sigma} = \frac{43 - 45}{4} = \frac{-2}{4} = -0.5$$

STEP 3 **Find the appropriate proportion in the unit normal table.**   Ignoring the negative size, locate $z = -0.50$ in column A. In this case, the proportion we want corresponds to the body of the distribution and the value is found in column B. For this example,

$$p(X > 43) = p(z > -0.50) = 0.6915$$

DEMONSTRATION 6.2

PROBABILITY AND THE BINOMIAL DISTRIBUTION

Suppose that you completely forgot to study for a quiz and now must guess on every question. It is a true/false quiz with $n = 40$ questions. What is the probability that you will get at least 26 questions correct just by chance? Stated in symbols,

$$p(X \geq 26) = ?$$

STEP 1 **Identify p and q.**   This problem is a binomial situation, in which

$$p = \text{probability of guessing correctly} = 0.50$$

$$q = \text{probability of guessing incorrectly} = 0.50$$

With $n = 40$ quiz items, both pn and qn are greater than 10. Thus, the criteria for the normal approximation to the binomial distribution are satisfied:

$$pn = 0.50(40) = 20$$

$$qn = 0.50(40) = 20$$

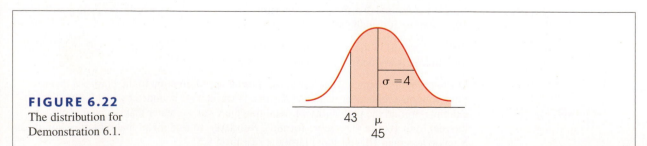

FIGURE 6.22
The distribution for
Demonstration 6.1.

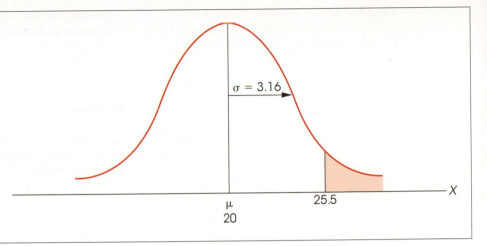

FIGURE 6.23
The normal approximation of a binomial distribution with $\mu = 20$ and $\sigma = 3.16$. The proportion corresponding to scores that are equal to or greater than 26 is shaded. Notice that the lower real limit (25.5) for $X = 26$ is used.

STEP 2   **Identify the parameters, and sketch the binomial distribution.**   For a true/false quiz, correct and incorrect guesses are equally likely, $p = q = \frac{1}{2}$. With pn and qn both greater than 10, the normal approximation is appropriate and will have a mean and a standard deviation as follows:

$$\mu = pn = 0.5(40) = 20$$
$$\sigma = \sqrt{npq} = \sqrt{10} = 3.16$$

Figure 6.23 shows the distribution. We are looking for the probability of getting $X = 26$ or more questions correct, so we will use the lower real limit for 26, which is 25.5.

STEP 3   The z-score for $X = 25.5$ is calculated as follows:

$$z = \frac{X - pn}{\sqrt{npq}} = \frac{25.5 - 20}{3.16} = +1.74$$

According to the unit normal table, the proportion we want is 0.0409. Thus, the probability of getting at least 26 questions right just by guessing is

$$p(X \geq 26) = 0.0409 \text{ (or } 4.09\%)$$

PROBLEMS

1. What are the two requirements for a random sample?

2. Define sampling with replacement and explain why is it used?

3. Around Halloween each year the grocery store sells three-pound bags of candy contains a mixture of three different mini-bars: Snickers, Milky Way, and Twix. If the bag has an equal number of each of the three bars, then what are the probabilities for each of the following?
   a. Randomly selecting a Milky Way bar.
   b. Randomly selecting either a Snickers or Twix bar.
   c. Randomly selecting something other than a Twix bar.

4. A psychology class consists of 28 males and 52 females. If the professor selects names from the class list using *random sampling*,
   a. What is the probability that the first student selected will be a female?
   b. If a random sample of $n = 3$ students is selected and the first two are both females, what is the probability that the third student selected will be a male?

5. Draw a vertical line through a normal distribution for each of the following z-score locations. Determine whether the tail is on the right or left side of the line and find the proportion in the tail.
   a. $z = 1.00$
   b. $z = 0.50$
   c. $z = -1.25$
   d. $z = -0.40$

6. Draw a vertical line through a normal distribution for each of the following z-score locations. Determine whether the body is on the right or left side of the line and find the proportion in the body.
   a. $z = 2.50$
   b. $z = 0.80$
   c. $z = -0.50$
   d. $z = -0.77$

7. Find each of the following probabilities for a normal distribution.
   a. $p(z > 1.25)$
   b. $p(z > -0.60)$
   c. $p(z < 0.70)$
   d. $p(z < -1.30)$

8. What proportion of a normal distribution is located between each of the following z-score boundaries?
   a. $z = -0.25$ and $z = +0.25$
   b. $z = -0.67$ and $z = +0.67$
   c. $z = -1.20$ and $z = +1.20$

9. Find each of the following probabilities for a normal distribution.
   a. $p(-0.80 < z < 0.80)$
   b. $p(-0.50 < z < 1.00)$
   c. $p(0.20 < z < 1.50)$
   d. $p(-1.20 < z < -0.80)$

10. Find the z-score location of a vertical line that separates a normal distribution as described in each of the following.
    a. 5% in the tail on the left
    b. 30% in the tail on the right
    c. 65% in the body on the left
    d. 80% in the body on the right

11. Find the z-score boundaries that separate a normal distribution as described in each of the following.
    a. The middle 30% from the 70% in the tails.
    b. The middle 40% from the 60% in the tails.
    c. The middle 50% from the 50% in the tails.
    d. The middle 60% from the 40% in the tails.

12. A normal distribution has a mean of $\mu = 70$ and a standard deviation of $\sigma = 8$. For each of the following scores, indicate whether the tail is to the right or left of the score and find the proportion of the distribution located in the tail.
    a. $X = 72$
    b. $X = 76$
    c. $X = 66$
    d. $X = 60$

13. A normal distribution has a mean of $\mu = 30$ and a standard deviation of $\sigma = 12$. For each of the following scores, indicate whether the body is to the right or

left of the score and find the proportion of the distribution located in the body.
    a. $X = 33$
    b. $X = 18$
    c. $X = 24$
    d. $X = 39$

14. For a normal distribution with a mean of $\mu = 60$ and a standard deviation of $\sigma = 10$, find the proportion of the population corresponding to each of the following.
    a. Scores greater than 65.
    b. Scores less than 68.
    c. Scores between 50 and 70.

15. In 2014, the New York Yankees had a team batting average of $\mu = 245$ (actually 0.245 but we will avoid the decimals). Of course, the batting average varies from game to game, but assuming that the distribution of batting averages for 162 games is normal with a standard deviation is $\sigma = 40$ points, answer each of the following.
    a. If you randomly select one game from 2014, what is the probability that the team batting average was over 300?
    b. If you randomly select one game from 2014, what is the probability that the team batting average was under 200?

16. IQ test scores are standardized to produce a normal distribution with a mean of $\mu = 100$ and a standard deviation of $\sigma = 15$. Find the proportion of the population in each of the following IQ categories.
    a. Genius or near genius: IQ over 140
    b. Very superior intelligence: IQ from 120–140
    c. Average or normal intelligence: IQ from 90–109

17. The distribution of scores on the SAT is approximately normal with a mean of $\mu = 500$ and a standard deviation of $\sigma = 100$. For the population of students who have taken the SAT,
    a. What proportion have SAT scores less than 400?
    b. What proportion have SAT scores greater than 650?
    c. What is the minimum SAT score needed to be in the highest 20% of the population?
    d. If the state college only accepts students from the top 40% of the SAT distribution, what is the minimum SAT score needed to be accepted?

18. According to a recent report, people smile an average of $\mu = 62$ time per day. Assuming that the distribution of smiles is approximately normal with a standard deviation of $\sigma = 18$, find each of the following values.
    a. What proportion of people smile more than 80 times a day?
    b. What proportion of people smile at least 50 times a day?

19. A report in 2010 indicates that Americans between the ages of 8 and 18 spend an average of $\mu = 7.5$ hours per day using some sort of electronic device such as smart phones, computers, or tablets. Assume that the distribution of times is normal with a standard deviation of $\sigma = 2.5$ hours and find the following values.
   a. What is the probability of selecting an individual who uses electronic devices more than 12 hours a day?
   b. What proportion of 8- to 18-year-old Americans spend between 5 and 10 hours per day using electronic devices? In symbols, $p(5 < X < 10) = ?$

20. Rochester, New York, averages $\mu = 21.9$ inches of snow for the month of December. The distribution of snowfall amounts is approximately normal with a standard deviation of $\sigma = 6.5$ inches. A local jewelry store advertised a refund of 50% off all purchases made in December, if we have more than 3 feet (36 inches) during the month. What is the probability that the jewelry store will have to pay off on its promise?

21. A multiple-choice test has 32 questions, each with four response choices. If a student is simply guessing at the answers,
   a. What is the probability of guessing correctly for any individual question?
   b. On average, how many questions would a student answer correctly for the entire test?
   c. What is the probability that a student would get more than 12 answers correct simply by guessing?

22. A true/false test has 20 questions. If a student is simply guessing at the answers,
   a. On average, how many questions would a student answer correctly for the entire test?
   b. What is the probability that a student would answer more than 15 questions correctly simply by guessing?
   c. What is the probability that a student would answer fewer than 7 questions correctly simply by guessing?

23. A roulette wheel has alternating red and black, numbered slots into which the ball finally stops to determine the winner. If a gambler always bets on black to win, then
   a. What is the probability of winning at least 40 times in a series of 64 spins? (Note that *at least* 40 wins means 40 or more.)
   b. What is the probability of winning more than 40 times in a series of 64 spins?
   c. Based on your answers to a and b, what is the probability of winning exactly 40 times?

24. A test developed to measure ESP involves using Zener cards. Each card shows one of five equally likely symbols (square, circle, star, cross, wavy lines), and the person being tested has to predict the shape on each card before it is selected. Find each of the probabilities requested for a person who has no ESP and is just guessing.
   a. What is the probability of correctly predicting exactly 20 cards in a series of 100 trials?
   b. What is the probability of correctly predicting 10 or more cards in a series of 36 trials?
   c. What is the probability of correctly predicting more than 16 cards in a series of 64 trials?

25. The student health clinic reports that only 30% of students got a flu shot this year. If a researcher surveys a sample of $n = 84$ students attending a basketball game,
   a. What is the probability that any individual student has had a flu shot?
   b. What is the probability that more than 30 students have had flu shots?
   c. What is the probability that 20 or fewer students have had shots?

Probability and Samples: The Distribution of Sample Means

© Deborah Batt

Tools You Will Need

The following items are considered essential background material for this chapter. If you doubt your knowledge of any of these items, you should review the appropriate chapter and section before proceeding.

- Random sampling (Chapter 6)
- Probability and the normal distribution (Chapter 6)
- z-Scores (Chapter 5)

In this chapter we extend the topic of probability to cover larger samples; specifically, samples that have more than one score. Fortunately, you already know the one basic fact governing probability for samples: Samples tend to be similar to the populations from which they are taken.

For example, if you take a sample from a population that consists of 75% females and only 25% males, you probably will get a sample that has more females than males. Or, if you select a sample from a population for which the average age is $\mu = 21$ years, you probably will get a sample with an average age around 21 years. We are confident that you already know this basic fact because research indicates that even 8-month-old infants understand this basic law of sampling.

Xu and Garcia (2008) began one experiment by showing 8-month-old infants a large box filled with ping-pong balls. The box was brought onto a puppet stage and the front panel was opened to reveal the balls inside. The box contained either mostly red with a few white balls or mostly white with a few red balls. The experimenter alternated between the two boxes until the infants had seen both displays several times. After the infants were familiar with the boxes, the researchers began a series of test trials. On each trial, the box was brought on stage with the front panel closed. The researcher reached in the box and, one at a time, drew out a sample of five balls. The balls were placed in a transparent container next to the box. On half of the trials, the sample was rigged to have 1 red and 4 white balls. For the other half, the sample had 1 white and 4 red balls. The researchers then removed the front panel to reveal the contents of the box and recorded how long the infants continued to look at the box.

The contents of the box were either consistent with the sample, and therefore expected, or inconsistent with the sample and therefore unexpected. An *expected* outcome, for example, means that a sample with 4 red and 1 white ball should come from a box with mostly red balls. This same sample is *unexpected* from a box with mostly white balls. The results showed that the infants stared consistently longer at an unexpected outcome ($M = 9.9$ seconds) than at an expected outcome ($M = 7.5$ seconds), indicating that the infants considered the unexpected outcome surprising and more interesting than the expected outcome.

Xu and Garcia's results strongly suggest that even 8-month-old infants understand the basic principles that identify which samples have high probability and which have low probability. Nevertheless, whenever you are picking ping pong balls from a box or recruiting people to participate in a research study, it usually is possible to obtain thousands or even millions of different samples from the same population. Under these circumstances, how can we determine the probability for obtaining any specific sample?

In this chapter we introduce the *distribution of sample means*, which allows us to find the exact probability of obtaining a specific sample mean from a specific population. This distribution describes the entire set of all the possible sample means for any sized sample. Because we can describe the entire set, we can find probabilities associated with specific sample means. (Recall from Chapter 6 that probabilities are equivalent to proportions of the entire distribution.) Also, because the distribution of sample means tends to be normal, it is possible to find probabilities using z-scores and the unit normal table. Although it is impossible to predict exactly which sample will be obtained, the probabilities allow researchers to determine which samples are likely and which are very unlikely.

7.1 | Samples, Populations, and the Distribution of Sample Means

LEARNING OBJECTIVE   1. Define the distribution of sample means and describe the logically predictable characteristics of the distribution.

The preceding two chapters presented the topics of z-scores and probability. Whenever a score is selected from a population, you should be able to compute a z-score that describes exactly where the score is located in the distribution. If the population is normal, you also should be able to determine the probability value for obtaining any individual score. In a normal distribution, for example, any score located in the tail of the distribution beyond $z = +2.00$ is an extreme value, and a score this large has a probability of only $p = 0.0228$.

However, the *z*-scores and probabilities that we have considered so far are limited to situations in which the sample consists of a single score. Most research studies involve much larger samples such as $n = 25$ preschool children or $n = 100$ American Idol contestants. In these situations, the sample mean, rather than a single score, is used to answer questions about the population. In this chapter we extend the concepts of *z*-scores and probability to cover situations with larger samples. In particular, we introduce a procedure for transforming a sample mean into a *z*-score. Thus, a researcher is able to compute a *z*-score that describes an entire sample. As always, a *z*-score value near zero indicates a central, representative sample; a *z*-value beyond $+2.00$ or -2.00 indicates an extreme sample. Thus, it is possible to describe how any specific sample is related to all the other possible samples. In addition, we can use the *z*-score values to look up probabilities for obtaining certain samples, no matter how many scores the sample contains.

In general, the difficulty of working with samples is that a sample provides an incomplete picture of the population. Suppose, for example, a researcher randomly selects a sample of $n = 25$ students from the state college. Although the sample should be representative of the entire student population, there are almost certainly some segments of the population that are not included in the sample. In addition, any statistics that are computed for the sample will not be identical to the corresponding parameters for the entire population. For example, the average IQ for the sample of 25 students will not be the same as the overall mean IQ for the entire population. This difference, or *error* between sample statistics and the corresponding population parameters, is called *sampling error* and was illustrated in Figure 1.2 (page 7).

DEFINITION

> **Sampling error** is the natural discrepancy, or amount of error, between a sample statistic and its corresponding population parameter.

Furthermore, samples are variable; they are not all the same. If you take two separate samples from the same population, the samples will be different. They will contain different individuals, they will have different scores, and they will have different sample means. How can you tell which sample gives the best description of the population? Can you even predict how well a sample will describe its population? What is the probability of selecting a sample with specific characteristics? These questions can be answered once we establish the rules that relate samples and populations.

■ The Distribution of Sample Means

As noted, two separate samples probably will be different even though they are taken from the same population. The samples will have different individuals, different scores, different means, and so on. In most cases, it is possible to obtain thousands of different samples from one population. With all these different samples coming from the same population, it may seem hopeless to try to establish some simple rules for the relationships between samples and populations. Fortunately, however, the huge set of possible samples forms a relatively simple and orderly pattern that makes it possible to predict the characteristics of a sample with some accuracy. The ability to predict sample characteristics is based on the *distribution of sample means*.

DEFINITION

> **The distribution of sample means** is the collection of sample means for all the possible random samples of a particular size (*n*) that can be obtained from a population.

Notice that the distribution of sample means contains *all the possible samples*. It is necessary to have all the possible values to compute probabilities. For example, if the entire set contains exactly 100 samples, then the probability of obtaining any specific sample is 1 out of 100: $p = \frac{1}{100}$ (Box 7.1).

Also, you should notice that the distribution of sample means is different from distributions we have considered before. Until now we always have discussed distributions of scores; now the values in the distribution are not scores, but statistics (sample means). Because statistics are obtained from samples, a distribution of statistics is referred to as a *sampling distribution*.

DEFINITION

> **A sampling distribution** is a distribution of statistics obtained by selecting all the possible samples of a specific size from a population.

Thus, the distribution of sample means is an example of a sampling distribution. In fact, it often is called the sampling distribution of *M*.

If you actually wanted to construct the distribution of sample means, you would first select a random sample of a specific size (*n*) from a population, calculate the sample mean, and place the sample mean in a frequency distribution. Then you select another random sample with the same number of scores. Again, you calculate the sample mean and add it to your distribution. You continue selecting samples and calculating means, over and over, until you have the complete set of all the possible random samples. At this point, your frequency distribution will show the distribution of sample means.

BOX 7.1 Probability and the Distribution of Sample Means

I have a bad habit of losing playing cards. This habit is compounded by the fact that I always save the old deck in the hope that someday I will find the missing cards. As a result, I have a drawer filled with partial decks of playing cards. Suppose that I take one of these almost-complete decks, shuffle the cards carefully, and then randomly select one card. What is the probability that I will draw a king?

You should realize that it is impossible to answer this probability question. To find the probability of selecting a king, you must know how many cards are in the deck and exactly which cards are missing. (It is crucial that you know whether or not any kings are missing.) The point of this simple example is that any probability question requires that you have complete information about the population from which the sample is being selected. In this case, you must know all the possible cards in the deck before you can find the probability for selecting any specific card.

In this chapter, we are examining probability and sample means. To find the probability for any specific sample mean, you first must know *all the possible sample means*. Therefore, we begin by defining and describing the set of all possible sample means that can be obtained from a particular population. Once we have specified the complete set of all possible sample means (i.e., the distribution of sample means), we will be able to find the probability of selecting any specific sample means.

■ Characteristics of the Distribution of Sample Means

We demonstrate the process of constructing a distribution of sample means in Example 7.1, but first we use common sense and a little logic to predict the general characteristics of the distribution.

1. The sample means should pile up around the population mean. Samples are not expected to be perfect but they are representative of the population.

As a result, most of the sample means should be relatively close to the population mean.

2. The pile of sample means should tend to form a normal-shaped distribution. Logically, most of the samples should have means close to μ, and it should be relatively rare to find sample means that are substantially different from μ. As a result, the sample means should pile up in the center of the distribution (around μ) and the frequencies should taper off as the distance between M and μ increases. This describes a normal-shaped distribution.

3. In general, the larger the sample size, the closer the sample means should be to the population mean, μ. Logically, a large sample should be a better representative than a small sample. Thus, the sample means obtained with a large sample size should cluster relatively close to the population mean; the means obtained from small samples should be more widely scattered.

As you will see, each of these three commonsense characteristics is an accurate description of the distribution of sample means. The following example demonstrates the process of constructing the distribution of sample means by repeatedly selecting samples from a population.

EXAMPLE 7.1   Consider a population that consists of only 4 scores: 2, 4, 6, 8. This population is pictured in the frequency distribution histogram in Figure 7.1.   ■

Remember that random sampling requires sampling with replacement.   We are going to use this population as the basis for constructing the distribution of sample means for $n = 2$. Remember: This distribution is the collection of sample means from all the possible random samples of $n = 2$ from this population. We begin by looking at all the possible samples. For this example, there are 16 different samples, and they are all listed in Table 7.1. Notice that the samples are listed systematically. First, we list all the possible samples with $X = 2$ as the first score, then all the possible samples with $X = 4$ as the first score, and so on. In this way, we are sure that we have all of the possible random samples.

Next, we compute the mean, M, for each of the 16 samples (see the last column of Table 7.1). The 16 means are then placed in a frequency distribution histogram in Figure 7.2. This is the distribution of sample means. Note that the distribution in Figure 7.2 demonstrates two of the characteristics that we predicted for the distribution of sample means.

1. The sample means pile up around the population mean. For this example, the population mean is $\mu = 5$, and the sample means are clustered around a value of 5. It should not surprise you that the sample means tend to approximate the population mean. After all, samples are supposed to be representative of the population.

FIGURE 7.1
Frequency distribution histogram for a population of 4 scores: 2, 4, 6, 8.

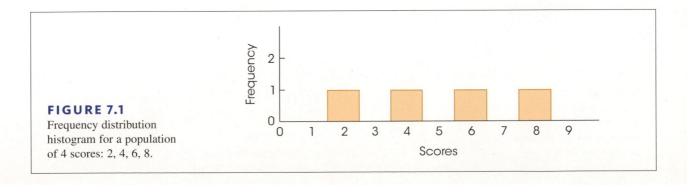

TABLE 7.1

All the possible samples of $n = 2$ scores that can be obtained from the population presented in Figure 7.1. Notice that the table lists *random samples*. This requires sampling with replacement, so is possible to select the same score twice.

| Sample | Scores | | Sample Mean |
	First	Second	(M)
1	2	2	2
2	2	4	3
3	2	6	4
4	2	8	5
5	4	2	3
6	4	4	4
7	4	6	5
8	4	8	6
9	6	2	4
10	6	4	5
11	6	6	6
12	6	8	7
13	8	2	5
14	8	4	6
15	8	6	7
16	8	8	8

   2. The distribution of sample means is approximately normal in shape. This is a characteristic that is discussed in detail later and is extremely useful because we already know a great deal about probabilities and the normal distribution (Chapter 6).

Remember that our goal in this chapter is to answer probability questions about samples with $n > 1$.

   Finally, you should realize that we can use the distribution of sample means to achieve the goal for this chapter, which is to answer probability questions about sample means. For example, if you take a sample of $n = 2$ scores from the original population, what is the probability of obtaining a sample with a mean greater than 7? In symbols,

$$p(M > 7) = ?$$

   Because probability is equivalent to proportion, the probability question can be restated as follows: Of all the possible sample means, what proportion has values greater than 7? In this form, the question is easily answered by looking at the distribution of sample means. All the possible sample means are pictured (see Figure 7.2), and only 1 out of the 16 means has a value greater than 7. The answer, therefore, is 1 out of 16, or $\frac{1}{16}$.

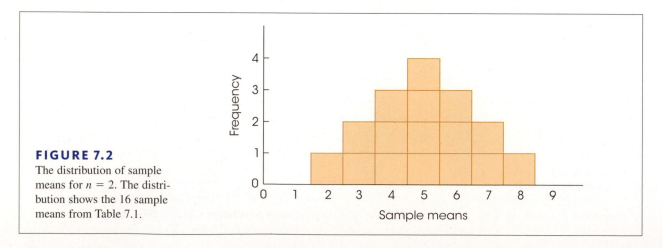

FIGURE 7.2

The distribution of sample means for $n = 2$. The distribution shows the 16 sample means from Table 7.1.

LEARNING CHECK

1. If all the possible random samples of size $n = 5$ are selected from a population with $\mu = 50$ and $\sigma = 10$ and the mean is computed for each sample, then what value will be obtained for the mean of all the sample means?

   a. 50

   b. $\frac{50}{5} = 10$

   c. $50(5) = 2500$

   d. Around 50 but probably not equal to 50.

2. All the possible random samples of size $n = 2$ are selected from a population with $\mu = 40$ and $\sigma = 10$ and the mean is computed for each sample. Then all the possible samples of size $n = 25$ are selected from the same population and the mean is computed for each sample. How will the distribution of sample means for $n = 2$ compare with the distribution for $n = 25$?

   a. The two distributions will have the same variances.

   b. The variance for $n = 25$ will be larger than the variance for $n = 2$.

   c. The variance for $n = 25$ will be smaller than the variance for $n = 2$.

   d. The variance for $n = 25$ will be two times larger than the variance for $n = 2$.

3. If all the possible random samples of size $n = 25$ are selected from a population with $\mu = 80$ and $\sigma = 10$ and the mean is computed for each sample, then what shape is expected for the distribution of sample means?

   a. The sample means tend to form a normal-shaped distribution whether the population is normal or not.

   b. The sample means tend to form a normal distribution only if the population distribution is normal.

   c. The sample means tend to be distributed evenly across the scale, forming a rectangular-shaped distribution.

   d. There are thousands of possible samples and it is impossible to predict the shape of the distribution.

ANSWERS   1. A, 2. C, 3. A

7.2 | The Distribution of Sample Means for any Population and any Sample Size

LEARNING OBJECTIVES

2. Explain how the central limit theorem specifies the shape, central tendency, and variability for the distribution of sample means.

3. Describe how the standard error of M is calculated, explain what it measures, and describe how it is related to the standard deviation for the population.

■ **The Central Limit Theorem**

Example 7.1 demonstrates the construction of the distribution of sample means for an overly simplified situation with a very small population and samples that each contain only $n = 2$ scores. In more realistic circumstances, with larger populations and larger samples,

the number of possible samples increases dramatically and it is virtually impossible to actually obtain every possible random sample. Fortunately, it is possible to determine exactly
what the distribution of sample means looks like without taking hundreds or thousands
of samples. Specifically, a mathematical proposition known as the *central limit theorem*
provides a precise description of the distribution that would be obtained if you selected
every possible sample, calculated every sample mean, and constructed the distribution of
the sample mean. This important and useful theorem serves as a cornerstone for much of
inferential statistics. Following is the essence of the theorem.

Central Limit Theorem   For any population with mean μ and standard deviation
σ, the distribution of sample means for sample size n will have a mean of μ and a
standard deviation of $\sigma/\sqrt{n}$ and will approach a normal distribution as n approaches
infinity.

   The value of this theorem comes from two simple facts. First, it describes the distribution of sample means for *any population*, no matter what shape, mean, or standard deviation. Second, the distribution of sample means "approaches" a normal distribution very
rapidly. By the time the sample size reaches $n = 30$, the distribution is almost perfectly
normal.

   Note that the central limit theorem describes the distribution of sample means by identifying the three basic characteristics that describe any distribution: shape, central tendency,
and variability. We will examine each of these.

■ The Shape of the Distribution of Sample Means

It has been observed that the distribution of sample means tends to be a normal distribution.
In fact, this distribution is almost perfectly normal if either of the following two conditions
is satisfied:

   1. The population from which the samples are selected is a normal distribution.

   2. The number of scores (n) in each sample is relatively large, around 30 or more.

   (As n gets larger, the distribution of sample means will closely approximate a normal
distribution. When $n > 30$, the distribution is almost normal regardless of the shape of the
original population.)

   As we noted earlier, the fact that the distribution of sample means tends to be normal is not surprising. Whenever you take a sample from a population, you expect the
sample mean to be near to the population mean. When you take lots of different samples, you expect the sample means to "pile up" around μ, resulting in a normal-shaped
distribution. You can see this tendency emerging (although it is not yet normal) in
Figure 7.2.

■ The Mean of the Distribution of Sample Means: The Expected Value of *M*

In Example 7.1, the distribution of sample means is centered at the mean of the population from which the samples were obtained. In fact, the average value of all the sample
means is exactly equal to the value of the population mean. This fact should be intuitively
reasonable; the sample means are expected to be close to the population mean, and they do
tend to pile up around μ. The formal statement of this phenomenon is that the mean of the
distribution of sample means always is identical to the population mean. This mean value
is called the *expected value of* M.

In commonsense terms, a sample mean is "expected" to be near its population mean. When all of the possible sample means are obtained, the average value is identical to μ.

The fact that the average value of M is equal to μ was first introduced in Chapter 4 (page 120) in the context of *biased* versus *unbiased* statistics. The sample mean is an example of an unbiased statistic, which means that on average the sample statistic produces a value that is exactly equal to the corresponding population parameter. In this case, the average value of all the sample means is exactly equal to μ.

DEFINITION

> The mean of the distribution of sample means is equal to the mean of the population of scores, μ, and is called the **expected value of M**.

■ The Standard Error of *M*

So far, we have considered the shape and the central tendency of the distribution of sample means. To completely describe this distribution, we need one more characteristic, variability. The value we will be working with is the standard deviation for the distribution of sample means. This standard deviation is identified by the symbol σ_M and is called the *standard error of* M.

When the standard deviation was first introduced in Chapter 4, we noted that this measure of variability serves two general purposes. First, the standard deviation describes the distribution by telling whether the individual scores are clustered close together or scattered over a wide range. Second, the standard deviation measures how well any individual score represents the population by providing a measure of how much distance is reasonable to expect between a score and the population mean. The standard error serves the same two purposes for the distribution of sample means.

1. The standard error describes the distribution of sample means. It provides a measure of how much difference is expected from one sample to another. When the standard error is small, all the sample means are close together and have similar values. If the standard error is large, the sample means are scattered over a wide range and there are big differences from one sample to another.

2. Standard error measures how well an individual sample mean represents the entire distribution. Specifically, it provides a measure of how much distance is reasonable to expect between a sample mean and the overall mean for the distribution of sample means. However, because the overall mean is equal to μ, the standard error also provides a measure of how much distance to expect between a sample mean (M) and the population mean, μ.

Remember that a sample is not expected to provide a perfectly accurate reflection of its population. Although a sample mean should be representative of the population mean, there typically is some error between the sample and the population. The standard error measures exactly how much difference is expected on average between a sample mean, M and the population mean, μ.

DEFINITION

> The standard deviation of the distribution of sample means, σ_M, is called the **standard error of M**. The standard error provides a measure of how much distance is expected on average between a sample mean (M) and the population mean (μ).

Once again, the symbol for the standard error is σ_M. The σ indicates that this value is a standard deviation, and the subscript M indicates that it is the standard deviation for the distribution of sample means. Similarly, it is common to use the symbol μ_M to represent the mean of the distribution of sample means. However, μ_M is always equal to μ and our primary interest in inferential statistics is to compare sample means (M) with their population means (μ). Therefore, we simply use the symbol μ to refer to the mean of the distribution of sample means.

The standard error is an extremely valuable measure because it specifies precisely how well a sample mean estimates its population mean—that is, how much error you should expect, on the average, between M and μ. Remember that one basic reason for taking samples is to use the sample data to answer questions about the population. However, you do not expect a sample to provide a perfectly accurate picture of the population. There always is some discrepancy or error between a sample statistic and the corresponding population parameter. Now we are able to calculate exactly how much error to expect. For any sample size (n), we can compute the standard error, which measures the average distance between a sample mean and the population mean.

The magnitude of the standard error is determined by two factors: (1) the size of the sample and (2) the standard deviation of the population from which the sample is selected. We will examine each of these factors.

The Sample Size   Earlier we predicted, based on commonsense, that the size of a sample should influence how accurately the sample represents its population. Specifically, a large sample should be more accurate than a small sample. In general, as the sample size increases, the error between the sample mean and the population mean should decrease. This rule is also known as the *law of large numbers*.

DEFINITION

> The **law of large numbers** states that the larger the sample size (n), the more probable it is that the sample mean will be close to the population mean.

The Population Standard Deviation   As we noted earlier, there is an inverse relationship between the sample size and the standard error: bigger samples have smaller error, and smaller samples have bigger error. At the extreme, the smallest possible sample (and the largest standard error) occurs when the sample consists of $n = 1$ score. At this extreme, each sample is a single score and the distribution of sample means is identical to the original distribution of scores. In this case, the standard deviation for the distribution of sample means, which is the standard error, is identical to the standard deviation for the distribution of scores. In other words, when $n = 1$, the standard error $= \sigma_M$ is identical to the standard deviation $= \sigma$.

When $n = 1$, $\sigma_M = \sigma$ (standard error = standard deviation).

You can think of the standard deviation as the "starting point" for standard error. When $n = 1$, the standard error and the standard deviation are the same: $\sigma_M = \sigma$. As sample size increases beyond $n = 1$, the sample becomes a more accurate representative of the population, and the standard error decreases. The formula for standard error expresses this relationship between standard deviation and sample size (n).

This formula is contained in the central limit theorem.

$$\text{standard error} = \sigma_M = \frac{\sigma}{\sqrt{n}} \qquad (7.1)$$

Note that the formula satisfies all of the requirements for the standard error. Specifically,

a. As sample size (n) increases, the size of the standard error decreases. (Larger samples are more accurate.)

b. When the sample consists of a single score ($n = 1$), the standard error is the same as the standard deviation ($\sigma_M = \sigma$).

In Equation 7.1 and in most of the preceding discussion, we have defined standard error in terms of the population standard deviation. However, the population standard deviation (σ) and the population variance (σ^2) are directly related, and it is easy to substitute variance into the equation for standard error. Using the simple equality $\sigma = \sqrt{\sigma^2}$, the equation for standard error can be rewritten as follows:

$$\text{standard error} = \sigma_M = \frac{\sigma}{\sqrt{n}} = \sqrt{\frac{\sigma^2}{n}} = \frac{\sigma^2}{n} \tag{7.2}$$

Throughout the rest of this chapter (and in Chapter 8), we will continue to define standard error in terms of the standard deviation (Equation 7.1). However, in later chapters (starting in Chapter 9) the formula based on variance (Equation 7.2) will become more useful.

Figure 7.3 illustrates the general relationship between standard error and sample size. (The calculations for the data points in Figure 7.3 are presented in Table 7.2.) Again, the basic concept is that the larger a sample is, the more accurately it represents its population. Also note that the standard error decreases in relation to the *square root* of the sample size. As a result, researchers can substantially reduce error by increasing sample size up to around $n = 30$. However, increasing sample size beyond $n = 30$ does not produce much additional improvement in how well the sample represents the population.

The following example is an opportunity for you to test your understanding of the standard error by computing it for yourself.

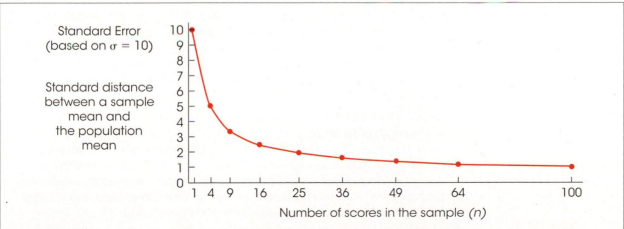

FIGURE 7.3

The relationship between standard error and sample size. As the sample size is increased, there is less error between the sample mean and the population mean.

TABLE 7.2

Calculations for the points shown in Figure 7.3. Again, notice that the size of the standard error decreases as the size of the sample increases.

Sample Size (n)	Standard Error	
1	$\sigma_M = \dfrac{10}{\sqrt{1}}$	= 10.00
4	$\sigma_M = \dfrac{10}{\sqrt{4}}$	= 5.00
9	$\sigma_M = \dfrac{10}{\sqrt{9}}$	= 3.33
16	$\sigma_M = \dfrac{10}{\sqrt{16}}$	= 2.50
25	$\sigma_M = \dfrac{10}{\sqrt{25}}$	= 2.00
49	$\sigma_M = \dfrac{10}{\sqrt{49}}$	= 1.43
64	$\sigma_M = \dfrac{10}{\sqrt{64}}$	= 1.25
100	$\sigma_M = \dfrac{10}{\sqrt{100}}$	= 1.00

EXAMPLE 7.2

If samples are selected from a population with $\mu = 50$ and $\sigma = 12$, then what is the standard error of the distribution of sample means for $n = 4$ and for $n = 16$? You should obtain answers of $\sigma_M = 6$ for $n = 4$ and $\sigma_M = 3$ for $n = 16$. ∎

■ Three Different Distributions

Before we move forward with our discussion of the distribution of sample means, we will pause for a moment to emphasize the idea that we are now dealing with three different but interrelated distributions.

1. First, we have the original population of scores. This population contains the scores for thousands or millions of individual people, and it has its own shape, mean, and standard deviation. For example, the population of IQ scores consists of millions of individual IQ scores that form a normal distribution with a mean of $\mu = 100$ and a standard deviation of $\sigma = 15$. An example of a population is shown in Figure 7.4(a).

2. Next, we have a sample that is selected from the population. The sample consists of a small set of scores for a few people who have been selected to represent the entire population. For example, we could select a sample of $n = 25$ people and measure each individual's IQ score. The 25 scores could be organized in a frequency distribution and we could calculate the sample mean and the sample standard deviation. Note that the sample also has its own shape, mean, and standard deviation. An example of a sample is shown in Figure 7.4(b).

3. The third distribution is the distribution of sample means. This is a theoretical distribution consisting of the sample means obtained from all the possible random samples of a specific size. For example, the distribution of sample means for samples of $n = 25$ IQ scores would be normal with a mean (expected value) of

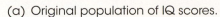

(a) Original population of IQ scores.

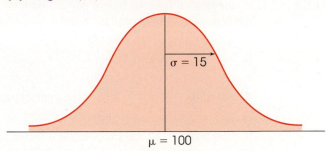

$\sigma = 15$

$\mu = 100$

(b) A sample of $n = 25$ IQ scores.

$\sigma = 11.5$

| 80 | 90 | 100 | 110 | 120 | 130 |

$M = 101.2$

(c) The distribution of sample means. Sample means for all the possible random samples of $n = 25$ IQ scores.

FIGURE 7.4

Three distributions. Part (a) shows the population of IQ scores. Part (b) shows a sample of $n = 25$ IQ scores. Part (c) shows the distribution of sample means for samples of $n = 25$ IQ scores. Note that the mean for the sample in part (b) is one of the thousands of sample means in the distribution shown in part (c).

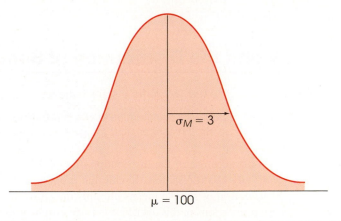

$\sigma_M = 3$

$\mu = 100$

$\mu = 100$ and a standard deviation (standard error) of $\sigma_M = \frac{15}{\sqrt{25}} = 3$. This distribution, shown in Figure 7.4(c), also has its own shape, mean, and standard deviation.

Note that the scores for the sample (Figure 7.4(b)) were taken from the original population (Figure 7.4(a)) and that the mean for the sample is one of the values contained in the distribution of sample means (Figure 7.4(c)). Thus, the three distributions are all connected, but they are all distinct.

LEARNING CHECK

1. If samples are selected from a population with $\mu = 80$ and $\sigma = 12$, then which of the following samples will have the largest expected value for M?

   a. $n = 10$ scores with $M = 82$

   b. $n = 20$ scores with $M = 84$

   c. $n = 30$ scores with $M = 86$

   d. All of the samples will have the same expected value.

2. If random samples, each with $n = 4$ scores are selected from a population with $\mu = 80$ and $\sigma = 12$, then how much distance is expected on average between the sample means and the population mean?

   a. $4(12) = 48$ points

   b. 12 points

   c. $\frac{12}{4} = 3$ points

   d. $\frac{12}{\sqrt{4}} = 6$ points

3. The standard distance between a sample mean and the population mean is 6 points for samples of $n = 16$ scores selected from a population with a mean of $\mu = 50$. What is the standard deviation for the population?

   a. 48

   b. 24

   c. 6

   d. 3

ANSWERS   1. D, 2. D, 3. B

7.3 | Probability and the Distribution of Sample Means

LEARNING OBJECTIVES

4. Calculate the z-score for a sample mean.

5. Describe the circumstances in which the distribution of sample means is normal and, in these circumstances, find the probability associated with a specific sample.

The primary use of the distribution of sample means is to find the probability associated with any specific sample. Recall that probability is equivalent to proportion. Because the distribution of sample means presents the entire set of all possible sample means, we can use proportions of this distribution to determine probabilities. The following example demonstrates this process.

EXAMPLE 7.3

The population of scores on the SAT forms a normal distribution with $\mu = 500$ and $\sigma = 100$. If you take a random sample of $n = 16$ students, what is the probability that the sample mean will be greater than $M = 525$?

First, you can restate this probability question as a proportion question: Out of all the possible sample means, what proportion has values greater than 525? You know about "all the possible sample means"; this is the distribution of sample means. The problem is to find a specific portion of this distribution.

Caution: Whenever you have a probability question about a sample mean, you must use the distribution of sample means.

Although we cannot construct the distribution of sample means by repeatedly taking samples and calculating means (as in Example 7.1), we know exactly what the distribution looks like based on the information from the central limit theorem. Specifically, the distribution of sample means has the following characteristics:

1. The distribution is normal because the population of SAT scores is normal.

2. The distribution has a mean of 500 because the population mean is $\mu = 500$.

3. For $n = 16$, the distribution has a standard error of $\sigma_M = 25$:

$$\sigma_M = \frac{\sigma}{\sqrt{n}} = \frac{100}{\sqrt{16}} = \frac{100}{4} = 25$$

This distribution of sample means is shown in Figure 7.5.

We are interested in sample means greater than 525 (the shaded area in Figure 7.5), so the next step is to use a z-score to locate the exact position of $M = 525$ in the distribution. The value 525 is located above the mean by 25 points, which is exactly 1 standard deviation (in this case, exactly 1 standard error). Thus, the z-score for $M = 525$ is $z = +1.00$.

Because this distribution of sample means is normal, you can use the unit normal table to find the probability associated with $z = +1.00$. The table indicates that 0.1587 of the distribution is located in the tail of the distribution beyond $z = +1.00$. Our conclusion is that it is relatively unlikely, $p = 0.1587$ (15.87%), to obtain a random sample of $n = 16$ students with an average SAT score greater than 525. ∎

■ A z-Score for Sample Means

As demonstrated in Example 7.3, it is possible to use a z-score to describe the exact location of any specific sample mean within the distribution of sample means. The z-score tells exactly where the sample mean is located in relation to all the other possible sample means that could have been obtained. As defined in Chapter 5, a z-score identifies the location with a signed number so that

1. The sign tells whether the location is above ($+$) or below ($-$) the mean.

2. The number tells the distance between the location and the mean in terms of the number of standard deviations.

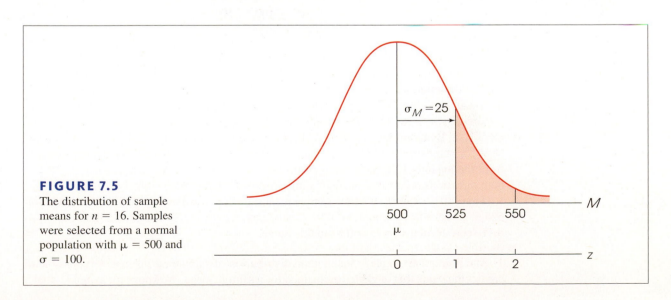

FIGURE 7.5

The distribution of sample means for $n = 16$. Samples were selected from a normal population with $\mu = 500$ and $\sigma = 100$.

However, we are now finding a location within the distribution of sample means. Therefore, we must use the notation and terminology appropriate for this distribution. First, we are finding the location for a sample mean (M) rather than a score (X). Second, the standard deviation for the distribution of sample means is the standard error, σ_M. With these changes, the z-score formula for locating a sample mean is

$$z = \frac{M - \mu}{\sigma_M} \tag{7.3}$$

Caution: **When computing z for a single score, use the standard deviation, σ. When computing z for a sample mean, you must use the standard error, σ_M.**

Just as every score (X) has a z-score that describes its position in the distribution of scores, every sample mean (M) has a z-score that describes its position in the distribution of sample means. Specifically, the sign of the z-scores tells whether the sample mean is above ($+$) or below ($-$) μ and the numerical value of the z–score is the distance between the sample mean and μ in terms of the number of standard errors. When the distribution of sample means is normal, it is possible to use z-scores and the unit normal table to find the probability associated with any specific sample mean (as in Example 7.3). The following example is an opportunity for you to test your understanding of z-scores and probability for sample means.

EXAMPLE 7.4

A sample of $n = 4$ scores is selected from a normal distribution with a mean of $\mu = 40$ and a standard deviation of $\sigma = 16$. The sample mean is $M = 42$. Find the z-score for this sample mean and determine the probability of obtaining a sample mean larger than $M = 42$. This is, find $p(M > 42)$ for $n = 4$. You should obtain $z = 0.25$ and $p = 0.4013$. ∎

The following example demonstrates that it also is possible to make quantitative predictions about the kinds of samples that should be obtained from any population.

EXAMPLE 7.5

Once again, the distribution of SAT scores forms a normal distribution with a mean of $\mu = 500$ and a standard deviation of $\sigma = 100$. For this example, we are going to determine what kind of sample mean is likely to be obtained as the average SAT score for a random sample of $n = 25$ students. Specifically, we will determine the exact range of values that is expected for the sample mean 80% of the time.

We begin with the distribution of sample means for $n = 25$. This distribution is normal with an expected value of $\mu = 500$ and, with $n = 25$, the standard error is

$$\sigma_M = \frac{\sigma}{\sqrt{n}} = \frac{100}{\sqrt{25}} = \frac{100}{5} = 20$$

(Figure 7.6). Our goal is to find the range of values that make up the middle 80% of the distribution. Because the distribution is normal we can use the unit normal table. First, the 80% in the middle is split in half, with 40% (0.4000) on each side of the mean. Looking up 0.4000 in column D (the proportion between the mean and z), we find a corresponding z-score of $z = 1.28$. Thus, the z-score boundaries for the middle 80% are $z = +1.28$ and $z = -1.28$. By definition, a z-score of 1.28 represents a location that is 1.28 standard deviations (or standard errors) from the mean. With a standard error of 20 points, the distance from the mean is $1.28(20) = 25.6$ points. The mean is $\mu = 500$, so a distance of 25.6 in both directions produces a range of values from 474.4–525.6.

Thus, 80% of all the possible sample means are contained in a range between 474.4 and 525.6. If we select a sample of $n = 25$ students, we can be 80% confident that the mean SAT score for the sample will be in this range. ∎

The point of Example 7.5 is that the distribution of sample means makes it possible to predict the value that ought to be obtained for a sample mean. We know, for example, that a

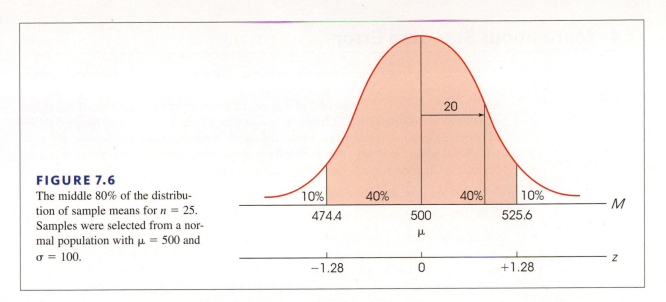

FIGURE 7.6
The middle 80% of the distribution of sample means for $n = 25$. Samples were selected from a normal population with $\mu = 500$ and $\sigma = 100$.

sample of $n = 25$ students ought to have a mean SAT score around 500. More specifically, we are 80% confident that the value of the sample mean will be between 474.4 and 525.6. The ability to predict sample means in this way will be a valuable tool for the inferential statistics that follow.

LEARNING CHECK

1. A normal population has $\mu = 50$ and $\sigma = 8$. A random sample of $n = 16$ scores from this population has a mean of 54. What is the z-score for this sample mean?
   a. $+0.50$
   b. $+1.00$
   c. $+2.00$
   d. $+4.00$

2. A random sample of $n = 9$ scores is obtained from a population with $\mu = 50$ and $\sigma = 9$. If the sample mean is $M = 53$, what is the z-score corresponding to the sample mean?
   a. $z = 0.33$
   b. $z = 1.00$
   c. $z = 3.00$
   d. cannot determine without additional information

3. A random sample of $n = 25$ scores is selected from a normally distributed population with $\mu = 500$ and $\sigma = 100$. What is the probability that the sample mean will be less than 490?
   a. 0.4602
   b. 0.3085
   c. 0.1587
   d. 0.0062

ANSWERS   1. C, 2. B, 3. B

7.4 | More about Standard Error

6. Describe how the magnitude of the standard error is related to the size of the sample.

In Chapter 5, we introduced the idea of z-scores to describe the exact location of individual scores within a distribution. In Chapter 6, we introduced the idea of finding the probability of obtaining any individual score, especially scores from a normal distribution. By now, you should realize that most of this chapter is simply repeating the same things that were covered in Chapters 5 and 6, but with two adjustments:

1. We are now using the distribution of sample means instead of a distribution of scores.

2. We are now using the standard error instead of the standard deviation.

Of these two adjustments, the primary new concept in Chapter 7 is the standard error and the single rule that you need to remember is: Whenever you are working with a sample mean, you must use the standard error.

This single rule encompasses essentially all of the new content of Chapter 7. Therefore, this section will focus on the concept of standard error to ensure that you have a good understanding of this new concept.

■ Sampling Error and Standard Error

At the beginning of this chapter, we introduced the idea that it is possible to obtain thousands of different samples from a single population. Each sample will have its own individuals, its own scores, and its own sample mean. The distribution of sample means provides a method for organizing all of the different sample means into a single picture. Figure 7.7 shows a prototypical distribution of sample means. To emphasize the fact that the distribution contains many different samples, we have constructed this figure so that the distribution is made up of hundreds of small boxes, each box representing a single

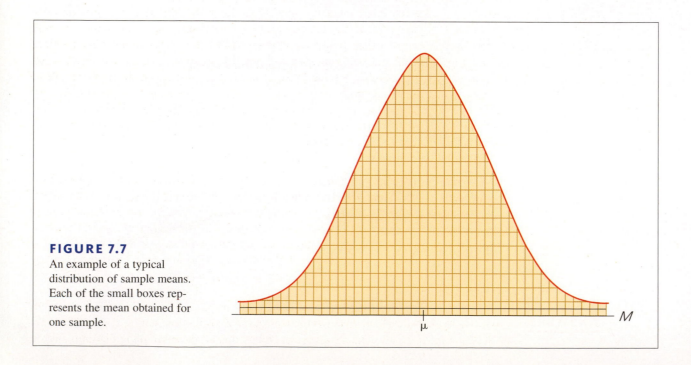

FIGURE 7.7

An example of a typical distribution of sample means. Each of the small boxes represents the mean obtained for one sample.

sample mean. Also, notice that the sample means tend to pile up around the population mean (μ), forming a normal-shaped distribution as predicted by the central limit theorem.

The distribution shown in Figure 7.7 provides a concrete example for reviewing the general concepts of sampling error and standard error. Although the following points may seem obvious, they are intended to provide you with a better understanding of these two statistical concepts.

1. **Sampling Error**   The general concept of sampling error is that a sample typically will not provide a perfectly accurate representation of its population. More specifically, there typically is some discrepancy (or error) between a statistic computed for a sample and the corresponding parameter for the population. As you look at Figure 7.7, notice that the individual sample means are not exactly equal to the population mean. In fact, 50% of the samples have means that are smaller than μ (the entire left-hand side of the distribution). Similarly, 50% of the samples produce means that overestimate the true population mean. In general, there will be some discrepancy, or *sampling error*, between the mean for a sample and the mean for the population from which the sample was obtained.

2. **Standard Error**   Again, looking at Figure 7.7, notice that most of the sample means are relatively close to the population mean (those in the center of the distribution). These samples provide a fairly accurate representation of the population. On the other hand, some samples produce means that are out in the tails of the distribution, relatively far from the population mean. These extreme sample means do not accurately represent the population. For each individual sample, you can measure the error (or distance) between the sample mean and the population mean. For some samples, the error will be relatively small, but for other samples, the error will be relatively large. The *standard error* provides a way to measure the "average", or standard, distance between a sample mean and the population mean.

Thus, the standard error provides a method for defining and measuring sampling error. Knowing the standard error gives researchers a good indication of how accurately their sample data represent the populations they are studying. In most research situations, for example, the population mean is unknown, and the researcher selects a sample to help obtain information about the unknown population. Specifically, the sample mean provides information about the value of the unknown population mean. The sample mean is not expected to give a perfectly accurate representation of the population mean; there will be some error, and the standard error tells *exactly how much error*, on average, should exist between the sample mean and the unknown population mean. The following example demonstrates the use of standard error and provides additional information about the relationship between standard error and standard deviation.

EXAMPLE 7.6   A recent survey of students at a local college included the following question: How many minutes do you spend each day watching electronic video (online, TV, cell phone, tablet, etc.). The average response was $\mu = 80$ minutes, and the distribution of viewing times was approximately normal with a standard deviation of $\sigma = 20$ minutes. Next, we take a sample from this population and examine how accurately the sample mean represents the population mean. More specifically, we will examine how sample size affects accuracy by considering three different samples: one with $n = 1$ student, one with $n = 4$ students, and one with $n = 100$ students.

Figure 7.8 shows the distributions of sample means based on samples of $n = 1$, $n = 4$, and $n = 100$. Each distribution shows the collection of all possible sample means that could be obtained for that particular sample size. Notice that all three sampling distributions are

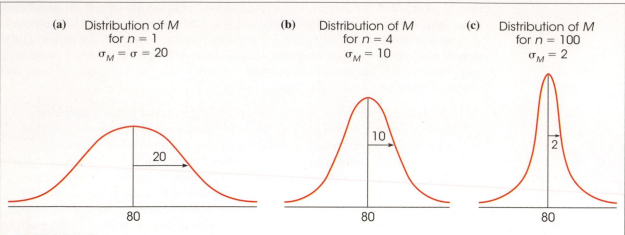

FIGURE 7.8

The distribution of sample means for (a) $n = 1$, (b) $n = 4$, and (c) $n = 100$ obtained from a normal population with $\mu = 80$ and $\sigma = 20$. Notice that the size of the standard error decreases as the sample size increases.

normal (because the original population is normal), and all three have the same mean, $\mu = 80$, which is the expected value of M. However, the three distributions differ greatly with respect to variability. We will consider each one separately.

The smallest sample size is $n = 1$. When a sample consists of a single student, the mean for the sample equals the score for the student, $M = X$. Thus, when $n = 1$, the distribution of sample means is identical to the original population of scores. In this case, the standard error for the distribution of sample means is equal to the standard deviation for the original population. Equation 7.1 confirms this observation.

$$\sigma_M = \frac{\sigma}{\sqrt{n}} = \frac{20}{\sqrt{1}} = 20$$

When the sample consists of a single student, you expect, on average, a 20-point difference between the sample mean and the mean for the population. As we noted earlier, the population standard deviation is the "starting point" for the standard error. With the smallest possible sample, $n = 1$, the standard error is equal to the standard deviation (see Figure 7.8(a)).

As the sample size increases, however, the standard error gets smaller. For a sample of $n = 4$ students, the standard error is

$$\sigma_M = \frac{\sigma}{\sqrt{n}} = \frac{20}{\sqrt{4}} = \frac{20}{2} = 10$$

That is, the typical (or standard) distance between M and μ is 10 points. Figure 7.8(b) illustrates this distribution. Notice that the sample means in this distribution approximate the population mean more closely than in the previous distribution where $n = 1$.

With a sample of $n = 100$, the standard error is still smaller.

$$\sigma_M = \frac{\sigma}{\sqrt{n}} = \frac{20}{\sqrt{100}} = \frac{20}{10} = 2$$

A sample of $n = 100$ students should produce a sample mean that represents the population much more accurately than a sample of $n = 4$ or $n = 1$. As shown in Figure 7.8(c), there is very little error between M and μ when $n = 100$. Specifically, you would expect on average only a 2-point difference between the population mean and the sample mean. ■

In summary, this example illustrates that with the smallest possible sample ($n = 1$), the standard error and the population standard deviation are the same. As sample size increases, the standard error gets smaller, and the sample means tend to approximate μ more closely. Thus, standard error defines the relationship between sample size and the accuracy with which M represents μ.

IN THE LITERATURE

Reporting Standard Error

As we will see in future chapters, the standard error plays a very important role in inferential statistics. Because of its crucial role, the standard error for a sample mean, rather than the sample standard deviation, is often reported in scientific papers. Scientific journals vary in how they refer to the standard error, but frequently the symbols *SE* and *SEM* (for standard error of the mean) are used. The standard error is reported in two ways. Much like the standard deviation, it may be reported in a table along with the sample means (Table 7.3). Alternatively, the standard error may be reported in graphs.

Figure 7.9 illustrates the use of a bar graph to display information about the sample mean and the standard error. In this experiment, two samples (groups A and B) are given different treatments, and then the subjects' scores on a dependent variable are recorded. The mean for group A is $M = 15$, and for group B, it is $M = 30$. For both samples, the standard error of M is $\sigma_M = 4$. Note that the mean is represented by the height of the bar, and the standard error is depicted by brackets at the top of each bar. Each bracket extends 1 standard error above and 1 standard error below the sample mean. Thus, the graph illustrates the mean for each group plus or minus 1 standard error ($M \pm SE$). When you glance at Figure 7.9, not only do you get a "picture" of the sample means, but also you get an idea of how much error you should expect for those means.

TABLE 7.3

The mean self-consciousness scores for participants who were working in front of a video camera and those who were not (controls)

	n	Mean	*SE*
Control	17	32.23	2.31
Camera	15	45.17	2.78

Figure 7.10 shows how sample means and standard error are displayed in a line graph. In this study, two samples representing different age groups are tested on a task for four trials. The number of errors committed on each trial is recorded for all participants. The graph shows the mean (M) number of errors committed for each group on each trial. The brackets show the size of the standard error for each sample mean. Again, the brackets extend 1 standard error above and below the value of the mean.

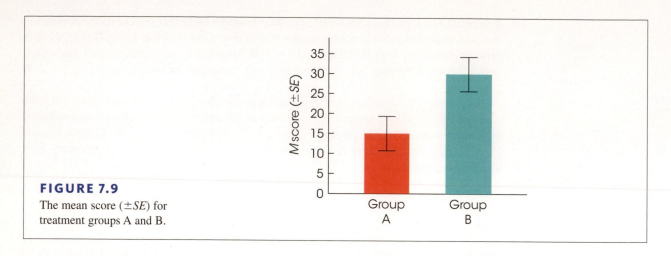

FIGURE 7.9

The mean score ($\pm SE$) for treatment groups A and B.

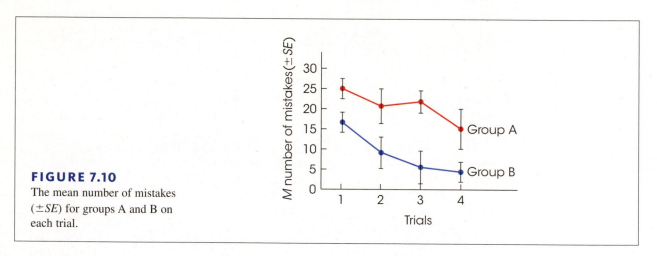

FIGURE 7.10

The mean number of mistakes ($\pm SE$) for groups A and B on each trial.

LEARNING CHECK

1. For samples selected from a population with $\mu = 100$ and $\sigma = 15$, which of the following has the smallest standard error?

   a. a sample of $n = 4$ scores

   b. a sample of $n = 16$ scores

   c. a sample of $n = 25$ scores

   d. The three samples all have the same standard error.

2. For samples selected from a population with $\mu = 40$ and $\sigma = 20$, what sample size is necessary to make the standard distance between the sample mean and the population mean equal to 2 points?

   a. $n = 2$

   b. $n = 10$

   c. $n = 25$

   d. $n = 100$

3. For a particular population, the standard distance between a sample mean and the population mean is 5 points for samples of $n = 4$ scores. What would the standard distance be for samples of $n = 16$ scores?

  a. 5 points

  b. 4 points

  c. 2.5 points

  d. 1 point

ANSWERS   **1.** C, **2.** D, **3.** C

7.5 | Looking Ahead to Inferential Statistics

LEARNING OBJECTIVE   **7.** Explain how the distribution of sample means can be used to evaluate a treatment effect by identifying likely and very unlikely samples.

Inferential statistics are methods that use sample data as the basis for drawing general conclusions about populations. However, we have noted that a sample is not expected to give a perfectly accurate reflection of its population. In particular, there will be some error or discrepancy between a sample statistic and the corresponding population parameter. In this chapter, we focused on sample means and observed that a sample mean will not be exactly equal to the population mean. The standard error of M specifies how much difference is expected on average between the mean for a sample and the mean for the population.

The natural differences that exist between samples and populations introduce a degree of uncertainty and error into all inferential processes. Specifically, there is always a margin of error that must be considered whenever a researcher uses a sample mean as the basis for drawing a conclusion about a population mean. Remember that the sample mean is not perfect. In the next seven chapters we introduce a variety of statistical methods that all use sample means to draw inferences about population means.

In each case, the distribution of sample means and the standard error will be critical elements in the inferential process. Before we begin this series of chapters, we pause briefly to demonstrate how the distribution of sample means, along with z-scores and probability, can help us use sample means to draw inferences about population means.

EXAMPLE 7.7   Suppose that a psychologist is planning a research study to evaluate the effect of a new growth hormone. It is known that regular, adult rats (with no hormone) weigh an average of $\mu = 400$ g. Of course, not all rats are the same size, and the distribution of their weights is normal with $\sigma = 20$. The psychologist plans to select a sample of $n = 25$ newborn rats, inject them with the hormone, and then measure their weights when they become adults. The structure of this research study is shown in Figure 7.11.

The psychologist will make a decision about the effect of the hormone by comparing the sample of treated rats with the regular untreated rats in the original population. If the treated rats in the sample are noticeably different from untreated rats, then the researcher has evidence that the hormone has an effect. The problem is to determine exactly how much difference is necessary before we can say that the sample is *noticeably different*.

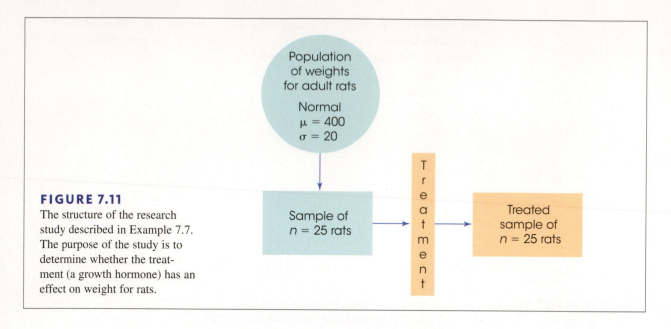

FIGURE 7.11

The structure of the research study described in Example 7.7. The purpose of the study is to determine whether the treatment (a growth hormone) has an effect on weight for rats.

The distribution of sample means and the standard error can help researchers make this decision. In particular, the distribution of sample means can be used to show exactly what would be expected for a sample of rats that do not receive any hormone injections. This allows researchers to make a simple comparison between

a. The sample of treated rats (from the research study) asdsd

b. Samples of untreated rats (from the distribution of sample means)

If our treated sample is noticeably different from the untreated samples, then we have evidence that the treatment has an effect. On the other hand, if our treated sample still looks like one of the untreated samples, then we must conclude that the treatment does not appear to have any effect.

We begin with the original population of untreated rats and consider the distribution of sample means for all the possible samples of $n = 25$ rats. The distribution of sample means has the following characteristics:

1. It is a normal distribution, because the population of rat weights is normal.

2. It has an expected value of 400, because the population mean for untreated rats is $\mu = 400$.

3. It has a standard error of $\sigma_M = \frac{20}{\sqrt{25}} = \frac{20}{5} = 4$, because the population standard deviation is $\sigma = 20$ and the sample size is $n = 25$.

The distribution of sample means is shown in Figure 7.12. Notice that a sample of $n = 25$ untreated rats (without the hormone) should have a mean weight around 400 grams. To be more precise, we can use z-scores to determine the middle 95% of all the possible sample means. As demonstrated in Chapter 6 (page 185), the middle 95% of a normal distribution is located between z-score boundaries of $z = +1.96$ and $z = -1.96$ (check the unit normal table). These z-score boundaries are shown in Figure 7.12. With a standard error of $\sigma_M = 4$ points, a z-score of $z = 1.96$ corresponds to a distance of 1.96(4) = 7.84 points from the mean. Thus, the z-score boundaries of ± 1.96 correspond to sample means of 392.16 and 407.84.

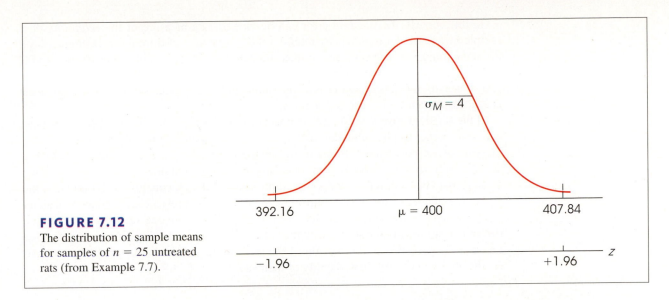

FIGURE 7.12

The distribution of sample means for samples of $n = 25$ untreated rats (from Example 7.7).

We determined that 95% of all the possible samples of untreated rats will have sample means between 392.16 and 407.84. At the same time, it is very unlikely (probability of 5% or less) that a sample of untreated rats would have a mean in the tails beyond these two boundaries. In other words, if the treatment has no effect, then it is very unlikely to obtain a sample mean greater than 407.84 or less than 392.16. If our sample produces one of these extreme means, then we have evidence of a treatment effect. Specifically, it is very unlikely that such an extreme outcome would occur without a treatment effect. ■

In Example 7.7 we used the distribution of sample means, together with z-scores and probability, to provide a description of what is reasonable to expect for an untreated sample. Then, we evaluated the effect of a treatment by determining whether the treated sample was noticeably different from an untreated sample. This procedure forms the foundation for the inferential technique known as *hypothesis testing* that is introduced in Chapter 8 and repeated throughout the remainder of this book.

■ Standard Error as a Measure of Reliability

The research situation shown in Figure 7.11 introduces one final issue concerning sample means and standard error. In Figure 7.11, as in most research studies, the researcher must rely on a *single* sample to provide an accurate representation of the population being investigated. As we have noted, however, if you take two different samples from the same population, you will get different individuals with different scores and different sample means. Thus, every researcher must face the nagging question, "If I had taken a different sample, would I have obtained different results?"

The importance of this question is directly related to the degree of similarity among all the different samples. For example, if there is a high level of consistency from one sample to another, then a researcher can be reasonably confident that the specific sample being studied will provide a good measurement of the population. That is, when all the samples are similar, then it does not matter which one you have selected. On the other hand, if there are big differences from one sample to another, then the researcher is left with some doubts about the accuracy of his/her specific sample. In this case, a different sample could have produced vastly different results.

In this context, the standard error can be viewed as a measure of the *reliability* of a sample mean. The term *reliability* refers to the consistency of different measurements of the same thing. More specifically, a measurement procedure is said to be reliable if you make two different measurements of the same thing and obtain identical (or nearly identical) values. If you view a sample as a "measurement" of a population, then a sample mean is a "measurement" of a population mean.

If the standard error is small, all the possible sample means are clustered close together and a researcher can be confident that any individual sample mean will provide a reliable measure of the population. On the other hand, a large standard error indicates that there are relatively large differences from one sample mean to another, and a researcher must be concerned that a different sample could produce a different conclusion. Fortunately, the size of the standard error can be controlled. In particular, if a researcher is concerned about a large standard error and the potential for big differences from one sample to another, the researcher has the option of reducing the standard error by selecting a larger sample (see Figure 7.3). Thus, the ability to compute the value of the standard error provides researchers the ability to control the reliability of their samples.

LEARNING CHECK

1. If a sample is selected from a normal population with $\mu = 50$ and $\sigma = 20$, which of the following samples is extreme and very unlikely to be obtained?
   a. $M = 45$ for a sample of $n = 4$ scores.
   b. $M = 45$ for a sample of $n = 25$ scores.
   c. $M = 45$ for a sample of $n = 100$ scores.
   d. The three samples are equally likely to be obtained.

2. A random sample is obtained from a population with $\mu = 80$ and $\sigma = 10$ and a treatment is administered to the sample. Which of the following outcomes would be considered noticeably different from a typical sample that did not receive the treatment?
   a. $n = 25$ with $M = 81$
   b. $n = 25$ with $M = 83$
   c. $n = 100$ with $M = 81$
   d. $n = 100$ with $M = 83$

3. If a sample of $n = 25$ scores is selected from a normal population with $\mu = 80$ and $\sigma = 10$, then what sample means form the boundaries that separate the middle 95% of all sample means from the extreme 5% in the tails?
   a. 76.08 and 83.92
   b. 76.70 and 83.30
   c. 77.44 and 82.56
   d. 78 and 82

ANSWERS   **1.** C, **2.** D, **3.** A

SUMMARY

1. The distribution of sample means is defined as the set of *M*s for all the possible random samples for a specific sample size (*n*) that can be obtained from a given population. According to the *central limit theorem*, the parameters of the distribution of sample means are as follows:

   a. *Shape.* The distribution of sample means is normal if either one of the following two conditions is satisfied:

      (1) The population from which the samples are selected is normal.

      (2) The size of the samples is relatively large (around *n* = 30 or more).

   b. *Central Tendency.* The mean of the distribution of sample means is identical to the mean of the population from which the samples are selected. The mean of the distribution of sample means is called the expected value of *M*.

   c. *Variability.* The standard deviation of the distribution of sample means is called the standard error of *M* and is defined by the formula

   $$\sigma_M = \frac{\sigma}{\sqrt{n}} \quad \text{or} \quad \sigma_M = \sqrt{\frac{\sigma^2}{n}}$$

   Standard error measures the standard distance between a sample mean (*M*) and the population mean (μ).

2. One of the most important concepts in this chapter is standard error. The standard error tells how much error to expect if you are using a sample mean to represent a population mean.

3. The location of each *M* in the distribution of sample means can be specified by a *z*-score:

   $$z = \frac{M - \mu}{\sigma_M}$$

   Because the distribution of sample means tends to be normal, we can use these *z*-scores and the unit normal table to find probabilities for specific sample means. In particular, we can identify which sample means are likely and which are very unlikely to be obtained from any given population. This ability to find probabilities for samples is the basis for the inferential statistics in the chapters ahead.

4. In general terms, the standard error measures how much discrepancy you should expect, between a sample statistic and a population parameter. Statistical inference involves using sample statistics to make a general conclusion about a population parameter. Thus, standard error plays a crucial role in inferential statistics.

KEY TERMS

sampling error (195)

distribution of sample means (195)

sampling distribution (196)

central limit theorem (200)

expected value of *M* (200)

standard error of *M* (201)

law of large numbers (202)

SPSS®

The statistical computer package SPSS is not structured to compute the standard error or a *z*-score for a sample mean. In later chapters, however, we introduce new inferential statistics that are included in SPSS. When these new statistics are computed, SPSS typically includes a report of standard error that describes how accurately, on average, the sample represents its population.

FOCUS ON PROBLEM SOLVING

1. Whenever you are working probability questions about sample means, you must use the distribution of sample means. Remember that every probability question can be restated as a proportion question. Probabilities for sample means are equivalent to proportions of the distribution of sample means.

2. When computing probabilities for sample means, the most common error is to use standard deviation (σ) instead of standard error (σ_M) in the z-score formula. Standard deviation measures the typical deviation (or "error") for a single score. Standard error measures the typical deviation (or error) for a sample. Remember: The larger the sample is, the more accurately the sample represents the population. Thus, sample size (n) is a critical part of the standard error.

$$\text{Standard error} = \sigma_M = \frac{\sigma}{\sqrt{n}}$$

3. Although the distribution of sample means is often normal, it is not always a normal distribution. Check the criteria to be certain the distribution is normal before you use the unit normal table to find probabilities (see item 1a of the Summary). Remember that all probability problems with a normal distribution are easier if you sketch the distribution and shade in the area of interest.

DEMONSTRATION 7.1

PROBABILITY AND THE DISTRIBUTION OF SAMPLE MEANS

A population forms a normal distribution with a mean of $\mu = 60$ and a standard deviation of $\sigma = 12$. For a sample of $n = 36$ scores from this population, what is the probability of obtaining a sample mean greater than 63?

$$p(M > 63) = ?$$

STEP 1   **Rephrase the probability question as a proportion question.**   Out of all the possible sample means for $n = 36$, what proportion will have values greater than 63? *All the possible sample means* is simply the distribution of sample means, which is normal, with a mean of $\mu = 60$ and a standard error of

$$\sigma_M = \frac{\sigma}{\sqrt{n}} = \frac{12}{\sqrt{36}} = \frac{12}{6} = 2$$

The distribution is shown in Figure 7.13.

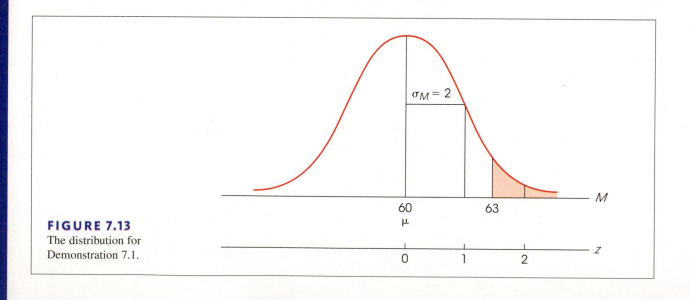

FIGURE 7.13

The distribution for
Demonstration 7.1.

STEP 2   **Compute the z-score for the sample mean.**   A sample mean of $M = 63$ corresponds to a z-score of

$$z = \frac{M - \mu}{\sigma_M} = \frac{63 - 60}{2} = \frac{3}{2} = 1.50$$

Therefore, $p(M > 63) = p(z > 1.50)$

STEP 3   **Look up the proportion in the unit normal table.**   Find $z = 1.50$ in column A and read across the row to find $p = 0.0668$ in column C. This is the answer.

$$p(M > 63) = p(z > 1.50) = 0.0668 \text{ (or 6.68\%)}$$

PROBLEMS

1. Briefly define each of the following:
   a. Distribution of sample means
   b. Expected value of M
   c. Standard error of M

2. A sample is selected from a population with a mean of $\mu = 40$ and a standard deviation of $\sigma = 8$.
   a. If the sample has $n = 4$ scores, what is the expected value of M and the standard error of M?
   b. If the sample has $n = 16$ scores, what is the expected value of M and the standard error of M?

3. Describe the distribution of sample means (shape, mean, standard error) for samples of $n = 64$ selected from a population with a mean of $\mu = 90$ and a standard deviation of $\sigma = 32$.

4. The distribution of sample means is not always a normal distribution. Under what circumstances is the distribution of sample means *not* be normal?

5. A population has a standard deviation of $\sigma = 24$.
   a. On average, how much difference should there be between the sample mean and the population mean for a random sample of $n = 4$ scores from this population?
   b. On average, how much difference should there be for a sample of $n = 9$ scores?
   c. On average, how much difference should there be for a sample of $n = 16$ scores?

6. For a population with a mean of $\mu = 45$ and a standard deviation of $\sigma = 10$, what is the standard error of the distribution of sample means for each of the following sample sizes?
   a. $n = 4$ scores
   b. $n = 25$ scores

7. For a population with $\sigma = 12$, how large a sample is necessary to have a standard error that is:
   a. less than 4 points?
   b. less than 3 points?
   c. less than 2 points?

8. If the population standard deviation is $\sigma = 10$, how large a sample is necessary to have a standard error that is
   a. less than 5 points?
   b. less than 2 points?
   c. less than 1 point?

9. For a sample of $n = 16$ scores, what is the value of the population standard deviation (σ) necessary to produce each of the following a standard error values?
   a. $\sigma_M = 8$ points?
   b. $\sigma_M = 4$ points?
   c. $\sigma_M = 1$ point?

10. For a population with a mean of $\mu = 40$ and a standard deviation of $\sigma = 8$, find the z-score corresponding to each of the following samples.
    a. $X = 36$ for a sample of $n = 1$ score
    b. $M = 36$ for a sample of $n = 4$ scores
    c. $M = 36$ for a sample of $n = 16$ scores

11. A sample of $n = 25$ scores has a mean of $M = 68$. Find the z-score for this sample:
    a. If it was obtained from a population with $\mu = 60$ and $\sigma = 10$.
    b. If it was obtained from a population with $\mu = 60$ and $\sigma = 20$.
    c. If it was obtained from a population with $\mu = 60$ and $\sigma = 40$.

12. A population forms a normal distribution with a mean of $\mu = 55$ and a standard deviation of $\sigma = 12$. For each of the following samples, compute the z-score for the sample mean.
    a. $M = 58$ for $n = 4$ scores
    b. $M = 58$ for $n = 16$ scores
    c. $M = 58$ for $n = 36$ scores

13. Scores on a standardized reading test for 4th-grade students form a normal distribution with $\mu = 60$ and

$\sigma = 20$. What is the probability of obtaining a sample mean greater than $M = 65$ for each of the following:
a. a sample of $n = 16$ students
b. a sample of $n = 25$ students
c. a sample of $n = 100$ students

14. IQ scores form a normal distribution with a mean of $\mu = 100$ and a standard deviation of $\sigma = 15$. What is the probability of obtaining a sample mean greater than $M = 103$,
    a. for a random sample of $n = 9$ people?
    b. for a random sample of $n = 25$ people?
    c. for a random sample of $n = 100$ people?

15. A normal distribution has a mean of $\mu = 54$ and a standard deviation of $\sigma = 6$.
    a. What is the probability of randomly selecting a score less than $X = 51$?
    b. What is the probability of selecting a sample of $n = 4$ scores with a mean less than $M = 51$?
    c. What is the probability of selecting a sample of $n = 36$ scores with a mean less than $M = 51$?

16. A population has a mean of $\mu = 30$ and a standard deviation of $\sigma = 8$
    a. If the population distribution is normal, what is the probability of obtaining a sample mean greater than $M = 32$ for a sample of $n = 4$?
    b. If the population distribution is positively skewed, what is the probability of obtaining a sample mean greater than $M = 32$ for a sample of $n = 4$?
    c. If the population distribution is normal, what is the probability of obtaining a sample mean greater than $M = 32$ for a sample of $n = 64$?
    d. If the population distribution is positively skewed, what is the probability of obtaining a sample mean greater than $M = 32$ for a sample of $n = 64$?

17. For random samples of size $n = 25$ selected from a normal distribution with a mean of mean of $\mu = 50$ and a standard deviation of $\sigma = 20$, find each of the following:
    a. The range of sample means that defines the middle 95% of the distribution of sample means.
    b. The range of sample means that defines the middle 99% of the distribution of sample means.

18. The distribution ages for students at the state college is positively skewed with a mean of $\mu = 21.5$ and a standard deviation of $\sigma = 3$.
    a. What is the probability of selecting a random sample of $n = 4$ students with an average age greater than 23? (Careful: This is a trick question.)
    b. What is the probability of selecting a random sample of $n = 36$ students with an average age greater than 23?
    c. For a sample of $n = 36$ students, what is the probability that the average age is between 21 and 22?

19. Jumbo shrimp are those that require 10–15 shrimp to make a pound. Suppose that the number of jumbo shrimp in a 1-pound bag averages $\mu = 12.5$ with a standard deviation of $\sigma = 1$, and forms a normal distribution. What is the probability of randomly picking a sample of $n = 25$ 1-pound bags that average more than $M = 13$ shrimp per bag?

20. Callahan (2009) conducted a study to evaluate the effectiveness of physical exercise programs for individuals with chronic arthritis. Participants with doctor-diagnosed arthritis either received a Tai Chi course immediately or were placed in a control group to begin the course 8 weeks later. At the end of the 8-week period, self-reports of pain were obtained for both groups. Data similar to the results obtained in the study are shown in the following table.

Self-Reported Level of Pain		
	Mean	SE
Tai Chi course	3.7	1.2
No Tai Chi course	7.6	1.7

a. Construct a bar graph that incorporates all of the information in the table.
b. Looking at your graph, do you think that participation in the Tai Chi course reduces arthritis pain?

21. A normal distribution has a mean of $\mu = 60$ and a standard deviation of $\sigma = 18$. For each of the following samples, compute the z-score for the sample mean and determine whether the sample mean is a typical, representative value or an extreme value for a sample of this size.
    a. $M = 67$ for $n = 4$ scores
    b. $M = 67$ for $n = 36$ scores

22. A random sample is obtained from a normal population with a mean of $\mu = 95$ and a standard deviation of $\sigma = 40$. The sample mean is $M = 86$.
    a. Is this a representative sample mean or an extreme value for a sample of $n = 16$ scores?
    b. Is this a representative sample mean or an extreme value for a sample of $n = 100$ scores?

23. A normal distribution has a mean of $\mu = 65$ and a standard deviation of $\sigma = 20$. For each of the following samples, compute the z-score for the sample mean and determine whether the sample mean is a typical, representative value or an extreme value for a sample of its size.
    a. $M = 74$ for a sample of 4 scores
    b. $M = 74$ for a sample of 25 scores

Introduction to Hypothesis Testing

Tools You Will Need

The following items are considered essential background material for this chapter. If you doubt your knowledge of any of these items, you should review the appropriate chapter or section before proceeding.

- *z*-Scores (Chapter 5)
- Distribution of sample means (Chapter 7)
  - Expected value
  - Standard error
  - Probability and sample means

© Deborah Batt

Most of us spend more time looking down at our mobile devices than we do looking up at the clouds. But if you do watch the clouds and have a little imagination, you occasionally will see them form into familiar shapes. Figure 8.1 is a photograph of a cloud formation seen over Kansas City around Christmas in 2008. Do you recognize a familiar image?

The cloud pattern shown in Figure 8.1 was formed simply by chance. Specifically, it was the random forces of wind and air currents that produced a portrait of Santa Claus. The clouds did not conspire to form the image, and it was not deliberately created by a team of professional skywriters. The point we would like to make is that what appear to be meaningful patterns can be produced by random chance.

Researchers often find meaningful patterns in the sample data obtained in research studies. The problem is deciding whether the patterns found in a sample reflect real patterns that exist in the population or are simply random, chance occurrences. To differentiate between real, systematic patterns and random, chance occurrences, researchers rely on a statistical technique known as *hypothesis testing*, which is introduced in this chapter. As you will see, a hypothesis test first determines the

probability that the pattern could have been produced by chance alone. If this probability is large enough, we conclude that the pattern can reasonably be explained by chance. However, if the probability is extremely small, we can rule out chance as a plausible explanation and conclude that some meaningful, systematic force has created the pattern. For example, it is reasonable, once in a lifetime, to see a cloud formation that resembles Santa Claus. However, it would be extremely unlikely if the clouds also included the words "Merry Christmas" spelled out beneath Santa's face. If this happened, we would conclude that the pattern was not produced by the random forces of chance, but rather was created by a deliberate, systematic act.

In this chapter we introduce the statistical procedure known as a hypothesis test, which is one of the fundamental techniques of inferential statistics. In general terms, a hypothesis test uses the limited information from a sample to reach a general conclusion about a population. Specifically, a hypothesis test helps researchers differentiate between real and random patterns in the data. We present the logic underlying a hypothesis test, the terminology used in a hypothesis test, and the general procedure for conducting a hypothesis test.

FIGURE 8.1
A cloud formation seen over Kansas City.

8.1 | The Logic of Hypothesis Testing

LEARNING OBJECTIVES

1. Describe the purpose of a hypothesis test and explain how the test accomplishes its goal.

2. Define the null hypothesis and the alternative hypothesis for a hypothesis test.

3. Define the alpha level (level of significance) and the critical region for a hypothesis test.

4. Conduct a hypothesis test and make a statistical decision about the effect of a treatment.

It is usually impossible or impractical for a researcher to observe every individual in a population. Therefore, researchers usually collect data from a sample and then use the sample data to help answer questions about the population. Hypothesis testing is a statistical procedure that allows researchers to use sample data to draw inferences about the population of interest.

Hypothesis testing is one of the most commonly used inferential procedures. In fact, most of the remainder of this book examines hypothesis testing in a variety of different situations and applications. Although the details of a hypothesis test change from one situation to another, the general process remains constant. In this chapter, we introduce the general procedure for a hypothesis test. You should notice that we use the statistical techniques that have been developed in the preceding three chapters—that is, we combine the concepts of z-scores, probability, and the distribution of sample means to create a new statistical procedure known as a *hypothesis test*.

DEFINITION

> A **hypothesis test** is a statistical method that uses sample data to evaluate a hypothesis about a population.

In very simple terms, the logic underlying the hypothesis-testing procedure is as follows:

1. First, we state a hypothesis about a population. Usually the hypothesis concerns the value of a population parameter. For example, we might hypothesize that American adults gain an average of $\mu = 7$ pounds between Thanksgiving and New Year's Day each year.

2. Before we select a sample, we use the hypothesis to predict the characteristics that the sample should have. For example, if we predict that the average weight gain for the population is $\mu = 7$ pounds, then we would predict that our sample should have a mean *around* 7 pounds. Remember: The sample should be similar to the population, but you always expect a certain amount of error.

3. Next, we obtain a random sample from the population. For example, we might select a sample of $n = 200$ American adults and measure the average weight change for the sample between Thanksgiving and New Year's Day.

4. Finally, we compare the obtained sample data with the prediction that was made from the hypothesis. If the sample mean is consistent with the prediction, we conclude that the hypothesis is reasonable. But if there is a big discrepancy between the data and the prediction, we decide that the hypothesis is wrong.

A hypothesis test is typically used in the context of a research study. That is, a researcher completes a research study and then uses a hypothesis test to evaluate the results. Depending on the type of research and the type of data, the details of the hypothesis test change from one research situation to another. In later chapters, we examine different versions of hypothesis testing that are used for different kinds of research. For now, however, we focus on the basic elements that are common to all hypothesis tests. To accomplish this general goal, we will examine a hypothesis test as it applies to the simplest possible situation—using a sample mean to test a hypothesis about a population mean.

In the six chapters that follow, we consider hypothesis testing in more complex research situations involving sample means and mean differences. In Chapters 15 and 16, we look at correlational research and examine how the relationships obtained for sample data are used to evaluate hypotheses about relationships in the population. Finally, in Chapters 17 and 18, we examine how the proportions that exist in a sample are used to test hypotheses about the corresponding proportions in the population.

Once again, we introduce hypothesis testing with a situation in which a researcher is using one sample mean to evaluate a hypothesis about one unknown population mean.

The Unknown Population Figure 8.2 shows the general research situation that we will use to introduce the process of hypothesis testing. Notice that the researcher begins with a known population. This is the set of individuals as they exist *before treatment*. For this example, we are assuming that the original set of scores forms a normal distribution with $\mu = 80$ and $\sigma = 20$. The purpose of the research is to determine the effect of a treatment on the individuals in the population. That is, the goal is to determine what happens to the population *after the treatment is administered.*

To simplify the hypothesis-testing situation, one basic assumption is made about the effect of the treatment: if the treatment has any effect, it is simply to add a constant amount to (or subtract a constant amount from) each individual's score. You should recall from Chapters 3 and 4 that adding (or subtracting) a constant changes the mean but does not change the shape of the population, nor does it change the standard deviation. Thus, we assume that the population after treatment has the same shape as the original population and the same standard deviation as the original population. This assumption is incorporated into the situation shown in Figure 8.2.

Note that the unknown population, after treatment, is the focus of the research question. Specifically, the purpose of the research is to determine what would happen if the treatment were administered to every individual in the population.

The Sample in the Research Study The goal of the hypothesis test is to determine whether the treatment has any effect on the individuals in the population (see Figure 8.2). Usually, however, we cannot administer the treatment to the entire population so the actual research study is conducted using a sample. Figure 8.3 shows the structure of the research

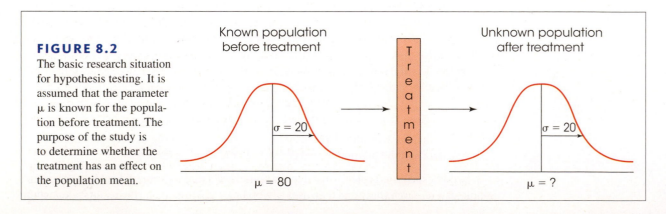

FIGURE 8.2

The basic research situation for hypothesis testing. It is assumed that the parameter μ is known for the population before treatment. The purpose of the study is to determine whether the treatment has an effect on the population mean.

Known population before treatment

$\sigma = 20$

$\mu = 80$

Treatment

Unknown population after treatment

$\sigma = 20$

$\mu = ?$

FIGURE 8.3

From the point of view of the hypothesis test, the entire population receives the treatment and then a sample is selected from the treated population. In the actual research study, however, a sample is selected from the original population and the treatment is administered to the sample. From either perspective, the result is a treated sample that represents the treated population.

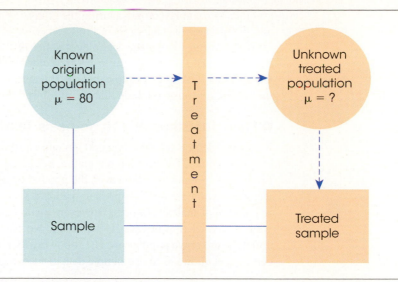

study from the point of view of the hypothesis test. The original population, before treatment, is shown on the left-hand side. The unknown population, after treatment, is shown on the right-hand side. Note that the unknown population is actually *hypothetical* (the treatment is never administered to the entire population). Instead, we are asking *what would happen if* the treatment were administered to the entire population. The research study involves selecting a sample from the original population, administering the treatment to the sample, and then recording scores for the individuals in the treated sample. Notice that the research study produces a treated sample. Although this sample was obtained indirectly, it is equivalent to a sample that is obtained directly from the unknown treated population. The hypothesis test uses the treated sample on the right-hand side of Figure 8.3 to evaluate a hypothesis about the unknown treated population on the right side of the figure.

A hypothesis test is a formalized procedure that follows a standard series of operations. In this way, researchers have a standardized method for evaluating the results of their research studies. Other researchers will recognize and understand exactly how the data were evaluated and how conclusions were reached. To emphasize the formal structure of a hypothesis test, we will present hypothesis testing as a four-step process that is used throughout the rest of the book. The following example provides a concrete foundation for introducing the hypothesis-testing procedure.

EXAMPLE 8.1   Previous research indicates that men rate women as being more attractive when they are wearing red (Elliot & Niesta, 2008). Based on these results, Guéguen and Jacob (2012) reasoned that the same phenomenon might influence the way that men react to waitresses wearing red. In their study, waitresses in five different restaurants wore the same T-shirt in six different colors (red, blue, green, yellow, black, and white) on different days during a six-week period. Except for the T-shirts, the waitresses were instructed to act normally and to record each customer's gender and how much was left as a tip. The results show that male customers gave significantly bigger tips to waitresses wearing red but that color had no effect on tipping for female customers.

A researcher decided to test this result by repeating the basic study at a local restaurant. Waitresses (and waiters) at the restaurant routinely wear white shirts with black pants and restaurant records indicate that the waitress' tips from male customers average $\mu = 15.8$ percent with a standard deviation of $\sigma = 2.4$ percentage points. The distribution of tip amounts is roughly normal. During the study, the waitresses are asked to wear red shirts and the researcher plans to record tips for a sample of $n = 36$ male customers.

If the mean tip for the sample is noticeably different from the baseline mean, the researcher can conclude that wearing the color red does appear to have an effect on tipping. On the other hand, if the sample mean is still around 15.8 percent (the same as the baseline), the researcher must conclude that the red shirt does not appear to have any effect. ■

■ The Four Steps of a Hypothesis Test

Figure 8.3 depicts the same general structure as the research situation described in the preceding example. The original population before treatment (before the red shirt) has a mean tip of $\mu = 15.8$ percent. However, the population after treatment is unknown. Specifically, we do not know what will happen to the mean score if the waitresses wear red for the entire population of male customers. However, we do have a sample of $n = 36$ participants who were served when waitresses wore red and we can use this sample to help draw inferences about the unknown population. The following four steps outline the hypothesis-testing procedure that allows us to use sample data to answer questions about an unknown population.

STEP 1   **State the hypothesis.**   As the name implies, the process of hypothesis testing begins by stating a hypothesis about the unknown population. Actually, we state two opposing hypotheses. Notice that both hypotheses are stated in terms of population parameters.

The first and most important of the two hypotheses is called the *null hypothesis*. The null hypothesis states that the treatment has no effect. In general, the null hypothesis states that there is no change, no effect, no difference—nothing happened, hence the name *null*. The null hypothesis is identified by the symbol H_0 (The H stands for *hypothesis*, and the zero subscript indicates that this is the *zero-effect* hypothesis.) For the study in Example 8.1, the null hypothesis states that the red shirt has no effect on tipping behavior for the population of male customers. In symbols, this hypothesis is

$$H_0: \mu_{\text{with red shirt}} = 15.8 \quad \text{(Even with a red shirt, the mean tip is still 15.8 percent.)}$$

> The goal of inferential statistics is to make general statements about the population by using sample data. Therefore, when testing hypotheses, we make our predictions about the population parameters.

DEFINITION

> The **null hypothesis** (H_0) states that in the general population there is no change, no difference, or no relationship. In the context of an experiment, H_0 predicts that the independent variable (treatment) *has no effect* on the dependent variable (scores) for the population.

The second hypothesis is simply the opposite of the null hypothesis, and it is called the *scientific,* or *alternative, hypothesis* (H_1). This hypothesis states that the treatment has an effect on the dependent variable.

DEFINITION

> The **alternative hypothesis** (H_1) states that there is a change, a difference, or a relationship for the general population. In the context of an experiment, H_1 predicts that the independent variable (treatment) *does have an effect* on the dependent variable.

> The null hypothesis and the alternative hypothesis are mutually exclusive and exhaustive. They cannot both be true. The data will determine which one should be rejected.

For this example, the alternative hypothesis states that the red shirt does have an effect on tipping for the population and will cause a change in the mean score. In symbols, the alternative hypothesis is represented as

$$H_1: \mu_{\text{with red shirt}} \neq 15.8 \quad \text{(with a red shirt, the mean tip will be different from 15.8 percent)}$$

Notice that the alternative hypothesis simply states that there will be some type of change. It does not specify whether the effect will be increased or decreased tips. In some circumstances, it is appropriate for the alternative hypothesis to specify the direction of the effect. For example, the researcher might hypothesize that a red shirt will increase tips ($\mu > 15.8$). This type of hypothesis results in a directional hypothesis test, which is examined in detail later in this chapter. For now we concentrate on nondirectional tests, for which the hypotheses simply state that the treatment has no effect (H_0) or has some effect (H_1).

STEP 2   **Set the criteria for a decision.**   Eventually the researcher will use the data from the sample to evaluate the credibility of the null hypothesis. The data will either provide support for the null hypothesis or tend to refute the null hypothesis. In particular, if there is a big discrepancy between the data and the hypothesis, we will conclude that the hypothesis is wrong.

To formalize the decision process, we use the null hypothesis to predict the kind of sample mean that ought to be obtained. Specifically, we determine exactly which sample means are consistent with the null hypothesis and which sample means are at odds with the null hypothesis.

For our example, the null hypothesis states that the red shirts have no effect and the population mean is still $\mu = 15.8$ percent, the same as the population mean for waitresses wearing white shirts. If this is true, then the sample mean should have a value around 15.8. Therefore, a sample mean near 15.8 is consistent with the null hypothesis. On the other hand, a sample mean that is very different from 15.8 is not consistent with the null hypothesis. To determine exactly which values are "near" 15.8 and which values are "very different from" 15.8, we will examine all of the possible sample means that could be obtained if the null hypothesis is true. For our example, this is the distribution of sample means for $n = 36$. According to the null hypothesis, this distribution is centered at $\mu = 15.8$. The distribution of sample means is then divided into two sections.

1. Sample means that are likely to be obtained if H_0 is true; that is, sample means that are close to the null hypothesis

2. Sample means that are very unlikely to be obtained if H_0 is true; that is, sample means that are very different from the null hypothesis

Figure 8.4 shows the distribution of sample means divided into these two sections. Notice that the high-probability samples are located in the center of the distribution and have sample means close to the value specified in the null hypothesis. On the other hand, the low-probability samples are located in the extreme tails of the distribution. After the distribution has been divided in this way, we can compare our sample data with the values in the distribution. Specifically, we can determine whether our sample mean is consistent with the null hypothesis (like the values in the center of the distribution) or whether our sample mean is very different from the null hypothesis (like the values in the extreme tails).

The Alpha Level   To find the boundaries that separate the high-probability samples from the low-probability samples, we must define exactly what is meant by "low" probability and "high" probability. This is accomplished by selecting a specific probability value, which is known as the *level of significance,* or the *alpha level,* for the hypothesis test. The alpha (α) value is a small probability that is used to identify the low-probability samples. By convention, commonly used alpha levels are $\alpha = .05$ (5%), $\alpha = .01$ (1%), and $\alpha = .001$ (0.1%). For example, with $\alpha = .05$, we separate the most unlikely 5% of the sample means (the extreme values) from the most likely 95% of the sample means (the central values).

With rare exceptions, an alpha level is never larger than .05.

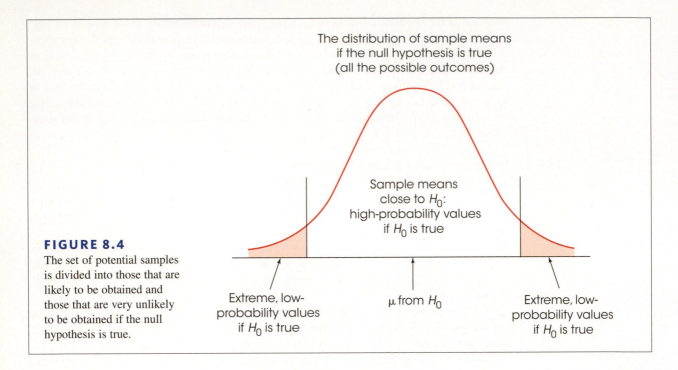

The distribution of sample means
if the null hypothesis is true
(all the possible outcomes)

Sample means
close to H_0:
high-probability values
if H_0 is true

Extreme, low-
probability values
if H_0 is true

μ from H_0

Extreme, low-
probability values
if H_0 is true

FIGURE 8.4

The set of potential samples is divided into those that are likely to be obtained and those that are very unlikely to be obtained if the null hypothesis is true.

The extremely unlikely values, as defined by the alpha level, make up what is called the *critical region*. These extreme values in the tails of the distribution define outcomes that are not consistent with the null hypothesis; that is, they are very unlikely to occur if the null hypothesis is true. Whenever the data from a research study produce a sample mean that is located in the critical region, we conclude that the data are not consistent with the null hypothesis, and we reject the null hypothesis.

DEFINITIONS

> The **alpha level**, or the **level of significance**, is a probability value that is used to define the concept of "very unlikely" in a hypothesis test.
>
> The **critical region** is composed of the extreme sample values that are very unlikely (as defined by the alpha level) to be obtained if the null hypothesis is true. The boundaries for the critical region are determined by the alpha level. If sample data fall in the critical region, the null hypothesis is rejected.

Technically, the critical region is defined by sample outcomes that are *very unlikely* to occur if the treatment has no effect (that is, if the null hypothesis is true). Reversing the point of view, we can also define the critical region as sample values that provide *convincing evidence* that the treatment really does have an effect. For our example, the regular population of male customers leaves a mean tip of $\mu = 15.8$ percent. We selected a sample from this population and administered a treatment (the red shirt) to the individuals in the sample. What kind of sample mean would convince you that the treatment has an effect? It should be obvious that the most convincing evidence would be a sample mean that is really different from $\mu = 15.8$ percent. In a hypothesis test, the critical region is determined by sample values that are "really different" from the original population.

The Boundaries for the Critical Region   To determine the exact location for the boundaries that define the critical region, we use the alpha-level probability and the unit normal table. In most cases, the distribution of sample means is normal, and the unit normal table provides the precise z-score location for the critical region boundaries. With $\alpha = .05$, for example, the boundaries separate the extreme 5% from the middle 95%. Because the extreme 5% is split between two tails of the distribution, there is exactly 2.5% (or 0.0250) in each tail. In the unit normal table, you can look up a proportion of 0.0250 in column C (the tail) and find that the z-score boundary is $z = 1.96$. Thus, for any normal distribution, the extreme 5% is in the tails of the distribution beyond $z = +1.96$ and $z = -1.96$. These values define the boundaries of the critical region for a hypothesis test using $\alpha = .05$ (Figure 8.5).

Similarly, an alpha level of $\alpha = .01$ means that 1% or .0100 is split between the two tails. In this case, the proportion in each tail is .0050, and the corresponding z-score boundaries are $z = \pm 2.58$ (± 2.57 is equally good). For $\alpha = .001$, the boundaries are located at $z = \pm 3.30$. You should verify these values in the unit normal table and be sure that you understand exactly how they are obtained.

STEP 3   **Collect data and compute sample statistics.**   At this time, we begin recording tips for male customers while the waitresses are wearing red. Notice that the data are collected *after* the researcher has stated the hypotheses and established the criteria for a decision. This sequence of events helps ensure that a researcher makes an honest, objective evaluation of the data and does not tamper with the decision criteria after the experimental outcome is known.

Next, the raw data from the sample are summarized with the appropriate statistics: For this example, the researcher would compute the sample mean. Now it is possible for the researcher to compare the sample mean (the data) with the null hypothesis. This is the heart of the hypothesis test: comparing the data with the hypothesis.

The comparison is accomplished by computing a z-score that describes exactly where the sample mean is located relative to the hypothesized population mean from H_0. In Step 2, we constructed the distribution of sample means that would be expected if the null

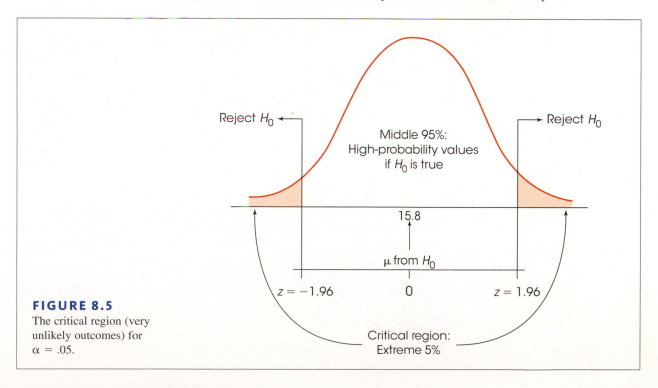

FIGURE 8.5

The critical region (very unlikely outcomes) for $\alpha = .05$.

hypothesis were true—that is, the entire set of sample means that could be obtained if the treatment has no effect (see Figure 8.5). Now we calculate a z-score that identifies where our sample mean is located in this hypothesized distribution. The z-score formula for a sample mean is

$$z = \frac{M - \mu}{\sigma_M}$$

In the formula, the value of the sample mean (M) is obtained from the sample data, and the value of μ is obtained from the null hypothesis. Thus, the z-score formula can be expressed in words as follows:

$$z = \frac{\text{sample mean} - \text{hypothesized population mean}}{\text{standard error between } M \text{ and } \mu}$$

Notice that the top of the z-score formula measures how much difference there is between the data and the hypothesis. The bottom of the formula measures the standard distance that ought to exist between a sample mean and the population mean.

STEP 4   **Make a decision.**   In the final step, the researcher uses the z-score value obtained in Step 3 to make a decision about the null hypothesis according to the criteria established in Step 2. There are two possible outcomes.

1. The sample data are located in the critical region. By definition, a sample value in the critical region is very unlikely to occur if the null hypothesis is true. Therefore, we conclude that the sample is not consistent with H_0 and our decision is to *reject the null hypothesis*. Remember, the null hypothesis states that there is no treatment effect, so rejecting H_0 means we are concluding that the treatment did have an effect.

   For the example we have been considering, suppose the sample produced a mean tip of $M = 16.7$ percent. The null hypothesis states that the population mean is $\mu = 15.8$ percent and, with $n = 36$ and $\sigma = 2.4$, the standard error for the sample mean is

   $$\sigma_M = \frac{\sigma}{\sqrt{n}} = \frac{2.4}{6} = 0.4$$

   Thus, a sample mean of $M = 16.7$ produces a z-score of

   $$z = \frac{M - \mu}{\sigma_M} = \frac{16.7 - 15.8}{0.4} = \frac{0.9}{0.4} = 2.25$$

   With an alpha level of $\alpha = .05$, this z-score is far beyond the boundary of 1.96. Because the sample z-score is in the critical region, we reject the null hypothesis and conclude that the red shirt did have an effect on tipping behavior.

2. The sample data are not in the critical region. In this case, the sample mean is reasonably close to the population mean specified in the null hypothesis (in the center of the distribution). Because the data do not provide strong evidence that the null hypothesis is wrong, our conclusion is to *fail to reject the null hypothesis*. This conclusion means that the treatment does not appear to have an effect.

   For the research study examining the effect of a red shirt, suppose our sample produced a mean tip of $M = 16.1$ percent. As before, the standard error for a sample of $n = 36$ is $\sigma_M = 0.4$, and the null hypothesis states that $\mu = 15.8$ percent. These values produce a z-score of

   $$z = \frac{M - \mu}{\sigma_M} = \frac{16.1 - 15.8}{0.4} = \frac{0.3}{0.4} = 0.75$$

The z-score of 0.75 is not in the critical region. Therefore, we would fail to reject the null hypothesis and conclude that the red shirt does not appear to have an effect on male tipping behavior.

In general, the final decision is made by comparing our treated sample with the distribution of sample means that would be obtained for untreated samples. If our treated sample (red shirt) looks much the same as untreated samples (white shirt), we conclude that the treatment does not appear to have any effect. On the other hand, if the treated sample is noticeably different from the majority of untreated samples, we conclude that the treatment does have an effect.

The following example is an opportunity to test your understanding of the process of a hypothesis test.

EXAMPLE 8.2   A normal population has a mean $\mu = 40$ and a standard deviation of $\sigma = 8$. After a treatment is administered to a sample of $n = 16$ individuals from the population, the sample mean is found to be $M = 45$. Does this sample provide sufficient evidence to reject the null hypothesis and conclude that there is a significant treatment effect with an alpha level of $\alpha = .05$? You should obtain $z = 2.50$, which is in the critical region and rejects the null hypothesis. ■

An Analogy for Hypothesis Testing   It may seem awkward to phrase both of the two possible decisions in terms of rejecting the null hypothesis; either we reject H_0 or we fail to reject H_0. These two decisions may be easier to understand if you think of a research study as an attempt to gather evidence to prove that a treatment works. From this perspective, the process of conducting a hypothesis test is similar to the process that takes place during a jury trial. For example,

1. The test begins with a null hypothesis stating that there is no treatment effect. The trial begins with a null hypothesis that there is no crime (innocent until proven guilty).

2. The research study gathers evidence to show that the treatment actually does have an effect, and the police gather evidence to show that there really is a crime. Note that both are trying to refute the null hypothesis.

3. If there is enough evidence, the researcher rejects the null hypothesis and concludes that there really is a treatment effect. If there is enough evidence, the jury rejects the hypothesis and concludes that the defendant is guilty of a crime.

4. If there is not enough evidence, the researcher fails to reject the null hypothesis. Note that the researcher does not conclude that there is no treatment effect, simply that there is not enough evidence to conclude that there is an effect. Similarly, if there is not enough evidence, the jury fails to find the defendant guilty. Note that the jury does not conclude that the defendant is innocent, simply that there is not enough evidence for a guilty verdict.

■ A Closer Look at the z-Score Statistic

The z-score statistic that is used in the hypothesis test is the first specific example of what is called a *test statistic*. The term *test statistic* simply indicates that the sample data are converted into a single, specific statistic that is used to test the hypotheses. In the chapters that follow, we introduce several other test statistics that are used in a variety of different research situations. However, most of the new test statistics have the same basic structure and serve the same purpose as the z-score. We have already described the z-score equation as a formal method for comparing the sample data and the population hypothesis. In this section, we discuss the z-score from two other perspectives that may give you a better understanding of hypothesis testing and the role that z-scores play in this inferential technique. In each case, keep in mind that the z-score serves as a general model for other test statistics that will come in future chapters.

The z-Score Formula as a Recipe   The z-score formula, like any formula, can be viewed as a recipe. If you follow instructions and use all the right ingredients, the formula produces a z-score. In the hypothesis-testing situation, however, you do not have all the necessary ingredients. Specifically, you do not know the value for the population mean (μ), which is one component or ingredient in the formula.

This situation is similar to trying to follow a cake recipe where one of the ingredients is not clearly listed. For example, the recipe may call for flour but there is a grease stain that makes it impossible to read how much flour. Faced with this situation, you might try the following steps.

1. Make a hypothesis about the amount of flour. For example, hypothesize that the correct amount is 2 cups.

2. To test your hypothesis, add the rest of the ingredients along with the hypothesized flour and bake the cake.

3. If the cake turns out to be good, you can reasonably conclude that your hypothesis was correct. But if the cake is terrible, you conclude that your hypothesis was wrong.

In a hypothesis test with z-scores, we do essentially the same thing. We have a formula (recipe) for z-scores but one ingredient is missing. Specifically, we do not know the value for the population mean, μ. Therefore, we try the following steps.

1. Make a hypothesis about the value of μ. This is the null hypothesis.

2. Plug the hypothesized value in the formula along with the other values (ingredients).

3. If the formula produces a z-score near zero (which is where z-scores are supposed to be), we conclude that the hypothesis was correct. On the other hand, if the formula produces an extreme value (a very unlikely result), we conclude that the hypothesis was wrong.

The z-Score Formula as a Ratio   In the context of a hypothesis test, the z-score formula has the following structure:

$$z = \frac{M - \mu}{\sigma_M} = \frac{\text{sample mean} - \text{hypothesized population mean}}{\text{standard error between } M \text{ and } \mu}$$

Notice that the numerator of the formula involves a direct comparison between the sample data and the null hypothesis. In particular, the numerator measures the obtained difference between the sample mean and the hypothesized population mean. The standard error in the denominator of the formula measures the standard amount of distance that exists naturally between a sample mean and the population mean without any treatment effect that causes the sample to be different. Thus, the z-score formula (and many other test statistics) forms a ratio

$$z = \frac{\text{actual difference between the sample } (M) \text{ and the hypothesis } (\mu)}{\text{standard difference between } M \text{ and } \mu \text{ with no treatment effect}}$$

Thus, for example, a z-score of $z = 3.00$ means that the obtained difference between the sample and the hypothesis is 3 times bigger than would be expected if the treatment had no effect.

In general, a large value for a test statistic like the z-score indicates a large discrepancy between the sample data and the null hypothesis. Specifically, a large value indicates that the sample data are very unlikely to have occurred by chance alone. Therefore, when we obtain a large value (in the critical region), we conclude that it must have been caused by a treatment effect.

LEARNING CHECK

1. A sample of $n = 16$ individuals is selected from a population with $\mu = 40$ and $\sigma = 12$, and a treatment is administered to the sample. After treatment, the sample mean is $M = 42$. How does this sample compare to samples that should be obtained if the treatment has no effect?
   a. It is an extreme value that would be very unlikely if the treatment has no effect.
   b. It is much the same as samples that would be obtained if the treatment has no effect.
   c. The mean is noticeably larger than would be expected if the treatment has no effect but it is not quite an extreme sample.
   d. There is no way to compare this sample with those that would be obtained if the treatment had no effect.

2. A researcher selects a sample of $n = 20$ from a normal population with $\mu = 80$ and $\sigma = 20$, and a treatment is administered to the sample. If a hypothesis test is used to evaluate the effect of the treatment, what is the null hypothesis?
   a. $\mu = 80$
   b. $M = 80$
   c. $\mu \neq 80$
   d. $M \neq 80$

3. What is the correct decision in a hypothesis if the data produce a z-score that is in the critical region?
   a. Reject H_0
   b. Fail to reject H_0
   c. Reject H_1
   d. Fail to reject H_1

4. A researcher administers a treatment to a sample of $n = 25$ participants and uses a hypothesis test to evaluate the effect of the treatment. The hypothesis test produces a z-score of $z = 1.77$. Assuming that the researcher is using a two-tailed test,
   a. The researcher should reject the null hypothesis with $\alpha = .05$ but not with $\alpha = .01$.
   b. The researcher should reject the null hypothesis with either $\alpha = .05$ or $\alpha = .01$.
   c. The researcher should fail to reject H_0 with either $\alpha = .05$ or $\alpha = .01$.
   d. Cannot answer without additional information.

ANSWERS   1. B, 2. A, 3. A, 4. C

8.2 | Uncertainty and Errors in Hypothesis Testing

5. Define a Type I error and a Type II error and explain the consequences of each.

Hypothesis testing is an *inferential process,* which means that it uses limited information as the basis for reaching a general conclusion. Specifically, a sample provides only limited or incomplete information about the whole population, and yet a hypothesis test uses a sample to draw a conclusion about the population. In this situation, there is always the possibility that an incorrect conclusion will be made. Although sample data are usually representative of the population, there is always a chance that the sample is misleading and will cause a researcher to make the wrong decision about the research results. In a hypothesis test, there are two different kinds of errors that can be made.

■ Type I Errors

It is possible that the data will lead you to reject the null hypothesis when in fact the treatment has no effect. Remember: samples are not expected to be identical to their populations, and some extreme samples can be very different from the populations they are supposed to represent. If a researcher selects one of these extreme samples by chance, then the data from the sample may give the appearance of a strong treatment effect, even though there is no real effect. In the previous section, for example, we discussed a research study examining how the tipping behavior of male customers is influenced by a waitress wearing the color red. Suppose the researcher selects a sample of $n = 36$ men who already were good tippers. Even if the red shirt (the treatment) has no effect at all, these men will still leave higher than average tips. In this case, the researcher is likely to conclude that the treatment does have an effect, when in fact it really does not. This is an example of what is called a *Type I error.*

DEFINITION

> A **Type I error** occurs when a researcher rejects a null hypothesis that is actually true. In a typical research situation, a Type I error means the researcher concludes that a treatment does have an effect when in fact it has no effect.

You should realize that a Type I error is not a stupid mistake in the sense that a researcher is overlooking something that should be perfectly obvious. On the contrary, the researcher is looking at sample data that appear to show a clear treatment effect. The researcher then makes a careful decision based on the available information. The problem is that the information from the sample is misleading.

In most research situations, the consequences of a Type I error can be very serious. Because the researcher has rejected the null hypothesis and believes that the treatment has a real effect, it is likely that the researcher will report or even publish the research results. A Type I error, however, means that this is a false report. Thus, Type I errors lead to false reports in the scientific literature. Other researchers may try to build theories or develop other experiments based on the false results. A lot of precious time and resources may be wasted.

The Probability of a Type I Error   A Type I error occurs when a researcher unknowingly obtains an extreme, nonrepresentative sample. Fortunately, the hypothesis test is structured to minimize the risk that this will occur. Figure 8.5 shows the distribution of sample means and the critical region for the waitress-tipping study we have been discussing. This distribution contains all of the possible sample means for samples of $n = 36$ if the null hypothesis is true. Notice that most of the sample means are near the hypothesized population mean, $\mu = 15.8$, and that means in the critical region are very unlikely to occur.

With an alpha level of $\alpha = .05$, only 5% of the samples have means in the critical region. Therefore, there is only a 5% probability ($p = .05$) that one of these samples will be obtained. Thus, the alpha level determines the probability of obtaining a sample mean in the critical region when the null hypothesis is true. In other words, the alpha level determines the probability of a Type I error.

DEFINITION	The **alpha level** for a hypothesis test is the probability that the test will lead to a Type I error. That is, the alpha level determines the probability of obtaining sample data in the critical region even though the null hypothesis is true.

In summary, whenever the sample data are in the critical region, the appropriate decision for a hypothesis test is to reject the null hypothesis. Normally this is the correct decision because the treatment has caused the sample to be different from the original population; that is, the treatment effect has pushed the sample mean into the critical region. In this case, the hypothesis test has correctly identified a real treatment effect. Occasionally, however, sample data are in the critical region just by chance, without any treatment effect. When this occurs, the researcher will make a Type I error; that is, the researcher will conclude that a treatment effect exists when in fact it does not. Fortunately, the risk of a Type I error is small and is under the control of the researcher. Specifically, the probability of a Type I error is equal to the alpha level.

■ Type II Errors

Whenever a researcher rejects the null hypothesis, there is a risk of a Type I error. Similarly, whenever a researcher fails to reject the null hypothesis, there is a risk of a *Type II error*. By definition, a Type II error is the failure to reject a false null hypothesis. In more straightforward English, a Type II error means that a treatment effect really exists, but the hypothesis test fails to detect it.

DEFINITION	A **Type II error** occurs when a researcher fails to reject a null hypothesis that is really false. In a typical research situation, a Type II error means that the hypothesis test has failed to detect a real treatment effect.

A Type II error occurs when the sample mean is not in the critical region even though the treatment has an effect on the sample. Often this happens when the effect of the treatment is relatively small. In this case, the treatment does influence the sample, but the magnitude of the effect is not big enough to move the sample mean into the critical region. Because the sample is not substantially different from the original population (it is not in the critical region), the statistical decision is to fail to reject the null hypothesis and to conclude that there is not enough evidence to say there is a treatment effect.

The consequences of a Type II error are usually not as serious as those of a Type I error. In general terms, a Type II error means that the research data do not show the results that the researcher had hoped to obtain. The researcher can accept this outcome and conclude that the treatment either has no effect or has only a small effect that is not worth pursuing, or the researcher can repeat the experiment (usually with some improvement, such as a larger sample) and try to demonstrate that the treatment really does work.

Unlike a Type I error, it is impossible to determine a single, exact probability for a Type II error. Instead, the probability of a Type II error depends on a variety of factors and therefore is a function, rather than a specific number. Nonetheless, the probability of a Type II error is represented by the symbol β, the Greek letter *beta*.

In summary, a hypothesis test always leads to one of two decisions.

1. The sample data provide sufficient evidence to reject the null hypothesis and conclude that the treatment has an effect.

2. The sample data do not provide enough evidence to reject the null hypothesis. In this case, you fail to reject H_0 and conclude that the treatment does not appear to have an effect.

In either case, there is a chance that the data are misleading and the decision is wrong. The complete set of decisions and outcomes is shown in Table 8.1. The risk of an error is especially important in the case of a Type I error, which can lead to a false report. Fortunately, the probability of a Type I error is determined by the alpha level, which is completely under the control of the researcher. At the beginning of a hypothesis test, the researcher states the hypotheses and selects the alpha level, which immediately determines the risk of a Type I error.

TABLE 8.1

Possible outcomes of a statistical decision.

		Actual Situation	
		No Effect, H_0 True	Effect Exists, H_0 False
Experimenter's Decision	Reject H_0	Type I error	Decision correct
	Fail to Reject H_0	Decision correct	Type II error

■ Selecting an Alpha Level

As you have seen, the alpha level for a hypothesis test serves two very important functions. First, alpha helps determine the boundaries for the critical region by defining the concept of "very unlikely" outcomes. At the same time, alpha determines the probability of a Type I error. When you select a value for alpha at the beginning of a hypothesis test, your decision influences both of these functions.

The primary concern when selecting an alpha level is to minimize the risk of a Type I error. Thus, alpha levels tend to be very small probability values. By convention, the largest permissible value is $\alpha = .05$. When there is no treatment effect, an alpha level of .05 means that there is still a 5% risk, or a 1-in-20 probability, of rejecting the null hypothesis and committing a Type I error. Because the consequences of a Type I error can be relatively serious, many individual researchers and many scientific publications prefer to use a more conservative alpha level such as .01 or .001 to reduce the risk that a false report is published and becomes part of the scientific literature. (For more information on the origins of the .05 level of significance, see the excellent short article by Cowles and Davis, 1982.)

At this point, it may appear that the best strategy for selecting an alpha level is to choose the smallest possible value to minimize the risk of a Type I error. However, there is a different kind of risk that develops as the alpha level is lowered. Specifically, a lower alpha level means less risk of a Type I error, but it also means that the hypothesis test demands more evidence from the research results.

The trade-off between the risk of a Type I error and the demands of the test is controlled by the boundaries of the critical region. For the hypothesis test to conclude that the

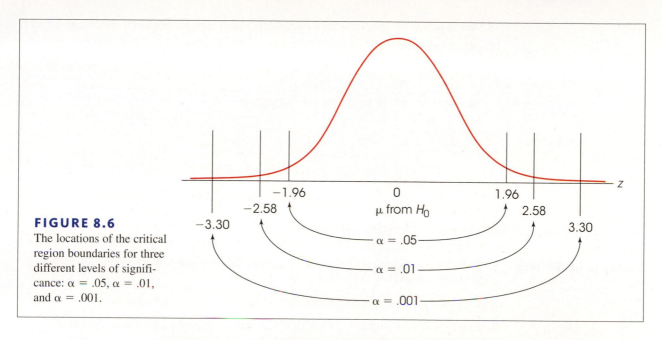

FIGURE 8.6
The locations of the critical region boundaries for three different levels of significance: $\alpha = .05$, $\alpha = .01$, and $\alpha = .001$.

treatment does have an effect, the sample data must be in the critical region. If the treatment really has an effect, it should cause the sample to be different from the original population; essentially, the treatment should push the sample into the critical region. However, as the alpha level is lowered, the boundaries for the critical region move farther out and become more difficult to reach. Figure 8.6 shows how the boundaries for the critical region move farther into the tails as the alpha level decreases. Notice that $z = 0$, in the center of the distribution, corresponds to the value of μ specified in the null hypothesis. The boundaries for the critical region determine how much distance between the sample mean and μ is needed to reject the null hypothesis. As the alpha level gets smaller, this distance gets larger.

Thus, an extremely small alpha level, such as .000001 (one in a million), would mean almost no risk of a Type I error but would push the critical region so far out that it would become essentially impossible to ever reject the null hypothesis; that is, it would require an enormous treatment effect before the sample data would reach the critical boundaries.

In general, researchers try to maintain a balance between the risk of a Type I error and the demands of the hypothesis test. Alpha levels of .05, .01, and .001 are considered reasonably good values because they provide a low risk of error without placing excessive demands on the research results.

LEARNING CHECK

1. When does a researcher risk a Type I error?
   a. anytime H_0 is rejected
   b. anytime H_1 is rejected
   c. anytime the decision is "fail to reject H_0"
   d. All of the other options are correct.

2. Which of the following defines a Type II error?
   a. rejecting a false null hypothesis
   b. rejecting a true null hypothesis
   c. failing to reject a false null hypothesis
   d. failing to reject a true null hypothesis

3. What is a likely outcome for a hypothesis test if a treatment has a very small effect?

    a. a Type I error

    b. a Type II error

    c. correctly reject the null hypothesis

    d. correctly fail to reject the null hypothesis

<u>**ANSWERS**</u>   **1.** A, **2.** C, **3.** B

8.3 | More about Hypothesis Tests

LEARNING OBJECTIVES

  6. Explain how the results of a hypothesis test with a z-score test statistic are reported in the literature.

  7. Describe the assumptions underlying a hypothesis test with a z-score test statistic.

  8. Explain how the outcome of a hypothesis test is influenced by the sample size, standard deviation, and difference between the sample mean and the hypothesized population mean.

■ A Summary of the Hypothesis Test

In Example 8.1 we presented a complete example of a hypothesis test evaluating the effect of waitresses wearing red on male customers tipping behavior. The 4-step process for that hypothesis test is summarized as follows.

STEP 1   **State the hypotheses and select an alpha level.**   For this example, the general population of men, left an average tip of $\mu = 15.8$ percent with $\sigma = 2.4$ percentage points when the waitress wore a white shirt. Therefore, the hypotheses are:

$$H_0: \mu_{\text{with red shirt}} = 15.8 \quad \text{(the red shirt has no effect)}$$
$$H_1: \mu_{\text{with red shirt}} \neq 15.8 \quad \text{(the red shirt does have an effect)}$$

We set $\alpha = .05$.

STEP 2   **Locate the critical region.**   For a normal distribution with $\alpha = .05$, the critical region consists of sample means that produce z-scores in the extreme tails of the distribution beyond $z = \pm 1.96$.

STEP 3   **Compute the test statistic (the z-score).**   We obtained an average tip of $M = 16.7$ percent for a sample of $n = 36$. With a standard error of $\sigma_M = 0.4$, we obtain $z = 2.25$.

STEP 4   **Make a decision.**   The z-score is in the critical region, which means that this sample mean is very unlikely if the null hypothesis is true. Therefore, we reject the null hypothesis and conclude that wearing a red shirt did have an effect on the men's tipping.

The following section explains how the results from this hypothesis test would be presented in a research report.

IN THE LITERATURE

Reporting the Results of the Statistical Test

A special jargon and notational system are used in published reports of hypothesis tests. When you are reading a scientific journal, for example, you typically will not be told explicitly that the researcher evaluated the data using a *z*-score as a test statistic with an alpha level of .05. Nor will you be told "the null hypothesis is rejected." Instead, you will see a statement such as

> Wearing a red shirt had a significant effect on the size of the tips left by male customers, $z = 2.25$, $p < .05$.

Let us examine this statement piece by piece. First, what is meant by the word *significant*? In statistical tests, a *significant* result means that the null hypothesis has been rejected, which means that the result is very unlikely to have occurred merely by chance. For this example, the null hypothesis stated that the red shirt has no effect, however the data clearly indicated that wearing a red shirt did have an effect. Specifically, it is very unlikely that the data would have been obtained if the red shirt did not have an effect.

DEFINITION

> A result is said to be **significant** or **statistically significant** if it is very unlikely to occur when the null hypothesis is true. That is, the result is sufficient to reject the null hypothesis. Thus, a treatment has a significant effect if the decision from the hypothesis test is to reject H_0.

Next, what is the meaning of $z = 2.25$? The *z* indicates that a *z*-score was used as the test statistic to evaluate the sample data and that its value is 2.25. Finally, what is meant by $p < .05$? This part of the statement is a conventional way of specifying the alpha level that was used for the hypothesis test. It also acknowledges the possibility (and the probability) of a Type I error. Specifically, the researcher is reporting that the treatment had an effect but admits that this could be a false report. That is, it is possible that the sample mean was in the critical region even though the red shirt had no effect. However, the probability (p) of obtaining a sample mean in the critical region is extremely small (less than .05) if there is no treatment effect.

In circumstances in which the statistical decision is to *fail to reject H_0*, the report might state that.

> There was no evidence that the red shirt had an effect on the size of the tips left by male customers, $z = 0.75$, $p > .05$.

The APA style does not use a leading zero in a probability value that refers to a level of significance.

In that case, we would be saying that the obtained result, $z = 0.75$, is not unusual (not in the critical region) and that it has a relatively high probability of occurring (greater than .05) even if the null hypothesis is true and there is no treatment effect.

Sometimes students become confused trying to differentiate between $p < .05$ and $p > .05$. Remember that you reject the null hypothesis with extreme, low-probability values, located in the critical region in the tails of the distribution. Thus, a significant result that rejects the null hypothesis corresponds to $p < .05$ (Figure 8.7).

When a hypothesis test is conducted using a computer program, the printout often includes not only a *z*-score value but also an exact value for *p*, the probability that the result occurred without any treatment effect. In this case, researchers are encouraged to report the exact *p* value instead of using the less-than or greater-than notation. For example, a research

FIGURE 8.7

Sample means that fall in the critical region (shaded area) have a probability *less than* alpha ($p < \alpha$). In this case, the null hypothesis should be rejected. Sample means that do not fall in the critical region have a probability *greater than* alpha ($p > \alpha$).

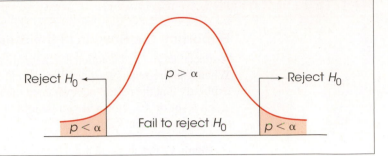

report might state that the treatment effect was significant, with $z = 2.25$, $p = .0244$. When using exact values for p, however, you must still satisfy the traditional criterion for significance; specifically, the p value must be smaller than .05 to be considered statistically significant. Remember: The p value is the probability that the result would occur if H_0 were true (without any treatment effect), which is also the probability of a Type I error. It is essential that this probability be very small.

■ Factors That Influence a Hypothesis Test

The final decision in a hypothesis test is determined by the value obtained for the z-score statistic. If the z-score is large enough to be in the critical region, we reject the null hypothesis and conclude that there is a significant treatment effect. Otherwise, we fail to reject H_0 and conclude that the treatment does not have a significant effect. The most obvious factor influencing the size of the z-score is the difference between the sample mean and the hypothesized population mean from H_0. A big mean difference indicates that the treated sample is noticeably different from the untreated population and usually supports a conclusion that the treatment effect is significant. In addition to the mean difference, however, the size of the z-score is also influenced by the standard error, which is determined by the variability of the scores (standard deviation or the variance) and the number of scores in the sample (n).

$$\sigma_M = \frac{\sigma}{\sqrt{n}}$$

Therefore, these two factors also help determine whether the z-score will be large enough to reject H_0. In this section we examine how the variability of the scores and the size of the sample can influence the outcome of a hypothesis test.

We will use the research study from Example 8.1 to examine each of these factors. The study used a sample of $n = 36$ male customers and concluded that wearing the color red has a significant effect on the tips that waitresses receive, $z = 2.25$, $p < .05$.

The Variability of the Scores   In Chapter 4 (page 124), we noted that high variability can make it very difficult to see any clear patterns in the results from a research study. In a hypothesis test, higher variability can reduce the chances of finding a significant treatment effect. For the study in Example 8.1, the standard deviation is $\sigma = 2.4$. With a sample of $n = 36$, this produced a standard error of $\sigma_M = 0.4$ points and a significant z-score of $z = 2.25$. Now consider what happens if the standard deviation is increased to $\sigma = 5.4$. With the increased variability, the standard error becomes $\sigma_M = \frac{5.4}{\sqrt{36}} = 0.9$ points. Using the same 0.9-point mean difference from the original example the new z-score becomes

$$z = \frac{M - \mu}{\sigma_M} = \frac{16.7 - 15.8}{0.9} = \frac{0.9}{0.9} = 1.00$$

The z-score is no longer beyond the critical boundary of 1.96, so the statistical decision is to fail to reject the null hypothesis. The increased variability means that the sample data are no longer sufficient to conclude that the treatment has a significant effect. In general, increasing the variability of the scores produces a larger standard error and a smaller value (closer to zero) for the z-score. If other factors are held constant, the larger the variability, the lower the likelihood of finding a significant treatment effect.

The Number of Scores in the Sample   The second factor that influences the outcome of a hypothesis test is the number of scores in the sample. The study in Example 8.1 used a sample of $n = 36$ male customers obtained a standard error of $\sigma_M = \frac{2.4}{\sqrt{36}} = 0.4$ points and a significant z-score of $z = 2.25$. Now consider what happens if we increase the sample size to $n = 144$ customers. With $n = 144$, the standard error becomes $\sigma_M = \frac{2.4}{\sqrt{144}} = 0.2$ points, and the z-score becomes

$$z = \frac{M - \mu}{\sigma_M} = \frac{16.7 - 15.8}{0.2} = \frac{0.9}{0.2} = 4.50$$

Increasing the sample size from $n = 36$ to $n = 144$ has doubled the size of the z-score. In general, increasing the number of scores in the sample produces a smaller standard error and a larger value for the z-score. If all other factors are held constant, the larger the sample size is, the greater the likelihood of finding a significant treatment effect. In simple terms, finding a 0.9-point treatment effect with large sample is more convincing than finding a 0.9-point effect with a small sample.

■ Assumptions for Hypothesis Tests with z-Scores

The mathematics used for a hypothesis test are based on a set of assumptions. When these assumptions are satisfied, you can be confident that the test produces a justified conclusion. However, if the assumptions are not satisfied, the hypothesis test may be compromised. In practice, researchers are not overly concerned with the assumptions underlying a hypothesis test because the tests usually work well even when the assumptions are violated. However, you should be aware of the fundamental conditions that are associated with each type of statistical test to ensure that the test is being used appropriately. The assumptions for hypothesis tests with z-scores are summarized as follows.

Random Sampling   It is assumed that the participants used in the study were selected randomly. Remember, we wish to generalize our findings from the sample to the population. Therefore, the sample must be representative of the population from which it has been drawn. Random sampling helps to ensure that it is representative.

Independent Observations   The values in the sample must consist of *independent* observations. In everyday terms, two observations are independent if there is no consistent, predictable relationship between the first observation and the second. More precisely, two events (or observations) are independent if the occurrence of the first event has no effect on the probability of the second event. Specific examples of independence and non-independence are examined in Box 8.1. Usually, this assumption is satisfied by using a *random* sample, which also helps ensure that the sample is representative of the population and that the results can be generalized to the population.

The Value of σ Is Unchanged by the Treatment   A critical part of the z-score formula in a hypothesis test is the standard error, σ_M. To compute the value for the standard error, we must know the sample size (n) and the population standard deviation (σ). In a hypothesis test, however, the sample comes from an *unknown* population (see Figure 8.3). If the population is really unknown, it would suggest that we do not know the standard deviation and, therefore, we cannot calculate the standard error. To solve this dilemma, we have made an assumption. Specifically, we assume that the standard deviation for the unknown population (after treatment) is the same as it was for the population before treatment.

Actually, this assumption is the consequence of a more general assumption that is part of many statistical procedures. This general assumption states that the effect of the treatment is to add a constant amount to (or subtract a constant amount from) every score in the population. You should recall that adding (or subtracting) a constant changes the mean but has no effect on the standard deviation. You also should note that this assumption is a theoretical ideal. In actual experiments, a treatment generally does not show a perfect and consistent additive effect.

Normal Sampling Distribution   To evaluate hypotheses with z-scores, we have used the unit normal table to identify the critical region. This table can be used only if the distribution of sample means is normal.

BOX 8.1 Independent Observations

Independent observations are a basic requirement for nearly all hypothesis tests. The critical concern is that each observation or measurement is not influenced by any other observation or measurement. An example of independent observations is the set of outcomes obtained in a series of coin tosses. Assuming that the coin is balanced, each toss has a 50–50 chance of coming up either heads or tails. More important, each toss is *independent* of the tosses that came before. On the fifth toss, for example, there is a 50% chance of heads no matter what happened on the previous four tosses; the coin does not remember what happened earlier and is not influenced by the past. (*Note:* Many people fail to believe in the independence of events. For example, after a series of four tails in a row, it is tempting to think that the probability of heads must increase because the coin is overdue to come up heads. This is a mistake, called the "gambler's fallacy." Remember that the coin does not know what happened on the preceding tosses and cannot be influenced by previous outcomes.)

In most research situations, the requirement for independent observations is typically satisfied by using a random sample of separate, unrelated individuals. Thus, the measurement obtained for each individual is not influenced by other participants in the study. The following two situations demonstrate circumstances in which the observations are *not* independent.

1. A researcher is interested in examining television preferences for children. To obtain a sample of $n = 20$ children, the researcher selects 4 children from family A, 3 children from family B, 5 children from family C, 2 children from family D, and 6 children from family E.

   It should be obvious that the researcher does *not* have 20 independent observations. Within each family, the children probably share television preference (at least, they watch the same shows). Thus, the response, for each child is likely to be related to the responses of his or her siblings.

2. The principle of independent observations is violated if the sample is obtained using *sampling without replacement*. For example, if you are selecting from a group of 20 potential participants, each individual has a 1 in 20 chance of being selected first. After the first person is selected, however, there are only 19 people remaining and the probability of being selected changes to 1 in 19. Because the probability of the second selection depends on the first, the two selections are not independent.

LEARNING CHECK

1. A research report summarizes the results of the hypothesis test by stating, "$z = 2.13$, $p < .05$." Which of the following is a correct interpretation of this report?

   a. The null hypothesis was rejected and the probability of a Type I error is less than .05.

   b. The null hypothesis was rejected and the probability of a Type II error is less than .05.

   c. The null hypothesis was not rejected and the probability of a Type I error is less than .05.

   d. The null hypothesis was not rejected and the probability of a Type II error is less than .05.

2. Which of the following is not an assumption for the z-score hypothesis test?

   a. The observations must be independent.

   b. The value for μ must be the same after treatment as it was before treatment.

   c. The value for σ must be the same after treatment as it was before treatment.

   d. The sample must be obtained by random sampling.

3. A sample of $n = 16$ individuals is selected from a population with $\mu = 80$ and $\sigma = 5$, and a treatment is administered to the sample. If the treatment really does have an effect, then what would be the effect of increasing the standard deviation to $\sigma = 25$?

   a. Increase the chances that the sample will produce an extreme z-score and increase the likelihood that you will conclude that a treatment effect exists.

   b. Increase the chances that the sample will produce an extreme z-score and increase the likelihood that you will conclude that a treatment effect does not exist.

   c. Increase the chances that the sample will produce a z-score near zero and increase the likelihood that you will conclude that a treatment effect exists.

   d. Increase the chances that the sample will produce a z-score near zero and increase the likelihood that you will conclude that a treatment effect does not exist.

ANSWERS   1. A, 2. B, 3. D

8.4 | Directional (One-Tailed) Hypothesis Tests

LEARNING OBJECTIVE   9. Describe the hypotheses and the critical region for a directional (one-tailed) hypothesis test.

The hypothesis-testing procedure presented in Sections 8.2 and 8.3 is the standard, or *two-tailed*, test format. The term *two-tailed* comes from the fact that the critical region is divided between the two tails of the distribution. This format is by far the most widely accepted procedure for hypothesis testing. Nonetheless, there is an alternative that is discussed in this section.

Usually a researcher begins an experiment with a specific prediction about the direction of the treatment effect. For example, a special training program is expected to *increase* student performance, or alcohol consumption is expected to *slow* reaction times. In these situations, it is possible to state the statistical hypotheses in a manner that incorporates the directional prediction into the statement of H_0 and H_1. The result is a directional test, or what commonly is called a *one-tailed test*.

<table>
<tr><td>DEFINITION</td><td>In a directional hypothesis test, or a one-tailed test, the statistical hypotheses (H_0 and H_1) specify either an increase or a decrease in the population mean. That is, they make a statement about the direction of the effect.</td></tr>
</table>

The following example demonstrates the elements of a one-tailed hypothesis test.

EXAMPLE 8.3   Earlier, in Example 8.1, we discussed a research study that examined the effect of waitresses wearing red on the tips given by male customers. In the study, each participant in a sample of $n = 36$ was served by a waitress wearing a red shirt and the size of the tip was recorded. For the general population of male customers (with waitresses wearing a white shirt), the distribution of tips was roughly normal with a mean of $\mu = 15.8$ percent and a standard deviation of $\sigma = 2.4$ percentage points. For this example, the expected effect is that the color red will increase tips. If the researcher obtains a sample mean of $M = 16.5$ percent for the $n = 36$ participants, is the result sufficient to conclude that the red shirt really increases tips?   ■

■ The Hypotheses for a Directional Test

Because a specific direction is expected for the treatment effect, it is possible for the researcher to perform a directional test. The first step (and the most critical step) is to state the statistical hypotheses. Remember that the null hypothesis states that there is no treatment effect and that the alternative hypothesis says that there is an effect. For this example, the predicted effect is that the red shirt will increase tips. Thus, the two hypotheses would state:

> H_0: Tips are not increased. (The treatment does not work.)
>
> H_1: Tips are increased. (The treatment works as predicted.)

To express directional hypotheses in symbols, it usually is easier to begin with the alternative hypothesis (H_1). Again, we know that the general population has an average of $\mu = 15.8$, and H_1 states that this value will be increased with the red shirt. Therefore, expressed in symbols, H_1 states,

> H_1: $\mu > 15.8$ (With the red shirt, the average tip is greater than 15.8 percent.)

The null hypothesis states that the red shirt does not increase tips. In symbols,

> H_0: $\mu \leq 15.8$ (With the red shirt, the average tip is not greater than 15.8.)

Note again that the two hypotheses are mutually exclusive and cover all of the possibilities. Also note that the two hypotheses concern the general population of male customers, not just the 36 men in the study. We are asking what would happen if all male customers were served by a waitress wearing a red shirt.

■ The Critical Region for Directional Tests

The critical region is defined by sample outcomes that are very unlikely to occur if the null hypothesis is true (that is, if the treatment has no effect). When the critical region was first defined (page 232), we noted that the critical region can also be defined in terms of sample values that provide *convincing evidence* that the treatment really does have an effect. For a directional test, the concept of "convincing evidence" is the simplest way to determine the location of the critical region. We begin with all the possible sample means that could be obtained if the null hypothesis is true. This is the distribution of sample means and it will be normal (because the population of test scores is normal), have an expected value of $\mu = 15.8$ (from H_0), and, for a sample of $n = 36$, will have a standard error of $\sigma_M = \frac{2.4}{\sqrt{36}} = 0.4$ points. The distribution is shown in Figure 8.8.

> **If the prediction is that the treatment will produce a *decrease* in scores, the critical region is located entirely in the left-hand tail of the distribution.**

For this example, the treatment is expected to increase test scores. If the regular population of male customers has an average tip of $\mu = 15.8$ percent, then a sample mean that is substantially more than 15.8 would provide convincing evidence that the red shirt worked. Thus, the critical region is located entirely in the right-hand tail of the distribution corresponding to sample means much greater than $\mu = 15.8$ (Figure 8.8). Because the critical region is contained in one tail of the distribution, a directional test is commonly called a *one-tailed* test. Also note that the proportion specified by the alpha level is not divided between two tails, but rather is contained entirely in one tail. Using $\alpha = .05$ for example, the whole 5% is located in one tail. In this case, the z-score boundary for the critical region is $z = 1.65$, which is obtained by looking up a proportion of .05 in column C (the tail) of the unit normal table.

Notice that a directional (one-tailed) test requires two changes in the step-by-step hypothesis-testing procedure.

1. In the first step of the hypothesis test, the directional prediction is incorporated into the statement of the hypotheses.

2. In the second step of the process, the critical region is located entirely in one tail of the distribution.

After these two changes, a one-tailed test continues exactly the same as a regular two-tailed test. Specifically, you calculate the z-score statistic and then make a decision about H_0 depending on whether the z-score is in the critical region.

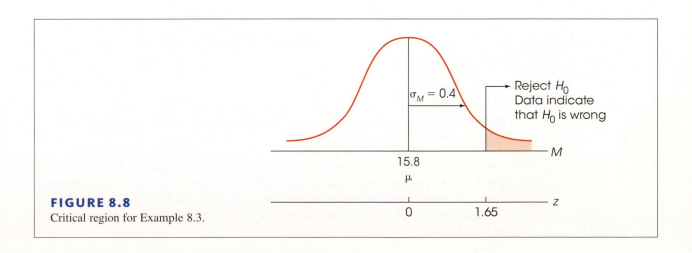

FIGURE 8.8
Critical region for Example 8.3.

For this example, the researcher obtained a mean of $M = 16.5$ percent for the 36 participants who were served by a waitress in a red shirt. This sample mean corresponds to a z-score of

$$z = \frac{M - \mu}{\sigma_M} = \frac{16.5 - 15.8}{0.4} = \frac{0.7}{0.4} = 1.75$$

A z-score of $z = 1.75$ is in the critical region for a one-tailed test (see Figure 8.8). This is a very unlikely outcome if H_0 is true. Therefore, we reject the null hypothesis and conclude that the red shirt produces a significant increase in tips from male customers. In the literature, this result would be reported as follows:

Wearing a red shirt produced a significant increase in tips, $z = 1.75$, $p < .05$, one tailed.

Note that the report clearly acknowledges that a one-tailed test was used.

■ Comparison of One-Tailed vs. Two-Tailed Tests

The general goal of hypothesis testing is to determine whether a particular treatment has any effect on a population. The test is performed by selecting a sample, administering the treatment to the sample, and then comparing the result with the original population. If the treated sample is noticeably different from the original population, then we conclude that the treatment has an effect, and we reject H_0. On the other hand, if the treated sample is still similar to the original population, then we conclude that there is no convincing evidence for a treatment effect, and we fail to reject H_0. The critical factor in this decision is the *size of the difference* between the treated sample and the original population. A large difference is evidence that the treatment worked; a small difference is not sufficient to say that the treatment has any effect.

The major distinction between one-tailed and two-tailed tests is in the criteria they use for rejecting H_0. A one-tailed test allows you to reject the null hypothesis when the difference between the sample and the population is relatively small, provided the difference is in the specified direction. A two-tailed test, on the other hand, requires a relatively large difference independent of direction. This point is illustrated in the following example.

EXAMPLE 8.4

Consider again the one-tailed test in Example 8.3 evaluating the effect of waitresses wearing red on the tips from male customers. If we had used a standard two-tailed test, the hypotheses would be

H_0: $\mu = 15.8$ (The red shirt has no effect on tips.)

H_1: $\mu \neq 15.8$ (The red shirt does have an effect on tips.)

For a two-tailed test with $\alpha = .05$, the critical region consists of z-scores beyond ± 1.96. The data from Example 8.3 produced a sample mean of $M = 16.5$ percent and $z = 1.75$. For the two-tailed test, this z-score is not in the critical region, and we conclude that the red shirt does not have a significant effect. ∎

With the two-tailed test in Example 8.4, the 0.7-point difference between the sample mean and the hypothesized population mean ($M = 16.5$ and $\mu = 15.8$) is not big enough to reject the null hypothesis. However, with the one-tailed test in Example 8.3, the same 0.7-point difference is large enough to reject H_0 and conclude that the treatment had a significant effect.

All researchers agree that one-tailed tests are different from two-tailed tests. However, there are several ways to interpret the difference. One group of researchers contends that a two-tailed test is more rigorous and, therefore, more convincing than a one-tailed test. Remember that the two-tailed test demands more evidence to reject H_0 and thus provides a stronger demonstration that a treatment effect has occurred.

Other researchers feel that one-tailed tests are preferable because they are more sensitive. That is, a relatively small treatment effect may be significant with a one-tailed test but fail to reach significance with a two-tailed test. Also, there is the argument that one-tailed tests are more precise because they test hypotheses about a specific directional effect instead of an indefinite hypothesis about a general effect.

In general, two-tailed tests should be used in research situations when there is no strong directional expectation or when there are two competing predictions. For example, a two-tailed test would be appropriate for a study in which one theory predicts an increase in scores but another theory predicts a decrease. One-tailed tests should be used only in situations when the directional prediction is made before the research is conducted and there is a strong justification for making the directional prediction. In particular, if a two-tailed test fails to reach significance, you should never follow up with a one-tailed test as a second attempt to salvage a significant result for the same data.

LEARNING CHECK

1. A researcher is predicting that a treatment will *increase* scores. If this treatment is evaluated using a directional hypothesis test, then the critical region for the test _____.
   a. would be entirely in the right-hand tail of the distribution
   b. would be entirely in the left-hand tail of the distribution
   c. would be divided equally between the two tails of the distribution
   d. cannot answer without knowing the value of the alpha level

2. A sample is selected from a population with a mean of $\mu = 45$ and a treatment is administered to the sample. If the treatment is expected to produce a *decrease* in the scores, then what would be the null hypothesis for a directional hypothesis test?
   a. $\mu \geq 45$
   b. $\mu \leq 45$
   c. $M \leq 45$
   d. $M \geq 45$

3. A researcher expects a treatment to produce an *increase* in the population mean. The treatment is evaluated using a one-tailed hypothesis test, and the test produces $z = -1.85$. Based on this result, what is the correct statistical decision?
   a. The researcher should reject the null hypothesis with $\alpha = .05$ but not with $\alpha = .01$.
   b. The researcher should reject the null hypothesis with either $\alpha = .05$ or $\alpha = .01$.
   c. The researcher should fail to reject H_0 with either $\alpha = .05$ or $\alpha = .01$.
   d. Cannot answer without additional information

ANSWERS   **1.** A, **2.** A, **3.** C

8.5 | Concerns about Hypothesis Testing: Measuring Effect Size

LEARNING OBJECTIVES

10. Explain why it is necessary to report a measure of effect size in addition to the outcome of a hypothesis test.

11. Calculate Cohen's *d* as a measure of effect size.

12. Explain how measures of effect size such as Cohen's *d* are influenced by the sample size and the standard deviation.

Although hypothesis testing is the most commonly used technique for evaluating and interpreting research data, a number of scientists have expressed a variety of concerns about the hypothesis testing procedure (for example, see Loftus, 1996; Hunter, 1997; and Killeen, 2005).

There are two serious limitations with using a hypothesis test to establish the significance of a treatment effect. The first concern is that the focus of a hypothesis test is on the data rather than the hypothesis. Specifically, when the null hypothesis is rejected, we are actually making a strong probability statement about the sample data, not about the null hypothesis. A significant result permits the following conclusion: "This specific sample mean is very unlikely ($p < .05$) if the null hypothesis is true." Note that the conclusion does not make any definite statement about the probability of the null hypothesis being true or false. The fact that the data are very unlikely *suggests* that the null hypothesis is also very unlikely, but we do not have any solid grounds for making a probability statement about the null hypothesis. Specifically, you cannot conclude that the probability of the null hypothesis being true is less than 5% simply because you rejected the null hypothesis with $\alpha = .05$.

A second concern is that demonstrating a *significant* treatment effect does not necessarily indicate a *substantial* treatment effect. In particular, statistical significance does not provide any real information about the absolute size of a treatment effect. Instead, the hypothesis test has simply established that the results obtained in the research study are very unlikely to have occurred if there is no treatment effect. The hypothesis test reaches this conclusion by (1) calculating the standard error, which measures how much difference is reasonable to expect between M and μ, and (2) demonstrating that the obtained mean difference is substantially bigger than the standard error.

Notice that the test is making a *relative* comparison: the size of the treatment effect is being evaluated relative to the standard error. If the standard error is very small, then the treatment effect can also be very small and still be large enough to be significant. Thus, a significant effect does not necessarily mean a big effect.

The idea that a hypothesis test evaluates the relative size of a treatment effect, rather than the absolute size, is illustrated in the following example.

EXAMPLE 8.5

We begin with a population of scores that forms a normal distribution with $\mu = 50$ and $\sigma = 10$. A sample is selected from the population and a treatment is administered to the sample. After treatment, the sample mean is found to be $M = 51$. Does this sample provide evidence of a statistically significant treatment effect?

Although there is only a 1-point difference between the sample mean and the original population mean, the difference may be enough to be significant. In particular, the outcome of the hypothesis test depends on the sample size. In particular, the outcome of the hypothesis test depends on the sample size.

For example, with a sample of $n = 25$ the standard error is

$$\sigma_M = \frac{\sigma}{\sqrt{n}} = \frac{10}{\sqrt{25}} = \frac{10}{5} = 2.00$$

and the z-score for $M = 51$ is

$$z = \frac{M - \mu}{\sigma_M} = \frac{51 - 50}{2} = \frac{1}{2} = 0.50$$

This z-score fails to reach the critical boundary of $z = 1.96$, so we fail to reject the null hypothesis. In this case, the 1-point difference between M and μ is not significant because it is being evaluated relative to a standard error of 2 points.

Now consider the outcome with a sample of $n = 400$. With a larger sample, the standard error is

$$\sigma_M = \frac{\sigma}{\sqrt{n}} = \frac{10}{\sqrt{400}} = \frac{10}{20} = 0.50$$

and the z-score for $M = 51$ is

$$z = \frac{M - \mu}{\sigma_M} = \frac{51 - 50}{0.5} = \frac{1}{0.5} = 2.00$$

Now the z-score is beyond the 1.96 boundary, so we reject the null hypothesis and conclude that there is a significant effect. In this case, the 1-point difference between M and μ is considered statistically significant because it is being evaluated relative to a standard error of only 0.5 points.   ■

The point of Example 8.5 is that a small treatment effect can still be statistically significant. If the sample size is large enough, any treatment effect, no matter how small, can be enough for us to reject the null hypothesis.

■ Measuring Effect Size

As noted in the previous section, one concern with hypothesis testing is that a hypothesis test does not really evaluate the absolute size of a treatment effect. To correct this problem, it is recommended that whenever researchers report a statistically significant effect, they also provide a report of the effect size (see the guidelines presented by *L*. Wilkinson and the APA Task Force on Statistical Inference, 1999). Therefore, as we present different hypothesis tests we will also present different options for measuring and reporting *effect size.* The goal is to measure and describe the absolute size of the treatment effect in a way that is not influenced by the number of scores in the sample.

DEFINITION

> A measure of **effect size** is intended to provide a measurement of the absolute magnitude of a treatment effect, independent of the size of the sample(s) being used.

One of the simplest and most direct methods for measuring effect size is *Cohen's d.* Cohen (1988) recommended that effect size can be standardized by measuring the mean difference in terms of the standard deviation. The resulting measure of effect size is computed as

$$\text{Cohen's } d = \frac{\text{mean difference}}{\text{standard deviation}} = \frac{\mu_{\text{treatment}} - \mu_{\text{no treatment}}}{\sigma} \tag{8.1}$$

For the z-score hypothesis test, the mean difference is determined by the difference between the population mean before treatment and the population mean after treatment. However, the population mean after treatment is unknown. Therefore, we must use the mean for the treated sample in its place. Remember, the sample mean is expected to be representative of the population mean and provides the best measure of the treatment effect. Thus, the actual calculations are really estimating the value of Cohen's *d* as follows:

Cohen's *d* measures the distance between two means and is typically reported as a positive number even when the formula produces a negative value.

$$\text{estimated Cohen's } d = \frac{\text{mean difference}}{\text{standard deviation}} = \frac{M_{\text{treatment}} - \mu_{\text{no treatment}}}{\sigma} \qquad (8.2)$$

The standard deviation is included in the calculation to standardize the size of the mean difference in much the same way that *z*-scores standardize locations in a distribution. For example, a 15-point mean difference can be a relatively large treatment effect or a relatively small effect depending on the size of the standard deviation. This phenomenon is demonstrated in Figure 8.9. The top portion of the figure (part a) shows the results of a treatment that produces a 15-point mean difference in SAT scores; before treatment, the average SAT score is $\mu = 500$, and after treatment the average is 515. Notice that the standard deviation for SAT scores is $\sigma = 100$, so the 15-point difference appears to be small. For this example, Cohen's *d* is

$$\text{Cohen's } d = \frac{\text{mean difference}}{\text{standard deviation}} = \frac{15}{100} = 0.15$$

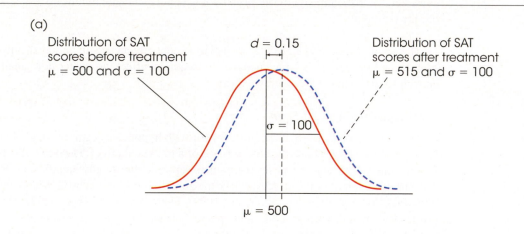

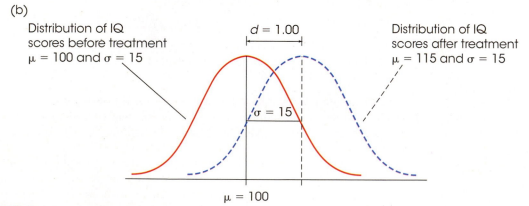

FIGURE 8.9

The appearance of a 15-point treatment effect in two different situations. In part (a), the standard deviation is $\sigma = 100$ and the 15-point effect is relatively small. In part (b), the standard deviation is $\sigma = 15$ and the 15-point effect is relatively large. Cohen's *d* uses the standard deviation to help measure effect size.

Now consider the treatment effect shown in Figure 8.9(b). This time, the treatment produces a 15-point mean difference in IQ scores; before treatment the average IQ is 100, and after treatment the average is 115. Because IQ scores have a standard deviation of $\sigma = 15$, the 15-point mean difference now appears to be large. For this example, Cohen's d is

$$\text{Cohen's } d = \frac{\text{mean difference}}{\text{standard deviation}} = \frac{15}{15} = 1.00$$

Notice that Cohen's d measures the size of the treatment effect in terms of the standard deviation. For example, a value of $d = 0.50$ indicates that the treatment changed the mean by half of a standard deviation; similarly, a value of $d = 1.00$ indicates that the size of the treatment effect is equal to one whole standard deviation. (Box 8.2.)

BOX 8.2 Overlapping Distributions

Figure 8.9(b) shows the results of a treatment with a Cohen's d of 1.00; that is, the effect of the treatment is to increase the mean by one full standard deviation. According to the guidelines in Table 8.2, a value of $d = 1.00$ is considered a large treatment effect. However, looking at the figure, you may get the impression that there really isn't that much difference between the distribution before treatment and the distribution after treatment. In particular, there is substantial overlap between the two distributions, so that many of the individuals who receive the treatment are not any different from the individuals who do not receive the treatment.

The overlap between distributions is a basic fact of life in most research situations; it is extremely rare for the scores after treatment to be *completely different* (no overlap) from the scores before treatment. Consider, for example, children's heights at different ages. Everyone knows that 8-year-old children are taller than 6-year-old children; on average, the difference is 3 or 4 inches. However, this does not mean that all 8-year-old children are taller than all 6-year-old children. In fact, there is considerable overlap between the two distributions, so that the tallest among the 6-year-old children are actually taller than most 8-year-old children. In fact, the height distributions for the two age groups would look a lot like the two distributions in Figure 8.9(b). Although there is a clear *mean difference* between the two distributions, there still can be substantial overlap.

Cohen's d measures the degree of separation between two distributions, and a separation of one standard deviation ($d = 1.00$) represents a large difference. Eight-year-old children really are bigger than 6-year-old children.

Cohen (1988) also suggested criteria for evaluating the size of a treatment effect as shown in Table 8.2.

TABLE 8.2
Evaluating effect size with Cohen's d.

Magnitude of d	Evaluation of Effect Size
$d = 0.2$	Small effect (mean difference around 0.2 standard deviation)
$d = 0.5$	Medium effect (mean difference around 0.5 standard deviation)
$d = 0.8$	Large effect (mean difference around 0.8 standard deviation)

As one final demonstration of Cohen's d, consider the two hypothesis tests in Example 8.5. For each test, the original population had a mean of $\mu = 50$ with a standard deviation of $\sigma = 10$. For each test, the mean for the treated sample was $M = 51$. Although one test used a sample of $n = 25$ and the other test used a sample of $n = 400$, the sample size is not considered when computing Cohen's d. Therefore, both of the hypothesis tests would produce the same value:

$$\text{Cohen's } d = \frac{\text{mean difference}}{\text{standard deviation}} = \frac{1}{10} = 0.10$$

Notice that Cohen's *d* simply describes the size of the treatment effect and is not influenced by the number of scores in the sample. For both hypothesis tests, the original population mean was $\mu = 50$ and, after treatment, the sample mean was $M = 51$. Thus, treatment appears to have increased the scores by 1 point, which is equal to one-tenth of a standard deviation (Cohen's $d = 0.1$).

LEARNING CHECK

1. Which of the following is most likely to reject the null hypothesis even if the treatment effect is very small?
   a. A small sample and a small value for σ.
   b. A small sample and a large value for σ.
   c. A large sample and a small value for σ.
   d. A large sample and a large value for σ.

2. A treatment is administered to a sample selected from a population with a mean of $\mu = 80$ and a standard deviation of $\sigma = 10$. After treatment, the sample mean is $M = 85$. Based on this information, the effect size as measured by Cohen's *d* is _____.
   a. $d = 5.00$
   b. $d = 2.00$
   c. $d = 0.50$
   d. impossible to calculate without more information

3. If other factors are held constant, then how does the size of the standard deviation affect the likelihood of rejecting the null hypothesis and the value for Cohen's *d*?
   a. A larger standard deviation increases the likelihood of rejecting the null hypothesis and increases the value of Cohen's *d*.
   b. A larger standard deviation increases the likelihood of rejecting the null hypothesis but decreases the value of Cohen's *d*.
   c. A larger standard deviation decreases the likelihood of rejecting the null hypothesis but increases the value of Cohen's *d*.
   d. A larger standard deviation decreases the likelihood of rejecting the null hypothesis and decreases the value of Cohen's *d*.

ANSWERS   1. C, 2. C, 3. D

8.6 | Statistical Power

LEARNING OBJECTIVES

13. Define the power of a hypothesis test and explain how power is related to the probability of a Type II error.

14. Identify the factors that influence power and explain how power is affected by each.

Instead of measuring effect size directly, an alternative approach to determining the size or strength of a treatment effect is to measure the power of the statistical test. The *power* of a test is defined as the probability that the test will reject the null hypothesis if the treatment really has an effect.

DEFINITION

> The **power** of a statistical test is the probability that the test will correctly reject a false null hypothesis. That is, power is the probability that the test will identify a treatment effect if one really exists.

Whenever a treatment has an effect, there are only two possible outcomes for a hypothesis test: either fail to reject H_0 or reject H_0. Because there are only two possible outcomes, the probability for the first and the probability for the second must add to 1.00. The first outcome, failing to reject H_0 when there is a real effect, was defined earlier (page 237) as a Type II error with a probability identified as $p = \beta$. Therefore, the second outcome must have a probability of $1 - \beta$. However, the second outcome, rejecting H_0 when there is a real effect, is the power of the test. Thus, the power of a hypothesis test is equal to $1 - \beta$. In the examples that follow, we demonstrate the calculation of power for a hypothesis test; that is, the probability that the test will correctly reject the null hypothesis. At the same time, however, we are computing the probability that the test will result in a Type II error. For example, if the power of the test is 70% $(1 - \beta)$ then the probability of a Type II error must be 30% (β).

Calculating Power   Researchers typically calculate power as a means of determining whether a research study is likely to be successful. Thus, researchers usually calculate the power of a hypothesis test *before* they actually conduct the research study. In this way, they can determine the probability that the results will be significant (reject H_0) before investing time and effort in the actual research. To calculate power, however, it is first necessary to make assumptions about a variety of factors that influence the outcome of a hypothesis test. Factors such as the sample size, the size of the treatment effect, and the value chosen for the alpha level can all influence a hypothesis test. The following example demonstrates the calculation of power for a specific research situation.

EXAMPLE 8.6

We start with a normal shaped population with a mean of $\mu = 80$ and a standard deviation of $\sigma = 10$. A researcher plans to select a sample of $n = 25$ individuals from this population and administer a treatment to each individual. It is expected that the treatment will have an 8-point effect; that is, the treatment will add 8 points to each individual's score.

The top half of Figure 8.10 shows the original population and two possible outcomes:

1. If the null hypothesis is true and there is no treatment effect.

2. If the researcher's expectation is correct and there is an 8-point effect.

The left-hand side of the figure shows what should happen according to the null hypothesis. In this case, the treatment has no effect and the population mean is still $\mu = 80$. On the right-hand side of the figure we show what would happen if the treatment has an 8-point effect. If the treatment adds 8 points to each person's score, the population mean after treatment will increase to $\mu = 88$.

The bottom half of Figure 8.10 shows the distribution of sample means for $n = 25$ for each of the two outcomes. According to the null hypothesis, the sample means are centered at $\mu = 80$ (bottom left). With an 8-point treatment effect, the sample means are centered at $\mu = 88$ (bottom right). Both distributions have a standard error of

$$\sigma_M = \frac{\sigma}{\sqrt{n}} = \frac{10}{\sqrt{25}} = \frac{10}{5} = 2$$

Notice that the distribution on the left shows all of the possible sample means if the null hypothesis is true. This is the distribution we use to locate the critical region for the hypothesis

FIGURE 8.10

A demonstration of measuring power for a hypothesis test. The left-hand side shows the distribution of sample means that would occur if the null hypothesis were true. The critical region is defined for this distribution. The right-hand side shows the distribution of sample means that would be obtained if there were an 8-point treatment effect. Notice that if there is an 8-point effect, then essentially all of the sample means would be in the critical region. Thus, the probability of rejecting the null hypothesis (the power of the test) is nearly 100% for an 8-point treatment effect.

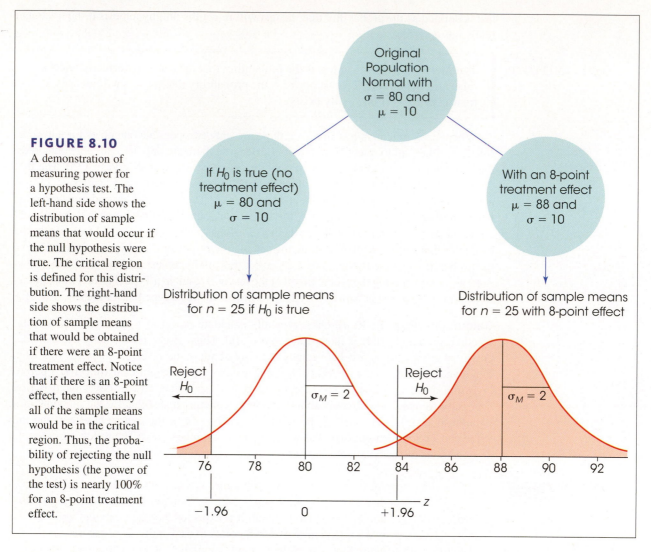

test. Using $\alpha = .05$, the critical region consists of extreme values in this distribution, specifically sample means beyond $z = 1.96$ or $z = -1.96$. These values are shown in Figure 8.10 and, for both distributions, we have shaded all the sample means located in the critical region.

Now turn your attention to the distribution on the right, which shows all of the possible sample means if there is an 8-point treatment effect. Notice that most of these sample means are located beyond the $z = 1.96$ boundary. This means that, if there is an 8-point treatment effect, you are almost guaranteed to obtain a sample mean in the critical region and reject the null hypothesis. Thus, the power of the test (the probability of rejecting H_0) is close to 100% if there is an 8-point treatment effect.

To calculate the exact value for the power of the test we must determine what portion of the distribution on the right-hand side is shaded. Thus, we must locate the exact boundary for the critical region, then find the probability value in the unit normal table. For the distribution on the left-hand side, the critical boundary of $z = +1.96$ corresponds to a location that is above the mean by 1.96 standard deviations. This distance is equal to

$$1.96\sigma_M = 1.96(2) = 3.92 \text{ points}$$

Thus, the critical boundary of $z = +1.96$ corresponds to a sample mean of $M = 80 + 3.92 = 83.92$. Any sample mean greater than $M = 83.92$ is in the critical region and would lead to rejecting the null hypothesis. Next, we determine what proportion of the treated samples are greater than $M = 83.92$. For the treated distribution (right-hand side), the population mean is $\mu = 88$ and a sample mean of $M = 83.92$ corresponds to a z-score of

$$z = \frac{M - \mu}{\sigma_M} = \frac{83.92 - 88}{2} = \frac{-4.08}{2} = -2.04$$

Finally, look up $z = -2.04$ in the unit normal table and determine that the shaded area ($z > -2.04$) corresponds to $p = 0.9793$ (or 97.93%). Thus, if the treatment has an 8-point effect, 97.93% of all the possible sample means will be in the critical region and we will reject the null hypothesis. In other words, the power of the test is 97.93%. In practical terms, this means that the research study is almost guaranteed to be successful. If the researcher selects a sample of $n = 25$ individuals, and if the treatment really does have an 8-point effect, then 97.93% of the time the hypothesis test will conclude that there is a significant effect. ∎

■ Power and Effect Size

Logically, it should be clear that power and effect size are related. Figure 8.10 shows the calculation of power for an 8-point treatment effect. Now consider what would happen if the treatment effect were only 4 points. With a 4-point treatment effect, the distribution on the right-hand side would shift to the left so that it is centered at $\mu = 84$. In this new position, only about 50% of the treated sample means would be beyond the $z = 1.96$ boundary. Thus, with a 4-point treatment effect, there is only a 50% probability of selecting a sample that leads to rejecting the null hypothesis. In other words, the power of the test is only about 50% for a 4-point effect compared to nearly 98% with an 8-point effect (Example 8.6). Again, it is possible to find the z-score corresponding to the exact location of the critical boundary and to look up the probability value for power in the unit normal table. For a 4-point treatment effect, you should that the critical boundary corresponds to $z = -0.04$ and the exact power of the test is $p = 0.5160$, or 51.60%

In general, as the effect size increases, the distribution of sample means on the right-hand side moves even farther to the right so that more and more of the samples are beyond the $z = 1.96$ boundary. Thus, as the effect size increases, the probability of rejecting H_0 also increases, which means that the power of the test increases. Therefore, measures of effect size such as Cohen's d and measures of power both provide an indication of the strength or magnitude of a treatment effect.

■ Other Factors that Affect Power

Although the power of a hypothesis test is directly influenced by the size of the treatment effect, power is not meant to be a pure measure of effect size. Instead, power is influenced by several factors, other than effect size, that are related to the hypothesis test. Some of these factors are considered in the following section.

Sample Size One factor that has a huge influence on power is the size of the sample. In Example 8.6 we demonstrated power for an 8-point treatment effect using a sample of $n = 25$. If the researcher decided to conduct the study using a sample of $n = 4$, then the power would be dramatically different. With $n = 4$, the standard error for the sample means would be

$$\sigma_M = \frac{\sigma}{\sqrt{n}} = \frac{10}{\sqrt{4}} = \frac{10}{2} = 5$$

Figure 8.11 shows the two distributions of sample means with $n = 4$ and a standard error of $\sigma_M = 5$ points. Again, the distribution on the left is centered at $\mu = 80$ and shows the set

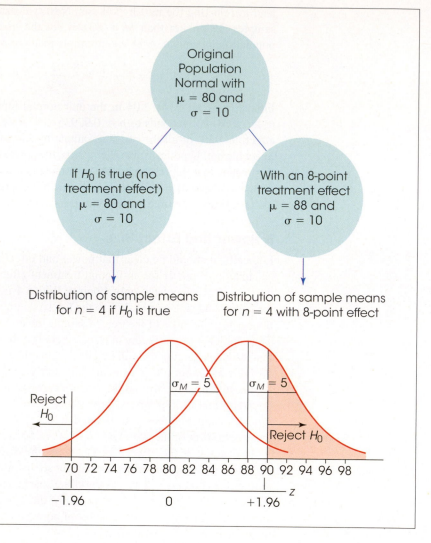

FIGURE 8.11

A demonstration of how sample size affects the power of a hypothesis test. The left-hand side shows the distribution of sample means if the null hypothesis were true. The critical region is defined in this distribution. The right-hand side shows the distribution of sample means if there were an 8-point treatment effect. Notice that reducing the sample size to $n = 4$ has reduced the power of the test to less than 50% compared to a power of nearly 100% with a sample of $n = 25$ in Figure 8.10.

of all possible sample means if H_0 is true. As always, this distribution is used to locate the critical boundaries for the hypothesis test, $z = -1.96$ and $z = +1.96$. The distribution on the right is centered at $\mu = 88$ and shows all possible sample means if there is an 8-point treatment effect. Note that less than half of the treated sample means in the right-hand distribution are now located beyond the 1.96 boundary. Thus, with a sample of $n = 4$, there is less than a 50% probability that the hypothesis test would reject H_0, even though the treatment has an 8-point effect. Earlier, in Example 8.6, we found power equal to 97.93% for a sample of $n = 25$. However, when the sample size is reduced to $n = 4$, power decreases to less than 50%. In general, a larger sample produces greater power for a hypothesis test.

The following example is an opportunity to test your understanding of statistical power.

EXAMPLE 8.7 Find the exact value of the power for the hypothesis test shown in Figure 8.11.

You should find that the critical boundary corresponds to a sample mean of $M = 89.8$ in the treatment distribution and power is $p = 0.3594$ or 35.94%. ∎

Because power is directly related to sample size, one of the primary reasons for computing power is to determine what sample size is necessary to achieve a reasonable probability for a successful research study. Before a study is conducted, researchers can compute power to determine the probability that their research will successfully reject the null hypothesis. If the probability (power) is too small, they always have the option of increasing sample size to increase power.

Alpha Level Reducing the alpha level for a hypothesis test also reduces the power of the test. For example, lowering α from .05 to .01 lowers the power of the hypothesis test. The effect of reducing the alpha level can be seen by looking at Figure 8.11. In this figure, the boundaries for the critical region are drawn using $\alpha = .05$. Specifically, the critical region on the right-hand side begins at $z = 1.96$. If a were changed to .01, the boundary would be moved farther to the right, out to $z = 2.58$. It should be clear that moving the critical boundary to the right means that a smaller portion of the treatment distribution (the distribution on the right-hand side) will be in the critical region. Thus, there would be a lower probability of rejecting the null hypothesis and a lower value for the power of the test.

One-Tailed vs. Two-Tailed Tests If the treatment effect is in the predicted direction, then changing from a regular two-tailed test to a one-tailed test increases the power of the hypothesis test. Again, this effect can be seen by referring to Figure 8.11. The figure shows the boundaries for the critical region using a two-tailed test with $\alpha = .05$ so that the critical region on the right-hand side begins at $z = 1.96$. Changing to a one-tailed test would move the critical boundary to the left to a value of $z = 1.65$. Moving the boundary to the left would cause a larger proportion of the treatment distribution to be in the critical region and, therefore, would increase the power of the test.

LEARNING CHECK

1. Which of the following defines the power of a hypothesis test?
   a. The probability of rejecting a true null hypothesis
   b. The probability of supporting true null hypothesis
   c. The probability of rejecting a false null hypothesis
   d. The probability of supporting a false null hypothesis

2. How does increasing the sample size influence the likelihood of rejecting the null hypothesis and the power of the hypothesis test?
   a. The likelihood of rejecting H_0 and the power of the test both increase.
   b. The likelihood of rejecting H_0 and the power of the test both decrease.
   c. The likelihood of rejecting H_0 increases but the power of the test is unchanged.
   d. The likelihood of rejecting H_0 decreases but the power of the test is unchanged.

3. How does an increase in the value of σ influence the likelihood of rejecting the null hypothesis and the power of the hypothesis test?
   a. The likelihood of rejecting H_0 and the power of the test both increase.
   b. The likelihood of rejecting H_0 and the power of the test both decrease.
   c. The likelihood of rejecting H_0 increases but the power of the test is unchanged.
   d. The likelihood of rejecting H_0 decreases but the power of the test is unchanged.

ANSWERS   1. C, 2. A, 3. B

SUMMARY

1. Hypothesis testing is an inferential procedure that uses the data from a sample to draw a general conclusion about a population. The procedure begins with a hypothesis about an unknown population. Then a sample is selected, and the sample data provide evidence that either supports or refutes the hypothesis.

2. In this chapter, we introduced hypothesis testing using the simple situation in which a sample mean is used to test a hypothesis about an unknown population mean. We begin with an unknown population, generally a population that has received a treatment. The question is to determine whether the treatment has an effect on the population mean (see Figure 8.2).

3. Hypothesis testing is structured as a four-step process that is used throughout the remainder of the book.
   a. State the null hypothesis (H_0), and select an alpha level. The null hypothesis states that there is no effect or no change. In this case, H_0 states that the mean for the treated population is the same as the mean before treatment. The alpha level, usually $\alpha = .05$ or $\alpha = .01$, provides a definition of the term *very unlikely* and determines the risk of a Type I error. Also state an alternative hypothesis (H_1), which is the exact opposite of the null hypothesis.
   b. Locate the critical region. The critical region is defined as sample outcomes that would be very unlikely to occur if the null hypothesis is true. The alpha level defines "very unlikely." For example, with $\alpha = .05$, the critical region is defined as sample means in the extreme 5% of the distribution of sample means. When the distribution is normal, the extreme 5% corresponds to z-scores beyond $z = \pm 1.96$.
   c. Collect the data, and compute the test statistic. The sample mean is transformed into a z-score by the formula

$$z = \frac{M - \mu}{\sigma_M}$$

   The value of μ is obtained from the null hypothesis. The z-score test statistic identifies the location of the sample mean in the distribution of sample means. Expressed in words, the z-score formula is

$$z = \frac{\text{sample mean} - \begin{array}{c}\text{hypothesized}\\\text{population mean}\end{array}}{\text{standard error}}$$

   d. Make a decision. If the obtained z-score is in the critical region, reject H_0 because it is very unlikely that these data would be obtained if H_0 were true. In this case, conclude that the treatment has changed the population mean. If the z-score is not in the critical region, fail to reject H_0 because the data are not significantly different from the null hypothesis. In this case, the data do not provide sufficient evidence to indicate that the treatment has had an effect.

4. Whatever decision is reached in a hypothesis test, there is always a risk of making the incorrect decision. There are two types of errors that can be committed.

   A Type I error is defined as rejecting a true H_0. This is a serious error because it results in falsely reporting a treatment effect. The risk of a Type I error is determined by the alpha level and therefore is under the experimenter's control.

   A Type II error is defined as the failure to reject a false H_0. In this case, the experiment fails to detect an effect that actually occurred. The probability of a Type II error cannot be specified as a single value and depends in part on the size of the treatment effect. It is identified by the symbol β (beta).

5. When a researcher expects that a treatment will change scores in a particular direction (increase or decrease), it is possible to do a directional, or one-tailed, test. The first step in this procedure is to incorporate the directional prediction into the hypotheses. For example, if the prediction is that a treatment will increase scores, the null hypothesis says that there is no increase and the alternative hypothesis states that there is an increase. To locate the critical region, you must determine what kind of data would refute the null hypothesis by demonstrating that the treatment worked as predicted. These outcomes will be located entirely in one tail of the distribution, so the entire critical region (5% or 1% depending on α) will be in one tail.

6. A one-tailed test is used when there is prior justification for making a directional prediction. These *a priori* reasons may be previous reports and findings or theoretical considerations. In the absence of the *a priori* basis, a two-tailed test is appropriate. In this situation, you might be unsure of what to expect in the study, or you might be testing competing theories.

7. In addition to using a hypothesis test to evaluate the *significance* of a treatment effect, it is recommended that you also measure and report the *effect size*. One measure of effect size is Cohen's d, which is a standardized measure of the mean difference. Cohen's d is computed as

$$\text{Cohen's } d = \frac{\text{mean difference}}{\text{standard deviation}}$$

8. The power of a hypothesis test is defined as the probability that the test will correctly reject the null hypothesis.

9. To determine the power for a hypothesis test, you must first identify the treatment and null distributions. Also, you must specify the magnitude of the treatment effect. Next, you locate the critical region in the null distribution. The power of the hypothesis test is the portion of the treatment distribution that is located beyond the boundary (critical value) of the critical region.

10. As the size of the treatment effect increases, statistical power increases. Also, power is influenced by several factors that can be controlled by the experimenter:
   a. Increasing the alpha level increases power.
   b. A one-tailed test has greater power than a two-tailed test.
   c. A large sample results in more power than a small sample.

KEY TERMS

hypothesis testing (225)

null hypothesis (228)

alternative hypothesis (228)

level of significance (230)

alpha level (230)

critical region (230)

test statistic (233)

Type I error (236)

Type II error (237)

beta (238)

significant (241)

directional test (246)

one-tailed test (246)

effect size (251)

Cohen's d (251)

power (255)

SPSS®

The statistical computer package SPSS is not structured to conduct hypothesis tests using z-scores. In truth, the z-score test presented in this chapter is rarely used in actual research situations. The problem with the z-score test is that it requires that you know the value of the population standard deviation, and this information is usually not available. Researchers rarely have detailed information about the populations that they wish to study. Instead, they must obtain information entirely from samples. In the following chapters we introduce new hypothesis-testing techniques that are based entirely on sample data. These new techniques are included in SPSS.

FOCUS ON PROBLEM SOLVING

1. Hypothesis testing involves a set of logical procedures and rules that enable us to make general statements about a population when all we have are sample data. This logic is reflected in the four steps that have been used throughout this chapter. Hypothesis-testing problems will become easier to tackle when you learn to follow the steps.

STEP 1 State the hypotheses and set the alpha level.

STEP 2 Locate the critical region.

STEP 3 Compute the test statistic (in this case, the z-score) for the sample.

STEP 4 Make a decision about H_0 based on the result of Step 3.

2. Students often ask, "What alpha level should I use?" Or a student may ask, "Why is an alpha of .05 used?" as opposed to something else. There is no single correct answer to either of these questions. Keep in mind the aim of setting an alpha level in the first place: *to reduce the risk of committing a Type I error*. Therefore, the maximum acceptable value is $\alpha = .05$. However, some researchers prefer to take even less risk and use alpha levels of .01 or smaller.

Most statistical tests are now done with computer programs that provide an exact probability (p value) for a Type I error. Because an exact value is available, most researchers simply report the p value from the computer printout rather than setting an alpha level at the beginning of the test. However, the same criterion still applies: A result is not significant unless the p value is less than .05.

3. Take time to consider the implications of your decision about the null hypothesis. The null hypothesis states that there is no effect. Therefore, if your decision is to reject H_0, you should conclude that the sample data provide evidence for a treatment effect. However, it is an entirely different matter if your decision is to fail to reject H_0. Remember that when you fail to reject the null hypothesis, the results are inconclusive. It is impossible to *prove* that H_0 is correct; therefore, you cannot state with certainty that "there is no effect" when H_0 is not rejected. At best, all you can state is that "there is insufficient evidence for an effect."

4. It is very important that you understand the structure of the z-score formula (page 234). It will help you understand many of the other hypothesis tests that are covered later.

5. When you are doing a directional hypothesis test, read the problem carefully, and watch for key words (such as increase or decrease, raise or lower, and more or less) that tell you which direction the researcher is predicting. The predicted direction will determine the alternative hypothesis (H_1) and the critical region. For example, if a treatment is expected to *increase* scores, H_1 would contain a *greater than* symbol, and the critical region would be in the tail associated with high scores.

DEMONSTRATION 8.1

HYPOTHESIS TEST WITH Z

A researcher begins with a known population—in this case, scores on a standardized test that are normally distributed with $\mu = 65$ and $\sigma = 15$. The researcher suspects that special training in reading skills will produce a change in the scores for the individuals in the population. Because it is not feasible to administer the treatment (the special training) to everyone in the population, a sample of $n = 25$ individuals is selected, and the treatment is given to this sample. Following treatment, the average score for this sample is $M = 70$. Is there evidence that the training has an effect on test scores?

STEP 1   **State the hypothesis and select an alpha level**   The null hypothesis states that the special training has no effect. In symbols,

$$H_0\text{: } \mu = 65 \text{ (After special training, the mean is still 65.)}$$

The alternative hypothesis states that the treatment does have an effect.

$$H_1\text{: } \mu \neq 65 \text{ (After training, the mean is different from 65.)}$$

At this time you also select the alpha level. For this demonstration, we will use $\alpha = .05$. Thus, there is a 5% risk of committing a Type I error if we reject H_0.

STEP 2   **Locate the critical region**   With $\alpha = .05$, the critical region consists of sample means that correspond to z-scores beyond the critical boundaries of $z = \pm 1.96$.

STEP 3   **Obtain the sample data, and compute the test statistic**   For this example, the distribution of sample means, according to the null hypothesis, will be normal with an expected value of $\mu = 65$ and a standard error of

$$\sigma_M = \frac{\sigma}{\sqrt{n}} = \frac{15}{\sqrt{25}} = \frac{15}{5} = 3$$

In this distribution, our sample mean of $M = 70$ corresponds to a z-score of

$$z = \frac{M - \mu}{\sigma_M} = \frac{70 - 65}{3} = \frac{5}{3} = +1.67$$

STEP 4   Make a decision about H_0, and state the conclusion   The z-score we obtained is not in the critical region. This indicates that our sample mean of $M = 70$ is not an extreme or unusual value to be obtained from a population with $\mu = 65$. Therefore, our statistical decision is to *fail to reject H_0*. Our conclusion for the study is that the data do not provide sufficient evidence that the special training changes test scores.

DEMONSTRATION 8.2

EFFECT SIZE USING COHEN'S *d*

We will compute Cohen's *d* using the research situation and the data from Demonstration 8.1. Again, the original population mean was $\mu = 65$ and, after treatment (special training), the sample mean was $M = 70$. Thus, there is a 5-point mean difference. Using the population standard deviation, $\sigma = 15$, we obtain an effect size of

$$\text{Cohen's } d = \frac{\text{mean difference}}{\text{standard deviation}} = \frac{5}{15} = 0.33$$

According to Cohen's evaluation standards (see Table 8.2), this is a medium treatment effect.

PROBLEMS

1. Identify the four steps of a hypothesis test as presented in this chapter.

2. Define the alpha level and the critical region for a hypothesis test.

3. Define a Type I error and a Type II error and explain the consequences of each.

4. If the alpha level is changed from $\alpha = .05$ to $\alpha = .01$,
   a. What happens to the boundaries for the critical region?
   b. What happens to the probability of a Type I error?

5. The value of the z-score in a hypothesis test is influenced by a variety of factors. Assuming that all other variables are held constant, explain how the value of z is influenced by each of the following:
   a. Increasing the difference between the sample mean and the original population mean.
   b. Increasing the population standard deviation.
   c. Increasing the number of scores in the sample.

6. Although there is a popular belief that herbal remedies such as Ginkgo biloba and Ginseng may improve learning and memory in healthy adults, these effects are usually not supported by well-controlled research (Persson, Bringlov, Nilsson, and Nyberg, 2004). In a typical study, a researcher obtains a sample of $n = 16$ participants and has each person take the herbal supplements every day for 90 days. At the end of the 90 days, each person takes a standardized memory test. For the general population, scores from the test form a normal distribution with a mean of $\mu = 50$ and a standard deviation of $\sigma = 12$. The sample of research participants had an average of $M = 54$.
   a. Assuming a two-tailed test, state the null hypothesis in a sentence that includes the two variables being examined.
   b. Using the standard 4-step procedure, conduct a two-tailed hypothesis test with $\alpha = .05$ to evaluate the effect of the supplements.

7. Babcock and Marks (2010) reviewed survey data from 2003–2005, and obtained an average of $\mu = 14$ hours per week spent studying by full-time students at 4-year colleges in the United States. To determine whether this average has changed in the past 10 years, researcher selected a sample of $n = 64$ of today's college students and obtained an average of $M = 12.5$ hours. If the standard deviation for the distribution is $\sigma = 4.8$ hours per week, does this sample indicate a significant change in the number of hours spent studying? Use a two-tailed test with $\alpha = .05$.

8. Childhood participation in sports, cultural groups, and youth groups appears to be related to improved self-esteem for adolescents (McGee, Williams, Howden-Chapman, Martin, & Kawachi, 2006). In a representative study, a sample of $n = 100$ adolescents with a history of group participation is given a standardized self-esteem questionnaire. For the general population of adolescents, scores on this questionnaire form a normal distribution with a mean of $\mu = 50$ and a standard deviation of $\sigma = 15$. The sample of group-participation adolescents had an average of $M = 53.8$.
   a. Does this sample provide enough evidence to conclude that self-esteem scores for these adolescents are significantly different from those of the general population? Use a two-tailed test with $\alpha = .05$.
   b. Compute Cohen's d to measure the size of the difference.
   c. Write a sentence describing the outcome of the hypothesis test and the measure of effect size as it would appear in a research report.

9. The psychology department is gradually changing its curriculum by increasing the number of online course offerings. To evaluate the effectiveness of this change, a random sample of $n = 36$ students who registered for Introductory Psychology is placed in the online version of the course. At the end of the semester, all students take the same final exam. The average score for the sample is $M = 76$. For the general population of students taking the traditional lecture class, the final exam scores form a normal distribution with a mean of $\mu = 71$.
   a. If the final exam scores for the population have a standard deviation of $\sigma = 12$, does the sample provide enough evidence to conclude that the new online course is significantly different from the traditional class? Use a two-tailed test with $\alpha = .05$.
   b. If the population standard deviation is $\sigma = 18$, is the sample sufficient to demonstrate a significant difference? Again, use a two-tailed test with $\alpha = .05$.
   c. Comparing your answers for parts a and b, explain how the magnitude of the standard deviation influences the outcome of a hypothesis test.

10. A random sample is selected from a normal population with a mean of $\mu = 30$ and a standard deviation of $\sigma = 8$. After a treatment is administered to the individuals in the sample, the sample mean is found to be $M = 33$.
   a. If the sample consists of $n = 16$ scores, is the sample mean sufficient to conclude that the treatment has a significant effect? Use a two-tailed test with $\alpha = .05$.
   b. If the sample consists of $n = 64$ scores, is the sample mean sufficient to conclude that the

treatment has a significant effect? Use a two-tailed test with $\alpha = .05$.
   c. Comparing your answers for parts a and b, explain how the size of the sample influences the outcome of a hypothesis test.

11. A random sample of $n = 25$ scores is selected from a normal population with a mean of $\mu = 40$. After a treatment is administered to the individuals in the sample, the sample mean is found to be $M = 44$.
   a. If the population standard deviation is $\sigma = 5$, is the sample mean sufficient to conclude that the treatment has a significant effect? Use a two-tailed test with $\alpha = .05$.
   b. If the population standard deviation is $\sigma = 15$, is the sample mean sufficient to conclude that the treatment has a significant effect? Use a two-tailed test with $\alpha = .05$.
   c. Comparing your answers for parts a and b, explain how the magnitude of the standard deviation influences the outcome of a hypothesis test.

12. Brunt, Rhee, and Zhong (2008) surveyed 557 undergraduate college students to examine their weight status, health behaviors, and diet. Using body mass index (BMI), they classified the students into four categories: underweight, healthy weight, overweight, and obese. They also measured dietary variety by counting the number of different foods each student ate from several food groups. Note that the researchers are not measuring the amount of food eaten, but rather the number of different foods eaten (variety, not quantity). Nonetheless, it was somewhat surprising that the four the four weight groups all ate essentially the same number of fatty and/or sugary snacks.
   Suppose a researcher conducting a follow up study obtains a sample of $n = 25$ students classified as healthy weight and a sample of $n = 36$ students classified as overweight. Each student completes the food variety questionnaire, and the healthy-weight group produces a mean of $M = 4.01$ for the fatty, sugary snack category compared to a mean of $M = 4.48$ for the overweight group. The results from the Brunt, Rhee, and Zhong study showed an overall mean score of $\mu = 4.22$ for the sweets or fats food group. Assume that the distribution of scores is approximately normal with a standard deviation of $\sigma = 0.60$.
   a. Does the sample of $n = 36$ indicate that number of fatty, sugary snacks eaten by overweight students is significantly different from the overall population mean? Use a two-tailed test with $\alpha = .05$.
   b. Based on the sample of $n = 25$ healthy-weight students, can you conclude that healthy-weight students eat significantly fewer fatty, sugary snacks than the overall population? Use a one-tailed test with $\alpha = .05$.

13. A random sample is selected from a normal population with a mean of $\mu = 100$ and a standard deviation of $\sigma = 20$. After a treatment is administered to the individuals in the sample, the sample mean is found to be $M = 96$.
    a. How large a sample is necessary for this sample mean to be statistically significant? Assume a two-tailed test with $\alpha = .05$.
    b. If the sample mean were $M = 98$, what sample size is needed to be significant for a two-tailed test with $\alpha = .05$?

14. In a study examining the effect of alcohol on reaction time, Liguori and Robinson (2001) found that even moderate alcohol consumption significantly slowed response time to an emergency situation in a driving simulation. In a similar study, researchers measured reaction time 30 minutes after participants consumed one 6-ounce glass of wine. Again, they used a standardized driving simulation task for which the regular population averages $\mu = 400$ msec. The distribution of reaction times is approximately normal with $\sigma = 40$. Assume that the researcher obtained a sample mean of $M = 422$ for the $n = 25$ participants in the study.
    a. Are the data sufficient to conclude that the alcohol has a significant effect on reaction time? Use a two-tailed test with $\alpha = .01$.
    b. Do the data provide evidence that the alcohol significantly increased (slowed) reaction time? Use a one-tailed test with $\alpha = .01$.
    c. Compute Cohen's d to estimate the size of the effect.

15. The researchers cited in the previous problem (Liguori and Robinson, 2001) also examined the effect of caffeine on response time in the driving simulator. In a similar study, researchers measured reaction time 30 minutes after participants consumed one 6-ounce cup of coffee. Using the same driving simulation task, for which the distribution of reaction times is normal with $\mu = 400$ msec and $\sigma = 40$, they obtained a mean of $M = 392$ for a sample of $n = 36$ participants.
    a. Are the data sufficient to conclude that caffeine has a significant effect on reaction time? Use a two-tailed test with $\alpha = .05$.
    b. Compute Cohen's d to estimate the size of the effect.
    c. Write a sentence describing the outcome of the hypothesis test and the measure of effect size as it would appear in a research report.

16. Researchers at a National Weather Center in the northeastern United States recorded the number of 90° days each year since records first started in 1875. The numbers form a normal shaped distribution with a mean of $\mu = 9.6$ and a standard deviation of $\sigma = 1.9$.

To see if the data showed any evidence of global warming, they also computed the mean number of 90° days for the most recent $n = 4$ years and obtained $M = 12.25$. Do the data indicate that the past four years have had significantly more 90° days than would be expected for a random sample from this population? Use a one-tailed test with $\alpha = .05$.

17. A high school teacher has designed a new course intended to help students prepare for the mathematics section of the SAT. A sample of $n = 20$ students is recruited to for the course and, at the end of the year, each student takes the SAT. The average score for this sample is $M = 562$. For the general population, scores on the SAT are standardized to form a normal distribution with $\mu = 500$ and $\sigma = 100$.
    a. Can the teacher conclude that students who take the course score significantly higher than the general population? Use a one-tailed test with $\alpha = .01$.
    b. Compute Cohen's d to estimate the size of the effect.
    c. Write a sentence demonstrating how the results of the hypothesis test and the measure of effect size would appear in a research report.

18. Researchers have noted a decline in cognitive functioning as people age (Bartus, 1990). However, the results from other research suggest that the antioxidants in foods such as blueberries can reduce and even reverse these age-related declines, at least in laboratory rats (Joseph et al., 1999). Based on these results, one might theorize that the same antioxidants might also benefit elderly humans. Suppose a researcher is interested in testing this theory. The researcher obtains a sample of $n = 16$ adults who are older than 65, and gives each participant a daily dose of a blueberry supplement that is very high in antioxidants. After taking the supplement for 6 months, the participants are given a standardized cognitive skills test and produce a mean score of $M = 50.2$. For the general population of elderly adults, scores on the test average $\mu = 45$ and form a normal distribution with $\sigma = 9$.
    a. Can the researcher conclude that the supplement has a significant effect on cognitive skill? Use a two-tailed test with $\alpha = .05$.
    b. Compute Cohen's d for this study.
    c. Write a sentence demonstrating how the outcome of the hypothesis test and the measure of effect size would appear in a research report.

19. A researcher is evaluating the influence of a treatment using a sample selected from a normally distributed population with a mean of $\mu = 40$ and a standard deviation of $\sigma = 12$. The researcher expects a 6-point

treatment effect and plans to use a two-tailed hypothesis test with $\alpha = .05$.

a. Compute the power of the test if the researcher uses a sample of $n = 9$ individuals. (See Example 8.6.)

b. Compute the power of the test if the researcher uses a sample of $n = 16$ individuals.

20. A researcher plans to conduct an experiment evaluating the effect of a treatment. A sample of $n = 9$ participants is selected and each person receives the treatment before being tested on a standardized dexterity task. The treatment is expected to lower scores on the test by an average of 30 points. For the regular population, scores on the dexterity task form a normal distribution with $\mu = 240$ and $\sigma = 30$.

a. If the researcher uses a two-tailed test with $\alpha = .05$, what is the power of the hypothesis test?

b. Again assuming a two-tailed test with $\alpha = .05$, what is the power of the hypothesis test if the sample size is increased to $n = 25$?

21. Research has shown that IQ scores have been increasing for years (Flynn, 1984, 1999). The phenomenon is called the Flynn effect and the data indicate that the increase appears to average about 7 points per decade. To examine this effect, a researcher obtains an IQ test with instructions for scoring from 10 years ago and plans to administers the test to a sample of $n = 25$ of today's high school students. Ten years ago, the scores on this IQ test produced a standardized distribution with a mean of $\mu = 100$ and a standard deviation $\sigma = 15$. If there actually has been a 7-point increase in the average IQ during the past 10 years, then find the power of the hypothesis test for each of the following.

a. The researcher uses a two-tailed hypothesis test with $\alpha = .05$ to determine if the data indicate a significant change in IQ over the past 10 years.

b. The researcher uses a one-tailed hypothesis test with $\alpha = .05$ to determine if the data indicate a significant increase in IQ over the past 10 years.

22. Briefly explain how increasing sample size influences each of the following. Assume that all other factors are held constant.

a. The size of the z-score in a hypothesis test.

b. The size of Cohen's d.

c. The power of a hypothesis test.

23. Explain how the power of a hypothesis test is influenced by each of the following. Assume that all other factors are held constant.

a. Increasing the alpha level from .01 to .05.

b. Changing from a one-tailed test to a two-tailed test.

Introduction to the *t* Statistic

© Deborah Batt

Tools You Will Need

The following items are considered essential background material for this chapter. If you doubt your knowledge of any of these items, you should review the appropriate chapter or section before proceeding.

- Sample standard deviation (Chapter 4)
- Degrees of freedom (Chapter 4)
- Standard error (Chapter 7)
- Hypothesis testing (Chapter 8)

In studies examining the effect of humor on interpersonal attractions, McGee and Shevlin (2009) found that a man's sense of humor had a significant effect on how he was perceived by women. In one part of the study, female college students were given brief descriptions of a potential romantic partner. The fictitious male was described positively as being single, ambitious, and having good job prospects. For one group of participants, the description also said that he had a great sense of humor. For another group, it said that he had no sense of humor. After reading the description, each participant was asked to rate the attractiveness of the man on a seven-point scale from 1 (very unattractive) to 7 (very attractive). A score of 4 indicates a neutral rating.

The females who read the "great sense of humor" description gave the potential partner an average attractiveness score of $M = 4.53$, which was found to be significantly higher than a neutral rating of $\mu = 4.0$. Those who read the description saying "no sense of humor" gave the potential partner an average attractiveness score of $M = 3.30$, which is significantly lower than neutral.

These results clearly indicate that a sense of humor is an important factor in interpersonal attraction. You also should recognize that the researchers used hypothesis tests to evaluate the significance of the results. For each group of participants, the null hypothesis would state that humor has no effect on rating of attractiveness and the average rating should be $\mu = 4.0$ (neutral) for each treatment condition. Thus, the researchers have a hypothesis and a sample mean for each of the treatment conditions. However, they do not know the population standard deviation (σ). This value is needed to compute the standard error for the sample mean (σ_M), which is the denominator of the z-score equation. Recall that the standard error measures how much difference is reasonable to expect between a sample mean (M) and the population mean (μ). This value is critical for determining whether the sample data are consistent with the null hypothesis or refute the null hypothesis. Without the standard error it is impossible to conduct a z-score hypothesis test.

Because it is impossible to compute the standard error, a z-score cannot be used for the hypothesis test. However, it is possible to *estimate* the standard error using the sample data. The estimated standard error can then be used to compute a new statistic that is similar to the z-score. The new statistic is called a *t statistic* and it can be used to conduct a new kind of hypothesis test. In this chapter we introduce the *t* statistic and demonstrate how it is used for hypothesis testing in situation for which the population standard deviation is unknown.

9.1 | The *t* Statistic: An Alternative to *z*

LEARNING OBJECTIVES

1. Describe the circumstances in which a *t* statistic is used for hypothesis testing instead of a z-score and explain the fundamental difference between a *t* statistic and a z-score.

2. Explain the relationship between the *t* distribution and the normal distribution.

In the previous chapter, we presented the statistical procedures that permit researchers to use a sample mean to test hypotheses about an unknown population mean. These statistical procedures were based on a few basic concepts, which we summarize as follows.

1. A sample mean (M) is expected to approximate its population mean (μ). This permits us to use the sample mean to test a hypothesis about the population mean.

2. The standard error provides a measure of how well a sample mean approximates the population mean. Specifically, the standard error determines how much difference is reasonable to expect between a sample mean (M) and the population mean (μ).

$$\sigma_M = \frac{\sigma}{\sqrt{n}} \quad \text{or} \quad \sigma_M = \sqrt{\frac{\sigma^2}{n}}$$

3. To quantify our inferences about the population, we compare the obtained sample mean (M) with the hypothesized population mean (μ) by computing a z-score test statistic.

$$z = \frac{M - \mu}{\sigma_M} = \frac{\text{obtained difference between data and hypothesis}}{\text{standard distance between } M \text{ and } \mu}$$

The goal of the hypothesis test is to determine whether the obtained difference between the data and the hypothesis is significantly greater than would be expected by chance. When the z-scores form a normal distribution, we are able to use the unit normal table (Appendix B) to find the critical region for the hypothesis test.

■ The Problem with *z*-Scores

The shortcoming of using a z-score for hypothesis testing is that the z-score formula requires more information than is usually available. Specifically, a z-score requires that we know the value of the population standard deviation (or variance), which is needed to compute the standard error. In most situations, however, the standard deviation for the population is not known. In fact, the whole reason for conducting a hypothesis test is to gain knowledge about an *unknown* population. This situation appears to create a paradox: You want to use a z-score to find out about an unknown population, but you must know about the population before you can compute a z-score. Fortunately, there is a relatively simple solution to this problem. When the variance (or standard deviation) for the population is not known, we use the corresponding sample value in its place.

■ Introducing the *t* Statistic

In Chapter 4, the sample variance was developed specifically to provide an unbiased estimate of the corresponding population variance. Recall that the formulas for sample variance and sample standard deviation are as follows:

The concept of degrees of freedom, *df* = *n* − 1, was introduced in Chapter 4 (page 115) and is discussed later in this chapter (page 271).

$$\text{sample variance} = s^2 = \frac{SS}{n - 1} = \frac{SS}{df}$$

$$\text{sample standard deviation} = s = \sqrt{\frac{SS}{n - 1}} = \sqrt{\frac{SS}{df}}$$

Using the sample values, we can now *estimate* the standard error. Recall from Chapters 7 and 8 that the value of the standard error can be computed using either standard deviation or variance:

$$\text{standard error} = \sigma_M = \frac{\sigma}{\sqrt{n}} \quad \text{or} \quad \sigma_M = \sqrt{\frac{\sigma^2}{n}}$$

Now we estimate the standard error by simply substituting the sample variance or standard deviation in place of the unknown population value:

$$\text{estimated standard error} = s_M = \frac{s}{\sqrt{n}} \quad \text{or} \quad s_M = \sqrt{\frac{s^2}{n}} \tag{9.1}$$

Notice that the symbol for the *estimated standard error of M* is s_M instead of σ_M, indicating that the estimated value is computed from sample data rather than from the actual population parameter.

DEFINITION

> The **estimated standard error** (s_M) is used as an estimate of the real standard error σ_M when the value of σ is unknown. It is computed from the sample variance or sample standard deviation and provides an estimate of the standard distance between a sample mean M and the population mean μ.

Finally, you should recognize that we have shown formulas for standard error (actual or estimated) using both the standard deviation and the variance. In the past (Chapters 7 and 8), we concentrated on the formula using the standard deviation. At this point, however, we shift our focus to the formula based on variance. Thus, throughout the remainder of this chapter, and in following chapters, the estimated standard error of M typically is presented and computed using

$$s_M = \sqrt{\frac{s^2}{n}}$$

There are two reasons for making this shift from standard deviation to variance:

1. In Chapter 4 (page 118) we saw that the sample variance is an *unbiased* statistic; on average, the sample variance (s^2) provides an accurate and unbiased estimate of the population variance (σ^2). Therefore, the most accurate way to estimate the standard error is to use the sample variance to estimate the population variance.

2. In future chapters we will encounter other versions of the *t* statistic that require variance (instead of standard deviation) in the formulas for estimated standard error. To maximize the similarity from one version to another, we will use variance in the formula for *all* of the different *t* statistics. Thus, whenever we present a *t* statistic, the estimated standard error will be computed as

$$\text{estimated standard error} = \sqrt{\frac{\text{sample variance}}{\text{sample size}}}$$

Now we can substitute the estimated standard error in the denominator of the z-score formula. The result is a new test statistic called a *t statistic*:

$$t = \frac{M - \mu}{s_M} \tag{9.2}$$

DEFINITION

> The *t* **statistic** is used to test hypotheses about an unknown population mean, μ, when the value of σ is unknown. The formula for the *t* statistic has the same structure as the z-score formula, except that the *t* statistic uses the estimated standard error in the denominator.

The only difference between the *t* formula and the z-score formula is that the z-score uses the actual population variance, σ^2 (or the standard deviation), and the *t* formula uses the corresponding sample variance (or standard deviation) when the population value is not known.

$$z = \frac{M - \mu}{\sigma_M} = \frac{M - \mu}{\sqrt{\sigma^2/n}} \qquad t = \frac{M - \mu}{s_M} = \frac{M - \mu}{\sqrt{s^2/n}}$$

■ Degrees of Freedom and the *t* Statistic

In this chapter, we have introduced the *t* statistic as a substitute for a z-score. The basic difference between these two is that the *t* statistic uses sample variance (s^2) and the z-score uses the population variance (σ^2). To determine how well a *t* statistic approximates a *z*-score, we must determine how well the sample variance approximates the population variance.

In Chapter 4, we introduced the concept of degrees of freedom (page 115). Reviewing briefly, you must know the sample mean before you can compute sample variance. This places a restriction on sample variability such that only $n - 1$ scores in a sample are independent and free to vary. The value $n - 1$ is called the *degrees of freedom* (or *df*) for the sample variance.

$$\text{degrees of freedom} = df = n - 1 \qquad \qquad \textbf{(9.3)}$$

DEFINITION
> **Degrees of freedom** describe the number of scores in a sample that are independent and free to vary. Because the sample mean places a restriction on the value of one score in the sample, there are $n - 1$ degrees of freedom for a sample with *n* scores (see Chapter 4).

The greater the value of *df* for a sample, the better the sample variance, s^2, represents the population variance, σ^2, and the better the *t* statistic approximates the *z*-score. This should make sense because the larger the sample (*n*) is, the better the sample represents its population. Thus, the degrees of freedom associated with s^2 also describe how well *t* represents *z*.

The following example is an opportunity for you to test your understanding of the estimated standard error for a *t* statistic.

EXAMPLE 9.1
For a sample of $n = 9$ scores with $SS = 288$, compute the sample variance and the estimated standard error for the sample mean. You should obtain $s^2 = 36$ and $s_M = 2$. ■

■ The *t* Distribution

Every sample from a population can be used to compute a *z*-score or a *t* statistic. If you select all the possible samples of a particular size (*n*), and compute the *z*-score for each sample mean, then the entire set of *z*-scores will form a *z*-score distribution. In the same way, you can compute the *t* statistic for every sample and the entire set of *t* values will form a *t* distribution. As we saw in Chapter 7, the distribution of *z*-scores for sample means tends to be a normal distribution. Specifically, if the sample size is large (around $n = 30$ or more) or if the sample is selected from a normal population, then the distribution of sample means is a nearly perfect normal distribution. In these same situations, the *t* distribution approximates a normal distribution, just as a *t* statistic approximates a *z*-score. How well a *t* distribution approximates a normal distributor is determined by degrees of freedom. In general, the greater the sample size (*n*) is, the larger the degrees of freedom ($n - 1$) are, and the better the *t* distribution approximates the normal distribution. This fact is demonstrated in Figure 9.1, which shows a normal distribution and two *t* distributions with $df = 5$ and $df = 20$.

DEFINITION
> A *t* **distribution** is the complete set of *t* values computed for every possible random sample for a specific sample size (*n*) or a specific degrees of freedom (*df*). The *t* distribution approximates the shape of a normal distribution.

■ The Shape of the *t* Distribution

The exact shape of a *t* distribution changes with degrees of freedom. In fact, statisticians speak of a "family" of *t* distributions. That is, there is a different sampling distribution of

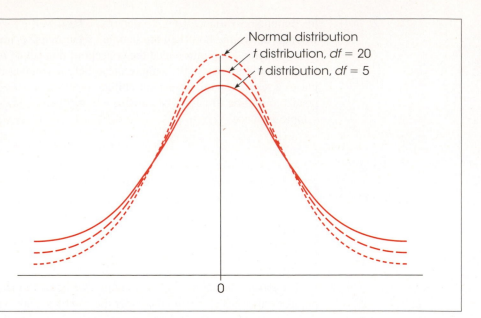

FIGURE 9.1

Distributions of the *t* statistic for different values of degrees of freedom are compared to a normal *z*-score distribution. Like the normal distribution, *t* distributions are bell-shaped and symmetrical and have a mean of zero. However, *t* distributions are more variable than the normal distribution as indicated by the flatter and more spread-out shape. The larger the value of *df* is, the more closely the *t* distribution approximates a normal distribution.

t (a distribution of all possible sample *t* values) for each possible number of degrees of freedom. As *df* gets very large, the *t* distribution gets closer in shape to a normal *z*-score distribution. A quick glance at Figure 9.1 reveals that distributions of *t* are bell-shaped and symmetrical and have a mean of zero. However, the *t* distribution has more variability than a normal *z* distribution, especially when *df* values are small (see Figure 9.1). The *t* distribution tends to be flatter and more spread out, whereas the normal *z* distribution has more of a central peak.

The reason that the *t* distribution is flatter and more variable than the normal *z*-score distribution becomes clear if you look at the structure of the formulas for *z* and *t*. For both *z* and *t*, the top of the formula, $M - \mu$, can take on different values because the sample mean (M) varies from one sample to another. For *z*-scores, however, the bottom of the formula does not vary, provided that all of the samples are the same size and are selected from the same population. Specifically, all the *z*-scores have the same standard error in the denominator, $\sigma_M = \sqrt{\sigma^2/n}$, because the population variance and the sample size are the same for every sample. For *t* statistics, on the other hand, the bottom of the formula varies from one sample to another. Specifically, the sample variance (s^2) changes from one sample to the next, so the estimated standard error also varies, $s_M = \sqrt{s^2/n}$. Thus, only the numerator of the *z*-score formula varies, but both the numerator and the denominator of the *t* statistic vary. As a result, *t* statistics are more variable than are *z*-scores, and the *t* distribution is flatter and more spread out. As sample size and *df* increase, however, the variability in the *t* distribution decreases, and it more closely resembles a normal distribution.

■ Determining Proportions and Probabilities for *t* Distributions

Just as we used the unit normal table to locate proportions associated with *z*-scores, we use a *t* distribution table to find proportions for *t* statistics. The complete *t* distribution table is presented in Appendix B, page 703, and a portion of this table is reproduced in Table 9.1. The two rows at the top of the table show proportions of the *t* distribution contained in either one or two tails, depending on which row is used. The first column of the table lists degrees of freedom for the *t* statistic. Finally, the numbers in the body of the table are the *t* values that mark the boundary between the tails and the rest of the *t* distribution.

For example, with $df = 3$, exactly 5% of the *t* distribution is located in the tail beyond $t = 2.353$ (Figure 9.2). The process of finding this value is highlighted in Table 9.1. Begin by locating $df = 3$ in the first column of the table. Then locate a proportion of 0.05 (5%) in the one-tail proportion row. When you line up these two values in the table, you should find $t = 2.353$. Because the distribution is symmetrical, 5% of the *t* distribution is also located in the tail beyond $t = -2.353$ (see Figure 9.2). Finally, notice that a total of 10% (or 0.10) is contained in the two tails beyond $t = \pm2.353$ (check the proportion value in the "two-tails combined" row at the top of the table).

TABLE 9.1   A portion of the *t*-distribution table. The numbers in the table are the values of *t* that separate the tail from the main body of the distribution. Proportions for one or two tails are listed at the top of the table, and *df* values for *t* are listed in the first column.

	Proportion in One Tail					
	0.25	0.10	0.05	0.025	0.01	0.005
	Proportion in Two Tails Combined					
df	0.50	0.20	0.10	0.05	0.02	0.01
1	1.000	3.078	6.314	12.706	31.821	63.657
2	0.816	1.886	2.920	4.303	6.965	9.925
3	0.765	1.638	2.353	3.182	4.541	5.841
4	0.741	1.533	2.132	2.776	3.747	4.604
5	0.727	1.476	2.015	2.571	3.365	4.032
6	0.718	1.440	1.943	2.447	3.143	3.707

A close inspection of the *t* distribution table in Appendix B will demonstrate a point we made earlier: as the value for *df* increases, the *t* distribution becomes more similar to a normal distribution. For example, examine the column containing *t* values for a 0.05 proportion in two tails. You will find that when $df = 1$, the *t* values that separate the extreme 5% (0.05) from the rest of the distribution are $t = \pm12.706$. As you read down the column, however, you should find that the critical *t* values become smaller and smaller, ultimately reaching ±1.96. You should recognize ±1.96 as the *z*-score values that separate the extreme 5% in a normal distribution. Thus, as *df* increases, the proportions in a *t* distribution become more like the proportions in a normal distribution. When the sample size (and degrees of freedom) is sufficiently large, the difference between a *t* distribution and the normal distribution becomes negligible.

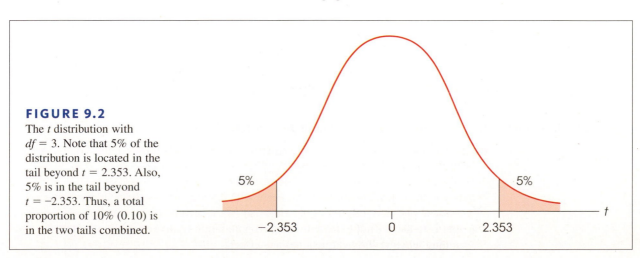

FIGURE 9.2
The *t* distribution with $df = 3$. Note that 5% of the distribution is located in the tail beyond $t = 2.353$. Also, 5% is in the tail beyond $t = -2.353$. Thus, a total proportion of 10% (0.10) is in the two tails combined.

5%   5%

-2.353 0   2.353   *t*

Caution: The *t* distribution table printed in this book has been abridged and does not include entries for every possible *df* value. For example, the table lists *t* values for *df* = 40 and for *df* = 60, but does not list any entries for *df* values between 40 and 60. Occasionally, you will encounter a situation in which your *t* statistic has a *df* value that is not listed in the table. In these situations, you should look up the critical *t* for both of the surrounding *df* values listed and then use the *larger* value for *t*. If, for example, you have *df* = 53 (not listed), look up the critical *t* value for both *df* = 40 and *df* = 60 and *then use the larger t value*. If your sample *t* statistic is greater than the larger value listed, you can be certain that the data are in the critical region, and you can confidently reject the null hypothesis.

LEARNING CHECK

1. Which of the following is a fundamental difference between the *t* statistic and a *z*-score?
    a. The *t* statistic uses the sample mean in place of the population mean.
    b. The *t* statistic uses the sample variance in place of the population variance.
    c. The *t* statistic computes the standard error by dividing the standard deviation by *n* − 1 instead of dividing by *n*.
    d. All of the above are differences between *t* and *z*.

2. Which of the following terms is *not* required when using the *t* statistic?
    a. *n*
    b. σ
    c. *df*
    d. *s* or s^2 or *SS*

3. How does the shape of the *t* distribution compare to a normal distribution?
    a. The *t* distribution is flatter and more spread out, especially when *n* is small.
    b. The *t* distribution is flatter and more spread out, especially when *n* is large.
    c. The *t* distribution is taller and less spread out, especially when *n* is small.
    d. The *t* distribution is taller and less spread out, especially when *n* is small.

ANSWERS   1. B, 2. B, 3. A

9.2 | Hypothesis Tests with the *t* Statistic

LEARNING OBJECTIVES

3. Conduct a hypothesis test using the *t* statistic.

4. Explain how the likelihood of rejecting the null hypothesis for a *t* test is influenced by sample size and sample variance.

In the hypothesis-testing situation, we begin with a population with an unknown mean and an unknown variance, often a population that has received some treatment (Figure 9.3). The goal is to use a sample from the treated population (a treated sample) as the basis for determining whether the treatment has any effect.

As always, the null hypothesis states that the treatment has no effect; specifically, H_0 states that the population mean is unchanged. Thus, the null hypothesis provides a specific value for the unknown population mean. The sample data provide a value for the sample mean. Finally, the variance and estimated standard error are computed from the sample data. When these values are used in the *t* formula, the result becomes

$$t = \frac{\substack{\text{sample mean} \\ \text{(from the data)}} - \substack{\text{population mean} \\ \text{(hypothesized from } H_0)}}{\substack{\text{estimated standard error} \\ \text{(computed from the sample data)}}}$$

As with the *z*-score formula, the *t* statistic forms a ratio. The numerator measures the actual difference between the sample data (M) and the population hypothesis (μ). The estimated standard error in the denominator measures how much difference is reasonable to expect between a sample mean and the population mean. When the obtained difference between the data and the hypothesis (numerator) is much greater than expected (denominator), we obtain a large value for *t* (either large positive or large negative). In this case, we conclude that the data are not consistent with the hypothesis, and our decision is to "reject H_0." On the other hand, when the difference between the data and the hypothesis is small relative to the standard error, we obtain a *t* statistic near zero, and our decision is "fail to reject H_0."

The Unknown Population   As mentioned earlier, the hypothesis test often concerns a population that has received a treatment. This situation is shown in Figure 9.3. Note that the value of the mean is known for the population before treatment. The question is whether the treatment influences the scores and causes the mean to change. In this case, the unknown population is the one that exists after the treatment is administered, and the null hypothesis simply states that the value of the mean is not changed by the treatment.

Although the *t* statistic can be used in the "before and after" type of research shown in Figure 9.3, it also permits hypothesis testing in situations for which you do not have a known population mean to serve as a standard. Specifically, the *t* test does not require any prior knowledge about the population mean or the population variance. All you need to compute a *t* statistic is a null hypothesis and a sample from the unknown population. Thus, a *t* test can be used in situations for which the null hypothesis is obtained from

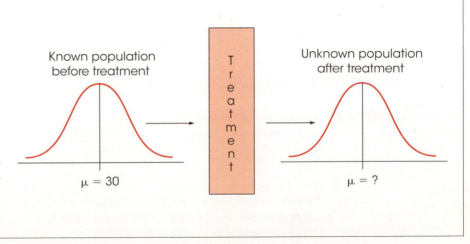

FIGURE 9.3
The basic research situation for the *t* statistic hypothesis test. It is assumed that the parameter μ is known for the population before treatment. The purpose of the research study is to determine whether the treatment has an effect. Note that the population after treatment has unknown values for the mean and the variance. We will use a sample to test a hypothesis about the population mean.

Known population before treatment

Treatment

Unknown population after treatment

$\mu = 30$

$\mu = ?$

a theory, a logical prediction, or just wishful thinking. For example, many studies use rating-scale questions to measure perceptions or attitudes (see the sense-of-humor study discussed in the Chapter Preview). Participants are presented with a statement and asked to express their opinion on a scale from 1–7, with 1 indicating "strongly negative" and 7 indicating "strongly positive." A score of 4 indicates a neutral position, with no strong opinion one way or the other. In this situation, the null hypothesis would state that there is no preference, or no strong opinion, in the population, and use a null hypothesis of H_0: $\mu = 4$. The data from a sample is then used to evaluate the hypothesis. Note that the researcher has no prior knowledge about the population mean and states a hypothesis that is based on logic.

■ Hypothesis Testing Example

The following research situation demonstrates the procedures of hypothesis testing with the *t* statistic. Note that this is another example of a null hypothesis that is founded in logic rather than prior knowledge of a population mean.

EXAMPLE 9.2   Infants, even newborns, prefer to look at attractive faces compared to less attractive faces (Slater et al., 1998). In the study, infants from 1–6 days old were shown two photographs of women's faces. Previously, a group of adults had rated one of the faces as significantly more attractive than the other. The babies were positioned in front of a screen on which the photographs were presented. The pair of faces remained on the screen until the baby accumulated a total of 20 seconds of looking at one or the other. The number of seconds looking at the attractive face was recorded for each infant. Suppose that the study used a sample of $n = 9$ infants and the data produced an average of $M = 13$ seconds for the attractive face with $SS = 72$. Note that all the available information comes from the sample. Specifically, we do not know the population mean or the population standard deviation.

STEP 1   **State the hypotheses and select an alpha level.**   Although we have no information about the population of scores, it is possible to form a logical hypothesis about the value of μ. In this case, the null hypothesis states that the infants have no preference for either face. That is, they should average half of the 20 seconds looking at each of the two faces. In symbols, the null hypothesis states

$$H_0: \mu_{attractive} = 10 \text{ seconds}$$

The alternative hypothesis states that there is a preference and one of the faces is preferred over the other. A directional, one-tailed test would specify which of the two faces is preferred, but the nondirectional alternative hypothesis is expressed as follows:

$$H_1: \mu_{attractive} \neq 10 \text{ seconds}$$

We will set the level of significance at $\alpha = .05$ for two tails.

STEP 2   **Locate the critical region.**   The test statistic is a *t* statistic because the population variance is not known. Therefore, the value for degrees of freedom must be determined before the critical region can be located. For this sample

$$df = n - 1 = 9 - 1 = 8$$

For a two-tailed test at the .05 level of significance and with 8 degrees of freedom, the critical region consists of *t* values greater than $+2.306$ or less than -2.306. Figure 9.4 depicts the critical region in this *t* distribution.

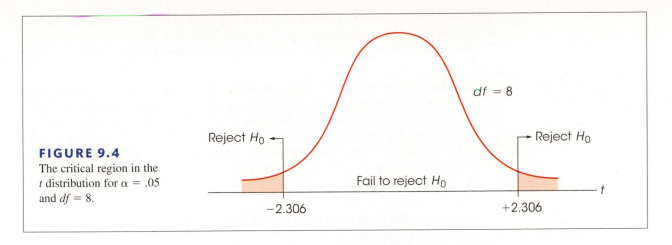

FIGURE 9.4
The critical region in the
t distribution for $\alpha = .05$
and $df = 8$.

STEP 3 **Calculate the test statistic.** The *t* statistic typically requires more computation than is necessary for a *z*-score. Therefore, we recommend that you divide the calculations into a three-stage process as follows.

a. First, calculate the sample variance. Remember that the population variance is unknown, and you must use the sample value in its place. (This is why we are using a *t* statistic instead of a *z*-score.)

$$s^2 = \frac{SS}{n-1} = \frac{SS}{df} = \frac{72}{8} = 9$$

b. Next, use the sample variance (s^2) and the sample size (n) to compute the estimated standard error. This value is the denominator of the *t* statistic and measures how much difference is reasonable to expect by chance between a sample mean and the corresponding population mean.

$$s_M = \sqrt{\frac{s^2}{n}} = \sqrt{\frac{9}{9}} = \sqrt{1} = 1$$

c. Finally, compute the *t* statistic for the sample data.

$$t = \frac{M - \mu}{s_M} = \frac{13 - 10}{1} = 3.00$$

STEP 4 **Make a decision regarding H_0.** The obtained *t* statistic of 3.00 falls into the critical region on the right-hand side of the *t* distribution (see Figure 9.4). Our statistical decision is to reject H_0 and conclude that babies do show a preference when given a choice between an attractive and an unattractive face. Specifically, the average amount of time that the babies spent looking at the attractive face was significantly different from the 10 seconds that would be expected if there were no preference. As indicated by the sample mean, there is a tendency for the babies to spend more time looking at the attractive face. ■

■ Assumptions of the *t* Test

Two basic assumptions are necessary for hypothesis tests with the *t* statistic.

1. The values in the sample must consist of *independent* observations.
   In everyday terms, two observations are independent if there is no consistent, predictable relationship between the first observation and the second. More

precisely, two events (or observations) are independent if the occurrence of the first event has no effect on the probability of the second event. We examined specific examples of independence and non-independence in Box 8.1 (page 244).

2. The population sampled must be normal.

   This assumption is a necessary part of the mathematics underlying the development of the *t* statistic and the *t* distribution table. However, violating this assumption has little practical effect on the results obtained for a *t* statistic, especially when the sample size is relatively large. With very small samples, a normal population distribution is important. With larger samples, this assumption can be violated without affecting the validity of the hypothesis test. If you have reason to suspect that the population distribution is not normal, use a large sample to be safe.

■ The Influence of Sample Size and Sample Variance

As we noted in Chapter 8 (page 242), a variety of factors can influence the outcome of a hypothesis test. In particular, the number of scores in the sample and the magnitude of the sample variance both have a large effect on the *t* statistic and thereby influence the statistical decision. The structure of the *t* formula makes these factors easier to understand.

$$t = \frac{M - \mu}{s_M} \quad \text{where } s_M = \sqrt{\frac{s^2}{n}}$$

Because the estimated standard error, s_M, appears in the denominator of the formula, a larger value for s_M produces a smaller value (closer to zero) for *t*. Thus, any factor that influences the standard error also affects the likelihood of rejecting the null hypothesis and finding a significant treatment effect. The two factors that determine the size of the standard error are the sample variance, s^2, and the sample size, *n*.

The estimated standard error is directly related to the sample variance so that the larger the variance, the larger the error. Thus, large variance means that you are less likely to obtain a significant treatment effect. In general, large variance is bad for inferential statistics. Large variance means that the scores are widely scattered, which makes it difficult to see any consistent patterns or trends in the data. In general, high variance reduces the likelihood of rejecting the null hypothesis.

On the other hand, the estimated standard error is inversely related to the number of scores in the sample. The larger the sample, the smaller the error. If all other factors are held constant, large samples tend to produce bigger *t* statistics and therefore are more likely to produce significant results. For example, a 2-point mean difference with a sample of $n = 4$ may not be convincing evidence of a treatment effect. However, the same 2-point difference with a sample of $n = 100$ is much more compelling.

LEARNING CHECK

1. If a *t* statistic is computed for a sample of $n = 4$ scores with $SS = 300$, then what are the sample variance and the estimated standard error for the sample mean?

   a. $s^2 = 10$ and $s_M = 5$

   b. $s^2 = 100$ and $s_M = 20$

   c. $s^2 = 10$ and $s_M = 20$

   d. $s^2 = 100$ and $s_M = 5$

2. A sample is selected from a population with $\mu = 40$, and a treatment is administered to the sample. If the sample variance is $s^2 = 96$, which set of sample characteristics has the greatest likelihood of rejecting the null hypothesis?
   a. $M = 37$ for a sample of $n = 16$
   b. $M = 37$ for a sample of $n = 64$
   c. $M = 34$ for a sample of $n = 16$
   d. $M = 34$ for a sample of $n = 64$

3. A sample is selected from a population and a treatment is administered to the sample. If there is a 2-point difference between the sample mean and the original population mean, which set of sample characteristics has the greatest likelihood of rejecting the null hypothesis?
   a. $s^2 = 12$ for a sample with $n = 25$
   b. $s^2 = 12$ for a sample with $n = 9$
   c. $s^2 = 32$ for a sample with $n = 25$
   d. $s^2 = 32$ for a sample with $n = 9$

ANSWERS   **1.** D, **2.** D, **3.** A

9.3 | Measuring Effect Size for the *t* Statistic

LEARNING OBJECTIVES

5. Calculate Cohen's *d* or the percentage of variance accounted for (r^2) to measure effect size for a hypothesis test with a *t* statistic.

6. Explain how measures of effect size for a *t* test are influenced by sample size and sample variance.

7. Explain how a confidence interval can be used to describe the size of a treatment effect and describe the factors that affect with width of a confidence interval.

8. Describe how the results from a hypothesis test using a *t* statistic are reported in the literature.

In Chapter 8 we noted that one criticism of a hypothesis test is that it does not really evaluate the size of the treatment effect. Instead, a hypothesis test simply determines whether the treatment effect is greater than chance, where "chance" is measured by the standard error. In particular, it is possible for a very small treatment effect to be "statistically significant," especially when the sample size is very large. To correct for this problem, it is recommended that the results from a hypothesis test be accompanied by a report of effect size such as Cohen's *d*.

■ Estimated Cohen's *d*

When Cohen's *d* was originally introduced (page 251), the formula was presented as

$$\text{Cohen's } d = \frac{\text{mean difference}}{\text{standard deviation}} = \frac{\mu_{\text{treatment}} - \mu_{\text{no treatment}}}{\sigma}$$

Cohen defined this measure of effect size in terms of the population mean difference and the population standard deviation. However, in most situations the population values are not known and you must substitute the corresponding sample values in their place. When this is done, many researchers prefer to identify the calculated value as an "*estimated* d" or name the value after one of the statisticians who first substituted sample statistics into Cohen's formula (e.g., Glass's *g* or Hedges's *g*). For hypothesis tests using the *t* statistic, the population mean with no treatment is the value specified by the null hypothesis. However, the population mean with treatment and the standard deviation are both unknown. Therefore, we use the mean for the treated sample and the standard deviation for the sample after treatment as estimates of the unknown parameters. With these substitutions, the formula for estimating Cohen's *d* becomes

$$\text{estimated } d = \frac{\text{mean difference}}{\text{sample standard deviation}} = \frac{M - \mu}{s} \qquad (9.4)$$

The numerator measures that magnitude of the treatment effect by finding the difference between the mean for the treated sample and the mean for the untreated population (μ from H_0). The sample standard deviation in the denominator standardizes the mean difference into standard deviation units. Thus, an estimated *d* of 1.00 indicates that the size of the treatment effect is equivalent to one standard deviation. The following example demonstrates how the estimated *d* is used to measure effect size for a hypothesis test using a *t* statistic.

EXAMPLE 9.3   For the infant face-preference study in Example 9.2, the babies averaged $M = 13$ out of 20 seconds looking at the attractive face. If there was no preference (as stated by the null hypothesis), the population mean would be $\mu = 10$ seconds. Thus, the results show a 3-second difference between the obtained mean ($M = 13$) and the hypothesized mean if there is no preference ($\mu = 10$). Also, for this study the sample standard deviation is

$$s = \sqrt{\frac{SS}{df}} = \sqrt{\frac{72}{8}} = \sqrt{9} = 3$$

Thus, Cohen's *d* for this example is estimated to be

$$\text{Cohen's } d = \frac{M - \mu}{s} = \frac{13 - 10}{3} = 1.00$$

According to the standards suggested by Cohen (Table 8.2, page 253), this is a large treatment effect. ∎

To help you visualize what is measured by Cohen's *d*, we have constructed a set of $n = 9$ scores with a mean of $M = 13$ and a standard deviation of $s = 3$ (the same values as in Examples 9.2 and 9.3). The set of scores is shown in Figure 9.5. Notice that the figure also includes an arrow that locates $\mu = 10$. Recall that $\mu = 10$ is the value specified by the null hypothesis and identifies what the mean ought to be if the treatment has no effect. Clearly, our sample is not centered at $\mu = 10$. Instead, the scores have been shifted to the right so that the sample mean is $M = 13$. This shift, from 10 to 13, is the 3-point mean difference that was caused by the treatment effect. Also notice that the 3-point mean difference is exactly equal to the standard deviation. Thus, the size of the treatment effect is equal to 1 standard deviation. In other words, Cohen's $d = 1.00$.

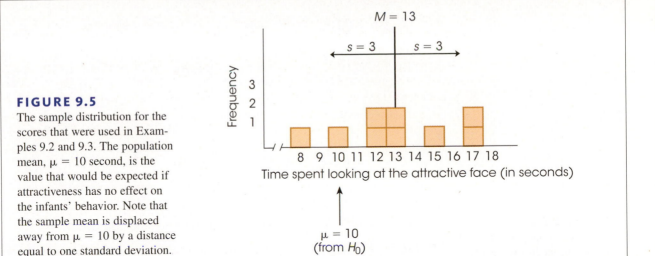

FIGURE 9.5

The sample distribution for the scores that were used in Examples 9.2 and 9.3. The population mean, $\mu = 10$ second, is the value that would be expected if attractiveness has no effect on the infants' behavior. Note that the sample mean is displaced away from $\mu = 10$ by a distance equal to one standard deviation.

The following example is an opportunity for you to test your understanding of hypothesis testing and effect size with the *t* statistic.

EXAMPLE 9.4 A sample of $n = 16$ individuals is selected from a population with a mean of $\mu = 40$. A treatment is administered to the individuals in the sample and, after treatment, the sample has a mean of $M = 44$ and a variance of $s^2 = 16$. Use a two-tailed test with $\alpha = .05$ to determine whether the treatment effect is significant and compute Cohen's *d* to measure the size of the treatment effect. You should obtain $t = 4.00$ with $df = 15$, which is large enough to reject H_0 with Cohen's $d = 1.00$. ∎

■ Measuring the Percentage of Variance Explained, r^2

An alternative method for measuring effect size is to determine how much of the variability in the scores is explained by the treatment effect. The concept behind this measure is that the treatment causes the scores to increase (or decrease), which means that the treatment is causing the scores to vary. If we can measure how much of the variability is explained by the treatment, we will obtain a measure of the size of the treatment effect.

To demonstrate this concept we will use the data from the hypothesis test in Example 9.2. Recall that the null hypothesis stated that the treatment (the attractiveness of the faces) has no effect on the infants' behavior. According to the null hypothesis, the infants should show no preference between the two photographs, and therefore should spend an average of $\mu = 10$ out of 20 seconds looking at the attractive face.

However, if you look at the data in Figure 9.5, the scores are not centered at $\mu = 10$. Instead, the scores are shifted to the right so that they are centered around the sample mean, $M = 13$. This shift is the treatment effect. To measure the size of the treatment effect we calculate deviations from the mean and the sum of squared deviations, *SS*, two different ways.

Figure 9.6(a) shows the original set of scores. For each score, the deviation from $\mu = 10$ is shown as a colored line. Recall that $\mu = 10$ is from the null hypothesis and represents the population mean if the treatment has no effect. Note that almost all of the scores are located on the right-hand side of $\mu = 10$. This shift to the right is the treatment effect. Specifically, the preference for the attractive face has caused the infants to spend more time looking

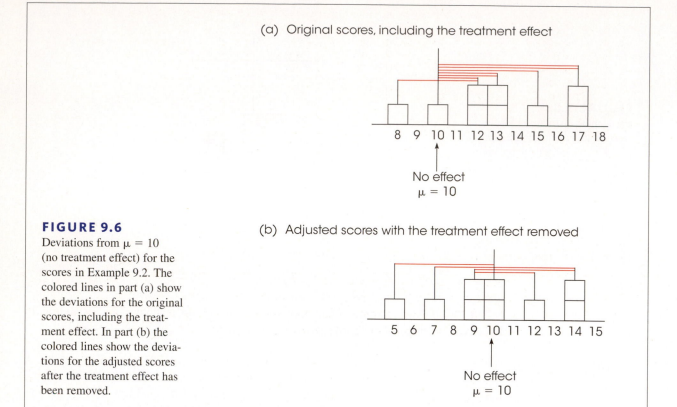

FIGURE 9.6

Deviations from $\mu = 10$ (no treatment effect) for the scores in Example 9.2. The colored lines in part (a) show the deviations for the original scores, including the treatment effect. In part (b) the colored lines show the deviations for the adjusted scores after the treatment effect has been removed.

at the attractive photograph, which means that their scores are generally greater than 10. Thus, the treatment has pushed the scores away from $\mu = 10$ and has increased the size of the deviations.

Next, we will see what happens if the treatment effect is removed. In this example, the treatment has a 3-point effect (the average increases from $\mu = 10$ to $M = 13$). To remove the treatment effect, we simply subtract 3 points from each score. The adjusted scores are shown in Figure 9.6(b) and, once again, the deviations from $\mu = 10$ are shown as colored lines. First, notice that the adjusted scores are centered at $\mu = 10$, indicating that there is no treatment effect. Also notice that the deviations, the colored lines, are noticeably smaller when the treatment effect is removed.

To measure how much the variability is reduced when the treatment effect is removed, we compute the sum of squared deviations, *SS*, for each set of scores. The left-hand columns of Table 9.2 show the calculations for the original scores (Figure 9.6(a)), and the right-hand columns show the calculations for the adjusted scores (Figure 9.6(b)). Note that the total variability, including the treatment effect, is $SS = 153$. However, when the treatment effect is removed, the variability is reduced to $SS = 72$. The difference between these two values, $153 - 72 = 81$ points, is the amount of variability that is accounted for by the treatment effect. This value is usually reported as a proportion or percentage of the total variability:

$$\frac{\text{variability accounted for}}{\text{total variability}} = \frac{81}{153} = 0.5294 \quad (52.94\%)$$

TABLE 9.2

Calculation of *SS*, the sum of squared deviations, for the data in Figure 9.6. The first three columns show the calculations for the original scores, including the treatment effect. The last three columns show the calculations for the adjusted scores after the treatment effect has been removed.

	Calculation of *SS* including the treatment effect			Calculation of *SS* after the treatment effect is removed		
Score	Deviation from $\mu = 10$	Squared Deviation	Adjusted Score	Deviation from $\mu = 10$	Squared Deviation
8	−2	4	8 − 3 = 5	−5	25
10	0	0	10 − 3 = 7	−3	9
12	2	4	12 − 3 = 9	−1	1
12	2	4	12 − 3 = 9	−1	1
13	3	9	13 − 3 = 10	0	0
13	3	9	13 − 3 = 10	0	0
15	5	25	15 − 3 = 12	2	4
17	7	49	17 − 3 = 14	4	16
17	7	49	17 − 3 = 14	4	16
		SS = 153			*SS* = 72

Thus, removing the treatment effect reduces the variability by 52.94%. This value is called the *percentage of variance accounted for by the treatment* and is identified as r^2.

Rather than computing r^2 directly by comparing two different calculations for *SS*, the value can be found from a single equation based on the outcome of the *t* test.

$$r^2 = \frac{t^2}{t^2 + df} \qquad (9.5)$$

The letter *r* is the traditional symbol used for a correlation, and the concept of r^2 is discussed again when we consider correlations in Chapter 15. Also, in the context of *t* statistics, the percentage of variance that we are calling r^2 is often identified by the Greek letter omega squared (ω^2).

For the hypothesis test in Example 9.2, we obtained $t = 3.00$ with $df = 8$. These values produce

$$r^2 = \frac{3^2}{3^2 + 8} = \frac{9}{17} = 0.5294 \quad (52.94\%)$$

Note that this is exactly the same value we obtained with the direct calculation of the percentage of variability accounted for by the treatment.

Interpreting r^2 In addition to developing the Cohen's *d* measure of effect size, Cohen (1988) also proposed criteria for evaluating the size of a treatment effect that is measured by r^2. The criteria were actually suggested for evaluating the size of a correlation, *r*, but are easily extended to apply to r^2. Cohen's standards for interpreting r^2 are shown in Table 9.3.

According to these standards, the data we constructed for Examples 9.1 and 9.2 show a very large effect size with $r^2 = .5294$.

TABLE 9.3

Criteria for interpreting the value of r^2 as proposed by Cohen (1988).

Percentage of Variance Explained, r^2	
$r^2 = 0.01$	Small effect
$r^2 = 0.09$	Medium effect
$r^2 = 0.25$	Large effect

As a final note, we should remind you that, although sample size affects the hypothesis test, this factor has little or no effect on measures of effect size. In particular, estimates of Cohen's *d* are not influenced at all by sample size, and measures of r^2 are only slightly affected by changes in the size of the sample. The sample variance, on the other hand, influences hypothesis tests and measures of effect size. Specifically, high variance reduces the likelihood of rejecting the null hypothesis and it reduces measures of effect size.

■ Confidence Intervals for Estimating μ

An alternative technique for describing the size of a treatment effect is to compute an estimate of the population mean after treatment. For example, if the mean before treatment is known to be $\mu = 80$ and the mean after treatment is estimated to be $\mu = 86$, then we can conclude that the size of the treatment effect is around 6 points.

Estimating an unknown population mean involves constructing a *confidence interval*. A confidence interval is based on the observation that a sample mean tends to provide a reasonably accurate estimate of the population mean. The fact that a sample mean tends to be near to the population mean implies that the population mean should be near to the sample mean. Thus, if we obtain a sample mean of $M = 86$, we can be reasonably confident that the population mean is around 86. Thus, a confidence interval consists of an interval of values around a sample mean, and we can be reasonably confident that the unknown population mean is located somewhere in the interval.

DEFINITION

> A **confidence interval** is an interval, or range of values centered around a sample statistic. The logic behind a confidence interval is that a sample statistic, such as a sample mean, should be relatively near to the corresponding population parameter.

Therefore, we can confidently estimate that the value of the parameter should be located in the interval.

■ Constructing a Confidence Interval

The construction of a confidence interval begins with the observation that every sample mean has a corresponding *t* value defined by the equation

$$t = \frac{M - \mu}{s_M}$$

Although the values for M and s_M are available from the sample data, we do not know the values for *t* or for μ. However, we can *estimate* the *t* value. For example, if the sample has $n = 9$ scores, then the *t* statistic has $df = 8$, and the distribution of all possible *t* values can be pictured as in Figure 9.7. Notice that the *t* values pile up around $t = 0$, so we can estimate that the *t* value for our sample should have a value around 0. Furthermore, the *t* distribution table lists a variety of different *t* values that correspond to specific proportions of the *t* distribution. With $df = 8$, for example, 80% of the *t* values are located between $t = +1.397$ and $t = -1.397$. To obtain these value, simply look up a two-tailed proportion of 0.20 (20%) for $df = 8$. Because 80% of all the possible *t* values are located between ±1.397, we can be 80% confident that our sample mean

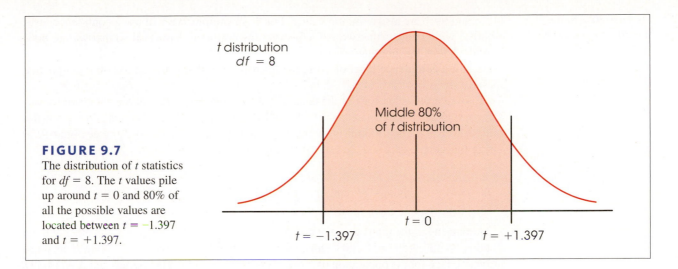

FIGURE 9.7

The distribution of *t* statistics for $df = 8$. The *t* values pile up around $t = 0$ and 80% of all the possible values are located between $t = -1.397$ and $t = +1.397$.

t distribution
$df = 8$

Middle 80%
of *t* distribution

$t = -1.397$ $t = 0$ $t = +1.397$

corresponds to a *t* value in this interval. Similarly, we can be 95% confident that the mean for a sample of $n = 9$ scores corresponds to a *t* value between ± 2.306. Notice that we are able to estimate the value of *t* with a specific level of confidence. To construct a confidence interval for μ, we plug the estimated *t* value into the *t* equation, and then we can calculate the value of μ.

Before we demonstrate the process of constructing a confidence interval for an unknown population mean, we simplify the calculations by regrouping the terms in the *t* equation. Because the goal is to compute the value of μ, we use simple algebra to solve the equation for μ. The result is $\mu = M - ts_M$. However, we estimate that the *t* value is in an interval around 0, with one end at $+t$ and the other end at $-t$. The $\pm t$ can be incorporated in the equation to produce

$$\mu = M \pm ts_M \tag{9.6}$$

This is the basic equation for a confidence interval. Notice that the equation produces an interval around the sample mean. One end of the interval is located at $M + ts_M$ and the other end is at $M - ts_M$. The process of using this equation to construct a confidence interval is demonstrated in the following example.

EXAMPLE 9.5   Example 9.2 describes a study in which infants displayed a preference for the more attractive face by looking at it, instead of the less attractive face, for the majority of a 20-second viewing period. Specifically, a sample of $n = 9$ infants spent an average of $M = 13$ seconds, out of a total 20-second period, looking at the more attractive face. The data produced an estimated standard error of $s_M = 1$. We will use this sample to construct a confidence interval to estimate the mean amount of time that the population of infants spends looking at the more attractive face. That is, we will construct an interval of values that is likely to contain the unknown population mean.

Again, the estimation formula is

$$\mu = M \pm t(s_M)$$

In the equation, the value of $M = 13$ and $s_M = 1$ are obtained from the sample data. The next step is to select a level of confidence that will determine the value of *t* in the equation. The most commonly used confidence level is probably 95% but values of 80%,

To have 80% in the middle there must be 20% (or .20) in the tails. To find the *t* values, look under two tails, .20 in the *t* table.

90%, and 99% are also common values. For this example, we will use a confidence level of 80%, which means that we will construct the confidence interval so that we are 80% confident that the population mean is actually contained in the interval. Because we are using a confidence level of 80%, the resulting interval is called the 80% *confidence interval for* μ.

To obtain the value for *t* in the equation, we simply estimate that the *t* statistic for our sample is located somewhere in the middle 80% of the *t* distribution. With $df = n - 1 = 8$, the middle 80% of the distribution is bounded by *t* values of $+1.397$ and -1.397 (see Figure 9.7). Using the sample data and the estimated range of *t* values, we obtain

$$\mu = M \pm t(s_M) = 13 \pm 1.397(1.00) = 13 \pm 1.397$$

At one end of the interval, we obtain $\mu = 13 + 1.397 = 14.397$, and at the other end we obtain $\mu = 13 - 1.397 = 11.603$. Our conclusion is that the average time looking at the more attractive face for the population of infants is between $\mu = 11.603$ seconds and $\mu = 14.397$ seconds, and we are 80% confident that the true population mean is located within this interval. The confidence comes from the fact that the calculation was based on only one assumption. Specifically, we assumed that the *t* statistic was located between $+1.397$ and -1.397, and we are 80% confident that this assumption is correct because 80% of all the possible *t* values are located in this interval. Finally, note that the confidence interval is constructed around the sample mean. As a result, the sample mean, $M = 13$, is located exactly in the center of the interval. ■

■ Factors Affecting the Width of a Confidence Interval

Two characteristics of the confidence interval should be noted. First, notice what happens to the width of the interval when you change the level of confidence (the percent confidence). To gain more confidence in your estimate, you must increase the width of the interval. Conversely, to have a smaller, more precise interval, you must give up confidence. In the estimation formula, the percentage of confidence influences the value of *t*. A larger level of confidence (the percentage), produces a larger the *t* value and a wider interval. This relationship can be seen in Figure 9.7. In the figure, we identified the middle 80% of the *t* distribution in order to find an 80% confidence interval. It should be obvious that if we were to increase the confidence level to 95%, it would be necessary to increase the range of *t* values and thereby increase the width of the interval.

Second, note what happens to the width of the interval if you change the sample size. This time the basic rule is as follows: The bigger the sample (*n*), the smaller the interval. This relationship is straightforward if you consider the sample size as a measure of the amount of information. A bigger sample gives you more information about the population and allows you to make a more precise estimate (a narrower interval). The sample size controls the magnitude of the standard error in the estimation formula. As the sample size increases, the standard error decreases, and the interval gets smaller. Because confidence intervals are influenced by sample size, they do not provide an unqualified measure of absolute effect size and are not an adequate substitute for Cohen's *d* or r^2. Nonetheless, they can be used in a research report to provide a description of the size of the treatment effect.

IN THE LITERATURE

Reporting the Results of a *t* Test

In Chapter 8, we noted the conventional style for reporting the results of a hypothesis test, according to APA format. First, recall that a scientific report typically uses the term *significant* to indicate that the null hypothesis has been rejected and the term *not significant* to indicate failure to reject H_0. Additionally, there is a prescribed format for reporting the calculated value of the test statistic, degrees of freedom, and alpha level for a *t* test. This format parallels the style introduced in Chapter 8 (page 241).

In Example 9.2 we calculated a *t* statistic of 3.00 with $df = 8$, and we decided to reject H_0 with alpha set at .05. Using the same data, we obtained $r^2 = 0.5294$ (52.94%) for the percentage of variance explained by the treatment effect. In a scientific report, this information is conveyed in a concise statement, as follows:

> The infants spent an average of $M = 13$ out of 20 seconds looking at the attractive face, with $SD = 3.00$. Statistical analysis indicates that the time spent looking at the attractive face was significantly greater than would be expected if there were no preference, $t(8) = 3.00$, $p < .05$, $r^2 = 0.5294$.

The first statement reports the descriptive statistics, the mean ($M = 13$) and the standard deviation ($SD = 3$), as previously described (Chapter 4, page 121). The next statement provides the results of the inferential statistical analysis. Note that the degrees of freedom are reported in parentheses immediately after the symbol *t*. The value for the obtained *t* statistic follows (3.00), and next is the probability of committing a Type I error (less than 5%). Finally, the effect size is reported, $r^2 = 52.94\%$. If the 80% confidence interval from Example 9.5 were included in the report as a description of effect size, it would be added after the results of the hypothesis test as follows:

$$t(8) = 3.00, p < .05, 80\% \text{ CI } [11.603, 14.397].$$

Often, researchers use a computer to perform a hypothesis test like the one in Example 9.2. In addition to calculating the mean, standard deviation, and the *t* statistic for the data, the computer usually calculates and reports the *exact probability* (or α level) associated with the *t* value. In Example 9.2 we determined that any *t* value beyond ± 2.306 has a probability of less than .05 (see Figure 9.4). Thus, the obtained *t* value, $t = 3.00$, is reported as being very unlikely, $p < .05$. A computer printout, however, would have included an exact probability for our specific *t* value.

Whenever a specific probability value is available, you are encouraged to use it in a research report. For example, the computer analysis of these data reports an exact *p* value of $p = .017$, and the research report would state "$t(8) = 3.00$, $p = .017$" instead of using the less specific "$p < .05$." As one final caution, we note that occasionally a *t* value is so extreme that the computer reports $p = 0.000$. The zero value does not mean that the probability is literally zero; instead, it means that the computer has rounded off the probability value to three decimal places and obtained a result of 0.000. In this situation, you do not know the exact probability value, but you can report $p < .001$.

LEARNING CHECK

1. A researcher uses a sample of $n = 10$ to evaluate a treatment effect and obtains $t = 3.00$. If effect size is measured using the percentage of variance accounted for, then what value will be obtained for r^2?
   a. $\frac{9}{18} = 0.50$
   b. $\frac{9}{17} = 0.53$
   c. $\frac{3}{12} = 0.25$
   d. $\frac{3}{11} = 0.27$

2. What happens to measures of effect size such as r^2 and Cohen's *d* as sample size increases?
   a. They also tend to increase.
   b. They tend to decrease.
   c. Sample size does not have any great influence on measures of effect size.
   d. The effect of sample size depends on other factors such as sample variance.

3. Which set of factors would produce the narrowest confidence interval for estimating a population mean?
   a. A large sample and a large percentage of confidence
   b. A large sample and a small percentage of confidence
   c. A small sample and a large percentage of confidence
   d. A small sample and a small percentage of confidence

4. The results of a hypothesis test are reported as follows: "$t(20) = 2.70, p < .05$." Based on this report, how many individuals were in the sample?
   a. 19
   b. 20
   c. 21
   d. cannot be determined from the information provided

ANSWERS   1. A, 2. C, 3. B, 4. C

9.4 | Directional Hypotheses and One-Tailed Tests

LEARNING OBJECTIVE

9. Conduct a directional (one-tailed) hypothesis test using the *t* statistic.

As noted in Chapter 8, the nondirectional (two-tailed) test is more commonly used than the directional (one-tailed) alternative. On the other hand, a directional test may be used in some research situations, such as exploratory investigations or pilot studies or when there is *a priori* justification (for example, a theory or previous findings). The following example demonstrates a directional hypothesis test with a *t* statistic, using the same experimental situation presented in Example 9.2.

EXAMPLE 9.6

The research question is whether attractiveness affects the behavior of infants looking at photographs of women's faces. The researcher is expecting the infants to prefer the more attractive face. Therefore, the researcher predicts that the infants will spend more than half of the 20-second period looking at the attractive face. For this example we will use the same sample data that were used in the original hypothesis test in Example 9.2. Specifically, the researcher tested a sample of $n = 9$ infants and obtained a mean of $M = 13$ seconds looking at the attractive face with $SS = 72$.

STEP 1   **State the hypotheses, and select an alpha level.**   With most directional tests, it is usually easier to state the hypothesis in words, including the directional prediction, and then convert the words into symbols. For this example, the researcher is predicting that attractiveness will cause the infants to increase the amount of time they spend looking at the attractive face; that is, more than half of the 20 seconds should be spent looking at the attractive face. In general, the null hypothesis states that the predicted effect will not happen. For this study, the null hypothesis states that the infants will not spend more than half of the 20 seconds looking at the attractive face. In symbols,

H_0: $\mu_{attractive} \leq 10$ seconds   **(Not more than half of the 20 seconds looking at the attractive face)**

Similarly, the alternative states that the treatment will work. In this case, H_1 states that the infants will spend more than half of the time looking at the attractive face. In symbols,

H_1: $\mu_{attractive} > 10$ seconds   **(More than half of the 20 seconds looking at the attractive face)**

This time, we will set the level of significance at $\alpha = .01$.

STEP 2   **Locate the critical region.**   In this example, the researcher is predicting that the sample mean (M) will be greater than 10 seconds. Thus, if the infants average more than 10 seconds looking at the attractive face, the data will provide support for the researcher's prediction and will tend to refute the null hypothesis. Also note that a sample mean greater than 10 will produce a positive value for the t statistic. Thus, the critical region for the one-tailed test will consist of positive t values located in the right-hand tail of the distribution. However, we must still determine exactly how large a value is necessary to reject the null hypothesis. To find the critical value, you must look in the t distribution table using the one-tail proportions. With a sample of $n = 9$, the t statistic will have $df = 8$; using $\alpha = .01$, you should find a critical value of $t = 2.896$. Therefore, if we obtain a sample mean greater than 10 seconds and the sample mean produces a t statistic greater than 2.896, we will reject the null hypothesis and conclude that the infants show a significant preference for the attractive face. Figure 9.8 shows the one-tailed critical region for this test.

STEP 3   **Calculate the test statistic.**   The computation of the t statistic is the same for either a one-tailed or a two-tailed test. Earlier (in Example 9.2), we found that the data for this experiment produce a test statistic of $t = 3.00$.

STEP 4   **Make a decision.**   The test statistic is in the critical region, so we reject H_0. In terms of the experimental variables, we have decided that the infants show a preference and spend significantly more time looking at the attractive face than they do looking at the unattractive face. In a research report the results would be presented as follows:

> The time spent looking at the attractive face was significantly greater than would be expected if there were no preference, $t(8) = 3.00$, $p < .01$, one tailed.

Note that the report clearly acknowledges that a one-tailed test was used.   ■

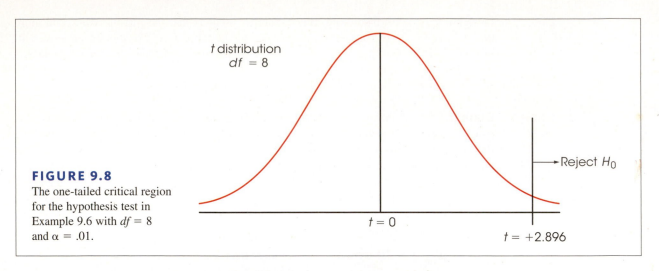

FIGURE 9.8

The one-tailed critical region for the hypothesis test in Example 9.6 with *df* = 8 and α = .01.

■ The Critical Region for a One-Tailed Test

In Step 2 of Example 9.6, we determined that the critical region is in the right-hand tail of the distribution. However, it is possible to divide this step into two stages that eliminate the need to determine which tail (right or left) should contain the critical region. The first stage in this process is simply to determine whether the sample mean is in the direction predicted by the original research question. For this example, the researcher predicted that the infants would prefer the attractive face and spend more time looking at it. Specifically, the researcher expects the infants to spend more than 10 out of 20 seconds focused on the attractive face. The obtained sample mean, *M* = 13 seconds, is in the correct direction. This first stage eliminates the need to determine whether the critical region is in the left- or right-hand tail. Because we already have determined that the effect is in the correct direction, the sign of the *t* statistic (+ or −) no longer matters. The second stage of the process is to determine whether the effect is large enough to be significant. For this example, the requirement is that the sample produces a *t* statistic greater than 2.896. If the magnitude of the *t* statistic, independent of its sign, is greater than 2.896, the result is significant and H_0 is rejected.

LEARNING CHECK

1. A researcher selects a sample of *n* = 16 individuals from a population with a mean of μ = 75 and administers a treatment to the sample. If the research predicts that the treatment will increase scores, then what is the correct statement of the null hypothesis for a directional (one-tailed) test?

   a. μ ≥ 75

   b. μ > 75

   c. μ ≤ 75

   d. μ < 75

2. A researcher is conducting a directional (one-tailed) test with a sample of *n* = 25 to evaluate the effect of a treatment that is predicted to decrease scores. If the researcher obtains *t* = −1.700, then what decision should be made?

   a. The treatment has a significant effect with either α = .05 or α = .01.

   b. The treatment does not have a significant effect with either α = .05 or α = .01.

   c. The treatment has a significant effect with α = .05 but not with α = .01.

   d. The treatment has a significant effect with α = .01 but not with α = .05.

3. A researcher rejects the null hypothesis with a regular two-tailed test using $\alpha = .05$. If the researcher used a directional (one-tailed) test with the same data, then what decision would be made?

   a. Definitely reject the null hypothesis.

   b. Definitely reject the null hypothesis if the treatment effect is in the predicted direction.

   c. Possibly not reject the null hypothesis even if the treatment effect is in the predicted direction.

   d. It is impossible to determine without more information.

ANSWERS   **1.** C, **2.** B, **3.** B

SUMMARY

1. The t statistic is used instead of a z-score for hypothesis testing when the population standard deviation (or variance) is unknown.

2. To compute the t statistic, you must first calculate the sample variance (or standard deviation) as a substitute for the unknown population value.

$$\text{sample variance} = s^2 = \frac{SS}{df}$$

Next, the standard error is *estimated* by substituting s^2 in the formula for standard error. The estimated standard error is calculated in the following manner:

$$\text{estimated standard error} = s_M = \sqrt{\frac{s^2}{n}}$$

Finally, a t statistic is computed using the estimated standard error. The t statistic is used as a substitute for a z-score that cannot be computed when the population variance or standard deviation is unknown.

$$t = \frac{M - \mu}{s_M}$$

3. The structure of the t formula is similar to that of the z-score in that

$$z \text{ or } t = \frac{\text{sample mean} - \text{population mean}}{\text{(estimated) standard error}}$$

For a hypothesis test, you hypothesize a value for the unknown population mean and plug the hypothesized value into the equation along with the sample mean and the estimated standard error, which are computed from the sample data. If the hypothesized mean produces an extreme value for t, you conclude that the hypothesis was wrong.

4. The t distribution is an approximation of the normal z distribution. To evaluate a t statistic that is obtained for a sample mean, the critical region must be located in a t distribution. There is a family of t distributions, with the exact shape of a particular distribution of t values depending on degrees of freedom $(n - 1)$. Therefore, the critical t values depend on the value for df associated with the t test. As df increases, the shape of the t distribution approaches a normal distribution.

5. When a t statistic is used for a hypothesis test, Cohen's d can be computed to measure effect size. In this situation, the sample standard deviation is used in the formula to obtain an estimated value for d:

$$\text{estimated } d = \frac{\text{mean difference}}{\text{standard deviation}} = \frac{M - \mu}{s}$$

6. A second measure of effect size is r^2, which measures the percentage of the variability that is accounted for by the treatment effect. This value is computed as follows:

$$r^2 = \frac{t^2}{t^2 + df}$$

7. An alternative method for describing the size of a treatment effect is to use a confidence interval for μ. A confidence interval is a range of values that estimates the unknown population mean. The confidence interval uses the *t* equation, solved for the unknown mean:

$$\mu = M \pm t(s_M)$$

First, select a level of confidence and then look up the corresponding *t* values to use in the equation. For example, for 95% confidence, use the range of *t* values that determine the middle 95% of the distribution.

KEY TERMS

estimated standard error (270)	*t* distribution (271)	percentage of variance accounted for by the treatment (r^2) (283)
t statistic (270)	estimated *d* (280)	
degrees of freedom (271)		confidence interval (284)

SPSS®

General instructions for using SPSS are presented in Appendix D. Following are detailed instructions for using SPSS to perform the *t* **Test** presented in this chapter.

Data Entry

1. Enter all of the scores from the sample in one column of the data editor, probably VAR00001.

Data Analysis

1. Click Analyze on the tool bar, select **Compare Means**, and click on **One-Sample T Test**.
2. Highlight the column label for the set of scores (VAR0001) in the left box and click the arrow to move it into the **Test Variable**(s) box.
3. In the **Test Value** box at the bottom of the One-Sample *t* Test window, enter the hypothesized value for the population mean from the null hypothesis. *Note*: The value is automatically set at zero until you type in a new value.
4. In addition to performing the hypothesis test, the program will compute a confidence interval for the population mean difference. The confidence level is automatically set at 95% but you can select **Options** and change the percentage.
5. Click **OK**.

SPSS Output

We used the SPSS program to analyze the data from the infants-and-attractive-faces study in Example 9.2 and the program output is shown in Figure 9.9. The output includes a table of sample statistics with the mean, standard deviation, and standard error for the sample mean. A second table shows the results of the hypothesis test, including the values for *t*, *df*, and the level of significance (the *p* value for the test), as well as the mean difference from the hypothesized value of $\mu = 10$ and a 95% confidence interval for the mean difference. To obtain a 95% confidence interval for the mean, simply add $\mu = 10$ points to the values in the table.

One-Sample Statistics

	N	Mean	Std. Deviation	Std. Error Mean
VAR00001	9	13.0000	3.00000	1.00000

One-Sample Test

	Test Value = 10					
					95% Confidence Interval of the Difference	
	t	df	Sig. (2-tailed)	Mean Difference	Lower	Upper
VAR00001	3.000	8	.017	3.00000	.6940	5.3060

FIGURE 9.9
The SPSS output for the hypothesis test in Example 9.2.

FOCUS ON PROBLEM SOLVING

1. The first problem we confront in analyzing data is determining the appropriate statistical test. Remember that you can use a z-score for the test statistic only when the value for σ is known. If the value for σ is not provided, then you must use the t statistic.

2. For the t test, the sample variance is used to find the value for estimated standard error. Remember that when computing the sample variance, use $n - 1$ in the denominator (see Chapter 4). When computing estimated standard error, use n in the denominator.

DEMONSTRATION 9.1

A HYPOTHESIS TEST WITH THE t STATISTIC

A psychologist has prepared an "Optimism Test" that is administered yearly to graduating college seniors. The test measures how each graduating class feels about its future—the higher the score, the more optimistic the class. Last year's class had a mean score of $\mu = 15$. A sample of $n = 9$ seniors from this year's class was selected and tested. The scores for these seniors are 7, 12, 11, 15, 7, 8, 15, 9, and 6, which produce a sample mean of $M = 10$ with $SS = 94$.

On the basis of this sample, can the psychologist conclude that this year's class has a different level of optimism than last year's class?

Note that this hypothesis test will use a t statistic because the population variance (σ^2) is not known.

STEP 1 State the hypotheses, and select an alpha level   The statements for the null hypothesis and the alternative hypothesis follow the same form for the t statistic and the z-score test.

$$H_0: \mu = 15 \quad \text{(There is no change.)}$$

$$H_1: \mu \neq 15 \quad \text{(This year's mean is different.)}$$

For this demonstration, we will use $\alpha = .05$, two tails.

STEP 2   **Locate the critical region**   With a sample of $n = 9$ students, the *t* statistic has $df = n - 1 = 8$. For a two-tailed test with $\alpha = .05$ and $df = 8$, the critical *t* values are $t = \pm 2.306$. These critical *t* values define the boundaries of the critical region. The obtained *t* value must be more extreme than either of these critical values to reject H_0.

STEP 3   **Compute the test statistic**   As we have noted, it is easier to separate the calculation of the *t* statistic into three stages.

Sample Variance

$$s^2 = \frac{SS}{n-1} = \frac{94}{8} = 11.75$$

Estimated Standard Error The estimated standard error for these data is

$$s_M = \sqrt{\frac{s^2}{n}} = \sqrt{\frac{11.75}{9}} = 1.14$$

The t Statistic. Now that we have the estimated standard error and the sample mean, we can compute the *t* statistic. For this demonstration,

$$t = \frac{M - \mu}{s_M} = \frac{10 - 15}{1.14} = \frac{-5}{1.14} = -4.39$$

STEP 4   **Make a decision about H_0, and state a conclusion**   The *t* statistic we obtained ($t = -4.39$) is in the critical region. Thus, our sample data are unusual enough to reject the null hypothesis at the .05 level of significance. We can conclude that there is a significant difference in level of optimism between this year's and last year's graduating classes, $t(8) = -4.39$, $p < .05$, two-tailed.

DEMONSTRATION 9.2

EFFECT SIZE: ESTIMATING COHEN'S *d* AND COMPUTING r^2

We will estimate Cohen's *d* for the same data used for the hypothesis test in Demonstration 9.1. The mean optimism score for the sample from this year's class was 5 points lower than the mean from last year ($M = 10$ vs. $\mu = 15$). In Demonstration 9.1 we computed a sample variance of $s^2 = 11.75$, so the standard deviation is $\sqrt{11.75} = 3.43$. With these values,

$$\text{estimated } d = \frac{\text{mean difference}}{\text{standard deviation}} = \frac{5}{3.43} = 1.46$$

To calculate the percentage of variance explained by the treatment effect, r^2, we need the value of *t* and the *df* value from the hypothesis test. In Demonstration 9.1 we obtained $t = -4.39$ with $df = 8$. Using these values in Equation 9.5, we obtain

$$r^2 = \frac{t^2}{t^2 + df} = \frac{(-4.39)^2}{(-4.39)^2 + 8} = \frac{19.27}{27.27} = 0.71$$

PROBLEMS

1. Under what circumstances is a t statistic used instead of a z-score for a hypothesis test?

2. A sample of $n = 16$ scores has a mean of $M = 56$ and a standard deviation of $s = 12$.
   a. Explain what is measured by the sample standard deviation.
   b. Compute the estimated standard error for the sample mean and explain what is measured by the standard error.

3. Find the estimated standard error for the sample mean for each of the following samples.
   a. $n = 9$ with $SS = 1152$
   b. $n = 16$ with $SS = 540$
   c. $n = 25$ with $SS = 600$

4. Explain why t distributions tend to be flatter and more spread out than the normal distribution.

5. Find the t values that form the boundaries of the critical region for a two-tailed test with $\alpha = .05$ for each of the following sample sizes:
   a. $n = 4$
   b. $n = 15$
   c. $n = 24$

6. Find the t value that forms the boundary of the critical region in the right-hand tail for a one-tailed test with $\alpha = .01$ for each of the following sample sizes.
   a. $n = 10$
   b. $n = 20$
   c. $n = 30$

7. The following sample of $n = 4$ scores was obtained from a population with unknown parameters.
   Scores: 2, 2, 6, 2
   a. Compute the sample mean and standard deviation. (Note that these are descriptive values that summarize the sample data.)
   b. Compute the estimated standard error for M. (Note that this is an inferential value that describes how accurately the sample mean represents the unknown population mean.)

8. The following sample was obtained from a population with unknown parameters.
   Scores: 13, 7, 6, 12, 0, 4
   a. Compute the sample mean and standard deviation. (Note that these are descriptive values that summarize the sample data.)
   b. Compute the estimated standard error for M. (Note that this is an inferential value that describes how accurately the sample mean represents the unknown population mean.)

9. A random sample of $n = 12$ individuals is selected from a population with $\mu = 70$, and a treatment is administered to each individual in the sample. After treatment, the sample mean is found to be $M = 74.5$ with $SS = 297$.
   a. How much difference is there between the mean for the treated sample and the mean for the original population? (*Note*: In a hypothesis test, this value forms the numerator of the t statistic.)
   b. How much difference is expected just by chance between the sample mean and its population mean? That is, find the standard error for M. (*Note*: In a hypothesis test, this value is the denominator of the t statistic.)
   c. Based on the sample data, does the treatment have a significant effect? Use a two-tailed test with $\alpha = .05$.

10. A random sample of $n = 25$ individuals is selected from a population with $\mu = 20$, and a treatment is administered to each individual in the sample. After treatment, the sample mean is found to be $M = 22.2$ with $SS = 384$.
    a. How much difference is there between the mean for the treated sample and the mean for the original population? (*Note*: In a hypothesis test, this value forms the numerator of the t statistic.)
    b. If there is no treatment effect, how much difference is expected between the sample mean and its population mean? That is, find the standard error for M. (*Note*: In a hypothesis test, this value is the denominator of the t statistic.)
    c. Based on the sample data, does the treatment have a significant effect? Use a two-tailed test with $\alpha = .05$.

11. To evaluate the effect of a treatment, a sample is obtained from a population with a mean of $\mu = 30$, and the treatment is administered to the individuals in the sample. After treatment, the sample mean is found to be $M = 31.3$ with a standard deviation of $\sigma = 3$.
    a. If the sample consists of $n = 16$ individuals, are the data sufficient to conclude that the treatment has a significant effect using a two-tailed test with $\alpha = .05$?
    b. If the sample consists of $n = 36$ individuals, are the data sufficient to conclude that the treatment has a significant effect using a two-tailed test with $\alpha = .05$?
    c. Comparing your answer for parts a and b, how does the size of the sample influence the outcome of a hypothesis test?

12. To evaluate the effect of a treatment, a sample of $n = 8$ is obtained from a population with a mean of $\mu = 40$, and the treatment is administered to the individuals in

the sample. After treatment, the sample mean is found to be $M = 35$.

a. If the sample variance is $s^2 = 32$, are the data sufficient to conclude that the treatment has a significant effect using a two-tailed test with $\alpha = .05$?

b. If the sample variance is $s^2 = 72$, are the data sufficient to conclude that the treatment has a significant effect using a two-tailed test with $\alpha = .05$?

c. Comparing your answer for parts *a* and *b*, how does the variability of the scores in the sample influence the outcome of a hypothesis test?

13. The spotlight effect refers to overestimating the extent to which others notice your appearance or behavior, especially when you commit a social faux pas. Effectively, you feel as if you are suddenly standing in a spotlight with everyone looking. In one demonstration of this phenomenon, Gilovich, Medvec, and Savitsky (2000) asked college students to put on a Barry Manilow T-shirt that fellow students had previously judged to be embarrassing. The participants were then led into a room in which other students were already participating in an experiment. After a few minutes, the participant was led back out of the room and was allowed to remove the shirt. Later, each participant was asked to estimate how many people in the room had noticed the shirt. The individuals who were in the room were also asked whether they noticed the shirt. In the study, the participants significantly overestimated the actual number of people who had noticed.

a. In a similar study using a sample of $n = 9$ participants, the individuals who wore the shirt produced an average estimate of $M = 6.4$ with $SS = 162$. The average number who said they noticed was 3.1. Is the estimate from the participants significantly different from the actual number? Test the null hypothesis that the true mean is $\mu = 3.1$ using a two-tailed test with $\alpha = .05$.

b. Is the estimate from the participants significantly higher than the actual number ($\mu = 3.1$)? Use a one-tailed test with $\alpha = .05$.

14. Many animals, including humans, tend to avoid direct eye contact and even patterns that look like eyes. Some insects, including moths, have evolved eye-spot patterns on their wings to help ward off predators. Scaife (1976) reports a study examining how eye-spot patterns affect the behavior of birds. In the study, the birds were tested in a box with two chambers and were free to move from one chamber to another. In one chamber, two large eye-spots were painted on one wall. The other chamber had plain walls. The researcher recorded the amount of time each bird spent in the plain chamber during a 60-minute session. Suppose the study produced a mean of $M = 34.5$ minutes on the plain chamber with $SS = 210$ for a sample

of $n = 15$ birds. (*Note*: If the eye spots have no effect, then the birds should spend an average of $\mu = 30$ minutes in each chamber.)

a. Is this sample sufficient to conclude that the eye-spots have a significant influence on the birds' behavior? Use a two-tailed test with $\alpha = .05$.

b. Compute the estimated Cohen's *d* to measure the size of the treatment effect.

c. Construct the 90% confidence interval to estimate the mean amount of time spent on the plain side for the population of birds.

15. Standardized measures seem to indicate that the average level of anxiety has increased gradually over the past 50 years (Twenge, 2000). In the 1950s, the average score on the Child Manifest Anxiety Scale was $\mu = 15.1$. A sample of $n = 16$ of today's children produces a mean score of $M = 23.3$ with $SS = 240$.

a. Based on the sample, has there been a significant change in the average level of anxiety since the 1950s? Use a two-tailed test with $\alpha = .01$.

b. Make a 90% confidence interval estimate of today's population mean level of anxiety.

c. Write a sentence that demonstrates how the outcome of the hypothesis test and the confidence interval would appear in a research report.

16. Weinstein, McDermott, and Roediger (2010) report that students who were given questions to be answered while studying new material had better scores when tested on the material compared to students who were simply given an opportunity to reread the material. In a similar study, an instructor in a large psychology class gave one group of students questions to be answered while studying for the final exam. The overall average for the exam was $\mu = 73.4$ but the $n = 16$ students who answered questions had a mean of $M = 78.3$ with a standard deviation of $\sigma = 8.4$. For this study, did answering questions while studying produce significantly higher exam scores? Use a one-tailed test with $\alpha = .01$.

17. Ackerman and Goldsmith (2011) found that students who studied text from printed hardcopy had better test scores than students who studied from text presented on a screen. In a related study, a professor noticed that several students in a large class had purchased the e-book version of the course textbook. For the final exam, the overall average for the entire class was $\mu = 81.7$ but the $n = 9$ students who used e-books had a mean of $M = 77.2$ with a standard deviation of $s = 5.7$.

a. Is the sample sufficient to conclude that scores for students using e-books were significantly different from scores for the regular class? Use a two-tailed test with $\alpha = .05$.

b. Construct the 90% confidence interval to estimate the mean exam score if the entire population used e-books.

c. Write a sentence demonstrating how the results from the hypothesis test and the confidence interval would appear in a research report.

18. A random sample of $n = 16$ scores is obtained from a population with a mean of $\mu = 45$. After a treatment is administered to the individuals in the sample, the sample mean is found to be $M = 49.2$.

a. Assuming that the sample standard deviation is $s = 8$, compute r^2 and the estimated Cohen's d to measure the size of the treatment effect.

b. Assuming that the sample standard deviation is $s = 20$, compute r^2 and the estimated Cohen's d to measure the size of the treatment effect.

c. Comparing your answers from parts a and b, how does the variability of the scores in the sample influence the measures of effect size?

19. A random sample is obtained from a population with a mean of $\mu = 45$. After a treatment is administered to the individuals in the sample, the sample mean is $M = 49$ with a standard deviation of $s = 12$.

a. Assuming that the sample consists of $n = 9$ scores, compute r^2 and the estimated Cohen's d to measure the size of treatment effect.

b. Assuming that the sample consists of $n = 16$ scores, compute r^2 and the estimated Cohen's d to measure the size of treatment effect.

c. Comparing your answers from parts a and b, how does the number of scores in the sample influence the measures of effect size?

20. An example of the vertical-horizontal illusion is shown in the figure. Although the two lines are exactly the same length, the vertical line appears to be much longer. To examine the strength of this illusion, a researcher prepared an example in which both lines were exactly 10 inches long. The example was shown to individual participants who were told that the horizontal line was 10 inches long and then were asked to estimate the length of the vertical line. For a sample of $n = 25$ participants, the average estimate was $M = 12.2$ inches with a standard deviation of $s = 1.00$.

a. Use a one-tailed hypothesis test with $\alpha = .01$ to demonstrate that the individuals in the sample significantly overestimate the true length of the line. (*Note*: Accurate estimation would produce a mean of $\mu = 10$ inches.)

b. Calculate the estimated d and r^2, the percentage of variance accounted for, to measure the size of this effect.

c. Construct a 95% confidence interval for the population mean estimated length of the vertical line.

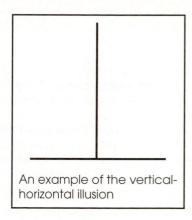

An example of the vertical-horizontal illusion

21. In the Preview for this Chapter, we discussed a study by McGee and Shevlin (2009) demonstrating that an individual's sense of humor had a significant effect on how the individual was perceived by others. In one part of the study, female college students were given brief descriptions of a potential romantic partner. The fictitious male was described positively and, for one group of participants, the description also said that he had a great sense of humor. Another group of female students read the same description except it now said that he has no sense of humor. After reading the description, each participant was asked to rate the attractiveness of the man on a seven-point scale from 1 (very unattractive) to 7 (very attractive) with a score of 4 indicating a neutral rating.

a. The females who read the "great sense of humor" description gave the potential partner an average attractiveness score of $M = 4.53$ with a standard deviation of $s = 1.04$. If the sample consisted of $n = 16$ participants, is the average rating significantly higher than neutral ($\mu = 4$)? Use a one-tailed test with $\alpha = .05$.

b. The females who read the description saying "no sense of humor" gave the potential partner an average attractiveness score of $M = 3.30$ with a standard deviation of $s = 1.18$. If the sample consisted of $n = 16$ participants, is the average rating significantly lower than neutral ($\mu = 4$)? Use a one-tailed test with $\alpha = .05$.

22. Oishi and Schimmack (2010) report that people who move from home to home frequently as children tend to have lower than average levels of well-being as adults. To further examine this relationship, a psychologist obtains a sample of $n = 12$ young adults who each experienced 5 or more different homes before they were 16 years old. These participants were given a standardized well-being questionnaire for which the general population has an average score of $\mu = 40$. The well-being scores for this sample are as follows: 38, 37, 41, 35, 42, 40, 33, 33, 36, 38, 32, 39.

   a. On the basis of this sample, is well-being for frequent movers significantly different from well-being in the general population? Use a two-tailed test with $\alpha = .05$.

   b. Compute the estimated Cohen's *d* to measure the size of the difference.

   c. Write a sentence showing how the outcome of the hypothesis test and the measure of effect size would appear in a research report.

23. Research examining the effects of preschool child-care has found that children who spent time in day care, especially high-quality day care, perform better on math and language tests than children who stay home with their mothers (Broberg, Wessels, Lamb, & Hwang, 1997). In a typical study, a researcher obtains a sample of $n = 10$ children who attended day care before starting school. The children are given a standardized math test for which the population mean is $\mu = 50$. The scores for the sample are as follows: 53, 57, 61, 49, 52, 56, 58, 62, 51, 56.

   a. Is this sample sufficient to conclude that the children with a history of preschool day care are significantly different from the general population? Use a two-tailed test with $\alpha = .01$.

   b. Compute Cohen's *d* to measure the size of the preschool effect.

   c. Write a sentence showing how the outcome of the hypothesis test and the measure of effect size would appear in a research report.

The *t* Test for Two Independent Samples

Tools You Will Need

The following items are considered essential background material for this chapter. If you doubt your knowledge of any of these items, you should review the appropriate chapter or section before proceeding.

- Sample variance (Chapter 4)
- Standard error formulas (Chapter 7)
- The *t* statistic (Chapter 9)
  - Distribution of *t* values
  - *df* for the *t* statistic
  - Estimated standard error

© Deborah Batt

PREVIEW

Does money in your hands make you feel better? There is actually some evidence that suggests that handling money really does reduce pain (Zhou, Vohs, & Baumeister, 2009). In their experiment, a group of college students was told that they were participating in a manual dexterity study. The researchers then created two treatment conditions by manipulating the kind of material that each participant would be handling. Half of the students were given a stack of money to count and the other half got a stack of blank pieces of paper. After the counting task, the participants were asked to dip their hands into bowls of very hot water (122°F) and rate how uncomfortable it was. Participants who had counted money consistently rated the pain lower than those who had counted paper.

Although the data show a mean difference between the two groups, you cannot automatically conclude that the difference was caused by the materials used in the counting task. Specifically, the two groups consist of different people with different backgrounds, different skills, different IQs, and so on. Because the two different groups consist of different individuals, you should expect them to have different scores and different means. This issue was first presented in Chapter 1 when we introduced the concept of sampling error (see Figure 1.2 on page 7). Thus, there are two possible explanations for the difference between the two groups.

1. It is possible that there really is a difference between the two treatment conditions so that counting money results in less pain than counting paper.

2. It is possible that there is no difference between the two treatment conditions and the mean difference obtained in the experiment is simply the result of sampling error.

A hypothesis test is necessary to determine which of the two explanations is most plausible. However, the hypothesis tests we have examined thus far are intended to evaluate the data from only one sample. In this study there are two separate samples.

In this chapter we introduce the *independent-measures t test,* which is a solution to the problem. The independent-measures *t* test is a hypothesis test that uses two separate samples to evaluate the mean difference between two treatment conditions or between two different populations. Like the *t* test introduced in Chapter 9, the independent-measures test uses the sample variance to compute an estimated standard error. This test, however, combines the variance from the two separate samples to evaluate the difference between two separate sample means.

10.1 | Introduction to the Independent-Measures Design

LEARNING OBJECTIVE   1. Define independent-measures designs and repeated-measures designs.

Until this point, all the inferential statistics we have considered involve using one sample as the basis for drawing conclusions about one population. Although these *single-sample* techniques are used occasionally in real research, most research studies require the comparison of two (or more) sets of data. For example, a social psychologist may want to compare men and women in terms of their political attitudes, an educational psychologist may want to compare two methods for teaching mathematics, or a clinical psychologist may want to evaluate a therapy technique by comparing depression scores for patients before therapy with their scores after therapy. When the scores are numerical values, the research question concerns a mean difference between two sets of data. The research designs that are used to obtain the two sets of data can be classified in two general categories:

1. The two sets of data could come from two completely separate groups of participants. For example, the study could involve a sample of men compared with a sample of women. Or, the study could compare grades for one group of freshmen who are given laptop computers with grades for a second group who are not given computers.

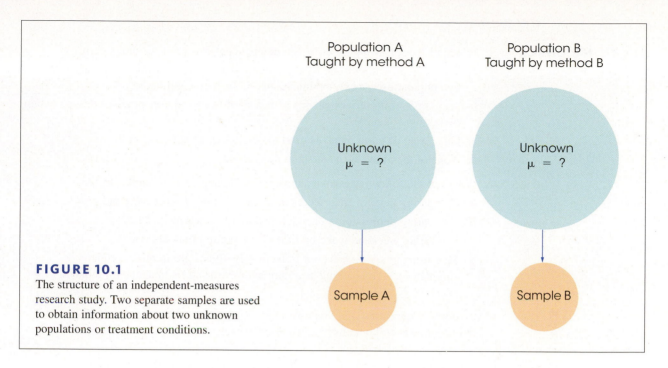

Population A
Taught by method A

Population B
Taught by method B

Unknown
$\mu = ?$

Unknown
$\mu = ?$

Sample A

Sample B

FIGURE 10.1

The structure of an independent-measures research study. Two separate samples are used to obtain information about two unknown populations or treatment conditions.

2. The two sets of data could come from the same group of participants. For example, the researcher could obtain one set of scores by measuring depression for a sample of patients before they begin therapy and then obtain a second set of data by measuring the same individuals after 6 weeks of therapy.

The first research strategy, using completely separate groups, is called an *independent-measures* research design or a *between-subjects* design. These terms emphasize the fact that the design involves separate and independent samples and makes a comparison between two groups of individuals. The structure of an independent-measures research design is shown in Figure 10.1. Notice that the research study uses two separate samples to represent the two different populations (or two different treatments) being compared.

DEFINITION

A research design that uses a separate group of participants for each treatment condition (or for each population) is called an **independent-measures** research design or a **between-subjects** design.

In this chapter, we examine the statistical techniques used to evaluate the data from an independent-measures design. More precisely, we introduce the hypothesis test that allows researchers to use the data from two separate samples to evaluate the mean difference between two populations or between two treatment conditions.

The second research strategy, in which the two sets of data are obtained from the same group of participants, is called a *repeated-measures* research design or a *within-subjects* design. The statistics for evaluating the results from a repeated-measures design are introduced in Chapter 11. Also, at the end of Chapter 11, we discuss some of the advantages and disadvantages of independent-measures and repeated-measures designs.

LEARNING CHECK

1. Which of the following describes an independent-measures research study?
   a. Two separate samples are used to obtain two groups of scores to represent the two populations or two treatment conditions being compared.
   b. The same group of participants is measured in two treatment conditions to obtain two groups of scores.
   c. One group of participants is used to obtain one group of scores.
   d. None of the above is an accurate description

2. Which of the following is an example of an independent-measures design?
   a. Comparing performance scores for 6-year-old boys and 6-year-old girls.
   b. Comparing depression scores before and after therapy.
   c. Measuring weekend wake-up times for a group of adolescents.
   d. Measuring smoking behavior immediately after a group of participants completes a stop-smoking program and measuring them again six months later.

3. A repeated-measures study comparing two treatment conditions uses ___ group(s) of participants and obtains ___ score(s) for each participant.
   a. 1, 1
   b. 1, 2
   c. 2, 1
   d. 2, 2

ANSWERS   1. A, 2. A, 3. B

10.2 | The Null Hypothesis and the Independent-Measures *t* Statistic

LEARNING OBJECTIVES

2. Describe the hypotheses for an independent-measures *t* test.

3. Describe the structure of the independent-measures *t* statistic and explain how it is related to the single-sample *t*.

4. Calculate the estimated standard error for a sample mean difference, $s_{(M_1-M_2)}$, including the pooled variance, which is part of the equation.

5. Calculate the complete independent-measures *t* statistic and its degrees of freedom.

Because an independent-measures study involves two separate samples, we need some special notation to help specify which data go with which sample. This notation involves the use of subscripts, which are small numbers written beside a sample statistic. For example, the number of scores in the first sample would be identified by n_1; for the second sample, the number of scores is n_2. The sample means would be identified by M_1 and M_2. The sums of squares would be SS_1 and SS_2.

■ The Hypotheses for an Independent-Measures Test

The goal of an independent-measures research study is to evaluate the mean difference between two populations (or between two treatment conditions). Using subscripts to differentiate the two populations, the mean for the first population is μ_1, and the second population mean is μ_2. The difference between means is simply $\mu_1 - \mu_2$. As always, the null hypothesis states that there is no change, no effect, or, in this case, no difference. Thus, in symbols, the null hypothesis for the independent-measures test is

$$H_0: \quad \mu_1 - \mu_2 = 0 \qquad \text{(No difference between the population means)}$$

You should notice that the null hypothesis could also be stated as $\mu_1 = \mu_2$. However, the first version of H_0 produces a specific numerical value (zero) that is used in the calculation of the *t* statistic. Therefore, we prefer to phrase the null hypothesis in terms of the difference between the two population means.

The alternative hypothesis states that there is a mean difference between the two populations,

$$H_1: \quad \mu_1 - \mu_2 \neq 0 \qquad \text{(There is a mean difference.)}$$

Equivalently, the alternative hypothesis can simply state that the two population means are not equal: $\mu_1 \neq \mu_2$.

■ The Formulas for an Independent-Measures Hypothesis Test

The independent-measures hypothesis test uses another version of the *t* statistic. The formula for this new *t* statistic has the same general structure as the *t* statistic formula that was introduced in Chapter 9. To help distinguish between the two *t* formulas, we refer to the original formula (Chapter 9) as the *single-sample* t *statistic* and we refer to the new formula as the *independent-measures* t *statistic*. Because the new independent-measures *t* includes data from two separate samples and hypotheses about two populations, the formulas may appear to be a bit overpowering. However, the new formulas are easier to understand if you view them in relation to the single-sample *t* formulas from Chapter 9. In particular, there are two points to remember.

1. The basic structure of the *t* statistic is the same for both the independent-measures and the single-sample hypothesis tests. In both cases,

$$t = \frac{\text{sample statistic} - \text{hypothesized population parameter}}{\text{estimated standard error}}$$

2. The independent-measures *t* is basically a *two-sample* t *that doubles all the elements of the single-sample* t *formulas.*

To demonstrate the second point, we examine the two *t* formulas piece by piece.

The Overall *t* Formula   The single-sample *t* uses one sample mean to test a hypothesis about one population mean. The sample mean and the population mean appear in the numerator of the *t* formula, which measures how much difference there is between the sample data and the population hypothesis.

$$t = \frac{\text{sample mean} - \text{population mean}}{\text{estimated standard error}} = \frac{M - \mu}{s_M}$$

The independent-measures *t* uses the difference between *two* sample means to evaluate a hypothesis about the difference between *two* population means. Thus, the independent-measures *t* formula is

$$t = \frac{\text{sample mean difference} - \text{population mean difference}}{\text{estimated standard error}} = \frac{(M_1 - M_2) - (\mu_1 - \mu_2)}{s_{(M_1 - M_2)}}$$

In this formula, the value of $M_1 - M_2$ is obtained from the sample data and the value for $\mu_1 - \mu_2$ comes from the null hypothesis. In a hypothesis test, the null hypothesis sets the population mean difference equal to zero, so the independent measures *t* formula can be simplified further,

$$t = \frac{\text{sample mean difference}}{\text{estimated standard error}}$$

In this form, the *t* statistic is a simple ratio comparing the actual mean difference (numerator) with the difference that is expected by chance (denominator).

The Estimated Standard Error In each of the *t*-score formulas, the standard error in the denominator measures how accurately the sample statistic represents the population parameter. In the single-sample *t* formula, the standard error measures the amount of error expected for a sample mean and is represented by the symbol s_M. For the independent-measures *t* formula, the standard error measures the amount of error that is expected when you use a sample mean difference $(M_1 - M_2)$ to represent a population mean difference $(\mu_1 - \mu_2)$. The standard error for the sample mean difference is represented by the symbol $s_{(M_1 - M_2)}$.

Caution: Do not let the notation for standard error confuse you. In general, standard error measures how accurately a statistic represents a parameter. The symbol for standard error takes the form $s_{\text{statistic}}$. When the statistic is a sample mean, *M*, the symbol for standard error is s_M. For the independent-measures test, the statistic is a sample mean difference $(M_1 - M_2)$, and the symbol for standard error is $s_{(M_1 - M_2)}$. In each case, the standard error tells how much discrepancy is reasonable to expect between the sample statistic and the corresponding population parameter.

Interpreting the Estimated Standard Error The *estimated standard error of* $M_1 - M_2$ that appears in the bottom of the independent-measures *t* statistic can be interpreted in two ways. First, the standard error is defined as a measure of the standard or average distance between a sample statistic $(M_1 - M_2)$ and the corresponding population parameter $(\mu_1 - \mu_2)$. As always, samples are not expected to be perfectly accurate and the standard error measures how much difference is reasonable to expect between a sample statistic and the population parameter.

When the null hypothesis is true, however, the population mean difference is zero. In this case, the standard error is measuring how far, on average, the sample mean difference is from zero. However, measuring how far it is from zero is the same as measuring how big it is. Therefore, when the null hypothesis is true, the standard error is measuring how big, on average, the sample mean difference is. Thus, there are two ways to interpret the estimated standard error of $(M_1 - M_2)$.

1. It measures the standard distance between $(M_1 - M_2)$ and $(\mu_1 - \mu_2)$.

2. It measures the standard, or average size of $(M_1 - M_2)$ if the null hypothesis is true. That is, it measures how much difference is reasonable to expect between the two sample means.

■ Calculating the Estimated Standard Error

To develop the formula for $s_{(M_1-M_2)}$ we consider the following three points.

1. Each of the two sample means represents it own population mean, but in each case there is some error.

$$M_1 \text{ approximates } \mu_1 \text{ with some error}$$
$$M_2 \text{ approximates } \mu_2 \text{ with some error}$$

   Thus, there are two sources of error.

2. The amount of error associated with each sample mean is measured by the estimated standard error of *M*. Using Equation 9.1 (page 269), the estimated standard error for each sample mean is computed as follows:

$$\text{For } M_1 \ s_M = \sqrt{\frac{s_1^2}{n_1}} \qquad \text{For } M_2 \ s_M = \sqrt{\frac{s_2^2}{n_2}}$$

3. For the independent-measures *t* statistic, we want to know the total amount of error involved in using *two* sample means to approximate *two* population means. To do this, we will find the error from each sample separately and then add the two errors together. The resulting formula for standard error is

$$s_{(M_1-M_2)} = \sqrt{\frac{s_1^2}{n_1} + \frac{s_2^2}{n_2}} \tag{10.1}$$

Because the independent-measures *t* statistic uses two sample means, the formula for the estimated standard error simply combines the error for the first sample mean and the error for the second sample mean (Box 10.1).

BOX 10.1 The Variability of Difference Scores

It may seem odd that the independent-measures *t* statistic *adds* together the two sample errors when it *subtracts* to find the difference between the two sample means. The logic behind this apparently unusual procedure is demonstrated here.

We begin with two populations, I and II (Figure 10.2). The scores in population I range from a high of 70 to a low of 50. The scores in population II range from 30 to 20. We will use the range as a measure of how spread out (variable) each population is:

For population I, the scores cover a range of 20 points. For population II, the scores cover a range of 10 points.

If we randomly select one score from population I and one score from population II and compute the difference between these two scores $(X_1 - X_2)$, what range of values is possible for these differences? To answer this question, we need to find the biggest possible difference and the smallest possible difference. Look at Figure 10.2 where the biggest difference occurs when $X_1 = 70$ and $X_2 = 20$. This is a difference of $X_1 - X_2 = 50$ points. The smallest difference occurs when $X_1 = 50$ and $X_2 = 30$. This is a difference of $X_1 - X_2 = 20$ points. Notice that the differences go from a high of 50 to a low of 20. This is a range of 30 points:

   range for population I (X_1 scores) = 20 points
   range for population II (X_2 scores) = 10 points
   range for the differences ($X_1 - X_2$) = 30 points

The variability for the difference scores is found by *adding* together the variability for each of the two populations.

In the independent-measures *t* statistics, we are computing the variability (standard error) for a sample mean difference. To compute this value, we add together the variability for each of the two sample means.

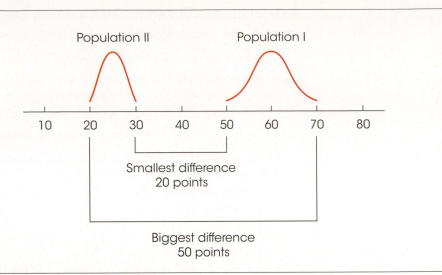

FIGURE 10.2
Two population distributions. The scores in population I range from 50–70 (a 20-point spread) and the scores in population II range from 20–30 (a 10-point spread). If you select one score from each of these two populations, the closest two values are $X_1 = 50$ and $X_2 = 30$. The two values that are farthest apart are $X_1 = 70$ and $X_2 = 20$.

■ Pooled Variance

An alternative to computing pooled variance is presented in Box 10.2, page 315.

Although Equation 10.1 accurately presents the concept of standard error for the independent-measures *t* statistic, this formula is limited to situations in which the two samples are exactly the same size (that is, $n_1 = n_2$). For situations in which the two sample sizes are different, the formula is *biased* and, therefore, inappropriate. The bias comes from the fact that Equation 10.1 treats the two sample variances equally. However, when the sample sizes are different, the two sample variances are not equally good and should not be treated equally. In Chapter 7, we introduced the law of large numbers, which states that statistics obtained from large samples tend to be better (more accurate) estimates of population parameters than statistics obtained from small samples. This same fact holds for sample variances: The variance obtained from a large sample is a more accurate estimate of σ^2 than the variance obtained from a small sample.

One method for correcting the bias in the standard error is to combine the two sample variances into a single value called the *pooled variance*. The pooled variance is obtained by averaging or "pooling" the two sample variances using a procedure that allows the bigger sample to carry more weight in determining the final value.

You should recall that when there is only one sample, the sample variance is computed as

$$s^2 = \frac{SS}{df}$$

For the independent-measures *t* statistic, there are two *SS* values and two *df* values (one from each sample). The values from the two samples are combined to compute what is called the *pooled variance*. The pooled variance is identified by the symbol s_p^2 and is computed as

$$\text{pooled variance} = s_p^2 = \frac{SS_1 + SS_2}{df_1 + df_2} \qquad \text{(10.2)}$$

With one sample, the variance is computed as *SS* divided by *df*. With two samples, the pooled variance is computed by combining the two *SS* values and then dividing by the combination of the two *df* values.

As we mentioned earlier, the pooled variance is actually an average of the two sample variances, but the average is computed so that the larger sample carries more weight in determining the final value. The following examples demonstrate this point.

Equal Samples Sizes   We begin with two samples that are exactly the same size. The first sample has $n = 6$ scores with $SS = 50$, and the second sample has $n = 6$ scores with $SS = 30$. Individually, the two sample variances are

$$\text{Variance for sample 1:} \quad s^2 = \frac{SS}{df} = \frac{50}{5} = 10$$

$$\text{Variance for sample 2:} \quad s^2 = \frac{SS}{df} = \frac{30}{5} = 6$$

The pooled variance for these two samples is

$$s_p^2 = \frac{SS_1 + SS_2}{df_1 + df_2} = \frac{50 + 30}{5 + 5} = \frac{80}{10} = 8.00$$

Note that the pooled variance is exactly halfway between the two sample variances. Because the two samples are exactly the same size, the pooled variance is simply the average of the two sample variances.

Unequal Samples Sizes   Now consider what happens when the samples are not the same size. This time the first sample has $n = 3$ scores with $SS = 20$, and the second sample has $n = 9$ scores with $SS = 48$. Individually, the two sample variances are

$$\text{Variance for sample 1:} \quad s^2 = \frac{SS}{df} = \frac{20}{2} = 10$$

$$\text{Variance for sample 2:} \quad s^2 = \frac{SS}{df} = \frac{48}{8} = 6$$

The pooled variance for these two samples is

$$s_p^2 = \frac{SS_1 + SS_2}{df_1 + df_2} = \frac{20 + 48}{2 + 8} = \frac{68}{10} = 6.80$$

This time the pooled variance is not located halfway between the two sample variances. Instead, the pooled value is closer to the variance for the larger sample ($n = 9$ and $s^2 = 6$) than to the variance for the smaller sample ($n = 3$ and $s^2 = 10$). The larger sample carries more weight when the pooled variance is computed.

When computing the pooled variance, the weight for each of the individual sample variances is determined by its degrees of freedom. Because the larger sample has a larger *df* value, it carries more weight when averaging the two variances. This produces an alternative formula for computing pooled variance.

$$\text{pooled variance} = s_p^2 = \frac{df_1 s_1^2 + df_2 s_2^2}{df_1 + df_2} \tag{10.3}$$

For example, if the first sample has $df_1 = 3$ and the second sample has $df_2 = 7$, then the formula instructs you to take 3 of the first sample variance and 7 of the second sample variance for a total of 10 variances. You then divide by 10 to obtain the average. The alternative formula is especially useful if the sample data are summarized as means and variances. Finally, you should note that because the pooled variance is an average of the two sample variances, the value obtained for the pooled variance is always located between the two sample variances.

■ Estimated Standard Error

Using the pooled variance in place of the individual sample variances, we can now obtain an unbiased measure of the standard error for a sample mean difference. The resulting formula for the independent-measures estimated standard error is

$$\text{estimated standard error of } M_1 - M_2 = s_{(M_1-M_2)} = \sqrt{\frac{s_p^2}{n_1} + \frac{s_p^2}{n_2}} \qquad (10.4)$$

Conceptually, this standard error measures how accurately the difference between two sample means represents the difference between the two population means. In a hypothesis test, H_0 specifies that $\mu_1 - \mu_2 = 0$, and the standard error also measures how much difference is expected on average between the two sample means. In either case, the formula combines the error for the first sample mean with the error for the second sample mean. Also note that the pooled variance from the two samples is used to compute the standard error for the two samples.

The following example is an opportunity to test your understanding of the pooled variance and the estimated standard error.

EXAMPLE 10.1   One sample from an independent-measures study has $n = 5$ with $SS = 100$. The other sample has $n = 10$ and $SS = 290$. For these data, compute the pooled variance and the estimated standard error for the mean difference. You should find that the pooled variance is $390/13 = 30$ and the estimated standard error is 3.   ■

■ The Final Formula and Degrees of Freedom

The complete formula for the independent-measures *t* statistic is as follows:

$$t = \frac{(M_1 - M_2) - (\mu_1 - \mu_2)}{s_{(M_1-M_2)}}$$

$$= \frac{\text{sample mean difference} - \text{population mean difference}}{\text{estimated standard error}} \qquad (10.5)$$

In the formula, the estimated standard error in the denominator is calculated using Equation 10.4, and requires calculation of the pooled variance using either Equation 10.2 or 10.3.

The degrees of freedom for the independent-measures *t* statistic are determined by the *df* values for the two separate samples:

$$\begin{aligned} df \text{ for the } t \text{ statistic} &= df \text{ for the first sample} + df \text{ for the second sample} \\ &= df_1 + df_2 \\ &= (n_1 - 1) + (n_2 - 1) \qquad (10.6) \end{aligned}$$

Equivalently, the *df* value for the independent-measures *t* statistic can be expressed as

$$df = n_1 + n_2 - 2. \qquad (10.7)$$

Note that the *df* formula subtracts 2 points from the total number of scores; 1 point for the first sample and 1 for the second.

The independent-measures *t* statistic is used for hypothesis testing. Specifically, we use the difference between two sample means $(M_1 - M_2)$ as the basis for testing hypotheses about the difference between two population means $(\mu_1 - \mu_2)$. In this context, the overall structure of the *t* statistic can be reduced to the following:

$$t = \frac{\text{data} - \text{hypothesis}}{\text{error}}$$

This same structure is used for both the single-sample *t* from Chapter 9 and the new independent-measures *t* that was introduced in the preceding pages. Table 10.1 identifies each component of these two *t* statistics and should help reinforce the point that we made earlier in the chapter; that is, the independent-measures *t* statistic simply doubles each aspect of the single-sample *t* statistic.

TABLE 10.1   The basic elements of a *t* statistic for the single-sample *t* and the independent-measures *t*.

	Sample Data	Hypothesized Population Parameter	Estimated Standard Error	Sample Variance
Single-sample *t* statistic	M	μ	$\sqrt{\dfrac{s^2}{n}}$	$s^2 = \dfrac{SS}{df}$
Independent-measures	$(M_1 - M_2)$	$(\mu_1 - \mu_2)$	$\sqrt{\dfrac{s_p^2}{n_1} + \dfrac{s_p^2}{n_2}}$	$s_p^2 = \dfrac{SS_1 + SS_2}{df_1 + df_2}$

LEARNING CHECK

1. When the null hypothesis is true, what is measured by the estimated standard error in the denominator of the independent-measures *t* statistic?
   a. The standard distance between a sample mean and its population mean
   b. The standard distance between two sample means selected from the same population
   c. The actual distance between a sample mean and its population mean
   d. The actual distance between two sample means

2. Which of the following is the correct null hypothesis for an independent-measures *t* test?
   a. $M_1 - M_2 = 0$
   b. $M_1 - M_2 \neq 0$
   c. $\mu_1 - \mu_2 = 0$
   d. $\mu_1 - \mu_2 \neq 0$

3. One sample from an independent-measures study has $n = 4$ with $SS = 100$. The other sample has $n = 8$ and $SS = 140$. What is the pooled variance?
   a. 53.33
   b. 42.5
   c. 24
   d. 20

4. One sample from an independent-measures study has $n = 9$ with a variance of $s^2 = 35$. The other sample has $n = 3$ and $s^2 = 40$. What is the *df* value for the *t* statistic?
   a. 36
   b. 12
   c. 11
   d. 10

ANSWERS   **1.** B, **2.** C, **3.** C, **4.** D

10.3 | Hypothesis Tests with the Independent-Measures *t* Statistic

LEARNING OBJECTIVES

6. Use the data from two samples to conduct an independent-measures *t* test evaluating the significance of the difference between two population means.

7. Conduct a directional (one-tailed) hypothesis test using the independent-measures *t* statistic.

8. Describe the basic assumptions underlying the independent-measures *t* hypothesis test, especially the homogeneity of variance assumption, and explain how the homogeneity assumption can be tested.

The independent-measures *t* statistic uses the data from two separate samples to help decide whether there is a significant mean difference between two populations or between two treatment conditions. A complete example of a hypothesis test with two independent samples follows.

EXAMPLE 10.2

Research has shown that people are more likely to show dishonest and self-interested behaviors in darkness than in a well-lit environment (Zhong, Bohns, & Gino, 2010). In one experiment, participants were given a set of 20 puzzles and were paid $0.50 for each one solved in a 5-minute period. However, the participants reported their own performance and there was no obvious method for checking their honesty. Thus, the task provided a clear opportunity to cheat and receive undeserved money. One group of participants was tested in a room with dimmed lighting and a second group was tested in a well-lit room. The reported number of solved puzzles was recorded for each individual. The following data represent results similar to those obtained in the study.

Number of Solved Puzzles			
Well-Lit Room		Dimly Lit Room	
11	6	7	9
9	7	13	11
4	12	14	15
5	10	16	11
$n = 8$		$n = 8$	
$M = 8$		$M = 12$	
$SS = 60$		$SS = 66$	

STEP 1 **State the hypotheses and select the alpha level.**

$$H_0: \quad \mu_1 - \mu_2 = 0 \qquad \text{(No difference.)}$$
$$H_1: \quad \mu_1 - \mu_2 \neq 0 \qquad \text{(There is a difference.)}$$

We will set $\alpha = .05$.

Directional hypotheses could be used and would specify whether the students who were tested in a dimly lit room should have higher or lower scores.

STEP 2 **This is an independent-measures design.**   The *t* statistic for these data has degrees of freedom determined by

$$df = df_1 + df_2$$
$$= (n_1 - 1) + (n_2 - 1)$$
$$= 7 + 7$$
$$= 14$$

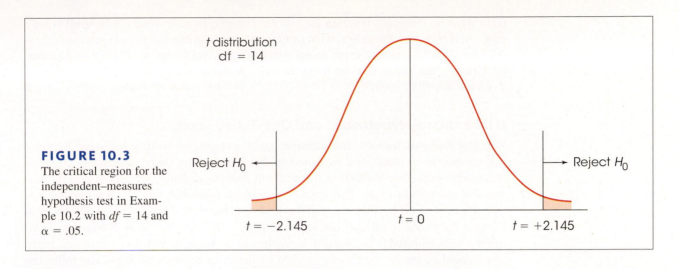

FIGURE 10.3
The critical region for the independent–measures hypothesis test in Example 10.2 with *df* = 14 and α = .05.

The *t* distribution for *df* = 14 is presented in Figure 10.3. For α = .05, the critical region consists of the extreme 5% of the distribution and has boundaries of *t* = +2.145 and *t* = −2.145.

STEP 3 **Obtain the data and compute the test statistic.**   The data are as given, so all that remains is to compute the *t* statistic. As with the single-sample *t* test in Chapter 9, we recommend that the calculations be divided into three parts.

First, find the pooled variance for the two samples:

Caution: **The pooled variance combines the two samples to obtain a single estimate of variance. In the formula, the two samples are combined in a single fraction.**

$$s_p^2 = \frac{SS_1 + SS_2}{df_1 + df_2}$$

$$= \frac{60 + 66}{7 + 7} = \frac{126}{14} = 9$$

Second, use the pooled variance to compute the estimated standard error:

Caution: **The standard error adds the errors from two separate samples. In the formula, these two errors are added as two separate fractions. In this case, the two errors are equal because the sample sizes are the same.**

$$s_{(M_1 - M_2)} = \sqrt{\frac{s_1^2}{n_1} + \frac{s_2^2}{n_2}} = \sqrt{\frac{9}{8} + \frac{9}{8}}$$

$$= \sqrt{2.25}$$

$$= 1.50$$

Third, compute the *t* statistic:

$$t = \frac{(M_1 - M_2) - (\mu_1 - \mu_2)}{s_{(M_1 - M_2)}} = \frac{(8 - 12) - 0}{1.5}$$

$$= \frac{-4}{1.5} = -2.67$$

STEP 4 **Make a decision.**   The obtained value (*t* = −2.67) is in the critical region. In this example, the obtained sample mean difference is 2.67 times greater than would be expected if there were no difference between the two populations. In other words, this result is very unlikely if H_0 is true. Therefore, we reject H_0 and conclude that there is a significant difference between the reported scores in the dimly lit room and the scores in the well-lit room. Specifically, the students in the dimly lit room reported significantly higher scores than those in the well-lit room.   ∎

■ Directional Hypotheses and One-Tailed Tests

When planning an independent-measures study, a researcher usually has some expectation or specific prediction for the outcome. For the cheating study in Example 10.2, the researchers expect the students in the dimly lit room to claim higher scores than the students in the well-lit room. This kind of directional prediction can be incorporated into the statement of the hypotheses, resulting in a directional, or one-tailed, test. Recall from Chapter 8 that one-tailed tests can lead to rejecting H_0 when the mean difference is relatively small compared to the magnitude required by a two-tailed test. As a result, one-tailed tests should be used when clearly justified by theory or previous findings. The following example demonstrates the procedure for stating hypotheses and locating the critical region for a one-tailed test using the independent-measures *t* statistic.

EXAMPLE 10.3   We will use the same research situation that was described in Example 10.2. The researcher is using an independent-measures design to examine the relationship between lighting and dishonest behavior. The prediction is that students in a dimly lit room are more likely to cheat (report higher scores) than are students in a well-lit room.

STEP 1   **State the hypotheses and select the alpha level.**   As always, the null hypothesis says that there is no effect, and the alternative hypothesis says that there is an effect. For this example, the predicted effect is that the students in the dimly lit room will claim to have higher scores. Thus, the two hypotheses are as follows:

$$H_0: \quad \mu_{\text{Dimly Lit}} \leq \mu_{\text{Well Lit}} \qquad \text{\textbf{(Reported scores are not higher in the dimly lit room)}}$$

$$H_1: \quad \mu_{\text{Dimly Lit}} > \mu_{\text{Well Lit}} \qquad \text{\textbf{(Reported scores are higher in the dimly lit room)}}$$

Note that it is usually easier to state the hypotheses in words before you try to write them in symbols. Also, it usually is easier to begin with the alternative hypothesis (H_1), which states that the treatment works as predicted. Also note that the equal sign goes in the null hypothesis, indicating *no difference* between the two treatment conditions. The idea of zero difference is the essence of the null hypothesis, and the numerical value of zero is used for ($\mu_1 - \mu_2$) during the calculation of the *t* statistic. For this test we will use $\alpha = .01$.

STEP 2   **Locate the critical region.**   For a directional test, the critical region is located entirely in one tail of the distribution. Rather than trying to determine which tail, positive or negative, is the correct location, we suggest you identify the criteria for the critical region in a two-step process as follows. First, look at the data and determine whether the sample mean difference is in the direction that was predicted. If the answer is no, then the data obviously do not support the predicted treatment effect, and you can stop the analysis. On the other hand, if the difference is in the predicted direction, then the second step is to determine whether the difference is large enough to be significant. To test for significance, simply find the one-tailed critical value in the *t* distribution table. If the calculated *t* statistic is more extreme (either positive or negative) than the critical value, then the difference is significant.

For this example, the students in the dimly lit room reported higher scores, as predicted. With $df = 14$, the one-tailed critical value for $\alpha = .01$ is $t = 2.624$.

STEP 3 **Collect the data and calculate the test statistic.** The details of the calculations were shown in Example 10.2. The data produce a *t* statistic of $t = -2.67$.

STEP 4 **Make a decision.** The *t* statistic of $t = -2.67$ is beyond the critical boundary of $t = 2.624$. Therefore, we reject the null hypothesis and conclude that the reported scores for students in the dimly lit room are significantly higher than the scores for students in the well-lit room. In a research report, the one-tailed test would be clearly noted:

> Reported scores were significantly higher for students in the dimly lit room, $t(14) = -2.67$, $p < .01$, one tailed. ∎

■ Assumptions Underlying the Independent-Measures *t* Formula

There are three assumptions that should be satisfied before you use the independent-measures *t* formula for hypothesis testing:

1. The observations within each sample must be independent (see page 244).
2. The two populations from which the samples are selected must be normal.
3. The two populations from which the samples are selected must have equal variances.

The first two assumptions should be familiar from the single-sample *t* hypothesis test presented in Chapter 9. As before, the normality assumption is the less important of the two, especially with large samples. When there is reason to suspect that the populations are far from normal, you should compensate by ensuring that the samples are relatively large.

The third assumption is referred to as *homogeneity of variance* and states that the two populations being compared must have the same variance. You may recall a similar assumption for the *z*-score hypothesis test in Chapter 8. For that test, we assumed that the effect of the treatment was to add a constant amount to (or subtract a constant amount from) each individual score. As a result, the population standard deviation after treatment was the same as it had been before treatment. We now are making essentially the same assumption but phrasing it in terms of variances.

Recall that the pooled variance in the *t*-statistic formula is obtained by averaging together the two sample variances. It makes sense to average these two values only if they both are estimating the same population variance—that is, if the homogeneity of variance assumption is satisfied. If the two sample variances are estimating different population variances, then the average is meaningless. (*Note:* If two people are asked to estimate the same thing—for example, what you weigh—it is reasonable to average the two estimates. However, it is not meaningful to average estimates of two different things. If one person estimates your weight and another estimates the number of beans in a pound of whole-bean coffee, it is meaningless to average the two numbers.)

Homogeneity of variance is most important when there is a large discrepancy between the sample sizes. With equal (or nearly equal) sample sizes, this assumption is less critical, but still important. Violating the homogeneity of variance assumption can negate any meaningful interpretation of the data from an independent-measures experiment. Specifically, when you compute the *t* statistic in a hypothesis test, all the numbers in the formula come from the data except for the population mean difference, which you get from H_0. Thus, you are sure of all the numbers in the formula except one. If you obtain an extreme result for the *t* statistic (a value in the critical region), you conclude that the hypothesized value was wrong. But consider what happens when you violate the homogeneity of variance assumption. In this case, you have two questionable values in the formula (the hypothesized population value and the meaningless average of the two variances). Now if you obtain an extreme *t* statistic, you do not know which of these two values is responsible. Specifically, you cannot reject the hypothesis because it may have been the pooled variance that produced the extreme *t* statistic. Without satisfying the homogeneity of variance requirement, you cannot accurately interpret a *t* statistic, and the hypothesis test becomes meaningless.

■ Hartley's *F*-Max Test

How do you know whether the homogeneity of variance assumption is satisfied? One simple test involves just looking at the two sample variances. Logically, if the two population variances are equal, then the two sample variances should be very similar. When the two sample variances are reasonably close, you can be reasonably confident that the homogeneity assumption has been satisfied and proceed with the test. However, if one sample variance is more than three or four times larger than the other, then there is reason for concern. A more objective procedure involves a statistical test to evaluate the homogeneity assumption. Although there are many different statistical methods for determining whether the homogeneity of variance assumption has been satisfied, Hartley's *F*-max test is one of the simplest to compute and to understand. An additional advantage is that this test can also be used to check homogeneity of variance with more than two independent samples. Later, in Chapter 12, we examine statistical methods for comparing several different samples, and Hartley's test will be useful again. The following example demonstrates the *F*-max test for two independent samples.

EXAMPLE 10.4

The *F*-max test is based on the principle that a sample variance provides an unbiased estimate of the population variance. The null hypothesis for this test states that the population variances are equal, therefore, the sample variances should be very similar. The procedure for using the *F*-max test is as follows:

1. Compute the sample variance, $s^2 = \dfrac{SS}{df}$, for each of the separate samples.

2. Select the largest and the smallest of these sample variances and compute

$$F\text{-max} = \frac{s^2(\text{largest})}{s^2(\text{smallest})}$$

   A relatively large value for *F*-max indicates a large difference between the sample variances. In this case, the data suggest that the population variances are different and that the homogeneity assumption has been violated. On the other hand, a small value of *F*-max (near 1.00) indicates that the sample variances are similar and that the homogeneity assumption is reasonable.

3. The *F*-max value computed for the sample data is compared with the critical value found in Table B.3 (Appendix B). If the sample value is larger than the table value, you conclude that the variances are different and that the homogeneity assumption is not valid.

   To locate the critical value in the table, you need to know

   a. k = number of separate samples. (For the independent-measures *t* test, $k = 2$.)

   b. $df = n - 1$ for each sample variance. The Hartley test assumes that all samples are the same size.

   c. The alpha level. The table provides critical values for $\alpha = .05$ and $\alpha = .01$. Generally, a test for homogeneity would use the larger alpha level.

Suppose, for example, that two independent samples each have $n = 10$ with sample variances of 12.34 and 9.15. For these data,

$$F\text{-max} = \frac{s^2(\text{largest})}{s^2(\text{smallest})} = \frac{12.34}{9.15} = 1.35$$

With $\alpha = .05$, $k = 2$, and $df = n - 1 = 9$, the critical value from the table is 4.03. Because the obtained F-max is smaller than this critical value, you conclude that the data do not provide evidence that the homogeneity of variance assumption has been violated.   ■

The goal for most hypothesis tests is to reject the null hypothesis to demonstrate a significant difference or a significant treatment effect. However, when testing for homogeneity of variance, the preferred outcome is to fail to reject H_0. Failing to reject H_0 with the F-max test means that there is no significant difference between the two population variances and the homogeneity assumption is satisfied. In this case, you may proceed with the independent-measures t test using pooled variance.

If the F-max test rejects the hypothesis of equal variances, or if you simply suspect that the homogeneity of variance assumption is not justified, you should not compute an independent-measures t statistic using pooled variance. However, there is an alternative formula for the t statistic that does not pool the two sample variances and does not require the homogeneity assumption. The alternative formula is presented in Box 10.2.

BOX 10.2   An Alternative to Pooled Variance

Computing the independent-measures t statistic using pooled variance requires that the data satisfy the homogeneity of variance assumption. Specifically, the two distributions from which the samples are obtained must have equal variances. To avoid this assumption, many statisticians recommend an alternative formula for computing the independent-measures t statistic that does not require pooled variance or the homogeneity assumption. The alternative procedure consists of two steps.

1. The standard error is computed using the two separate sample variances as in Equation 10.1.
2. The value of degrees of freedom for the t statistic is adjusted using the following equation:

$$df = \frac{(V_1 + V_2)^2}{\dfrac{V_1^2}{n_1 - 1} + \dfrac{V_2^2}{n_2 - 1}} \quad \text{where } V_1 = \frac{s_1^2}{n_1} \text{ and } V_2 = \frac{s_2^2}{n_2}$$

Decimal values for df should be rounded down to the next lower integer.

The adjustment to degrees of freedom lowers the value of df, which pushes the boundaries for the critical region farther out. Thus, the adjustment makes the test more demanding and therefore corrects for the same bias problem that the pooled variance attempts to avoid.

Note: Many computer programs that perform statistical analysis (such as SPSS) report two versions of the independent-measures t statistic; one using pooled variance (with equal variances assumed) and one using the adjustment shown here (with equal variances not assumed).

LEARNING CHECK

1. A researcher obtains an independent-measures t statistic of $t = 2.38$ for a study comparing two treatments with a sample of $n = 14$ in each treatment. What is the correct decision for a regular two-tailed hypothesis test?
   a. Reject the null hypothesis with $\alpha = .05$ but not with $\alpha = .01$.
   b. Reject the null hypothesis with $\alpha = .05$ or with $\alpha = .01$.
   c. Reject the null hypothesis with $\alpha = .01$ but not with $\alpha = .05$.
   d. Fail to reject the null hypothesis with either alpha level.

2. A researcher study comparing performance for males and females predicts that females will have higher scores. If a one-tailed test is used to evaluate the results, then what is the correct statement for the alternative hypothesis (H_1)?
   a. $\mu_{females} > \mu_{males}$
   b. $\mu_{females} \geq \mu_{males}$
   c. $\mu_{females} < \mu_{males}$
   d. $\mu_{females} \leq \mu_{males}$

3. A researcher is using an F_{max} test to evaluate the homogeneity of variance assumption and an independent-measures t test with pooled variance to evaluate the mean difference between two treatment conditions. What are the preferred outcomes for the two tests?
   a. Reject H_0 for both.
   b. Fail to reject H_0 for both.
   c. Reject H_0 for F_{max} but fail to reject H_0 for the *t* test.
   d. Fail to reject H_0 for F_{max} but reject H_0 for the *t* test.

ANSWERS   1. A, 2. A, 3. D

10.4 | Effect Size and Confidence Intervals for the Independent-Measures *t*

LEARNING OBJECTIVES

9. Measure effect size for an independent-measures *t* test using either Cohen's *d* or r^2, the percentage of variance accounted for.

10. Use the data from two separate samples to compute a confidence interval describing the size of the mean difference between two treatment conditions or two populations.

11. Describe the relationship between a hypothesis test with an independent-measures t statistic using $\alpha = .05$ and the corresponding 95% confidence interval for the mean difference.

12. Describe how the results of an independent-measures *t* test and measures of effect size are reported in the scientific literature.

As noted in Chapters 8 and 9, a hypothesis test is usually accompanied by a report of effect size to provide an indication of the absolute magnitude of the treatment effect. One technique for measuring effect size is Cohen's *d*, which produces a standardized measure of mean difference. In its general form, Cohen's *d* is defined as

$$d = \frac{\text{mean difference}}{\text{standard deviation}} = \frac{\mu_1 - \mu_2}{\sigma}$$

In the context of an independent-measures research study, the difference between the two sample means ($M_1 - M_2$) is used as the best estimate of the mean difference between the two populations, and the pooled standard deviation (the square root of the pooled variance) is used to estimate the population standard deviation. Thus, the formula for estimating Cohen's d becomes

$$\text{estimated } d = \frac{\text{estimated mean difference}}{\text{estimated standard deviation}} = \frac{M_1 - M_2}{\sqrt{s_p^2}} \qquad (10.8)$$

For the data from Example 10.2, the two sample means are 8 and 12, and the pooled variance is 9. The estimated d for these data is

$$d = \frac{M_1 - M_2}{\sqrt{s_p^2}} = \frac{8 - 12}{\sqrt{9}} = \frac{-4}{3} = -1.33$$

Note: Cohen's d is typically reported as a positive value; in this case $d = 1.33$. Using the criteria established to evaluate Cohen's d (see Table 8.2 on page 253), this value indicates a very large treatment effect.

The independent-measures t hypothesis test also allows for measuring effect size by computing the percentage of variance accounted for, r^2. As we saw in Chapter 9, r^2 measures how much of the variability in the scores can be explained by the treatment effects. For example, some of the variability in the reported scores for the cheating study can be explained by knowing the room in which particular student was tested; students in the dimly lit room tend to report higher scores and students in the well-lit room tend to report lower scores. By measuring exactly how much of the variability can be explained, we can obtain a measure of how big the treatment effect actually is. The calculation of r^2 for the independent-measures t is exactly the same as it was for the single-sample t in Chapter 9:

$$r^2 = \frac{t^2}{t^2 + df} \qquad (10.9)$$

For the data in Example 10.2, we obtained $t = -2.67$ with $df = 14$. These values produce an r^2 of

$$r^2 = \frac{-2.67^2}{-2.67^2 + 14} = \frac{7.13}{7.13 + 14} = \frac{7.13}{21.13} = 0.337$$

According to the standards used to evaluate r^2 (see Table 9.3 on page 283), this value also indicates a large treatment effect.

Although the value of r^2 is usually obtained by using Equation 10.9, it is possible to determine the percentage of variability directly by computing SS values for the set of scores. The following example demonstrates this process using the data from the cheating study in Example 10.2.

EXAMPLE 10.5   The cheating study described in Example 10.2 examined dishonest behavior for two groups of students; one group who were tested in a dimly lit room and one group tested in a well-lit room. If we assume that the null hypothesis is true and there is no difference between the two populations of students, there should be no systematic difference between the two samples. In this case, the two samples can be combined to form a single set of $n = 16$ scores with an overall mean of $M = 10$. The two samples are shown as a single distribution in Figure 10.4(a).

For this example, however, the conclusion from hypothesis test is that there is a real difference between the two groups. The students tested in the dimly lit room have a mean score of $M = 12$, which is 2 points above the overall average. Similarly, the students tested in the well-lit room had a mean score of $M = 8$, 2 points below the overall average. Thus,

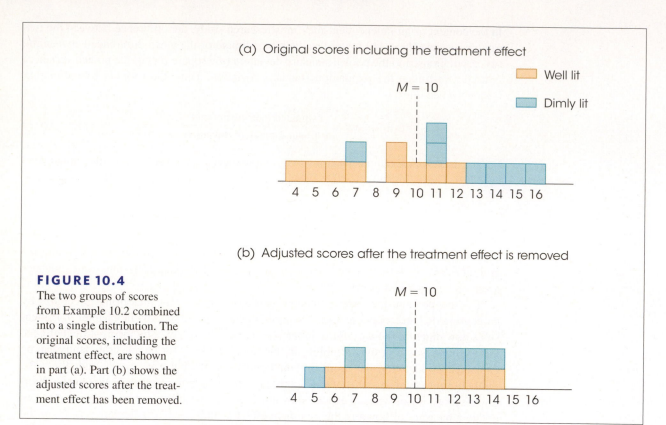

FIGURE 10.4

The two groups of scores from Example 10.2 combined into a single distribution. The original scores, including the treatment effect, are shown in part (a). Part (b) shows the adjusted scores after the treatment effect has been removed.

the lighting effect causes one group of scores to move toward the right of the distribution, away from the middle, and causes the other group to move toward the left. The result is that the lighting effect causes the scores to spread out and increases the variability.

To determine how much the treatment effect has increased the variability, we remove the treatment effect and examine the resulting scores. To remove the effect, we add 2 points to the score for each student tested in the well-lit room and we subtract 2 points for each student tested in the dimly lit room. This adjustment causes both groups to have a mean of $M = 10$ so there is no longer any mean difference between the two groups. The adjusted scores are shown in Figure 10.4(b).

It should be clear that the adjusted scores in Figure 10.4(b) are less variable (more closely clustered) than the original scores in Figure 10.4(a). That is, removing the treatment effect has reduced the variability. To determine exactly how much the treatment influences variability, we have computed SS, the sum of squared deviations, for each set of scores. For the scores in Figure 10.4(a), including the treatment effect, we obtain $SS = 190$. When the treatment effect is removed, in Figure 10.4(b), the variability is reduced to $SS = 126$. The difference between these two values is 64 points. Thus, the treatment effect accounts for 64 points of the total variability in the original scores. When expressed as a proportion of the total variability, we obtain

$$\frac{\text{variability explained by the treatment}}{\text{total variability}} = \frac{64}{190} = 0.337$$

You should recognize that this is exactly the same value we obtained for r^2 using Equation 10.9. ∎

The following example is an opportunity to test you understanding of Cohen's d and r^2 for the independent-measures t statistic.

EXAMPLE 10.6

In an independent-measures study with $n = 16$ scores in each treatment, one sample has $M = 89.5$ with $SS = 1005$ and the second sample has $M = 82.0$ with $SS = 1155$. The data produce $t(30) = 2.50$. Use these data to compute Cohen's d and r^2 for these data. You should find that $d = 0.883$ and $r^2 = 0.172$.   ■

■ Confidence Intervals for Estimating $\mu_1 - \mu_2$

As noted in chapter 9, it is possible to compute a confidence interval as an alternative method for measuring and describing the size of the treatment effect. For the single-sample t, we used a single sample mean, M, to estimate a single population mean. For the independent-measures t, we use a sample mean difference, $M_1 - M_2$, to estimate the population mean difference, $\mu_1 - \mu_2$. In this case, the confidence interval literally estimates the size of the population mean difference between the two populations or treatment conditions.

As with the single-sample t, the first step is to solve the t equation for the unknown parameter. For the independent-measures t statistic, we obtain

$$\mu_1 - \mu_2 = M_1 - M_2 \pm ts_{(M_1-M_2)} \tag{10.10}$$

In the equation, the values for $M_1 - M_2$ and for $s_{(M_1-M_2)}$ are obtained from the sample data. Although the value for the t statistic is unknown, we can use the degrees of freedom for the t statistic and the t distribution table to estimate the t value. Using the estimated t and the known values from the sample, we can then compute the value of $\mu_1 - \mu_2$. The following example demonstrates the process of constructing a confidence interval for a population mean difference.

EXAMPLE 10.7

Earlier we presented a research study comparing puzzle-solving scores for students who were tested in a dimly lit room scores for students tested in a well-lit room (p. 310). The results of the hypothesis test indicated a significant mean difference between the two populations of students. Now, we will construct a 95% confidence interval to estimate the size of the population mean difference.

The data from the study produced a mean grade of $M = 12$ for the group in the dimly lit room and a mean of $M = 8$ for the group in the well-lit room The estimated standard error for the mean difference was $s_{(M_1-M_2)} = 1.5$. With $n = 8$ scores in each sample, the independent-measures t statistic has $df = 14$. To have 95% confidence, we simply estimate that the t statistic for the sample mean difference is located somewhere in the middle 95% of all the possible t values. According to the t distribution table, with $df = 14$, 95% of the t values are located between t = +2.145 and $t = -2.145$. Using these values in the estimation equation, we obtain

$$\mu_1 - \mu_2 = M_1 - M_2 \pm ts_{(M_1-M_2)}$$
$$= 12 - 8 \pm 2.145(1.5)$$
$$= 4 \pm 3.218$$

This produces an interval of values ranging from $4 - 3.218 = 0.782$ to $4 + 3.218 = 7.218$. Thus, our conclusion is that students who were tested in the dimly lit room had higher scores that those who were tested in a well-lit room, and the mean difference between the two populations is somewhere between 0.782 points and 7.218 points. Furthermore, we are 95% confident that the true mean difference is in this interval because the only value estimated during the calculations was the t statistic, and we are 95% confident that the t value is located in the middle 95% of the distribution. Finally, note that the confidence interval is constructed around the sample mean difference. As a result, the sample mean difference, $M_1 - M_2 = 12 - 8 = 4$ points, is located exactly in the center of the interval.   ■

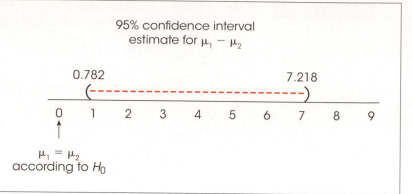

FIGURE 10.5

The 95% confidence interval for the population mean difference $(\mu_1 = \mu_2)$ from Example 10.7. Note that $\mu_1 - \mu_2 = 0$ is excluded from the confidence interval, indicating that a zero difference is not an acceptable value (H_0 would be rejected in a hypothesis test with $\alpha = .05$).

As with the confidence interval for the single-sample *t* (p. 286), the confidence interval for an independent-measures *t* is influenced by a variety of factors other than the actual size of the treatment effect. In particular, the width of the interval depends on the percentage of confidence used so that a larger percentage produces a wider interval. Also, the width of the interval depends on the sample size, so that a larger sample produces a narrower interval. Because the interval width is related to sample size, the confidence interval is not a pure measure of effect size like Cohen's *d* or r^2.

■ Confidence Intervals and Hypothesis Tests

The hypothesis test for these data was conducted in Example 10.2 (p. 310) and the decision was to reject H_0 with $\alpha = .05$.

In addition to describing the size of a treatment effect, estimation can be used to get an indication of the *significance* of the effect. Example 10.7 presented an independent-measures research study examining the effect of room lighting on performance scores (cheating). Based on the results of the study, the 95% confidence interval estimated that the population mean difference for the two groups of students was between 0.782 and 7.218 points. The confidence interval estimate is shown in Figure 10.5. In addition to the confidence interval for $\mu_1 - \mu_2$, we have marked the spot where the mean difference is equal to zero. You should recognize that a mean difference of zero is exactly what would be predicted by the null hypothesis if we were doing a hypothesis test. You also should realize that a zero difference ($\mu_1 - \mu_2 = 0$) is *outside* the 95% confidence interval. In other words, $\mu_1 - \mu_2 = 0$ is not an acceptable value if we want 95% confidence in our estimate. To conclude that a value of zero is *not acceptable* with 95% confidence is equivalent to concluding that a value of zero is *rejected* with 95% confidence. This conclusion is equivalent to rejecting H_0 with $\alpha = .05$. On the other hand, if a mean difference of zero were included within the 95% confidence interval, then we would have to conclude that $\mu_1 - \mu_2 = 0$ is an acceptable value, which is the same as failing to reject H_0.

IN THE LITERATURE

Reporting the Results of an Independent-Measures *t* Test

A research report typically presents the descriptive statistics followed by the results of the hypothesis test and measures of effect size (inferential statistics). In Chapter 4 (page 121), we demonstrated how the mean and the standard deviation are reported in APA format. In Chapter 9 (page 287), we illustrated the APA style for reporting the results of a *t* test. Now we use the APA format to report the results of Example 10.2, an independent-measures *t* test. A concise statement might read as follows:

> The students who were tested in a dimly lit room reported higher performance scores ($M = 12$, $SD = 2.93$) than the students who were tested in the well-lit room ($M = 8$, $SD = 3.07$). The mean difference was significant, $t(14) = 2.67$, $p < .05$, $d = 1.33$.

You should note that standard deviation is not a step in the computations for the independent-measures *t* test, yet it is useful when providing descriptive statistics for each treatment group. It is easily computed when doing the *t* test because you need SS and *df* for both groups to determine the pooled variance. Note that the format for reporting *t* is exactly the same as that described in Chapter 9 (page 287) and that the measure of effect size is reported immediately after the results of the hypothesis test.

Also, as we noted in Chapter 9, if an exact probability is available from a computer analysis, it should be reported. For the data in Example 10.2, the computer analysis reports a probability value of $p = .018$ for $t = 2.67$ with $df = 14$. In the research report, this value would be included as follows:

> The difference was significant, $t(14) = 2.67$, $p = .018$, $d = 1.33$.

Finally, if a confidence interval is reported to describe effect size, it appears immediately after the results from the hypothesis test. For the cheating behavior examples (Example 10.2 and Example 10.7) the report would be as follows:

> The difference was significant, $t(14) = 2.67$, $p = .018$, 95% CI [0.782, 7.218].

LEARNING CHECK

1. An independent-measures study with $n = 8$ in each treatment produces $M = 86$ for the first treatment and $M = 82$ for the second treatment with a pooled variance of 16. What is Cohen's *d* for these data?
   a. 0.25
   b. 0.50
   c. 1.00
   d. 2.00

2. An independent-measures study with $n = 12$ in each treatment produces $M = 34$ with $SS = 44$ for the first treatment and $M = 42$ with $SS = 66$ for the second treatment. If the data are used to construct a 95% confidence interval for the population mean difference, then what value will be at the center of the interval?
   a. 0
   b. 2.074
   c. 8
   d. 22

3. An independent-measures study with $n = 10$ in each treatment produces $M = 35$ for the first treatment and $M = 30$ for the second treatment. If the data are evaluated with a hypothesis test and the decision is to reject the null hypothesis with $\alpha = .05$, then what value cannot be inside the 95% confidence interval for the population mean difference?
   a. 0
   b. 2.101
   c. 5
   d. Impossible to determine without more information.

4. A researcher completes a hypothesis test with an independent-measures t statistic and obtains $t = 2.36$ with $df = 18$, which is significant with $\alpha = .05$, and computes Cohen's $d = 0.36$. How would this result appear in a scientific report?

   a. $t(18) = 2.36$, $d = 0.36$, $p > .05$

   b. $t(18) = 2.36$, $d = 0.36$, $p < .05$

   c. $t(18) = 2.36$, $p > .05$, $d = 0.36$

   d. $t(18) = 2.36$, $p < .05$, $d = 0.36$

ANSWERS   **1.** C, **2.** C, **3.** A, **4.** D

10.5 | The Role of Sample Variance and Sample Size in the Independent-Measures t Test

LEARNING OBJECTIVE **13.** Describe how sample size and sample variance influence the outcome of a hypothesis test and measures of effect size for the independent-measures t statistic.

In Chapter 9 (page 278) we identified several factors that can influence the outcome of a hypothesis test. Two factors that play important roles are the variability of the scores and the size of the samples. Both factors influence the magnitude of the estimated standard error in the denominator of the t statistic. The standard error is directly related to sample variance so that larger variance leads to larger error. As a result, larger variance produces a smaller value for the t statistic (closer to zero) and reduces the likelihood of finding a significant result. By contrast, the standard error is inversely related to sample size (larger size leads to smaller error). Thus, a larger sample produces a larger value for the t statistic (farther from zero) and increases the likelihood of rejecting H_0.

Although variance and sample size both influence the hypothesis test, only variance has a large influence on measures of effect size such as Cohen's d and r^2; larger variance produces smaller measures of effect size. Sample size, on the other hand, has no effect on the value of Cohen's d and only a small influence on r^2.

The following example provides a visual demonstration of how large sample variance can obscure a mean difference between samples and lower the likelihood of rejecting H_0 for an independent-measures study.

EXAMPLE 10.8   We will use the data in Figure 10.6 to demonstrate the influence of sample variance. The figure shows the results from a research study comparing two treatments. Notice that the study uses two separate samples, each with $n = 9$, and there is a 5-point mean difference between the two samples: $M = 8$ for treatment 1 and $M = 13$ for treatment 2. Also notice that there is a clear difference between the two distributions; the scores for treatment 2 are clearly higher than the scores for treatment 1.

For the hypothesis test, the data produce a pooled variance of 1.50 and an estimated standard error of 0.58. The t statistic is

$$t = \frac{\text{mean difference}}{\text{estimated standard error}} = \frac{5}{0.58} = 8.62$$

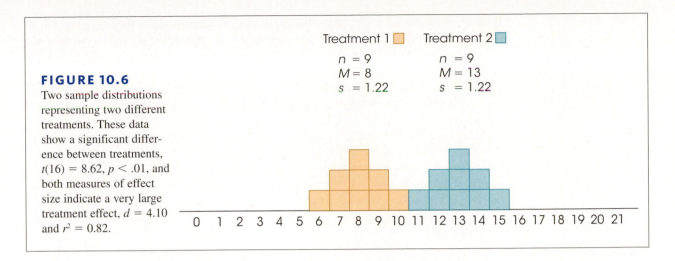

FIGURE 10.6

Two sample distributions representing two different treatments. These data show a significant difference between treatments, $t(16) = 8.62$, $p < .01$, and both measures of effect size indicate a very large treatment effect, $d = 4.10$ and $r^2 = 0.82$.

With $df = 16$, this value is far into the critical region (for $\alpha = .05$ or $\alpha = .01$), so we reject the null hypothesis and conclude that there is a significant difference between the two treatments.

Now consider the effect of increasing sample variance. Figure 10.7 shows the results from a second research study comparing two treatments. Notice that there are still $n = 9$ scores in each sample, and the two sample means are still $M = 8$ and $M = 13$. However, the sample variances have been greatly increased: each sample now has $s^2 = 44.25$ as compared with $s^2 = 1.5$ for the data in Figure 10.6. Notice that the increased variance means that the scores are now spread out over a wider range, with the result that the two samples are mixed together without any clear distinction between them.

The absence of a clear difference between the two samples is supported by the hypothesis test. The pooled variance is 44.25, the estimated standard error is 3.14, and the independent-measures *t* statistic is

$$ t = \frac{\text{mean difference}}{\text{estimated standard error}} = \frac{5}{3.14} = 1.59 $$

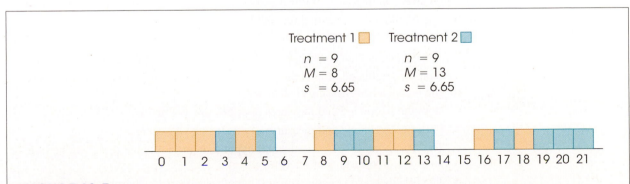

FIGURE 10.7

Two sample distributions representing two different treatments. These data show exactly the same mean difference as the scores in Figure 10.6, however the variance has been greatly increased. With the increased variance, there is no longer a significant difference between treatments, $t(16) = 1.59$, $p > .05$, and both measures of effect size are substantially reduced, $d = 0.75$ and $r^2 = 0.14$.

With $df = 16$ and $\alpha = .05$, this value is not in the critical region. Therefore, we fail to reject the null hypothesis and conclude that there is no significant difference between the two treatments. Although there is still a 5-point difference between sample means (as in Figure 10.6), the 5-point difference is not significant with the increased variance. In general, large sample variance can obscure any mean difference that exists in the data and reduces the likelihood of obtaining a significant difference in a hypothesis test. ■

Finally, we should note that the problems associated with high variance often can be minimized by transforming the original scores to ranks and then conducting an alternative statistical analysis known as the *Mann-Whitney test*, which is designed specifically for ordinal data. The Mann-Whitney test is presented in Appendix E, which also discusses the general purpose and process of converting numerical scores into ranks. The Mann-Whitney test also can be used if the data violate one of the assumptions for the independent-measures *t* test (pp. 313–314).

LEARNING CHECK

1. Which of the following sets of data is most likely to reject the null hypothesis in a test with the independent-measures *t* statistic. Assume that other factors are held constant.
   a. $n = 20$ and $SS = 210$ for both samples
   b. $n = 10$ and $SS = 210$ for both samples
   c. $n = 20$ and $SS = 410$ for both samples
   d. $n = 10$ and $SS = 410$ for both samples

2. If other factors are held constant, what effect does increasing the sample sizes have on the likelihood of rejecting the null hypothesis and measures of effect size for the independent-measures *t* statistic?
   a. Both the likelihood and measures of effect size will increase.
   b. Both the likelihood and measures of effect size will decrease.
   c. The likelihood of rejecting the null hypothesis increases but there will be little or no effect on measures of effect size.
   d. The likelihood of rejecting the null hypothesis decreases but there will be little or no effect on measures of effect size.

3. If other factors are held constant, what effect does increasing the sample variance have on the likelihood of rejecting the null hypothesis and measures of effect size for the independent-measures *t* statistic?
   a. Both the likelihood and measures of effect size will increase.
   b. Both the likelihood and measures of effect size will decrease.
   c. The likelihood of rejecting the null hypothesis increases but there will be little or no effect on measures of effect size.
   d. The likelihood of rejecting the null hypothesis decreases but there will be little or no effect on measures of effect size.

ANSWERS   **1.** A, **2.** C, **3.** B

SUMMARY

1. The independent-measures t statistic uses the data from two separate samples to draw inferences about the mean difference between two populations or between two different treatment conditions.

2. The formula for the independent-measures t statistic has the same structure as the original z-score or the single-sample t:

$$t = \frac{\text{sample statistic} - \text{population parameter}}{\text{estimated standard error}}$$

   For the independent-measures t, the sample statistic is the sample mean difference $(M_1 - M_2)$. The population parameter is the population mean difference, $(\mu_1 - \mu_2)$. The estimated standard error for the sample mean difference is computed by combining the errors for the two sample means. The resulting formula is

$$t = \frac{(M_1 - M_2) - (\mu_1 - \mu_2)}{s_{(M_1 - M_2)}}$$

   where the estimated standard error is

$$s_{(M_1 - M_2)} = \sqrt{\frac{s_p^2}{n_1} + \frac{s_p^2}{n_2}}$$

   The pooled variance in the formula, s_p^2, is the weighted mean of the two sample variances:

$$s_p^2 = \frac{SS_1 + SS_2}{df_1 + df_2}$$

   This t statistic has degrees of freedom determined by the sum of the df values for the two samples:

$$\begin{aligned} df &= df_1 + df_2 \\ &= (n_1 - 1) + (n_2 - 1) \end{aligned}$$

3. For hypothesis testing, the null hypothesis states that there is no difference between the two population means:

$$H_0: \mu_1 = \mu_2 \text{ or } \mu_1 - \mu_2 = 0$$

4. When a hypothesis test with an independent-measures t statistic indicates a significant difference, it is recommended that you also compute a measure of the effect size. One measure of effect size is Cohen's d, which is a standardized measured of the mean difference. For the independent-measures t statistic, Cohen's d is estimated as follows:

$$\text{estimated } d = \frac{M_1 - M_2}{\sqrt{s_p^2}}$$

   A second common measure of effect size is the percentage of variance accounted for by the treatment effect. This measure is identified by r^2 and is computed as

$$r^2 = \frac{t^2}{t^2 + df}$$

5. An alternative method for describing the size of the treatment effect is to construct a confidence interval for the population mean difference, $\mu_1 - \mu_2$. The confidence interval uses the independent-measures t equation, solved for the unknown mean difference:

$$\mu_1 - \mu_2 = M_1 - M_2 \pm ts_{(M_1 - M_2)}$$

   First, select a level of confidence and then look up the corresponding t values. For example, for 95% confidence, use the range of t values that determine the middle 95% of the distribution. The t values are then used in the equation along with the values for the sample mean difference and the standard error, which are computed from the sample data.

6. Appropriate use and interpretation of the t statistic using pooled variance require that the data satisfy the homogeneity of variance assumption. This assumption stipulates that the two populations have equal variances. An informal test of the assumption can be made by verifying that the two sample variances are approximately equal. Hartley's F-max test provides a statistical technique for determining whether the data satisfy the homogeneity assumption. An alternative technique that avoids pooling variances and eliminates the need for the homogeneity assumption is presented in Box 10.2.

KEY TERMS

independent-measures research design (301)

between-subjects research design (301)

repeated-measures research design (301)

within-subjects research design (301)

independent-measures t statistic (303)

estimated standard error of $M_1 - M_2$ (304)

pooled variance (306)

Mann-Whitney test (324)

homogeneity of variance (313)

General instructions for using SPSS are presented in Appendix D. Following are detailed instructions for using SPSS to perform the **Independent-Measures *t* Test** presented in this chapter.

Data Entry

1. The scores are entered in what is called *stacked format*, which means that all the scores from *both samples* are entered in one column of the data editor (probably VAR00001). Enter the scores for sample #2 directly beneath the scores from sample #1 with no gaps or extra spaces.

2. Values are then entered into a second column (VAR00002) to identify the sample or treatment condition corresponding to each of the scores. For example, enter a 1 beside each score from sample #1 and enter a 2 beside each score from sample #2.

Data Analysis

1. Click **Analyze** on the tool bar, select **Compare Means**, and click on **Independent-Samples T Test**.

2. Highlight the column label for the set of scores (VAR0001) in the left box and click the arrow to move it into the **Test Variable(s)** box.

3. Highlight the label from the column containing the sample numbers (VAR0002) in the left box and click the arrow to move it into the **Group Variable** box.

4. Click on **Define Groups**.

5. Assuming that you used the numbers 1 and 2 to identify the two sets of scores, enter the values 1 and 2 into the appropriate group boxes.

6. Click **Continue**.

7. In addition to performing the hypothesis test, the program will compute a confidence interval for the population mean difference. The confidence level is automatically set at 95% but you can select **Options** and change the percentage.

8. Click **OK**.

SPSS Output

We used the SPSS program to analyze the data from Example 10.2, which examined the relationship between cheating and room lighting. The program output is shown in Figure 10.8. The output includes a table of sample statistics with the mean, standard deviation, and standard error of the mean for each group. A second table, which is split into two sections in Figure 10.8, begins with the results of Levene's test for homogeneity of variance. This test should *not* be significant (you do not want the two variances to be different), so you want the reported Sig. value to be greater than .05. Next, the results of the independent-measures *t* test are presented using two different assumptions. The top row shows the outcome assuming equal variances, using the pooled variance to compute *t*. The second row does not assume equal variances and computes the *t* statistic using the alternative method presented in Box 10.2. Each row reports the calculated *t* value, the degrees of freedom, the level of significance (the *p* value for the test), the size of the mean difference and the standard error for the mean difference (the denominator of the *t* statistic). Finally, the output includes a 95% confidence interval for the mean difference.

Group Statistics

VAR00002		N	Mean	Std. Deviation	Std. Error Mean
VAR00001	1.00	8	8.0000	2.92770	1.03510
	2.00	8	12.0000	3.07060	1.08562

Independent Samples Test

		Levene's Test for Equality of Variances		t-test for Equality of Means	
		F	Sig.	t	df
VAR00001	Equal variances assumed	.000	1.000	−2.667	14
	Equal variances not assumed			−2.667	13.968

Independent Samples Test

		t-test for Equality of Means				
					95% Confidence Interval of the Difference	
		Sig. (2-tailed)	Mean Difference	Std. Error Difference	Lower	Upper
VAR00001	Equal variances assumed	.018	−4.00000	1.50000	−7.21718	−.78282
	Equal variances not assumed	.018	−4.00000	1.50000	−7.21786	−.78214

FIGURE 10.8
The SPSS output for the independent-measures hypothesis test in Example 10.2.

FOCUS ON PROBLEM SOLVING

1. As you learn more about different statistical methods, one basic problem will be deciding which method is appropriate for a particular set of data. Fortunately, it is easy to identify situations in which the independent-measures t statistic is used. First, the data will always consist of two separate samples (two ns, two Ms, two SSs, and so on). Second, this t statistic is always used to answer questions about a mean difference: On the average, is one group different (better, faster, smarter) than the other group? If you examine the data and identify the type of question that a researcher is asking, you should be able to decide whether an independent-measures t is appropriate.

2. When computing an independent-measures t statistic from sample data, we suggest that you routinely divide the formula into separate stages rather than trying to do all the calculations at once. First, find the pooled variance. Second, compute the standard error. Third, compute the t statistic.

3. One of the most common errors for students involves confusing the formulas for pooled variance and standard error. When computing pooled variance, you are "pooling" the two samples together into a single variance. This variance is computed as a *single fraction*, with two SS values in the numerator and two df values in the denominator. When computing the standard error, you are adding the error from the first sample and the error from the second sample. These two separate errors are added as *two separate fractions* under the square root symbol.

DEMONSTRATION 10.1

THE INDEPENDENT-MEASURES *t* TEST

In a study of jury behavior, two samples of participants were provided details about a trial in which the defendant was obviously guilty. Although group 2 received the same details as group 1, the second group was also told that some evidence had been withheld from the jury by the judge. Later the participants were asked to recommend a jail sentence. The length of term suggested by each participant is presented here. Is there a significant difference between the two groups in their responses?

Group 1	Group 2	
4	3	
4	7	
3	8	for Group 1: $M = 3$ and $SS = 16$
2	5	
5	4	for Group 2: $M = 6$ and $SS = 24$
1	7	
1	6	
4	8	

There are two separate samples in this study. Therefore, the analysis will use the independent-measures *t* test.

STEP 1 **State the hypothesis, and select an alpha level**

H_0: $\mu_1 - \mu_2 = 0$ **(For the population, knowing evidence has been withheld has no effect on the suggested sentence.)**

H_1: $\mu_1 - \mu_2 \neq 0$ **(For the population, knowledge of withheld evidence has an effect on the jury's response.)**

We will set the level of significance to $\alpha = .05$, two tails.

STEP 2 **Identify the critical region**   For the independent-measures *t* statistic, degrees of freedom are determined by

$$df = n_1 + n_2 - 2$$
$$= 8 + 8 - 2$$
$$= 14$$

The *t* distribution table is consulted, for a two-tailed test with $\alpha = .05$ and $df = 14$. The critical *t* values are $+2.145$ and -2.145.

STEP 3 **Compute the test statistic**   As usual, we recommended that the calculation of the *t* statistic be separated into three stages.

Pooled Variance   For these data, the pooled variance equals

$$s_p^2 = \frac{SS_1 + SS_2}{df_1 + df_2} = \frac{16 + 24}{7 + 7} = \frac{40}{14} = 2.86$$

Estimated Standard Error   Now we can calculate the estimated standard error for mean differences.

$$s_{(M_1 - M_2)} = \sqrt{\frac{s_p^2}{n_1} + \frac{s_p^2}{n_2}} = \sqrt{\frac{2.86}{8} + \frac{2.86}{8}} = \sqrt{0.358 + 0.358} = \sqrt{0.716} = 0.85$$

The t Statistic Finally, the t statistic can be computed.

$$t = \frac{(M_1 - M_2) - (\mu_1 - \mu_2)}{s_{(M_1-M_2)}} = \frac{(3-6)-0}{0.85} = \frac{-3}{0.85} = -3.53$$

STEP 4 **Make a decision about H_0, and state a conclusion** The obtained t value of -3.53 falls in the critical region of the left tail (critical $t = \pm 2.145$). Therefore, the null hypothesis is rejected. The participants who were informed about the withheld evidence gave significantly longer sentences, $t(14) = -3.53$, $p < .05$, two tails.

DEMONSTRATION 10.2

EFFECT SIZE FOR THE INDEPENDENT-MEASURES t

We will estimate Cohen's d and compute r^2 for the jury decision data in Demonstration 10.1. For these data, the two sample means are $M_1 = 3$ and $M_2 = 6$, and the pooled variance is 2.86. Therefore, our estimate of Cohen's d is

$$\text{estimated } d = \frac{M_1 - M_2}{\sqrt{s_p^2}} = \frac{3-6}{\sqrt{2.86}} = \frac{3}{1.69} = 1.78$$

With a t value of $t = 3.53$ and $df = 14$, the percentage of variance accounted for is

$$r^2 = \frac{t^2}{t^2 + df} = \frac{(3.53)^2}{(3.53)^2 + 14} = \frac{12.46}{26.46} = 0.47 \quad (\text{or } 47\%)$$

PROBLEMS

1. Describe the basic characteristics of an independent-measures, or a between-subjects, research study.

2. Describe what is measured by the estimated standard error in the bottom of the independent-measures t statistic.

3. One sample has $SS = 36$ and a second sample has $SS = 18$.
   a. If $n = 4$ for both samples, find each of the sample variances and compute the pooled variance. Because the samples are the same size, you should find that the pooled variance is exactly halfway between the two sample variances.
   b. Now assume that $n = 4$ for the first sample and $n = 7$ for the second. Again, calculate the two sample variances and the pooled variance. You should find that the pooled variance is closer to the variance for the larger sample.

4. One sample has $SS = 60$ and a second sample has $SS = 48$.
   a. If $n = 7$ for both samples, find each of the sample variances, and calculate the pooled variance. Because the samples are the same size, you should find that the pooled variance is exactly halfway between the two sample variances.

   b. Now assume that $n = 7$ for the first sample and $n = 5$ for the second. Again, calculate the two sample variances and the pooled variance. You should find that the pooled variance is closer to the variance for the larger sample.

5. Two separate samples, each with $n = 15$ individuals, receive different treatments. After treatment, the first sample has $SS = 1740$ and the second has $SS = 1620$.
   a. Find the pooled variance for the two samples.
   b. Compute the estimated standard error for the sample mean difference.
   c. If the sample mean difference is 8 points, is this enough to reject the null hypothesis and conclude that there is a significant difference for a two-tailed test at the .05 level?

6. Two separate samples receive different treatments. After treatment, the first sample has $n = 9$ with $SS = 462$, and the second has $n = 7$ with $SS = 420$.
   a. Compute the pooled variance for the two samples.
   b. Calculate the estimated standard error for the sample mean difference.
   c. If the sample mean difference is 10 points, is this enough to reject the null hypothesis using a two-tailed test with $\alpha = .05$?

7. Research results suggest a relationship between the TV viewing habits of 5-year-old children and their future performance in high school. For example, Anderson, Huston, Wright, and Collins (1998) report that high school students who regularly watched Sesame Street as children had better grades in high school than their peers who did not watch Sesame Street. Suppose that a researcher intends to examine this phenomenon using a sample of 20 high school students.

The researcher first surveys the students' parents to obtain information on the family's TV viewing habits during the time that the students were 5 years old. Based on the survey results, the researcher selects a sample of $n = 10$ students with a history of watching "Sesame Street" and a sample of $n = 10$ students who did not watch the program. The average high school grade is recorded for each student and the data are as follows:

Average High School Grade			
Watched Sesame Street		Did Not Watch Sesame Street	
86	99	90	79
87	97	89	83
91	94	82	86
97	89	83	81
98	92	85	92
$n = 10$		$n = 10$	
$M = 93$		$M = 85$	
$SS = 200$		$SS = 160$	

Use an independent-measures *t* test with $\alpha = .01$ to determine whether there is a significant difference between the two types of high school student.

8. It appears that there is some truth to the old adage "That which doesn't kill us makes us stronger." Seery, Holman, and Silver (2010) found that individuals with some history of adversity report better mental health and higher well-being compared to people with little or no history of adversity. In an attempt to examine this phenomenon, a researcher surveys a group of college students to determine the negative life events that they experienced in the past 5 years and their current feeling of well-being. For $n = 18$ participants with 2 or fewer negative experiences, the average well-being score is $M = 42$ with $SS = 398$, and for $n = 16$ participants with 5 to 10 negative experiences the average score is $M = 48.6$ with $SS = 370$.
   a. Is there a significant difference between the two populations represented by these two samples? Use a two-tailed test with $\alpha = .01$.

   b. Compute Cohen's *d* to measure the size of the effect.
   c. Write a sentence demonstrating how the outcome of the hypothesis test and the measure of effect size would appear in a research report.

9. Does posting calorie content for menu items affect people's choices in fast food restaurants? According to results obtained by Elbel, Gyamfi, and Kersh (2011), the answer is no. The researchers monitored the calorie content of food purchases for children and adolescents in four large fast food chains before and after mandatory labeling began in New York City. Although most of the adolescents reported noticing the calorie labels, apparently the labels had no effect on their choices. Data similar to the results obtained show an average of $M = 786$ calories per meal with $s = 85$ for $n = 100$ children and adolescents before the labeling, compared to an average of $M = 772$ calories with $s = 91$ for a similar sample of $n = 100$ after the mandatory posting.
   a. Use a two-tailed test with $\alpha = .05$ to determine whether the mean number of calories after the posting is significantly different than before calorie content was posted.
   b. Calculate r^2 to measure effect size for the mean difference.

10. In 1974, Loftus and Palmer conducted a classic study demonstrating how the language used to ask a question can influence eyewitness memory. In the study, college students watched a film of an automobile accident and then were asked questions about what they saw. One group was asked, "About how fast were the cars going when they smashed into each other?" Another group was asked the same question except the verb was changed to "hit" instead of "smashed into." The "smashed into" group reported significantly higher estimates of speed than the "hit" group. Suppose a researcher repeats this study with a sample of today's college students and obtains the following results.

Estimated Speed	
Smashed into	Hit
$n = 15$	$n = 15$
$M = 40.8$	$M = 34.0$
$SS = 510$	$SS = 414$

   a. Do the results indicate a significantly higher estimated speed for the "smashed into" group? Use a one-tailed test with $\alpha = .01$.
   b. Compute the estimated value for Cohen's *d* to measure the size of the effect.

c. Write a sentence demonstrating how the results of the hypothesis test and the measure of effect size would appear in a research report.

11. Recent research has shown that creative people are more likely to cheat than their less creative counterparts (Gino and Ariely, 2012). Participants in the study first completed creativity assessment questionnaires and then returned to the lab several days later for a series of tasks. One task was a multiple-choice general knowledge test for which the participants circled their answers on the test sheet. Afterward, they were asked to transfer their answers to a bubble sheets for computer scoring. However, the experimenter admitted that the wrong bubble sheet had been copied so that the correct answers were still faintly visible. Thus, the participants had an opportunity to cheat and inflate their test scores. Higher scores were valuable because participants were paid based on the number of correct answers. However, the researchers had secretly coded the original tests and the bubble sheets so that they could measure the degree of cheating for each participant. Assuming that the participants were divided into two groups based on their creativity scores, the following data are similar to the cheating scores obtained in the study.

High Creativity Participants	Low Creativity Participants
$n = 27$	$n = 27$
$M = 7.41$	$M = 4.78$
$SS = 749.5$	$SS = 830$

a. Use a one-tailed test with $\alpha = .05$ to determine whether these data are sufficient to conclude that high creativity people are more likely to cheat than people with lower levels of creativity.
b. Compute Cohen's d to measure the size of the effect.
c. Write a sentence demonstrating how the results from the hypothesis test and the measure of effect size would appear in a research report.

12. Recent research has demonstrated that music-based physical training for elderly people can improve balance, walking efficiency, and reduce the risk of falls (Trombetti et al., 2011). As part of the training, participants walked in time to music and responded to changes in the music's rhythm during a 1-hour per week exercise program. After 6 months, participants in the training group increased their walking speed and their stride length compared to individuals in the

control group. The following data are similar to the results obtained in the study.

Exercise Group Stride Length				Control Group Stride Length			
24	25	22	24	26	23	20	23
26	17	21	22	20	16	21	17
22	19	24	23	18	23	16	20
23	28	25	23	25	19	17	16

Do the results indicate a significant difference in the stride length for the two groups? Use a two-tailed test with $\alpha = .05$.

13. McAllister et al. (2012) compared varsity football and hockey players with varsity athletes from noncontact sports to determine whether exposure to head impacts during one season have an effect on cognitive performance. In the study, tests of new learning performance were significantly poorer for the contact sport athletes compared to the noncontact sport athletes. The following table presents data similar to the results obtained in the study.

Noncontact Athletes	Contact Athletes
10	7
8	4
7	9
9	3
13	7
7	6
6	10
12	2

a. Are the test scores significantly lower for the contact sport athletes than for the noncontact athletes? Use a one-tailed test with $\alpha = .05$.
b. Compute the value of r^2 (percentage of variance accounted for) for these data.

14. In the Chapter Preview we presented a study showing that handling money reduces the perception pain (Zhou, Vohs, & Baumeister, 2009). In the experiment, a group of college students was told that they were participating in a manual dexterity study. Half of the students were given a stack of money to count and the other half got a stack of blank pieces of paper. After the counting task, the participants were asked to dip their hands into bowls of very hot water (122°F) and rate how uncomfortable it was. The following data show ratings of pain similar to the results obtained in the study.

Counting Money	Counting Paper
7	9
8	11
10	13
6	10
8	11
5	9
7	15
12	14
5	10

a. Is there a significant difference in reported pain between the two conditions? Use a two-tailed test with $\alpha = .01$.

b. Compute Cohen's *d* to estimate the size of the treatment effect.

15. In a classic study in the area of problem solving, Katona (1940) compared the effectiveness of two methods of instruction. One group of participants was shown the exact, step-by-step procedure for solving a problem and was required to memorize the solution. Participants in a second group were encouraged to study the problem and find the solution on their own. They were given helpful hints and clues, but the exact solution was never explained. The study included the problem in the following figure showing a pattern of five squares made of matchsticks. The problem is to change the pattern into exactly four squares by moving only three matches. (All matches must be used, none can be removed, and all the squares must be the same size.) After 3 weeks, both groups returned to be tested again. The two groups did equally well on the matchstick problem they had learned earlier. But when they were given new problems (similar to the matchstick problem), the memorization group had much lower scores than the group who explored and found the solution on their own. The following data demonstrate this result.

Memorization of the Solution	Find a Solution on Your Own
$n = 8$	$n = 8$
$M = 10.50$	$M = 6.16$
$SS = 108$	$SS = 116$

a. Is there a significant difference in performance on new problems for these two groups? Use a two-tailed test with $\alpha = .05$.

b. Construct a 90% confidence interval to estimate the size of the mean difference.

Incidentally, if you still have not discovered the solution to the matchstick problem, keep trying. According to Katona's results, it would be very poor teaching strategy for us to give you the answer. If you still have not discovered the solution, however, check Appendix C at the beginning of the problem solutions for Chapter 10 and we will show you how it is done.

16. A researcher conducts an independent-measures study comparing two treatments and reports the *t* statistic as $t(25) = 2.071$.

a. How many individuals participated in the entire study?

b. Using a two-tailed test with $\alpha = .05$, is there a significant difference between the two treatments?

c. Compute r^2 to measure the percentage of variance accounted for by the treatment effect.

17. In a recent study, Piff, Kraus, Côté, Cheng, and Keitner (2010) found that people from lower social economic classes tend to display greater prosocial behavior than their higher class counterparts. In one part of the study, participants played a game with an anonymous partner. Part of the game involved sharing points with the partner. The lower economic class participants were significantly more generous with their points compared with the upper class individuals. Results similar to those found in the study, show that $n = 12$ lower class participants shared an average of $M = 5.2$ points with $SS = 11.91$, compared to an average of $M = 4.3$ with $SS = 9.21$ for the $n = 12$ upper class participants.

a. Are the data sufficient to conclude that there is a significant mean difference between the two economic populations? Use a two-tailed test with $\alpha = .05$

b. Construct an 80% confidence interval to estimate the size of the population mean difference.

18. Describe the homogeneity of variance assumption and explain why it is important for the independent-measures *t* test.

19. If other factors are held constant, explain how each of the following influences the value of the independent-measures t statistic, the likelihood of rejecting the null hypothesis, and the magnitude of measures of effect size.
   a. Increasing the number of scores in each sample.
   b. Increasing the variance for each sample.

20. As noted on page 304, when the two population means are equal, the estimated standard error for the independent-measures t test provides a measure of how much difference to expect between two sample means. For each of the following situations, assume that $\mu_1 = \mu_2$ and calculate how much difference should be expected between the two sample means.
   a. One sample has $n = 6$ scores with $SS = 75$ and the second sample has $n = 10$ scores with $SS = 135$.
   b. One sample has $n = 6$ scores with $SS = 310$ and the second sample has $n = 10$ scores with $SS = 530$.
   c. In part b, the samples have larger variability (bigger SS values) than in part a, but the sample sizes are unchanged. How does larger variability affect the magnitude of the standard error for the sample mean difference?

21. Two samples are selected from the same population. For each of the following, calculate how much difference is expected, on average, between the two sample means.
   a. One sample has $n = 4$, the second has $n = 6$, and the pooled variance is 60.
   b. One sample has $n = 12$, the second has $n = 15$, and the pooled variance is 60.
   c. In part b, the sample sizes are larger but the pooled variance is unchanged. How does larger sample size affect the magnitude of the standard error for the sample mean difference?

22. For each of the following, assume that the two samples are obtained from populations with the same mean, and calculate how much difference should be expected, on average, between the two sample means.
   a. Each sample has $n = 4$ scores with $s^2 = 68$ for the first sample and $s^2 = 76$ for the second. (*Note:* Because the two samples are the same size, the pooled variance is equal to the average of the two sample variances.)
   b. Each sample has $n = 16$ scores with $s^2 = 68$ for the first sample and $s^2 = 76$ for the second.
   c. In part b, the two samples are bigger than in part a, but the variances are unchanged. How does sample size affect the size of the standard error for the sample mean difference?

23. For each of the following, calculate the pooled variance and the estimated standard error for the sample mean difference
   a. The first sample has $n = 4$ scores and a variance of $s^2 = 17$, and the second sample has $n = 8$ scores and a variance of $s^2 = 27$.
   b. Now the sample variances are increased so that the first sample has $n = 4$ scores and a variance of $s^2 = 68$, and the second sample has $n = 8$ scores and a variance of $s^2 = 108$.
   c. Comparing your answers for parts a and b, how does increased variance influence the size of the estimated standard error?

The *t* Test for Two Related Samples

Tools You Will Need

The following items are considered essential background material for this chapter. If you doubt your knowledge of any of these items, you should review the appropriate chapter or section before proceeding.

- Introduction to the *t* statistic (Chapter 9)
  - Estimated standard error
  - Degrees of freedom
  - *t* Distribution
  - Hypothesis tests with the *t* statistic
- Independent-measures design (Chapter 10)

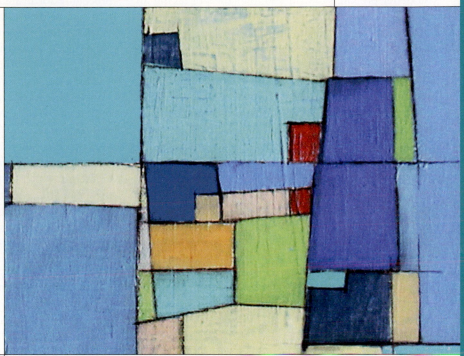

© Deborah Batt

It's the night before an exam. It is getting late and you are trying to decide whether to study or to sleep. There are obvious advantages to studying, especially if you feel that you do not have a good grasp of the material to be tested. On the other hand, a good night's sleep will leave you better prepared to deal with the stress of taking an exam. "To study or to sleep?" was the question addressed by a recent research study (Gillen-O'Neel, Huynh, & Fuligni, 2013). The researchers started with a sample of 535 9th-grade students and followed up when the students were in the 10th and 12th grades. Each year the students completed a diary every day for two weeks, recording how much time they spent studying outside of school and how much time they slept the night before. The students also reported the occurrence of academic problems each day such as "did not understand something taught in class" and "did poorly on a test, quiz, or homework." The data showed a general trade-off between study time and sleep time, especially for the older students. The primary result, however is that the students reported more academic problems following nights with less than average sleep than they did after nights with more than average sleep. Notice that the researchers recorded academic performance for each student in two conditions, after nights with more sleep and after nights with less sleep, and the goal is to compare these two sets of scores.

In the previous chapter, we introduced a statistical procedure for evaluating the mean difference between two sets of data (the independent-measures t statistic). However, the independent-measures t statistic is intended for research situations involving two separate and independent samples. You should realize that the two sets of scores in the sleep-or-study example are not independent samples. In fact, the same group individuals participated in both of the treatment conditions. What is needed is a new statistical analysis for comparing two means that are both obtained from the same group of participants.

In this chapter, we introduce the *repeated-measures* t *statistic*, which is used for hypothesis tests evaluating the mean difference between two sets of scores obtained from the same group of individuals. As you will see, however, this new t statistic is very similar to the original t statistic that was introduced in Chapter 9.

11.1 | Introduction to Repeated-Measures Designs

LEARNING OBJECTIVES

1. Define a repeated-measures design and explain how it differs from an independent-measures design.

2. Define a matched-subjects design and explain how it differs from repeated-measures and independent-measures designs.

In the previous chapter, we introduced the independent-measures research design as one strategy for comparing two treatment conditions or two populations. The independent-measures design is characterized by the fact that two separate samples are used to obtain the two sets of scores that are to be compared. In this chapter, we examine an alternative strategy known as a *repeated-measures design*, or a *within-subjects design*. With a repeated-measures design, one group of participants is measured in two different treatment conditions so there are two separate scores for each individual in the sample. For example, a group of patients could be measured before therapy and then measured again after therapy. Or, response time could be measured in a driving simulation task for a group of individuals who are first tested when they are sober and then tested again after two alcoholic drinks. In each case, the same variable is being measured twice for the same set of individuals; that is, we are literally repeating measurements on the same sample.

DEFINITION

A **repeated-measures design**, or a **within-subject design**, is one in which the dependent variable is measured two or more times for each individual in a single sample. The same group of subjects is used in all of the treatment conditions.

The main advantage of a repeated-measures study is that it uses exactly the same individuals in all treatment conditions. Thus, there is no risk that the participants in one treatment are substantially different from the participants in another. With an independent-measures design, on the other hand, there is always a risk that the results are biased because the individuals in one sample are systematically different (smarter, faster, more extroverted, and so on) than the individuals in the other sample. At the end of this chapter, we present a more detailed comparison of repeated-measures studies and independent-measures studies, considering the advantages and disadvantages of both types of research.

■ The Matched-Subjects Design

Occasionally, researchers try to approximate the advantages of a repeated-measures design by using a technique known as *matched subjects*. A matched-subjects design involves two separate samples, but each individual in one sample is matched one-to-one with an individual in the other sample. Typically, the individuals are matched on one or more variables that are considered to be especially important for the study. For example, a researcher studying verbal learning might want to be certain that the two samples are matched in terms of IQ. In this case, a participant with an IQ of 120 in one sample would be matched with another participant with an IQ of 120 in the other sample. Although the participants in one sample are not *identical* to the participants in the other sample, the matched-subjects design at least ensures that the two samples are equivalent (or matched) with respect to a specific variable.

DEFINITION

> In a **matched-subjects** study, each individual in one sample is matched with an individual in the other sample. The matching is done so that the two individuals are equivalent (or nearly equivalent) with respect to a specific variable that the researcher would like to control.

Of course, it is possible to match participants on more than one variable. For example, a researcher could match pairs of subjects on age, gender, race, and IQ. In this case, for example, a 22-year-old white female with an IQ of 115 who was in one sample would be matched with another 22-year-old white female with an IQ of 115 in the second sample. The more variables that are used, however, the more difficult it is to find matching pairs. The goal of the matching process is to simulate a repeated-measures design as closely as possible. In a repeated-measures design, the matching is perfect because the same individual is used in both conditions. In a matched-subjects design, however, the best you can get is a degree of match that is limited to the variable(s) that are used for the matching process.

In a repeated-measures design or a matched-subjects design comparing two treatment conditions, the data consist of two sets of scores, which are grouped into sets of two, corresponding to the two scores obtained for each individual or each matched pair of subjects (Table 11.1). Because the scores in one set are directly related, one-to-one, with the scores in the second set, the two research designs are statistically equivalent and share the common name *related-samples* designs (or correlated-samples designs). In this chapter, we focus our discussion on repeated-measures designs because they are overwhelmingly the more common example of related-samples designs. However, you should realize that the statistical techniques used for repeated-measures studies also can be applied directly to data from a matched-subjects study. We should also note that a matched-subjects study occasionally is called a *matched samples design*, but the subjects in the samples must be matched one-to-one before you can use the statistical techniques in this chapter.

TABLE 11.1   An example of the data from a repeated-measures or a matched-subjects study using $n = 5$ participants (or matched pairs).

Participant or Matched Pair	First Score	Second Score	
#1	12	15	←the 2 scores for
#2	10	14	one participant or
#3	15	17	one matched pair
#4	17	17	
#5	12	18	

Now we will examine the statistical techniques that allow a researcher to use the sample data from a repeated-measures study to draw inferences about the general population.

LEARNING CHECK

1. A researcher conducts an experiment comparing two treatment conditions and obtains 20 scores in each treatment. Which design would require the smallest number of subjects?
   a. a repeated-measures design
   b. an independent-measures design
   c. a matched-subjects design
   d. either an independent-measures or a matched-subjects design

2. For an experiment comparing two treatment conditions, an independent-measures design would obtain _____ score(s) for each subject and a repeated-measures design would obtain _____ score(s) for each subject.
   a. 1, 1
   b. 1, 2
   c. 2, 1
   d. 2, 2

3. For a research study comparing two treatments with $n = 10$ scores in each treatment, which design would require the largest number of participants?
   a. repeated-measures
   b. independent-measures
   c. matched-subjects
   d. independent-measures or match-subjects (both use the same number)

ANSWERS   1. A, 2. B, 3. D

11.2 | The *t* Statistic for a Repeated-Measures Research Design

3. Describe the data (difference scores) that are used for the repeated-measures *t* statistic.

4. Describe the hypotheses for a repeated-measures *t* test.

5. Describe the structure of the repeated-measures *t* statistic, including the estimated standard error and the degrees of freedom, and explain how the formula is related to the single-sample *t*.

The *t* statistic for a repeated-measures design is structurally similar to the other *t* statistics we have examined. As we will see, it is essentially the same as the single-sample *t* statistic covered in Chapter 9. The major distinction of the related-samples *t* is that it is based on difference scores rather than raw scores (*X* values). In this section, we examine difference scores and develop the *t* statistic for related samples.

■ Difference Scores: The Data for a Repeated-Measures Study

Many over-the-counter cold medications include the warning "may cause drowsiness." Table 11.2 shows an example of data from a study that examines this phenomenon. Note that there is one sample of $n = 4$ participants, and that each individual is measured twice. The first score for each person (X_1) is a measurement of reaction time before the medication was administered. The second score (X_2) measures reaction time 1 hour after taking the medication. Because we are interested in how the medication affects reaction time, we have computed the difference between the first score and the second score for each individual. The *difference scores*, or *D* values, are shown in the last column of the table. Notice that the difference scores measure the amount of change in reaction time for each person. Typically, the difference scores are obtained by subtracting the first score (before treatment) from the second score (after treatment) for each person:

$$\text{difference score} = D = X_2 - X_1 \tag{11.1}$$

Note that the sign of each *D* score tells you the direction of the change. Person A, for example, shows a decrease in reaction time after taking the medication (a negative change), but person B shows an increase (a positive change).

The sample of difference scores (*D* values) serves as the sample data for the hypothesis test and all calculations are done using the *D* scores. To compute the *t* statistic, for example, we use the number of *D* scores (*n*) as well as the mean for the sample of *D* scores (M_D) and the value of *SS* for the sample of *D* scores.

TABLE 11.2

Reaction time measurements taken before and after taking an over-the-counter cold medication.

Person	Before Medication (X_1)	After Medication (X_2)	Difference D
A	215	210	−5
B	221	242	21
C	196	219	23
D	203	228	25

$$\Sigma D = 64$$

Note that M_D is the mean for the sample of *D* scores.

$$M_D = \frac{\Sigma D}{n} = \frac{64}{4} = 16$$

■ The Hypotheses for a Related-Samples Test

The researcher's goal is to use the sample of difference scores to answer questions about the general population. In particular, the researcher would like to know whether there is any difference between the two treatment conditions for the general population. Note that we are interested in a population of *difference scores*. That is, we would like to know what would happen if every individual in the population were measured in two treatment conditions (X_1 and X_2) and a difference score (D) were computed for everyone. Specifically, we are interested in the mean for the population of difference scores. We identify this population mean difference with the symbol μ_D (using the subscript letter D to indicate that we are dealing with D values rather than X scores).

As always, the null hypothesis states that for the general population there is no effect, no change, or no difference. For a repeated-measures study, the null hypothesis states that the mean difference for the general population is zero. In symbols,

$$H_0: \mu_D = 0$$

Again, this hypothesis refers to the mean for the entire population of difference scores. Although the population mean is zero, the individual scores in the population are not all equal to zero. Thus, even when the null hypothesis is true, we still expect some individuals to have positive difference scores and some to have negative difference scores. However, the positives and negatives are unsystematic and, in the long run, balance out to $\mu_D = 0$. Also note that a sample selected from this population will probably not have a mean exactly equal to zero. As always, there will be some error between a sample mean and the population mean, so even if $\mu_D = 0$ (H_0 is true), we do not expect M_D to be exactly equal to zero.

The alternative hypothesis states that there is a treatment effect that causes the scores in one treatment condition to be systematically higher (or lower) than the scores in the other condition. In symbols,

$$H_1: \mu_D \neq 0$$

According to H_1, the difference scores for the individuals in the population tend to be systematically positive (or negative), indicating a consistent, predictable difference between the two treatments. See Box 11.1 for further discussion of H_0 and H_1.

BOX 11.1 Analogies for H_0 and H_1 in the Repeated-Measures Test

An Analogy for H_0: Intelligence is a fairly stable characteristic; that is, you do not get noticeably smarter or dumber from one day to the next. However, if we gave you an IQ test every day for a week, we probably would get seven different numbers. The day-to-day changes in your IQ score are caused by random factors such as your health, your mood, and your luck at guessing answers you do not know. Some days your IQ score is slightly higher, and some days it is slightly lower. On average, the day-to-day changes in IQ should balance out to zero. This is the situation that is predicted by the null hypothesis for a repeated-measures test. According to H_0, any changes that occur either for an individual or for a sample are just due to chance, and in the long run, they will average out to zero.

An Analogy for H_1: On the other hand, suppose we evaluate your performance on a new video game by measuring your score every day for a week. Again, we probably will find small differences in your scores from one day to the next, just as we did with the IQ scores. However, the day-to-day changes in your game score will not be random. Instead, there should be a general trend toward higher scores as you gain more experience with the new game. Thus, most of the day-to-day changes should show an increase. This is the situation that is predicted by the alternative hypothesis for the repeated-measures test. According to H_1, the changes that occur are systematic and predictable and will not average out to zero.

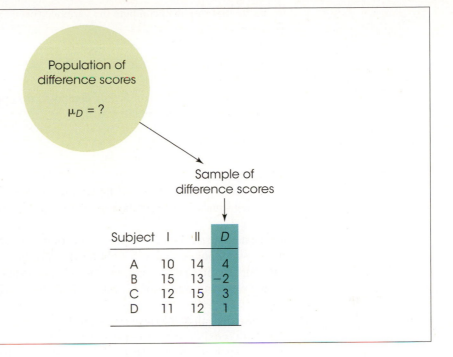

FIGURE 11.1

A sample of $n = 4$ people is selected from the population. Each individual is measured twice, once in treatment I and once in treatment II, and a difference score, D is computed for each individual. This sample of difference scores is intended to represent the population. Note that we are using a sample of difference scores to represent a population of difference scores. Note that the mean for the population of difference scores is unknown. The null hypothesis states that there is no consistent or systematic difference between the two treatment conditions, so the population mean difference is $\mu_D = 0$.

Population of difference scores

$\mu_D = ?$

Sample of difference scores

Subject	I	II	D
A	10	14	4
B	15	13	−2
C	12	15	3
D	11	12	1

■ The *t* Statistic for Related Samples

Figure 11.1 shows the general situation that exists for a repeated-measures hypothesis test. You may recognize that we are facing essentially the same situation that we encountered in Chapter 9. In particular, we have a population for which the mean and the standard deviation are unknown, and we have a sample that will be used to test a hypothesis about the unknown population. In Chapter 9, we introduced the single-sample *t* statistic, which allowed us to use a sample mean as a basis for testing hypotheses about an unknown population mean. This *t*-statistic formula will be used again here to develop the repeated-measures *t* test. To refresh your memory, the single-sample *t* statistic (Chapter 9) is defined by the formula

$$t = \frac{M - \mu}{s_M}$$

In this formula, the sample mean, M, is calculated from the data, and the value for the population mean, μ, is obtained from the null hypothesis. The estimated standard error, s_M, is also calculated from the data and provides a measure of how much difference it is reasonable to expect between a sample mean and the population mean.

For the repeated-measures design, the sample data are difference scores and are identified by the letter D, rather than X. Therefore, we will use Ds in the formula to emphasize that we are dealing with difference scores instead of X values. Also, the population mean that is of interest to us is the population mean difference (the mean amount of change for the entire population), and we identify this parameter with the symbol μ_D. With these simple changes, the *t* formula for the repeated-measures design becomes

As noted, the repeated-measures t formula is also used for matched-subjects designs.

$$t = \frac{M_D - \mu_D}{s_{M_D}} \tag{11.2}$$

In this formula, the *estimated standard error*, s_{M_D}, is computed in exactly the same way as it is computed for the single-sample *t* statistic. To calculate the estimated standard error, the first step is to compute the variance (or the standard deviation) for the sample of *D* scores.

$$s^2 = \frac{SS}{n-1} = \frac{SS}{df} \quad \text{or} \quad s = \sqrt{\frac{SS}{df}}$$

The estimated standard error is then computed using the sample variance (or sample standard deviation) and the sample size, *n*.

$$s_{M_D} = \sqrt{\frac{s^2}{n}} \quad \text{or} \quad s_{M_D} = \frac{s}{\sqrt{n}} \tag{11.3}$$

The following example is an opportunity to test your understanding of variance and estimated standard error for the repeated-measures *t* statistic.

EXAMPLE 11.1   A repeated-measures study with a sample of $n = 8$ participants produces a mean difference of $M_D = 5.5$ points with $SS = 224$ for the difference scores. For these data, find the variance for the difference scores and the estimated standard error for the sample mean. You should obtain a variance of 32 and an estimated standard error of 2.   ∎

   Notice that all of the calculations are done using the difference scores (the *D* scores) and that there is only one *D* score for each subject. With a sample of *n* participants, the number of *D* scores is *n*, and the *t* statistic has $df = n - 1$. Remember that *n* refers to the number of *D* scores, not the number of *X* scores in the original data.
   You should also note that the *repeated-measures t statistic* is conceptually similar to the *t* statistics we have previously examined:

$$t = \frac{\text{sample statistic} - \text{population parameter}}{\text{estimated standard error}}$$

In this case, the sample data are represented by the sample mean of the difference scores (M_D), the population parameter is the value predicted by H_0 $(\mu_D = 0)$, and the amount of sampling error is measured by the standard error of the sample mean (s_{M_D}).

LEARNING CHECK   1. For the following data from a repeated-measures study, what is the sum of the squared deviations (*SS*) for the difference scores?

		I	II
a.	54	4	8
b.	18	9	3
c.	12	6	7
d.	0	8	9

2. For a repeated-measures hypothesis test, which of the following is the correct statement of the alternative hypothesis (H_1)?

   a. For the individuals in the entire population there is no consistent difference between the two treatments.

   b. For the individuals in the entire population there is a consistent difference between the two treatments.

c. For the individuals in the sample there is no consistent difference between the two treatments.

d. For the individuals in the sample there is a consistent difference between the two treatments.

3. For a repeated-measures study comparing two treatments with 10 scores in each treatment, what is the *df* value for the *t* statistic?

a. 8

b. 9

c. 18

d. 19

ANSWERS   **1.** A, **2.** B, **3.** B

11.3 | Hypothesis Tests for the Repeated-Measures Design

LEARNING OBJECTIVES

6. Conduct a repeated-measures *t* test to evaluate the significance of the population mean difference using the data from a repeated-measures or a matched-subjects study comparing two treatment conditions.

7. Conduct a directional (one-tailed) hypothesis test using the repeated-measures *t* statistic.

In a repeated-measures study, each individual is measured in two different treatment conditions and we are interested in whether there is a systematic difference between the scores in the first treatment condition and the scores in the second treatment condition. A difference score (*D* value) is computed for each person and the hypothesis test uses the difference scores from the sample to evaluate the overall mean difference, μ_D, for the entire population. The hypothesis test with the repeated-measures *t* statistic follows the same four-step process that we have used for other tests. The complete hypothesis-testing procedure is demonstrated in Example 11.2.

EXAMPLE 11.2

Swearing is a common, almost reflexive, response to pain. Whether you knock your shin into the edge of a coffee table or smash your thumb with a hammer, most of us respond with a streak of obscenities. One question, however, is whether swearing focuses attention on the pain and, thereby, increases its intensity, or serves as a distraction that reduces pain. To address this issue, Stephens, Atkins, and Kingston (2009) conducted an experiment comparing swearing with other responses to pain. In the study, participants were asked to place one hand in icy cold water for as long as they could bear the pain. Half of the participants were told to repeat their favorite swear word over and over for as long as their hands were in the water. The other half repeated a neutral word. The participants rated the intensity of the pain and the researchers recorded how long each participant was able to tolerate the ice water. After a brief rest, the two groups switched words and repeated the ice water plunge. Thus, all the participants experienced both conditions (swearing and neutral) with half swearing on their first plunge and half on their second. The results clearly showed that swearing significantly increased pain tolerance and decreased the perceived level of pain. The data in Table 11.3 are representative of the results obtained in the study and represented the reports of pain level of $n = 9$ participants.

TABLE 11.3
Ratings of pain level on a scale from 1 to 10 for participants who were swearing or repeating a neutral word.

Participant	Neutral Word	Swearing	D	D²
A	9	7	−2	4
B	8	7	−1	1
C	7	3	−4	16
D	7	8	+1	1
E	8	6	−2	4
F	9	4	−5	25
G	7	6	−1	1
H	7	7	0	0
I	8	4	−4	16

$$\Sigma D = -18 \qquad \Sigma D^2 = 68$$

$$M_D = \frac{-18}{9} = -2 \qquad SS = \Sigma D^2 - \frac{(\Sigma D)^2}{N} = 68 - \frac{(-18)^2}{9} = 68 - 36 = 32$$

STEP 1　**State the hypotheses, and select the alpha level.**

$$H_0: \mu_D = 0 \qquad \textbf{(There is no difference between the two conditions.)}$$
$$H_1: \mu_D \neq 0 \qquad \textbf{(There is a difference.)}$$

For this test, we use $\alpha = .05$

STEP 2　**Locate the critical region.**　For this example, $n = 9$, so the *t* statistic has $df = n - 1 = 8$. For $\alpha = .05$, the critical value listed in the *t* distribution table is ±2.306. The critical region is shown in Figure 11.2.

STEP 3　**Calculate the *t* statistic.**　Table 11.3 shows the sample data and the calculations of $M_D = -2$ and $SS = 32$. Note that all calculations are done with the difference scores. As we have done with the other *t* statistics, we present the calculation of the *t* statistic as a three-step process.

First, compute the sample variance.

$$s^2 = \frac{SS}{n-1} = \frac{32}{8} = 4$$

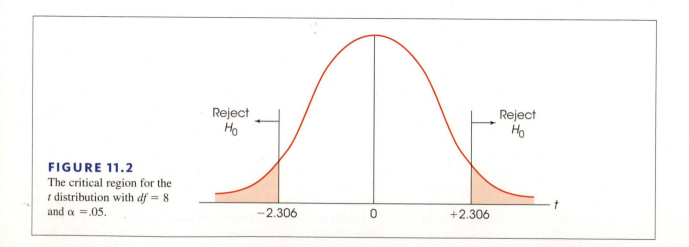

FIGURE 11.2
The critical region for the *t* distribution with $df = 8$ and $\alpha = .05$.

Next, use the sample variance to compute the estimated standard error.

$$s_{M_D} = \sqrt{\frac{s^2}{n}} = \sqrt{\frac{4}{9}} = 0.667$$

Finally, use the sample mean (M_D) and the hypothesized population mean (μ_D) along with the estimated standard error to compute the value for the t statistic.

$$t = \frac{M_D - \mu_D}{s_{M_D}} = \frac{-2 - 0}{0.667} = -3.00$$

STEP 4   **Make a decision.**   The t value we obtained falls in the critical region (see Figure 11.2). The researcher rejects the null hypothesis and concludes that cursing, as opposed to repeating a neutral work, has a significant effect on pain perception.   ■

■ Directional Hypotheses and One-Tailed Tests

In many repeated-measures and matched-subjects studies, the researcher has a specific prediction concerning the direction of the treatment effect. For example, in the study described in Example 11.2, the researcher could predict that the level of perceived pain will be lower when the participant is cursing. This kind of directional prediction can be incorporated into the statement of the hypotheses, resulting in a directional, or one-tailed, hypothesis test. The following example demonstrates how the hypotheses and critical region are determined for a directional test.

EXAMPLE 11.3   We will reexamine the experiment presented in Example 11.2. The researcher is using a repeated-measures design to investigate the effect of swearing on perceived pain. The researcher predicts that the pain ratings for the ice water will decrease when the participants are swearing compared to repeating a neutral word.

STEP 1   **State the hypotheses and select the alpha level.**   For this example, the researcher predicts that pain ratings will decrease when the participants are swearing. The null hypothesis, on the other hand, says that the pain ratings will not decrease but rather will be unchanged or even increased with the swearing. In symbols,

$$H_0: \mu_D \geq 0 \qquad \textbf{\color{blue}(There is no decrease with swearing.)}$$

The alternative hypothesis says that the treatment does work. For this example, H_1 says that swearing will decrease the pain ratings.

$$H_1: \mu_D < 0 \qquad \textbf{\color{blue}(The ratings are decreased.)}$$

We use $\alpha = .01$.

STEP 2   **Locate the critical region.**   As we demonstrated with the independent-measures t statistic (p. 312), the critical region for a one-tailed test can be located using a two-stage process. Rather than trying to determine which tail of the distribution contains the critical region, you first look at the sample mean difference to verify that it is in the predicted direction. If not, then the treatment clearly did not work as expected and you can stop the test. If the change is in the correct direction, then the question is whether it is large enough to be significant. For this example, change is in the predicted direction (the researcher predicted lower ratings and the sample mean shows a decrease.) With $n = 9$, we obtain $df = 8$ and a critical value of $t = 2.896$ for a one-tailed test with $\alpha = .01$. Thus, any t statistic beyond 2.896 (positive or negative) is sufficient to reject the null hypothesis.

STEP 3   **Compute the *t* statistic.**   We calculated the t statistic in Example 11.2, and obtained $t = -3.00$.

STEP 4   **Make a decision.**   The obtained t statistic is beyond the critical boundary. Therefore, we reject the null hypothesis and conclude that swearing significantly reduced the pain ratings. ■

■ Assumptions of the Related-Samples *t* Test

The related-samples t statistic requires two basic assumptions.

1. The observations within each treatment condition must be independent (see p. 244). Notice that the assumption of independence refers to the scores *within* each treatment. Inside each treatment, the scores are obtained from different individuals and should be independent of one another.

2. The population distribution of difference scores (D values) must be normal.

As before, the normality assumption is not a cause for concern unless the sample size is relatively small. In the case of severe departures from normality, the validity of the t test may be compromised with small samples. However, with relatively large samples ($n > 30$), this assumption can be ignored.

If there is reason to suspect that one of the assumptions for the repeated-measures t test has been violated, an alternative analysis known as the *Wilcoxon test* is presented in Appendix E. The Wilcoxon test requires that the original scores be transformed into ranks before evaluating the difference between the two treatment conditions.

LEARNING CHECK

1. For a repeated-measures study comparing two treatments with a sample of $n = 9$ participants, a researcher obtains a sample mean difference is $M_D = 5$ with $SS = 288$ for the difference scores. What is the repeated-measures t statistic for these data?

   a. 5
   b. 2.5
   c. 1.25
   d. 0.83

2. For a repeated-measures study comparing two treatments with $n = 15$ scores in each treatment, the data produce $t = 1.78$. If the mean difference is in the direction that is predicted by the researcher, then which of the following is the correct decision for a hypothesis test with $\alpha = .05$?

   a. Reject H_0 for a one-tailed test but not for two-tailed.
   b. Reject H_0 for a two-tailed test but not for one-tailed.
   c. Reject H_0 for either a one-tailed test or a two-tailed test.
   d. Fail to reject H_0 for both a one-tailed test and a two-tailed test.

ANSWERS   **1.** B, **2.** A

11.4 | Effect Size and Confidence Intervals for the Repeated-Measures *t*

LEARNING OBJECTIVES

8. Measure effect size for a repeated-measures *t* test using either Cohen's *d* or r^2, the percentage of variance accounted for.

9. Use the data from a repeated-measures study to compute a confidence interval describing the size of the population mean difference.

10. Describe how the results of a repeated-measures *t* test and measures of effect size are reported in the scientific literature.

11. Describe how the outcome of a hypothesis test and measures of effect size using the repeated-measures *t* statistic are influenced by sample size and sample variance.

12. Describe how the consistency of the treatment effect is reflected in the variability of the difference scores and explain how this influences the outcome of a hypothesis test.

As we noted with other hypothesis tests, whenever a treatment effect is found to be statistically significant, it is recommended that you also report a measure of the absolute magnitude of the effect. The most commonly used measures of effect size are Cohen's *d* and r^2, the percentage of variance accounted for. The size of the treatment effect also can be described with a confidence interval estimating the population mean difference, μ_D. Using the data from Example 11.2, we will demonstrate how these values are calculated to measure and describe effect size.

Cohen's *d* In Chapters 8 and 9 we introduced Cohen's *d* as a standardized measure of the mean difference between treatments. The standardization simply divides the population mean difference by the standard deviation. For a repeated-measures study, Cohen's *d* is defined as

$$d = \frac{\text{population mean difference}}{\text{standard deviation}} = \frac{\mu_D}{\sigma_D}$$

Because the population mean and standard deviation are unknown, we use the sample values instead. The sample mean, M_D, is the best estimate of the actual mean difference, and the sample standard deviation (square root of sample variance) provides the best estimate of the actual standard deviation. Thus, we are able to estimate the value of *d* as follows:

Because we are measuring the size of the effect and not the direction, it is customary to ignore the minus sign and report Cohen's *d* as a positive value.

$$\text{estimated } d = \frac{\text{sample mean difference}}{\text{sample mean deviation}} = \frac{M_D}{s} \qquad (11.4)$$

For the repeated-measures study in Example 11.2, the sample mean difference is $M_D = -2$, or simply 2 points, and the sample variance is $s^2 = 4.00$, so the data produce

$$\text{estimated } d = \frac{M_D}{s} = \frac{2}{\sqrt{4.00}} = \frac{2}{2} = 1.00$$

Any value greater than 0.80 is considered to be a large effect, and these data are clearly in that category (see Table 8.2 on p. 253).

The Percentage of Variance Accounted for, *r²*   Percentage of variance is computed using the obtained *t* value and the *df* value from the hypothesis test, exactly as was done for the single-sample *t* (see p. 283) and for the independent-measures *t* (see p. 317). For the data in Example 11.2, we obtain

$$r^2 = \frac{t^2}{t^2 + df} = \frac{(3.00)^2}{(3.00)^2 + 8} = \frac{9}{17} = 0.529 \text{ or } 52.9\%$$

For these data, 52.9% of the variance in the scores is explained by the effect of cursing. More specifically, swearing caused the estimated pain ratings to be consistently negative. Thus, the deviations from zero are largely explained by the treatment.

The following example is an opportunity to test your understanding of Cohen's *d* and *r²* to measure effect size for the repeated-measures *t* statistic.

EXAMPLE 11.4

A repeated-measures study with *n* = 16 participants produces a mean difference of M_D = 6 points, *SS* = 960 for the difference scores, and *t* = 3.00. Calculate Cohen's *d* and *r²* to measure the effect size for this study. You should obtain $d = \frac{6}{8} = 0.75$ and $r^2 = \frac{9}{24} = 0.375$.   ∎

Confidence Intervals for Estimating μ_D   As noted in the previous two chapters, it is possible to compute a confidence interval as an alternative method for measuring and describing the size of the treatment effect. For the repeated-measures *t*, we use a sample mean difference, M_D, to estimate the population mean difference, μ_D. In this case, the confidence interval literally estimates the size of the treatment effect by estimating the population mean difference between the two treatment conditions.

As with the other *t* statistics, the first step is to solve the *t* equation for the unknown parameter. For the repeated-measures t statistic, we obtain

$$\mu_D = M_D \pm ts_{M_D} \tag{11.5}$$

In the equation, the values for M_D and for s_{M_D}, are obtained from the sample data. Although the value for the *t* statistic is unknown, we can use the degrees of freedom for the *t* statistic and the *t* distribution table to estimate the *t* value. Using the estimated *t* and the known values from the sample, we can then compute the value of μ_D. The following example demonstrates the process of constructing a confidence interval for a population mean difference.

EXAMPLE 11.5

In Example 11.2 we presented a research study demonstrating how swearing influenced the perception of pain. In the study, a sample of *n* = 9 participants rated their level of pain significantly lower when they were repeating a swear word than when they repeated a neutral word. The mean difference between treatments was M_D = −2 points and the estimated standard error for the mean difference was s_{M_D} = 0.667. Now, we construct a 95% confidence interval to estimate the size of the population mean difference.

With a sample of *n* = 9 participants, the repeated-measures *t* statistic has *df* = 8. To have 95% confidence, we simply estimate that the *t* statistic for the sample mean difference is located somewhere in the middle 95% of all the possible *t* values. According to the *t* distribution table, with *df* = 8, 95% of the *t* values are located between *t* = +2.306 and *t* = −2.306. Using these values in the estimation equation, together with the values for the sample mean and the standard error, we obtain

$$\mu_D = M_D \pm ts_{M_D}$$
$$= -2 \pm 2.306(0.667)$$
$$= -2 \pm 1.538$$

This produces an interval of values ranging from $-2 - 1.538 = -3.538$ to $-2 + 1.538 = -0.462$. Our conclusion is that for the general population, swearing instead of repeating a neutral word decreases the perceived pain between 0.462 and 3.538 points. We are 95% confident that the true mean difference is in this interval because the only value estimated during the calculations was the *t* statistic, and we are 95% confident that the *t* value is located in the middle 95% of the distribution. Finally note that the confidence interval is constructed around the sample mean difference. As a result, the sample mean difference, $M_D = -2$ points, is located exactly in the center of the interval.   ■

As with the other confidence intervals presented in Chapters 9 and 10, the confidence interval for a repeated-measures *t* is influenced by a variety of factors other than the actual size of the treatment effect. In particular, the width of the interval depends on the percentage of confidence used so that a larger percentage produces a wider interval. Also, the width of the interval depends on the sample size, so that a larger sample produces a narrower interval. Because the interval width is related to sample size, the confidence interval is not a pure measure of effect size like Cohen's *d* or r^2.

Finally, we should note that the 95% confidence interval computed in Example 11.5 does not include the value $\mu_D = 0$. In other words, we are 95% confident that the population mean difference is not $\mu_D = 0$. This is equivalent to concluding that a null hypothesis specifying that $\mu_D = 0$ would be rejected with a test using $\alpha = .05$. If $\mu_D = 0$ were included in the 95% confidence interval, it would indicate that a hypothesis test would fail to reject H_0 with $\alpha = .05$.

IN THE LITERATURE

Reporting the Results of a Repeated-Measures *t* Test

As we have seen in Chapters 9 and 10, the APA format for reporting the results of *t* tests consists of a concise statement that incorporates the *t* value, degrees of freedom, and alpha level. One typically includes values for means and standard deviations, either in a statement or a table (Chapter 4). For Example 11.2, we observed a mean difference of $M_D = -2.00$ with $s = 2.00$. Also, we obtained a *t* statistic of $t = -3.00$ with $df = 8$, and our decision was to reject the null hypothesis at the .05 level of significance. Finally, we measured effect size by computing the percentage of variance explained and obtained $r^2 = 0.529$. A published report of this study might summarize the results as follows:

> Changing from a neutral word to a swear word reduced the perceived level of pain by an average of $M = 2.00$ points with $SD = 2.00$. The treatment effect was statistically significant, $t(8) = -3.00$, $p < .05$, $r^2 = 0.529$.

When the hypothesis test is conducted with a computer program, the printout typically includes an exact probability for the level of significance. The *p*-value from the printout is then stated as the level of significance in the research report. For example, the data from Example 11.2 produced a significance level of $p = .017$, and the results would be reported as "statistically significant, $t(8) = -3.00$, $p = .017$, $r^2 = 0.529$." Occasionally, a probability is so small that the computer rounds it off to 3 decimal points and produces a value of zero. In this situation you do not know the exact probability value and should report $p < .001$.

If the confidence interval from Example 11.5 is reported as a description of effect size together with the results from the hypothesis test, it would appear as follows:

> Changing from a neutral word to a swear word significantly reduced the perceived level of pain, $t(8) = -3.00$, $p < .05$, 95% CI $[-0.462, -3.538]$.

■ Descriptive Statistics and the Hypothesis Test

Often, a close look at the sample data from a research study makes it easier to see the size of the treatment effect and to understand the outcome of the hypothesis test. In Example 11.2, we obtained a sample of $n = 9$ participants who produce a mean difference of $M_D = -2.00$ with a standard deviation of $s = 2.00$ points. The sample mean and standard deviation describe a set of scores centered at $M_D = -2.00$ with most of the scores located within 2.00 points of the mean. Figure 11.3 shows the actual set of difference scores that were obtained in Example 11.2. In addition to showing the scores in the sample, we have highlighted the position of $\mu_D = 0$; that is, the value specified in the null hypothesis. Notice that the scores in the sample are displaced away from zero. Specifically, the data are not consistent with a population mean of $\mu_D = 0$, which is why we rejected the null hypothesis. In addition, note that the sample mean is located one standard deviation below zero. This distance corresponds to the effect size measured by Cohen's $d = 1.00$. For these data, the picture of the sample distribution (see Figure 11.3) should help you to understand the measure of effect size and the outcome of the hypothesis test.

■ Sample Variance and Sample Size in the Repeated-Measures *t* Test

In previous chapters we identified sample variability and sample size as two factors that affect the estimated standard error and the outcome of the hypothesis test. The standard error is inversely related to sample size (larger size leads to smaller error) and is directly related to sample variance (larger variance leads to larger error). As a result, a larger sample produces a larger value for the *t* statistic (farther from zero) and increases the likelihood of rejecting H_0. Larger variance, on the other hand, produces a smaller value for the *t* statistic (closer to zero) and reduces the likelihood of finding a significant result.

Although variance and sample size both influence the hypothesis test, only variance has a large influence on measures of effect size such as Cohen's d and r^2; larger variance produces smaller measures of effect size. Sample size, on the other hand, has no effect on the value of Cohen's d and only a small influence on r^2.

Variability as a Measures of Consistency for the Treatment Effect In a repeated-measures study, the variability of the difference scores becomes a relatively concrete and easy-to-understand concept. In particular, the sample variability describes the *consistency* of the treatment effect. For example, if a treatment consistently adds a few

---

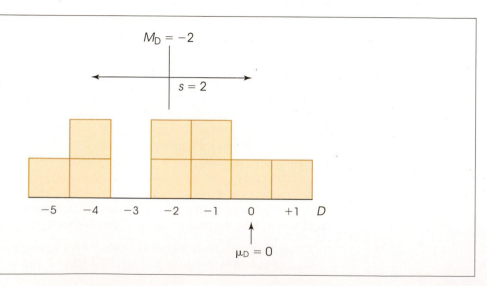

FIGURE 11.3

The sample of difference scores from Example 11.2. The mean is $M_D = -2$ and the standard deviation is $s = 2$. The difference scores are consistently negative, indicating a decrease in perceived pain, suggest that $\mu_D = 0$ (no effect) is not a reasonable hypothesis.

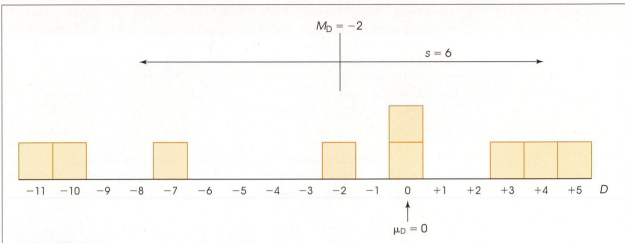

FIGURE 11.4

A sample of difference scores with a mean of $M_D = -2$ and standard deviation of $s = 6$. The data do not show a consistent increase or decrease in scores. Because there is no consistent treatment effect, $\mu_D = 0$ is a reasonable hypothesis.

points to each individual's score, then the set of difference scores will be clustered together with relatively small variability. This is the situation that we observed in Example 11.2 (see Figure 11.3) in which nearly all of the participants had lower ratings of pain in the swearing condition. In this situation, with small variability, it is easy to see the treatment effect and it is likely to be significant.

Now consider what happens when the variability is large. Suppose that the swearing study in Example 11.2 produced a sample of $n = 9$ difference scores consisting of -11, -10, -7, -2, 0, 0, $+3$, $+4$, and $+5$. These difference scores also have a mean of $M_D = -2.00$, but now the variability is substantially increased so that $SS = 288$ and the standard deviation is $s = 6.00$. Figure 11.4 shows the new set of difference scores. Again, we have highlighted the position of $\mu_D = 0$, which is the value specified in the null hypothesis. Notice that the high variability means that there is no consistent treatment effect. Some participants rate the pain higher while swearing (the positive differences) and some rate it lower (the negative differences). In the hypothesis test, the high variability increases the size of the estimated standard error and results in a hypothesis test that produces $t = -1.00$, which is not in the critical region. With these data, we would fail to reject the null hypothesis and conclude that swearing has no effect on the perceived level of pain.

With small variability (see Figure 11.3), the 2-point treatment effect is easy to see and is statistically significant. With large variability (see Figure 11.4), the 2-point effect is not easy to see and is not significant. As we have noted several times in the past, large variability can obscure patterns in the data and reduces the likelihood of finding a significant treatment effect.

LEARNING CHECK

1. A repeated-measures study with 9 scores in each treatment produces a mean difference of $M_D = 4.0$ and $SS = 288$ for the difference scores. If the effect size is described using Cohen's d, then what is the value for d?

   a. 0.111

   b. 0.667

   c. 1.00

   d. 2.00

2. A researcher obtains a sample of $n = 9$ individuals and tests each person in two different treatment conditions. The sample mean difference is $M_D = 12$ points with $SS = 72$. Which of the following is the correct equation for the 80% confidence interval for the population mean difference?

   a. $\mu_D = 0 \pm 3(1.397)$

   b. $\mu_D = 0 \pm 1(1.397)$

   c. $\mu_D = 12 \pm 3(1.397)$

   d. $\mu_D = 12 \pm 1(1.397)$

3. The results of a hypothesis test with a repeated-measures *t* statistic are reported as follows: $t(18) = 2.25, p < .05$. Which of the following is consistent with the report?

   a. The study used a total of 20 participants and the mean difference was not significant.

   b. The study used a total of 20 participants and the mean difference was significant.

   c. The study used a total of 19 participants and the mean difference was not significant.

   d. The study used a total of 19 participants and the mean difference was significant.

4. Which of the following would have little or no influence on effect size as measured by Cohen's *d* or by r^2?

   a. Increasing the sample size

   b. Increasing the size of the sample mean difference

   c. Increasing the sample variance

   d. All of the other three options would influence the magnitude of effect size.

5. A researcher is using a repeated-measures study to evaluate the difference between two treatments. If there is a consistent difference between the treatments then the data should produce _____.

   a. a small variance for the difference scores and a small standard error

   b. a small variance for the difference scores and a large standard error

   c. a large variance for the difference scores and a small standard error

   d. a large variance for the difference scores and a large standard error

ANSWERS   **1.** B, **2.** D, **3.** D, **4.** A, **5.** A

11.5 | Comparing Repeated- and Independent-Measures Designs

LEARNING OBJECTIVE   **13.** Describe the advantages and disadvantages of choosing a repeated-measures design instead of an independent-measures design to compare two treatment conditions.

In many research situations, it is possible to use either a repeated-measures design or an independent-measures design to compare two treatment conditions. The independent-measures design would use two separate samples (one in each treatment condition) and the repeated-measures design would use only one sample with the same individuals participating in both treatments. The decision about which design to use is often made by considering the advantages and disadvantages of the two designs. In general, the repeated-measures design has most of the advantages.

Number of Subjects   A repeated-measures design typically requires fewer subjects than an independent-measures design. The repeated-measures design uses the subjects more efficiently because each individual is measured in both of the treatment conditions. This can be especially important when there are relatively few subjects available (for example, when you are studying a rare species or individuals in a rare profession).

Study Changes Over Time   The repeated-measures design is especially well suited for studying learning, development, or other changes that take place over time. Remember that this design often involves measuring individuals at one time and then returning to measure the same individuals at a later time. In this way, a researcher can observe behaviors that change or develop over time.

Individual Differences   The primary advantage of a repeated-measures design is that it reduces or eliminates problems caused by individual differences. *Individual differences* are characteristics such as age, IQ, gender, and personality that vary from one individual to another. These individual differences can influence the scores obtained in a research study, and they can affect the outcome of a hypothesis test. Consider the data in Table 11.4. The first set of data represents the results from a typical independent-measures study and the second set represents a repeated-measures study. Note that we have identified each participant by name to help demonstrate the effects of individual differences.

TABLE 11.4   Hypothetical data showing the results from an independent-measures study and a repeated-measures study. The two sets of data use exactly the same numerical scores and they both show the same 5-point mean difference between treatments.

Independent-Measures Study (2 Separate Samples)		Repeated-Measures Study (Same Sample in Both Treatments)		
Treatment 1	Treatment 2	Treatment 1	Treatment 2	D
(John) $X = 18$	(Sue) $X = 15$	(John) $X = 18$	(John) $X = 15$	-3
(Mary) $X = 27$	(Tom) $X = 20$	(Mary) $X = 27$	(Mary) $X = 20$	-7
(Bill) $X = 33$	(Dave) $X = 28$	(Bill) $X = 33$	(Bill) $X = 28$	-5
$M = 26$	$M = 21$			$M_D = -5$
$SS = 114$	$SS = 86$			$SS = 8$

For the independent-measures data, note that every score represents a different person. For the repeated-measures study, on the other hand, the same participants are measured in both of the treatment conditions. This difference between the two designs has some important consequences.

1. We have constructed the data so that both research studies have exactly the same scores and they both show the same 5-point mean difference between treatments. In each case, the researcher would like to conclude that the 5-point difference was caused by the treatments. However, with the independent-measures design, there is always the possibility that the participants in treatment 1 have different characteristics than those in treatment 2. For example, the three participants in treatment

1 may be more intelligent than those in treatment 2 and their higher intelligence caused them to have higher scores. Note that this problem disappears with the repeated-measures design. Specifically, with repeated measures there is no possibility that the participants in one treatment are different from those in another treatment because the same participants are used in all the treatments.

2. Although the two sets of data contain exactly the same scores and have exactly the same 5-point mean difference, you should realize that they are very different in terms of the variance used to compute standard error. For the independent-measures study, you calculate the *SS* or variance for the scores in each of the two separate samples. Note that in each sample there are big differences between participants. In treatment 1, for example, Bill has a score of 33 and John's score is only 18. These individual differences produce a relatively large sample variance and a large standard error. For the independent-measures study, the standard error is 5.77, which produces a *t* statistic of $t = 0.87$. For these data, the hypothesis test concludes that there is no significant difference between treatments.

In the repeated-measures study, the *SS* and variance are computed for the difference scores. If you examine the repeated-measures data in Table 11.4, you will see that the big differences between John and Bill that exist in treatment 1 and in treatment 2 are eliminated when you get to the difference scores. Because the individual differences are eliminated, the variance and standard error are dramatically reduced. For the repeated-measures study, the standard error is 1.15 and the *t* statistic is $t = -4.35$. With the repeated-measures *t*, the data show a significant difference between treatments. Thus, one big advantage of a repeated-measures study is that it reduces variance by removing individual differences, which increases the chances of finding a significant result.

■ Time-Related Factors and Order Effects

The primary disadvantage of a repeated-measures design is that the structure of the design allows for factors other than the treatment effect to cause a participant's score to change from one treatment to the next. Specifically, in a repeated-measures design, each individual is measured in two different treatment conditions, often *at two different times*. In this situation, outside factors that change over time may be responsible for changes in the participants' scores. For example, a participant's health or mood may change over time and cause a difference in the participant's scores. Outside factors such as the weather can also change and may have an influence on participants' scores. Because a repeated-measures study typically takes place over time, it is possible that time-related factors (other than the two treatments) are responsible for causing changes in the participants' scores.

Also, it is possible that participation in the first treatment influences the individual's score in the second treatment. If the researcher is measuring individual performance, for example, the participants may gain experience during the first treatment condition, and this extra practice helps their performance in the second condition. In this situation, the researcher would find a mean difference between the two conditions; however, the difference would not be caused by the treatments, instead it would be caused by practice effects. Changes in scores that are caused by participation in an earlier treatment are called *order effects* and can distort the mean differences found in repeated-measures research studies.

Counterbalancing One way to deal with time-related factors and order effects is to counterbalance the order of presentation of treatments. That is, the participants are randomly divided into two groups, with one group receiving treatment 1 followed by treatment 2, and the other group receiving treatment 2 followed by treatment 1. The goal of counter-balancing is to distribute any outside effects evenly over the two treatments. For example, if practice effects are a problem, then half of the participants will gain experience in treatment 1, which then helps their performance in treatment 2. However, the other half will gain experience in treatment 2, which helps their performance in treatment 1. Thus, prior experience helps the two treatments equally.

Finally, if there is reason to expect strong time-related effects or strong order effects, your best strategy is not to use a repeated-measures design. Instead, use independent-measures (or a matched-subjects design) so that each individual participates in only one treatment and is measured only one time.

LEARNING CHECK

1. An advantage of a repeated-measured design (compared to an independent-measures design) is that it reduces the contribution of error variability due to _____.
   a. M_D
   b. degrees of freedom
   c. the effect of the treatment
   d. individual differences

2. Compared to a repeated-measures design, which of the following is an advantage of an independent-measures research design?
   a. It usually requires fewer participants.
   b. It eliminates the concern that performance in one treatment condition may be influenced by experience in the other treatment condition.
   c. It eliminates the concern that the participants in one treatment may be smarter than those in the other treatment.
   d. It tends to have a smaller variance and a smaller standard error.

3. For which of the following situations would a repeated-measures design have the maximum advantage over an independent-measures design?
   a. when many subjects are available and individual differences are small
   b. when very few subjects are available and individual differences are small
   c. when many subjects are available and individual differences are large
   d. when very few subjects are available and individual differences are large

ANSWERS   1. D, 2. B, 3. D

SUMMARY

1. In a related-samples research study, the individuals in one treatment condition are directly related, one-to-one, with the individuals in the other treatment condition(s). The most common related-samples study is a repeated-measures design, in which the same sample of individuals is tested in all of the treatment conditions. This design literally repeats measurements on the same subjects. An alternative is a matched-subjects design, in which the individuals in one sample are matched one-to-one with individuals in another sample. The matching is based on a variable relevant to the study.

2. The repeated-measures *t* test begins by computing a difference between the first and second measurements for each subject (or the difference for each matched pair). The difference scores, or *D* scores, are obtained by

$$D = X_2 - X_1$$

The sample mean, M_D, and sample variance, s^2, are used to summarize and describe the set of difference scores.

3. The formula for the repeated-measures *t* statistic is

$$t = \frac{M_D - \mu_D}{s_{M_D}}$$

In the formula, the null hypothesis specifies $\mu_D = 0$, and the estimated standard error is computed by

$$s_{M_D} = \sqrt{\frac{s^2}{n}}$$

4. For a repeated-measures design, effect size can be measured using either r^2 (the percentage of variance accounted for) or Cohen's *d* (the standardized mean difference). The value of r^2 is computed the same for both independent and repeated-measures designs,

$$r^2 = \frac{t^2}{t^2 + df}$$

Cohen's *d* is defined as the sample mean difference divided by standard deviation for both repeated- and independent-measures designs. For repeated-measures studies, Cohen's *d* is estimated as

$$\text{estimated } d = \frac{M_D}{s}$$

5. An alternative method for describing the size of the treatment effect is to construct a confidence interval for the population mean difference, μ_D. The confidence interval uses the repeated-measures *t* equation, solved for the unknown mean difference:

$$\mu_D = M_D \pm ts_{M_D}$$

First, select a level of confidence and then look up the corresponding *t* values. For example, for 95% confidence, use the range of *t* values that determine the middle 95% of the distribution. The *t* values are then used in the equation along with the values for the sample mean difference and the standard error, which are computed from the sample data.

6. A repeated-measures design may be preferred to an independent-measures study when one wants to observe changes in behavior in the same subjects, as in learning or developmental studies. An important advantage of the repeated-measures design is that it removes or reduces individual differences, which in turn lowers sample variability and tends to increase the chances for obtaining a significant result.

KEY TERMS

repeated-measures design (336)

within-subjects design (336)

matched-subjects design (337)

related-samples design (337)

difference scores (339)

estimated standard error for M_D (341)

repeated-measures *t* statistic (342)

individual differences (353)

Wilcoxon test (346)

order effects (354)

SPSS®

General instructions for using SPSS are presented in Appendix D. Following are detailed instructions for using SPSS to perform the **Repeated-Measures *t* Test** presented in this chapter.

Data Entry

1. Enter the data into two columns (VAR0001 and VAR0002) in the data editor with the first score for each participant in the first column and the second score in the second column. The two scores for each participant must be in the same row.

Data Analysis

1. Click **Analyze** on the tool bar, select **Compare Means**, and click on **Paired-Samples T Test**.
2. One at a time, highlight the column labels for the two data columns and click the arrow to move them into the **Paired Variables** box.
3. In addition to performing the hypothesis test, the program will compute a confidence interval for the population mean difference. The confidence level is automatically set at 95% but you can select **Options** and change the percentage.
4. Click **OK**.

SPSS Output

We used the SPSS program to analyze the data from the cursing and pain experiment in Example 11.2 and the program output is shown in Figure 11.5. The output includes a table of sample statistics with the mean and standard deviation for each treatment. A second table shows the correlation between the two sets of scores (correlations are presented in

Paired Samples Statistics

		Mean	N	Std. Deviation	Std. Error Mean
Pair 1	VAR00001	7.7778	9	.83333	.27778
	VAR00002	5.7778	9	1.71594	.57198

Paired Samples Correlations

		N	Correlation	Sig.
Pair 1	VAR00001 & VAR00002	9	−.126	.746

Paired Samples Test

		Paired Differences		
		Mean	Std. Deviation	Std. Error Mean
Pair 1	VAR00001 - VAR00002	2.00000	2.00000	.66667

Paired Samples Test

		Paired Differences 95% Confidence Interval of the Difference				
		Lower	Upper	t	df	Sig. (2-tailed)
Pair 1	VAR00001 - VAR00002	.46266	3.53734	3.000	8	.017

FIGURE 11.5

The SPSS output for the repeated-measures hypothesis test in Example 11.2.

Chapter 15). The final table, which is split into two sections in Figure 11.5, shows the results of the hypothesis test, including the mean and standard deviation for the difference scores, the standard error for the mean, a 95% confidence interval for the mean difference, and the values for *t*, *df*, and the level of significance (the *p* value for the test).

FOCUS ON PROBLEM SOLVING

1. Once data have been collected, we must then select the appropriate statistical analysis. How can you tell whether the data call for a repeated-measures *t* test? Look at the experiment carefully. Is there only one sample of subjects? Are the same subjects tested a second time? If your answers are yes to both of these questions, then a repeated-measures *t* test should be done. There is only one situation in which the repeated-measures *t* can be used for data from two samples, and that is for *matched-subjects* studies (p. 337).

2. The repeated-measures *t* test is based on difference scores. In finding difference scores, be sure you are consistent with your method. That is, you may use either $X_2 - X_1$ or $X_1 - X_2$ to find *D* scores, but you must use the same method for all subjects.

DEMONSTRATION 11.1

A REPEATED-MEASURES *t* TEST

A major oil company would like to improve its tarnished image following a large oil spill. Its marketing department develops a short television commercial and tests it on a sample of $n = 7$ participants. People's attitudes about the company are measured with a short questionnaire, both before and after viewing the commercial. The data are as follows:

Person	X_1 (Before)	X_2 (After)	D (Difference)	
A	15	15	0	
B	11	13	+2	$\Sigma D = 21$
C	10	18	+8	
D	11	12	+1	$M_D = 21/7 = 3.00$
E	14	16	+2	
F	10	10	0	$SS = 74$
G	11	19	+8	

Was there a significant change? Note that participants are being tested twice—once before and once after viewing the commercial. Therefore, we have a repeated-measures design.

STEP 1 **State the hypotheses, and select an alpha level** The null hypothesis states that the commercial has no effect on people's attitude, or in symbols,

$$H_0: \mu_D = 0 \quad \text{(The mean difference is zero.)}$$

The alternative hypothesis states that the commercial does alter attitudes about the company, or

$$H_1: \mu_D \neq 0 \quad \text{(There is a mean change in attitudes.)}$$

For this demonstration, we will use an alpha level of .05 for a two-tailed test.

STEP 2   **Locate the critical region**   Degrees of freedom for the repeated-measures t test are obtained by the formula

$$df = n - 1$$

For these data, degrees of freedom equal

$$df = 7 - 1 = 6$$

The t distribution table is consulted for a two-tailed test with $\alpha = .05$ for $df = 6$. The critical t values for the critical region are $t = \pm2.447$.

STEP 3   **Compute the test statistic**   Once again, we suggest that the calculation of the t statistic be divided into a three-part process.

Variance for the D Scores The variance for the sample of D scores is

$$s^2 = \frac{SS}{n - 1} = \frac{74}{6} = 12.33$$

Estimated Standard Error for M_D The estimated standard error for the sample mean difference is computed as follows:

$$s_{M_D} = \sqrt{\frac{s^2}{n}} = \sqrt{\frac{12.33}{7}} = \sqrt{1.76} = 1.33$$

The Repeated-Measures t Statistic Now we have the information required to calculate the t statistic.

$$t = \frac{M_D - \mu_D}{s_{M_D}} = \frac{3 - 0}{1.33} = 2.26$$

STEP 4   **Make a decision about H_0, and state the conclusion**   The obtained t value is not extreme enough to fall in the critical region. Therefore, we fail to reject the null hypothesis. We conclude that there is no evidence that the commercial will change people's attitudes, $t(6) = 2.26$, $p > .05$, two-tailed. (Note that we state that p is *greater than* .05 because we failed to reject H_0.)

DEMONSTRATION 11.2

EFFECT SIZE FOR THE REPEATED-MEASURES t

We will estimate Cohen's d and calculate r^2 for the data in Demonstration 11.1. The data produced a sample mean difference of $M_D = 3.00$ with a sample variance of $s^2 = 12.33$. Based on these values, Cohen's d is

$$\text{estimated } d = \frac{\text{mean difference}}{\text{standard deviation}} = \frac{M_D}{s} = \frac{3.00}{\sqrt{12.33}} = \frac{3.00}{3.51} = 0.86$$

The hypothesis test produced $t = 2.26$ with $df = 6$. Based on these values,

$$r^2 = \frac{t^2}{t^2 + df} = \frac{(2.26)^2}{(2.26)^2 + 6} = \frac{5.11}{11.11} = 0.46 \quad (\text{or } 46\%)$$

PROBLEMS

1. For the each of the following studies determine whether a repeated-measures *t* test is the appropriate analysis. Explain your answers.
   a. A researcher is examining the effect of violent video games on behavior by comparing aggressive behaviors for one group who just finished playing a violent game with another group who played a neutral game.
   b. A researcher is examining the effect on humor on memory by presenting a group of participants with a series of humorous and not humorous sentences and them recording how many of each type of sentence is recalled by each participant.
   c. A researcher is evaluating the effectiveness of a new cholesterol medication by recording the cholesterol level for each individual in a sample before they start taking the medication and again after 8 weeks with the medication.

2. What is the defining characteristic of a repeated-measures or within-subjects research design?

3. Explain the difference between a matched-subjects design and a repeated-measures design.

4. A researcher conducts an experiment comparing two treatment conditions with 20 scores in each treatment condition.
   a. If an independent-measures design is used, how many subjects are needed for the experiment?
   b. If a repeated-measures design is used, how many subjects are needed for the experiment?
   c. If a matched-subjects design is used, how many subjects are needed for the experiment?

5. A sample of $n = 9$ individuals participates in a repeated-measures study that produces a sample mean difference of $M_D = 4.25$ with $SS = 128$ for the difference scores.
   a. Calculate the standard deviation for the sample of difference scores. Briefly explain what is measured by the standard deviation.
   b. Calculate the estimated standard error for the sample mean difference. Briefly explain what is measured by the estimated standard error.

6. In the Preview for Chapter 2, we presented a study showing that a woman shown in a photograph was judged as less attractive when the photograph showed a visible tattoo compared to the same photograph with the tattoo removed (Resenhoeft, Villa, & Wiseman, 2008). Suppose a similar experiment is conducted as a repeated-measures study. A sample of $n = 12$ males looks at a set of 30 photographs of women and rates the attractiveness of each woman using a 5-point scale (5 = most positive). One photograph appears twice in the set, once with a tattoo and once with the tattoo removed. For each participant, the researcher records the difference between the two ratings of the same photograph. On average, the photograph without the tattoo is rated $M_D = 1.2$ points higher than the photograph with the tattoo, with $SS = 33$ for the difference scores. Does the presence of a visible tattoo have a significant effect on the attractiveness ratings? Use a two-tailed test with $\alpha = .05$.

7. The following data are from a repeated-measures study examining the effect of a treatment by measuring a group of $n = 6$ participants before and after they receive the treatment.
   a. Calculate the difference scores and M_D.
   b. Compute SS, sample variance, and estimated standard error.
   c. Is there a significant treatment effect? Use $\alpha = .05$, two tails.

Participant	Before Treatment	After Treatment
A	7	8
B	2	9
C	4	6
D	5	7
E	5	6
F	3	8

8. When you get a surprisingly low price on a product do you assume that you got a really good deal or that you bought a low-quality product? Research indicates that you are more likely to associate low price and low quality if someone else makes the purchase rather than yourself (Yan and Sengupta, 2011). In a similar study, $n = 16$ participants were asked to rate the quality of low-priced items under two scenarios: purchased by a friend or purchased yourself. The results produced a mean difference of $M_D = 2.6$ and $SS = 135$, with self-purchases rated higher.
   a. Is the judged quality of objects significantly different for self-purchases than for purchases made by others? Use a two-tailed test with $\alpha = .05$.
   b. Compute Cohen's d to measure the size of the treatment effect.

9. Masculine-themed words (such as competitive, independent, analyze, strong) are commonly used in job recruitment materials, especially for job advertisements in male-dominated areas (Gaucher, Friesen, & Kay, 2010). The same study found that these

words also make the jobs less appealing to women. In a similar study, female participants were asked to read a series of job advertisements and then rate how interesting or appealing the job appeared to be. Half of the advertisements were constructed to include several masculine-themed words and the others were worded neutrally. The average rating for each type of advertisement was obtained for each participant. For $n = 25$ participants, the mean difference between the two types of advertisements is $M_D = 1.32$ points (neutral ads rated higher) with $SS = 150$ for the difference scores.

a. Is this result sufficient to conclude that there is a significant difference in the ratings for two types of advertisements? Use a two-tailed test with $\alpha = .05$.

b. Compute r^2 to measure the size of the treatment effect.

c. Write a sentence describing the outcome of the hypothesis test and the measure of effect size as it would appear in a research report.

10. The stimulant Ritalin has been shown to increase attention span and improve academic performance in children with ADHD (Evans et al., 2001). To demonstrate the effectiveness of the drug, a researcher selects a sample of $n = 20$ children diagnosed with the disorder and measures each child's attention span before and after taking the drug. The data show an average increase of attention span of $M_D = 4.8$ minutes with a variance of $s^2 = 125$ for the sample of difference scores.

a. Is this result sufficient to conclude that Ritalin significantly improves attention span? Use a one-tailed test with $\alpha = .05$.

b. Compute the 80% confidence interval for the mean change in attention span for the population.

11. College athletes, especially males, are often perceived as having very little interest in the academic side of their college experience. One common problem is class attendance. To address the problem of class attendance, a group of researchers developed and demonstrated a relatively simple but effective intervention (Bicard, Lott, Mills, Bicard, & Baylot-Casey, 2012). The researchers asked each athlete to text his academic counselor "in class" as soon as he arrived at the classroom. The researchers found significantly better attendance after the students began texting. In a similar study, a researcher monitored class attendance for a sample of $n = 16$ male athletes during the first 3 weeks of the semester and recorded the number of minutes that each student was late to class. The athletes were then asked to begin texting their arrival at the classroom and the researcher continued to monitor attendance for another 3 weeks. For each athlete, the average lateness for the first three weeks

and for the second three weeks were calculated, and the difference score was recorded. The data showed that lateness to class decreased by an average of $M_D = 21$ minutes with $SS = 2940$ when the students were texting.

a. Use a two-tailed test with $\alpha = .01$ to determine whether texting produced a significant change in attendance.

b. Compute a 95% confidence interval to estimate the mean change in attendance for the population.

12. Callahan (2009) demonstrated that Tai Chi can significantly reduce symptoms for individuals with arthritis. Participants were 18 years old or older with doctor-diagnosed arthritis. Self-reports of pain and stiffness were measured at the beginning of an 8-week Tai Chi course and again at the end. Suppose that the data produced an average decrease in pain and stiffness of $M_D = 8.5$ points with a standard deviation of 21.5 for a sample of $n = 40$ participants.

a. Use a two-tailed test with $\alpha = .05$ to determine whether the Tai Chi had a significant effect on pain and stiffness.

b. Compute Cohen's d to measure the size of the treatment effect.

13. Research results indicate that physically attractive people are also perceived as being more intelligent (Eagly, Ashmore, Makhijani, & Longo, 1991). As a demonstration of this phenomenon, a researcher obtained a set of 10 photographs, 5 showing men who were judged to be attractive and 5 showing men who were judged to be unattractive. The photographs were shown to a sample of $n = 25$ college students and the students were asked to rate the intelligence of the person in the photo on a scale from 1 to 10. For each student, the researcher determined the average rating for the 5 attractive photos and the average for the 5 unattractive photos, and then computed the difference between the two scores. For the entire sample, the average difference was $M_D = 2.7$ (attractive photos rated higher) with $s = 2.00$. Are the data sufficient to conclude that there was a significant difference in perceived intelligence for the two sets of photos? Use a two-tailed test with $\alpha = .05$.

14. There is some evidence suggesting that you are likely to improve your test score if you rethink and change answers on a multiple-choice exam (Johnston, 1975). To examine this phenomenon, a teacher gave the same final exam to two sections of a psychology course. The students in one section were told to turn in their exams immediately after finishing, without changing any of their answers. In the other section, students were encouraged to reconsider each question and to change answers whenever they felt it was appropriate. Before

the final exam, the teacher had matched 9 students in the first section with 9 students in the second section based on their midterm grades. For example, a student in the no-change section with an 89 on the midterm exam was matched with student in the change section who also had an 89 on the midterm. The difference between the two final exam grades for each matched pair was computed and the data showed that the students who were allowed to change answers scoring higher by an average of $M_D = 7$ points with $SS = 288$.

a. Do the data indicate a significant difference between the two conditions? Use a two-tailed test with $\alpha = .05$.

b. Construct a 95% confidence interval to estimate the size of the population mean difference.

c. Write a sentence demonstrating how the results of the hypothesis test and the confidence interval would appear in a research report.

15. Research indicates that the color red increases men's attraction to women (Elliot and Niesta, 2008). In the original study, men were shown women's photographs presented on either a white or a red background. Photographs presented on red were rated significantly more attractive than the same photographs mounted on white. In a similar study, a researcher prepares a set of 30 women's photographs, with 15 mounted on a white background and 15 mounted on red. One picture is identified as the test photograph, and appears twice in the set, once on white and once on red. Each male participant looks through the entire set of photographs and rates the attractiveness of each woman on a 10-point scale. The following table summarizes the ratings of the test photograph for a sample of $n = 9$ men. Are the ratings for the test photograph significantly different when it is presented on a red background compared to a white background? Use a two-tailed test with $\alpha = .01$.

Participant	White Background	Red Background
A	4	7
B	6	7
C	5	8
D	5	9
E	6	9
F	4	7
G	3	9
H	8	9
I	6	9

16. Example 11.2 in this chapter presented a repeated-measures research study demonstrating that swearing

can help reduce ratings of pain (Stephens, Atkins, & Kingston, 2009). In the study, each participant was asked to plunge a hand into icy water and keep it there as long as the pain would allow. In one condition, the participants repeated their favorite curse words while their hands were in the water. In the other condition, the participants repeated a neutral word. In addition to lowering the participants' perception of pain, swearing also increased the amount of time that they were able to tolerate the pain. Data similar to the results obtained in the study are shown in the following table.

a. Do these data indicate a significant difference in pain tolerance between the two conditions? Use a two-tailed test with $\alpha = .05$.

b. Compute r^2, the percentage of variance accounted for, to measure the size of the treatment effect.

c. Write a sentence demonstrating how the results of the hypothesis test and the measure of effect size would appear in a research report.

| | Amount of Time (in Seconds) | |
Participant	Swear words	Neutral words
1	94	59
2	70	61
3	52	47
4	83	60
5	46	35
6	117	92
7	69	53
8	39	30
9	51	56
10	73	61

17. a. A repeated-measures study with a sample of $n = 16$ participants produces a mean difference of $M_D = 3$ with a standard deviation of $s = 4$. Use a two-tailed hypothesis test with $\alpha = .05$ to determine whether this sample provides evidence of a significant treatment effect.

b. Now assume that the sample standard deviation is $s = 12$ and repeat the hypothesis test.

c. Explain how the size of the sample standard deviation influences the likelihood of finding a significant mean difference.

18. a. A repeated-measures study with a sample of $n = 16$ participants produces a mean difference of $M_D = 4$ with a standard deviation of $s = 8$. Use a two-tailed hypothesis test with $\alpha = .05$ to determine whether it is likely that this sample came from a population with $\mu_D = 0$.

b. Now assume that the sample mean difference is $M_D = 10$, and once again visualize the sample distribution. Use a two-tailed hypothesis test with $\alpha = .05$ to determine whether it is likely that this sample came from a population with $\mu_D = 0$.

c. Explain how the size of the sample mean difference influences the likelihood of finding a significant mean difference.

19. A sample of difference scores from a repeated-measures experiment has a mean of $M_D = 3$ with a standard deviation of $s = 4$.

a. If $n = 4$, is this sample sufficient to reject the null hypothesis using a two-tailed test with $\alpha = .05$?

b. Would you reject H_0 if $n = 16$? Again, assume a two-tailed test with $\alpha = .05$.

c. Explain how the size of the sample influences the likelihood of finding a significant mean difference.

20. Participants enter a research study with unique characteristics that produce different scores from one person to another. For an independent-measures study, these individual differences can cause problems. Identify the problems and briefly explain how they are eliminated or reduced with a repeated-measures study.

21. In the Chapter Preview we described a study showing that students had more academic problems following nights with less than average sleep compared to nights with more than average sleep (Gillen-O'Neel, Huynh, & Fuligni, 2013). Suppose a researcher is attempting to replicate this study using a sample of $n = 8$ college freshmen. Each student records the amount of study time and amount of sleep each night and reports the number of academic problems each day. The following data show the results from the study.

Number of Academic Problems

Student	Following Nights with Above Average Sleep	Following Nights with Below Average Sleep
A	10	13
B	8	6
C	5	9
D	5	6
E	4	6
F	10	14
G	11	13
H	3	5

a. Treat the data as if the scores are from an independent-measures study using two separate samples, each with $n = 8$ participants. Compute the pooled variance, the estimated standard error for the mean difference, and the independent-measures t statistic. Using a two-tailed test with $\alpha = .05$, is there a significant difference between the two sets of scores?

b. Now assume that the data are from a repeated-measures study using the same sample of $n = 8$ participants in both treatment conditions. Compute the variance for the sample of difference scores, the estimated standard error for the mean difference and the repeated-measures t statistic. Using a two-tailed test with $\alpha = .05$, is there a significant difference between the two sets of scores? (You should find that the repeated-measures design substantially reduces the variance and increases the likelihood of rejecting H_0.)

22. The previous problem demonstrates that removing individual differences can substantially reduce variance and lower the standard error. However, this benefit only occurs if the individual differences are consistent across treatment conditions. In problem 21, for example, the participants with the highest scores in the more-sleep condition also had the highest scores in the less-sleep condition. Similarly, participants with the lowest scores in the first condition also had the lowest scores in the second condition. To construct the following data, we started with the scores in problem 21 and scrambled the scores in treatment 1 to eliminate the consistency of the individual differences.

Number of Academic Problems

Student	Following Nights with Above Average Sleep	Following Nights with Below Average Sleep
A	10	13
B	8	14
C	5	13
D	5	5
E	4	9
F	10	6
G	11	6
H	3	6

a. Treat the data as if the scores are from an independent-measures study using two separate samples, each with $n = 8$ participants. Compute the pooled variance, the estimated standard error for the mean difference, and the independent-measures *t* statistic. Using a two-tailed test with $\alpha = .05$, is there a significant difference between the two sets of scores? Note: The scores in each treatment are the same as in Problem 21. Nothing has changed.

b. Now assume that the data are from a repeated-measures study using the same sample of $n = 8$ participants in both treatment conditions. Compute the variance for the sample of difference scores, the estimated standard error for the mean difference and the repeated-measures *t* statistic. Using a two-tailed test with $\alpha = .05$, is there a significant difference between the two sets of scores? (You should find that removing the individual differences with a repeated-measures *t* no longer reduces the variance because there are no consistent individual differences.)

Introduction to Analysis of Variance

Tools You Will Need

The following items are considered essential background material for this chapter. If you doubt your knowledge of any of these items, you should review the appropriate chapter or section before proceeding.

- Variability (Chapter 4)
  - Sum of squares
  - Sample variance
  - Degrees of freedom
- Introduction to hypothesis testing (Chapter 8)
  - The logic of hypothesis testing
- Independent-measures t statistic (Chapter 10)

© Deborah Batt

PREVIEW

Your job interview is tomorrow morning and suddenly a huge zit appears on your chin. Will it make a difference? A recent report suggests that it will. In two studies, Madera and Hebl (2012) compared interviews involving job applicants who either had a facial stigma or did not have any facial blemish. Each study included three treatment conditions: the applicant had a scar on the right cheek, a port-wine birthmark on the right cheek, or no facial blemish. The same applicants appeared in all three conditions and the facial stigmas were cosmetic. In one study, undergraduate students watched a computer recording of a job interview while their eye movements were recorded. The students were then tested on their memory of the applicant. A second study involved face-to-face interviews between applicants and full-time managers in an MBA program. The results showed that participants spent more time looking at the right cheek for the stigmatized applicants than for the clear skin applicants, they recalled less information about the applicants with the facial stigmas, and the managers gave the stigmatized applicants lower ratings. The researchers theorize that the facial stigma is a distraction that disrupts attention to the interview, which leads to poorer memory and lower applicant ratings.

In terms of job interviews, the Madera and Hebl studies are notable because they demonstrate a reaction to facial stigma that is difficult to control and can lead to workplace discrimination. In terms of statistics, however, this study is notable because it compares three different treatment conditions in a single experiment. We now have three different means and need a hypothesis test to evaluate the mean differences. Unfortunately, the t tests introduced in Chapter 10 and 11 are limited to comparing only two treatments. A new hypothesis test is needed for this kind of data.

In this chapter we introduce a hypothesis test known as *analysis of variance*, which is designed to evaluate the mean differences from research studies producing two or more sample means. Although "two or more" may seem like a small step from "two," this new hypothesis testing procedure provides researchers with a tremendous gain in experimental sophistication. In this chapter, and the two that follow, we examine some of the many applications of analysis of variance.

12.1 | Introduction (An Overview of Analysis of Variance)

LEARNING OBJECTIVES
1. Describe the terminology that is used for ANOVA, especially the terms *factor* and *level*, and identify the hypotheses for this test.

2. Identify the circumstances in which you should use ANOVA instead of t tests to evaluate mean differences, and explain why.

3. Describe the F-ratio that is used in ANOVA and explain how it is related to the t statistic.

Analysis of variance (ANOVA) is a hypothesis-testing procedure that is used to evaluate mean differences between two or more treatments (or populations). As with all inferential procedures, ANOVA uses sample data as the basis for drawing general conclusions about populations. It may appear that ANOVA and t tests are simply two different ways of doing exactly the same job: testing for mean differences. In some respects, this is true—both tests use sample data to test hypotheses about population means. However, ANOVA has a tremendous advantage over t tests. Specifically, t tests are limited to situations in which there are only two treatments to compare. The major advantage of ANOVA is that it can be used to compare *two or more treatments*. Thus, ANOVA provides researchers with much greater flexibility in designing experiments and interpreting results.

Figure 12.1 shows a typical research situation for which ANOVA would be used. Note that the study involves three samples representing three populations. The goal of the analysis is to determine whether the mean differences observed among the samples provide

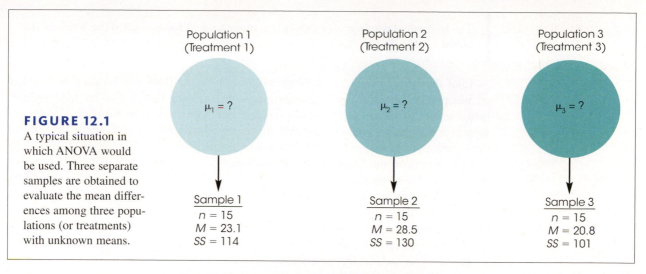

FIGURE 12.1

A typical situation in which ANOVA would be used. Three separate samples are obtained to evaluate the mean differences among three populations (or treatments) with unknown means.

Population 1 (Treatment 1)
$\mu_1 = ?$

Population 2 (Treatment 2)
$\mu_2 = ?$

Population 3 (Treatment 3)
$\mu_3 = ?$

Sample 1
$n = 15$
$M = 23.1$
$SS = 114$

Sample 2
$n = 15$
$M = 28.5$
$SS = 130$

Sample 3
$n = 15$
$M = 20.8$
$SS = 101$

enough evidence to conclude that there are mean differences among the three populations. Specifically, we must decide between two interpretations:

1. There really are no differences between the populations (or treatments). The observed differences between the sample means are caused by random, unsystematic factors (sampling error) that differentiate one sample from another.

2. The populations (or treatments) really do have different means, and these population mean differences are responsible for causing systematic differences between the sample means.

You should recognize that these two interpretations correspond to the two hypotheses (null and alternative) that are part of the general hypothesis-testing procedure.

■ Terminology in Analysis of Variance

Before we continue, it is necessary to introduce some of the terminology that is used to describe the research situation shown in Figure 12.1. Recall (from Chapter 1) that when a researcher manipulates a variable to create the treatment conditions in an experiment, the variable is called an independent variable. For example, Figure 12.1 could represent a study examining driving performance under three different telephone conditions: driving with no phone, talking on a hands-free phone, and talking on a hand-held phone. Note that the three conditions are created by the researcher. On the other hand, when a researcher uses a non-manipulated variable to designate groups, the variable is called a *quasi-independent variable*. For example, the three groups in Figure 12.1 could represent 6-year-old, 8-year-old, and 10-year-old children. In the context of ANOVA, an independent variable or a quasi-independent variable is called a *factor*. Thus, Figure 12.1 could represent an experimental study in which the telephone condition is the factor being evaluated or it could represent a nonexperimental study in which age is the factor being examined.

DEFINITION

> In analysis of variance, the variable (independent or quasi-independent) that designates the groups being compared is called a **factor**.

In addition, the individual groups or treatment conditions that are used to make up a factor are called the *levels* of the factor. For example, a study that examined performance under three different telephone conditions would have three levels of the factor.

DEFINITION	The individual conditions or values that make up a factor are called the **levels** of the factor.

Like the *t* tests presented in Chapters 10 and 11, ANOVA can be used with either an independent-measures or a repeated-measures design. Recall that an independent-measures design means that there is a separate group of participants for each of the treatments (or populations) being compared. In a repeated-measures design, on the other hand, the same group is tested in all of the different treatment conditions. In addition, ANOVA can be used to evaluate the results from a research study that involves more than one factor. For example, a researcher may want to compare two different therapy techniques, examining their immediate effectiveness as well as the persistence of their effectiveness over time. In this situation, the research study could involve two different groups of participants, one for each therapy, and measure each group at several different points in time. The structure of this design is shown in Figure 12.2. Notice that the study uses two factors, one independent-measures factor and one repeated-measures factor:

1. Factor 1: Therapy technique. A separate group is used for each technique (independent measures).

2. Factor 2: Time. Each group is tested at three different times (repeated measures).

In this case, the ANOVA would evaluate mean differences between the two therapies as well as mean differences between the scores obtained at different times. A study that combines two factors, like the one in Figure 12.2, is called a *two-factor design* or a *factorial design*.

The ability to combine different factors and to mix different designs within one study provides researchers with the flexibility to develop studies that address scientific questions that could not be answered by a single design using a single factor.

Although ANOVA can be used in a wide variety of research situations, this chapter introduces ANOVA in its simplest form. Specifically, we consider only *single-factor* designs. That is, we examine studies that have only one independent variable (or only one quasi-independent variable). Second, we consider only *independent-measures* designs; that is, studies that use

FIGURE 12.2

A research design with two factors. One factor uses two levels of therapy technique (I vs. II) and the second factor uses three levels of time (before after, and 6 months after). Also notice that the therapy factor uses two separate groups (independent measures) and the time factor uses the same group for all three levels (repeated measures).

			TIME	
		Before Therapy	**After Therapy**	**6 Months After Therapy**
THERAPY TECHNIQUE	Therapy I (Group 1)	Scores for group 1 measured before Therapy I	Scores for group 1 measured after Therapy I	Scores for group 1 measured 6 months after Therapy I
	Therapy II (Group 2)	Scores for group 2 measured before Therapy II	Scores for group 2 measured after Therapy II	Scores for group 2 measured 6 months after Therapy II

a separate group of participants for each treatment condition. The basic logic and procedures that are presented in this chapter form the foundation for more complex applications of ANOVA. For example, in Chapter 13 we extend the analysis to single-factor, repeated-measures designs and in Chapter 14 we introduce two-factor designs. But for now, in this chapter, we limit our discussion of ANOVA to *single-factor, independent-measures* research studies.

■ Statistical Hypotheses for ANOVA

The following example introduces the statistical hypotheses for ANOVA. Suppose that a researcher examined driving performance under three different telephone conditions: no phone, a hands-free phone, and a hand-held phone. Three samples of participants are selected, one sample for each treatment condition. The purpose of the study is to determine whether using a telephone affects driving performance. In statistical terms, we want to decide between two hypotheses: the null hypothesis (H_0), which states that the telephone condition has no effect, and the alternative hypothesis (H_1), which states that the telephone condition does affect driving. In symbols, the null hypothesis states

$$H_0: \mu_1 = \mu_2 = \mu_3$$

In other words, the null hypothesis states that the telephone condition has no effect on driving performance. That is, the population means for the three telephone conditions are all the same. In general, H_0 states that there is no treatment effect.

The alternative hypothesis states that the population means are not all the same:

H_1: There is at least one mean difference among the populations.

In general, H_1 states that the treatment conditions are not all the same; that is, there is a real treatment effect. As always, the hypotheses are stated in terms of population parameters, even though we use sample data to test them.

Notice that we are not stating a specific alternative hypothesis. This is because many different alternatives are possible, and it would be tedious to list them all. One alternative, for example, would be that the first two populations are identical, but that the third is different. Another alternative states that the last two means are the same, but that the first is different. Other alternatives might be

$$H_1: \mu_1 \neq \mu_2 \neq \mu_3 \quad \textbf{(all three means are different.)}$$
$$H_1: \mu_1 = \mu_3, \text{ but } \mu_2 \text{ is different.}$$

We should point out that a researcher typically entertains only one (or at most a few) of these alternative hypotheses. Usually a theory or the outcomes of previous studies will dictate a specific prediction concerning the treatment effect. For the sake of simplicity, we will state a general alternative hypothesis rather than try to list all possible specific alternatives.

■ Type I Errors and Multiple-Hypothesis Tests

If we already have *t* tests for comparing mean differences, you might wonder why ANOVA is necessary. Why create a whole new hypothesis-testing procedure that simply duplicates what the *t* tests can already do? The answer to this question is based in a concern about Type I errors.

Remember that each time you do a hypothesis test, you select an alpha level that determines the risk of a Type I error. With $\alpha = .05$, for example, there is a 5%, or a 1-in-20, risk of a Type I error, whenever your decision is to reject the null hypothesis. Often a single experiment requires several hypothesis tests to evaluate all the mean differences. However, each test has a risk of a Type I error, and the more tests you do, the greater the risk.

For this reason, researchers often make a distinction between the *testwise alpha level* and the *experimentwise alpha level*. The testwise alpha level is simply the alpha level you select for each individual hypothesis test. The experimentwise alpha level is the total probability of a Type I error accumulated from all of the separate tests in the experiment. As the number of separate tests increases, so does the experimentwise alpha level.

DEFINITIONS

> The **testwise alpha level** is the risk of a Type I error, or alpha level, for an individual hypothesis test.
>
> When an experiment involves several different hypothesis tests, the **experimentwise alpha level** is the total probability of a Type I error that is accumulated from all of the individual tests in the experiment. Typically, the experimentwise alpha level is substantially greater than the value of alpha used for any one of the individual tests.

For example, an experiment involving three treatments would require three separate *t* tests to compare all of the mean differences:

Test 1 compares treatment I vs. treatment II.

Test 2 compares treatment I vs. treatment III.

Test 3 compares treatment II vs. treatment III.

If all tests use $\alpha = .05$, then there is a 5% risk of a Type I error for the first test, a 5% risk for the second test, and another 5% risk for the third test. The three separate tests accumulate to produce a relatively large experimentwise alpha level. The advantage of ANOVA is that it performs all three comparisons simultaneously in one hypothesis test. Thus, no matter how many different means are being compared, ANOVA uses one test with one alpha level to evaluate the mean differences and thereby avoids the problem of an inflated experimentwise alpha level.

■ The Test Statistic for ANOVA

The test statistic for ANOVA is very similar to the *t* statistics used in earlier chapters. For the *t* statistic, we first computed the standard error, which measures the difference between two sample means that is reasonable to expect if there is no treatment effect (that is, if H_0 is true). Then we computed the *t* statistic with the following structure:

$$t = \frac{\text{obtained difference between two sample means}}{\text{standard error (the difference expected with no treatment effect)}}$$

For ANOVA, however, we want to compare differences among two *or more* sample means. With more than two samples, the concept of "difference between sample means" becomes difficult to define or measure. For example, if there are only two samples and they have means of $M = 20$ and $M = 30$, then there is a 10-point difference between the sample means. Suppose, however, that we add a third sample with a mean of $M = 35$. Now how much difference is there between the sample means? It should be clear that we have a problem. The solution to this problem is to use variance to define and measure the size of the differences among the sample means. Consider the following two sets of sample means:

Set 1	Set 2
$M_1 = 20$	$M_1 = 28$
$M_2 = 30$	$M_2 = 30$
$M_3 = 35$	$M_3 = 31$

If you compute the variance for the three numbers in each set, then the variance is $s^2 = 58.33$ for set 1 and the variance is $s^2 = 2.33$ for set 2 is. Notice that the two variances provide an accurate representation of the size of the differences. In set 1 there are relatively large differences between sample means and the variance is relatively large. In set 2 the mean differences are small and the variance is small.

Thus, we can use variance to measure sample mean differences when there are two or more samples. The test statistic for ANOVA uses this fact to compute an F-*ratio* with the following structure:

$$F = \frac{\text{variance (differences) between sample means}}{\text{variance (differences) expected with no treatment effect}}$$

Note that the F-ratio has the same basic structure as the t statistic but is based on *variance* instead of sample mean *difference*. The variance in the numerator of the F-ratio provides a single number that measures the differences among all of the sample means. The variance in the denominator of the F-ratio, like the standard error in the denominator of the t statistic, measures the mean differences that would be expected if there is no treatment effect. Thus, the t statistic and the F-ratio provide the same basic information. In each case, a large value for the test statistic provides evidence that the sample mean differences (numerator) are larger than would be expected if there were no treatment effects (denominator).

LEARNING CHECK

1. Which of the following is the correct description for a research study comparing problem-solving scores obtained for 3 different age groups of children?
   a. single-factor design
   b. two-factor design
   c. three-factor design
   d. factorial design

2. ANOVA is necessary to evaluate mean differences among three or more treatments, in order to _____ .
   a. minimize risk of Type I error
   b. maximize risk of Type I error
   c. minimize risk of Type II error
   d. maximize risk of Type II error

3. The numerator of the F-ratio measures the size of the sample mean differences. How is this number obtained, especially when there are more than two sample means?
   a. by computing the sum of the sample mean differences
   b. by computing the average of the sample mean differences
   c. by the largest of the sample mean differences.
   d. by computing the variance for the sample means

ANSWERS   **1.** A, **2.** A, **3.** D

12.2 | The Logic of Analysis of Variance

LEARNING OBJECTIVE

4. Identify the sources that contribute to the variance between-treatments and the variance within-treatments, and describe how these two variances are compared in the *F*-ratio to evaluate the null hypothesis.

The formulas and calculations required in ANOVA are somewhat complicated, but the logic that underlies the whole procedure is fairly straightforward. Therefore, this section gives a general picture of ANOVA before we start looking at the details. We will introduce the logic of ANOVA with the help of the hypothetical data in Table 12.1. These data represent the results of an independent-measures experiment using three separate samples, each with $n = 5$ participants, to compare performance in a driving simulator under three telephone conditions.

TABLE 12.1

Hypothetical data from an experiment examining driving performance under three telephone conditions.

Treatment 1 No Phone (Sample 1)	Treatment 2 Hand-Held (Sample 2)	Treatment 3 Hands-Free (Sample 3)
4	0	1
3	1	2
6	3	2
3	1	0
4	0	0
$M = 4$	$M = 1$	$M = 1$

One obvious characteristic of the data in Table 12.1 is that the scores are not all the same. In everyday language, the scores are different; in statistical terms, the scores are variable. Our goal is to measure the amount of variability (the size of the differences) and to explain why the scores are different.

The first step is to determine the total variability for the entire set of data. To compute the total variability, we combine all the scores from all the separate samples to obtain one general measure of variability for the complete experiment. Once we have measured the total variability, we can begin to break it apart into separate components. The word *analysis* means dividing into smaller parts. Because we are going to analyze variability, the process is called *analysis of variance*. This analysis process divides the total variability into two basic components.

1. **Between-Treatments Variance.** Looking at the data in Table 12.1, we clearly see that much of the variability in the scores results from general differences between treatment conditions. For example, the scores in the no-phone condition tend to be much higher ($M = 4$) than the scores in the hand-held condition ($M = 1$). We will calculate the variance between treatments to provide a measure of the overall differences between treatment conditions. Notice that the variance between treatments is really measuring the differences between sample means.

2. **Within-Treatment Variance.** In addition to the general differences between treatment conditions, there is variability within each sample. Looking again at Table 12.1, we see that the scores in the no-phone condition are not all the same; they are variable. The within-treatments variance provides a measure of the variability inside each treatment condition.

Analyzing the total variability into these two components is the heart of ANOVA. We will now examine each of the components in more detail.

■ Between-Treatments Variance

Remember that calculating variance is simply a method for measuring how big the differences are for a set of numbers. When you see the term *variance*, you can automatically translate it into the term *differences*. Thus, the *between-treatments variance* simply measures how much difference exists between the treatment conditions. There are two possible explanations for these between-treatment differences:

1. The differences between treatments are not caused by any treatment effect but are simply the naturally occurring, random and unsystematic differences that exist between one sample and another. That is, the differences are the result of sampling error.

2. The differences between treatments have been caused by the *treatment effects*. For example, if using a telephone really does interfere with driving performance, then scores in the telephone conditions should be systematically lower than scores in the no-phone condition.

Thus, when we compute the between-treatments variance, we are measuring differences that could be caused by a systematic treatment effect or could simply be random and unsystematic mean differences caused by sampling error. To demonstrate that there really is a treatment effect, we must establish that the differences between treatments are bigger than would be expected by sampling error alone. To accomplish this goal, we determine how big the differences are when there is no systematic treatment effect; that is, we measure how much difference (or variance) can be explained by random and unsystematic factors. To measure these differences, we compute the variance within treatments.

■ Within-Treatments Variance

Inside each treatment condition, we have a set of individuals who all receive exactly the same treatment; that is, the researcher does not do anything that would cause these individuals to have different scores. In Table 12.1, for example, the data show that five individuals were tested while talking on a hand-held phone (sample 2). Although these five individuals all received exactly the same treatment, their scores are different. Why are the scores different? The answer is that there is no specific cause for the differences. Instead, the differences that exist within a treatment represent random and unsystematic differences that occur when there are no treatment effects causing the scores to be different. Thus, the *within-treatments variance* provides a measure of how big the differences are when H_0 is true.

Figure 12.3 shows the overall ANOVA and identifies the sources of variability that are measured by each of the two basic components.

■ The *F*-Ratio: The Test Statistic for ANOVA

Once we have analyzed the total variability into two basic components (between treatments and within treatments), we simply compare them. The comparison is made by computing an F-*ratio*. For the independent-measures ANOVA, the *F*-ratio has the following structure:

$$F = \frac{\text{variance between treatments}}{\text{variance within treatments}} = \frac{\text{differences including any treatment effects}}{\text{differences with no treatment effects}} \quad \text{(12.1)}$$

When we express each component of variability in terms of its sources (see Figure 12.3), the structure of the *F*-ratio is

$$F = \frac{\text{systematic treatment effects} + \text{random, unsystematic differences}}{\text{random, unsystematic differences}} \quad \text{(12.2)}$$

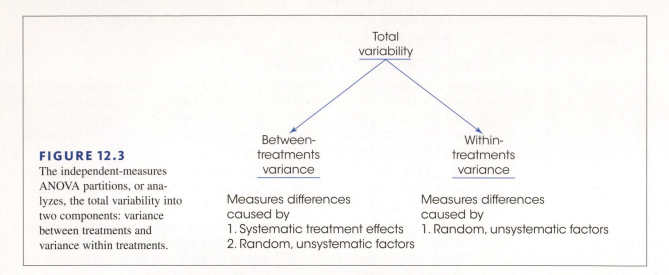

FIGURE 12.3

The independent-measures ANOVA partitions, or analyzes, the total variability into two components: variance between treatments and variance within treatments.

The value obtained for the *F*-ratio helps determine whether any treatment effects exist. Consider the following two possibilities:

1. When there are no systematic treatment effects, the differences between treatments (numerator) are entirely caused by random, unsystematic factors. In this case, the numerator and the denominator of the *F*-ratio are both measuring random differences and should be roughly the same size. With the numerator and denominator roughly equal, the *F*-ratio should have a value around 1.00. In terms of the formula, when the treatment effect is zero, we obtain

$$F = \frac{0 + \text{random, unsystematic differences}}{\text{random, unsystematic differences}}$$

    Thus, an *F*-ratio near 1.00 indicates that the differences between treatments (numerator) are random and unsystematic, just like the differences in the denominator. With an *F*-ratio near 1.00, we conclude that there is no evidence to suggest that the treatment has any effect.

2. When the treatment does have an effect, causing systematic differences between samples, then the combination of systematic and random differences in the numerator should be larger than the random differences alone in the denominator. In this case, the numerator of the *F*-ratio should be noticeably larger than the denominator, and we should obtain an *F*-ratio noticeably larger than 1.00. Thus, a large *F*-ratio is evidence for the existence of systematic treatment effects; that is, there are significant differences between treatments.

Because the denominator of the *F*-ratio measures only random and unsystematic variability, it is called the *error term*. The numerator of the *F*-ratio always includes the same unsystematic variability as in the error term, but it also includes any systematic differences caused by the treatment effect. The goal of ANOVA is to find out whether a treatment effect exists.

DEFINITION

For ANOVA, the denominator of the *F*-ratio is called the **error term**. The error term provides a measure of the variance caused by random, unsystematic differences. When the treatment effect is zero (H_0 is true), the error term measures the same sources of variance as the numerator of the *F*-ratio, so the value of the *F*-ratio is expected to be nearly equal to 1.00.

1. In an analysis of variance, the primary effect of large mean differences from one sample to another is to increase the value for _____.
   a. the variance between treatments.
   b. the variance within treatments
   c. the total variance.
   d. large mean differences will not directly affect any of the three variances.

2. When the null hypothesis is true for an ANOVA, what is the expected value for the F-ratio?
   a. 0
   b. 1.00
   c. much greater than 1.00
   d. less than zero

ANSWERS   **1.** A, **2.** B

12.3 | ANOVA Notation and Formulas

5. Calculate the three SS values, the three df values, and the two mean squares (MS values) that are needed for the F-ratio and describe the relationships among them.

Because ANOVA typically is used to examine data from more than two treatment conditions (and more than two samples), we need a notational system to keep track of all the individual scores and totals. To help introduce this notational system, we use the hypothetical data from Table 12.1 again. The data are reproduced in Table 12.2 along with some of the notation and statistics that will be described.

1. The letter k is used to identify the number of treatment conditions—that is, the number of levels of the factor. For an independent-measures study, k also specifies the number of separate samples. For the data in Table 12.2, there are three treatments, so $k = 3$.

2. The number of scores in each treatment is identified by a lowercase letter n. For the example in Table 12.2, $n = 5$ for all the treatments. If the samples are of different sizes, you can identify a specific sample by using a subscript. For example, n_2 is the number of scores in treatment 2.

3. The total number of scores in the entire study is specified by a capital letter N. When all the samples are the same size (n is constant), $N = kn$. For the data in Table 12.2, there are $n = 5$ scores in each of the $k = 3$ treatments, so we have a total of $N = 3(5) = 15$ scores in the entire study.

4. The sum of the scores (ΣX) for each treatment condition is identified by the capital letter T (for treatment total). The total for a specific treatment can be identified by adding a numerical subscript to the T. For example, the total for the second treatment in Table 12.2 is $T_2 = 5$.

Because ANOVA formulas require ΣX for each treatment and ΣX for the entire set of scores, we have introduced new notation (T and G) to help identify which ΣX is being used. Remember: T stands for *treatment total* and G stands for *grand total*.

5. The sum of all the scores in the research study (the grand total) is identified by G. You can compute G by adding up all N scores or by adding up the treatment totals: $G = \Sigma T$.

6. Although there is no new notation involved, we also have computed SS and M for each sample, and we have calculated ΣX^2 for the entire set of $N = 15$ scores in the study. These values are given in Table 12.2 and are important in the formulas and calculations for ANOVA.

Finally, we should note that there is no universally accepted notation for ANOVA. Although we are using Gs and Ts, for example, you may find that other sources use other symbols.

TABLE 12.2

The same data that appeared in Table 12.1 with summary values and notation appropriate for an ANOVA.

	Telephone Conditions		
Treatment 1 No Phone (Sample 1)	Treatment 2 Hand-Held (Sample 2)	Treatment 3 Hands-Free (Sample 3)	
4	0	1	$\Sigma X^2 = 106$
3	1	2	$G = 30$
6	3	2	$N = 15$
3	1	0	$k = 3$
4	0	0	
$T_1 = 20$	$T_2 = 5$	$T_3 = 5$	
$SS_1 = 6$	$SS_2 = 6$	$SS_3 = 4$	
$n_1 = 5$	$n_2 = 5$	$n_3 = 5$	
$M_1 = 4$	$M_2 = 1$	$M_3 = 1$	

■ ANOVA Formulas

Because ANOVA requires extensive calculations and many formulas, one common problem for students is simply keeping track of the different formulas and numbers. Therefore, we will examine the general structure of the procedure and look at the organization of the calculations before we introduce the individual formulas.

1. The final calculation for ANOVA is the F-ratio, which is composed of two variances:

$$F = \frac{\text{variance between treatments}}{\text{variance within treatments}}$$

2. Each of the two variances in the F-ratio is calculated using the basic formula for sample variance.

$$\text{sample variance} = s^2 = \frac{SS}{df}$$

Therefore, we need to compute an SS and a df for the variance between treatments (numerator of F), and we need another SS and df for the variance within treatments (denominator of F). To obtain these SS and df values, we must go through two separate analyses: First, compute SS for the total study, and analyze it into two

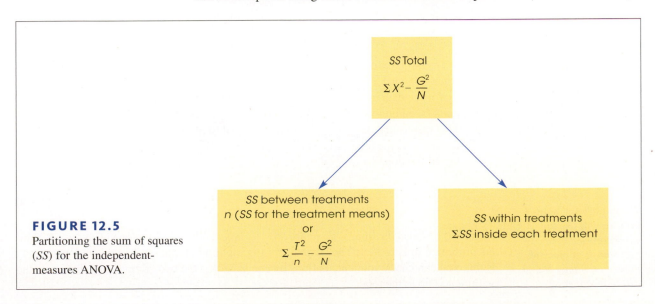

The final goal for the ANOVA is an F-ratio	$F = \dfrac{\text{Variance between treatments}}{\text{Variance within treatments}}$	
Each variance in the F-ratio is computed as *SS/df*	Variance between treatments $= \dfrac{SS\text{ between}}{df\text{ between}}$	Variance within treatments $= \dfrac{SS\text{ within}}{df\text{ within}}$
To obtain each of the *SS* and *df* values, the total variability is analyzed into the two components	*SS* total *SS* between *SS* within	*df* total *df* between *df* within

FIGURE 12.4

The structure and sequence of calculations for the ANOVA.

components (between and within). Then compute *df* for the total study, and analyze it into two components (between and within).

Thus, the entire process of ANOVA requires nine calculations: three values for *SS*, three values for *df*, two variances (between and within), and a final *F*-ratio. However, these nine calculations are all logically related and are all directed toward finding the final *F*-ratio. Figure 12.4 shows the logical structure of ANOVA calculations.

■ Analysis of Sum of Squares (*SS*)

The ANOVA requires that we first compute a total sum of squares and then partition this value into two components: between treatments and within treatments. This analysis is outlined in Figure 12.5. We will examine each of the three components separately.

1. **Total Sum of Squares, SS_{total}.** As the name implies, SS_{total} is the sum of squares for the entire set of *N* scores. As described in Chapter 4 (pp. 110–112), this *SS* value can be computed using either a definitional or a computational formula. However,

SS Total

$$\Sigma X^2 - \frac{G^2}{N}$$

SS between treatments
n (*SS* for the treatment means)
or
$$\Sigma \frac{T^2}{n} - \frac{G^2}{N}$$

SS within treatments
ΣSS inside each treatment

FIGURE 12.5

Partitioning the sum of squares (*SS*) for the independent-measures ANOVA.

ANOVA typically involves a large number of scores and the mean is often not a whole number. Therefore, it is usually much easier to calculate SS_{total} using the computational formula:

$$SS = \Sigma X^2 - \frac{(\Sigma X)^2}{N}$$

To make this formula consistent with the ANOVA notation, we substitute the letter G in place of ΣX and obtain

$$SS_{total} = \Sigma X^2 - \frac{G^2}{N} \qquad (12.3)$$

Applying this formula to the set of data in Table 12.2, we obtain

$$SS_{total} = 106 - \frac{30^2}{15}$$

$$= 106 - 60$$

$$= 46$$

2. **Within-Treatments Sum of Squares, $SS_{within\ treatments}$.** Now we are looking at the variability inside each of the treatment conditions. We already have computed the SS within each of the three treatment conditions (Table 12.2): $SS_1 = 6$, $SS_2 = 6$, and $SS_3 = 4$. To find the overall within-treatment sum of squares, we simply add these values together:

$$SS_{within\ treatments} = \Sigma SS_{inside\ each\ treatment} \qquad (12.4)$$

For the data in Table 12.2, this formula gives

$$SS_{within\ treatments} = 6 + 6 + 4$$

$$= 16$$

3. **Between-Treatments Sum of Squares, $SS_{between\ treatments}$.** Before we introduce any equations for $SS_{between\ treatments}$, consider what we have found so far. The total variability for the data in Table 12.2 is $SS_{total} = 46$. We intend to partition this total into two parts (see Figure 12.5). One part, $SS_{within\ treatments}$, has been found to be equal to 16. This means that $SS_{between\ treatments}$ must be equal to 30 so that the two parts (16 and 30) to add up to the total (46). Thus, the value for $SS_{between\ treatments}$ can be found simply by subtraction:

$$SS_{between} = SS_{total} - SS_{within} \qquad (12.5)$$

To simplify the notation we will use the subscripts *between* and *within* in place of *between treatments* and *within treatments*.

However, it is also possible to compute $SS_{between}$ independently, using one of the two formulas presented in Box 12.1. The advantage of computing all three SS values independently is that you can check your calculations by ensuring that the two components, between and within, add up to the total.

BOX 12.1 Alternative Formulas for $SS_{between}$

Recall that the variability between treatments is measuring the differences between treatment means. Conceptually, the most direct way of measuring the amount of variability among the treatment means is to compute the sum of squares for the set of sample means, SS_{means}. For the data in Table 12.2, the samples means are 4, 1, and 1. These three values produce $SS_{means} = 6$. However, each of the three means represents a group of $n = 5$ scores. Therefore, the final value for $SS_{between}$ is obtained by multiplying SS_{means} by n.

$$SS_{between} = n(SS_{means}) \qquad (12.6)$$

For the data in Table 12.2, we obtain

$$SS_{between} = n(SS_{means}) = 5(6) = 30$$

Unfortunately, Equation 12.6 can only be used when all of the samples are exactly the same size

(equal ns), and the equation can be very awkward, especially when the treatment means are not whole numbers. Therefore, we also present a computational formula for $SS_{between}$ that uses the treatment totals (T) instead of the treatment means.

$$SS_{between} = \Sigma \frac{T^2}{n} - \frac{G^2}{N} \qquad (12.7)$$

For the data in Table 12.2 this formula produces:

$$SS_{between} = \frac{20^2}{5} + \frac{5^2}{5} + \frac{5^2}{5} - \frac{30^2}{15}$$
$$= 80 + 5 + 5 - 60$$
$$= 90 - 60$$
$$= 30$$

Note that all three techniques (Equations 12.5, 12.6, and 12.7) produce the same result, $SS_{between} = 30$.

Computing $SS_{between}$ Including the two formulas in Box 12.1, we have presented three different equations for computing $SS_{between}$. Rather than memorizing all three, however, we suggest that you pick one formula and use it consistently. There are two reasonable alternatives to use. The simplest is Equation 12.5, which finds $SS_{between}$ simply by subtraction: First you compute SS_{total} and SS_{within}, then subtract:

$$SS_{between} = SS_{total} - SS_{within}$$

The second alternative is to use Equation 12.7, which computes $SS_{between}$ using the treatment totals (the T values). The advantage of this alternative is that it provides a way to check your arithmetic: Calculate SS_{total}, $SS_{between}$, and SS_{within} separately, and then check to be sure that the two components add up to equal SS_{total}.

Using Equation 12.6, which computes SS for the set of sample means, is usually not a good choice. Unless the sample means are all whole numbers, this equation can produce very tedious calculations. In most situations, one of the other two equations is a better alternative.

The following example is an opportunity for you to test your understanding of the analysis of SS in ANOVA.

EXAMPLE 12.1   Three samples, each with $n = 5$ participants, are used to evaluate the mean differences among three treatment conditions. The three sample means and SS values are $T_1 = 10$ with $SS = 16$, $T_2 = 25$ with $SS = 20$, and $T_3 = 40$ with $SS = 24$. If $SS_{total} = 150$, then what are the values for $SS_{between}$ and SS_{within}? You should find that $SS_{between} = 90$ and $SS_{within} = 60$. ■

■ The Analysis of Degrees of Freedom (*df*)

The analysis of degrees of freedom (*df*) follows the same pattern as the analysis of SS. First, we find *df* for the total set of N scores, and then we partition this value into two components: degrees of freedom between treatments and degrees of freedom within treatments. In computing degrees of freedom, there are two important considerations to keep in mind:

1. Each *df* value is associated with a specific *SS* value.

2. Normally, the value of *df* is obtained by counting the number of items that were used to calculate *SS* and then subtracting 1. For example, if you compute *SS* for a set of *n* scores, then $df = n - 1$.

With this in mind, we will examine the degrees of freedom for each part of the analysis.

1. **Total Degrees of Freedom, df_{total}.** To find the *df* associated with SS_{total}, you must first recall that this *SS* value measures variability for the entire set of *N* scores. Therefore, the *df* value is

$$df_{total} = N - 1 \tag{12.8}$$

   For the data in Table 12.2, the total number of scores is $N = 15$, so the total degrees of freedom are

$$df_{total} = 15 - 1$$
$$= 14$$

2. **Within-Treatments Degrees of Freedom, df_{within}.** To find the *df* associated with SS_{within}, we must look at how this *SS* value is computed. Remember, we first find *SS* inside of each of the treatments and then add these values together. Each of the treatment *SS* values measures variability for the *n* scores in the treatment, so each *SS* has $df = n - 1$. When all these individual treatment values are added together, we obtain

$$df_{within} = \Sigma(n - 1) = \Sigma df_{in\ each\ treatment} \tag{12.9}$$

   For the experiment we have been considering, each treatment has $n = 5$ scores. This means there are $n - 1 = 4$ degrees of freedom inside each treatment. Because there are three different treatment conditions, this gives a total of 12 for the within-treatments degrees of freedom. Notice that this formula for *df* simply adds up the number of scores in each treatment (the *n* values) and subtracts 1 for each treatment. If these two stages are done separately, you obtain

$$df_{within} = N - k \tag{12.10}$$

   (Adding up all the *n* values gives *N*. If you subtract 1 for each treatment, then altogether you have subtracted *k* because there are *k* treatments.) For the data in Table 12.2, $N = 15$ and $k = 3$, so

$$df_{within} = 15 - 3$$
$$= 12$$

3. **Between-Treatments Degrees of Freedom, $df_{between}$.** The *df* associated with $SS_{between}$ can be found by considering how the *SS* value is obtained. This *SS* formulas measure the variability for the set of treatments (totals or means). To find $df_{between}$, simply count the number of treatments and subtract 1. Because the number of treatments is specified by the letter *k*, the formula for *df* is

$$df_{between} = k - 1 \tag{12.11}$$

For the data in Table 12.2, there are three different treatment conditions (three *T* values or three sample means), so the between-treatments degrees of freedom are computed as follows:

$$df_{between} = 3 - 1$$
$$= 2$$

Notice that the two parts we obtained from this analysis of degrees of freedom add up to equal the total degrees of freedom:

$$df_{total} = df_{within} + df_{between}$$
$$14 = 12 + 2$$

The complete analysis of degrees of freedom is shown in Figure 12.6.

As you are computing the *SS* and *df* values for ANOVA, keep in mind that the labels that are used for each value can help you understand the formulas. Specifically,

1. The term **total** refers to the entire set of scores. We compute *SS* for the whole set of *N* scores, and the *df* value is simply $N - 1$.

2. The term **within treatments** refers to differences that exist inside the individual treatment conditions. Thus, we compute *SS* and *df* inside each of the separate treatments.

3. The term **between treatments** refers to differences from one treatment to another. With three treatments, for example, we are comparing three different means (or totals) and have $df = 3 - 1 = 2$.

■ Calculation of Variances (*MS*) and the *F*-Ratio

The next step in the ANOVA procedure is to compute the variance between treatments and the variance within treatments, which are used to calculate the *F*-ratio (see Figure 12.4).

In ANOVA, it is customary to use the term *mean square*, or simply *MS*, in place of the term *variance*. Recall (from Chapter 4) that variance is defined as the mean of the squared deviations. In the same way that we use *SS* to stand for the sum of the squared deviations,

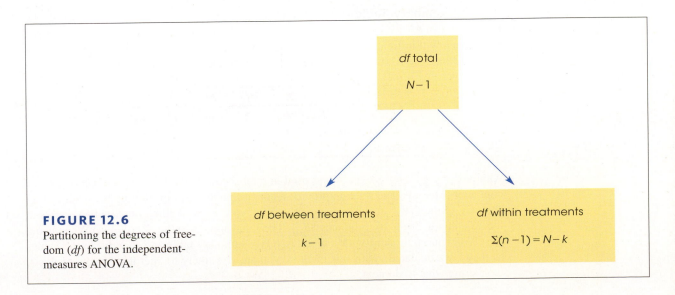

FIGURE 12.6
Partitioning the degrees of freedom (*df*) for the independent-measures ANOVA.

we now will use *MS* to stand for the mean of the squared deviations. For the final *F*-ratio we will need an *MS* (variance) between treatments for the numerator and an *MS* (variance) within treatments for the denominator. In each case

$$MS(\text{variance}) = s^2 = \frac{SS}{df} \quad \quad (12.12)$$

For the data we have been considering,

$$MS_{\text{between}} = s^2_{\text{between}} = \frac{SS_{\text{between}}}{df_{\text{between}}} = \frac{30}{2} = 15$$

and

$$MS_{\text{within}} = s^2_{\text{within}} = \frac{SS_{\text{within}}}{df_{\text{within}}} = \frac{16}{12} = 1.33$$

We now have a measure of the variance (or differences) between the treatments and a measure of the variance within the treatments. The *F*-ratio simply compares these two variances:

$$F = \frac{s^2_{\text{between}}}{s^2_{\text{within}}} = \frac{MS_{\text{between}}}{MS_{\text{within}}} \quad \quad (12.13)$$

For the experiment we have been examining, the data give an *F*-ratio of

$$F = \frac{15}{1.33} = 11.28$$

For this example, the obtained value of $F = 11.28$ indicates that the numerator of the *F*-ratio is substantially bigger than the denominator. If you recall the conceptual structure of the *F*-ratio as presented in Equations 12.1 and 12.2, the *F* value we obtained indicates that the differences between treatments are more than 11 times bigger than what would be expected if there is no treatment effect. Stated in terms of the experimental variables, using a telephone while driving does appear to have an effect on driving performance. However, to properly evaluate the *F*-ratio, we must select an α level and consult the *F*-distribution table that is discussed in the next section.

ANOVA Summary Tables It is useful to organize the results of the analysis in one table called an *ANOVA summary table*. The table shows the source of variability (between treatments, within treatments, and total variability), *SS*, *df*, *MS*, and *F*. For the previous computations, the ANOVA summary table is constructed as follows:

Source	SS	df	MS	
Between treatments	30	2	15	F = 11.28
Within treatments	16	12	1.33	
Total	46	14		

Although these tables are no longer used in published reports, they are a common part of computer printouts, and they do provide a concise method for presenting the results of an analysis. (Note that you can conveniently check your work: Adding the first two entries in the *SS* column, $30 + 16$, produces SS_{total}. The same applies to the *df* column.) When using ANOVA, you might start with a blank ANOVA summary table and then fill in the values as they are calculated. With this method, you are less likely to "get lost" in the analysis, wondering what to do next.

LEARNING CHECK

1. An analysis of variance produces $SS_{between\ treatments} = 40$ and $MS_{between\ treatments} = 10$. In this analysis, how many treatment conditions are being compared?

   a. 4

   b. 5

   c. 30

   d. 50

2. An analysis of variance is used to evaluate the mean differences for a research study comparing four treatment conditions with a separate sample of $n = 5$ in each treatment. The analysis produces $SS_{within\ treatments} = 32$, $SS_{between\ treatments} = 40$, and $SS_{total} = 72$. For this analysis, what is $MS_{within\ treatments}$?

   a. $\frac{32}{5}$

   b. $\frac{32}{4}$

   c. $\frac{32}{16}$

   d. $\frac{32}{20}$

3. In analysis of variance, an MS value is a measure of _____.

   a. variance

   b. average differences among means

   c. the total variability for the set of N scores

   d. the overall mean for the set of N scores

ANSWERS   1. B, 2. C, 3. A

12.4 | Examples of Hypothesis Testing and Effect Size with ANOVA

LEARNING OBJECTIVES

6. Define the *df* values for and *F*-ratio and use the *df* values, together with an alpha level, to locate the critical region in the distribution of *F*-ratios.

7. Conduct a complete ANOVA to evaluate the differences among a set of means and compute a measure of effect size to describe the mean differences.

8. Explain how the results from an ANOVA and measures of effect size are reported in the literature.

■ The Distribution of *F*-Ratios

In analysis of variance, the *F*-ratio is constructed so that the numerator and denominator of the ratio are measuring exactly the same variance when the null hypothesis is true (see Equation 12.2). In this situation, we expect the value of *F* to be around 1.00.

   If the null hypothesis is false, the *F*-ratio should be much greater than 1.00. The problem now is to define precisely which values are "around 1.00" and which are "much greater than 1.00." To answer this question, we need to look at all the possible F values that can be obtained when the null hypothesis is true—that is, the *distribution of F-ratios*.

Before we examine this distribution in detail, you should note two obvious characteristics:

1. Because F-ratios are computed from two variances (the numerator and denominator of the ratio), F values always are positive numbers. Remember that variance is always positive.

2. When H_0 is true, the numerator and denominator of the F-ratio are measuring the same variance. In this case, the two sample variances should be about the same size, so the ratio should be near 1. In other words, the distribution of F-ratios should pile up around 1.00.

With these two factors in mind, we can sketch the distribution of F-ratios. The distribution is cut off at zero (all positive values), piles up around 1.00, and then tapers off to the right (Figure 12.7). The exact shape of the F distribution depends on the degrees of freedom for the two variances in the F-ratio. You should recall that the precision of a sample variance depends on the number of scores or the degrees of freedom. In general, the variance for a large sample (large df) provides a more accurate estimate of the population variance. Because the precision of the MS values depends on df, the shape of the F distribution also depends on the df values for the numerator and denominator of the F-ratio. With very large df values, nearly all the F-ratios are clustered very near to 1.00. With the smaller df values, the F distribution is more spread out.

■ The *F* Distribution Table

For ANOVA, we expect F near 1.00 if H_0 is true. An F-ratio that is much larger than 1.00 is an indication that H_0 is not true. In the F distribution, we need to separate those values that are reasonably near 1.00 from the values that are significantly greater than 1.00. These critical values are presented in an F distribution table in Appendix B, page 647. A portion of the F distribution table is shown in Table 12.3. To use the table, you must know the df values for the F-ratio (numerator and denominator), and you must know the alpha level for the hypothesis test. It is customary for an F table to have the df

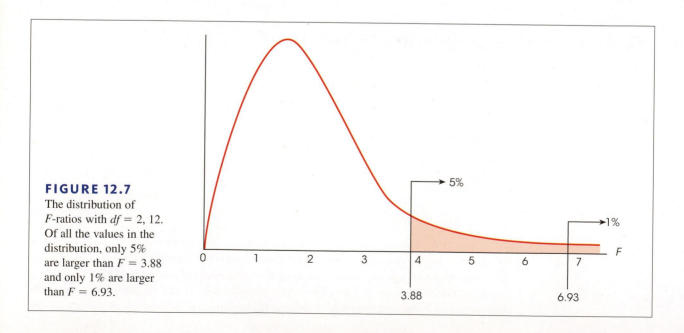

FIGURE 12.7

The distribution of F-ratios with $df = 2, 12$. Of all the values in the distribution, only 5% are larger than $F = 3.88$ and only 1% are larger than $F = 6.93$.

values for the numerator of the F-ratio printed across the top of the table. The df values for the denominator of F are printed in a column on the left-hand side. For the experiment we have been considering, the numerator of the F-ratio (between treatments) has $df = 2$, and the denominator of the F-ratio (within treatments) has $df = 12$. This F-ratio is said to have "degrees of freedom equal to 2 and 12." The degrees of freedom would be written as $df = 2, 12$. To use the table, you would first find $df = 2$ across the top of the table and $df = 12$ in the first column. When you line up these two values, they point to a pair of numbers in the middle of the table. These numbers give the critical cutoffs for $\alpha = .05$ and $\alpha = .01$. With $df = 2, 12$, for example, the numbers in the table are 3.88 and 6.93. Thus, only 5% of the distribution ($\alpha = .05$) corresponds to values greater than 3.88 and only 1% of the distribution ($\alpha = .01$) corresponds to values greater than 6.93 (see Figure 12.7).

TABLE 12.3

A portion of the F distribution table. Entries in roman type are critical values for the .05 level of significance, and bold type values are for the .01 level of significance. The critical values for $df = 2, 12$ have been highlighted (see text).

Degrees of Freedom: Denominator	Degrees of Freedom: Numerator					
	1	2	3	4	5	6
10	4.96	4.10	3.71	3.48	3.33	3.22
	10.04	**7.56**	**6.55**	**5.99**	**5.64**	**5.39**
11	4.84	3.98	3.59	3.36	3.20	3.09
	9.65	**7.20**	**6.22**	**5.67**	**5.32**	**5.07**
12	4.75	3.88	3.49	3.26	3.11	3.00
	9.33	**6.93**	**5.95**	**5.41**	**5.06**	**4.82**
13	4.67	3.80	3.41	3.18	3.02	2.92
	9.07	**6.70**	**5.74**	**5.20**	**4.86**	**4.62**
14	4.60	3.74	3.34	3.11	2.96	2.85
	8.86	**6.51**	**5.56**	**5.03**	**4.69**	**4.46**

In the experiment comparing driving performance under different telephone conditions, we obtained an F-ratio of 11.28. According to the critical cutoffs in Figure 12.7, this value is extremely unlikely (it is in the most extreme 1%). Therefore, we would reject H_0 with an α level of either .05 or .01, and conclude that the different telephone conditions significantly affect driving performance.

■ An Example of Hypothesis Testing and Effect Size with ANOVA

The Hypothesis Test   Although we have now seen all the individual components of ANOVA, the following example demonstrates the complete ANOVA process using the standard four-step procedure for hypothesis testing.

EXAMPLE 12.2

Over the years, students and teachers have developed a variety of strategies to help prepare for an upcoming test. But how do you know which is best? A partial answer to this question comes from a research study comparing three different strategies (Weinstein, McDermott, & Roediger, 2010). In the study, students read a passage knowing that they would be tested on the material. In one condition, participants simply reread the material to be tested. In a second condition, the students answered prepared comprehension questions about the material, and in a third condition, the students generated and answered their own questions. The results showed that answering comprehension questions significantly improved exam performance but it did not matter whether the students had to generate their own questions. The authors note that generating questions took twice as much time as simply answering questions but the extra time did not seem to have any effect on test performance.

The data in Table 12.4 are from an independent-measures study attempting to replicate the results from the Weinstein et al. study. We use an ANOVA is to determine whether there are any significant differences among the three study strategies.

Before we begin the hypothesis test, note that we have already computed several summary statistics for the data in Table 12.4. Specifically, the treatment totals (T) and SS values are shown for each sample, and the grand total (G) as well as N and ΣX^2 are shown for the entire set of data. Having these summary values simplifies the computations in the hypothesis test, and we suggest that you always compute these summary statistics before you begin an ANOVA.

TABLE 12.4

Test scores for students using three different study strategies.

Read and Reread	Read, then Answer Prepared Questions	Read, then Create and Answer Questions	
2	5	8	
3	9	6	$N = 18$
8	10	12	$G = 144$
6	13	11	$\Sigma X^2 = 1324$
5	8	11	
6	9	12	
$T = 30$	$T = 54$	$T = 60$	
$M = 5$	$M = 9$	$M = 10$	
$SS = 24$	$SS = 34$	$SS = 30$	

STEP 1   **State the hypotheses and select an alpha level.**

$$H_0: \mu_1 = \mu_2 = \mu_3 \quad \textbf{(There is no treatment effect.)}$$

$$H_1: \text{At least one of the treatment means is different.}$$

We will use $\alpha = .05$.

STEP 2   **Locate the critical region.**   We first must determine degrees of freedom for $MS_{\text{between treatments}}$ and $MS_{\text{within treatments}}$ (the numerator and denominator of the F-ratio), so we begin by analyzing the degrees of freedom. For these data, the total degrees of freedom are

$$df_{\text{total}} = N - 1 = 18 - 1 = 17$$

Often it is easier to post-pone finding the critical region until after Step 3, where you compute the df values as part of the calculations for the F-ratio.

Analyzing this total into two components, we obtain

$$df_{\text{between}} = k - 1 = 3 - 1 = 2$$

$$df_{\text{within}} = \Sigma df_{\text{inside each treatment}} = 5 + 5 + 5 = 15$$

The F-ratio for these data has $df = 2, 15$. The distribution of all the possible F-ratios with $df = 2, 15$ is presented in Figure 12.8. Note that F-ratios larger than 3.68 are extremely rare ($p < .05$) if H_0 is true and, therefore, form the critical region for the test.

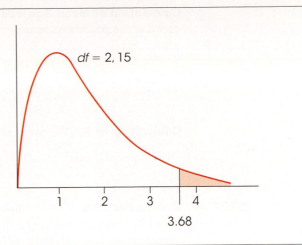

FIGURE 12.8
The distributions of *F*-ratios with $df = 2, 15$.
The critical boundary for $\alpha = .05$ is 3.68.

STEP 3 **Compute the *F*-ratio.**   The series of calculations for computing *F* is presented in Figure 12.4 and can be summarized as follows:

   a. Analyze the *SS* to obtain $SS_{between}$ and SS_{within}.

   b. Use the *SS* values and the *df* values (from step 2) to calculate the two variances, $MS_{between}$ and MS_{within}.

   c. Finally, use the two *MS* values (variances) to compute the *F*-ratio.

Analysis of *SS*.   First, we compute the total *SS* and then the two components, as indicated in Figure 12.5.

SS_{total} is simply the *SS* for the total set of $N = 18$ scores.

$$SS_{total} = \Sigma X^2 - \frac{G^2}{N}$$

$$= 1324 - 1152$$

$$= 172$$

SS_{within} combines the *SS* values from inside each of the treatment conditions.

$$SS_{within} = \Sigma SS_{inside\ each\ treatment} = 24 + 34 + 30 = 88$$

$SS_{between}$ measures the differences among the three treatment means (or treatment totals). Because we have already calculated SS_{total} and SS_{within}, the simplest way to obtain $SS_{between}$ is by subtraction (Equation 12.5).

$$SS_{between} = SS_{total} - SS_{within}$$

$$= 172 - 88$$

$$= 84$$

Calculation of Mean Squares.   Because we already found the *df* values (Step 2), we now can compute the variance or *MS* value for each of the two components.

$$MS_{between} = \frac{SS_{between}}{df_{between}} = \frac{84}{2} = 42$$

$$MS_{within} = \frac{SS_{within}}{df_{within}} = \frac{88}{15} = 5.87$$

Calculation of F.   We compute the *F*-ratio:

$$F = \frac{MS_{between}}{MS_{within}} = \frac{42}{5.87} = 7.16$$

STEP 4   **Make a decision.**   The *F* value we obtained, $F = 7.16$, is in the critical region (see Figure 12.8). It is very unlikely ($p < .05$) that we would obtain a value this large if H_0 is true. Therefore, we reject H_0 and conclude that there is a significant treatment effect.   ■

Example 12.2 demonstrated the complete, step-by-step application of the ANOVA procedure. There are two additional points that can be made using this example.

First, you should look carefully at the statistical decision. We have rejected H_0 and concluded that not all the treatments are the same. But we have not determined which ones are different. Is answering prepared questions different from making up and answering your own questions? Is answering prepared questions different from simply rereading? Unfortunately, these questions remain unanswered. We do know that at least one difference exists (we rejected H_0), but additional analysis is necessary to find out exactly where this difference is. We address this problem in Section 12.5.

Second, as noted earlier, all of the components of the analysis (the *SS*, *df*, *MS*, and *F*) can be presented together in one summary table. The summary table for the analysis in Example 12.2 is as follows:

Source	SS	df	MS	
Between treatments	84	2	42.00	$F = 7.16$
Within treatments	88	15	5.87	
Total	172	17		

Although these tables are very useful for organizing the components of an ANOVA, they are not commonly used in published reports. The current method for reporting the results from an ANOVA is presented on p. 389.

■ Measuring Effect Size for ANOVA

As we noted previously, a *significant* mean difference simply indicates that the difference observed in the sample data is very unlikely to have occurred just by chance. Thus, the term significant does not necessarily mean *large*, it simply means larger than expected by chance. To provide an indication of how large the effect actually is, it is recommended that researchers report a measure of effect size in addition to the measure of significance.

For ANOVA, the simplest and most direct way to measure effect size is to compute the percentage of variance accounted for by the treatment conditions. Like the r^2 value used to measure effect size for the *t* tests in Chapters 9, 10, and 11, this percentage measures how much of the variability in the scores is accounted for by the differences between treatments. For ANOVA, the calculation and the concept of the percentage of variance is extremely straightforward. Specifically, we determine how much of the total SS is accounted for by the $SS_{\text{between treatments}}$.

$$\text{The percentage of variance accounted for} = \frac{SS_{\text{between treatments}}}{SS_{\text{total}}} \qquad \text{(12.14)}$$

For the data in Example 12.2, we obtain

$$\text{The percentage of variance accounted for} = \frac{84}{172} = 0.488 \text{ (or 48.8\%)}$$

In published reports of ANOVA results, the percentage of variance accounted for by the treatment effect is usually called η^2 (the Greek letter *eta squared*) instead of using r^2. Thus, for the study in Example 12.2, $\eta^2 = 0.488$.

The following example is an opportunity for you to test your understanding of computing η^2 to measure effect size in ANOVA.

EXAMPLE 12.3   The ANOVA from an independent-measures study is summarized in the following table.

Source	SS	df	MS	
Between treatments	84	2	42	$F(2, 12) = 7.00$
Within treatments	72	12	6	
Total	156	14		

Compute η^2 to measure effect size for this study. You should find that $\eta^2 = 0.538$. ■

IN THE LITERATURE

Reporting the Results of Analysis of Variance

The APA format for reporting the results of ANOVA begins with a presentation of the treatment means and standard deviations in the narrative of the article, a table, or a graph. These descriptive statistics are not needed in the calculations of the actual ANOVA, but you can easily determine the treatment means from n and T ($M = T/n$) and the standard deviations from the SS values for each treatment [$s = \sqrt{SS/(n-1)}$]. Next, report the results of the ANOVA. For the study described in Example 12.1, the report might state the following:

The means and standard deviations are presented in Table 1. The analysis of variance indicates that there are significant differences among the three strategies for studying, $F(2, 15) = 7.16, p < .05, \eta^2 = 0.488$.

TABLE 1
Quiz scores following three different strategies for studying.

	Simply reread	Answer prepared questions	Create and answer your own questions
M	5.00	9.00	10.00
SD	2.19	2.61	2.45

Note how the F-ratio is reported. In this example, degrees of freedom for between and within treatments are $df = 2, 15$, respectively. These values are placed in parentheses immediately following the symbol F. Next, the calculated value for F is reported, followed by the probability of committing a Type I error (the alpha level) and the measure of effect size.

When an ANOVA is done using a computer program, the F-ratio is usually accompanied by an exact value for p. The data from Example 12.1 were analyzed using the SPSS program (see SPSS at the end of this chapter) and the computer output included a significance level of $p = .007$. Using the exact p value from the computer output, the research report would conclude, "The analysis of variance revealed significant differences among the three strategies for studying, $F(2, 15) = 7.16$, $p = .007$, $\eta^2 = 0.488$."

■ An Example with Unequal Sample Sizes

In the previous examples, all the samples were exactly the same size (equal ns). However, the formulas for ANOVA can be used when the sample size varies within an experiment. You also should note, however, that the general ANOVA procedure is most accurate when used to examine experimental data with equal sample sizes. Therefore, researchers generally try to plan experiments with equal ns. However, there are circumstances in which it is impossible or impractical to have an equal number of subjects in every treatment condition. In these situations, ANOVA still provides a valid test, especially when the samples are relatively large and when the discrepancy between sample sizes is not extreme.

The following example demonstrates an ANOVA with samples of different sizes.

EXAMPLE 12.4

A researcher is interested in the amount of homework required by different academic majors. Students were recruited from Biology, English, and Psychology to participant in the study. The researcher randomly selects one course that each student is currently taking and asks the student to record the amount of out-of-class work required each week for the course. The researcher used all of the volunteer participants, which resulted in unequal sample sizes. The data are summarized in Table 12.5.

TABLE 12.5

Average hours of homework per week for one course for students in three academic majors.

Biology	English	Psychology	
$n = 4$	$n = 10$	$n = 6$	$N = 20$
$M = 9$	$M = 13$	$M = 14$	$G = 250$
$T = 36$	$T = 130$	$T = 84$	$\Sigma X^2 = 3377$
$SS = 37$	$SS = 90$	$SS = 60$	

STEP 1 **State the hypotheses, and select the alpha level.**

$$H_0: \mu_1 = \mu_2 = \mu_3$$

H_1: At least one population is different.

$$\alpha = .05$$

STEP 2 **Locate the critical region.** To find the critical region, we first must determine the df values for the F-ratio:

$$df_{total} = N - 1 = 20 - 1 = 19$$
$$df_{between} = k - 1 = 3 - 1 = 2$$
$$df_{within} = N - k = 20 - 3 = 17$$

The F-ratio for these data has $df = 2, 17$. With $\alpha = .05$, the critical value for the F-ratio is 3.59.

STEP 3 **Compute the F-ratio.** First, compute the three SS values. As usual, SS_{total} is the SS for the total set of $N = 20$ scores, and SS_{within} combines the SS values from inside each of the treatment conditions.

$$SS_{total} = \Sigma X^2 - \frac{G^2}{N}$$

$$= 3377 - \frac{250^2}{20} \qquad SS_{within} = \Sigma SS_{\text{inside each treatment}}$$

$$= 3377 - 3125 \qquad\qquad = 37 + 90 + 60$$

$$= 252 \qquad\qquad\qquad = 187$$

$SS_{between}$ can be found by subtraction (Equation 12.5).

$$SS_{between} = SS_{total} - SS_{within}$$
$$= 252 - 187$$
$$= 65$$

Or, $SS_{between}$ can be calculated using the computation formula (Equation 12.7). If you use the computational formula, be careful to match each treatment total (T) with the appropriate sample size (n) as follows:

$$SS_{between} = \Sigma\frac{T^2}{n} - \frac{G^2}{N}$$

$$= \frac{36^2}{4} + \frac{130^2}{10} + \frac{84^2}{6} - \frac{250^2}{20}$$

$$= 324 + 1690 + 1176 - 3125$$

$$= 65$$

Finally, compute the MS values and the F-ratio:

$$MS_{between} = \frac{SS}{df} = \frac{65}{2} = 32.5$$

$$MS_{within} = \frac{SS}{df} = \frac{187}{17} = 11$$

$$F = \frac{MS_{between}}{MS_{within}} = \frac{32.5}{11} = 2.95$$

STEP 4 **Make a decision.** Because the obtained F-ratio is not in the critical region, we fail to reject the null hypothesis and conclude that there are no significant differences among the three populations of students in terms of the average amount of homework each week. ■

■ Assumptions for the Independent-Measures ANOVA

The independent-measures ANOVA requires the same three assumptions that were necessary for the independent-measures t hypothesis test:

1. The observations within each sample must be independent (see p. 244).
2. The populations from which the samples are selected must be normal.

3. The populations from which the samples are selected must have equal variances (homogeneity of variance).

Ordinarily, researchers are not overly concerned with the assumption of normality, especially when large samples are used, unless there are strong reasons to suspect the assumption has not been satisfied. The assumption of homogeneity of variance is an important one. If a researcher suspects it has been violated, it can be tested by Hartley's F-max test for homogeneity of variance (Chapter 10, p. 314).

Finally, if you suspect that one of the assumptions for the independent-measures ANOVA has been violated, you can still proceed by transforming the original scores to ranks and then using an alternative statistical analysis known as the Kruskal-Wallis test, which is designed specifically for ordinal data. The Kruskal-Wallis test is presented in Appendix E. The Kruskal-Wallis test also can be useful if large sample variance prevents the independent-measures ANOVA from producing a significant result.

LEARNING CHECK

1. A researcher reports an F-ratio with $df = 3, 36$ for an independent-measures experiment. How many treatment conditions were compared in this experiment?

   a. 3
   b. 4
   c. 36
   d. 39

2. The following table shows the results of an analysis of variance comparing three treatment conditions with a sample of $n = 7$ participants in each treatment. Note that several values are missing in the table. What is the missing value for SS_{total}?

   a. 16
   b. 23
   c. 29
   d. 56

Source	SS	df	MS	
Between	20	xx	xx	$F = $ xx
Within	xx	xx	2	
Total	xx	xx		

3. A research report concludes that there are significant differences among treatments, with "$F(2,27) = 8.62, p < .01, \eta^2 = 0.28$." If the same number of participants was used in all of the treatment conditions, then how many individuals were in each treatment?

   a. 7
   b. 9
   c. 10
   d. cannot determine without additional information

ANSWERS   **1.** B, **2.** D, **3.** C

12.5 | Post Hoc Tests

LEARNING OBJECTIVE   **9.** Describe the circumstances in which posttests are necessary and explain what the tests accomplish.

As noted earlier, the primary advantage of ANOVA (compared to *t* tests) is it allows researchers to test for significant mean differences when there are *more than two* treatment conditions. ANOVA accomplishes this feat by comparing all the individual mean differences simultaneously within a single test. Unfortunately, the process of combining several mean differences into a single test statistic creates some difficulty when it is time to interpret the outcome of the test. Specifically, when you obtain a significant *F*-ratio (reject H_0), it simply indicates that somewhere among the entire set of mean differences there is at least one that is statistically significant. In other words, the overall *F*-ratio only tells you that a significant difference exists; it does not tell exactly which means are significantly different and which are not.

In Example 12.2 we presented an independent-measures study using three samples to compare three strategies for studying in preparation for a test: rereading the material to be tested, answering prepared questions on the material, creating and answering your own questions. The three sample means were $M_1 = 5$, $M_2 = 9$, and $M_3 = 10$. In this study there are three mean differences:

1. There is a 4-point difference between M_1 and M_2.
2. There is a 1-point difference between M_2 and M_3.
3. There is a 5-point difference between M_1 and M_3.

The ANOVA used to evaluate these data produced a significant *F*-ratio indicating that at least one of the sample mean differences is large enough to satisfy the criterion of statistical significance. In this example, the 5-point difference is the biggest of the three and, therefore, it must indicate a significant difference between the first treatment and the third treatment ($\mu_1 \neq \mu_3$). But what about the 4-point difference? Is it also large enough to be significant? And what about the 1-point difference between M_2 and M_3? Is it also significant? The purpose of *post hoc tests* is to answer these questions.

DEFINITION | **Post hoc tests** (or **posttests**) are additional hypothesis tests that are done after an ANOVA to determine exactly which mean differences are significant and which are not.

As the name implies, post hoc tests are done after an ANOVA. More specifically, these tests are done after ANOVA when

1. You reject H_0 and
2. there are three or more treatments ($k \geq 3$).

Rejecting H_0 indicates that at least one difference exists among the treatments. If there are only two treatments, then there is no question about which means are different and, therefore, no need for posttests. However, with three or more treatments ($k \geq 3$), the problem is to determine exactly which means are significantly different.

■ Posttests and Type I Errors

In general, a post hoc test enables you to go back through the data and compare the individual treatments two at a time. In statistical terms, this is called making *pairwise comparisons*.

For example, with $k = 3$, we would compare μ_1 vs. μ_2, then μ_2 vs. μ_3, and then μ_1 vs. μ_3. In each case, we are looking for a significant mean difference. The process of conducting pairwise comparisons involves performing a series of separate hypothesis tests, and each of these tests includes the risk of a Type I error. As you do more and more separate tests, the risk of a Type I error accumulates and is called the *experimentwise alpha level* (see p. 370).

We have seen, for example, that a research study with three treatment conditions produces three separate mean differences, each of which could be evaluated using a post hoc test. If each test uses $\alpha = .05$, then there is a 5% risk of a Type I error for the first posttest, another 5% risk for the second test, and one more 5% risk for the third test. Although the probability of error is not simply the sum across the three tests, it should be clear that increasing the number of separate tests definitely increases the total, experimentwise probability of a Type I error.

Whenever you are conducting posttests, you must be concerned about the experimentwise alpha level. Statisticians have worked with this problem and have developed several methods for trying to control Type I errors in the context of post hoc tests. We will consider two alternatives.

■ Tukey's Honestly Significant Difference (HSD) Test

The first post hoc test we consider is *Tukey's HSD test*. We selected Tukey's HSD test because it is a commonly used test in psychological research. Tukey's test allows you to compute a single value that determines the minimum difference between treatment means that is necessary for significance. This value, called the *honestly significant difference*, or HSD, is then used to compare any two treatment conditions. If the mean difference exceeds Tukey's HSD, you conclude that there is a significant difference between the treatments. Otherwise, you cannot conclude that the treatments are significantly different. The formula for Tukey's HSD is

$$HSD = q\sqrt{\frac{MS_{within}}{n}} \tag{12.15}$$

The *q* value used in Tukey's HSD test is called a Studentized range statistic.

where the value of q is found in Table B.5 (Appendix B, p. 656), $MS_{within\ treatments}$ is the within-treatments variance from the ANOVA, and n is the number of scores in each treatment. Tukey's test requires that the sample size, n, be the same for all treatments. To locate the appropriate value of q, you must know the number of treatments in the overall experiment (k), the degrees of freedom for $MS_{within\ treatments}$ (the error term in the F-ratio), and you must select an alpha level (generally the same α used for the ANOVA).

EXAMPLE 12.5

To demonstrate the procedure for conducting post hoc tests with Tukey's HSD, we use the data from Example 12.2, which are summarized in Table 12.6. Note that the table displays summary statistics for each sample and the results from the overall ANOVA. With $k = 3$ treatments, $n = 6$, and $\alpha = .05$, you should find that the value of q for the test is $q = 3.67$ (see Table B.5). Therefore, Tukey's HSD is

$$HSD = q\sqrt{\frac{MS_{within}}{n}} = 3.67\sqrt{\frac{5.87}{6}} = 3.63$$

Thus, the mean difference between any two samples must be at least 3.63 to be significant. Using this value, we can make the following conclusions:

1. Treatment A is significantly different from treatment B ($M_A - M_B = 4.00$).
2. Treatment A is also significantly different from treatment C ($M_A - M_C = 5.00$).
3. Treatment B is not significantly different from treatment C ($M_B - M_C = 1.00$). ■

TABLE 12.6

Results from the research study in Example 12.2. Summary statistics are presented for each treatment along with the outcome from the ANOVA.

Treatment A Reread	Treatment B Prepared Questions	Treatment C Create Questions
$n = 6$	$n = 6$	$n = 6$
$T = 30$	$T = 54$	$T = 60$
$M = 5.00$	$M = 9.00$	$M = 10.00$

Source	SS	df	MS
Between	84	2	42
Within	88	15	5.87
Total	172	17	
Overall $F(2, 15) = 7.16$			

■ **The Scheffè Test**

Because it uses an extremely cautious method for reducing the risk of a Type I error, the *Scheffé test* has the distinction of being one of the safest of all possible post hoc tests (smallest risk of a Type I error). The Scheffé test uses an *F*-ratio to evaluate the significance of the difference between any two treatment conditions. The numerator of the *F*-ratio is an *MS* between treatments that is calculated using *only the two treatments you want to compare*. The denominator is the same MS_{within} that was used for the overall ANOVA. The "safety factor" for the Scheffé test comes from the following two considerations:

1. Although you are comparing only two treatments, the Scheffé test uses the value of *k* from the original experiment to compute *df* between treatments. Thus, *df* for the numerator of the *F*-ratio is $k - 1$.

2. The critical value for the Scheffé *F*-ratio is the same as was used to evaluate the *F*-ratio from the overall ANOVA. Thus, Scheffé requires that every posttest satisfy the same criterion that was used for the complete ANOVA. The following example uses the data from Example 12.2 (p. 385) to demonstrate the Scheffé posttest procedure.

EXAMPLE 12.6

Remember that the Scheffé procedure requires a separate $SS_{between}$, $MS_{between}$, and *F*-ratio for each comparison being made. Although Scheffé computes $SS_{between}$ using the regular computational formula (Equation 12.7), you must remember that all the numbers in the formula are entirely determined by the two treatment conditions being compared. We begin with the smallest mean difference, which involves comparing treatment B (with $T = 54$ and $n = 6$) and treatment C (with $T = 60$ and $n = 6$). The first step is to compute $SS_{between}$ for these two groups. In the formula for *SS*, notice that the grand total for the two groups is $G = 54 + 60 = 114$, and the total number of scores for the two groups is $N = 6 + 6 = 12$.

$$SS_{between} = \Sigma \frac{T^2}{n} - \frac{G^2}{N}$$
$$= \frac{(54)^2}{6} + \frac{(60)^2}{6} - \frac{(114)^2}{12}$$
$$= 486 + 600 - 1083$$
$$= 3$$

Although we are comparing only two groups, these two were selected from a study consisting of $k = 3$ samples. The Scheffé test uses the overall study to determine the degrees of freedom between treatments. Therefore, $df_{between} = 3 - 1 = 2$, and the MS between treatments is

$$MS_{between} = \frac{SS_{between}}{df_{between}} = \frac{3}{2} = 1.5$$

Finally, the Scheffé procedure uses the error term from the overall ANOVA to compute the F-ratio. In this case, $MS_{within} = 5.87$ with $df_{within} = 15$. Thus, the Scheffé test produces an F-ratio of

$$F = \frac{MS_{between}}{MS_{within}} = \frac{1.5}{5.87} = 0.26$$

With $df = 2$, 15 and $\alpha = .05$, the critical value for F is 3.68 (see Table B.4). Therefore, our obtained F-ratio is not in the critical region, and we conclude that these data show no significant difference between treatment B and treatment C.

The second largest mean difference involves treatment A ($T = 30$) vs. treatment B ($T = 54$). This time the data produce $SS_{between} = 48$, $MS_{between} = 24$, and $F(2, 15) = 4.09$ (check the calculations for yourself). Once again the critical value for F is 3.68, so we conclude that there is a significant difference between treatment A and treatment B.

The final comparison is treatment A ($M = 5$) vs. treatment C ($M = 10$). We have already found that the 4-point mean difference between A and B is significant, so the 5-point difference between A and C also must be significant. Thus, the Scheffé posttest indicates that both B and C (answering prepared questions and creating and answering your own questions) are significantly different from treatment A (simply rereading), but there is no significant difference between B and C. ∎

In this case, the two post-test procedures, Tukey's HSD and Scheffé, produce exactly the same results. You should be aware, however, that there are situations in which Tukey's test will find a significant difference but Scheffé will not. Again, the Scheffé test is one of the safest of the posttest techniques because it provides the greatest protection from Type I errors. To provide this protection, the Scheffé test simply requires a larger difference between sample means before you may conclude that the difference is significant.

LEARNING CHECK

1. Under what circumstances are posttests necessary?
    a. reject the null hypothesis with $k = 2$ treatments.
    b. reject the null hypothesis with $k > 2$ treatments.
    c. fail to reject the null hypothesis with $k = 2$ treatments.
    d. fail to reject the null hypothesis with $k > 2$ treatments.

2. Which of the following accurately describes the purpose of posttests?
    a. They determine which treatments are different.
    b. They determine how much difference there is between treatments.
    c. They determine whether a Type I Error was made in the ANOVA.
    d. They determine whether a complete ANOVA is justified.

3. An analysis of variance is used to compare three treatment conditions with a group of 12 participants in each treatment. If a Scheffé test is used to evaluate the mean difference between the first two treatments, then what are the *df* values for the Scheffé *F*-ratio?
   a. 1, 33
   b. 1, 22
   c. 2, 33
   d. 2, 22

ANSWERS   **1.** B, **2.** A, **3.** C

12.6 | More about ANOVA

10. Explain how the outcome of an ANOVA and measures of effect size are influenced by sample size, sample variance, and sample mean differences.

11. Explain the relationship between the independent-measures *t* test and an ANOVA when evaluating the difference between two means from an independent-measures study.

■ A Conceptual View of ANOVA

Because analysis of variance requires relatively complex calculations, students encountering this statistical technique for the first time often tend to be overwhelmed by the formulas and arithmetic and lose sight of the general purpose for the analysis. The following two examples are intended to minimize the role of the formulas and shift attention back to the conceptual goal of the ANOVA process.

EXAMPLE 12.7

The following data represent the outcome of an experiment using two separate samples to evaluate the mean difference between two treatment conditions. Take a minute to look at the data and, without doing any calculations, try to predict the outcome of an ANOVA for these values. Specifically, predict what values should be obtained for the between-treatments variance (*MS*) and the *F*-ratio. If you do not "see" the answer after 20 or 30 seconds, try reading the hints that follow the data.

Treatment I	Treatment II	
4	2	$N = 8$
0	1	$G = 16$
1	0	$\Sigma X^2 = 56$
3	5	
$T = 8$	$T = 8$	
$SS = 10$	$SS = 14$	

If you are having trouble predicting the outcome of the ANOVA, read the following hints, and then go back and look at the data.

Hint 1:   Remember: $SS_{between}$ and $MS_{between}$ provide a measure of how much difference there is between treatment conditions.

Hint 2:   Find the mean or total (T) for each treatment, and determine how much difference there is between the two treatments.

You should realize by now that the data have been constructed so that there is zero difference between treatments. The two sample means (and totals) are identical, so $SS_{between} = 0$, $MS_{between} = 0$, and the F-ratio is zero.   ■

Conceptually, the numerator of the F-ratio always measures how much difference exists between treatments. In Example 12.7, we constructed an extreme set of scores with zero difference. However, you should be able to look at any set of data and quickly compare the means (or totals) to determine whether there are big differences between treatments or small differences between treatments.

Being able to estimate the magnitude of between-treatment differences is a good first step in understanding ANOVA and should help you to predict the outcome of an ANOVA. However, the *between-treatment* differences are only one part of the analysis. You must also understand the *within-treatment* differences that form the denominator of the F-ratio. The following example is intended to demonstrate the concepts underlying SS_{within} and MS_{within}. In addition, the example should give you a better understanding of how the between-treatment differences and the within-treatment differences act together within the ANOVA.

EXAMPLE 12.8   The purpose of this example is to present a visual image for the concepts of between-treatments variability and within-treatments variability. In this example, we compare two hypothetical outcomes for the same experiment. In each case, the experiment uses two separate samples to evaluate the mean difference between two treatments. The following data represent the two outcomes, which we call experiment A and experiment B.

Experiment A Treatment		Experiment B Treatment	
I	II	I	II
8	12	4	12
8	13	11	9
7	12	2	20
9	11	17	6
8	13	0	16
9	12	8	18
7	11	14	3
$M = 8$	$M = 12$	$M = 8$	$M = 12$
$s = 0.82$	$s = 0.82$	$s = 6.35$	$s = 6.35$

The data from experiment A are displayed in a frequency distribution graph in Figure 12.9(a). Notice that there is a 4-point difference between the treatment means ($M_1 = 8$ and $M_2 = 12$). This is the *between-treatments* difference that contributes to the numerator of the F-ratio. Also notice that the scores in each treatment are clustered close around the mean, indicating that the variance inside each treatment is relatively small. This is the *within-treatments* variance that contributes to the denominator of the F-ratio. Finally, you should realize that it is easy to see the mean difference between the two samples. The fact that there is a clear mean difference between the two treatments is confirmed by computing the F-ratio for experiment A.

$$F = \frac{\text{between-treatments difference}}{\text{within-treatments differences}} = \frac{MS_{between}}{MS_{within}} = \frac{56}{0.667} = 83.96$$

An F-ratio of $F = 83.96$ is sufficient to reject the null hypothesis, so we conclude that there is a significant difference between the two treatments.

Now consider the data from experiment B, which are shown in Figure 12.9(b) and present a very different picture. This experiment has the same 4-point difference between treatment means that we found in experiment A ($M_1 = 8$ and $M_2 = 12$). However, for these data the scores in each treatment are scattered across the entire scale, indicating relatively large variance inside each treatment. In this case, the large variance within treatments overwhelms the relatively small mean difference between treatments. In the figure it is almost impossible to see the mean difference between treatments. The within-treatments variance appears in the bottom of the F-ratio and, for these data, the F-ratio confirms that there is no clear mean difference between treatments.

$$F = \frac{\text{between-treatments difference}}{\text{within-treatments differences}} = \frac{MS_{between}}{MS_{within}} = \frac{56}{40.33} = 1.39$$

For experiment B, the F-ratio is not large enough to reject the null hypothesis, so we conclude that there is no significant difference between the two treatments. Once again, the

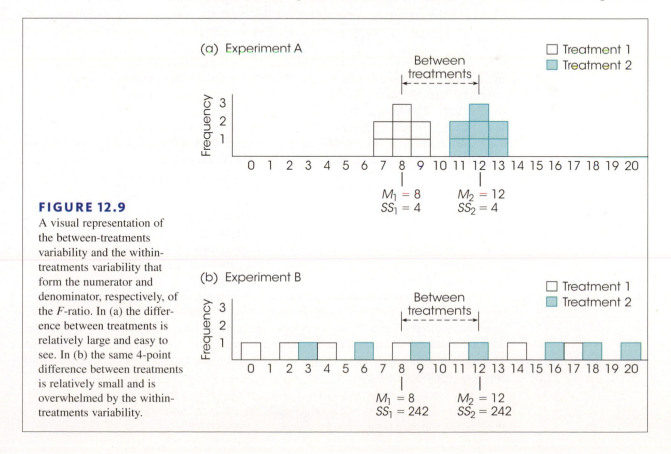

FIGURE 12.9

A visual representation of the between-treatments variability and the within-treatments variability that form the numerator and denominator, respectively, of the F-ratio. In (a) the difference between treatments is relatively large and easy to see. In (b) the same 4-point difference between treatments is relatively small and is overwhelmed by the within-treatments variability.

statistical conclusion is consistent with the appearance of the data in Figure 12.9(b). Looking at the figure, we see that the scores from the two samples appear to be intermixed randomly with no clear distinction between treatments.

As a final point, note that the denominator of the F-ratio, MS_{within}, is a measure of the variability (or variance) within each of the separate samples. As we have noted in previous chapters, high variability makes it difficult to see any patterns in the data. In Figure 12.9(a), the 4-point mean difference between treatments is easy to see because the sample variability is small. In Figure 12.9(b), the 4-point difference gets lost because the sample variability is large. In general, you can think of variance as measuring the amount of "noise" or "confusion" in the data. With large variance there is a lot of noise and confusion and it is difficult to see any clear patterns. ■

Although Examples 12.7 and 12.8 present somewhat simplified demonstrations with exaggerated data, the general point of the examples is to help you *see* what happens when you perform an ANOVA. Specifically:

1. The numerator of the F-ratio ($MS_{between}$) measures how much difference exists between the treatment means. The bigger the mean differences, the bigger the F-ratio.

2. The denominator of the F-ratio (MS_{within}) measures the variance of the scores inside each treatment; that is, the variance for each of the separate samples. In general, larger sample variance produces a smaller F-ratio.

We should note that the number of scores in the samples also influences the outcome of an ANOVA. As with most other hypothesis tests, if other factors are held constant, increasing the sample size tends to increase the likelihood of rejecting the null hypothesis. However, changes in sample size have little or no effect on measures of effect size such as η^2.

Finally, we should note that the problems associated with high variance often can be minimized by transforming the original scores to ranks and then conducting an alternative statistical analysis known as the Kruskal-Wallis test, which is designed specifically for ordinal data. The Kruskal-Wallis test is presented in Appendix E, which also discusses the general purpose and process of converting numerical scores into ranks. The Kruskal-Wallis test also can be used if the data violate one of the assumptions for the independent-measures *ANOVA*, which are outlined at the end of Section 12.4.

■ MS_{within} and Pooled Variance

You may have recognized that the two research outcomes presented in Example 12.8 are similar to those presented earlier in Example 10.8 in Chapter 10. Both examples are intended to demonstrate the role of variance in a hypothesis test. Both examples show that large values for sample variance can obscure any patterns in the data and reduce the potential for finding significant differences between means.

For the independent-measures t statistic in Chapter 10, the sample variance contributes directly to the standard error in the bottom of the t formula. Now, the sample variance contributes directly to the value of MS_{within} in the bottom of the F-ratio. In the t-statistic and in the F-ratio the variances from the separate samples are pooled together to create one average value for sample variance. For the independent-measures t statistic, we pooled two samples together to compute

$$\text{pooled variance} = s_p^2 = \frac{SS_1 + SS_2}{df_1 + df_2}$$

Now, in ANOVA, we are combining two or more samples to calculate

$$MS_{within} = \frac{SS_{within}}{df_{within}} = \frac{\Sigma SS}{\Sigma df} = \frac{SS_1 + SS_2 + SS_3 + \ldots}{df_1 + df_2 + df_3 + \ldots}$$

Notice that the concept of pooled variance is the same whether you have exactly two samples or more than two samples. In either case, you simply add the SS values and divide by the sum of the df values. The result is an average of all the different sample variances.

■ The Relationship between ANOVA and *t* Tests

The relationship between pooled variance and MS_{within} is only part of the general relationship between the independent-measures t test and the corresponding ANOVA. When you are evaluating the mean difference from an independent-measures study comparing only two treatments (two separate samples), you can use either an independent-measures t test (Chapter 10) or the ANOVA presented in this chapter. In practical terms, it makes no difference which you choose. These two statistical techniques always result in the same statistical decision. In fact the two methods use many of the same calculations and are very closely related in several other respects. The basic relationship between t statistics and F-ratios can be stated in an equation:

$$F = t^2$$

This relationship can be explained by first looking at the structure of the formulas for F and t. The t statistic compares *distances*: the distance between two sample means (numerator) and the distance computed for the standard error (denominator). The F-ratio, on the other hand, compares *variances*. You should recall that variance is a measure of squared distance. Hence, the relationship: $F = t^2$.

There are several other points to consider in comparing the t statistic to the F-ratio.

1. It should be obvious that you will be testing the same hypotheses whether you choose a t test or an ANOVA. With only two treatments, the hypotheses for either test are

$$H_0: \mu_1 = \mu_2$$

$$H_1: \mu_1 \neq \mu_2$$

2. The degrees of freedom for the t statistic and the df for the denominator of the F-ratio (df_{within}) are identical. For example, if you have two samples, each with six scores, the independent-measures t statistic will have $df = 10$, and the F-ratio will have $df = 1, 10$. In each case, you are adding the df from the first sample ($n - 1$) and the df from the second sample ($n - 1$).

3. The distribution of t and the distribution of F-ratios match perfectly if you take into consideration the relationship $F = t^2$. Consider the t distribution with $df = 18$ and the corresponding F distribution with $df = 1, 18$ that are presented in Figure 12.10. Notice the following relationships:

    a. If each of the t values is squared, then all of the negative values become positive. As a result, the whole left-hand side of the t distribution (below zero) will be flipped over to the positive side. This creates an asymmetrical, positively skewed distribution—that is, the F distribution.

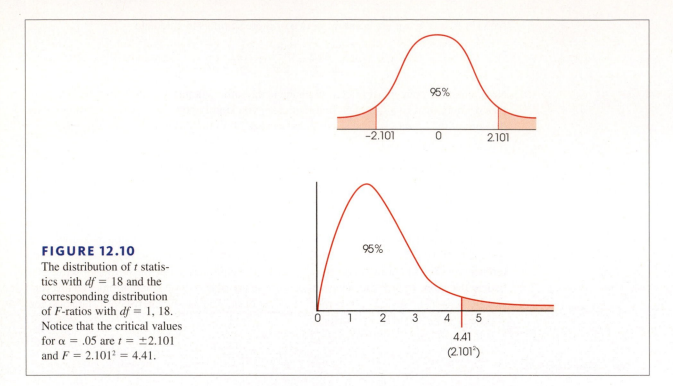

FIGURE 12.10

The distribution of t statistics with $df = 18$ and the corresponding distribution of F-ratios with $df = 1, 18$. Notice that the critical values for $\alpha = .05$ are $t = \pm 2.101$ and $F = 2.101^2 = 4.41$.

b. For $\alpha = .05$, the critical region for t is determined by values greater than $+2.101$ or less than -2.101. When these boundaries are squared, you get $\pm 2.1012 = 4.41$

Notice that 4.41 is the critical value for $\alpha = .05$ in the F distribution. Any value that is in the critical region for t will end up in the critical region for F-ratios after it is squared.

LEARNING CHECK

1. If an analysis of variance is used for the following data, what would be the effect of changing the value of SS_2 to 100?
   a. increase SS_{within} and increase the size of the F-ratio
   b. increase SS_{within} and decrease the size of the F-ratio
   c. decrease SS_{within} and increase the size of the F-ratio
   d. decrease SS_{within} and decrease the size of the F-ratio

Sample Data	
$M_1 = 15$	$M_2 = 25$
$SS_1 = 90$	$SS_2 = 70$

2. For an ANOVA, how does an increase in the sample sizes influence the likelihood of rejecting the null hypothesis and measures of effect size?
   a. both will increase.
   b. both will decrease
   c. the likelihood of rejecting H_0 will increase but there will be little or no effect on measures of effect size.
   d. the likelihood of rejecting H_0 will decrease but there will be little or no effect on measures of effect size.

3. An independent-measures t test produced a t statistic with $df = 20$. If the same data had been evaluated with an analysis of variance, what would be the df values for the F-ratio?

   a. 1, 19
   b. 1, 20
   c. 2, 19
   d. 2, 20

ANSWERS   **1.** B, **2.** C, **3.** B

SUMMARY

1. Analysis of variance (ANOVA) is a statistical technique that is used to test for mean differences among two or more treatment conditions. The null hypothesis for this test states that in the general population there are no mean differences among the treatments. The alternative states that at least one mean is different from another.

2. The test statistic for ANOVA is a ratio of two variances called an F-ratio. The variances in the F-ratio are called mean squares, or MS values. Each MS is computed by

$$MS = \frac{SS}{df}$$

3. For the independent-measures ANOVA, the F-ratio is

$$F = \frac{MS_{between}}{MS_{within}}$$

The $MS_{between}$ measures differences between the treatments by computing the variability of the treatment means or totals. These differences are assumed to be produced by
   a. treatment effects (if they exist) or
   b. differences resulting from chance.

The MS_{within} measures variability inside each of the treatment conditions. Because individuals inside a treatment condition are all treated exactly the same, any differences within treatments cannot be caused by treatment effects. Thus, the within-treatments MS is produced only by differences caused by chance. With these factors in mind, the F-ratio has the following structure:

$$F = \frac{\text{treatment effect} + \text{differences due to chance}}{\text{differences due to chance}}$$

When there is no treatment effect (H_0 is true), the numerator and the denominator of the F-ratio are measuring the same variance, and the obtained ratio should be near 1.00. If there is a significant treatment effect, the numerator of the ratio should be larger than the denominator, and the obtained F value should be much greater than 1.00.

4. The formulas for computing each SS, df, and MS value are presented in Figure 12.11, which also shows the general structure for the ANOVA.

5. The F-ratio has two values for degrees of freedom, one associated with the MS in the numerator and one associated with the MS in the denominator. These df values are used to find the critical value for the F-ratio in the F distribution table.

6. Effect size for the independent-measures ANOVA is measured by computing eta squared, the percentage of variance accounted for by the treatment effect.

$$\eta^2 = \frac{SS_{between}}{SS_{between} + SS_{within}}$$
$$= \frac{SS_{between}}{SS_{total}}$$

7. When the decision from an ANOVA is to reject the null hypothesis and when the experiment has more than two treatment conditions, it is necessary to continue the analysis with a post hoc test, such as Tukey's HSD test or the Scheffé test. The purpose of these tests is to determine exactly which treatments are significantly different and which are not.

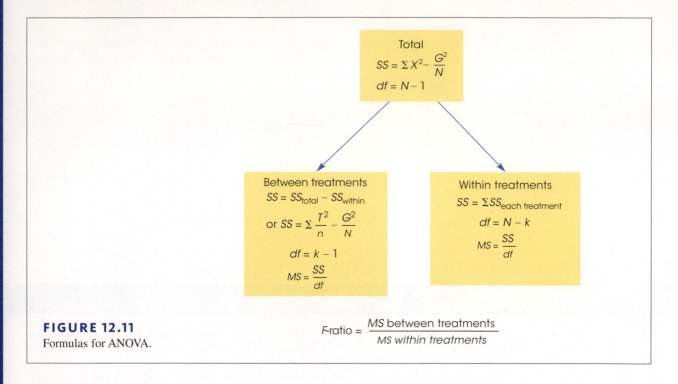

FIGURE 12.11
Formulas for ANOVA.

$$F\text{-ratio} = \frac{MS \text{ between treatments}}{MS \text{ within treatments}}$$

KEY TERMS

analysis of variance (ANOVA) (366)

factor (367)

levels (368)

experimentwise alpha level (370)

between-treatments variance (373)

treatment effect (373)

within-treatments variance (373)

F-ratio (373)

error term (374)

ANOVA summary table (380)

mean square (MS) (381)

distribution of F-ratios (384)

eta squared (η^2) (389)

post hoc tests (393)

pairwise comparisons (393)

Tukey's HSD test (394)

Scheffé test (395)

Kruskal-Wallis test (400)

SPSS®

General instructions for using SPSS are presented in Appendix D. Following are detailed instructions for using SPSS to perform the **Single-Factor, Independent-Measures Analysis of Variance (ANOVA)** presented in this chapter.

Data Entry

1. The scores are entered in a *stacked format* in the data editor, which means that all the scores from all of the different treatments are entered in a single column (VAR00001). Enter the scores for treatment #2 directly beneath the scores from treatment #1 with no gaps or extra spaces. Continue in the same column with the scores from treatment #3, and so on.

2. In the second column (VAR00002), enter a number to identify the treatment condition for each score. For example, enter a 1 beside each score from the first treatment, enter a 2 beside each score from the second treatment, and so on.

Data Analysis

1. Click **Analyze** on the tool bar, select **Compare Means**, and click on **One-Way ANOVA**.
2. Highlight the column label for the set of scores (VAR0001) in the left box and click the arrow to move it into the **Dependent List** box.
3. Highlight the label for the column containing the treatment numbers (VAR0002) in the left box and click the arrow to move it into the **Factor** box.
4. If you want descriptive statistics for each treatment, click on the **Options** box, select **Descriptives**, and click **Continue**.
5. Click **OK**.

SPSS Output

We used the SPSS program to analyze the data from the study comparing studying strategies in Example 12.2 and the program output is shown in Figure 12.12. The output begins with a table showing descriptive statistics (number of scores, mean, standard deviation, standard error for the mean, a 95% confidence interval for the mean, maximum, and minimum scores) for each sample. The second part of the output presents a summary table showing the results from the ANOVA.

Descriptives

VAR00001

	N	Mean	Std. Deviation	Std. Error	95% Confidence Interval for Mean Lower Bound	95% Confidence Interval for Mean Upper Bound	Minimum	Maximum
1.00	6	5.0000	2.19089	.89443	2.7008	7.2992	2.00	8.00
2.00	6	9.0000	2.60768	1.06458	6.2634	11.7366	5.00	13.00
3.00	6	10.0000	2.44949	1.00000	7.4294	12.5706	6.00	12.00
Total	18	8.0000	3.18082	.74973	6.4182	9.5818	2.00	13.00

ANOVA

VAR00001

	Sum of Squares	df	Mean Square	F	Sig.
Between Groups	84.000	2	42.000	7.159	.007
Within Groups	88.000	15	5.867		
Total	172.000	17			

FIGURE 12.12
The SPSS output from the ANOVA for the studying strategy experiment in Example 12.2.

FOCUS ON PROBLEM SOLVING

1. It can be helpful to compute all three SS values separately, and then check to verify that the two components (between and within) add up to the total. However, you can greatly simplify the calculations if you simply find SS_{total} and $SS_{within \, treatments}$, then obtain $SS_{between \, treatments}$ by subtraction.

2. Remember that an F-ratio has two separate values for df: a value for the numerator and one for the denominator. Properly reported, the $df_{between}$ value is stated first. You will need both df values when consulting the F distribution table for the critical F value. You should recognize immediately that an error has been made if you see an F-ratio reported with a single value for df.

3. When you encounter an F-ratio and its df values reported in the literature, you should be able to reconstruct much of the original experiment. For example, if you see "$F(2, 36) = 4.80$," you should realize that the experiment compared $k = 3$ treatment groups (because $df_{between} = k - 1 = 2$), with a total of $N = 39$ subjects participating in the experiment (because $df_{within} = N - k = 36$).

DEMONSTRATION 12.1

ANALYSIS OF VARIANCE

A human factors psychologist studied three computer keyboard designs. Three samples of individuals were given material to type on a particular keyboard, and the number of errors committed by each participant was recorded. The data are as follows:

Keyboard A	Keyboard B	Keyboard C	
0	6	6	$N = 15$
4	8	5	$G = 60$
0	5	9	$\Sigma X^2 = 356$
1	4	4	
0	2	6	
$T = 5$	$T = 25$	$T = 30$	
$SS = 12$	$SS = 20$	$SS = 14$	

Are these data sufficient to conclude that there are significant differences in typing performance among the three keyboard designs?

STEP 1 **State the hypotheses, and specify the alpha level**   The null hypothesis states that there is no difference among the keyboards in terms of number of errors committed. In symbols, we would state

$$H_0: \mu_1 = \mu_2 = \mu_3 \quad \text{(type of keyboard used has no effect.)}$$

As noted previously in this chapter, there are a number of possible statements for the alternative hypothesis. Here we state the general alternative hypothesis:

$$H_1: \quad \text{At least one of the treatment means is different.}$$

We will set alpha at $\alpha = .05$.

STEP 2   **Locate the critical region**   To locate the critical region, we must obtain the values for $df_{between}$ and df_{within}.

$$df_{between} = k - 1 = 3 - 1 = 2$$

$$df_{within} = N - k = 15 - 3 = 12$$

The F-ratio for this problem has $df = 2, 12$. Consult the F-distribution table for $df = 2$ in the numerator and $df = 12$ in the denominator. The critical F value for $\alpha = .05$ is $F = 3.88$. The obtained F-ratio must exceed this value to reject H_0.

STEP 3   **Perform the analysis**   The analysis involves the following steps:

1. Perform the analysis of SS.

2. Perform the analysis of df.

3. Calculate mean squares.

4. Calculate the F-ratio.

Perform the Analysis of SS   We will compute SS_{total} followed by its two components.

$$SS_{total} = \Sigma X^2 - \frac{G^2}{N} = 356 - \frac{60^2}{15} = 356 - \frac{3600}{15}$$

$$= 356 - 240 = 116$$

$$SS_{within} = \Sigma SS_{inside\ each\ treatment}$$
$$= 12 + 20 + 14$$
$$= 46$$

$$SS_{between} = \Sigma \frac{T^2}{n} - \frac{G^2}{N}$$
$$= \frac{5^2}{5} + \frac{25^2}{5} + \frac{30^2}{5} - \frac{60^2}{15}$$
$$= \frac{25}{5} + \frac{625}{5} + \frac{900}{5} - \frac{3600}{15}$$
$$= 5 + 125 + 180 - 240$$
$$= 70$$

Analyze Degrees of Freedom   We will compute df_{total}. Its components, $df_{between}$ and df_{within}, were previously calculated (step 2).

$$df_{total} = N - 1 = 15 - 1 = 14$$
$$df_{between} = 2$$
$$df_{within} = 12$$

Calculate the MS Values   The values for $MS_{between}$ and MS_{within} are determined.

$$MS_{between} = \frac{SS_{between}}{df_{between}} = \frac{70}{2} = 35$$

$$MS_{within} = \frac{SS_{within}}{df_{within}} = \frac{46}{12} = 3.83$$

Compute the *F*-Ratio   Finally, we can compute *F*.

$$F = \frac{MS_{between}}{MS_{within}} = \frac{35}{3.83} = 9.14$$

STEP 4   **Make a decision about H_0, and state a conclusion**   The obtained *F* of 9.14 exceeds the critical value of 3.88. Therefore, we can reject the null hypothesis. The type of keyboard used has a significant effect on the number of errors committed, $F(2, 12) = 9.14, p < .05$. The following table summarizes the results of the analysis:

Source	SS	df	MS	
Between treatments	70	2	35	F = 9.14
Within treatments	46	12	3.83	
Total	116	14		

DEMONSTRATION 12.2

COMPUTING EFFECT SIZE FOR ANALYSIS OF VARIANCE

We will compute eta squared (η^2), the percentage of variance explained, for the data that were analyzed in Demonstration 12.1. The data produced a between-treatments *SS* of 70 and a total *SS* of 116. Thus,

$$\eta^2 = \frac{SS_{between}}{SS_{total}} = \frac{70}{116} = 0.60 \ (\text{or } 60\%)$$

PROBLEMS

1. Explain why the *F*-ratio is expected to be near 1.00 when the null hypothesis is true.

2. Describe the similarities between an *F*-ratio and a *t* statistic.

3. Why should you use ANOVA instead of several *t* tests to evaluate mean differences when an experiment consists of three or more treatment conditions?

4. Calculate SS_{total}, $SS_{between}$, and SS_{within} for the following set of data:

Treatment 1	Treatment 2	Treatment 3	
$n = 10$	$n = 10$	$n = 10$	$N = 30$
$T = 10$	$T = 20$	$T = 30$	$G = 60$
$SS = 27$	$SS = 16$	$SS = 23$	$\Sigma X^2 = 206$

5. A researcher uses an ANOVA to compare three treatment conditions with a sample of $n = 8$ in each treatment. For this analysis, find df_{total}, $df_{between}$, and df_{within}.

6. A researcher reports an *F*-ratio with $df_{between} = 2$ and $df_{within} = 30$ for an independent-measures ANOVA.
   a. How many treatment conditions were compared in the experiment?
   b. How many subjects participated in the experiment?

7. A researcher reports an *F*-ratio with $df = 2, 27$ from an independent-measures research study.
   a. How many treatment conditions were compared in the study?
   b. What was the total number of participants in the study?

8. A research report from an independent-measures study states that there are significant differences between treatments, $F(3, 48) = 2.95, p < .05$.
   a. How many treatment conditions were compared in the study?
   b. What was the total number of participants in the study?

9. The following values are from an independent-measures study comparing three treatment conditions.

Treatment		
I	II	III
$n = 10$	$n = 10$	$n = 10$
$SS = 63$	$SS = 66$	$SS = 87$

a. Compute the variance for each sample.
b. Compute MS_{within}, which would be the denominator of the F-ratio for an ANOVA. Because the samples are all the same size, you should find that MS_{within} is equal to the average of the three sample variances.

10. A researcher conducts an experiment comparing four treatment conditions with a separate sample of $n = 6$ in each treatment. An ANOVA is used to evaluate the data, and the results of the ANOVA are presented in the following table. Complete all missing values in the table. *Hint:* Begin with the values in the df column.

Source	SS	df	MS	
Between treatments	___	___	___	$F =$ ___
Within treatments	___	___	2	
Total	58	___		

11. The following summary table presents the results from an ANOVA comparing four treatment conditions with $n = 10$ participants in each condition. Complete all missing values. (*Hint:* Start with the df column.)

Source	SS	df	MS	
Between treatments	___	___	10	$F =$ ___
Within treatments	___	___	___	
Total	174	___		

12. A developmental psychologist is examining the development of language skills from age 2 to age 4. Three different groups of children are obtained, one for each age, with $n = 18$ children in each group. Each child is given a language-skills assessment test. The resulting data were analyzed with an ANOVA to test for mean differences between age groups. The results of the ANOVA are presented in the following table. Fill in all missing values.

Source	SS	df	MS	
Between treatments	48	___	___	$F =$ ___
Within treatments	___	___	___	
Total	252	___		

13. The following data were obtained from an independent-measures research study comparing three treatment conditions. Use an ANOVA with $\alpha = .05$ to determine whether there are any significant mean differences among the treatments.

Treatment		
I	II	III
5	2	7
1	6	3
2	2	2
3	3	4
0	5	5
1	3	2
2	0	4
2	3	5

14. The following data were obtained from an independent-measures research study comparing three treatment conditions. Use an ANOVA with $\alpha = .05$ to determine whether there are any significant mean differences among the treatments.

Treatment			
I	II	III	
$n = 8$	$n = 6$	$n = 4$	$N = 18$
$T = 16$	$T = 24$	$T = 32$	$G = 72$
$SS = 40$	$SS = 24$	$SS = 16$	$\Sigma X^2 = 464$

15. A research study comparing three treatment conditions produces $T = 20$ with $n = 4$ for the first treatment, $T = 10$ with $n = 5$ for the second treatment, and $T = 30$ with $n = 6$ for the third treatment. Calculate $SS_{between\ treatments}$ for these data.

16. Several factors influence the size of the F-ratio. For each of the following, indicate whether it would influence the numerator or the denominator of the F-ratio, and indicate whether the size of the F-ratio would increase or decrease.
a. Increase the differences between the sample means.
b. Increase the sample variances.

17. A researcher used ANOVA and computed an F–ratio for the following data.

Treatments		
I	II	III
$n = 10$	$n = 10$	$n = 10$
$M = 20$	$M = 28$	$M = 35$
$SS = 105$	$SS = 191$	$SS = 180$

a. If the mean for treatment III were changed to $M = 25$, what would happen to the size of the F-ratio (increase or decrease)? Explain your answer.

b. If the SS for treatment I were changed to $SS = 1400$, what would happen to the size of the F-ratio (increase or decrease)? Explain your answer.

18. The following data were obtained from an independent-measures study comparing three treatment conditions.

	Treatment		
I	**II**	**III**	
4	3	8	$N = 12$
3	1	4	$G = 48$
5	3	6	$\Sigma X^2 = 238$
4	1	6	
$M = 4$	$M = 2$	$M = 6$	
$T = 16$	$T = 8$	$T = 24$	
$SS = 2$	$SS = 4$	$SS = 8$	

a. Calculate the sample variance for each of the three samples.

b. Use an ANOVA with $\alpha = .05$ to determine whether there are any significant differences among the three treatment means.

19. For the preceding problem you should find that there are significant differences among the three treatments. One reason for the significance is that the sample variances are relatively small. To create the following data, we kept the same sample means that appeared in problem 8 but increased the SS values within each sample.

	Treatment		
I	**II**	**III**	
4	4	9	$N = 12$
2	0	3	$G = 48$
6	3	6	$\Sigma X^2 = 260$
4	1	6	
$M = 4$	$M = 2$	$M = 6$	
$T = 16$	$T = 8$	$T = 24$	
$SS = 8$	$SS = 10$	$SS = 18$	

a. Calculate the sample variance for each of the three samples. Describe how these sample variances compare with those from problem 8.

b. Predict how the increase in sample variance should influence the outcome of the analysis. That is, how will the F-ratio for these data compare with the value obtained in problem 8?

c. Use an ANOVA with $\alpha = .05$ to determine whether there are any significant differences

among the three treatment means. (Does your answer agree with your prediction in part b?)

20. The following data summarize the results from an independent-measures study comparing three treatment conditions.

$M = 2$	$M = 3$	$M = 4$
$n = 10$	$n = 10$	$n = 10$
$T = 20$	$T = 30$	$T = 40$
$s^2 = 2.67$	$s^2 = 2.00$	$s^2 = 1.33$

a. Use an ANOVA with $\alpha = .05$ to determine whether there are any significant differences among the three treatment means. Note: Because the samples are all the same size, MS_{within} is the average of the three sample variances.

b. Calculate η^2 to measure the effect size for this study.

21. To create the following data we started with the same sample means and variances that appeared in problem 20 but increased the sample size to $n = 25$.

$M = 2$	$M = 3$	$M = 4$
$n = 25$	$n = 25$	$n = 25$
$T = 50$	$T = 75$	$T = 100$
$s^2 = 2.67$	$s^2 = 2.00$	$s^2 = 1.33$

a. Predict how the increase in sample size should affect the F-ratio for these data compared to the F-ratio in problem 20. Use an ANOVA to check your prediction. Note: Because the samples are all the same size, MS_{within} is the average of the three sample variances.

b. Predict how the increase in sample size should affect the value of η^2 for these data compared to the η^2 in problem 10. Calculate η^2 to check your prediction.

22. The following values are from an independent-measures study comparing three treatment conditions.

	Treatment	
I	**II**	**III**
$n = 8$	$n = 8$	$n = 8$
$SS = 42$	$SS = 28$	$SS = 98$

a. Compute the variance for each sample.

b. Compute MS_{within}, which would be the denominator of the F-ratio for an ANOVA. Because the samples are all the same size, you should find that MS_{within} is equal to the average of the three sample variances.

23. An ANOVA produces an F-ratio with $df = 1, 34$. Could the data have been analyzed with a t test? What would be the degrees of freedom for the t statistic?

24. The following scores are from an independent-measures study comparing two treatment conditions.
    a. Use an independent-measures t test with $\alpha = .05$ to determine whether there is a significant mean difference between the two treatments.

    b. Use an ANOVA with $\alpha = .05$ to determine whether there is a significant mean difference between the two treatments. You should find that $F = t^2$.

Treatment		
I	II	
1	12	
6	5	$N = 8$
6	6	$G = 48$
3	9	$\Sigma X^2 = 368$

Repeated-Measures Analysis of Variance

Tools You Will Need

The following items are considered essential background material for this chapter. If you doubt your knowledge of any of these items, you should review the appropriate chapter or section before proceeding.

- Independent-measures analysis of variance (Chapter 12)
- Repeated-measures designs (Chapter 11)
- Individual differences

© Deborah Batt

PREVIEW

Suppose that you were offered a choice between receiving $1000 in 5 years or a smaller amount today. How much would you be willing to take today to avoid waiting 5 years to get the full $1000 payment?

The general result for this kind of decision is that the longer the $1000 payment is delayed, the smaller the amount that people will accept today. For example, you may be willing to take $300 today rather than waiting 5 years for the $1000. However, you might be willing to settle for $100 today if you have to wait 10 years for the $1000. This phenomenon is known as delayed discounting because people discount the value of a future reward depending on how long it is delayed (Green, Fry, & Myerson, 1994). In a typical study examining delayed discounting, people are asked to place a value on a future reward for several different delay periods. For example, how much would you accept today instead of waiting for a future reward of $1000 if you had to wait 1 month before receiving the payment? How about waiting 6 months? 12 months? 24 months? 60 months?

Typical results for a sample of college students are shown in Figure 13.1. Note that the average value declines regularly as the delay period increases. The statistical question is whether the mean differences from one delay period to another are significant.

You should recognize that evaluating mean differences for more than two sample means is a job for analysis of variance. However, the discounting study is a repeated-measures design with five scores for each individual, and the ANOVA introduced in Chapter 12 is intended for independent-measures studies. Once again, a new hypothesis test is needed.

In this chapter we introduce the *repeated-measures ANOVA*. As the name implies, this new procedure is used to evaluate the differences between two or more sample means obtained from a repeated-measures research study. As you will see, many of the notational symbols and computations are the same as those used for the independent-measures ANOVA. In fact, your best preparation for this chapter is a good understanding of the basic ANOVA procedure presented in Chapter 12.

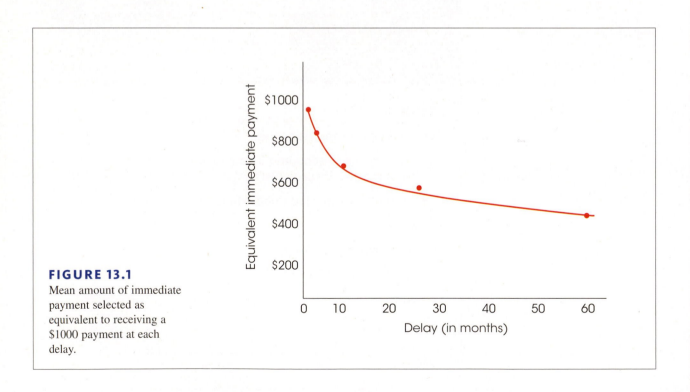

FIGURE 13.1

Mean amount of immediate payment selected as equivalent to receiving a $1000 payment at each delay.

13.1 | Overview of the Repeated-Measures ANOVA

1. Describe the *F*-ratio for the repeated-measures ANOVA and explain how it is related to the *F*-ratio for an independent-measures ANOVA.

In the preceding chapter we introduced analysis of variance (ANOVA) as a hypothesis-testing procedure for evaluating differences among two or more sample means. The specific advantage of ANOVA, especially in contrast to *t* tests, is that ANOVA can be used to evaluate the significance of mean differences in situations in which there are more than two sample means being compared. However, the presentation of ANOVA in Chapter 12 was limited to single-factor, independent-measures research designs. Recall that *single factor* indicates that the research study involves only one independent variable (or only one quasi-independent variable), and the term *independent-measures* indicates that the study uses a separate sample for each of the different treatment conditions being compared.

In this chapter we extend the ANOVA procedure to single-factor, repeated-measures designs. The defining characteristic of a repeated-measures design is that one group of individuals participates in all of the different treatment conditions. The repeated-measures ANOVA is used to evaluate mean differences in two general research situations:

1. An experimental study in which the researcher manipulates an independent variable to create two or more treatment conditions, with the same group of individuals tested in all of the conditions.

2. A nonexperimental study in which the same group of individuals is simply observed at two or more different times.

Examples of these two research situations are presented in Table 13.1. Table 13.1(a) shows data from a study in which the researcher changes the type of distraction to create three treatment conditions. One group of participants is then tested in all three conditions. In this study, the factor being examined is the type of distraction.

TABLE 13.1

Two sets of data representing typical examples of single-factor, repeated-measures research designs.

(a) Data from an experimental study evaluating the effects of different types of distraction on the performance of a visual detection task.

	Visual Detection Scores		
Participant	No Distraction	Visual Distraction	Auditory Distraction
A	47	22	41
B	57	31	52
C	38	18	40
D	45	32	43

(b) Data from a nonexperimental design evaluating the effectiveness of a clinical therapy for treating depression.

	Depression Scores		
Participant	Before Therapy	Immediately After Therapy	6-Month Follow-Up
A	71	53	55
B	62	45	44
C	82	56	61
D	77	50	46
E	81	54	55

Table 13.1(b) shows a study in which a researcher observes depression scores for the same group of individuals at three different times. In this study, the time of measurement is the factor being examined. Another common example of this type of design is found in developmental psychology when the participants' age is the factor being studied. For example, a researcher could study the development of vocabulary skill by measuring vocabulary for a sample of 3-year-old children, then measuring the same children again at ages 4 and 5.

■ The Hypotheses and Logic for the Repeated-Measures ANOVA

The hypotheses for the repeated-measures ANOVA are exactly the same as those for the independent-measures ANOVA presented in Chapter 12. Specifically, the null hypothesis states that for the general population there are no mean differences among the treatment conditions being compared. In symbols,

$$H_0: \mu_1 = \mu_2 = \mu_3 = \dots$$

The null hypothesis states that, on average, all of the treatments have exactly the same effect. According to the null hypothesis, any differences that may exist among the sample means are not caused by systematic treatment effects but rather are the result of random and unsystematic factors.

The alternative hypothesis states that there are mean differences among the treatment conditions. Rather than specifying exactly which treatments are different, we use a generic version of H_1, which simply states that differences exist:

$$H_1: \text{At least one treatment mean } (\mu) \text{ is different from another.}$$

Notice that the alternative says that, on average, the treatments do have different effects. Thus, the treatment conditions may be responsible for causing mean differences among the samples. As always, the goal of the ANOVA is to use the sample data to determine which of the two hypotheses is more likely to be correct.

■ The *F*-Ratio for Repeated-Measures ANOVA

The *F*-ratio for the repeated-measures ANOVA has the same structure that was used for the independent-measures ANOVA in Chapter 12. In each case, the *F*-ratio compares the actual mean differences between treatments with the amount of difference that would be expected if there were no treatment effect. The numerator of the *F*-ratio measures the actual mean differences between treatments. The denominator measures how big the differences should be if there is no treatment effect. As always, the *F*-ratio uses variance to measure the size of the differences. Thus, the *F*-ratio for both ANOVAs has the general structure

$$F = \frac{\text{variance (differences) between treatments}}{\text{variance (differences) expected if there is no treatment effect}}$$

A large value for the *F*-ratio indicates that the differences between treatments are greater than would be expected without any treatment effect. If the *F*-ratio is larger than the critical value in the *F* distribution table, we can conclude that the differences between treatments are *significantly* larger than would be caused by chance.

Individual Differences in the *F*-Ratio   Although the structure of the *F*-ratio is the same for independent-measures and repeated-measures designs, there is a fundamental difference between the two designs that produces a corresponding difference in the two *F*-ratios. Specifically, individual differences are a part of the independent-measures *F*-ratio but are eliminated from the repeated-measures *F*-ratio.

You should recall that the term *individual differences* refers to participant characteristics such as age, personality, and gender that vary from one person to another and may influence the measurements that you obtain for each person. Suppose, for example, that you are measuring reaction time. The first participant in your study is a 19-year-old female with an IQ of 136 who is on the college varsity volleyball team. The next participant is a 42-year-old male with an IQ of 111 who returned to college after losing his job and comes to the research study with a head cold. Would you expect to obtain the same reaction time score for these two individuals?

Individual differences are a part of the variance in the numerator and in the denominator of the *F*-ratio for the independent-measures ANOVA. However, individual difference are eliminated or removed from the variances in the *F*-ratio for the repeated measures ANOVA. The idea of removing individual differences was first presented in Chapter 11 when we introduced the repeated-measures design (p. 353), but we will repeat it briefly now.

In a repeated-measures study exactly the same individuals participate in all of the treatment conditions. Therefore, if there are any mean differences between treatments, they cannot be explained by individual differences. Thus, individual differences are automatically eliminated from the numerator of the repeated-measures *F*-ratio.

A repeated-measures design also allows you to remove individual differences from the variance in the denominator of the *F*-ratio. Because the same individuals are measured in every treatment condition, it is possible to measure the size of the individual differences. In Table 13.1(a), for example, participant A has scores that are consistently 10 points lower than the scores for participant B. Because the individual differences are systematic and predictable, they can be measured and separated from the random, unsystematic differences in the denominator of the *F*-ratio.

Thus, individual differences are automatically eliminated from the numerator of the repeated-measures *F*-ratio. In addition, they can be measured and removed from the denominator. As a result, the structure of the final *F*-ratio is as follows:

$$F = \frac{\text{variance/differences between treatments}\,(\text{without individual differences})}{\text{variance/differences with no treatment effect}\,(\text{with individual differences removed})}$$

The process of removing individual differences is an important part of the procedure for a repeated-measures ANOVA.

■ The Logic of the Repeated-Measures ANOVA

The general purpose of the repeated-measures ANOVA is to determine whether the differences that are found between treatment conditions are significantly greater than would be expected if there is no treatment effect. In the numerator of the *F*-ratio, the *between-treatments variance* measures the actual mean differences between the treatment conditions. The variance in the denominator is intended to measure how much difference is reasonable to expect if there are no systematic treatment effects and no systematic individual differences. In other words, the denominator measures variability caused entirely by random and unsystematic factors. For this reason, the variance in the denominator is called the *error variance*. In this section we examine the elements that make up the two variances in the repeated-measures *F*-ratio.

The Numerator of the *F*-Ratio: Between-Treatments Variance   Logically, any differences that are found between treatments can be explained by only two factors:

1. **Systematic Differences Caused by the Treatments**. It is possible that the different treatment conditions really do have different effects and, therefore, cause the individuals' scores in one condition to be higher (or lower) than in another. Remember that the purpose for the research study is to determine whether or not a *treatment effect* exists.

2. **Random, Unsystematic Differences**. Even if there is no treatment effect, it is possible for the scores in one treatment condition to be different from the scores in another. For example, suppose that I measure your IQ score on a Monday morning. A week later I come back and measure your IQ again under exactly the same conditions. Will you get exactly same IQ score both times? In fact, minor differences between the two measurement situations will probably cause you to end up with two different scores. For example, for one of the IQ tests you might be more tired, or hungry, or worried, or distracted than you were on the other test. These differences can cause your scores to vary. The same thing can happen in a repeated-measures research study. The same individuals are being measured two or more different times and, even though there may be no difference between the two treatment conditions, you can still end up with different scores. However, these differences are random and unsystematic and are classified as error variance.

Thus, it is possible that any differences (or variance) found between treatments could be caused by treatment effects, and it is possible that the differences could simply be the result of chance. On the other hand, it is *impossible* that the differences between treatments are caused by individual differences. Because the repeated-measures design uses exactly the same individuals in every treatment condition, individual differences are *automatically eliminated* from the variance between treatments in the numerator of the *F*-ratio.

The Denominator of the *F*-Ratio: Error Variance   The goal of the ANOVA is to determine whether the differences that are observed in the data are greater than would be expected without any systematic treatment effects. To accomplish this goal, the denominator of the *F*-ratio is intended to measure how much difference (or variance) is reasonable to expect from random and unsystematic factors. This means that we must measure the variance that exists when there are no treatment effects or any other systematic differences.

We begin exactly as we did with the independent-measures *F*-ratio; specifically, we calculate the variance that exists within treatments. Recall from Chapter 12 that within each treatment all of the individuals are treated exactly the same. Therefore, any differences that exist within treatments cannot be caused by treatment effects.

In a repeated-measures design, however, it is possible that individual differences can cause systematic differences between the scores within treatments. For example, one individual may score consistently higher than another. To eliminate the individual differences from the denominator of the *F*-ratio, we measure the individual differences and then subtract them from the rest of the variability. The variance that remains is a measure of pure *error* without any systematic differences that can be explained by treatment effects or by individual differences.

In summary, the *F*-ratio for a repeated-measures ANOVA has the same basic structure as the *F*-ratio for independent measures (Chapter 12) except that it includes no variability caused by individual differences. The individual differences are automatically eliminated from the variance between treatments (numerator) because the repeated-measures design uses the same individuals in all treatments. In the denominator, the individual differences are subtracted out during the analysis. As a result, the repeated-measures F-ratio has the following structure:

$$F = \frac{\text{between} - \text{treatments variance}}{\text{error variance}}$$

$$= \frac{\text{treatment effects} + \text{random, unsystematic differences}}{\text{random, unsystematic differences}} \tag{13.1}$$

Note that this F-ratio is structured so that there are no individual differences contributing to either the numerator or the denominator. When there is no treatment effect, the F-ratio is balanced because the numerator and denominator are both measuring exactly the same variance. In this case, the F-ratio should have a value near 1.00. When research results produce an F-ratio near 1.00, we conclude that there is no evidence of a treatment effect and we fail to reject the null hypothesis. On the other hand, when a treatment effect does exist, it contributes only to the numerator and should produce a large value for the F-ratio. Thus, a large value for F indicates that there is a real treatment effect and therefore we should reject the null hypothesis.

LEARNING CHECK

1. In an independent-measures ANOVA, individual differences contribute to the variance in the numerator and in the denominator of the F-ratio. For a repeated-measures ANOVA, what happens to the individual differences in the numerator of the F-ratio?

   a. They do not exist because the same individuals participate in all of the treatments.

   b. They are measured and subtracted out during the analysis.

   c. Individual differences contribute to the variance in the numerator.

   d. None of the other options accurately describes individual differences in the numerator.

2. In an independent-measures ANOVA, individual differences contribute to the variance in the numerator and in the denominator of the F-ratio. For a repeated-measures ANOVA, what happens to the individual differences in the denominator of the F-ratio?

   a. They do not exist because the same individuals participate in all of the treatments.

   b. They are measured and subtracted out during the analysis.

   c. Individual differences contribute to the variance in the numerator.

   d. None of the other options accurately describes individual differences in the numerator.

ANSWERS   **1.** A, **2.** B

13.2 | Hypothesis Testing and Effect Size with the Repeated-Measures ANOVA

LEARNING OBJECTIVES

2. Describe the two-stage structure of the repeated-measures ANOVA and explain what happens in each stage.

3. Calculate all of the *SS*, *df*, and *MS* values needed for a repeated-measures ANOVA and explain the relationships among them.

4. Conduct a complete repeated-measures ANOVA and a measure of effect size.

5. Describe how the results of a repeated-measures ANOVA are reported in the literature and explain how to determine the number of treatments and the number of participants from the reported *df* values for the *F*-ratio.

The overall structure of the repeated-measures ANOVA is shown in Figure 13.2. Note that the ANOVA can be viewed as a two-stage process. In the first stage, the total variance is partitioned into two components: *between-treatments variance* and *within-treatments variance*. This stage is identical to the analysis that we conducted for the independent-measures design in Chapter 12.

The second stage of the analysis is intended to remove the individual differences from the denominator of the *F*-ratio. In the second stage, we begin with the variance within treatments and then measure and subtract out the *between-subject variance*, which measures the size of the individual differences. The remaining variance, often called the *residual variance*, or *error variance*, provides a measure of how much variance is reasonable to expect

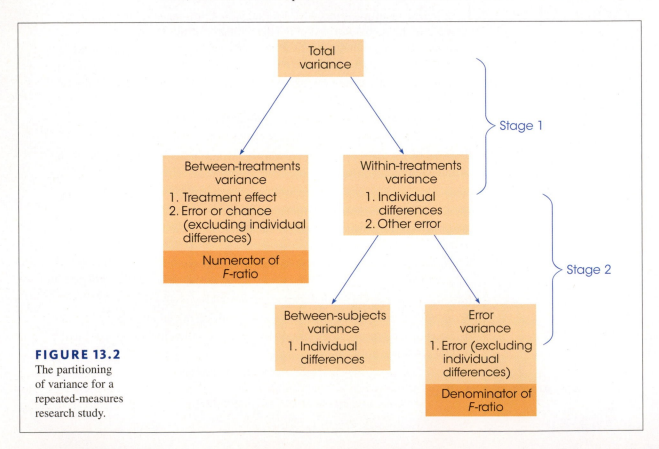

FIGURE 13.2

The partitioning of variance for a repeated-measures research study.

after the treatment effects and individual differences have been removed. The second stage of the analysis is what differentiates the repeated-measures ANOVA from the independent-measures ANOVA. Specifically, the repeated-measures design requires that the individual differences be removed.

DEFINITION

> In a repeated-measures ANOVA, the denominator of the F-ratio is called the **residual variance**, or the **error variance**, and measures how much variance is expected if there are no systematic treatment effects and no individual differences contributing to the variability of the scores.

■ Notation for the Repeated-Measures ANOVA

We will use the data in Table 13.2 to introduce the notation for the repeated-measures ANOVA. The data are similar to the results of a study by Weinstein, McDermott, and Roediger (2010) comparing three strategies for studying in preparation for a test. In the study, students read a passage knowing that they would be tested on the material. In one condition, participants simply reread the material to be tested. In a second condition, the students answered prepared comprehension questions about the material, and in a third condition, the students generated and answered their own questions. Each participant was tested in all three conditions and the order of the conditions was balanced across the participants. The data are the three test scores for each student. You may notice that this research study and the numerical values in the table are identical to those used to demonstrate the independent-measures ANOVA in the previous chapter (Example 12.2, p. 385). In this case, however, the data represent a repeated-measures study in which the same group of $n = 6$ students is tested in all three treatment conditions.

TABLE 13.2

Test scores for six students using three different study strategies.

Student	Reread	Answer Prepared Questions	Create and Answer Questions	Person Totals	
A	2	5	8	$P = 15$	$n = 6$
B	3	9	6	$P = 18$	$k = 3$
C	8	10	12	$P = 30$	$N = 18$
D	6	13	11	$P = 30$	$G = 144$
E	5	8	11	$P = 24$	$\Sigma X^2 = 1324$
F	6	9	12	$P = 27$	
	$T = 30$	$T = 54$	$T = 60$		
	$M = 5$	$M = 9$	$M = 10$		
	$SS = 24$	$SS = 34$	$SS = 30$		

You also should recognize that most of the notation in Table 13.2 is identical to the notation used in an independent-measures analysis (Chapter 12). For example, there are $n = 6$ participants who are tested in $k = 3$ treatment conditions, producing a total of $N = 18$ scores that add up to a grand total of $G = 144$. Note, however, that $N = 18$ now refers to the total number of scores in the study, not the number of participants.

The repeated-measures ANOVA introduces only one new notational symbol. The letter P is used to represent the total of all the scores for each individual in the study. You can think of the P values as "Person totals" or "Participant totals." In Table 13.2, for example, participant A had scores of 2, 5, and 8 for a total of $P = 15$. The P values are used to define and measure the magnitude of the individual differences in the second stage of the analysis.

EXAMPLE 13.1

We use the data in Table 13.2 to demonstrate the repeated-measures ANOVA. Again, the goal of the test is to determine whether there are any significant differences among the three strategies being compared. Specifically, are any of the mean differences in the data greater than would be expected if there were no systematic differences among the three strategies.

■ Stage 1 of the Repeated-Measures Analysis

The first stage of the repeated-measures analysis is identical to the independent-measures ANOVA that was presented in Chapter 12. Specially, the SS and df for the total set of scores are analyzed into within-treatments and between-treatments components.

Because the numerical values in Table 13.2 are the same as the values used in Example 12.2 (p. 385), the computations for the first stage of the repeated-measures analysis are identical to those in Example 12.2. Rather than repeating the same arithmetic, the results of the first stage of the repeated-measures analysis can be summarized as follows:

Total:

$$SS_{total} = \Sigma X^2 - \frac{G^2}{N} = 1324 - 1152 = 172$$

$$= 172$$

$$df_{total} = N - 1 = 17$$

Within treatments:

$$SS_{within\ treatments} = \Sigma SS_{inside\ each\ treatment} = 24 + 34 + 30 = 88$$

$$df_{within\ treatments} = \Sigma df_{inside\ each\ treatment} = 5 + 5 + 5 = 15$$

Between treatments: For this example we will use the computational formula.

$$SS_{between\ treatments} = \Sigma \frac{T^2}{n} - \frac{G^2}{N} = \frac{30^2}{6} + \frac{54^2}{6} + \frac{60^2}{6} - \frac{144^2}{18} = 84$$

$$df_{between\ treatments} = k - 1 = 2$$

For more details on the formulas and calculations see Example 12.2, pp. 385–388.

This completes the first stage of the repeated-measures analysis. Note that the two components, between and within, add up to the total for the SS values and for the df values. Also note that the between-treatments SS and df values provide a measure of the mean differences between treatments and are used to compute the variance in the numerator of the final F-ratio.

■ Stage 2 of the Repeated-Measures Analysis

The second stage of the analysis involves measuring the individual differences and then removing them from the denominator of the F-ratio.

Measuring Individual Differences   Because the same individuals are used in every treatment, it is possible to measure the size of the individual differences. For the data in Table 13.2, for example, person A tends to have the lowest scores and participants C and D tend to have the highest scores. These individual differences are reflected in the P values, or person totals, in the right-hand column. We will use these P values to create a computational formula for $SS_{between\ subjects}$ in much the same way that we used the treatment totals, the T values, in the computational formula for $SS_{\text{between treatments}}$. Specifically, the formula for the between-subjects SS is

$$SS_{\text{between subjects}} = \Sigma \frac{P^2}{k} - \frac{G^2}{N} \tag{13.2}$$

Notice that the formula for the between-subjects SS has exactly the same structure as the computational formula for the between-treatments SS (see the calculation in Stage 1). In this case we use the person totals (P values) instead of the treatment totals (T values). Each P value is squared and divided by the number of scores that were added to obtain the total. In this case, each person has k scores, one for each treatment. Box 13.1 presents another demonstration of the similarity of the formulas for SS between subjects and SS between treatments. For the data in Table 13.2,

$$SS_{\text{between subjects}} = \Sigma \frac{P^2}{k} - \frac{G^2}{N} = \frac{15^2}{3} + \frac{18^2}{3} + \frac{30^2}{3} + \frac{30^2}{3} + \frac{24^2}{3} + \frac{27^2}{3} - \frac{144^2}{18} = 66$$

The value of $SS_{\text{between subjects}}$ provides a measure of the size of the individual differences—that is, the differences between subjects.

BOX 13.1 $SS_{\text{between subjects}}$ and $SS_{\text{between treatments}}$

The data for a repeated-measures study are normally presented in a matrix, with the treatment conditions determining the columns and the participants defining the rows. The data in Table 13.2 demonstrate this normal presentation. The calculation of $SS_{\text{between treatments}}$ provides a measure of the differences between treatment conditions—that is, a measure of the mean differences between the *columns* in the data matrix. For the data in Table 13.2, the column totals are 30, 54, and 60. These values are variable, and $SS_{\text{between treatments}}$ measures the amount of variability.

The following table reproduces the data from Table 13.2, but now we have turned the data matrix on its side so that the participants define the columns and the treatment conditions define the rows.

In this new format, the differences between the columns represent the between-subjects variability.

The column totals are now P values (instead of T values) and the number of scores in each column is now identified by k (instead of n). With these changes in notation, the formula for $SS_{\text{between subjects}}$ has exactly the same structure as the formula for $SS_{\text{between treatments}}$. If you examine the two equations, the similarity should be clear.

	Participant						
	A	B	C	D	E	F	
Reread	2	3	8	6	5	6	$T = 30$
Prepared Questions	5	9	10	13	8	9	$T = 54$
Create Questions	8	6	12	11	11	12	$T = 60$

$$P = 15 \;\; P = 18 \;\; P = 30 \;\; P = 30 \;\; P = 24 \;\; P = 27$$

Removing the Individual Differences $SS_{\text{between subjects}}$ measures the individual differences. The next step is to subtract out the individual differences to obtain the measure of error that forms the denominator of the F-ratio. Thus, the final step in the analysis of SS is

$$SS_{\text{error}} = SS_{\text{within treatments}} - SS_{\text{between subjects}} \tag{13.3}$$

We have already computed $SS_{\text{within treatments}} = 88$ and $SS_{\text{between subjects}} = 66$, therefore

$$SS_{\text{error}} = 88 - 66 = 22$$

The analysis of degrees of freedom follows exactly the same pattern that was used to analyze SS. First we measures the df for the individual differences, then subtract this value from the df we obtained for within treatments. Remember that we use the P values to calculate $SS_{\text{between subjects}}$. The number of P values corresponds to the number of subjects, n, so the corresponding df is

$$df_{\text{between subjects}} = n - 1 \qquad (13.4)$$

For the data in Table 13.2, there are $n = 6$ participants and

$$df_{\text{between subjects}} = 6 - 1 = 5$$

Next, we subtract the individual differences from the within-subjects component to obtain a measure of error. For degrees of freedom,

$$df_{\text{error}} = df_{\text{within treatments}} - df_{\text{between subjects}} \qquad (13.5)$$

For the data in Table 13.2,

$$df_{\text{error}} = 15 - 5 = 10$$

An algebraically equivalent formula for df_{error} uses only the number of treatment conditions (k) and the number of participants (n):

$$df_{\text{error}} = (k - 1)(n - 1) \qquad (13.6)$$

The usefulness of equation 13.6 is discussed in Box 13.2.

BOX 13.2 Using the Alternate Formula for df_{error}

The statistics presented in a research report not only describe the significance of the results but also typically provide enough information to reconstruct the research design. The alternative formula for df_{error} is particularly useful for this purpose. Suppose, for example, that a research report for a repeated-measures study includes an F-ratio with $df = 2, 10$. How many treatment conditions were compared in the study and how many individuals participated?

To answer these question, begin with the first df value, which is $df_{\text{between treatments}} = 2 = k - 1$. From this value, it is clear that $k = 3$ treatments. Next, use the second df value, which is $df_{\text{error}} = 10$. Using this value and the fact that $k - 1 = 2$, use Equation 13.6 to find the number of participants.

$$df_{\text{error}} = 10 = (k - 1)(n - 1) = 2(n - 1)$$

If $2(n - 1) = 10$, then $n - 1$ must equal 5. Therefore, $n = 6$.

Therefore, we conclude that a repeated-measures study producing an F-ratio with $df = 2, 10$ must have compared 3 treatment conditions using a sample of 6 participants.

Remember: The purpose for the second stage of the analysis is to measure the individual differences and then remove the individual differences from the denominator of the F-ratio. This goal is accomplished by computing SS and df between subjects (the individual differences) and then subtracting these values from the within-treatments values. The result is a measure of variability with the individual differences removed. This error variance (SS and df) is used in the denominator of the F-ratio.

■ Calculation of the Variances (*MS* Values) and the *F*-Ratio

The final calculation in the analysis is the *F*-ratio, which is a ratio of two variances. Each variance is called a *mean square,* or *MS*, and is obtained by dividing the appropriate *SS* by its corresponding *df* value. The *MS* in the numerator of the *F*-ratio measures the size of the differences between treatments and is calculated as

$$MS_{\text{between treatments}} = \frac{SS_{\text{between treatments}}}{df_{\text{between treatments}}} \tag{13.7}$$

For the data in Table 13.2,

$$MS_{\text{between treatments}} = \frac{SS_{\text{between treatments}}}{df_{\text{between treatments}}} = \frac{84}{2} = 42$$

The denominator of the *F*-ratio measures how much difference is reasonable to expect if there are no systematic treatment effects and the individual differences have been removed. This is the error variance or the residual obtained in stage 2 of the analysis.

$$MS_{\text{error}} = \frac{SS_{\text{error}}}{df_{\text{error}}} \tag{13.8}$$

For the data in Table 13.2,

$$MS_{\text{error}} = \frac{SS_{\text{error}}}{df_{\text{error}}} = \frac{22}{10} = 2.20$$

Finally, the *F*-ratio is computed as

$$F = \frac{MS_{\text{between treatments}}}{MS_{\text{error}}} \tag{13.9}$$

For the data in Table 13.2,

$$F = \frac{MS_{\text{between treatments}}}{MS_{\text{error}}} = \frac{42}{2.20} = 19.09$$

Once again, notice that the repeated-measures ANOVA uses MS_{error} in the denominator of the *F*-ratio. This *MS* value is obtained in the second stage of the analysis, after the individual differences have been removed. As a result, individual differences are completely eliminated from the repeated-measures *F*-ratio, so that the general structure is

$$F = \frac{\text{treatment effects} + \text{unsystematic differences (without individual diff's)}}{\text{unsystematic differences (without individual diff's)}}$$

For the data we have been examining, the *F*-ratio is $F = 19.09$, indicating that the differences between treatments (numerator) are almost 20 times bigger than you would expect without any treatment effects (denominator). A ratio this large provides clear evidence that there is a real treatment effect. To verify this conclusion you must consult the *F* distribution table to determine the appropriate critical value for the test. The degrees of freedom for the *F*-ratio are determined by the two variances that form the numerator and the denominator. For a repeated-measures ANOVA, the *df* values for the *F*-ratio are reported as

$$df = df_{\text{between treatments}}, df_{\text{error}}$$

For the example we are considering, the F-ratio has $df = 2, 10$ ("degrees of freedom equal two and ten"). Using the F distribution table (p. 653) with $\alpha = .05$, the critical value is $F = 4.10$, and with $\alpha = .01$ the critical value is $F = 7.56$. Our obtained F-ratio, $F = 19.09$, is well beyond either of the critical values, so we can conclude that the differences between treatments are *significantly* greater than expected by chance using either $\alpha = .05$ or $\alpha = .01$. ∎

The summary table for the repeated-measures ANOVA from Example 13.1 is presented in Table 13.3. Although these tables are no longer commonly used in research reports, they provide a concise format for displaying all of the elements of the analysis.

TABLE 13.3

A summary table for the repeated-measures ANOVA for the data from Example 13.1.

Source	SS	df	MS	F
Between treatments	84	2	42.00	$F(2, 10) = 19.09$
Within treatments	88	15		
Between subjects	66	5		
Error	22	10	2.20	
Total	172	17		

The following example is an opportunity to test your understanding of the calculation of SS values for the repeated-measures ANOVA.

EXAMPLE 13.2

For the following data, compute $SS_{\text{between treatments}}$ and $SS_{\text{between subjects}}$.

		Treatment				
Subject	1	2	3	4		
A	2	2	2	2	$G = 32$	
B	4	0	0	4	$\Sigma X^2 = 96$	
C	2	0	2	0		
D	4	2	2	4		
	$T = 12$	$T = 4$	$T = 6$	$T = 10$		

You should find that $SS_{\text{between treatments}} = 10$ and $SS_{\text{between subjects}} = 8$. ∎

■ Measuring Effect Size for the Repeated-Measures ANOVA

The most common method for measuring effect size with ANOVA is to compute the percentage of variance that is explained by the treatment differences. In the context of ANOVA, the percentage of variance is commonly identified as η^2 (eta squared). In Chapter 12, for the independent-measures analysis, we computed η^2 as

$$\eta^2 = \frac{SS_{\text{between treatments}}}{SS_{\text{between treatments}} + SS_{\text{within treatments}}} = \frac{SS_{\text{between treatments}}}{SS_{\text{total}}}$$

The intent is to measure how much of the total variability is explained by the differences between treatments. With a repeated-measures design, however, there is another component that can explain some of the variability in the data. Specifically, part of the variability is caused by differences between individuals. In Table 13.2, for example, person C consistently scored higher than person A or B. This consistent difference explains some of the variability in the data. When computing the size of the treatment effect, it is customary to remove any variability that can be explained by other factors, and then compute the percentage of the remaining variability that can be explained by the treatment effects. Thus, for a repeated-measures ANOVA, the variability from the individual differences is removed before computing η^2. As a result, η^2 is computed as

$$\eta^2 = \frac{SS_{\text{between treatments}}}{SS_{\text{total}} - SS_{\text{between subjects}}} \qquad (13.10)$$

Because Equation 13.10 computes a percentage that is not based on the total variability of the scores (one part, $SS_{\text{between subjects}}$, is removed), the result is often called a *partial* eta squared.

The general goal of Equation 13.10 is to calculate a percentage of the variability that has not already been explained by other factors. Thus, the denominator of Equation 13.9 is limited to variability from the treatment differences and variability that is exclusively from random, unsystematic factors. With this in mind, an equivalent version of the η^2 formula is

$$\eta^2 = \frac{SS_{\text{between treatments}}}{SS_{\text{between treatments}} + SS_{\text{error}}} \qquad (13.11)$$

In this new version of the eta-squared formula, the denominator consists of the variability that is explained by the treatment differences plus the other *unexplained* variability. Using either formula, the data from Example 13.1 produce

$$\eta^2 = \frac{84}{106} = 0.79$$

This result means that 79% of the variability in the data (except for the individual differences) is accounted for by the differences between treatments.

The following example is an opportunity to test your understanding of effect size for the repeated-measures ANOVA.

EXAMPLE 13.3   The results from a repeated-measures ANOVA are presented in the following summary table. Compute η^2 to measure the size of the treatment effect. You should obtain $\eta^2 = 0.579$.

Source	SS	df	MS	F
Between treatments	44	2	22	$F(2, 16) = 11$
Within treatments	98	24		
Between subjects	66	8		
Error	32	16	2	
Total	142	26		

IN THE LITERATURE

Reporting the Results of a Repeated-Measures ANOVA

As described in Chapter 12 (p. 389), the format for reporting ANOVA results in journal articles consists of

1. A summary of descriptive statistics (at least treatment means and standard deviations, and tables or graphs as needed)
2. A concise statement of the outcome of the ANOVA

For the study in Example 13.1, the report could state:

The means and variances for the three strategies are shown in Table 1. A repeated-measures analysis of variance indicated significant mean differences in the participants' test scores for the three study strategies, $F(2, 10) = 19.09$, $p < .01$, $\eta^2 = 0.79$.

TABLE 1

Test scores using three different study strategies

	Reread	Answer Prepared Questions	Create and Answer Questions
M	5.00	9.00	10.00
SD	2.19	2.61	2.45

■ Post Hoc Tests with Repeated Measures

Recall that ANOVA provides an overall test of significance for the mean differences between treatments. When the null hypothesis is rejected, it indicates only that there is a difference between at least two of the treatment means. If $k = 2$, it is obvious which two treatments are different. However, when k is greater than 2, the situation becomes more complex. To determine exactly which differences are significant, the researcher must follow the ANOVA with post hoc tests. In Chapter 12, we used Tukey's HSD and the Scheffé test to make these multiple comparisons among treatment means. These two procedures attempt to control the overall alpha level by making adjustments for the number of potential comparisons.

For a repeated-measures ANOVA, Tukey's HSD and the Scheffé test can be used in the exact same manner as was done for the independent-measures ANOVA, *provided* that you substitute MS_{error} in place of $MS_{\text{within treatments}}$ in the formulas and use df_{error} in place of $df_{\text{within treatments}}$ when locating the critical value in a statistical table.

■ Assumptions of the Repeated-Measures ANOVA

The basic assumptions for the repeated-measures ANOVA are identical to those required for the independent-measures ANOVA.

1. The observations within each treatment condition must be independent (see p. 244).

2. The population distribution within each treatment must be normal. (As before, the assumption of normality is important only with small samples.)

3. The variances of the population distributions for each treatment should be equivalent.

For the repeated-measures ANOVA, there is an additional assumption, called homogeneity of covariance. Basically, it refers to the requirement that the relative standing of each subject be maintained in each treatment condition. This assumption is violated if the effect of the treatment is not consistent for all of the subjects or if order effects exist for some, but not other, subjects. This issue is very complex and is beyond the scope of this book. However, methods do exist for dealing with violations of this assumption (for a discussion, see Keppel, 1973).

If there is reason to suspect that one of the assumptions for the repeated-measures ANOVA has been violated, an alternative analysis known as the Friedman test is presented in Appendix E. The Friedman test requires that the original scores be transformed into ranks before evaluating the differences between treatment conditions.

LEARNING CHECK

1. Which of the following accurately describes the two stages of a repeated-measures analysis of variance?

   a. The first stage is identical to the independent-measures analysis and the second stage removes individual differences from the numerator of the F-ratio.

   b. The first stage is identical to the independent-measures analysis and the second stage removes individual differences from the denominator of the F-ratio.

   c. The first stage removes individual differences from the numerator of the F-ratio and the second stage is identical to the independent-measures analysis.

   d. The first stage removes individual differences from the denominator of the F-ratio and the second stage is identical to the independent-measures analysis.

2. For the repeated-measures ANOVA, SS_{error} is found by _____.

   a. $SS_{total} - SS_{between treatments}$

   b. $SS_{between treatments} - SS_{within}$

   c. $SS_{between treatments} - SS_{between subjects}$

   d. $SS_{within} - SS_{between subjects}$

3. A researcher uses a repeated-measures ANOVA to test for mean differences among three treatment conditions using a sample of $n = 10$ participants. What are the df values for the F-ratio from this analysis?

   a. 3, 9

   b. 2, 9

   c. 3, 18

   d. 2, 18

4. The results of a repeated-measures ANOVA are reported as follows, $F(2, 24) = 1.12$, $p > .05$. How many individuals participated in the study?

   a. 25

   b. 13

   c. 12

   d. 7

ANSWERS   1. B, 2. D, 3. D, 4. B

13.3 | More about the Repeated-Measures Design

LEARNING OBJECTIVES

6. Describe the general advantages and disadvantages of repeated-measures vs. independent-measures designs and identify the circumstances in which each is more appropriate.

7. Explain the relationship between the repeated-measures t test and a repeated-measures ANOVA when evaluating the difference between two means from a repeated-measures design.

■ Advantages and Disadvantages of the Repeated-Measures Design

When we first encountered the repeated-measure design (Chapter 11), we noted that this type of research study has certain advantages and disadvantages compared to the independent-measures design (pp. 352–355). On the bright side, a repeated-measures study may be desirable if the supply of participants is limited. A repeated-measures study is economical in that the research requires relatively few participants. Also, a repeated-measures design eliminates or minimizes most of the problems associated with individual differences. However, disadvantages also exist. These take the form of order effects, such as fatigue, that can make the interpretation of the data difficult.

Now that we have introduced the repeated-measures ANOVA we can examine one of the primary advantages of this design—namely, the elimination of variability caused by individual differences. Consider the structure of the F-ratio for both the independent-and the repeated-measures designs.

$$F = \frac{\text{treatment effects} + \text{random, unsystematic differences}}{\text{random, unsystematic differences}}$$

In each case, the goal of the analysis is to determine whether the data provide evidence for a treatment effect. If there is no treatment effect, the numerator and denominator are both measuring the same random, unsystematic variance and the F-ratio should produce a value near 1.00. On the other hand, the existence of a treatment effect should make the numerator substantially larger than the denominator and result in a large value for the F-ratio.

For the independent-measures design, the unsystematic differences include individual differences as well as other random sources of error. Thus, for the independent-measures ANOVA, the F-ratio has the following structure:

$$F = \frac{\text{treatment effect} + \text{individual differences and other error}}{\text{individual differences and other error}}$$

For the repeated-measures design, the individual differences are eliminated or subtracted out, and the resulting F-ratio is structured as follows:

$$F = \frac{\text{treatment effects} + \text{error (excluding individual differences)}}{\text{error (excluding individual differences)}}$$

The removal of individual differences from the analysis becomes an advantage in situations in which very large individual differences exist among the participants being studied.

When individual differences are large, the presence of a treatment effect may be masked if an independent-measures study is performed. In this case, a repeated-measures design would be more sensitive in detecting a treatment effect because individual differences do not influence the value of the F-ratio.

This point will become evident in the following example. Suppose that we know how much variability is accounted for by the different sources of variance. For example,

$$\text{treatment effect} = 10 \text{ units of variance}$$

$$\text{individual differences} = 10 \text{ units of variance}$$

$$\text{other error} = 1 \text{ unit of variance}$$

Notice that a large amount of the variability in the experiment is due to individual differences. By comparing the F-ratios for an independent- and a repeated-measures analysis, we are able to see a fundamental difference between the two types of experimental designs. For an independent-measures experiment, we obtain

$$F = \frac{\text{treatment effect} + \text{individual differences} + \text{error}}{\text{individual differences} + \text{error}}$$

$$= \frac{10 + 10 + 1}{10 + 1} = \frac{21}{11} = 1.91$$

Thus, the independent-measures ANOVA produces an F-ratio of $F = 1.91$. Recall that the F-ratio is structured to produce $F = 1.00$ if there is no treatment effect whatsoever. In this case, the F-ratio is near to 1.00 and strongly suggests that there is little or no treatment effect. If you check the F-distribution table in Appendix B, you will find that it is almost impossible for an F-ratio as small as 1.91 to be significant. For the independent-measures ANOVA, the 10-point treatment effect is overwhelmed by all the other variance.

Now consider what happens with a repeated-measures analysis. With the individual differences removed, the F-ratio becomes:

$$F = \frac{\text{treatment effect} + \text{error}}{\text{error}}$$
$$= \frac{10 + 1}{1} = \frac{11}{1} = 11$$

For the repeated-measures analysis, the numerator of the F-ratio (which includes the treatment effect) is 11 times larger than the denominator (which has no treatment effect). This result strongly indicates that there is a substantial treatment effect. In this example, the F-ratio is much larger for the repeated-measures study because the individual differences, which are extremely large, have been removed. In the independent-measures ANOVA, the presence of a treatment effect is obscured by the influence of individual differences. This problem is eliminated by the repeated-measures design, in which variability due to individual differences is partitioned out of the analysis. When the individual differences are large, a repeated-measures experiment may provide a more sensitive test for a treatment effect. In statistical terms, a repeated-measures test has more *power* than an independent-measures test; that is, it is more likely to detect a real treatment effect.

■ Factors that Influence the Outcome of a Repeated-Measures ANOVA and Measures of Effect Size

In previous chapters addressing hypothesis testing, we have repeatedly noted that the outcome of a hypothesis test is influenced by three major factors: the size of the treatment effect, the size of the sample(s), and the variance of the scores. Obviously, the bigger the treatment effect, the more likely it is to be significant. Also, a larger treatment effect tends to produce a larger measure of effect size. Similarly, a treatment effect demonstrated with a large sample is more convincing that an effect obtained with a small sample. Thus, larger samples tend to increase the likelihood of rejecting the null hypothesis. However, sample size has little or no effect on measures of effect size. These factors also apply to the repeated-measures ANOVA. The role of variance in the repeated-measures ANOVA, however, is somewhat more complicated, as discussed in the following section.

Individual Differences and the Consistency of the Treatment Effects   As we have demonstrated, one major advantage of a repeated-measures design is that it removes individual differences from the denominator of the F-ratio, which usually increases the likelihood of obtaining a significant result. However, removing individual differences is an advantage only when the treatment effects are reasonably consistent for all of the participants. If the treatment effects are not consistent across participants, the individual differences tend to disappear and value in the denominator is not noticeably reduced by removing them. This phenomenon is demonstrated in the following example.

EXAMPLE 13.4

Table 13.4 presents hypothetical data from a repeated-measures research study. We constructed the data specifically to demonstrate the relationship between consistent treatment effects and large individual differences.

TABLE 13.4

Data from a repeated-measures study comparing three treatments. The data show consistent treatment effects from one participant to another, which produce consistent and relatively large differences in the individual P totals.

Person	Treatment I	II	III	
A	0	1	2	$P = 3$
B	1	2	3	$P = 6$
C	2	4	6	$P = 12$
D	3	5	7	$P = 15$
	$T = 6$	$T = 12$	$T = 18$	
	$SS = 5$	$SS = 10$	$SS = 17$	

First, notice the consistency of the treatment effects. Treatment II has the same effect on every participant, increasing everyone's score by 1 or 2 points compared to treatment I. Also, treatment III produces a consistent increase of 1 or 2 points compared to treatment II. One consequence of the consistent treatment effects is that the individual differences are maintained in all of the treatment conditions. For example, participant A has the lowest score in all three treatments, and participant D always has the highest score. The participant totals (P values) reflect the consistent differences. For example, participant D has the largest score in every treatment and, therefore, has the largest P value. Also notice that there are big differences between the P totals from one individual to the next. For these data, $SS_{between subjects} = 30$ points.

Now consider the data in Table 13.5. To construct these data we started with the same numbers within each treatment that were used in Table 13.4. However, we scrambled the numbers within each column to eliminate the consistency of the treatment effects. In Table 13.5, for example, two participants show an increase in scores as they go from treatment I to treatment II, and two show a decrease. The data also show an inconsistent treatment effect as the participants go from treatment II to treatment III. One consequence of the inconsistent treatment effects is that there are no consistent individual differences between participants. Participant C, for example, has the lowest score in treatment II and the highest score in treatment III. As a result, there are no longer consistent differences between the individual participants. All of the P totals are about the same. For these data, $SS_{between subjects} = 3.33$ points. Because the two sets of data (Tables 13.4 and 13.5) have the same treatment totals (T values) and SS values, they have the same $SS_{between treatments}$ and $SS_{within treatments}$. For both sets of data,

$$SS_{between treatments} = 18 \text{ and } SS_{within treatments} = 32$$

TABLE 13.5

Data from a repeated-measures study comparing three treatments. The data show treatment effects that are inconsistent from one participant to another and, as a result, produce relatively small differences in the individual P totals. Note that the data have exactly the same scores within each treatment as the data in Table 13.5, however, the scores have been scrambled to eliminate the consistency of the treatment effects.

Person	Treatment I	II	III	
A	0	4	3	$P = 7$
B	1	5	2	$P = 8$
C	2	1	7	$P = 10$
D	3	2	6	$P = 11$
	$T = 6$	$T = 12$	$T = 18$	
	$SS = 5$	$SS = 10$	$SS = 17$	

However, there is a huge difference between the two sets of data when you compute SS_{error} for the denominator of the F-ratio. For the data in Table 13.4, with consistent treatment effects and large individual differences,

$$SS_{error} = SS_{within\ treatments} - SS_{between\ subjects}$$
$$= 32 - 30$$
$$= 2$$

For the data in Table 13.5, with no consistent treatment effects and relatively small differences between the individual P totals,

$$SS_{error} = SS_{within\ treatments} - SS_{between\ subjects}$$
$$= 32 - 3.33$$
$$= 28.67$$

Thus, consistent treatment effects tend to produce a relatively small error term for the F-ratio. As a result, consistent treatment effects are more likely to be statistically significant (reject the null hypothesis). For the examples we have been considering, the data in Table 13.4 produce an F-ratio of $F = 27.0$. With $df = 2, 6$, this F-ratio is well into the critical region for $\alpha = .05$ or .01 and we conclude that there are significant differences among the three treatments. On the other hand, the same mean differences in Table 13.5 produce $F = 1.88$. With $df = 2, 6$, this value is not in the critical region for $\alpha = .05$ or .01 and we conclude that there are no significant differences.

In summary, when treatment effects are consistent from one individual to another, the individual differences also tend to be consistent and relatively large. The large individual differences get subtracted from the denominator of the F-ratio producing a larger value for F and increasing the likelihood that the F-ratio will be in the critical region. ■

■ Repeated-Measures ANOVA and Repeated-Measures *t*

As we noted in Chapter 12 (pp. 401-402), whenever you are evaluating the difference between two sample means, you can use either a *t* test or analysis of variance. In Chapter 12 we demonstrated that the two tests are related in many respects, including:

1. The two tests always reach the same conclusion about the null hypothesis.
2. The basic relationship between the two test statistics is $F = t^2$.
3. The *df* value for the *t* statistic is identical to the *df* value for the denominator of the F-ratio.
4. If you square the critical value for the two-tailed *t* test, you will obtain the critical value for the F-ratio. Again, the basic relationship is $F = t^2$.

In Chapter 12, these relationships were demonstrated for the independent-measures tests, but they are also true for repeated-measures designs comparing two treatment conditions. The following example demonstrates the relationships.

EXAMPLE 13.5   The following table shows the data from a repeated-measures study comparing two treatment conditions. We have structured the data in a format that is compatible with the repeated-measures *t* test. Note that the calculations for the *t* test are based on the difference scores (*D* values) in the final column.

Participant	Treatment		
	I	II	D
A	3	5	2
B	4	14	10
C	5	7	2
D	4	6	2

$$M_D = 4$$
$$SS_D = 48$$

The Repeated-Measures *t* Test   The null hypothesis for the *t* test states that for the general population there is no mean difference between the two treatment conditions.

$$H_0: \mu_D = 0$$

With $n = 4$ participants, the test has $df = 3$ and the critical boundaries for a two-tailed test with $\alpha = .05$ are $t = \pm 3.182$.

For these data, the sample mean difference is $M_D = 4$, the variance for the difference scores is $s^2 = 16$, and the standard error is $S_{M_D} = 2$ points. These values produce a *t* statistic of

$$t = \frac{M_D - \mu_D}{S_{M_D}}$$

$$= \frac{4 - 0}{2} = 2.00$$

The *t* value is not in the critical region so we fail to reject H_0 and conclude that there is no significant difference between the two treatments.

The Repeated-Measures ANOVA   Now we will reorganize the data into a format that is compatible with a repeated-measures ANOVA. Notice that the ANOVA uses the original scores (not the difference scores) and requires the *P* totals for each participant.

Participant	Treatment			
	I	II	P	
A	3	5	8	$G = 48$
B	4	14	18	$\Sigma X^2 = 372$
C	5	7	12	$N = 8$
D	4	6	10	

Again, the null hypothesis states that for the general population there is no mean difference between the two treatment conditions.

$$H_0: \mu_1 - \mu_2 = 0$$

For this study, $df_{\text{between treatments}} = 1$, $df_{\text{within treatments}} = 6$, $df_{\text{between subjects}} = 3$, which produce $df_{\text{error}} = (6 - 3) = 3$. Thus, the F-ratio has $df = 1, 3$ and the critical value for $\alpha = .05$ is $F = 10.13$. Note that the denominator of the F-ratio has the same df value as the *t* statistic ($df = 3$) and that the critical value for F is equal to the squared critical value for t ($10.13 = 3.182^2$).

For these data, $SS_{total} = 84$,

$$SS_{within} = 52$$

$$SS_{between\ treatments} = (84 - 52) = 32$$

$$SS_{between\ subjects} = 28$$

$$SS_{error} = (52 - 28) = 24$$

The two variances in the F-ratio are

$$MS_{between\ treatments} = \frac{SS_{between\ treatments}}{df_{between\ treatments}} = \frac{32}{1} = 32$$

and $MS_{error} = \dfrac{SS_{error}}{df_{error}} = \dfrac{24}{3} = 8$

and the F-ratio is $F = \dfrac{MS_{between\ treatments}}{MS_{error}} = \dfrac{32}{8} = 4.00$

Notice that the F-ratio and the t statistic are related by the equation $F = t^2$ ($4 = 2^2$). The F-ratio (like the t statistic) is not in the critical region so, once again, we fail to reject H_0 and conclude that there is no significant difference between the two treatments. ■

LEARNING CHECK

1. Under what circumstances would a study using a repeated-measures ANOVA have a distinct advantage over a study using an independent-measures ANOVA?

   a. When there are many participants available and individual differences are large.

   b. When there are few participants available and individual differences are large.

   c. When there are many participants available and individual differences are small.

   d. When there are few participants available and individual differences are small.

2. A repeated-measures ANOVA produced an F-ratio of $F = 9.00$ with $df = 1, 14$. If the same data were analyzed with a repeated-measures t test, what value would be obtained for the t statistic?

   a. 3

   b. 9

   c. 81

   d. Cannot determine without more information.

ANSWERS   1. B, 2. A

SUMMARY

1. The repeated-measures ANOVA is used to evaluate the mean differences obtained in a research study comparing two or more treatment conditions using the same sample of individuals in each condition. The test statistic is an F-ratio, where the numerator measures the variance (differences) between treatments and the denominator measures the variance (differences) that are expected without any treatment effects or individual differences.

$$F = \frac{MS_{\text{between treatments}}}{MS_{\text{error}}}$$

2. The first stage of the repeated-measures ANOVA is identical to the independent-measures analysis and separates the total variability into two components: between-treatments and within-treatments. Because a repeated-measures design uses the same subjects in every treatment condition, the differences between treatments cannot be caused by individual differences. Thus, individual differences are automatically eliminated from the between-treatments variance in the numerator of the F-ratio.

3. In the second stage of the repeated-measures analysis, individual differences are computed and removed from the denominator of the F-ratio. To remove the individual differences, you first compute the variability between subjects (SS and df) and then subtract these values from the corresponding within-treatments values. The residual provides a measure of error excluding individual differences, which is the appropriate denominator for the repeated-measures F-ratio. The equations for analyzing SS and df for the repeated-measures ANOVA are presented in Figure 13.3.

4. Effect size for the repeated-measures ANOVA is measured by computing eta squared, the percentage of variance accounted for by the treatment effect. For the repeated-measures ANOVA

$$\eta^2 = \frac{SS_{\text{between treatments}}}{SS_{\text{total}} - SS_{\text{between subjects}}}$$

$$= \frac{SS_{\text{between treatments}}}{SS_{\text{between treatments}} + SS_{\text{error}}}$$

Because part of the variability (the SS due to individual differences) is removed before computing η^2, this measure of effect size is often called a partial eta squared.

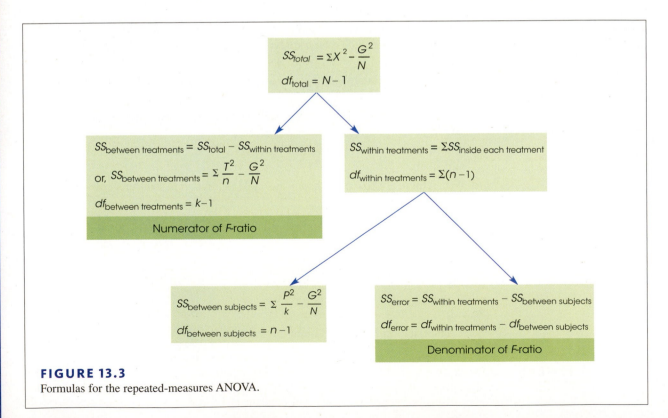

FIGURE 13.3
Formulas for the repeated-measures ANOVA.

5. When the obtained F-ratio is significant (that is, H_0 is rejected), it indicates that a significant difference lies between at least two of the treatment conditions. To determine exactly where the difference lies, post hoc comparisons may be made. Post hoc tests, such as Tukey's HSD, use MS_{error} and df_{error} rather than $MS_{within\ treatments}$ and $df_{within\ treatments}$.

6. A repeated-measures ANOVA eliminates the influence of individual differences from the analysis. If individual differences are extremely large, a treatment effect might be masked in an independent-measures experiment. In this case, a repeated-measures design might be a more sensitive test for a treatment effect.

KEY TERMS

individual differences (417)

between-treatments variance (417)

between-subjects variability (420)

error variance (421)

SPSS®

General instructions for using SPSS are presented in Appendix D. Following are detailed instructions for using SPSS to perform the Single-Factor, **Repeated-Measures Analysis of Variance (ANOVA)** presented in this chapter.

Data Entry

1. Enter the scores for each treatment condition in a separate column, with the scores for each individual in the same row. All the scores for the first treatment go in the VAR00001 column, the second treatment scores in the VAR00002 column, and so on.

Data Analysis

1. Click **Analyze** on the tool bar, select **General Linear Model**, and click on **Repeated-Measures**.
2. SPSS will present a box titled **Repeated-Measures Define Factors**. Within the box, the Within-Subjects Factor Name should already contain **Factor 1**. If not, type in Factor 1.
3. Enter the **Number of levels** (number of different treatment conditions) in the next box.
4. Click on **Add**.
5. Click **Define**.
6. One by one, move the column labels for your treatment conditions into the **Within Subjects Variables** box. (Highlight the column label on the left and click the arrow to move it into the box.)
7. If you want descriptive statistics for each treatment, click on the **Options** box, select **Descriptives**, and click **Continue**.
8. Click **OK**.

SPSS Output

We used the SPSS program to analyze the data from Example 13.1 comparing three strategies for studying before an exam. Portions of the program output are shown in Figure 13.4. Note that large portions of the SPSS output are not relevant for our purposes and are not included. The first item of interest is the table of **Descriptive Statistics**, which presents the mean, standard deviation, and number of scores for each treatment. Next, we skip to the table showing **Tests of Within-Subjects Effects**. The top line of the factor1 box (Sphericity Assumed) shows the between-treatments sum of squares, degrees of freedom, and mean square that form the numerator of the F-ratio. The same line reports the value of the F-ratio and the level of significance (the p value or alpha level). Similarly, the top line of the Error (factor1) box shows the sum of squares, the degrees of freedom, and the mean square for the error term (the

Descriptive Statistics

	Mean	Std. Deviation	N
VAR00001	5.0000	2.19089	6
VAR00002	9.0000	2.60768	6
VAR00003	10.0000	2.44949	6

Tests of Within-Subjects Effects

Measure: MEASURE_1

Source		Type III Sum of Squares	df	Mean Square	F	Sig.
factor 1	Sphericity Assumed	84.000	2	42.000	19.091	.000
	Greenhouse-Geisser	84.000	1.228	68.380	19.091	.004
	Huynh-Feldt	84.000	1.424	58.991	19.091	.002
	Lower-bound	84.000	1.000	84.000	19.091	.007
Error (factor 1)	Sphericity Assumed	22.000	10	2.200		
	Greenhouse-Geisser	22.000	6.142	3.582		
	Huynh-Feldt	22.000	7.120	3.090		
	Lower-bound	22.000	5.000	4.400		

FIGURE 13.4
Portions of the SPSS output for the repeated-measures ANOVA for the study in Example 13.1.

denominator of the F-ratio). The final box in the output (not shown in Figure 13.4) is labeled **Tests of Between-Subjects Effects** and the bottom line (Error) reports the between-subjects sum of squares and degrees of freedom (ignore the Mean Square and F-ratio which are not part of the repeated-measures ANOVA).

FOCUS ON PROBLEM SOLVING

1. Before you begin a repeated-measures ANOVA, complete all the preliminary calculations needed for the ANOVA formulas. This requires that you find the total for each treatment (Ts), the total for each person (Ps), the grand total (G), the SS for each treatment condition, and ΣX^2 for the entire set of N scores. As a partial check on these calculations, be sure that the T values add up to G and that the P values have a sum of G.

2. To help remember the structure of repeated-measures ANOVA, keep in mind that a repeated-measures experiment eliminates the contribution of individual differences. There are no individual differences contributing to the numerator of the F-ratio ($MS_{\text{between treatments}}$) because the same individuals are used for all treatments. Therefore,

you must also eliminate individual differences in the denominator. This is accomplished by partitioning within-treatments variability into two components: between-subjects variability and error variability. It is the *MS* value for error variability that is used in the denominator of the *F*-ratio.

DEMONSTRATION 13.1

REPEATED-MEASURES ANOVA

The following data were obtained from a research study examining the effect of sleep deprivation on motor-skills performance. A sample of five participants was tested on a motor-skills task after 24 hours of sleep deprivation, tested again after 36 hours, and tested once more after 48 hours. The dependent variable is the number of errors made on the motor-skills task. Do these data indicate that the number of hours of sleep deprivation has a significant effect on motor skills performance?

Participant	24 Hours	36 Hours	48 Hours	P totals	
A	0	0	6	6	$N = 15$
B	1	3	5	9	$G = 45$
C	0	1	5	6	$\Sigma X^2 = 245$
D	4	5	9	18	
E	0	1	5	6	
	$T = 5$	$T = 10$	$T = 30$		
	$SS = 12$	$SS = 16$	$SS = 12$		

STEP 1 **State the hypotheses, and specify alpha** The null hypothesis states that for the general population there are no differences among the three deprivation conditions. Any differences that exist among the samples are simply the result of chance or error. In symbols,

$$H_0: \mu_1 = \mu_2 = \mu_3$$

The alternative hypothesis states that there are differences among the conditions.

$$H_1: \text{At least one of the treatment means is different.}$$

We will use $\alpha = .05$.

STEP 2 **The repeated-measures analysis** Rather than compute the *df* values and look for a critical value for *F* at this time, we proceed directly to the ANOVA.

STAGE 1 The first stage of the analysis is identical to the independent-measures ANOVA presented in Chapter 13.

$$SS_{total} = \Sigma X^2 - \frac{G^2}{N} = 245 - \frac{45^2}{15} = 110$$

$$SS_{within} = \Sigma SS_{inside\ each\ treatment} = 12 + 16 + 12 = 40$$

$$SS_{between} = \Sigma \frac{T^2}{n} - \frac{G^2}{N} = \frac{5^2}{5} + \frac{10^2}{5} + \frac{30^2}{5} - \frac{45^2}{15} = 70$$

and the corresponding degrees of freedom are

$$df_{total} = N - 1 = 14$$

$$df_{within} = \Sigma df = 4 + 4 + 4 = 12$$

$$df_{between} = k - 1 = 2$$

STAGE 2   The second stage of the repeated-measures analysis measures and removes the individual differences from the denominator of the F-ratio.

$$SS_{between\ subjects} = \Sigma \frac{P^2}{k} - \frac{G^2}{N}$$

$$= \frac{6^2}{3} + \frac{9^2}{3} + \frac{6^2}{3} + \frac{18^2}{3} + \frac{6^2}{3} - \frac{45^2}{15}$$

$$= 36$$

$$SS_{error} = SS_{within} - SS_{between\ subjects}$$

$$= 40 - 36$$

$$= 4$$

and the corresponding df values are

$$df_{between\ subjects} = n - 1 = 4$$

$$df_{error} = df_{within} - df_{between\ subjects}$$

$$= 12 - 4$$

$$= 8$$

The mean square values that form the F-ratio are as follows:

$$MS_{between} = \frac{SS_{between}}{df_{between}} = \frac{70}{2} = 35$$

$$MS_{error} = \frac{SS_{error}}{df_{error}} = \frac{4}{8} = 0.50$$

Finally, the F-ratio is

$$F = \frac{MS_{between}}{MS_{error}} = \frac{35}{0.50} = 70.00$$

STEP 3   **Make a decision and state a conclusion**   With $df = 2, 8$ and $\alpha = .05$, the critical value is $F = 4.46$. Our obtained F-ratio ($F = 70.00$) is well into the critical region, so our decision is to reject the null hypothesis and conclude that there are significant differences among the three levels of sleep deprivation.

DEMONSTRATION 13.2

EFFECT SIZE FOR THE REPEATED-MEASURES ANOVA

We will compute η^2, the percentage of variance explained by the treatment differences, for the data in Demonstration 13.1. Using Equation 13.11 we obtain

$$\eta^2 = \frac{SS_{between\ treatments}}{SS_{between\ treatments} + SS_{error}} = \frac{70}{70 + 4} = \frac{70}{74} = 0.95 \quad (or\ 95\%)$$

PROBLEMS

1. How does the denominator of the F-ratio (the error term) for a repeated-measures ANOVA compare to the denominator for an independent-measures ANOVA?

2. The repeated-measures ANOVA can be viewed as a two-stage process. What is the purpose for the second stage?

3. A researcher conducts an experiment comparing four treatment conditions with $n = 12$ scores in each condition.
   a. If the researcher uses an independent-measures design, how many individuals are needed for the study and what are the df values for the F-ratio?
   b. If the researcher uses a repeated-measures design, how many individuals are needed for the study and what are the df values for the F-ratio?

4. A researcher conducts a repeated-measures experiment using a sample of $n = 15$ subjects to evaluate the differences among three treatment conditions. If the results are examined with an ANOVA, what are the df values for the F-ratio?

5. The following data were obtained from a repeated-measures study comparing three treatment conditions. Use a repeated-measures ANOVA with $\alpha = .05$ to determine whether there are significant mean differences among the three treatments.

		Treatments		Person
Person	I	II	III	Totals
A	0	2	4	$P = 6$
B	0	3	6	$P = 9$
C	3	7	8	$P = 18$
D	0	7	5	$P = 12$
E	2	6	7	$P = 15$

$N = 15$
$G = 60$
$\Sigma X^2 = 350$

$M = 1 \quad M = 5 \quad M = 6$
$T = 5 \quad T = 25 \quad T = 30$
$SS = 8 \quad SS = 22 \quad SS = 10$

6. The following data represent the results of a repeated-measures study comparing different viewing distances for a 42-inch high-definition television. Four viewing distances were evaluated, 9 feet, 12 feet, 15 feet, and 18 feet. Each participant was free to move back and forth among the four distances while watching a 30-minute video on the television. The only restriction was that each person had to spend at least 2 minutes watching from each of the four distances. At the end of the video, each participant rated the all of the viewing distances on a scale from 1 (Very Bad, definitely need to move closer or farther away) to 7 (excellent, perfect viewing distance).
   a. Use a repeated-measures ANOVA with $\alpha = .05$ to determine whether there are significant difference among the four viewing distances.
   b. Compute η^2 to measure the size of the treatment effect.

		Viewing Distance			Person
Person	9 Feet	12 Feet	15 Feet	18 Feet	Totals
A	3	4	7	6	$P = 20$
B	0	3	6	3	$P = 12$
C	2	1	5	4	$P = 12$
D	0	1	4	3	$P = 8$
E	0	1	3	4	$P = 8$

$n = 5$
$k = 4$
$N = 20$
$G = 60$
$\Sigma X^2 = 262$

$T = 5 \quad T = 10 \quad T = 25 \quad T = 20$
$SS = 8 \quad SS = 8 \quad SS = 10 \quad SS = 6$

7. The following data were obtained from a repeated-measures study comparing two treatment conditions. Use a repeated-measures ANOVA with $\alpha = .05$ to determine whether there are significant mean differences between the two treatments.

	Treatments		Person
Person	I	II	Totals
A	3	5	$P = 8$
B	5	9	$P = 14$
C	1	5	$P = 6$
D	1	7	$P = 8$
E	5	9	$P = 14$
F	3	7	$P = 10$
G	2	6	$P = 8$
H	4	8	$P = 12$

$N = 16$
$G = 80$
$\Sigma X_2 = 500$

$M = 3 \quad M = 7$
$T = 24 \quad T = 56$
$SS = 18 \quad SS = 18$

8. The following data were obtained from a repeated-measures study comparing three treatment conditions.
   a. Use a repeated-measures ANOVA with $\alpha = .05$ to determine whether there are significant mean differences among the three treatments.
   b. Compute η^2, the percentage of variance accounted for by the mean differences, to measures the size of the treatment effects.
   c. Write a sentence demonstrating how a research report would present the results of the hypothesis test and the measure of effect size.

Person	Treatments			Person Totals
	I	II	III	
A	1	1	4	$P = 6$
B	3	4	8	$P = 15$ $N = 15$
C	0	2	7	$P = 9$ $G = 45$
D	0	0	6	$P = 6$ $\Sigma X^2 = 231$
E	1	3	5	$P = 9$

$$M = 1 \quad M = 2 \quad M = 6$$
$$T = 5 \quad T = 10 \quad T = 30$$
$$SS = 6 \quad SS = 10 \quad SS = 10$$

9. For the data in problem 8,
   a. Compute SS_{total} and $SS_{between\ treatments}$.
   b. Eliminate the mean differences between treatments by adding 2 points to each score in treatment I, adding 1 point to each score in treatment II, and subtracting 3 points from each score in treatment II. (All three treatments should end up with $M = 3$ and $T = 15$.)
   c. Calculate SS_{total} for the modified scores. (Caution: You first must find the new value for ΣX^2.)
   d. Because the treatment effects were eliminated in part b, you should find that SS_{total} for the modified scores is smaller than SS_{total} for the original scores. The difference between the two SS values should be exactly equal to the value of $SS_{between\ treatments}$ for the original scores.

10. In the Preview section for this chapter, we presented an example of a delayed discounting study in which people are willing to settle for a smaller reward today in exchange for a larger reward in the future. The following data represent the typical results from one of these studies. The participants are asked how much they would take today instead of waiting for a specific delay period to receive $1000. Each participant responds to all 5 of the delay periods. Use a repeated-measures ANOVA with $\alpha = .01$ to determine whether there are significant differences among the 5 delay periods for the following data:

Participant	1 month	6 months	1 year	2 years	5 years
A	950	850	800	700	550
B	800	800	750	700	600
C	850	750	650	600	500
D	750	700	700	650	550
E	950	900	850	800	650
F	900	900	850	750	650

11. The endorphins released by the brain act as natural painkillers. For example, Gintzler (1970) monitored endorphin activity and pain thresholds in pregnant rats during the days before they gave birth. The data showed an increase in pain threshold as the pregnancy progressed. The change was gradual until 1 or 2 days before birth, at which point there was an abrupt increase in pain threshold. Apparently a natural painkilling mechanism was preparing the animals for the stress of giving birth. The following data represent pain-threshold scores similar to the results obtained by Gintzler. Do these data indicate a significant change in pain threshold? Use a repeated-measures ANOVA with $\alpha = .01$.

Subject	Days Before Giving Birth			
	7	5	3	1
A	39	40	49	52
B	38	39	44	55
C	44	46	50	60
D	40	42	46	56
E	34	33	41	52

12. A researcher is evaluating customer satisfaction with the service and coverage of two phone carriers. Each individual in a sample of $n = 16$ uses one carrier for two weeks and then switches to the other. Each participant then rates two carriers. The following table presents the results from the repeated-measures ANOVA comparing the average ratings. Fill in the missing values in the table. (*Hint:* Start with the df values.)

Source	SS	df	MS	
Between treatments	___	___	8	$F = $ ___
Within treatments	___	___		
Between subjects	___	___		
Error	60	___	___	
Total	103	___		

13. The following summary table presents the results from a repeated-measures ANOVA comparing three treatment conditions with a sample of $n = 8$ participants. Fill in the missing values in the table. (*Hint:* Start with the *df* values.)

Source	SS	df	MS	
Between treatments	___	___	___	$F = 6.50$
Within treatments	70	___		
Between subjects	___	___		
Error	28	___	___	
Total	___	___		

14. The following summary table presents the results from a repeated-measures ANOVA comparing four treatment conditions, each with a sample of $n = 20$ participants. Fill in the missing values in the table. (*Hint:* Start with the *df* values.)

Source	SS	df	MS	
Between treatments	33	___	___	$F = $ ___
Within treatments	___	___		
Between subjects	___	___		
Error	___	___	3	
Total	263	___		

15. A researcher uses a repeated-measures ANOVA to evaluate the results from a research study and reports an *F*-ratio with $df = 3, 24$.
   a. How many treatment conditions were compared in the study?
   b. How many individuals participated in the study?

16. A published report of a repeated-measures research study includes the following description of the statistical analysis. "The results show significant differences among the treatment conditions, $F(2, 26) = 4.87, p < .05$."
   a. How many treatment conditions were compared in the study?
   b. How many individuals participated in the study?

17. The following data are from a repeated-measures study comparing three treatment conditions.
   a. Use a repeated-measures ANOVA with $\alpha = .05$ to determine whether there are significant differences among the treatments and compute η^2 to measure the size of the treatment effect.
   b. Double the number of scores in each treatment by simply repeated the original scores in each treatment a second time. For example, the $n = 8$ scores in treatment I become 1, 4, 2, 1, 1, 4, 2, 1.

Note that this will not change the treatment means but it will double $SS_{\text{between treatments}}$, $SS_{\text{between subjects}}$, and the *SS* value for each treatment. For the new data, use a repeated-measures ANOVA with $\alpha = .05$ to determine whether there are significant differences among the treatments and compute η^2 to measure the size of the treatment effect.
   c. Describe how doubling the sample size affected the value of the F-ratio and the value of η^2.

		Treatments		Person
Person	I	II	III	Totals
A	1	4	7	$P = 12$
B	4	8	6	$P = 18$
C	2	7	9	$P = 18$
D	1	5	6	$P = 12$

$M = 2$ $M = 6$ $M = 7$
$T = 8$ $T = 24$ $T = 28$
$SS = 6$ $SS = 10$ $SS = 6$

$N = 12$
$G = 60$
$\Sigma X^2 = 378$

18. The following data were obtained from a repeated-measures study comparing three treatment conditions.

		Treatment		
Subject	I	II	III	P
A	6	8	10	24
B	5	5	5	15
C	1	2	3	6
D	0	1	2	3

$T = 12$ $T = 16$ $T = 20$
$SS = 26$ $SS = 30$ $SS = 38$

$G = 48$
$\Sigma X^2 = 294$

Use a repeated-measures ANOVA with $\alpha = .05$ to determine whether these data are sufficient to demonstrate significant differences between the treatments.

19. In Problem 18 the data show large and consistent differences between subjects. For example, subject A has the largest score in every treatment and subject D always has the smallest score. In the second stage of the ANOVA, the large individual differences get subtracted out of the denominator of the *F*-ratio, which results in a larger value for *F*.

The following data were created by using the same numbers that appeared in Problem 18. However, we eliminated the consistent individual differences by scrambling the scores within each treatment.

Subject	Treatment I	II	III	P	
A	6	2	3	11	$G = 48$
B	5	1	5	11	$\Sigma X^2 = 294$
C	0	5	10	15	
D	1	8	2	11	
	$T = 12$	$T = 16$	$T = 20$		
	$SS = 26$	$SS = 30$	$SS = 38$		

a. Use a repeated-measures ANOVA with $\alpha = .05$ to determine whether these data are sufficient to demonstrate significant differences between the treatments.

b. The data in problem 18 showed consistent differences between subjects and produced significant treatment effects. Explain how eliminating the consistent individual differences affected the results of this analysis compared with the results from Problem 18.

20. A recent study indicates that simply giving college students a pedometer can result in increased walking (Jackson & Howton, 2008). Students were given pedometers for a 12-week period, and asked to record the average number of steps per day during weeks 1, 6, and 12. The following data are similar to the results obtained in the study.

	Number of steps (×1000)				
Participant	Week 1	6	12	P	
A	6	8	10	24	
B	4	5	6	15	
C	5	5	5	15	$G = 72$
D	1	2	3	6	$\Sigma X^2 = 400$
E	0	1	2	3	
F	2	3	4	9	
	$T = 18$	$T = 24$	$T = 30$		
	$SS = 28$	$SS = 32$	$SS = 40$		

a. Use a repeated-measures ANOVA with $\alpha = .05$ to determine whether the mean number of steps changes significantly from one week to another.

b. Compute η^2 to measure the size of the treatment effect.

c. Write a sentence demonstrating how a research report would present the results of the hypothesis test and the measure of effect size.

21. One of the primary advantages of a repeated-measures design, compared to independent-measures, is that it reduces the overall variability by removing variance caused by individual differences. In the previous problem, the very large differences among the P totals indicate very large individual differences. For the repeated-measures ANOVA, removing these differences greatly reduces the variance and results in a significant F-ratio and a large value for η^2. Assume that the same data were obtained from an independent-measures study comparing three treatment and use an independent-measures ANOVA to evaluate the mean differences and then compute η^2. You should find that the F is no longer significant and η^2 is relatively small. (Note that most of the calculations were completed in Problem 20.)

22. One of the primary advantages of a repeated-measures design, compared to independent-measures, is that it reduces the overall variability by removing variance caused by individual differences. The following data are from a research study comparing three treatment conditions.

a. Assume that the data are from an independent-measures study using three separate samples, each with $n = 6$ participants. Ignore the column of P totals and use an independent-measures ANOVA with $\alpha = .05$ to test the significance of the mean differences.

b. Now assume that the data are from a repeated-measures study using the same sample of $n = 6$ participants in all three treatment conditions. Use a repeated-measures ANOVA with $\alpha = .05$ to test the significance of the mean differences.

c. Explain why the two analyses lead to different conclusions.

Treatment 1	Treatment 2	Treatment 3	P	
6	9	12	27	
8	8	8	24	$N = 18$
5	7	9	21	$G = 108$
0	4	8	12	$\Sigma X^2 = 800$
2	3	4	9	
3	5	7	15	
$M = 4$	$M = 6$	$M = 8$		
$T = 24$	$T = 36$	$T = 48$		
$SS = 42$	$SS = 28$	$SS = 34$		

23. The following data are from an experiment comparing three different treatment conditions:

A	B	C	
0	1	2	$N = 15$
2	5	5	$\Sigma X^2 = 354$
1	2	6	
5	4	9	
2	8	8	
$T = 10$	$T = 20$	$T = 30$	
$SS = 14$	$SS = 30$	$SS = 30$	

a. If the experiment uses an *independent-measures design,* can the researcher conclude that the treatments are significantly different? Test at the .05 level of significance.
b. If the experiment is done with a *repeated-measures design,* should the researcher conclude that the treatments are significantly different? Set alpha at .05 again.
c. Explain why the analyses in parts a and b lead to different conclusions.

24. The following data are from a repeated-measures study comparing two treatment conditions.
   a. Use a repeated-measures ANOVA with $\alpha = .05$ to determine whether the mean difference between treatments is significant.
   b. Now use a repeated-measures t test with $\alpha = .05$ to evaluate the mean difference between treatments. Comparing your answers from a and b, you should find that $F = t^2$.

	Treatment	
Participant	I	II
A	11	13
B	8	7
C	10	13
D	8	8
E	7	13
F	10	12

25. In the Preview section for Chapter 2 we presented a study showing that a visible tattoo can significantly lower the attractiveness rating of a woman shown in a photograph (Resenhoeft, Villa, & Wiseman, 2008). Suppose a similar experiment is conducted as a repeated-measures study. A sample of $n = 9$ males looks at a set of 30 photographs of women and rates

the attractiveness of each woman using a 10-point scale (10 = most positive). One photograph appears twice in the set, once with a tattoo and once with the tattoo removed. For each participant, the researcher records the difference between the two ratings of the same photograph. The results are shown in the following table.
   a. Use a repeated-measures ANOVA with $\alpha = .05$ to determine whether the mean difference between treatments is significant.
   b. Now use a repeated-measures t test with $\alpha = .05$ to evaluate the mean difference between treatments. Comparing your answers from a and b, you should find that $F = t^2$.

Participant	No Tattoo	With Tattoo
A	7	5
B	9	3
C	7	5
D	8	3
E	9	8
F	6	3
G	6	6
H	8	3
I	6	3

26. A repeated-measures experiment comparing only two treatments can be evaluated with either a t statistic or an ANOVA. As we found with the independent-measures design, the t test and the ANOVA produce equivalent conclusions, and the two test statistics are related by the equation $F = t^2$. The following data are from a repeated-measures study:

Subject	Treatment 1	Treatment 2	Difference
1	2	4	+2
2	1	3	+2
3	0	10	+10
4	1	3	+2

a. Use a repeated-measures t statistic with $\alpha = .05$ to determine whether the data provide evidence of a significant difference between the two treatments. (*Caution:* ANOVA calculations are done with the X values, but for t you use the difference scores.)
b. Use a repeated-measures ANOVA with $\alpha = .05$ to evaluate the data. (You should find $F = t^2$.)

27. For either independent-measures or repeated-measures designs comparing two treatments, the mean difference can be evaluated with either a *t* test or an ANOVA. The two tests are related by the equation $F = t^2$. For the following data,

a. Use a repeated-measures *t* test with $\alpha = .05$ to determine whether the mean difference between treatments is statistically significant.

b. Use a repeated-measures ANOVA with $\alpha = .05$ to determine whether the mean difference between treatments is statistically significant. (You should find that $F = t^2$.)

Person	Treatment 1	Treatment 2	Difference
A	4	7	3
B	2	11	9
C	3	6	3
D	7	10	3
	$M = 4$	$M = 8.5$	$M_D = 4.5$
	$T = 16$	$T = 34$	
	$SS = 14$	$SS = 17$	$SS = 27$

Two-Factor Analysis of Variance (Independent Measures)

Tools You Will Need

The following items are considered essential background material for this chapter. If you doubt your knowledge of any of these items, you should review the appropriate chapter or section before proceeding.

- Introduction to analysis of variance (Chapter 12)
  - The logic of analysis of variance
  - ANOVA notation and formulas
  - Distribution of *F*-ratios

© Deborah Batt

Does experiencing violence in video games have an effect on the players' behavior? One study suggests that the answer is yes and no. Bartholow and Anderson (2002) randomly assigned male and female undergraduate students to play a violent video game or a nonviolent game. After the game, each participant was asked to take part in a competitive reaction time game with another student who was actually part of the research team (a confederate). Both students were instructed to respond as quickly as possible to a stimulus tone and, on each trial, the loser was punished with a blast of white noise delivered through headphones. Part of the instructions allowed the participant to set the level of intensity for the punishment noise and the level selected was used as a measure of aggressive behavior for that participant, with higher levels indicating more aggressive behavior. The results of the study showed that the level of violence in the video game had essentially no effect on the behavior of the female participants but the males were significantly more aggressive after playing the violent game compared to the nonviolent game.

The Bartholow and Anderson study is an example of research that involves two independent variables in the same study. The independent variables are:

1. Level of violence in the video game (high or low)

2. Gender (male or female)

The results of the study indicate that the effect of one variable (violence) *depends on* another variable (gender).

You should realize that it is quite common to have two variables that interact in this way. For example, a drug may have profound effects on some patients and have no effect whatsoever on others. Some children survive abusive environments and live normal, productive lives, while others show serious difficulties. To observe how one variable interacts with another, it is necessary to study both variables simultaneously in one study. However, the analysis of variance (ANOVA) procedures introduced in Chapters 12 and 13 are limited to evaluating mean differences produced by one independent variable and are not appropriate for mean differences involving two (or more) independent variables.

Fortunately, ANOVA is a very flexible hypothesis testing procedure and can be modified again to evaluate the mean differences produced in a research study with two (or more) independent variables. In this chapter we introduce the *two-factor* ANOVA, which tests the significance of each independent variable acting alone as well as the interaction between variables.

14.1 | An Overview of the Two-Factor, Independent-Measures, ANOVA: Main Effects and Interactions

LEARNING OBJECTIVES

1. Describe the structure of a factorial research design, especially a two-factor independent-measures design, using the terms factor and level.

2. Define a main effect and an interaction and identify the patterns of data that produce main effects and interactions.

3. Identify the three F-ratios for a two-factor ANVOA and explain how they are related to each other.

In most research situations, the goal is to examine the relationship between two variables. Typically, the research study attempts to isolate the two variables to eliminate or reduce the influence of any outside variables that may distort the relationship being studied. A typical experiment, for example, focuses on one independent variable (which is expected to influence behavior) and one dependent variable (which is a measure of the behavior). In real life, however, variables rarely exist in isolation. That is, behavior usually is influenced by a variety of different variables acting and interacting simultaneously. To examine these more complex, real-life situations, researchers often design research studies that include more than one independent variable. Thus, researchers systematically change two (or more) variables and then observe how the changes influence another (dependent) variable.

An independent variable is a manipulated variable in an experiment. A quasi-independent variable is not manipulated but defines the groups of scores in a nonexperimental study.

In Chapters 12 and 13, we examined ANOVA for *single-factor* research designs—that is, designs that included only one independent variable or only one quasi-independent variable. When a research study involves more than one factor, it is called a *factorial design*. In this chapter, we consider the simplest version of a factorial design. Specifically, we examine ANOVA as it applies to research studies with exactly two factors. In addition, we limit our discussion to studies that use a separate sample for each treatment condition—that is, independent-measures designs. Finally, we consider only research designs for which the sample size (n) is the same for all treatment conditions. In the terminology of ANOVA, this chapter examines *two-factor, independent-measures, equal n designs.*

We will use the Bartholow and Anderson video game violence study described in the Chapter Preview to introduce the two-factor research design. Table 14.1 shows the structure of the study. Note that the study involves two separate factors: one factor is manipulated by the researcher, changing from a violent to a nonviolent game, and the second factor is gender, which varies from male to female. The two factors are used to create a *matrix* with the different genders defining the rows and the different levels of violence defining the columns. The resulting two-by-two matrix shows four different combinations of the variables, producing four different conditions. Thus, the research study would require four separate samples, one for each *cell*, or box, in the matrix. The dependent variable for the study is the level of aggressive behavior for the participants in each of the four conditions.

TABLE 14.1

The structure of a two-factor experiment presented as a matrix. The two factors are gender and level of violence in a video game, with two levels for each factor.

		Factor *B*: Level of Violence	
		Nonviolent	Violent
Factor *A*: Gender	Male	Scores for a group of males who play a nonviolent video game	Scores for a group of males who play a violent video game
	Female	Scores for a group of females who play a nonviolent video game	Scores for a group of females who play a violent video game

The two-factor ANOVA tests for mean differences in research studies that are structured like the gender-and-video-violence study in Table 14.1. For this example, the two-factor ANOVA evaluates three separate sets of mean differences.

1. What happens to the level of aggressive behavior when violence is added or taken away from the game?

2. Is there a difference in the aggressive behavior for male participants compared to females?

3. Is aggressive behavior affected by specific combinations of game violence and gender? (For example, a violent game may have a large effect on males but only a small effect for females.)

Thus, the two-factor ANOVA allows us to examine three types of mean differences within one analysis. In particular, we conduct three separate hypotheses tests for the same data, with a separate *F*-ratio for each test. The three *F*-ratios have the same basic structure:

$$F = \frac{\text{variance (differences) between treatments}}{\text{variance (differences) expected if there is no treatment effect}}$$

In each case, the numerator of the F-ratio measures the actual mean differences in the data, and the denominator measures the differences that would be expected if there is no treatment effect. As always, a large value for the F-ratio indicates that the sample mean differences are greater than would be expected by chance alone, and therefore provides evidence of a treatment effect. To determine whether the obtained F-ratios are *significant*, we need to compare each F-ratio with the critical values found in the F-distribution table in Appendix B.

■ Main Effects and Interactions

As noted in the previous section, a two-factor ANOVA actually involves three distinct hypothesis tests. In this section, we examine these three tests in more detail.

Traditionally, the two independent variables in a two-factor experiment are identified as factor A and factor B. For the study presented in Table 14.1, gender is factor A, and the level of violence in the game is factor B. The goal of the study is to evaluate the mean differences that may be produced by either of these factors acting independently or by the two factors acting together.

■ Main Effects

One purpose of the study is to determine whether differences in gender (factor A) result in differences in behavior. To answer this question, we compare the mean score for all the males with the mean for the females. Note that this process evaluates the mean difference between the top row and the bottom row in Table 14.1.

To make this process more concrete, we present a set of hypothetical data in Table 14.2. The table shows the mean score for each of the treatment conditions (cells) as well as the overall mean for each column (each level of violence) and the overall mean for each row (each gender group). These data indicate that the male participants (the top row) had an overall mean of $M = 8$. This overall mean was obtained by computing the average of the two means in the top row. In contrast, the female participants had an overall mean of $M = 4$ (the mean for the bottom row). The difference between these means constitutes what is called the *main effect* for gender, or the *main effect for factor* A.

TABLE 14.2   Hypothetical data for an experiment examining the effect of violence in a video game on the aggressive behavior of males and females.

	Nonviolent Game	Violent Game	
Males	$M = 7$	$M = 9$	$M = 8$
Females	$M = 3$	$M = 5$	$M = 4$
	$M = 5$	$M = 7$	

Similarly, the main effect for factor B (level of violence) is defined by the mean difference between the columns of the matrix. For the data in Table 14.2, the two groups of participants who played a nonviolent game had an overall mean score of $M = 5$. Participants who played a violent game had an overall average score of $M = 7$. The difference between these means constitutes the *main effect* for the level of game violence, or the *main effect for factor* B.

DEFINITION

> The mean differences among the levels of one factor are referred to as the **main effect** of that factor. When the design of the research study is represented as a matrix with one factor determining the rows and the second factor determining the columns, then the mean differences among the rows describe the main effect of one factor, and the mean differences among the columns describe the main effect for the second factor.

The mean differences between columns or rows simply *describe* the main effects for a two-factor study. As we have observed in earlier chapters, the existence of sample mean differences does not necessarily imply that the differences are *statistically significant*. In general, two samples are not expected to have exactly the same means. There will always be small differences from one sample to another, and you should not automatically assume that these differences are an indication of a systematic treatment effect. In the case of a two-factor study, any main effects that are observed in the data must be evaluated with a hypothesis test to determine whether they are statistically significant effects. Unless the hypothesis test demonstrates that the main effects are significant, you must conclude that the observed mean differences are simply the result of sampling error.

The evaluation of main effects accounts for two of the three hypothesis tests in a two-factor ANOVA. We state hypotheses concerning the main effect of factor A and the main effect of factor B and then calculate two separate F-ratios to evaluate the hypotheses.

For the example we are considering, factor A involves the comparison of two different genders. The null hypothesis would state that there is no difference between the two levels; that is, gender has no effect on aggressive behavior. In symbols,

$$H_0: \mu_{A_1} = \mu_{A_2}$$

The alternative hypothesis is that the two genders do produce different aggression scores:

$$H_1: \mu_{A_1} \neq \mu_{A_2}$$

To evaluate these hypotheses, we compute an F-ratio that compares the actual mean differences between the two genders vs. the amount of difference that would be expected without any systematic difference.

$$F = \frac{\text{variance (differences) between the means for factor } A}{\text{variance (differences) expected if there is no treatment effect}}$$

$$F = \frac{\text{variance (differences) between the row means}}{\text{variance (differences) expected if there is no treatment effect}}$$

Similarly, factor B involves the comparison of the two different violence conditions. The null hypothesis states that there is no difference in the mean level of aggression between the two conditions. In symbols,

$$H_0: \mu_{B_1} = \mu_{B_2}$$

As always, the alternative hypothesis states that the means are different:

$$H_1: \mu_{B_1} \neq \mu_{B_2}$$

Again, the F-ratio compares the obtained mean difference between the two violence conditions vs. the amount of difference that would be expected if there is no systematic treatment effect.

$$F = \frac{\text{variance (differences) between the means for factor } B}{\text{variance (differences) expected if there is no treatment effect}}$$

$$F = \frac{\text{variance (differences) between the column means}}{\text{variance (differences) expected if there is no treatment effect}}$$

■ Interactions

In addition to evaluating the main effect of each factor individually, the two-factor ANOVA allows you to evaluate other mean differences that may result from unique combinations of the two factors. For example, specific combinations of game violence and gender acting together may have effects that are different from the effects of gender or game violence acting alone. Any "extra" mean differences that are not explained by the main effects are called an *interaction*, or an *interaction between factors*. The real advantage of combining two factors within the same study is the ability to examine the unique effects caused by an interaction.

DEFINITION

> An **interaction** between two factors occurs whenever the mean differences between individual treatment conditions, or cells, are different from what would be predicted from the overall main effects of the factors.

To make the concept of an interaction more concrete, we reexamine the data shown in Table 14.2. For these data, there is no interaction; that is, there are no extra mean differences that are not explained by the main effects. For example, within each violence condition (each column of the matrix) the average level of aggression for the male participants is 4 points higher than the average for the female participants. This 4-point mean difference is exactly what is predicted by the overall main effect for gender.

Now consider a different set of data shown in Table 14.3. These new data show exactly the same main effects that existed in Table 14.2 (the column means and the row means have not been changed). There is still a 4-point mean difference between the two rows (the main effect for gender) and a 2-point mean difference between the two columns (the main effect for violence). But now there is an interaction between the two factors. For example, for the male participants (top row), there is a 4-point difference in the level of aggression after a violent game vs. a nonviolent game. This 4-point difference cannot be explained by the 2-point main effect for the violence factor. Also, for the female participants (bottom row), the data show no difference between the two game violence conditions. Again, the zero difference is not what would be expected based on the 2-point main effect for the game violence factor. Mean differences that are not explained by the main effects are an indication of an interaction between the two factors.

The data in Table 14.3 show the same pattern of results that was obtained in the Bartholow and Anderson research study.

TABLE 14.3

Hypothetical data for an experiment examining the effect of violence in a video game on the aggressive behavior of males and females. The data show the same main effects as the values in Table 14.2 but the individual treatment means have been modified to create an interaction.

	Nonviolent Game	Violent Game	
Male	$M = 6$	$M = 10$	$M = 8$
Female	$M = 4$	$M = 4$	$M = 4$
	$M = 5$	$M = 7$	

To evaluate the interaction, the two-factor ANOVA first identifies mean differences that are not explained by the main effects. The extra mean differences are then evaluated by an *F*-ratio with the following structure:

$$F = \frac{\text{variance (mean differences) not explained by main effects}}{\text{variance (differences) expected if there is no treatment effects}}$$

The null hypothesis for this *F*-ratio simply states that there is no interaction:

H_0:   There is no interaction between factors A and B. The mean differences between treatment conditions are explained by the main effects of the two factors.

The alternative hypothesis is that there is an interaction between the two factors:

H_1:   There is an interaction between factors. The mean differences between treatment conditions are not what would be predicted from the overall main effects of the two factors.

■ More about Interactions

In the previous section, we introduced the concept of an interaction as the unique effect produced by two factors working together. This section presents two alternative definitions of an interaction. These alternatives are intended to help you understand the concept of an interaction and to help you identify an interaction when you encounter one in a set of data. You should realize that the new definitions are equivalent to the original and simply present slightly different perspectives on the same concept.

The first new perspective on the concept of an interaction focuses on the notion of independence for the two factors. More specifically, if the two factors are independent, so that one factor does not influence the effect of the other, then there is no interaction. On the other hand, when the two factors are not independent, so that the effect of one factor *depends on* the other, then there is an interaction. The notion of dependence between factors is consistent with our earlier discussion of interactions. If one factor influences the effect of the other, then unique combinations of the factors produce unique effects.

DEFINITION

> When the effect of one factor depends on the different levels of a second factor, then there is an **interaction** between the factors.

This definition of an interaction should be familiar in the context of a "drug interaction." Your doctor and pharmacist are always concerned that the effect of one medication may be altered or distorted by a second medication that is being taken at the same time. Thus, the effect of one drug (factor *A*) depends on a second drug (factor *B*), and you have an interaction between the two drugs.

Returning to Table 14.2, you will notice that the size of the game-violence effect (first column vs. second column) *does not depend* on the gender of the participants. For these data, adding violence produces the same 2-point increase in aggressive behavior for both groups of participants. Thus, the effect of game violence does not depend on gender and there is no interaction. Now consider the data in Table 14.3. This time, the effect of adding violence *depends on* the gender of the participants. For example, there is a 4-point increase in aggressive behavior for the males but adding violence has no effect on aggression for the females. Thus, the effect of game violence depends on gender, which means that there is an interaction between the two factors.

The second alternative definition of an interaction is obtained when the results of a two-factor study are presented in a graph. In this case, the concept of an interaction can be defined in terms of the pattern displayed in the graph. Figure 14.1 shows the two sets of data we have been considering. The original data from Table 14.2, where there is no interaction, are presented in Figure 14.1(a). To construct this figure, we selected one of the factors to be displayed on the horizontal axis; in this case, the different levels of game violence are displayed. The dependent variable, the level of aggressive behavior, is shown on the vertical axis. Note that the figure actually contains two separate graphs: The top line shows the relationship between game violence and aggression for the males, and the bottom line shows the relationship for the females. In general, the picture in

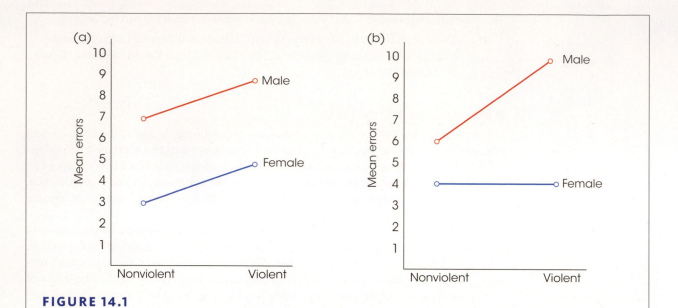

FIGURE 14.1
(a) Graph showing the treatment means for Table 14.2, for which there is no interaction. (b) Graph for Table 14.3, for which there is an interaction.

the graph matches the structure of the data matrix; the columns of the matrix appear as values along the X-axis, and the rows of the matrix appear as separate lines in the graph (see Box 14.1).

For the original set of data, Figure 14.1(a), note that the two lines are parallel; that is, the distance between lines is constant. In this case, the distance between lines reflects the 2-point difference in the mean aggression scores for males and females, and this 2-point difference is the same for both game violence conditions.

Now look at a graph that is obtained when there is an interaction in the data. Figure 14.1(b) shows the data from Table 14.3. This time, note that the lines in the graph are not parallel. The distance between the lines changes as you scan from left to right. For these data, the distance between the lines corresponds to the gender effect—that is, the mean difference in aggression for male vs. female participants. The fact that this difference depends on the level of game violence is an indication of an interaction between the two factors.

DEFINITION

When the results of a two-factor study are presented in a graph, the existence of nonparallel lines (lines that cross or converge) indicates an **interaction** between the two factors.

For many students, the concept of an interaction is easiest to understand using the perspective of interdependency; that is, an interaction exists when the effects of one variable *depend* on another factor. However, the easiest way to identify an interaction within a set of data is to draw a graph showing the treatment means. The presence of nonparallel lines is an easy way to spot an interaction.

BOX 14.1 Graphing Results from a Two-Factor Design

One of the best ways to get a quick overview of the results from a two-factor study is to present the data in a graph. Because the graph must display the means obtained for *two* independent variables (two factors), constructing the graph can be a bit more complicated than constructing the single-factor graphs we presented in Chapter 3 (pp. 90–91).

Figure 14.2 shows two possible graphs presenting the results from a two-factor study with 2 levels of factor *A* and 3 levels of factor *B*. With a 2×3 design, there are a total of 6 different treatment means that are shown in the following matrix.

		Factor *B*		
		B1	B2	B3
Factor *A*	A_1	$M = 10$	$M = 40$	$M = 20$
	A_2	$M = 30$	$M = 50$	$M = 30$

In the two graphs, note that values for the dependent variable (the treatment means) are shown on the vertical axis. Also note that the levels for one factor (we selected factor B) are displayed on the horizontal axis. Either factor can be used, but it is better to select one for which the different levels are measured on an interval or ratio scale. The reason for this suggestion is that an interval or ratio scale will permit you to construct a line graph as in Figure 14.2(a). If neither of the two factors is measured on an interval or ratio scale, you should use a bar graph as in Figure 14.2(b).

Line graphs: In Figure 14.2(a), we have assumed that factor *B* is an interval or a ratio variable, and the three levels for this factor are listed on the horizontal axis. Directly above the B_1 value on the horizontal axis, we have placed two dots corresponding to the two means in the B_1 column of the data matrix. Similarly, we have placed two dots above B_2 and another two dots above B_3. Finally, we have drawn a line connecting the three dots corresponding to level 1 of factor *A* (the three means in the top row of the data matrix). We have also drawn a second line that connects the three dots corresponding to level 2 of factor *A*. These lines are labeled A_1 and A_2 in the figure.

Bar graphs: Figure 14.2(b) also shows the three levels of factor *B* displayed on the horizontal axis. This time, however, we assume that factor *B* is measured on a nominal or ordinal scale, and the result is a bar graph. Directly above the B_1 value, we have drawn two bars so that the heights of the bars correspond to the two means in the B_1 column of the data matrix. Similarly, we have drawn two bars above B_2 and two more bars above B_3. Finally, the three bars corresponding to level 1 of factor *A* (the top row of the data matrix) are all colored (or shaded) to differentiate them from the three bars for level 2 of factor *A*.

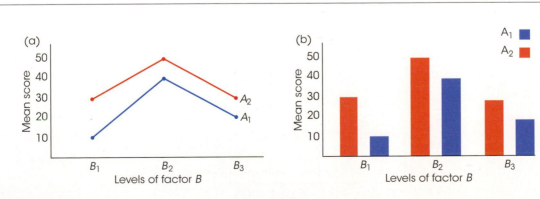

FIGURE 14.2

Two graphs showing the results from a two-factor study. A line graph is shown in (a) and a bar graph (b).

The $A \times B$ interaction typically is called "A by B" interaction. If there is an interaction between video game violence and gender, it may be called the "violence by gender" interaction.

■ Independence of Main Effects and Interactions

The two-factor ANOVA consists of three hypothesis tests, each evaluating specific mean differences: the A effect, the B effect, and the $A \times B$ interaction. As we have noted, these are three *separate* tests, but you should also realize that the three tests are *independent*. That is, the outcome for any one of the three tests is totally unrelated to the outcome for either of the other two. Thus, it is possible for data from a two-factor study to display any possible combination of significant and/or not significant main effects and interactions. The data sets in Table 14.4 show several possibilities.

Table 14.4(a) shows data with mean differences between levels of factor A (an A effect) but no mean differences for factor B and no interaction. To identify the A effect, notice that the overall mean for A_1 (the top row) is 10 points higher than the overall mean for A_2 (the bottom row). This 10-point difference is the main effect for factor A. To evaluate the B effect, notice that both columns have exactly the same overall mean, indicating no difference between levels of factor B; hence, there is no B effect. Finally, the absence of an interaction is indicated by the fact that the overall A effect (the 10-point difference) is constant within each column; that is, the A effect *does not depend* on the levels of factor B. (Alternatively, the data indicate that the overall B effect is constant within each row.)

TABLE 14.4

Three sets of data showing different combinations of main effects and interaction for a two-factor study. (The numerical value in each cell of the matrices represents the mean value obtained for the sample in that treatment condition.)

(a) Data showing a main effect for factor A but no B effect and no interaction

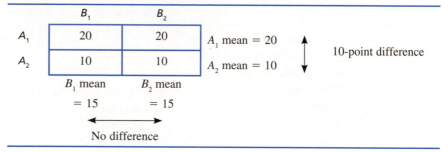

(b) Data showing main effects for both factor A and factor B but no interaction

(c) Data showing no main effect for either factor but an interaction

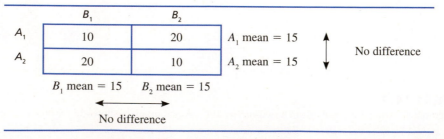

Table 14.4(b) shows data with an A effect and a B effect but no interaction. For these data, the A effect is indicated by the 10-point mean difference between rows, and the B effect is indicated by the 20-point mean difference between columns. The fact that the 10-point A effect is constant within each column indicates no interaction.

Finally, Table 14.4(c) shows data that display an interaction but no main effect for factor A or for factor B. For these data, there is no mean difference between rows (no A effect) and no mean difference between columns (no B effect). However, within each row (or within each column), there are mean differences. The "extra" mean differences within the rows and columns cannot be explained by the overall main effects and therefore indicate an interaction.

The following example is an opportunity to test your understanding of main effects and interactions.

EXAMPLE 14.1

The following matrix represents the outcome of a two-factor experiment. Describe the main effect for factor A and the main effect for factor B. Does there appear to be an interaction between the two factors?

	Experiment I	
	B_1	B_2
A_1	$M = 10$	$M = 20$
A_2	$M = 30$	$M = 40$

You should conclude that there is a main effect for factor A (the scores in A_2 average 20 points higher than in A_1) and there is a main effect for factor B (the scores in B_2 average 10 points higher than in B_1) but there is no interaction; there is a constant 20-point difference between A_1 and A_2 that does not depend on the levels of factor B. ∎

LEARNING CHECK

1. How many separate samples would be needed for a two-factor, independent-measures research study with 2 levels of factor A and 3 levels of factor B?
   a. 2
   b. 3
   c. 5
   d. 6

2. In a two-factor experiment with 2 levels of factor A and 2 levels of factor B, three of the treatment means are essentially identical and one is substantially different from the others. What result(s) would be produced by this pattern of treatment means?
   a. a main effect for factor A
   b. a main effect for factor B
   c. an interaction between A and B
   d. The pattern would produce main effects for both A and B, and an interaction.

3. In a two-factor ANOVA, what is the implication of a significant $A \times B$ interaction?
   a. At least one of the main effects must also be significant.
   b. Both of the main effects must also be significant.
   c. Neither of the two main effects can be significant.
   d. The significance of the interaction has no implications for the main effects.

ANSWERS   1. D, 2. D, 3. D

14.2 | An Example of the Two-Factor ANOVA and Effect Size

4. Describe the two-stage structure of a two-factor ANOVA and explain what happens in each stage.

5. Compute the *SS*, *df*, and *MS* values needed for a two-factor ANOVA and explain the relationships among them.

6. Conduct a two-factor ANOVA including measures of effect size for both main effects and the interaction.

The two-factor ANOVA is composed of three distinct hypothesis tests:

1. The main effect of factor *A* (often called the *A*-effect). Assuming that factor *A* is used to define the rows of the matrix, the main effect of factor *A* evaluates the mean differences between rows.

2. The main effect of factor *B* (called the *B*-effect). Assuming that factor *B* is used to define the columns of the matrix, the main effect of factor *B* evaluates the mean differences between columns.

3. The interaction (called the $A \times B$ interaction). The interaction evaluates mean differences between treatment conditions that are not predicted from the overall main effects from factor *A* or factor *B*.

For each of these three tests, we are looking for mean differences between treatments that are larger than would be expected if there are no treatment effects. In each case, the significance of the treatment effect is evaluated by an *F*-ratio. All three *F*-ratios have the same basic structure:

$$F = \frac{\text{variance (mean differences) between treatments}}{\text{variance (mean differences) expected if there are no treatment effects}} \qquad \text{(14.1)}$$

The general structure of the two-factor ANOVA is shown in Figure 14.3. Note that the overall analysis is divided into two stages. In the first stage, the total variability is separated into two components: between-treatments variability and within-treatments variability. This first stage is identical to the single-factor ANOVA introduced in Chapter 12 with each cell in the two-factor matrix viewed as a separate treatment condition. The within-treatments variability

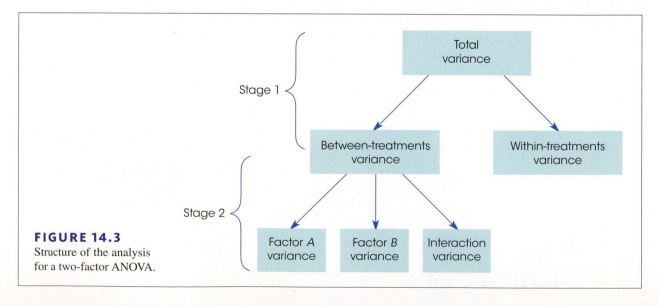

FIGURE 14.3
Structure of the analysis
for a two-factor ANOVA.

that is obtained in stage 1 of the analysis is used as the denominator for the F-ratios. As we noted in Chapter 12, within each treatment, all of the participants are treated exactly the same. Thus, any differences that exist within the treatments cannot be caused by treatment effects. As a result, the within-treatments variability provides a measure of the differences that exist when there are no systematic treatment effects influencing the scores (see Equation 14.1).

The between-treatments variability obtained in stage 1 of the analysis combines all the mean differences produced by factor A, factor B, and the interaction. The purpose of the second stage is to partition the differences into three separate components: differences attributed to factor A, differences attributed to factor B, and any remaining mean differences that define the interaction. These three components form the numerators for the three F-ratios in the analysis.

The goal of this analysis is to compute the variance values needed for the three F-ratios. We need three between-treatments variances (one for factor A, one for factor B, and one for the interaction), and we need a within-treatments variance. Each of these variances (or mean squares) is determined by a sum of squares value (SS) and a degree of freedom value (df):

Remember that in ANOVA a variance is called a mean square, or *MS*.

$$\text{mean square} = MS = \frac{SS}{df}$$

EXAMPLE 14.2

To demonstrate the two-factor ANOVA, we will use a research study based on previous work by Ackerman and Goldsmith (2011). Their study compared learning performance by students who studied text either from printed pages or from a computer screen. The results from the study indicate that students do much better studying from printed pages if their study time is self-regulated. However, when the researchers fixed the time spent studying, there was no difference between the two conditions. Apparently, students are less accurate predicting their learning performance or have trouble regulating study time when working with a computer screen compared to working with paper. Table 14.5 shows data from a two-factor study replicating the Ackerman and Goldsmith experiment. The two factors are mode of presentation (paper or computer screen) and time control (self-regulated or fixed). A separate group of $n = 5$ students was tested in each of the four conditions. The dependent variable is student performance on a quiz covering the text that was studied.

TABLE 14.5

Data for a two-factor study comparing two levels of time control (self-regulated or fixed by the researchers) and two levels of text presentation (paper and computer screen). The dependent variable is performance on a quiz covering the text that was presented. The study involves four treatment conditions with $n = 5$ participants in each treatment.

		Factor B: Text Presentation Mode		
		Paper	Computer Screen	
	Self-regulated	11 8 9 10 7 $M = 9$ $T = 45$ $SS = 10$	4 4 8 5 4 $M = 5$ $T = 25$ $SS = 12$	$T_{row} = 70$
Factor A Time Control	Fixed	10 7 10 6 7 $M = 8$ $T = 40$ $SS = 14$	10 6 10 10 9 $M = 9$ $T = 45$ $SS = 12$	$T_{row} = 85$
		$T_{col} = 85$	$T_{col} = 70$	

$N = 20$
$G = 155$
$\Sigma X^2 = 1303$

The data are displayed in a matrix with the two levels of time control (factor A) making up the rows and the two levels of presentation mode (factor B) making up the columns. Note that the data matrix has a total of four *cells* or treatment conditions with a separate sample of $n = 5$ participants in each condition. Most of the notation should be familiar from the single-factor ANOVA presented in Chapter 12. Specifically, the treatment totals are identified by T values, the total number of scores in the entire study is $N = 20$, and the grand total (sum) of all 20 scores is $G = 155$. In addition to these familiar values, we have included the totals for each row and for each column in the matrix. The goal of the ANOVA is to determine whether the mean differences observed in the data are significantly greater than would be expected if there were no treatment effects.

■ Stage 1 of the Two-Factor Analysis

The first stage of the two-factor analysis separates the total variability into two components: between-treatments and within-treatments. The formulas for this stage are identical to the formulas used in the single-factor ANOVA in Chapter 12 with the provision that each cell in the two-factor matrix is treated as a separate treatment condition. The formulas and the calculations for the data in Table 14.5 are as follows:

Total Variability

$$SS_{total} = \Sigma X^2 - \frac{G^2}{N} \tag{14.2}$$

For these data,

$$SS_{total} = 1303 - \frac{155^2}{20}$$
$$= 1303 - 1201.25$$
$$= 101.75$$

This SS value measures the variability for all $N = 20$ scores and has degrees of freedom given by

$$df_{total} = N - 1 \tag{14.3}$$

For the data in Table 14.5, $df_{total} = 19$.

Within-Treatments Variability   To compute the variance within treatments, we first compute SS and $df = n - 1$ for each of the individual treatment conditions. Then the within-treatments SS is defined as

$$SS_{within\ treatments} = \Sigma SS_{each\ treatment} \tag{14.4}$$

And the within-treatments df is defined as

$$df_{within\ treatments} = \Sigma df_{each\ treatment} \tag{14.5}$$

For the four treatment conditions in Table 14.5,

$$SS_{within\ treatments} = 10 + 12 + 14 + 12 = 48$$

$$df_{within\ treatments} = 4 + 4 + 4 + 4 = 16$$

Between-Treatments Variability   Because the two components in stage 1 must add up to the total, the easiest way to find $SS_{between\ treatments}$ is by subtraction.

$$SS_{between\ treatments} = SS_{total} - SS_{within} \tag{14.6}$$

For the data in Table 14.5, we obtain

$$SS_{\text{between treatments}} = 101.75 - 48 = 53.75$$

However, you can also use the computational formula to calculate $SS_{\text{between treatments}}$ directly.

$$SS_{\text{between treatments}} = \Sigma\frac{T^2}{n} - \frac{G^2}{N} \tag{14.7}$$

For the data in Table 14.5, there are four treatments (four T values), each with $n = 5$ scores, and the between-treatments SS is

$$SS_{\text{between treatments}} = \frac{45^2}{5} + \frac{25^2}{5} + \frac{40^2}{5} + \frac{45^2}{5} - \frac{155^2}{20}$$

$$= 405 + 125 + 320 + 405 - 1201.25$$

$$= 53.75$$

The between-treatments df value is determined by the number of treatments (or the number of T values) minus one. For a two-factor study, the number of treatments is equal to the number of cells in the matrix. Thus,

$$df_{\text{between treatments}} = \text{number of cells} - 1 \tag{14.8}$$

For these data, $df_{\text{between treatments}} = 3$.

This completes the first stage of the analysis. Note that the two components add to equal the total for both SS values and df values.

$$SS_{\text{between treatments}} + SS_{\text{within treatments}} = SS_{\text{total}}$$

$$53.75 + 48 = 101.75$$

$$df_{\text{between treatments}} + df_{\text{within treatments}} = df_{\text{total}}$$

$$3 + 16 = 19$$

■ Stage 2 of the Two-Factor Analysis

The second stage of the analysis determines the numerators for the three F-ratios. Specifically, this stage determines the between-treatments variance for factor A, factor B, and the interaction.

1. **Factor A** The main effect for factor A evaluates the mean differences between the levels of factor A. For this example, factor A defines the rows of the matrix, so we are evaluating the mean differences between rows. To compute the SS for factor A, we calculate a between-treatment SS using the row totals exactly the same as we computed $SS_{\text{between treatments}}$ using the treatment totals (T values) earlier. For factor A, the row totals are 70 and 85, and each total was obtained by adding 10 scores.

   Therefore,

   $$SS_A = \Sigma\frac{T_{\text{ROW}}^2}{n_{\text{ROW}}} - \frac{G^2}{N} \tag{14.9}$$

   For our data,

   $$SS_A = \frac{70^2}{10} + \frac{85^2}{10} - \frac{155^2}{20}$$

   $$= 490 + 722.5 - 1201.25$$

   $$= 11.25$$

Factor A involves two treatments (or two rows), easy and difficult, so the df value is

$$df_A = \text{number of rows} - 1 \tag{14.10}$$
$$= 2 - 1$$
$$= 1$$

2. **Factor B**   The calculations for factor B follow exactly the same pattern that was used for factor A, except for substituting columns in place of rows. The main effect for factor B evaluates the mean differences between the levels of factor B, which define the columns of the matrix.

$$SS_B = \Sigma \frac{T^2_{\text{COL}}}{n_{\text{COL}}} - \frac{G^2}{N} \tag{14.11}$$

For our data, the column totals are 85 and 70, and each total was obtained by adding 10 scores. Thus,

$$SS_B = \frac{85^2}{10} + \frac{70^2}{10} - \frac{155^2}{20}$$
$$= 722.5 + 490 - 1201.25$$
$$= 11.25$$

$$df_B = \text{number of columns} - 1 \tag{14.12}$$
$$= 2 - 1$$
$$= 1$$

3. **The $A \times B$ Interaction**   The $A \times B$ interaction is defined as the "extra" mean differences not accounted for by the main effects of the two factors. We use this definition to find the SS and df values for the interaction by simple subtraction. Specifically, the between-treatments variability is partitioned into three parts: the A effect, the B effect, and the interaction (see Figure 14.3). We have already computed the SS and df values for A and B, so we can find the interaction values by subtracting to find out how much is left. Thus,

$$SS_{A \times B} = SS_{\text{between treatments}} - SS_A - SS_B \tag{14.13}$$

For our data,

$$SS_{A \times B} = 53.75 - 11.25 - 11.25$$
$$= 31.25$$

Similarly,

$$df_{A \times B} = df_{\text{between treatments}} - df_A - df_B \tag{14.14}$$
$$= 3 - 1 - 1$$
$$= 1$$

An easy to remember alternative formula for $df_{A \times B}$ is

$$df_{A \times B} = df_A \times df_B \tag{14.15}$$
$$= 1 \times 1 = 1$$

■ Mean Squares and *F*-Ratios for the Two-Factor ANOVA

The two-factor ANOVA consists of three separate hypothesis tests with three separate *F*-ratios. The denominator for each *F*-ratio is intended to measure the variance (differences) that would be expected if there are no treatment effects. As we saw in Chapter 12, the within-treatments variance is the appropriate denominator for an independent-measures design (see p. 373). The within-treatments variance is called a *mean square*, or *MS*, and is computed as follows:

$$MS_{within\ treatments} = \frac{SS_{within\ treatments}}{df_{within\ treatments}}$$

For the data in Table 14.5,

$$MS_{within\ treatments} = \frac{48}{16} = 3$$

This value forms the denominator for all three *F*-ratios.

The numerators of the three *F*-ratios all measured variance or differences between treatments: differences between levels of factor *A*, differences between levels of factor *B*, and extra differences that are attributed to the $A \times B$ interaction. These three variances are computed as follows:

$$MS_A = \frac{SS_A}{df_A} \quad MS_B = \frac{SS_B}{df_B} \quad MS_{A \times B} = \frac{SS_{A \times B}}{df_{A \times B}}$$

For the data in Table 14.4, the three *MS* values are

$$MS_A = \frac{11.25}{1} = 11.25 \quad MS_B = \frac{11.25}{1} = 11.25 \quad MS_{A \times B} = \frac{31.25}{1} = 31.25$$

Finally, the three *F*-ratios are

$$F_A = \frac{MS_A}{MS_{within\ treatments}} = \frac{11.25}{3} = 3.75$$

$$F_B = \frac{MS_B}{MS_{within\ treatments}} = \frac{11.25}{3} = 3.75$$

$$F_{A \times B} = \frac{MS_{A \times B}}{MS_{within\ treatments}} = \frac{31.25}{3} = 10.42$$

To determine the significance of each *F*-ratio, we must consult the *F* distribution table using the *df* values for each of the individual *F*-ratios. For this example, all three *F*-ratios have $df = 1$ for the numerator and $df = 16$ for the denominator. Checking the table with $df = 1, 16$, we find a critical value of 4.49 for $\alpha = .05$ and a critical value of 8.53 for $\alpha = .01$. For both main effects, we obtained $F = 3.75$, so neither of the main effects is significant. For the interaction, we obtained $F = 10.42$, which exceeds both of the critical values, so we conclude that there is a significant interaction between the two factors. That is, the difference between the two modes of presentation depends on how studying time is controlled. ■

Table 14.6 is a summary table for the complete two-factor ANOVA from Example 14.2. Although these tables are no longer commonly used in research reports, they provide a concise format for displaying all of the elements of the analysis.

Source	SS	df	MS	F
Between treatments	53.75	3		
Factor A (time control)	11.25	1	11.25	$F(1, 16) = 3.75$
Factor B (paper/screen)	11.25	1	11.25	$F(1, 16) = 3.75$
$A \times B$	31.25	1	31.25	$F(1, 16) = 10.42$
Within treatments	48	16	3	
Total	101.75	19		

The following example is an opportunity to test your understanding of the calculations required for a two-factor ANOVA.

EXAMPLE 14.3   The following data summarize the results from a two-factor independent-measures experiment:

	Factor B		
	B_1	B_2	B_3
A_1	$n = 10$ $T = 0$ $SS = 30$	$n = 10$ $T = 10$ $SS = 40$	$n = 10$ $T = 20$ $SS = 50$
A_2	$n = 10$ $T = 40$ $SS = 60$	$n = 10$ $T = 30$ $SS = 50$	$n = 10$ $T = 20$ $SS = 40$

Factor A

Calculate the total for each level of factor A and compute SS for factor A, then calculate the totals for factor B, and compute SS for this factor. You should find that the totals for factor A are 30 and 90, and $SS_A = 60$. All three totals for factor B are equal to 40. Because they are all the same, there is no variability, and $SS_B = 0$. ∎

■ Measuring Effect Size for the Two-Factor ANOVA

The general technique for measuring effect size with an ANOVA is to compute a value for η^2, the percentage of variance that is explained by the treatment effects. For a two-factor ANOVA, we compute three separate values for eta squared: one measuring how much of the variance is explained by the main effect for factor A, one for factor B, and a third for the interaction. As we did with the repeated-measures ANOVA (p. 427) we remove any variability that can be explained by other sources before we calculate the percentage for each of the three specific treatment effects. Thus, for example, before we compute the η^2 for factor A, we remove the variability that is explained by factor B and the variability explained by the interaction. The resulting equation is

$$\text{for factor } A, \eta^2 = \frac{SS_A}{SS_{\text{total}} - SS_B - SS_{A \times B}} \qquad (14.16)$$

Note that the denominator of Equation 14.15 consists of the variability that is explained by factor A and the other *unexplained* variability. Thus, an equivalent version of the equation is,

$$\text{for factor } A, \eta^2 = \frac{SS_A}{SS_A + SS_{\text{within treatments}}} \qquad (14.17)$$

Similarly, the η^2 formulas for factor B and for the interaction are as follows:

$$\text{for factor } B, \eta^2 = \frac{SS_B}{SS_{\text{total}} - SS_A - SS_{A \times B}} = \frac{SS_B}{SS_B + SS_{\text{within treatments}}} \tag{14.18}$$

$$\text{for } A \times B, \eta^2 = \frac{SS_{A \times B}}{SS_{\text{total}} - SS_A - SS_B} = \frac{SS_{A \times B}}{SS_{A \times B} + SS_{\text{within treatments}}} \tag{14.19}$$

Because each of the η^2 equations computes a percentage that is not based on the total variability of the scores, the results are often called *partial* eta squares. For the data in Example 14.2, the equations produce the following values:

$$\eta^2 \text{ for factor } A \text{ (times control)} = \frac{11.25}{11.25 + 48} = 0.190$$

$$\eta^2 \text{ for factor } B \text{ (presentation mode)} = \frac{11.25}{11.25 + 48} = 0.190$$

$$\eta^2 \text{ for the } A \times B \text{ interaction} = \frac{31.25}{31.25 + 48} = 0.394$$

IN THE LITERATURE

Reporting the Results of a Two-Factor ANOVA

The APA format for reporting the results of a two-factor ANOVA follows the same basic guidelines as the single-factor report. First, the means and standard deviations are reported. Because a two-factor design typically involves several treatment conditions, these descriptive statistics often are presented in a table or a graph. Next, the results of all three hypothesis tests (F-ratios) are reported. The results for the study in Example 14.2 could be reported as follows:

The means and standard deviations for all treatment conditions are shown in Table 1. The two-factor analysis of variance showed no significant main effect for time control, $F(1, 16) = 3.75, p > .05, \eta^2 = 0.190$ or for presentation mode, $F(1, 16) = 3.75, p > .05, \eta^2 = 0.190$. However, the interaction between factors was significant, $F(1, 16) = 10.41, p < .01, \eta^2 = 0.394$.

TABLE 1

Mean quiz score for each treatment condition

		Presentation Mode	
		Paper	Computer Screen
Time Control	Self-regulated	$M = 9.00$ $SD = 1.58$	$M = 5.00$ $SD = 1.73$
	Fixed	$M = 8.00$ $SD = 1.87$	$M = 9.00$ $SD = 1.73$

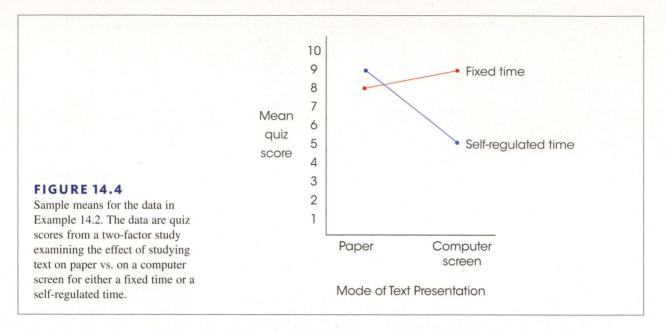

FIGURE 14.4

Sample means for the data in Example 14.2. The data are quiz scores from a two-factor study examining the effect of studying text on paper vs. on a computer screen for either a fixed time or a self-regulated time.

■ Interpreting the Results from a Two-Factor ANOVA

Because the two-factor ANOVA involves three separate tests, you must consider the overall pattern of results rather than focusing on the individual main effects or the interaction. In particular, whenever there is a significant interaction, you should be cautious about accepting the main effects at face value (whether they are significant or not). Remember, an interaction means that the effect of one factor *depends on* the level of the second factor. Because the effect changes from one level to the next, there is no consistent "main effect."

Figure 14.4 shows the sample means obtained from the paper vs. computer screen study. Recall that the analysis showed that both main effects were not significant but the interaction was significant. Although both main effects were too small to be significant, it would be incorrect to conclude that neither factor influenced behavior. For this example, the difference between studying text presented on paper vs. on a computer screen depends on how studying time is controlled. Specifically, there is little or no difference between paper and a computer screen when the time spent studying is fixed by the researchers. However, studying text from paper produces much higher quiz scores when participants regulate their own study time. Thus, the difference between studying from paper and studying from a computer screen *depends on* how the time spent studying is controlled. This interdependence between factors is the source of the significant interaction.

LEARNING CHECK   1. What happens in the second stage of a two-factor ANOVA?

    a. The total variability is divided into between treatments and within treatments.

    b. The variability within treatment is divided into between subjects and error.

    c. The variability within treatments is divided into the two main effects and the interaction.

    d. The variability between treatments is divided into the two main effects and the interaction.

2. In a two-factor ANOVA, which of the following is not computed directly but rather is found by subtraction?

   a. $SS_{between\ treatments}$

   b. SS_A

   c. SS_B

   d. $SS_{A \times B}$

3. A two-factor, independent-measures research study is evaluated using an analysis of variance. The F-ratio for factor A has $df = 1, 40$ and the F-ratio for factor B has $df = 3, 40$. Based on this information, what are the df values for the F-ratio for the $A \times B$ interaction?

   a. $df = 2, 40$

   b. $df = 3, 40$

   c. $df = 4, 40$

   d. $df = 7, 40$

ANSWERS   **1.** D, **2.** D, **3.** B

14.3 | More about the Two-Factor ANOVA

LEARNING OBJECTIVES

7. Explain how simple main effects can be used to analyze and describe the details of main effects and interactions.

8. Explain how adding a participant variable as a second factor can reduce variability caused by individual differences.

■ Testing Simple Main Effects

The existence of a significant interaction indicates that the effect (mean difference) for one factor depends on the levels of the second factor. When the data are presented in a matrix showing treatment means, a significant interaction indicates that the mean differences within one column (or row) show a different pattern than the mean differences within another column (or row). In this case, a researcher may want to perform a separate analysis for each of the individual columns (or rows). In effect, the researcher is separating the two-factor experiment into a series of separate single-factor experiments. The process of testing the significance of mean differences within one column (or one row) of a two-factor design is called testing *simple main effects*.

To demonstrate this process, we once again use the data from the paper vs. computer screen study (Example 14.2), which are summarized in Figure 14.4.

EXAMPLE 14.4

For this demonstration we test for significant mean differences within each row of the two-factor data matrix. That is, we test for significant mean differences between paper vs. screen for the self-regulated condition, and then repeat the test for the fixed-time condition. In terms of the two-factor notational system, we test the simple main effect of factor B for each level of factor A.

For the Self-Regulated Condition Because we are restricting the data to the first row of the data matrix, the data effectively have been reduced to a single-factor study comparing only two treatment conditions. Therefore, the analysis is essentially a single-factor ANOVA duplicating the procedure presented in Chapter 12. To facilitate the change from a two-factor to a single-factor analysis, the data for the self-regulated condition (first row of the matrix) are reproduced as follows using the notation for a single-factor study.

	Self-regulated time	
Paper	Computer screen	
$n = 5$	$n = 5$	$N = 10$
$M = 9$	$M = 5$	$G = 70$
$T = 45$	$T = 25$	

STEP 1 State the hypothesis. For this restricted set of the data, the null hypothesis would state that there is no difference between the mean for the paper condition and the mean for the computer screen condition. In symbols,

$$H_0: \mu_{paper} = \mu_{screen} \quad \text{for self-regulated study time}$$

STEP 2 To evaluate this hypothesis, we use an F-ratio for which the numerator, $MS_{\text{between treatments}}$, is determined by the mean differences between these two groups and the denominator consists of $MS_{\text{within treatments}}$ from the original ANOVA. Thus, the F-ratio has the structure

$$F = \frac{\text{variance (differences) for the means in row 1}}{\text{variance (differences) expected if there are no treatment effects}}$$

$$= \frac{MS_{\text{between treatments}} \text{ for the two treatments in row 1}}{MS_{\text{within treatments}} \text{ from the original ANOVA}}$$

To compute the $MS_{\text{between treatments}}$, we begin with the two treatment totals $T = 45$ and $T = 25$. Each of these totals is based on $n = 5$ scores, and the two totals add up to a grand total of $G = 70$. The $SS_{\text{between treatments}}$ for the two treatments is

$$SS_{\text{between treatments}} = \Sigma \frac{T^2}{n} - \frac{G^2}{N}$$

$$= \frac{45^2}{5} + \frac{25^2}{5} - \frac{70^2}{10}$$

$$= 405 + 125 - 490$$

$$= 40$$

Because this SS value is based on only two treatments, it has $df = 1$. Therefore,

$$MS_{\text{between treatments}} = \frac{40}{1} = 40$$

Using $MS_{\text{within treatments}} = 3$ with $df = 16$ from the original two-factor analysis, the final F-ratio is

$$F = \frac{MS_{\text{between treatments}}}{MS_{\text{within treatments}}} = \frac{40}{3} = 13.33$$

Note that this F-ratio has the same df values (1, 16) as the test for factor B main effects (paper vs. screen) in the original ANOVA. Therefore, the critical value for the F-ratio is the same as that in the original ANOVA. With $df = 1, 16$ the critical value is 4.49. In this case, our F-ratio far exceeds the critical value, so we conclude that there is a significant difference between the two modes of presentation, paper and screen, when study time is self-regulated.

For the Fixed-Time Condition   The test for the fixed-time condition follows the same process. The data for this condition are as follows:

Fixed-time condition		
	Computer	
Paper	Screen	
$n = 5$	$n = 5$	$N = 10$
$M = 8$	$M = 9$	$G = 85$
$T = 40$	$T = 45$	

For these data,

$$SS_{\text{between treatments}} = \Sigma \frac{T^2}{n} - \frac{G^2}{N}$$

$$= \frac{40^2}{5} + \frac{45^2}{5} - \frac{85^2}{10}$$

$$= 320 + 405 - 722.5$$

$$= 2.5$$

Again, we are comparing only two treatments, so $df = 1$ and,

$$MS_{\text{between treatments}} = \frac{2.5}{1} = 2.5$$

Using $MS_{\text{within treatments}} = 3$ from the original two-factor analysis, the F-ratio for the fixed-time condition is

$$F = \frac{MS_{\text{between treatments}}}{MS_{\text{within treatments}}} = \frac{2.5}{3} = 0.833$$

As before, this F-ratio has $df = 1, 16$ and is compared with the critical value $F = 4.49$. This time the F-ratio is not the critical region and we conclude that there is no significant difference between paper and computer when study time is self-regulated. ∎

As a final note, we should point out that the evaluation of simple main effects is used to account for the interaction as well as the overall main effect for one factor. In Example 14.2, the significant interaction indicates that the difference between paper vs. computer screen (factor B) depends on how time is controlled (factor A). The evaluation of the simple main effects demonstrates this dependency. Specifically, the mode of presentation has no significant effect on performance when study time is fixed, but does have a significant effect when time is self-controlled. Thus, the analysis of simple main effects provides a detailed evaluation of the effects of one factor *including its interaction with a second factor*.

The fact that the simple main effects for one factor encompass both the interaction and the overall main effect of the factor can be seen if you consider the SS values. For this demonstration,

Simple Main Effects for Factor B (Mode of Presentation)	Interaction and Main Effect for Factor B from the original ANOVA
$SS_{\text{self-regulated}} = 40$	$SS_{A \times B} = 31.25$
$SS_{\text{fixed}} = 2.5$	$SS_B = 11.25$
Total $SS = 42.5$	Total $SS = 42.5$

Notice that the total variability from the simple main effects of factor B (mode of presentation) completely accounts for the total variability of factor B and the $A \times B$ interaction.

■ Using a Second Factor to Reduce Variance Caused by Individual Differences

As we noted in Chapters 10 and 12, a concern for independent-measures designs is the variance that exists within each treatment condition. Specifically, large variance tends to reduce the size of the t statistic or F-ratio and, therefore, reduces the likelihood of finding significant mean differences. Much of the variance in an independent-measures study comes from individual differences. Recall that individual differences are the characteristics such as age or gender that differ from one participant to the next and can influence the scores obtained in the study.

Occasionally, there are consistent individual differences for one specific participant characteristic. For example, the males in a study may consistently have lower scores than the females. Or, the older participants may have consistently higher scores than the younger participants. For example, suppose that a researcher compares two treatment conditions using a separate group of children for each condition. Each group of participants contains a mix of boys and girls. Hypothetical data for this study are shown in Table 14.7(a), with each child's gender noted with an M or an F. While examining the results, the researcher notices that the girls tend to have higher scores than the boys, which produces big individual differences and high variance within each group. Fortunately, there is a relatively simple solution to the problem of high variance. The solution involves using the specific variable, in this case gender, as a second factor. Instead of one group in each treatment, the researcher divides the participants into two separate groups within each treatment: a group of boys and a group of girls. This process creates the two-factor study shown in Table 14.7(b), with one factor consisting of the two treatments (I and II) and the second factor consisting of the gender (male and female).

TABLE 14.7

A single-factor study comparing two treatments (a) can be transformed into a two-factor study (b) by using a participant characteristic (gender) as a second factor. This process creates smaller, more homogeneous groups, which reduces the variance within groups.

(a)

Treatment I	Treatment II
3 (M)	8 (F)
4 (F)	4 (F)
4 (F)	1 (M)
0 (M)	10 (F)
6 (F)	5 (M)
1 (M)	5 (M)
2 (F)	10 (F)
4 (M)	5 (M)
$M = 3$	$M = 6$
$SS = 50$	$SS = 68$

(b)

	Treatment I	Treatment II
Males	3	1
	0	5
	1	5
	4	5
	$M = 2$	$M = 4$
	$SS = 10$	$SS = 12$
Females	4	8
	4	4
	6	10
	2	10
	$M = 4$	$M = 8$
	$SS = 8$	$SS = 24$

By adding a second factor and creating four groups of participants instead of only two, the researcher has greatly reduced the individual differences (gender differences) within each group. This should produce a smaller variance within each group and, therefore, increase the likelihood of obtaining a significant mean difference. This process is demonstrated in the following example.

EXAMPLE 14.5

We will use the data in Table 14.7 to demonstrate how the variance caused by individual differences can be reduced by adding a participant characteristic, such as age or gender, as a second factor. For the single-factor study in Table 14.7(a), the two treatments produce $SS_{\text{within treatments}} = 50 + 68 = 118$. With $n = 8$ in each treatment, we obtain $df_{\text{within treatments}} = 7 + 7 = 14$. These values produce $MS_{\text{within treatments}} = \frac{118}{14} = 8.43$, which will be the denominator of the F-ratio evaluating the mean difference between treatments. For the two-factor study in Table 14.7(b), the four treatments produce $SS_{\text{within treatments}} = 10 + 12 + 8 + 24 = 54$. With $n = 4$ in each treatment, we obtain $df_{\text{within treatments}} = 3 + 3 + 3 + 3 = 12$. These value produce $MS_{\text{within treatments}} = \frac{54}{12} = 4.50$, which will be the denominator of the F-ratio evaluating the main effect for the treatments. Notice that the error term for the single-factor F is nearly twice as big as the error term for the two-factor F. Reducing the individual differences within each group has greatly reduced the within-treatment variance that forms the denominator of the F-ratio.

Both designs, single-factor and two-factor, will evaluate the difference between the two treatment means, $M = 3$ and $M = 6$, with $n = 8$ in each treatment. These values produce $SS_{\text{between treatments}} = 36$ and, with $k = 2$ treatments, we obtain $df_{\text{between treatments}} = 1$. Thus, $MS_{\text{between treatments}} = \frac{36}{1} = 36$. (For the two-factor design, this is the MS for the main effect of the treatment factor.) With different denominators, however, the two designs produce very different F-ratios. For the single-factor design, we obtain

$$F = \frac{MS_{\text{between treatments}}}{MS_{\text{within treatments}}} = \frac{36}{8.43} = 4.27$$

With $df = 1, 14$, the critical value for $\alpha = .05$ is $F = 4.60$. Our F-ratio is not in the critical region so we fail to reject the null hypothesis and must conclude that there is no significant difference between the two treatments.

For the two-factor design, however, we obtain

$$F = \frac{MS_{\text{between treatments}}}{MS_{\text{within treatments}}} = \frac{36}{4.50} = 8.00$$

With $df = 1, 12$, the critical value for $\alpha = .05$ is $F = 4.75$. Our F-ratio is well beyond this value so we reject the null hypothesis and conclude that there is a significant difference between the two treatments. ∎

For the single-factor study in Example 14.5, the individual differences caused by gender were part of the variance within each treatment condition. This increased variance reduced the F-ratio and resulted in a conclusion of no significant difference between treatments. In the two-factor analysis, the individual differences caused by gender are measured by the main effect for gender, which is a between-groups factor. Because the gender differences are now between-groups rather than within-groups, they no longer contribute to the variance.

The two-factor analysis has other advantages beyond reducing the variance. Specifically, it allows you to evaluate mean differences between genders as well as differences between treatments, and it reveals any interaction between treatment and gender.

■ Assumptions for the Two-Factor ANOVA

The validity of the ANOVA presented in this chapter depends on the same three assumptions we have encountered with other hypothesis tests for independent-measures designs (the *t* test in Chapter 10 and the single-factor ANOVA in Chapter 12):

1. The observations within each sample must be independent (see p. 244).
2. The populations from which the samples are selected must be normal.
3. The populations from which the samples are selected must have equal variances (homogeneity of variance).

As before, the assumption of normality generally is not a cause for concern, especially when the sample size is relatively large. The homogeneity of variance assumption is more important, and if it appears that your data fail to satisfy this requirement, you should conduct a test for homogeneity before you attempt the ANOVA. Hartley's *F*-max test (see p. 314) allows you to use the sample variances from your data to determine whether there is evidence for any differences among the population variances. Remember, for the two-factor ANOVA, there is a separate sample for each cell in the data matrix. The test for homogeneity applies to all these samples and the populations they represent.

LEARNING CHECK

1. A two-factor study has 2 levels of Factor *A* and 3 levels of Factor *B*. Because the ANOVA produces a significant interaction, the researcher decides to evaluate the simple mean effect of Factor *A* for each level of Factor *B*. How many *F*-ratios will this require?
   a. 1
   b. 2
   c. 3
   d. 6

2. Which of the following can often help reduce the variance caused by individual differences in a single-factor design?
   a. Counterbalance the order of treatments.
   b. Create a factorial design using a participant variable (such as age) as a second factor.
   c. Create a factorial design using the order of treatments as a second factor.
   d. The other three options are all methods for reducing variance.

ANSWERS   1. C, 2. B

SUMMARY

1. A research study with two independent variables is called a two-factor design. Such a design can be diagramed as a matrix with the levels of one factor defining the rows and the levels of the other factor defining the columns. Each cell in the matrix corresponds to a specific combination of the two factors.

2. Traditionally, the two factors are identified as factor A and factor B. The purpose of the ANOVA is to determine whether there are any significant mean differences among the treatment conditions or cells in the experimental matrix. These treatment effects are classified as follows:
   a. The A-effect: Overall mean differences among the levels of factor A.
   b. The B-effect: Overall mean differences among the levels of factor B.
   c. The A × B interaction: Extra mean differences that are not accounted for by the main effects.

3. The two-factor ANOVA produces three F-ratios: one for factor A, one for factor B, and one for the A × B interaction. Each F-ratio has the same basic structure:

$$F = \frac{MS_{\text{treatment effect}}(\text{either } A \text{ or } B \text{ or } A \times B)}{MS_{\text{within treatments}}}$$

The formulas for the SS, df, and MS values for the two-factor ANOVA are presented in Figure 14.5.

4. A significant interaction means that the effect of one factor depends on the levels of the second factor. In this case, evaluating the simple main effects for the factor can provide a complete description of the effect of that factor including its interaction with the second factor.

5. Large individual differences within treatment often can be reduced by using a participant variable, such as age or gender, as an additional factor.

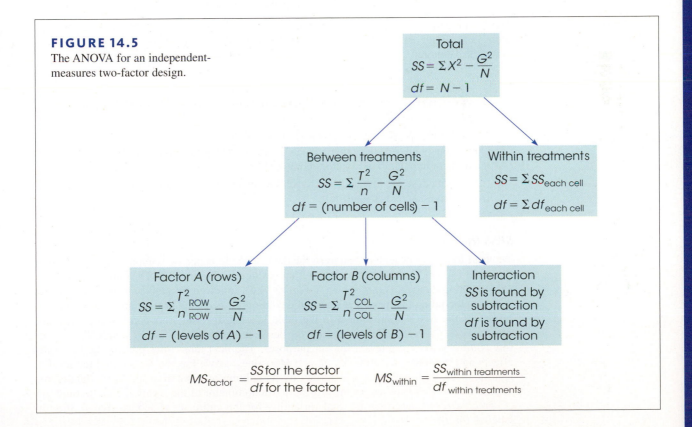

FIGURE 14.5
The ANOVA for an independent-measures two-factor design.

Total
$$SS = \Sigma X^2 - \frac{G^2}{N}$$
$$df = N - 1$$

Between treatments
$$SS = \Sigma \frac{T^2}{n} - \frac{G^2}{N}$$
$$df = (\text{number of cells}) - 1$$

Within treatments
$$SS = \Sigma SS_{\text{each cell}}$$
$$df = \Sigma df_{\text{each cell}}$$

Factor A (rows)
$$SS = \Sigma \frac{T^2_{\text{ROW}}}{n_{\text{ROW}}} - \frac{G^2}{N}$$
$$df = (\text{levels of } A) - 1$$

Factor B (columns)
$$SS = \Sigma \frac{T^2_{\text{COL}}}{n_{\text{COL}}} - \frac{G^2}{N}$$
$$df = (\text{levels of } B) - 1$$

Interaction
SS is found by subtraction
df is found by subtraction

$$MS_{\text{factor}} = \frac{SS \text{ for the factor}}{df \text{ for the factor}} \qquad MS_{\text{within}} = \frac{SS_{\text{within treatments}}}{df_{\text{within treatments}}}$$

KEY TERMS

two-factor design (449)   cells (449)   interaction (452)

matrix (449)   main effect (450)

SPSS®

General instructions for using SPSS are presented in Appendix D. Following are detailed instructions for using SPSS to perform the **Two-Factor, Independent-Measures Analysis of Variance (ANOVA)** presented in this chapter.

Data Entry

1. The scores are entered into the SPSS data editor in a *stacked format,* which means that all the scores from all the different treatment conditions are entered in a single column (VAR00001).

2. In a second column (VAR00002) enter a code number to identify the level of factor *A* for each score. If factor *A* defines the rows of the data matrix, enter a 1 beside each score from the first row, enter a 2 beside each score from the second row, and so on.

3. In a third column (VAR00003) enter a code number to identify the level of factor *B* for each score. If factor *B* defines the columns of the data matrix, enter a 1 beside each score from the first column, enter a 2 beside each score from the second column, and so on.

Thus, each row of the SPSS data editor will have one score and two code numbers, with the score in the first column, the code for factor *A* in the second column, and the code for factor *B* in the third column.

Data Analysis

1. Click **Analyze** on the tool bar, select **General Linear Model**, and click on **Univariant**.

2. Highlight the column label for the set of scores (VAR0001) in the left box and click the arrow to move it into the **Dependent Variable** box.

3. One by one, highlight the column labels for the two factor codes and click the arrow to move them into the **Fixed Factors** box.

4. If you want descriptive statistics for each treatment, click on the **Options** box, select **Descriptives**, and click **Continue**.

5. Click **OK**.

SPSS Output

We used the SPSS program to analyze the data from the paper vs. computer screen study in Example 14.2 and part of the program output is shown in Figure 14.6. The output begins with a table listing the factors (not shown in Figure 14.6), followed by a table showing descriptive statistics including the mean and standard deviation for each cell or treatment condition. The results of the ANOVA are shown in the table labeled **Tests of Between-Subjects Effects**. The top row (*Corrected Model*) presents the between-treatments *SS* and *df* values. The second row (*Intercept*) is not relevant for our purposes. The next three rows present the two main effects and the interaction (the *SS, df,* and *MS* values, as well as the *F*-ratio and the level of significance), with each factor identified by its column number from the SPSS data editor. The next row (*Error*) describes the error term (denominator of the *F*-ratio), and the final row (*Corrected Total*) describes the total variability for the entire set of scores. (Ignore the row labeled *Total*.)

Descriptive Statistics

Dependent Variable: VAR00001

VAR00002	VAR00003	Mean	Std. Deviation	N
1.00	1.00	9.0000	1.58114	5
	2.00	5.0000	1.73205	5
	Total	7.0000	2.62467	10
2.00	1.00	8.0000	1.87083	5
	2.00	9.0000	1.73205	5
	Total	8.5000	1.77951	10
Total	1.00	8.5000	1.71594	10
	2.00	7.0000	2.66667	10
	Total	7.7500	2.31414	20

Tests of Between-Subjects Effects

Dependent Variable: VAR00001

Source	Type III Sum of Squares	df	Mean Square	F	Sig.
Corrected Model	53.750[a]	3	17.917	5.972	.006
Intercept	1201.250	1	1201.250	400.417	.000
VAR00002	11.250	1	11.250	3.750	.071
VAR00003	11.250	1	11.250	3.750	.071
VAR00002 * VAR00003	31.250	1	31.250	10.417	.005
Error	48.000	16	3.000		
Total	1303.000	20			
Corrected Total	101.750	19			

FIGURE 14.6
Portions of the SPSS output for the two-factor ANOVA for the study in Example 14.2.

FOCUS ON PROBLEM SOLVING

1. Before you begin a two-factor ANOVA, take time to organize and summarize the data. It is best if you summarize the data in a matrix with rows corresponding to the levels of one factor and columns corresponding to the levels of the other factor. In each cell of the matrix, show the number of scores (n), the total and mean for the cell, and the SS within the cell. Also compute the row totals and column totals that are needed to calculate main effects.

2. For a two-factor ANOVA, there are three separate F-ratios. These three F-ratios use the same error term in the denominator (MS_{within}). On the other hand, these F-ratios have different numerators and may have different df values associated with each of these numerators. Therefore, you must be careful when you look up the critical F values in the table. The two factors and the interaction may have different critical F values.

DEMONSTRATION 14.1

TWO-FACTOR ANOVA

The following data are representative of the results obtained in a research study examining the relationship between eating behavior and body weight (Schachter, 1968). **The two factors in this study were:**

1. The participant's weight (normal or obese)
2. The participant's state of hunger (full stomach or empty stomach)

All participants were led to believe that they were taking part in a taste test for several types of crackers, and they were allowed to eat as many crackers as they wanted. The dependent variable was the number of crackers eaten by each participant. There were two specific predictions for this study. First, it was predicted that normal participants' eating behavior would be determined by their state of hunger. That is, people with empty stomachs would eat more and people with full stomachs would eat less. Second, it was predicted that eating behavior for obese participants would not be related to their state of hunger. Specifically, it was predicted that obese participants would eat the same amount whether their stomachs were full or empty. Note that the researchers are predicting an interaction: The effect of hunger will be different for the normal participants and the obese participants. The data are as follows.

		Factor B: Hunger			
		Empty stomach	Full stomach		
Factor A: Weight	Normal	$n = 20$ $M = 22$ $T = 440$ $SS = 1540$	$n = 20$ $M = 15$ $T = 300$ $SS = 1270$	$T_{normal} = 740$	$G = 1440$ $N = 80$ $\Sigma X^2 = 31,836$
	Obese	$n = 20$ $M = 17$ $T = 340$ $SS = 1320$	$n = 20$ $M = 18$ $T = 360$ $SS = 1266$	$T_{obese} = 700$	
		$T_{empty} = 780$	$T_{full} = 660$		

STEP 1 **State the hypotheses, and select alpha**   For a two-factor study, there are three separate hypotheses, the two main effects and the interaction.

For factor A, the null hypothesis states that there is no difference in the amount eaten for normal participants vs. obese participants. In symbols,

$$H_0: \mu_{normal} = \mu_{obese}$$

For factor B, the null hypothesis states that there is no difference in the amount eaten for full-stomach vs. empty-stomach conditions. In symbols,

$$H_0: \mu_{full} = \mu_{empty}$$

For the $A \times B$ interaction, the null hypothesis can be stated two different ways. First, the difference in eating between the full-stomach and empty-stomach conditions will be the same for normal and obese participants. Second, the difference in eating between the normal and obese participants will be the same for the full-stomach and empty-stomach conditions. In more general terms,

H_0:   The effect of factor A does not depend on the levels of factor B (and B does not depend on A).

We will use $\alpha = .05$ for all tests.

STEP 2   **The two-factor analysis**   Rather than compute the *df* values and look up critical values for *F* at this time, we will proceed directly to the ANOVA.

STAGE 1   The first stage of the analysis is identical to the independent-measures ANOVA presented in Chapter 13, where each cell in the data matrix is considered a separate treatment condition.

$$SS_{total} = \Sigma X^2 - \frac{G^2}{N}$$

$$= 31,836 - \frac{1440^2}{80} = 5916$$

$$SS_{within\ treatments} = \Sigma SS_{inside\ each\ treatment} = 1540 + 1270 + 1320 + 1266 = 5396$$

$$SS_{between\ treatments} = \Sigma \frac{T^2}{n} - \frac{G^2}{N}$$

$$= \frac{440^2}{20} + \frac{300^2}{20} + \frac{340^2}{20} + \frac{360^2}{20} - \frac{1440^2}{80}$$

$$= 520$$

The corresponding degrees of freedom are

$$df_{total} = N - 1 = 79$$

$$df_{within\ treatments} = \Sigma df = 19 + 19 + 19 + 19 = 76$$

$$df_{between\ treatments} = \text{number of treatments} - 1 = 3$$

STAGE 2   The second stage of the analysis partitions the between-treatments variability into three components: the main effect for factor *A*, the main effect for factor *B*, and the *A* × *B* interaction.

For factor *A* (normal/obese),

$$SS_A = \Sigma \frac{T^2_{ROWS}}{n_{ROWS}} - \frac{G^2}{N}$$

$$= \frac{740^2}{40} + \frac{700^2}{40} - \frac{1440^2}{80}$$

$$= 20$$

For factor *B* (full/empty),

$$SS_B = \Sigma \frac{T^2_{COLS}}{n_{COLS}} - \frac{G^2}{N}$$

$$= \frac{780^2}{40} + \frac{660^2}{40} - \frac{1440^2}{80}$$

$$= 180$$

For the *A* × *B* interaction,

$$SS_{A \times B} = SS_{between\ treatments} - SS_A - SS_B$$

$$= 520 - 20 - 180$$

$$= 320$$

The corresponding degrees of freedom are

$$df_A = \text{number of rows} - 1 = 1$$

$$df_B = \text{number of columns} - 1 = 1$$

$$df_{A \times B} = df_{between\ treatments} - df_A - df_B$$

$$= 3 - 1 - 1$$

$$= 1$$

The *MS* values needed for the *F*-ratios are

$$MS_A = \frac{SS_A}{df_A} = \frac{20}{1} = 20$$

$$MS_B = \frac{SS_B}{df_B} = \frac{180}{1} = 180$$

$$MS_{A\times B} = \frac{SS_{A\times B}}{df_{A\times B}} = \frac{320}{1} = 320$$

$$MS_{\text{within treatments}} = \frac{SS_{\text{within treatments}}}{df_{\text{within treatments}}} = \frac{5396}{76} = 71$$

Finally, the *F*-ratios are

$$F_A = \frac{MS_A}{MS_{\text{within treatments}}} = \frac{20}{71} = 0.28$$

$$F_B = \frac{MS_B}{MS_{\text{within treatments}}} = \frac{180}{71} = 2.54$$

$$F_{A\times B} = \frac{MS_{A\times B}}{MS_{\text{within treatments}}} = \frac{320}{71} = 4.51$$

STEP 3   **Make a decision and state a conclusion**   All three *F*-ratios have $df = 1, 76$. With $\alpha = .05$, the critical *F* value is 3.98 for all three tests.

For these data, factor *A* (weight) has no significant effect; $F(1, 76) = 0.28$. Statistically, there is no difference in the number of crackers eaten by normal vs. obese participants.

Similarly, factor *B* (fullness) has no significant effect; $F(1, 76) = 2.54$. Statistically, the number of crackers eaten by full participants is no different from the number eaten by hungry participants. (*Note:* This conclusion concerns the combined group of normal and obese participants. The interaction concerns these two groups separately.)

These data produce a significant interaction; $F(1, 76) = 4.51, p < .05$. This means that the effect of fullness does depend on weight. A closer look at the original data shows that the degree of fullness did affect the normal participants, but it had no effect on the obese participants.

DEMONSTRATION 14.2

MEASURING EFFECT SIZE FOR THE TWO-FACTOR ANOVA

Effect size for each main effect and for the interaction is measured by eta squared (η^2), the percentage of variance explained by the specific main effect or interaction. In each case, the variability that is explained by other sources is removed before the percentage is computed. For the two-factor ANOVA in Demonstration 14.1,

$$\text{For factor } A, \ \eta^2 = \frac{SS_A}{SS_{\text{total}} - SS_B - SS_{A\times B}} = \frac{20}{5916 - 180 - 320} = 0.004 \ (\text{or } 0.4\%)$$

$$\text{For factor } B, \ \eta^2 = \frac{SS_B}{SS_{\text{total}} - SS_A - SS_{A\times B}} = \frac{180}{5916 - 20 - 320} = 0.032 \ (\text{or } 3.2\%)$$

$$\text{For } A \times B, \ \eta^2 = \frac{SS_{A\times B}}{SS_{\text{total}} - SS_A - SS_B} = \frac{320}{5916 - 20 - 180} = 0.056 \ (\text{or } 5.6\%)$$

PROBLEMS

1. Define each of the following terms:
   a. Factor
   b. Level
   c. Two-factor study

2. The structure of a two-factor study can be presented as a matrix with the levels of one factor determining the rows and the levels of the second factor determining the columns. With this structure in mind, describe the mean differences that are evaluated by each of the three hypothesis tests that make up a two-factor ANOVA.

3. Briefly explain what happens during the second stage of the two-factor ANOVA.

4. For the data in the following matrix:

	No Treatment	Treatment	
Male	$M = 8$	$M = 14$	Overall $M = 11$
Female	$M = 4$	$M = 10$	Overall $M = 7$
	overall $M = 6$	overall $M = 12$	

   a. Which two means are compared to describe the treatment main effect?
   b. Which two means are compared to describe the gender main effect?
   c. Is there an interaction between gender and treatment? Explain your answer.

5. The following matrix presents the results from an independent-measures, two-factor study with a sample of $n = 10$ participants in each treatment condition. Note that one treatment mean is missing.

	Factor B	
	B_1	B_2
Factor A A_1	$M = 3$	$M = 7$
A_2	$M = 5$	

   a. What value for the missing mean would result in no main effect for factor A?
   b. What value for the missing mean would result in no main effect for factor B?
   c. What value for the missing mean would result in no interaction?

6. The following matrix presents the results of a two-factor study with $n = 10$ scores in each of the six treatment conditions. Note that one of the treatment means is missing.

		Factor B		
		B_1	B_2	B_3
Factor A	A_1	$M = 4$	$M = 6$	$M = 14$
	A_2	$M = 2$	$M = 4$	

   a. What value for the missing mean would result in no main effect for factor A?
   b. What value for the missing mean would result in no interaction?

7. For the data in the following graph:

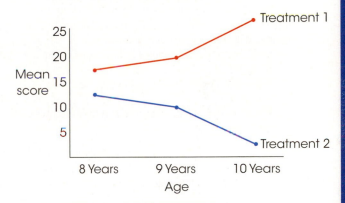

   a. Is there a main effect for the treatment factor?
   b. Is there a main effect for the age factor?
   c. Is there an interaction between age and treatment?

8. A researcher conducts an independent-measures, two-factor study with two levels of factor A and two levels of factor B, using a sample of $n = 8$ participants in each treatment condition.
   a. What are the df values for the F-ratio evaluating the main effect of factor A?
   b. What are the df values for the F-ratio evaluating the main effect of factor B?
   c. What are the df values for the F-ratio evaluating the interaction?

9. A researcher conducts an independent-measures, two-factor study with two levels of factor A and three levels of factor B, using a sample of $n = 10$ participants in each treatment condition.
   a. What are the df values for the F-ratio evaluating the main effect of factor A?
   b. What are the df values for the F-ratio evaluating the main effect of factor B?
   c. What are the df values for the F-ratio evaluating the interaction?

10. A researcher conducts an independent-measures, two-factor study using a separate sample of $n = 5$ participants in each treatment condition. The results are evaluated using an ANOVA and the researcher reports an F-ratio with $df = 1, 24$ for factor A, and an F-ratio with $df = 2, 24$ for factor B.
    a. How many levels of factor A were used in the study?
    b. How many levels of factor B were used in the study?
    c. What are the df values for the F-ratio evaluating the interaction?

11. A researcher conducts an independent-measures, two-factor study using a separate sample of $n = 10$ participants in each treatment condition. The results are evaluated using an ANOVA and the researcher reports an F-ratio with $df = 2, 108$ for factor A, and an F-ratio with $df = 3, 108$ for factor B.
    a. How many levels of factor A were used in the study?
    b. How many levels of factor B were used in the study?
    c. What are the df values for the F-ratio evaluating the interaction?

12. The following results are from an independent-measures, two-factor study with $n = 10$ participants in each treatment condition.

		Factor B	
		B_1	B_2
Factor A	A_1	T = 40 M = 4 SS = 50	T = 10 M = 1 SS = 30
	A_2	T = 50 M = 5 SS = 60	T = 20 M = 2 SS = 40

$$N = 40$$
$$G = 120$$
$$\Sigma X^2 = 640$$

    a. Use a two-factor ANOVA with $\alpha = .05$ to evaluate the main effects and the interaction.
    b. Compute η^2 to measure the effect size for each of the main effects and the interaction.

13. The following results are from an independent-measures, two-factor study with $n = 5$ participants in each treatment condition.

		Factor B		
		B_1	B_2	B_3
Factor A	A_1	T = 25 M = 5 SS = 30	T = 40 M = 8 SS = 38	T = 70 M = 14 SS = 46
	A_2	T = 15 M = 3 SS = 22	T = 20 M = 4 SS = 26	T = 40 M = 8 SS = 30

$$N = 30$$
$$G = 210$$
$$\Sigma X^2 = 2062$$

    a. Use a two-factor ANOVA with $\alpha = .05$ to evaluate the main effects and the interaction.
    b. Test the simple main effects using $\alpha = .05$ to evaluate the mean difference between treatment A_1 and A_2 for each level of factor B.

14. Most sports injuries are immediate and obvious, like a broken leg. However, some can be more subtle, like the neurological damage that may occur when soccer players repeatedly head a soccer ball. To examine effects of repeated heading, McAllister et al. (2012) examined a group of football and ice hockey players and a group of athletes in noncontact sports before and shortly after the season. The dependent variable was performance on a conceptual thinking task. Following are hypothetical data from an independent-measures study similar to the one by McAllister et al. The researchers measured conceptual thinking for contact and noncontact athletes at the beginning of their first season and for separate groups of athletes at the end of their second season.
    a. Use a two-factor ANOVA with $\alpha = .05$ to evaluate the main effects and interaction.
    b. Calculate the effects size (η^2) for the main effects and the interaction.
    c. Briefly describe the outcome of the study.

		Factor B: Time	
		Before the First Season	After the Second Season
Factor A: Sport	Contact Sport	n = 20 M = 9 T = 180 SS = 380	n = 20 M = 4 T = 80 SS = 390
	Non- contact Sport	n = 20 M = 9 T = 180 SS = 350	n = 20 M = 8 T = 160 SS = 400

$$\Sigma X^2 = 6360$$

15. Research results indicate that 5-year-old children who watched a lot of educational programming such as Sesame Street and Mr. Rogers had higher high-school grades than their peers (Anderson, Huston, Wright, & Collins, 1998). The same study reported that 5-year-old children who watched a lot of non-educational TV programs had relatively low high-school grades compared to their peers. A researcher attempting to replicate this result using an independent-measures study with four separate groups of high school students obtained the following data. The dependent variable is a rating of high school academic performance, with higher scores indicating higher levels of performance.

  a. Use a two-factor ANOVA with $\alpha = .01$ to evaluate the main effects and interaction.

  b. Calculate the effect size (η^2) for the main effects and the interaction.

  c. Briefly describe the outcome of the study.

	Little Watching	Substantial Watching
Educational TV	1 5 4 3 2 $M = 3$ $SS = 10$	6 6 4 5 4 $M = 5$ $SS = 4$
Non-educational TV	5 1 5 3 1 $M = 3$ $SS = 16$	0 0 1 3 1 $M = 1$ $SS = 6$
$N = 20$	$G = 60$	$\Sigma X^2 = 256$

16. Some people like to pour beer gently down the side of the glass to preserve bubbles. Others, splash it down the center to release the bubbles into a foamy head and free the aromas. Champagne, however, is best when the bubbles remain concentrated in the wine. A group of French scientists recently verified the difference between the two pouring methods by measuring the amount of bubbles in each glass of champagne poured two different ways and at three different temperatures (Liger-Belair et al., 2010). The following data present the pattern of results obtained in the study.

Champagne Temperature (°F)

	40°	46°	52°
Gentle Pour	$n = 10$ $M = 7$ $SS = 64$	$n = 10$ $M = 3$ $SS = 57$	$n = 10$ $M = 2$ $SS = 47$
Splashing Pour	$n = 10$ $M = 5$ $SS = 56$	$n = 10$ $M = 1$ $SS = 54$	$n = 10$ $M = 0$ $SS = 46$

  a. Use a two-factor ANOVA with $\alpha = .05$ to evaluate the mean differences.

  b. Briefly explain how temperature and pouring influence the bubbles in champagne according to this pattern of results.

17. The following table summarizes the results from a two-factor study with 2 levels of factor A and 4 levels of factor B using a separate sample of $n = 5$ participants in each treatment condition. Fill in the missing values. (*Hint:* Start with the *df* values.)

Source	SS	df	MS	
Between treatments	___	___		
Factor A	20	___	___	$F =$ ___
Factor B	___	___	12	$F =$ ___
A × B Interaction	___	___	6	$F =$ ___
Within treatments	64	___	___	
Total				

18. The following table summarizes the results from a two-factor study with 2 levels of factor A and 3 levels of factor B using a separate sample of $n = 8$ participants in each treatment condition. Fill in the missing values. (*Hint:* Start with the *df* values.)

Source	SS	df	MS	
Between treatments	72	___		
Factor A	___	___	___	$F =$ ___
Factor B	___	___	___	$F = 6.0$
A × B Interaction	___	___	12	$F =$ ___
Within treatments	126	___	___	
Total	___	___		

19. The following table summarizes the results from a two-factor study with 2 levels of factor A and 2 levels of factor B using a separate sample of $n = 15$ participants in each treatment condition. Fill in the missing values. (*Hint:* Start with the df values.)

Source	SS	df	MS	
Between treatments	_____	_____		
Factor A	16	_____	_____	$F = 8$
Factor B	_____	_____	_____	$F = 12$
$A \times B$ Interaction	_____	_____	_____	$F = 6$
Within treatments	_____	_____	_____	
Total	_____	_____		

20. The following data are from a two-factor study examining the effects of two treatment conditions on males and females.
    a. Use an ANOVA with $\alpha = .05$ for all tests to evaluate the significance of the main effects and the interaction.
    b. Compute η^2 to measure the size of the effect for each main effect and the interaction.

	Treatments		
	I	II	
male	3 8 9 4 $M = 6$ $T = 24$ $SS = 26$	2 8 7 7 $M = 6$ $T = 24$ $SS = 22$	$T_{male} = 48$
Female	0 0 2 6 $M = 2$ $T = 8$ $SS = 24$	12 6 9 13 $M = 10$ $T = 40$ $SS = 30$	$T_{female} = 48$

Factor A: Gender

$N = 16$
$G = 96$
$\Sigma X^2 = 806$

$T_I = 32 \quad T_{II} = 64$

21. The following data are from a two-factor study examining the effects of three treatment conditions on males and females.
    a. Use an ANOVA with $\alpha = .05$ for all tests to evaluate the significance of the main effects and the interaction.

b. Test the simple main effects using $\alpha = .05$ to evaluate the mean difference between males and females for each of the three treatments.

	Treatments			
	I	II	III	
Male	1 2 6 $M = 3$ $T = 9$ $SS = 14$	7 2 9 $M = 6$ $T = 18$ $SS = 26$	9 11 7 $M = 9$ $T = 27$ $SS = 8$	$T_{male} = 54$
Female	3 1 5 $M = 3$ $T = 9$ $SS = 8$	10 11 15 $M = 12$ $T = 36$ $SS = 14$	16 18 11 $M = 15$ $T = 45$ $SS = 26$	$T_{female} = 90$

Factor A: Gender

$N = 18$
$G = 144$
$\Sigma X^2 = 1608$

22. The following data are representative of many studies examining the relationship between arousal and performance. The general result of these studies is that increasing the level of arousal (or motivation) tends to improve the level of performance. (You probably have tried to "psych yourself up" to do well on a task.) For very difficult tasks, however, increasing arousal beyond a certain point tends to lower the level of performance. (Your friends have probably advised you to "calm down and stay focused" when you get overanxious about doing well.) This relationship between arousal and performance is known as the Yerkes-Dodson law.
    a. Use a two-factor ANOVA with $\alpha = .05$ to evaluate the significance of the main effects and the interaction.
    b. Calculate the η^2 values to measure the effect size for the two main effects.
    c. Describe the pattern of results. How does the level of arousal (motivation) affect performance?

	Factor B Arousal Level		
	Low	Medium	High

	Low	Medium	High	
Easy	3 1 1 6 4	1 4 8 6 6	10 10 14 7 9	$T_{ROW1} = 90$
	$M = 3$ $T = 15$ $SS = 18$	$M = 5$ $T = 25$ $SS = 28$	$M = 10$ $T = 50$ $SS = 26$	$N = 30$
Difficult	0 2 0 0 3	2 7 2 2 2	1 1 1 6 1	$G = 120$ $\Sigma X^2 = 860$ $T_{ROW2} = 30$
	$M = 1$ $T = 5$ $SS = 8$	$M = 3$ $T = 15$ $SS = 20$	$M = 2$ $T = 10$ $SS = 20$	
	$T_{COL1} = 20$	$T_{COL2} = 40$	$T_{COL3} = 60$	

Factor A Task Difficulty (row label)

23. Kirschner and Karpinski (2010) report that college students who are on Facebook (or have it running in the background) while studying had lower grades than students who did not use the social network. A researcher would like to know if the same result extends to students in lower grade levels. The researcher planned a two-factor study comparing Facebook users with non-users for middle school students, high school students, and college students. For consistency across groups, grades were converted into six categories, numbered 0–5 from low to high. The results are presented in the following matrix.

   a. Use a two-factor ANOVA with $\alpha = .05$ to evaluate the mean differences.
   b. Describe the pattern of results.

	Middle School	High School	College
User	3 5 5 3	5 5 2 4	5 4 2 5
Non-user	5 3 2 2	1 2 3 2	1 0 0 3

24. In Chapter 11 we described a research study in which the color red appeared to increase men's attraction to women (Elliot & Niesta, 2008). The same researchers have published other results showing that red also increases women's attraction to men but does

not appear to affect judgments of same sex individuals (Elliot et al., 2010). Combining these results into one study produces a two-factor design in which men judge photographs of both women and men, which are shown on both red and white backgrounds. The dependent variable is a rating of attractiveness for the person shown in the photograph. The study uses a separate group of participants for each condition. The following table presents data similar to the results from previous research.

		Person Shown in Photograph	
		Female	Male
Background Color for Photograph	White	$n = 10$ $M = 4.5$ $SS = 6$	$n = 10$ $M = 4.4$ $SS = 7$
	Red	$n = 10$ $M = 7.5$ $SS = 9$	$n = 10$ $M = 4.6$ $SS = 8$

   a. Use a two-factor ANOVA with $\alpha = .05$ to evaluate the main effects and the interaction.
   b. Describe the effect of background color on judgments of males and females.

25. In a classic research study, Shrauger (1972) examined the effect of an audience on the performance of two different personality types. Data from this experiment are as follows. The dependent variable is the number of errors made by each participant.

		Alone	Audience
Self-esteem	High	3 6 2 2 4 7	9 4 5 8 4 6
	Low	8 8 3 7 9 7	11 15 12 16 12 12

   a. Use an ANOVA with $\alpha = .05$ to evaluate the data. Describe the effect of the audience and the effect of self-esteem on performance.
   b. Calculate the effect size (η^2) for each main effect and for the interaction.

Correlation

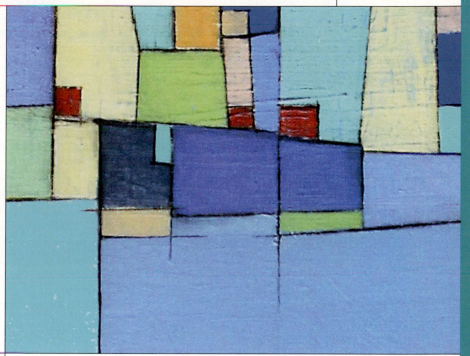

© Deborah Batt

Tools You Will Need

The following items are considered essential background material for this chapter. If you doubt your knowledge of any of these items, you should review the appropriate chapter or section before proceeding.

- Sum of squares (*SS*) (Chapter 4)
  - Computational formula
  - Definitional formula
- *z*-scores (Chapter 5)
- Hypothesis testing (Chapter 8)

A recent report describing the results of two separate studies, one with German participants and one with Americans, examined the relationship between income and weight for both men and women (Judge & Cable, 2011). Based on social standards, the authors predicted a negative relationship for women, with income decreasing as weight increased. For men, however, the expectation was for a generally positive relationship, with income increasing as weight increased, at least until obesity. These predictions came from the observation that being thin is seen as a sign of success and self-discipline for women. For men, however, a degree of extra weight is often viewed as a sign of success. For example, wealthy tycoons are typically portrayed in political cartoons as overweight men smoking cigars in tight-fitting suits. If these observations are correct, then it is possible that weight influences discrimination at some stage during employment from hiring and initial salary to training opportunities and promotion. Extra weight would lead to positive discrimination for men and negative discrimination for women, resulting in the predicted weight-income relationships.

The pattern of results from both the German and the American studies is shown in Figure 15.1. Note that the results support the researchers' predictions; income declined with increasing weight for women and income increased with additional weight for men.

Although the data in Figure 15.1 appear to show clear relationships, we need some procedure to measure relationships and a hypothesis test to determine whether they are significant. In the preceding five chapters, we described relationships between variables in terms of mean differences between two or more groups of scores, and we used hypothesis tests that evaluate the significance of mean differences. For the data in Figure 15.1, however, there is only one group of men and only one group of women. Calculating a mean weight and a mean income for the men is not going to help describe the relationship between the two variables. To evaluate these data, a completely different approach is needed for both descriptive and inferential statistics.

The two sets of data in Figure 15.1 are examples of the results from correlational research studies. In Chapter 1, the correlational design was introduced as a method for examining the relationship between two variables by measuring two different variables for each individual in one group of participants. The relationship obtained in a correlational study is typically described and evaluated with a statistical measure known as a *correlation*. Just as a sample mean provides a concise description of an entire sample, a correlation provides a description of a relationship. In this chapter, we introduce correlations and examine how correlations are used and interpreted.

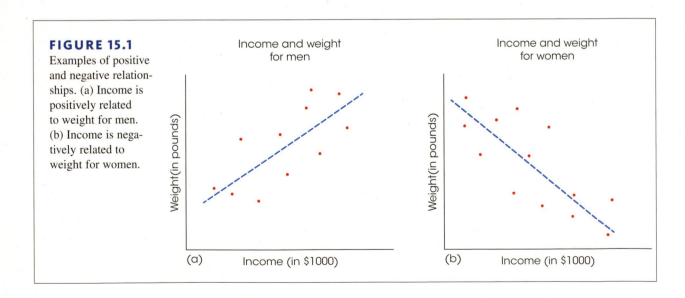

FIGURE 15.1
Examples of positive and negative relationships. (a) Income is positively related to weight for men. (b) Income is negatively related to weight for women.

15.1 | Introduction

LEARNING OBJECTIVE

1. Describe the information provided by the sign (+/−) and the numerical value of a correlation.

Correlation is a statistical technique that is used to measure and describe the relationship between two variables. Usually the two variables are simply observed as they exist naturally in the environment—there is no attempt to control or manipulate the variables. For example, a researcher could check high school records (with permission) to obtain a measure of each student's academic performance, and then survey each family to obtain a measure of income. The resulting data could be used to determine whether there is relationship between high school grades and family income. Notice that the researcher is not manipulating any student's grade or any family's income, but is simply observing what occurs naturally. You also should notice that a correlation requires two scores for each individual (one score from each of the two variables). These scores normally are identified as *X* and *Y*. The pairs of scores can be listed in a table, or they can be presented graphically in a scatter plot (Figure 15.2). In the scatter plot, the values for the *X* variable are listed on the horizontal axis and the *Y* values are listed on the vertical axis. Each individual is represented by a single point in the graph so that the horizontal position corresponds to the individual's *X* value and the vertical position corresponds to the *Y* value. The value of a scatter plot is that it allows you to see any patterns or trends that exist in the data. The scores in Figure 15.2, for example, show a clear relationship between family income and student grades: as income increases, grades also increase.

■ The Characteristics of a Relationship

A correlation is a numerical value that describes and measures three characteristics of the relationship between *X* and *Y*. These three characteristics are as follows

1. **The Direction of the Relationship**   The sign of the correlation, positive or negative, describes the direction of the relationship.

Person	Family Income (in $1000)	Student's Average Grade
A	31	72
B	38	86
C	42	81
D	44	78
E	49	85
F	56	80
G	58	91
H	65	89
I	70	94
J	90	83
K	92	90
L	106	97
M	135	89
N	174	95

FIGURE 15.2

Correlational data showing the relationship between family income (*X*) and student grades (*Y*) for a sample of *n* = 14 high school students. The scores are listed in order from lowest to highest family income and are shown in a scatter plot.

DEFINITIONS

In a **positive correlation**, the two variables tend to change in the same direction: as the value of the X variable increases from one individual to another, the Y variable also tends to increase; when the X variable decreases, the Y variable also decreases.

In a **negative correlation**, the two variables tend to go in opposite directions. As the X variable increases, the Y variable decreases. That is, it is an inverse relationship.

The two examples presented in the Preview provide examples of positive and negative relationships (see Figure 15.1). For the men, there is a positive relationship between weight and income, with income increasing as weight increases from person to person. For the women, the relationship is negative, with income decreasing as weight increases.

2. **The Form of the Relationship** In the preceding salary and weight examples (Figure 15.1), the relationships tend to have a linear form; that is, the points in the scatter plot tend to cluster around a straight line. We have drawn a line through the middle of the data points in each figure to help show the relationship. The most common use of correlation is to measure straight-line relationships. However, other forms of relationships do exist and there are special correlations used to measure them. (We examine alternatives in Section 15.5.)

3. **The Strength or Consistency of the Relationship** Finally, the correlation measures the consistency of the relationship. For a linear relationship, for example, the data points could fit perfectly on a straight line. Every time X increases by one point, the value of Y also changes by a consistent and predictable amount. Figure 15.3(a) shows an example of a perfect linear relationship. However, relationships are usually not perfect. Although there may be a tendency for the value of Y to increase whenever X increases, the amount that Y changes is not always the same, and occasionally, Y decreases when X increases. In this situation, the data points do not fall perfectly on a straight line. The consistency of the relationship is measured by the numerical value

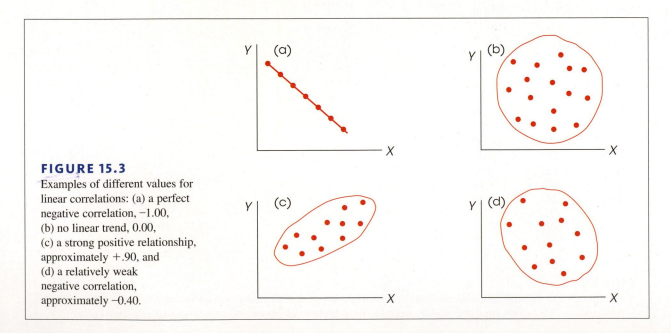

FIGURE 15.3
Examples of different values for linear correlations: (a) a perfect negative correlation, −1.00, (b) no linear trend, 0.00, (c) a strong positive relationship, approximately +.90, and (d) a relatively weak negative correlation, approximately −0.40.

of the correlation. A *perfect correlation* always is identified by a correlation of 1.00 and indicates a perfectly consistent relationship. For a correlation of 1.00 (or −1.00), each change in *X* is accompanied by a perfectly predictable change in *Y*. At the other extreme, a correlation of 0 indicates no consistency at all. For a correlation of 0, the data points are scattered randomly with no clear trend (see Figure 15.3(b)). Intermediate values between 0 and 1 indicate the degree of consistency.

Examples of different values for linear correlations are shown in Figure 15.3. In each example we have sketched a line around the data points. This line, called an *envelope* because it encloses the data, often helps you to see the overall trend in the data. As a rule of thumb, when the envelope is shaped roughly like a football, the correlation is around 0.7. Envelopes that are fatter than a football indicate correlations closer to 0, and narrower shapes indicate correlations closer to 1.00.

You should also note that the sign (+ or −) and the strength of a correlation are independent. For example, a correlation of 1.00 indicates a perfectly consistent relationship whether it is positive (+1.00) or negative (−1.00). Similarly, correlations of +0.80 and −0.80 are equally consistent relationships. Finally, you should notice that a correlation can never be larger than +1.00 or smaller than −1.00. If your calculations produce a value outside this range, then you should realize immediately that you have made a mistake.

LEARNING CHECK

1. A negative value for a correlation indicates _____.
   a. a much stronger relationship than if the correlation were positive
   b. a much weaker relationship than if the correlation were positive
   c. increases in *X* tend to be accompanied by increases in *Y*
   d. increases in *X* tend to be accompanied by decreases in *Y*

2. Which of the following is the correct order, from strongest and most consistent to weakest and least consistent, for the following correlations?
   a. −1.00, 0.85, −0.43, 0.02
   b. 0.85, 0.02, −0.43, −1.00
   c. −1.00, −0.43, 0.02. 0.85
   d. 0.85, −0.43, 0.02, −1.00

ANSWERS   1. D, 2. A

15.2 | The Pearson Correlation

LEARNING OBJECTIVES

2. Calculate the sum of products of deviations (*SP*) for a set of scores using the definitional and the computational formulas.

3. Calculate the Pearson correlation for a set of scores and explain what it measures.

4. Explain how the value of the Pearson correlation is affected when a constant is added to each of the *X* scores and/or the *Y* scores, and when the *X* and/or *Y* scores are all multiplied by a constant.

By far the most common correlation is the Pearson correlation (or the Pearson product–moment correlation), which measures the degree of straight-line relationship.

DEFINITION

> The **Pearson correlation** measures the degree and the direction of the linear relationship between two variables.

The Pearson correlation for a sample is identified by the letter r. The corresponding correlation for the entire population is identified by the Greek letter rho (ρ), which is the Greek equivalent of the letter r. Conceptually, this correlation is computed by

$$r = \frac{\text{degree to which } X \text{ and } Y \text{ vary together}}{\text{degree to which } X \text{ and } Y \text{ vary separately}}$$

$$= \frac{\text{covariability of } X \text{ and } Y}{\text{variability of } X \text{ and } Y \text{ separately}}$$

When there is a perfect linear relationship, every change in the X variable is accompanied by a corresponding change in the Y variable. In Figure 15.3(a), for example, every time the value of X increases, there is a perfectly predictable decrease in the value of Y. The result is a perfect linear relationship, with X and Y always varying together. In this case, the covariability (X and Y together) is identical to the variability of X and Y separately, and the formula produces a correlation with a magnitude of 1.00 or −1.00. At the other extreme, when there is no linear relationship, a change in the X variable does not correspond to any predictable change in the Y variable. In this case, there is no covariability, and the resulting correlation is zero.

■ The Sum of Products of Deviations

To calculate the Pearson correlation, it is necessary to introduce one new concept: the *sum of products* of deviations, or *SP*. This new value is similar to *SS* (the sum of squared deviations), which is used to measure variability for a single variable. Now, we use *SP* to measure the amount of covariability between two variables. The value for *SP* can be calculated with either a definitional formula or a computational formula.

The *definitional formula* for the sum of products is

$$SP = \Sigma(X - M_X)(Y - M_Y) \tag{15.1}$$

where M_X is the mean for the X scores and M_Y is the mean for the Ys.

The definitional formula instructs you to perform the following sequence of operations:

1. Find the X deviation and the Y deviation for each individual.
2. Find the product of the deviations for each individual.
3. Add the products.

Notice that this process literally *defines* the value being calculated; that is, the formula actually computes the sum of the products of the deviations.

The *computational formula* for the sum of products of deviations is

Caution: **The n in this formula refers to the number of pairs of scores.**

$$SP = \Sigma XY - \frac{\Sigma X \Sigma Y}{n} \tag{15.2}$$

Because the computational formula uses the original scores (X and Y values), it usually results in easier calculations than those required with the definitional formula, especially if M_X or M_Y is not a whole number. However, both formulas will always produce the same value for *SP*.

You may have noted that the formulas for *SP* are similar to the formulas you have learned for *SS* (sum of squares). The relationship between the two sets of formulas is described in Box 15.1. The following example demonstrates the calculation of *SP* with both formulas.

BOX 15.1 Comparing the *SP* and *SS* Formulas

It may help you to learn the formulas for *SP* if you note the similarity between the two *SP* formulas and the corresponding formulas for *SS* that were presented in Chapter 4. The definitional formula for *SS* is

$$SS = \Sigma(X - M)^2$$

In this formula, you must square each deviation, which is equivalent to multiplying it by itself. With this in mind, the formula can be rewritten as

$$SS = \Sigma(X - M)(X - M)$$

The similarity between the *SS* formula and the *SP* formula should be obvious—the *SS* formula uses squares and the *SP* formula uses products. This same relationship exists for the computational formulas. For *SS*, the computational formula is

$$SS = \Sigma X^2 - \frac{(\Sigma X)^2}{n}$$

As before, each squared value is equivalent to multiplying by itself, so the formula can be rewritten as

$$SS = \Sigma XX - \frac{\Sigma X \Sigma X}{n}$$

Again, note the similarity in structure between the *SS* formula and the *SP* formula: If you remember that the *SS* formulas use squared values and *SP* formulas use products, then the two new formulas for the sum of products should be easy to learn.

EXAMPLE 15.1

The same set of $n = 4$ pairs of scores are used to calculate *SP*, first using the definitional formula and then using the computational formula.

For the definitional formula, you need deviation scores for each of the *X* values and each of the *Y* values. Note that the mean for the *X*s is $M_X = 2.5$ and the mean for the *Y*s is $M_Y = 5$. The deviations and the products of deviations are shown in the following table:

Caution: The signs (+ and −) are critical in determining the sum of products, *SP*.

Scores		Deviations		Products
X	Y	$X - M_X$	$Y - M_Y$	$(X - M_X)(Y - M_Y)$
1	3	−1.5	−2	+3
2	6	−0.5	+1	−0.5
4	4	+1.5	−1	−1.5
3	7	+0.5	+2	+1
				+2 = SP

For these scores, the sum of the products of the deviations is $SP = +2$.

For the computational formula, you need the *X* value, the *Y* value, and the *XY* product for each individual. Then you find the sum of the *X*s, the sum of the *Y*s, and the sum of the *XY* products. These values are as follows:

X	Y	XY	
1	3	3	
2	6	12	
4	4	16	
3	7	21	
10	20	52	Totals

Substituting the totals in the formula gives

$$SP = \Sigma XY - \frac{\Sigma X \Sigma Y}{n}$$

$$= 52 - \frac{10(20)}{4}$$

$$= 52 - 50$$

$$= 2$$

Both formulas produce the same result, $SP = 2$. ∎

The following example is an opportunity to test your understanding of the calculation of SP (the sum of products of deviations).

| **EXAMPLE 15.2** | Calculate the sum of products of deviations (SP) for the following set of scores. Use the definitional formula and then the computational formula. You should obtain $SP = 5$ with both formulas. |

X	Y
0	1
3	3
2	3
5	2
0	1

∎

■ Calculation of the Pearson Correlation

As noted earlier, the Pearson correlation consists of a ratio comparing the covariability of X and Y (the numerator) with the variability of X and Y separately (the denominator). In the formula for the Pearson r, we use SP to measure the covariability of X and Y. The variability of X is measured by computing SS for the X scores and the variability of Y is measured by SS for the Y scores. With these definitions, the formula for the Pearson correlation becomes

$$r = \frac{SP}{\sqrt{SS_X SS_Y}} \tag{15.3}$$

Note that you *multiply* SS for X by SS for Y in the denominator of the Pearson formula.

The following example demonstrates the use of this formula with a simple set of scores.

| **EXAMPLE 15.3** | The Pearson correlation is computed for the set of $n = 5$ pairs of scores shown in the margin. |

X	Y
0	2
10	6
4	2
8	4
8	6

Before starting any calculations, it is useful to put the data in a scatter plot and make a preliminary estimate of the correlation. These data have been graphed in Figure 15.4. Looking at the scatter plot, it appears that there is a very good (but not perfect) positive correlation. You should expect an approximate value of $r = +.8$ or $+.9$. To find the Pearson correlation, we need SP, SS for X, and SS for Y. The calculations for each of these values, using the definitional formulas, are presented in Table 15.1. (Note that the mean for the X values is $M_X = 6$ and the mean for the Y scores is $M_Y = 4$.)

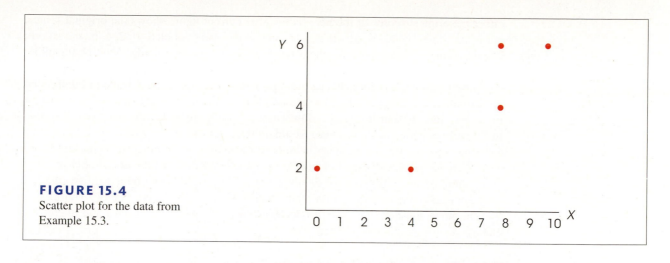

FIGURE 15.4
Scatter plot for the data from
Example 15.3.

TABLE 15.1
Calculation of SS_X, SS_Y, and SP for a sample of $n = 5$ pairs of scores.

Scores		Deviations		Squared Deviations		Products
X	Y	$X - M_X$	$Y - M_Y$	$(X - M_X)^2$	$(Y - M_Y)^2$	$(X - M_X)(Y - M_Y)$
0	2	−6	−2	36	4	+12
10	6	+4	+2	16	4	+8
4	2	−2	−2	4	4	+4
8	4	+2	0	4	0	0
8	6	+2	+2	4	4	+4
				$SS_X = 64$	$SS_Y = 16$	$SP = +28$

Using the values from Table 15.1, the Pearson correlation is

$$r = \frac{SP}{\sqrt{(SS_X)(SS_Y)}} = \frac{28}{\sqrt{(64)(16)}} = \frac{28}{32} = +0.875$$

■

■ Correlation and the Pattern of Data Points

Note that the value we obtained for the correlation in Example 15.3 is perfectly consistent with the pattern formed by the data points in Figure 15.4. The positive sign for the correlation indicates that the points are clustered around a line that slopes up to the right. Second, the high value for the correlation (near 1.00) indicates that the points are very tightly clustered close to the line. Thus, the value of the correlation describes the relationship that exists in the data.

Because the Pearson correlation describes the pattern formed by the data points, any factor that does not change the pattern also does not change the correlation. For example, if 5 points were added to each of the X values in Figure 15.4, then each data point would move to the right. However, because all of the data points shift to the right, the overall pattern is not changed, it is simply moved to a new location. Similarly, if 5 points were subtracted from each X value, the pattern would shift to the left. In either case, the overall pattern stays the same and the correlation is not changed. In the same way, adding a constant to (or subtracting a constant from) each Y value simply shifts the pattern up (or down) but does not change the pattern and, therefore, does not change the correlation. Similarly, multiplying each X and/or Y value by a positive constant also does not change the pattern formed by the data points and does not change the correlation. For example, if each of the X values in Figure 15.4 were multiplied by 2, then same scatter plot could be used to display either the original scores or the new scores. The current figure shows the original scores, but if the values on the X-axis (0, 1, 2, 3, and so on) were doubled (0, 2, 4, 6, and so on), then the same figure would show the pattern formed by the new scores. Multiplying either the X or the Y values by a negative number, however, does not change the numerical value of the correlation but it does change the sign. For example, if each X value in Figure 15.4 were multiplied by −1, then the current data points would be moved to the left-hand side of the Y-axis, forming a mirror image of the current pattern. Instead of the positive correlation in the current figure, the new pattern would produce a negative correlation with exactly the same numerical value. In summary, adding a constant to (or subtracting a constant from) each X and/or Y value does not change the pattern of data points and does not change the correlation. Also, multiplying (or dividing) each X or each Y value by a positive constant does not change the pattern and does not change the value of the correlation. Multiplying by a negative constant, however, produces a mirror image of the pattern and, therefore, changes the sign of the correlation.

■ The Pearson Correlation and z-Scores

The Pearson correlation measures the relationship between an individual's location in the X distribution and his or her location in the Y distribution. For example, a positive correlation means that individuals who score high on X also tend to score high on Y. Similarly, a negative correlation indicates that individuals with high X scores tend to have low Y scores.

Recall from Chapter 5 that z-scores identify the exact location of each individual score within a distribution. With this in mind, each X value can be transformed into a z-score, z_X, using the mean and standard deviation for the set of Xs. Similarly, each Y score can be transformed into z_Y. If the X and Y values are viewed as a sample, the transformation is completed using the sample formula for z (Equation 5.3). If the X and Y values form a complete population, the z-scores are computed using Equation 5.1. After the transformation, the formula for the Pearson correlation can be expressed entirely in terms of z-scores.

$$\text{For a sample, } r = \frac{\Sigma z_X z_Y}{(n-1)} \tag{15.4}$$

$$\text{For a population, } \rho = \frac{\Sigma z_X z_Y}{N} \tag{15.5}$$

Note that the population value is identified with a Greek letter rho.

LEARNING CHECK

1. What is the sum of products (SP) for the following data?

    a. 6

    b. −5

    c. 43

    d. None of the other 3 choices is correct.

X	Y
2	4
5	2
3	5
2	5

2. A set of $n = 5$ pairs of X and Y values has $SS_X = 5$, $SS_Y = 20$ and $SP = 8$. For these data, the Pearson correlation is _____.

    a. $r = \frac{8}{100} = 0.08$

    b. $r = \frac{8}{10} = 0.80$

    c. $r = \frac{8}{25} = 0.32$

    d. $r = \frac{8}{20} = 0.40$

3. A set of $n = 15$ pairs of X and Y values has a Pearson correlation of $r = 0.10$. If each of the X values were multiplied by 2, then what is the correlation for the resulting data?

   a. 0.10

   b. −0.10

   c. 0.20

   d. −0.20

ANSWERS   **1.** B, **2.** B, **3.** A

15.3 | Using and Interpreting the Pearson Correlation

LEARNING OBJECTIVES

5. Explain how a correlation can be influenced by a restricted range of scores or by outliers.

6. Define the coefficient of determination and explain what it measures.

7. Explain what is measured by a partial correlation.

■ Where and Why Correlation Are Used

Although correlations have a number of different applications, a few specific examples are presented next to give an indication of the value of this statistical measure.

1. **Prediction**   If two variables are known to be related in some systematic way, it is possible to use one of the variables to make accurate predictions about the other. For example, when you applied for admission to college, you were required to submit a great deal of personal information, including your scores on the Scholastic Achievement Test (SAT). College officials want this information so they can predict your chances of success in college. It has been demonstrated over several years that SAT scores and college grade point averages are correlated. Students who do well on the SAT tend to do well in college; students who have difficulty with the SAT tend to have difficulty in college. Based on this relationship, college admissions officers can make a prediction about the potential success of each applicant. You should note that this prediction is not perfectly accurate. Not everyone who does poorly on the SAT will have trouble in college. That is why you also submit letters of recommendation, high school grades, and other information with your application. The process of using relationships to make predictions is called *regression* and is discussed in the next chapter.

2. **Validity**   Suppose a psychologist develops a new test for measuring intelligence. How could you show that this test truly measures what it claims; that is, how could you demonstrate the validity of the test? One common technique for demonstrating validity is to use a correlation. If the test actually measures intelligence, then the scores on the test should be related to other measures of intelligence—for example, standardized IQ tests, performance on learning tasks, problem-solving ability, and so on. The psychologist could measure the correlation between the new test and each of these other measures of intelligence to demonstrate that the new test is valid.

3. **Reliability**   In addition to evaluating the validity of a measurement procedure, correlations are used to determine reliability. A measurement procedure is considered

reliable to the extent that it produces stable, consistent measurements. That is, a reliable measurement procedure will produce the same (or nearly the same) scores when the same individuals are measured twice under the same conditions. For example, if your IQ were measured as 113 last week, you would expect to obtain nearly the same score if your IQ were measured again this week. One way to evaluate reliability is to use correlations to determine the relationship between two sets of measurements. When reliability is high, the correlation between two measurements should be strong and positive. Further discussion of the concept of reliability is presented in Box 15.2.

4. **Theory Verification**   Many psychological theories make specific predictions about the relationship between two variables. For example, a theory may predict a relationship between brain size and learning ability; a developmental theory may predict a relationship between the parents' IQs and the child's IQ; a social psychologist may have a theory predicting a relationship between personality type and behavior in a social situation. In each case, the prediction of the theory could be tested by determining the correlation between the two variables.

BOX 15.2 Reliability and Error in Measurement

The idea of reliability of measurement is tied directly to the notion that each individual measurement includes an element of error. Expressed as an equation,

$$\text{measured score} = \text{true score} + \text{error}$$

For example, if I try to measure your intelligence with an IQ test, the score that I get is determined partially by your actual level of intelligence (your true score) but it also is influenced by a variety of other factors such as your current mood, your level of fatigue, your general health, and so on. These other factors are lumped together as *error,* and are typically a part of any measurement.

It is generally assumed that the error component changes randomly from one measurement to the next and this causes your score to change. For example, your IQ score is likely to be higher when you are well rested and feeling good compared to a measurement that is taken when you are tired and depressed. Although your actual intelligence hasn't changed, the error component causes your score to change from one measurement to another.

As long as the error component is relatively small, then your scores will be relatively consistent from one measurement to the next, and the measurements are said to be reliable. If you are feeling especially happy and well rested, it may affect your IQ score by a few points, but it is not going to boost your IQ from 110 to 170.

On the other hand, if the error component is relatively large, then you will find huge differences from one measurement to the next and the measurements are not reliable. Measurements of reaction time, for example, tend to be very unreliable. Suppose, for example, that you are seated at a desk with your finger on a button and a light bulb in front of you. Your job is to push the button as quickly as possible when the light goes on. On some trials, you are focused on the light with your finger tensed and ready to push. On other trials, you are distracted, or day dreaming, or blink when the light goes on so that time passes before you finally push the button. As a result, there is a huge error component to the measurement and your reaction time can change dramatically from one trial to the next. When measurements are unreliable you cannot trust any single measurement to provide an accurate indication of the individual's true score. To deal with this problem, researchers typically measure reaction time repeatedly and then average over a large number of measurements.

Correlations can be used to help researchers measure and describe reliability. By taking two measurements for each individual, it is possible to compute the correlation between the first score and the second score. A strong, positive correlation indicates a good level of reliability: people who scored high on the first measurement also scored high on the second. A weak correlation indicates that there is not a consistent relationship between the first score and the second score; that is, a weak correlation indicates poor reliability.

■ Interpreting Correlations

When you encounter correlations, there are four additional considerations that you should bear in mind.

1. Correlation simply describes a relationship between two variables. It does not explain why the two variables are related. Specifically, a correlation should not and cannot be interpreted as proof of a cause-and-effect relationship between the two variables.

2. The value of a correlation can be affected greatly by the range of scores represented in the data.

3. One or two extreme data points, often called *outliers,* can have a dramatic effect on the value of a correlation.

4. When judging how "good" a relationship is, it is tempting to focus on the numerical value of the correlation. For example, a correlation of +0.5 is halfway between 0 and 1.00 and therefore appears to represent a moderate degree of relationship. However, a correlation should not be interpreted as a proportion. Although a correlation of 1.00 does mean that there is a 100% perfectly predictable relationship between X and Y, a correlation of .5 does not mean that you can make predictions with 50% accuracy. To describe how accurately one variable predicts the other, you must square the correlation. Thus, a correlation of $r = .5$ means that one variable *partially* predicts the other, but the predictable portion is only $r^2 = .5^2 = 0.25$ (or 25%) of the total variability.

We now discuss each of these four points in detail.

■ Correlation and Causation

One of the most common errors in interpreting correlations is to assume that a correlation necessarily implies a cause-and-effect relationship between the two variables. (Even Pearson blundered by asserting causation from correlational data; Blum, 1978.) We constantly are bombarded with reports of relationships: cigarette smoking is related to heart disease; alcohol consumption is related to birth defects; carrot consumption is related to good eyesight. Do these relationships mean that cigarettes cause heart disease or carrots cause good eyesight? The answer is *no.* Although there may be a causal relationship, the simple existence of a correlation does not prove it. Earlier, for example, we discussed a study showing a relationship between high school grades and family income. However, this result does not mean that having a higher family income *causes* students to get better grades. For example, if mom gets an unexpected bonus at work, it is unlikely that her child's grades will also show a sudden increase. To establish a cause-and-effect relationship, it is necessary to conduct a true experiment (see p. 13) in which one variable is manipulated by a researcher and other variables are rigorously controlled. The fact that a correlation does not establish causation is demonstrated in the following example.

EXAMPLE 15.4

Suppose we select a variety of different cities and towns throughout the United States and measure the number of churches (X variable) and the number of serious crimes (Y variable) for each. A scatter plot showing hypothetical data for this study is presented in Figure 15.5. Notice that this scatter plot shows a strong, positive correlation between churches and crime. You also should note that these are realistic data. It is reasonable that the small towns would have less crime and fewer churches and that the large cities would have large values for both variables. Does this relationship mean that churches cause crime? Does it mean that crime causes churches? It should be clear that both answers are no. Although a strong correlation exists between churches and crime, the real cause of the relationship is the size of the population. ■

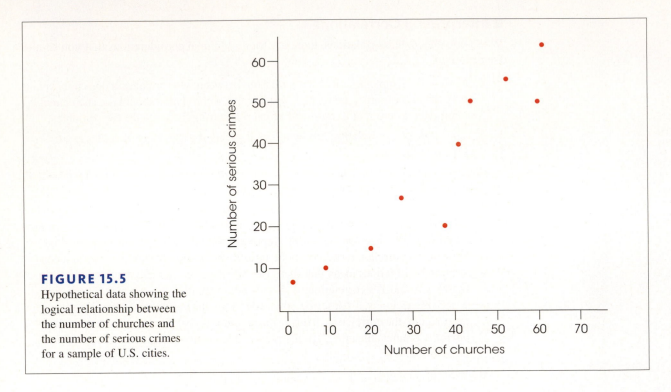

FIGURE 15.5

Hypothetical data showing the logical relationship between the number of churches and the number of serious crimes for a sample of U.S. cities.

■ Correlation and Restricted Range

Whenever a correlation is computed from scores that do not represent the full range of possible values, you should be cautious in interpreting the correlation. Suppose, for example, you are interested in the relationship between IQ and creativity. If you select a sample of your fellow college students, your data probably will represent only a limited range of IQ scores (most likely from 110–130). The correlation within this *restricted range* could be completely different from the correlation that would be obtained from a full range of IQ scores. For example, Figure 15.6 shows a strong positive relationship between X and Y when the entire range of scores is considered. However, this relationship is obscured when the data are limited to a restricted range.

To be safe, you should not generalize any correlation beyond the range of data represented in the sample. For a correlation to provide an accurate description for the general population, there should be a wide range of X and Y values in the data.

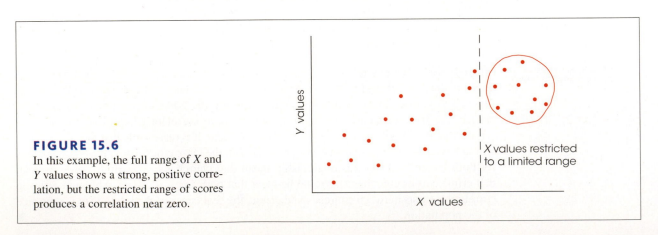

FIGURE 15.6

In this example, the full range of X and Y values shows a strong, positive correlation, but the restricted range of scores produces a correlation near zero.

■ Outliers

An outlier is an individual with X and/or Y values that are substantially different (larger or smaller) from the values obtained for the other individuals in the data set. The data point of a single outlier can have a dramatic influence on the value obtained for the correlation. This effect is illustrated in Figure 15.7. Figure 15.7(a) shows a set of $n = 5$ data points for which the correlation between the X and Y variables is nearly zero (actually $r = -0.08$). In Figure 15.7(b), one extreme data point (14, 12) has been added to the original data set. When this outlier is included in the analysis, a strong, positive correlation emerges (now $r = +0.85$). Note that the single outlier drastically alters the value for the correlation and thereby can affect one's interpretation of the relationship between variables X and Y. Without the outlier, one would conclude there is no relationship between the two variables. With the extreme data point, $r = +0.85$ implies a strong relationship with Y increasing consistently as X increases. The problem of outliers is a good reason for looking at a scatter plot, instead of simply basing your interpretation on the numerical value of the correlation. If you only "go by the numbers," you might overlook the fact that one extreme data point inflated the size of the correlation.

■ Correlation and the Strength of the Relationship

A correlation measures the degree of relationship between two variables on a scale from 0–1.00. Although this number provides a measure of the degree of relationship, many researchers prefer to square the correlation and use the resulting value to measure the strength of the relationship.

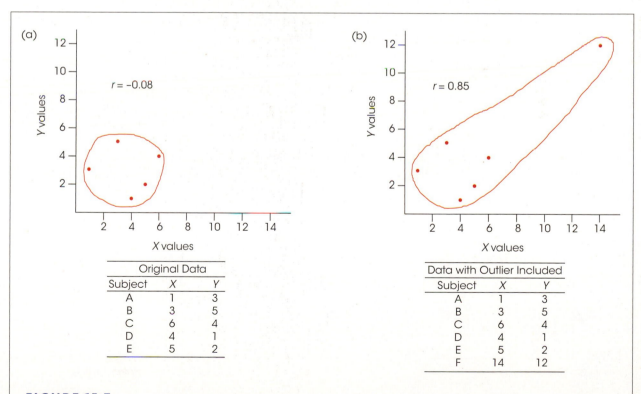

FIGURE 15.7

A demonstration of how one extreme data point (an outlier) can influence the value of a correlation.

One of the common uses of correlation is for prediction. If two variables are correlated, you can use the value of one variable to predict the other. For example, college admissions officers do not just guess which applicants are likely to do well; they use other variables (SAT scores, high school grades, and so on) to predict which students are most likely to be successful. These predictions are based on correlations. By using correlations, the admissions officers expect to make more accurate predictions than would be obtained by chance. In general, the squared correlation (r^2) measures the gain in accuracy that is obtained from using the correlation for prediction. The squared correlation measures the proportion of variability in the data that is explained by the relationship between X and Y. It is sometimes called the *coefficient of determination*.

DEFINITION

> The value r^2 is called the **coefficient of determination** because it measures the proportion of variability in one variable that can be determined from the relationship with the other variable. A correlation of $r = 0.80$ (or -0.80), for example, means that $r^2 = 0.64$ (or 64%) of the variability in the Y scores can be predicted from the relationship with X.

In earlier chapters (see pp. 281, 317 and 348) we introduced r^2 as a method for measuring effect size for research studies where mean differences were used to compare treatments. Specifically, we measured how much of the variance in the scores was accounted for by the differences between treatments. In experimental terminology, r^2 measures how much of the variance in the dependent variable is accounted for by the independent variable. Now we are doing the same thing, except that there is no independent or dependent variable. Instead, we simply have two variables, X and Y, and we use r^2 to measure how much of the variance in one variable can be determined from its relationship with the other variable. The following example demonstrates this concept.

EXAMPLE 15.5

Figure 15.8 shows three sets of data representing different degrees of linear relationship. The first set of data (Figure 15.8(a)) shows the relationship between IQ and shoe size. In this case, the correlation is $r = 0$ (and $r^2 = 0$), and you have no ability to predict a person's IQ based on his or her shoe size. Knowing a person's shoe size provides no information (0%) about the person's IQ. In this case, shoe size provides no help explaining why different people have different IQs.

Now consider the data in Figure 15.8(b). These data show a moderate, positive correlation, $r = +0.60$, between IQ scores and college grade point averages (GPA). Students with

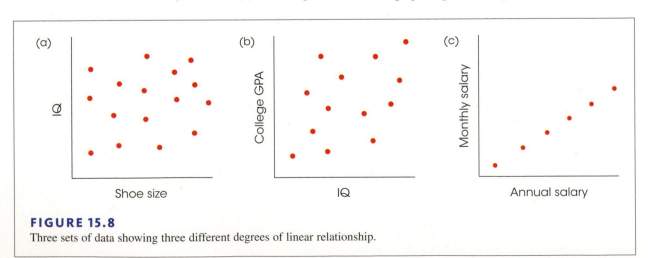

FIGURE 15.8
Three sets of data showing three different degrees of linear relationship.

high IQs tend to have higher grades than students with low IQs. From this relationship, it is possible to predict a student's GPA based on his or her IQ. However, you should realize that the prediction is not perfect. Although students with high IQs *tend* to have high GPAs, this is not always true. Thus, knowing a student's IQ provides some information about the student's grades, or knowing a student's grades provides some information about the student's IQ. In this case, IQ scores help explain the fact that different students have different GPAs. Specifically, you can say that *part* of the differences in GPA are accounted for by IQ. With a correlation of $r = +0.60$, we obtain $r^2 = 0.36$, which means that 36% of the variance in GPA can be explained by IQ.

Finally, consider the data in Figure 15.8(c). This time we show a perfect linear relationship ($r = +1.00$) between monthly salary and yearly salary for a group of college employees. With $r = 1.00$ and $r^2 = 1.00$, there is 100% predictability. If you know a person's monthly salary, you can predict the person's annual salary with perfect accuracy. If two people have different annual salaries, the difference can be completely explained (100%) by the difference in their monthly salaries. ■

Just as r^2 was used to evaluate effect size for mean differences in Chapters 9, 10, and 11, r^2 can now be used to evaluate the size or strength of the correlation. The same standards that were introduced in Table 9.3 (p. 283), apply to both uses of the r^2 measure. Specifically, an r^2 value of 0.01 indicates a small effect or a small correlation, an r^2 value of 0.09 indicates a medium correlation, and r^2 of 0.25 or larger indicates a large correlation.

More information about the coefficient of determination (r^2) is presented in Section 15.5 and in Chapter 16. For now, you should realize that whenever two variables are consistently related, it is possible to use one variable to predict values for the second variable. One final comment concerning the interpretation of correlations is presented in Box 15.3.

BOX 15.3   Regression Toward the Mean

Consider the following problem.

Explain why the rookie of the year in major-league baseball usually does not perform as well in his second season.

Notice that this question does not appear to be statistical or mathematical in nature. However, the answer to the question is directly related to the statistical concepts of correlation and regression (Chapter 16). Specifically, there is a simple observation about correlations known as *regression toward the mean*.

DEFINITION   When there is a less-than-perfect correlation between two variables, extreme scores (high or low) for one variable tend to be paired with the less extreme scores (more toward the mean) on the second variable. This fact is called **regression toward the mean**.

Figure 15.9 shows a scatter plot with a less-than-perfect correlation between two variables. The data

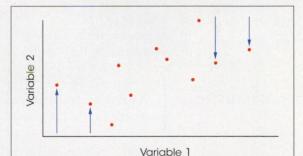

FIGURE 15.9
A demonstration of regression toward the mean. The figure shows a scatter plot for data with a less-than-perfect correlation. Notice that the highest scores on variable I (extreme right-hand points) are not the highest scores on variable 2, but are displaced downward toward the mean. Also, the lowest scores on variable 1 (extreme left-hand points) are not the lowest scores on variable 2, but are displaced upward toward the mean.

(continues)

BOX 15.3 Regression Toward the Mean *(continued)*

points in this figure might represent batting averages for baseball rookies in 2014 (variable 1) and batting averages for the same players in 2015 (variable 2). Because the correlation is less than perfect, the highest scores on variable 1 are generally *not* the highest scores on variable 2. In baseball terms, the rookies with the highest averages in 2014 do not have the highest averages in 2015.

Remember that a correlation does not explain *why* one variable is related to the other; it simply says that there is a relationship. The correlation cannot explain why the best rookie does not perform as well in his second year. But, because the correlation is

not perfect, it is a statistical fact that extremely high scores in one year generally will *not* be paired with extremely high scores in the next year.

Now try using the concept of regression toward the mean to explain the following phenomena.

1. You have a truly outstanding meal at a restaurant. However, when you go back with a group of friends, you find that the food is disappointing.

2. You have the highest score on exam I in your statistics class, but score only a little above average on exam II.

■ Partial Correlations

Occasionally a researcher may suspect that the relationship between two variables is being distorted by the influence of a third variable. Earlier in the chapter, for example, we found a strong positive relationship between the number of churches and the number of serious crimes for a sample of different towns and cities (see Example 15.4, p 497). However, it is unlikely that there is a direct relationship between churches and crime. Instead, both variables are influenced by population: Large cities have a lot of churches and high crime rates compared to smaller towns, which have fewer churches and less crime. If population is controlled, there probably would be no real correlation between churches and crime.

Fortunately, there is a statistical technique, known as *partial correlation,* that allows a researcher to measure the relationship between two variables while eliminating or holding constant the influence of a third variable. Thus, a researcher could use a partial correlation to examine the relationship between churches and crime without risk that the relationship is distorted by the size of the population.

DEFINITION

> **A partial correlation** measures the relationship between two variables while controlling the influence of a third variable by holding it constant.

In a situation with three variables, X, Y, and Z, it is possible to compute three individual Pearson correlations:

1. r_{XY} measuring the correlation between X and Y
2. r_{XZ} measuring the correlation between X and Z
3. r_{YZ} measuring the correlation between Y and Z

These three individual correlations can then be used to compute a partial correlation. For example, the partial correlation between X and Y, holding Z constant, is determined by the formula

$$r_{XY \cdot Z} = \frac{r_{XY} - (r_{XZ}r_{YZ})}{\sqrt{(1 - r_{XZ}^2)(1 - r_{YZ}^2)}}$$

(15.6)

The following example demonstrates the calculation and interpretation of a partial correlation.

EXAMPLE 15.6

We begin with the hypothetical data shown in Table 15.2. These scores have been constructed to simulate the church/crime/population situation for a sample of $n = 15$ cities. The X variable represents the number of churches, Y represents the number of crimes, and Z represents the population for each city. Note that the cities are grouped into three categories based on population (small cities, medium cities, large cities) with $n = 5$ cities in each group.

TABLE 15.2

Hypothetical data showing the relationship between the number of churches, the number of crimes, and the populations for a set of $n = 15$ cities.

Small Cities ($Z = 1$)		Medium Cities ($Z = 2$)		Large Cities ($Z = 3$)	
Churches (X)	Crimes (Y)	Churches (X)	Crimes (Y)	Churches (X)	Crimes (Y)
1	4	7	8	13	15
2	3	8	11	14	14
3	1	9	9	15	16
4	2	10	7	16	17
5	5	11	10	17	13

The data points for the 15 cities are shown in the scatter plot in Figure 15.10. Note that the population variable, Z, separates the scores into three distinct clusters: When $Z = 1$, the population is low and churches and crime (X and Y) are also low; when $Z = 2$, the population is moderate and churches and crime (X and Y) are also moderate; and when $Z = 3$, the population is large and churches and crime are both high. Thus, as the population increases from one city to another, the number of churches and crimes also increase, and the result is a strong positive correlation between churches and crime.

For the full set of 15 cities, the individual Pearson correlations are all large and positive:

a. The correlation between churches and crime is $r_{XY} = 0.923$.

b. The correlation between churches and population is $r_{XZ} = 0.961$.

c. The correlation between crime and population is $r_{YZ} = 0.961$.

Within each of the three population categories, however, there is no linear relationship between churches and crime. Specifically, within each group, the population variable is constant and the five data points for X and Y form a circular pattern, indicating no consistent linear relationship. The strong positive correlation between churches and crime appears to be caused by the differences in population. The partial correlation allows us to hold population constant across the entire sample and measure the underlying relationship between churches and crime without any influence from population. For these data, the partial correlation is

$$r_{XY-Z} = \frac{0.923 - 0.961(0.961)}{\sqrt{(1 - 0.961^2)(1 - 0.961^2)}}$$

$$= \frac{0}{0.076}$$

$$= 0$$

Thus, when the population differences are eliminated, there is no correlation remaining between churches and crime ($r = 0$). ■

In Example 15.6, the population differences, which correspond to the different values of the Z variable, were eliminated mathematically in the calculation of the partial correlation. However, it is possible to visualize how these differences are eliminated in the actual data. Looking at Figure 15.10, focus on the five points in the bottom left corner. These are

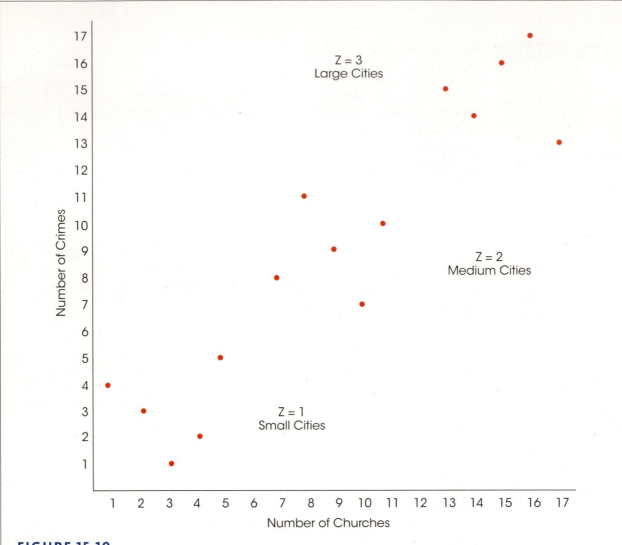

FIGURE 15.10

Hypothetical data showing the logical relationship between the number of churches and the number of crimes for three groups of cities: those with small populations ($Z = 1$), those with medium populations ($Z = 2$), and those with large populations ($Z = 3$).

the five cities with small populations, few churches, and little crime. The five points in the upper right corner represent the five cities with large populations, many churches and a lot of crime. The partial correlation controls population size by mathematically equalizing the populations for all 15 cities. Population is increased for the five small cities, which also increases churches and crime. Similarly, population is decreased for the five large cities, which also decreases churches and crime. In Figure 15.10, imagine the five points in the bottom left moving up and to the right so they overlap with the points in the center. At the same time, the five points in the upper right move down and to the left so they also overlap the points in the center. When population is equalized, the resulting set of 15 cities is shown in Figure 15.11. Note that controlling the population appears to have eliminated the relationship between churches and crime. This appearance is verified by the correlation for the 15 data points in Figure 15.11, which is $r = 0$, exactly the same as the partial correlation.

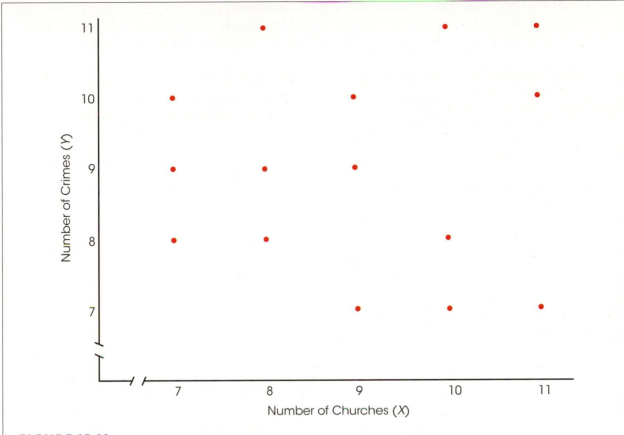

FIGURE 15.11

The relationship between the number of churches and the number of crimes for the same 15 cities shown in Figure 15.10 after the populations have been equalized.

LEARNING CHECK

1. The correlation for a set of X and Y scores is $r = 0.60$. The scores are separated into two groups, with one group consisting of individuals with X values that are equal to or above the median and the other group consisting of individuals with X values that are below the median. If the correlation is computed for the group with X values below the median, how will the correlation compare with the correlation for the full set of scores?

   a. It will still be $r = 0.60$.

   b. It will be greater than $r = 0.60$

   c. It will be smaller than $r = 0.60$.

   d. It is impossible to predict how the correlation for the smaller group will be related to the correlation for the entire group.

2. Suppose the correlation between height and weight for adults is $+0.80$. What proportion (or percent) of the variability in weight can be explained by the relationship with height?

   a. 80%

   b. 64%

   c. $100 - 80 = 20\%$

   d. 40%

3. What is measured by a partial correlation?

  a. It is the correlation obtained for a sample with missing scores (X or Y values).

  b. It is the correlation obtained for a restricted range of scores.

  c. It eliminates the influence of outliers (extreme scores) when computing a correlation.

  d. It measures the relationship between two variables while controlling the influence of a third variable.

ANSWERS   **1.** D, **2.** B, **3.** D

15.4 | Hypothesis Tests with the Pearson Correlation

LEARNING OBJECTIVE   **8.** Conduct a hypothesis test evaluating the significance of a correlation.

The Pearson correlation is generally computed for sample data. As with most sample statistics, however, a sample correlation is often used to answer questions about the corresponding population correlation. For example, a psychologist would like to know whether there is a relationship between IQ and creativity. This is a general question concerning a population. To answer the question, a sample would be selected, and the sample data would be used to compute the correlation value. You should recognize this process as an example of inferential statistics: using samples to draw inferences about populations. In the past, we have been concerned primarily with using sample means as the basis for answering questions about population means. In this section, we examine the procedures for using a sample correlation as the basis for testing hypotheses about the corresponding population correlation.

■ The Hypotheses

The basic question for this hypothesis test is whether a correlation exists in the population. The null hypothesis is "No. There is no correlation in the population." or "The population correlation is zero." The alternative hypothesis is "Yes. There is a real, nonzero correlation in the population." Because the population correlation is traditionally represented by ρ (the Greek letter rho), these hypotheses would be stated in symbols as

  $H_0: \rho = 0$ **(There is no population correlation.)**

  $H_1: \rho \neq 0$ **(There is a real correlation.)**

When there is a specific prediction about the direction of the correlation, it is possible to do a directional, or one-tailed test. For example, if a researcher is predicting a positive relationship, the hypotheses would be

  $H_0: \rho \leq 0$ **(The population correlation is not positive.)**

  $H_1: \rho > 0$ **(The population correlation is positive.)**

The correlation from the sample data is used to evaluate the hypotheses. For the regular, nondirectional test, a sample correlation near zero provides support for H_0 and a sample value far from zero tends to refute H_0. For a directional test, a positive value for the sample correlation would tend to refute a null hypothesis stating that the population correlation is not positive.

Although sample correlations are used to test hypotheses about population correlations, you should keep in mind that samples are not expected to be identical to the populations from which they come; there will be some discrepancy (sampling error) between a sample statistic and the corresponding population parameter. Specifically, you should always expect some error between a sample correlation and the population correlation it represents. One implication of this fact is that even when there is no correlation in the population ($\rho = 0$), you are still likely to obtain a nonzero value for the sample correlation. This is particularly true for small samples. Figure 15.12 illustrates how a small sample from a population with a near-zero correlation could result in a correlation that deviates from zero. The colored dots in the figure represent the entire population and the three circled dots represent a random sample. Note that the three sample points show a relatively good, positive correlation even through there is no linear trend ($\rho = 0$) for the population.

When you obtain a nonzero correlation for a sample, the purpose of the hypothesis test is to decide between the following two interpretations.

1. There is no correlation in the population ($\rho = 0$) and the sample value is the result of sampling error. Remember, a sample is not expected to be identical to the population. There always is some error between a sample statistic and the corresponding population parameter. This is the situation specified by H_0.

2. The nonzero sample correlation accurately represents a real, nonzero correlation in the population. This is the alternative stated in H_1.

The correlation from the sample will help to determine which of these two interpretations is more likely. A sample correlation near zero supports the conclusion that the population correlation is also zero. A sample correlation that is substantially different from zero supports the conclusion that there is a real, nonzero correlation in the population.

■ The Hypothesis Test

The hypothesis test evaluating the significance of a correlation can be conducted using either a t statistic or an F-ratio. The F-ratio is discussed later (pp. 541–543) and we focus

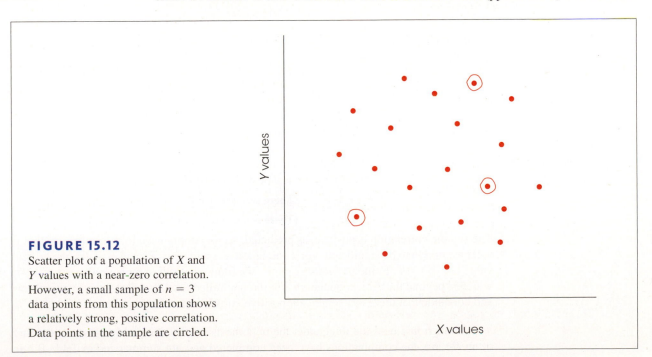

FIGURE 15.12

Scatter plot of a population of X and Y values with a near-zero correlation. However, a small sample of $n = 3$ data points from this population shows a relatively strong, positive correlation. Data points in the sample are circled.

on the t statistic here. The t statistic for a correlation has the same general structure as t statistics introduced in Chapters 9, 10, and 11.

$$t = \frac{\text{sample statistic} - \text{population parameter}}{\text{standard error}}$$

In this case, the sample statistic is the sample correlation (r) and the corresponding parameter is the population correlation (ρ). The null hypothesis specifies that the population correlation is $\rho = 0$. The final part of the equation is the standard error, which is determined by

$$\text{standard error for } r = s_r = \sqrt{\frac{1 - r^2}{n - 2}} \tag{15.7}$$

Thus, the complete t statistic is

$$t = \frac{r - \rho}{\sqrt{\frac{(1 - r^2)}{(n - 2)}}} \tag{15.8}$$

Degrees of Freedom for the t Statistic   The t statistic has degrees of freedom defined by $df = n - 2$. An intuitive explanation for this value is that a sample with only $n = 2$ data points has no degrees of freedom. Specifically, if there are only two points, they will fit perfectly on a straight line, and the sample produces a perfect correlation of $r = +1.00$ or $r = -1.00$. Because the first two points always produce a perfect correlation, the sample correlation is free to vary only when the data set contains more than two points. Thus, $df = n - 2$.

The following examples demonstrate the hypothesis test.

EXAMPLE 15.7   A researcher is using a regular, two-tailed test with $\alpha = .05$ to determine whether a non-zero correlation exists in the population. A sample of $n = 30$ individuals is obtained and produces a correlation of $r = 0.35$. The null hypothesis states that there is no correlation in the population.

$$H_0: \rho = 0$$

For this example, $df = 28$ and the critical values are $t = \pm 2.048$. With $r^2 = 0.35^2 = 0.1225$, the data produce

$$t = \frac{0.35 - 0}{\sqrt{(1 - 0.1225)/28}} = \frac{0.35}{0.177} = 1.97$$

The t value is not in the critical region so we fail to reject the null hypothesis. The sample correlation is not large enough to reject the null hypothesis.   ∎

EXAMPLE 15.8   With a sample of $n = 30$ and a correlation of $r = 0.35$, this time we use a directional, one-tailed test to determine whether there is a positive correlation in the population.

$$H_0: \rho \leq 0 \quad \textbf{(There is not a positive correlation.)}$$
$$H_1: \rho > 0 \quad \textbf{(There is a positive correlation.)}$$

The sample correlation is positive, as predicted, so we simply need to determine whether it is large enough to be significant. For a one-tailed test with $df = 28$ and $\alpha = .05$, the critical value is $t = 1.701$. In the previous example, we found that this sample produces $t = 1.97$, which is beyond the critical boundary. For the one-tailed test, we reject the null hypothesis and conclude that there is a significant positive correlation in the population.   ∎

Although it is possible to conduct the hypothesis test by computing either a t statistic or an F-ratio, the computations have been completed and are summarized in Table B.6 in

Appendix B. To use the table, you need to know the sample size (n) and the alpha level. In Examples 15.7 and 15.8, we used a sample of $n = 30$, a correlation of $r = 0.35$, and an alpha level of .05. In the table, you locate $df = n - 2 = 28$ in the left-hand column and the value .05 for either one tail or two tails across the top of the table. For $df = 28$ and $\alpha = .05$ for a two-tailed test, the table shows a critical value of 0.361. Because our sample correlation is not greater than this critical value, we fail to reject the null hypothesis (as in Example 15.7). For a one-tailed test, the table lists a critical value of 0.306. This time, our sample correlation is greater than the critical value so we reject the null hypothesis and conclude that the correlation is significantly greater than zero (as in Example 15.8).

As with most hypothesis tests, if other factors are held constant, the likelihood of finding a significant correlation increases as the sample size increases. For example, a sample correlation of $r = 0.50$ produces a nonsignificant $t(8) = 1.63$ for a sample of $n = 10$, but the same correlation produces a significant $t(18) = 2.45$ if the sample size is increased to $n = 20$.

The following example is an opportunity to test your understanding of the hypothesis test for the significance of a correlation. ∎

EXAMPLE 15.9

A researcher obtains a correlation of $r = -0.39$ for a sample of $n = 25$ individuals. For a two-tailed test with $\alpha = .05$, does this sample provide sufficient evidence to conclude that there is a significant, nonzero correlation in the population? Calculate the t statistic and then check your conclusion using the critical value in Table B6. You should obtain $t(23) = 2.03$. With a critical value of $t = 2.069$, the correlation is not significant. From Table B6 the critical value is 0.396. Again, the correlation is not significant. ∎

IN THE LITERATURE

Reporting Correlations

There is a standard APA format for reporting correlations. The report should include the sample size, the calculated value for the correlation, whether it is a statistically significant relationship, the probability level, and the type of test used (one- or two-tailed). For example, a correlation might be reported as follows:

> A correlation for the data revealed a significant relationship between amount of education and annual income, $r = +.65$, $n = 30$, $p < .01$, two tails.

Sometimes a study might look at several variables, and correlations between all possible variable pairings are computed. Suppose, for example, that a study measured people's annual income, amount of education, age, and intelligence. With four variables, there are six possible pairings leading to six different correlations. The results from multiple correlations are most easily reported in a table called a *correlation matrix,* using footnotes to indicate which correlations are significant. For example, the report might state:

> The analysis examined the relationships among income, amount of education, age, and intelligence for $n = 30$ participants. The correlations between pairs of variables are reported in Table 1. Significant correlations are noted in the table.

TABLE 1

Correlation matrix for income, amount of education, age, and intelligence

	Education	Age	IQ
Income	+.65*	+.41**	+.27
Education		+.11	+.38**
Age			−.02

$n = 30$
*$p < .01$, two tails
**$p < .05$, two tails

LEARNING CHECK

1. For a hypothesis test for the Pearson correlation, what is stated by the null hypothesis?
   a. There is a non-zero correlation for the general population.
   b. The population correlation is zero.
   c. There is a non-zero correlation for the sample.
   d. The sample correlation is zero.

2. A researcher evaluating the significance of a Pearson correlation, obtains $t = 2.099$ for a sample of $n = 20$ participants. For a two-tailed test, which of the following accurately describes the significance of the correlation?
   a. The correlation is significant with $\alpha = .05$ but not with $\alpha = .01$.
   b. The correlation is significant with either $\alpha = .05$ or $\alpha = .01$.
   c. The correlation is not significant with either $\alpha = .05$ or $\alpha = .01$.
   d. There is not enough information to evaluate the significance of the correlation.

ANSWERS   1. B, 2. C

15.5 | Alternatives to the Pearson Correlation

LEARNING OBJECTIVES

9. Explain how ranks are assigned to a set of scores, especially tied scores.

10. Compute the Spearman correlation for a set of data and explain what it measures.

11. Describe the circumstances in which the point-biserial correlation is used and explain what it measures.

12. Describe the circumstances in which the phi-coefficient is used and explain what it measures.

The Pearson correlation measures the degree of linear relationship between two variables when the data (X and Y values) consist of numerical scores from an interval or ratio scale of measurement. However, other correlations have been developed for nonlinear relationships and for other types of data. In this section we examine three additional correlations: the Spearman correlation, the point-biserial correlation, and the phi-coefficient. As you will see, all three can be viewed as special applications of the Pearson correlation.

■ The Spearman Correlation

When the Pearson correlation formula is used with data from an ordinal scale (ranks), the result is called the *Spearman correlation*. The Spearman correlation is used in two situations.

First, the Spearman correlation is used to measure the relationship between X and Y when both variables are measured on ordinal scales. Recall from Chapter 1 that an ordinal scale typically involves ranking individuals rather than obtaining numerical scores. Rank-order data are fairly common because they are often easier to obtain than interval or ratio scale data. For example, a teacher may feel confident about rank-ordering students' leadership abilities but would find it difficult to measure leadership on some other scale.

In addition to measuring relationships for ordinal data, the Spearman correlation can be used as a valuable alternative to the Pearson correlation, even when the original raw scores

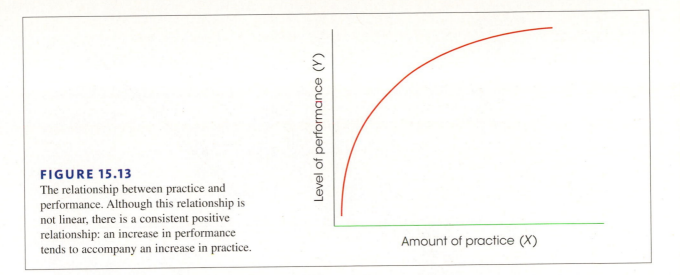

FIGURE 15.13
The relationship between practice and performance. Although this relationship is not linear, there is a consistent positive relationship: an increase in performance tends to accompany an increase in practice.

are on an interval or a ratio scale. As we have noted, the Pearson correlation measures the degree of *linear relationship* between two variables—that is, how well the data points fit on a straight line. However, a researcher often expects the data to show a consistently one-directional relationship but not necessarily a linear relationship. For example, Figure 15.13 shows the typical relationship between practice and performance. For nearly any skill, increasing amounts of practice tend to be associated with improvements in performance (the more you practice, the better you get). However, it is not a straight-line relationship. When you are first learning a new skill, practice produces large improvements in performance. After you have been performing a skill for several years, however, additional practice produces only minor changes in performance. Although there is a consistent relationship between the amount of practice and the quality of performance, it clearly is not linear. If the Pearson correlation were computed for these data, it would not produce a correlation of 1.00 because the data do not fit perfectly on a straight line. In a situation like this, the Spearman correlation can be used to measure the consistency of the relationship, independent of its form.

The reason that the Spearman correlation measures consistency, rather than form, comes from a simple observation: When two variables are consistently related, their ranks are linearly related. For example, a perfectly consistent positive relationship means that every time the *X* variable increases, the *Y* variable also increases. Thus, the smallest value of *X* is paired with the smallest value of *Y*, the second-smallest value of *X* is paired with the second smallest value of *Y*, and so on. Every time the rank for *X* goes up by 1 point, the rank for *Y* also goes up by 1 point. As a result, the ranks fit perfectly on a straight line. This phenomenon is demonstrated in the following example.

EXAMPLE 15.10

Table 15.3 presents *X* and *Y* scores for a sample of *n* = 4 people. Note that the data show a perfectly consistent relationship. Each increase in *X* is accompanied by an increase in *Y*. However the relationship is not linear, as can be seen in the graph of the data in Figure 15.14(a).

TABLE 15.3
Scores and ranks for Example 15.10.

Person	X	Y	X-Rank	Y-Rank
A	2	2	1	1
B	3	8	2	2
C	4	9	3	3
D	10	10	4	4

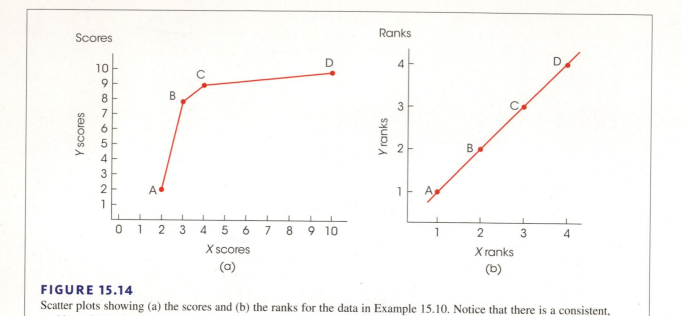

FIGURE 15.14

Scatter plots showing (a) the scores and (b) the ranks for the data in Example 15.10. Notice that there is a consistent, positive relationship between the X and Y scores, although it is not a linear relationship. Also notice that the scatter plot for the ranks shows a perfect linear relationship.

Next, we convert the scores to ranks. The lowest X is assigned a rank of 1, the next lowest a rank of 2, and so on. The Y scores are then ranked in the same way. The ranks are listed in Table 15.3 and shown in Figure 15.14(b). Note that the perfect consistency for the scores produces a perfect linear relationship for the ranks.   ■

The preceding example demonstrates that a consistent relationship among scores produces a linear relationship when the scores are converted to ranks. Thus, if you want to measure the consistency of a relationship for a set of scores, you can simply convert the scores to ranks and then use the Pearson correlation formula to measure the linear relationship for the ranked data. The degree of linear relationship for the ranks provides a measure of the degree of consistency for the original scores.

To summarize, the Spearman correlation measures the relationship between two variables when both are measured on ordinal scales (ranks). There are two general situations in which the Spearman correlation is used.

1. Spearman is used when the original data are ordinal; that is, when the X and Y values are ranks. In this case, you simply apply the Pearson correlation formula to the set of ranks.

2. Spearman is used when a researcher wants to measure the consistency of a relationship between X and Y, independent of the specific form of the relationship. In this case, the original scores are first converted to ranks; then the Pearson correlation formula is used with the ranks. Because the Pearson formula measures the degree to which the ranks fit on a straight line, it also measures the degree of consistency in the relationship for the original scores. Incidentally, when there is a consistently one-directional relationship between two variables, the relationship is said to be *monotonic*. Thus, the Spearman correlation measures the degree of monotonic relationship between two variables.

The word *monotonic* describes a sequence that is consistently increasing (or decreasing). Like the word *monotonous*, it means constant and unchanging.

In either case, the Spearman correlation is identified by the symbol r_S to differentiate it from the Pearson correlation. The complete process of computing the Spearman correlation, including ranking scores, is demonstrated in Example 15.11.

EXAMPLE 15.11

The following data show a nearly perfect monotonic relationship between X and Y. When X increases, Y tends to decrease, and there is only one reversal in this general trend. To compute the Spearman correlation, we first rank the X and Y values, and we then compute the Pearson correlation for the ranks.

We have listed the X values in order so that the trend is easier to recognize.

Original Data			Ranks		
X	Y		X	Y	XY
3	12		1	5	5
4	10		2	3	6
10	11		3	4	12
11	9		4	2	8
12	2		5	1	5
					$36 = \Sigma XY$

To compute the correlation, we need SS for X, SS for Y, and SP. Remember that all these values are computed with the ranks, not the original scores. The X ranks are simply the integers 1, 2, 3, 4, and 5. These values have $\Sigma X = 15$ and $\Sigma X^2 = 55$. The SS for the X ranks is

$$SS_X = \Sigma X^2 - \frac{(\Sigma X)^2}{n} = 55 - \frac{(15)^2}{5} = 10$$

Note that the ranks for Y are identical to the ranks for X; that is, they are the integers 1, 2, 3, 4, and 5. Therefore, the SS for Y is identical to the SS for X:

$$SS_Y = 10$$

To compute the SP value, we need ΣX, ΣY, and ΣXY for the ranks. The XY values are listed in the table with the ranks, and we already have found that both the Xs and the Ys have a sum of 15. Using these values, we obtain

$$SP = \Sigma XY - \frac{(\Sigma X)(\Sigma Y)}{n} = 36 - \frac{(15)(15)}{5} = -9$$

Finally, the Spearman correlation simply uses the Pearson formula for the ranks.

$$r_S = \frac{SP}{\sqrt{(SS_X)(SS_Y)}} = \frac{-9}{\sqrt{10(10)}} = -0.9$$

The Spearman correlation indicates that the data show a consistent (nearly perfect) negative trend. ∎

■ Ranking Tied Scores

When you are converting scores into ranks for the Spearman correlation, you may encounter two (or more) identical scores. Whenever two scores have exactly the same value, their ranks should also be the same. This is accomplished by the following procedure.

1. List the scores in order from smallest to largest. Include tied values in the list.

2. Assign a rank (first, second, etc.) to each position in the ordered list.

3. When two (or more) scores are tied, compute the mean of their ranked positions, and assign this mean value as the final rank for each score.

The process of finding ranks for tied scores is demonstrated here. These scores have been listed in order from smallest to largest.

Scores	Rank Position	Final Rank	
3	1	1.5	
3	2	1.5	Mean of 1 and 2
5	3	3	
6	4	5	
6	5	5	Mean of 4, 5, and 6
6	6	5	
12	7	7	

Note that this example has seven scores and uses all seven ranks. For $X = 12$, the largest score, the appropriate rank is 7. It cannot be given a rank of 6 because that rank has been used for the tied scores.

■ Special Formula for the Spearman Correlation

After the original X values and Y values have been ranked, the calculations necessary for SS and SP can be greatly simplified. First, you should note that the X ranks and the Y ranks are really just a set of integers: 1, 2, 3, 4, …, n. To compute the mean for these integers, you can locate the midpoint of the series by $M = (n + 1)/2$. Similarly, the SS for this series of integers can be computed by

$$SS = \frac{n(n^2 - 1)}{12} \quad \text{(Try it out.)}$$

Also, because the X ranks and the Y ranks are the same values, the SS for X is identical to the SS for Y.

Because calculations with ranks can be simplified and because the Spearman correlation uses ranked data, these simplifications can be incorporated into the final calculations for the Spearman correlation. Instead of using the Pearson formula after ranking the data, you can put the ranks directly into a simplified formula:

Caution: **In this formula, you compute the value of the fraction and then subtract from 1. The 1 is not part of the fraction.**

$$r_S = 1 - \frac{6\Sigma D^2}{n(n^2 - 1)} \tag{15.9}$$

where D is the difference between the X rank and the Y rank for each individual. This special formula produces the same result that would be obtained from the Pearson formula. However, note that this special formula can be used only after the scores have been converted to ranks and only when there are no ties among the ranks. If there are relatively few tied ranks, the formula still may be used, but it loses accuracy as the number of ties increases. The application of this formula is demonstrated in the following example.

EXAMPLE 15.12   To demonstrate the special formula for the Spearman correlation, we use the same data that were presented in Example 15.11. The ranks for these data are shown again here:

Ranks		Difference	
X	Y	D	D^2
1	5	4	16
2	3	1	1
3	4	1	1
4	2	-2	4
5	1	-4	16
			$38 = \Sigma D^2$

Using the special formula for the Spearman correlation, we obtain

$$r_S = 1 - \frac{6\Sigma D^2}{n(n^2 - 1)}$$

$$= 1 - \frac{6(38)}{5(25 - 1)}$$

$$= 1 - \frac{228}{120}$$

$$= 1 - 1.90$$

$$= -0.90$$

This is exactly the same answer that we obtained in Example 15.11, using the Pearson formula on the ranks. ■

The following example is an opportunity to test your understanding of the Spearman correlation.

EXAMPLE 15.13   Compute the Spearman correlation for the following set of scores:

X	Y
2	7
12	38
9	6
10	19

You should obtain $r_S = 0.80$.

■ Testing the Significance of the Spearman Correlation

Testing a hypothesis for the Spearman correlation is similar to the procedure used for the Pearson r. The basic question is whether a correlation exists in the population. The sample correlation could be due to chance, or perhaps it reflects an actual relationship between the variables in the population. For the Pearson correlation, the Greek letter rho (ρ) was used for the population correlation. For the Spearman, ρ_S is used for the population parameter. Note that this symbol is consistent with the sample statistic, r_S. The null hypothesis states that there is no correlation (no monotonic relationship) between the variables for the population, or in symbols:

$$H_0: \rho_S = 0 \quad \text{(The population correlation is zero.)}$$

The alternative hypothesis predicts that a nonzero correlation exists in the population, which can be stated in symbols as

$$H_1: \rho_S \neq 0 \quad \text{(There is a real correlation.)}$$

To determine whether the Spearman correlation is statistically significant (that is, H_0 should be rejected), consult Table B.7. This table is similar to the one used to determine the significance of Pearson's r (Table B.6); however, the first column is sample size (n) rather than degrees of freedom. To use the table, line up the sample size in the first column with the alpha level at the top. The values in the body of the table identify the magnitude of the Spearman correlation that is necessary to be significant. The table is built on the concept that a sample correlation should be representative of the corresponding population value. In particular, if the population correlation is $\rho_S = 0$ (as specified in H_0), then the sample correlation should be near zero. For each sample size and alpha level, the table identifies the minimum sample correlation that is significantly different from zero. The following example demonstrates the use of the table.

EXAMPLE 15.14 An industrial psychologist selects a sample of $n = 15$ employees. These employees are ranked in order of work productivity by their manager. They also are ranked by a peer. The Spearman correlation computed for these data revealed a correlation of $r_S = .60$. Using Table B.7 with $n = 15$ and $\alpha = .05$, a correlation of $\pm.521$ is needed to reject H_0. The observed correlation for the sample easily surpasses this critical value. The correlation between manager and peer ratings is statistically significant. ∎

■ The Point-Biserial Correlation and Measuring Effect Size with r^2

In Chapters 9, 10, and 11 we introduced r^2 as a measure of effect size that often accompanies a hypothesis test using the t statistic. The r^2 used to measure effect size and the r used to measure a correlation are directly related, and we now have an opportunity to demonstrate the relationship. Specifically, we compare the independent-measures t test (Chapter 10) and a special version of the Pearson correlation known as the *point-biserial correlation.*

The point-biserial correlation is used to measure the relationship between two variables in situations in which one variable consists of regular, numerical scores, but the second variable has only two values. A variable with only two values is called a *dichotomous variable* or a *binomial variable.* Some examples of dichotomous variables are:

1. male vs. female
2. college graduate vs. not a college graduate
3. first-born child vs. later-born child
4. success vs. failure on a particular task
5. older than 30 years old vs. younger than 30 years old

It is customary to use the numerical values 0 and 1, but any two different numbers would work equally well and would not affect the value of the correlation.

To compute the point-biserial correlation, the dichotomous variable is first converted to numerical values by assigning a value of zero (0) to one category and a value of one (1) to the other category. Then the regular Pearson correlation formula is used with the converted data.

To demonstrate the point-biserial correlation and its association with the r^2 measure of effect size, we use the data from Example 10.2 (p. 310). The original example compared cheating behavior in a dimly lit room compared to a well-lit room. The results showed that participants in the dimly lit room claimed to have solved significantly more puzzles than the participants in the well-lit room. The data from the independent-measures study are presented on the left side of Table 15.4. Notice that the data consist of two separate samples and the independent-measures t was used to determine whether there was a significant mean difference between the two populations represented by the samples.

On the right-hand side of Table 15.4 we have reorganized the data into a form that is suitable for a point-biserial correlation. Specifically, we used each participant's puzzle-solving score as the X value and we have created a new variable, Y, to represent the group or condition for each individual. In this case, we have used $Y = 0$ for individuals in the well-lit room and $Y = 1$ for participants in the dimly lit room.

TABLE 15.4

The same data are organized in two different formats. On the left-hand side, the data appear as two separate samples appropriate for an independent-measures *t* hypothesis test. On the right-hand side, the same data are shown as a single sample, with two scores for each individual: the number of puzzles solved and a dichotomous score (*Y*) that identifies the group in which the participant is located (Well-lit = 0 and Dimly lit = 1). The data on the right are appropriate for a point-biserial correlation.

Data for the Point-Biserial Correlation. Two scores *X* and *Y* for each of the *n* = 16 participants

Number of Solved Puzzles					Participant	Puzzles Solved *X*	Group *Y*
Well-Lit Room		**Dimly Lit Room**					
11	6	7	9		A	11	0
9	7	13	11		B	9	0
4	12	14	15		C	4	0
5	10	16	11		D	5	0
$n = 8$		$n = 8$			E	6	0
$M = 8$		$M = 12$			F	7	0
$SS = 60$		$SS = 66$			G	12	0
					H	10	0
					I	7	1
					J	13	1
					K	14	1
					L	16	1
					M	9	1
					N	11	1
					O	15	1
					P	11	1

When the data in Table 15.4 were originally presented in Chapter 10, we conducted an independent-measures *t* hypothesis test and obtained $t = -2.67$ with $df = 14$. We measured the size of the treatment effect by calculating r^2, the percentage of variance accounted for, and obtained $r^2 = 0.337$.

Calculating the point-biserial correlation for these data also produces a value for *r*. Specifically, the *X* scores produce $SS = 190$; the *Y* values produce $SS = 4.00$, and the sum of the products of the *X* and *Y* deviations produces $SP = 16$. The point-biserial correlation is

$$r = \frac{SP}{\sqrt{SS_X SS_Y}} = \frac{16}{\sqrt{(190)(4)}} = \frac{16}{27.57} = 0.58$$

Notice that squaring the value of the point-biserial correlation produces $r^2 = (0.58)^2 = 0.336$, which, within rounding error, is the same as the value of r^2 we obtained measuring effect size.

In some respects, the point-biserial correlation and the independent-measures hypothesis test are evaluating the same thing. Specifically, both are examining the relationship between room lighting and cheating behavior.

1. The correlation is measuring the *strength* of the relationship between the two variables. A large correlation (near 1.00 or −1.00) would indicate that there is a consistent, predictable relationship between cheating and the amount of light in the room. In particular, the value of r^2 measures how much of the variability in cheating can be predicted by knowing whether the participants were tested in a will-lit or a dimly lit room.

2. The *t* test evaluates the *significance* of the relationship. The hypothesis test determines whether the mean difference in grades between the two groups is greater than can be reasonably explained by chance alone.

As we noted in Chapter 10 (pp. 316–317), the outcome of the hypothesis test and the value of r^2 are often reported together. The *t* value measures statistical significance and r^2

measures the effect size. Also, as we noted in Chapter 10, the values for t and r^2 are directly related. In fact, either can be calculated from the other by the equations

$$r^2 = \frac{t^2}{t^2 + df} \quad \text{and} \quad t^2 = \frac{r^2}{(1 - r^2)/df}$$

where df is the degrees of freedom for the t statistic.

However, you should note that r^2 is determined entirely by the size of the correlation, whereas t is influenced by the size of the correlation and the size of the sample. For example, a correlation of $r = 0.30$ produces $r^2 = 0.09$ (9%) no matter how large the sample may be. On the other hand, a point-biserial correlation of $r = 0.30$ for a total sample of 10 people ($n = 5$ in each group) produces a nonsignificant value of $t = 0.791$. If the sample is increased to 50 people ($n = 25$ in each group), the same correlation produces a significant t value of $t = 4.75$. Although t and r are related, they are measuring different things.

■ The Phi-Coefficient

When both variables (X and Y) measured for each individual are dichotomous, the correlation between the two variables is called the *phi-coefficient*. To compute phi (Φ), you follow a two-step procedure.

1. Convert each of the dichotomous variables to numerical values by assigning a 0 to one category and a 1 to the other category for each of the variables.

2. Use the regular Pearson formula with the converted scores.

This process is demonstrated in the following example.

EXAMPLE 15.15 A researcher is interested in examining the relationship between birth-order position and personality. A random sample of $n = 8$ individuals is obtained, and each individual is classified in terms of birth-order position as first-born or only child vs. later-born. Then each individual's personality is classified as either introvert or extrovert.

The original measurements are then converted to numerical values by the following assignments:

Birth Order	Personality
1st or only child = 0	Introvert = 0
Later-born child = 1	Extrovert = 1

The original data and the converted scores are as follows:

Original Data		Converted Scores	
Birth Order X	Personality Y	Birth Order X	Personality Y
1st	Introvert	0	0
3rd	Extrovert	1	1
Only	Extrovert	0	1
2nd	Extrovert	1	1
4th	Extrovert	1	1
2nd	Introvert	1	0
Only	Introvert	0	0
3rd	Extrovert	1	1

The Pearson correlation formula is then used with the converted data to compute the phi-coefficient.

Because the assignment of numerical values is arbitrary (either category could be designated 0 or 1), the sign of the resulting correlation is meaningless. As with most correlations, the *strength* of the relationship is best described by the value of r^2, the coefficient of determination, which measures how much of the variability in one variable is predicted or determined by the association with the second variable.

We also should note that although the phi-coefficient can be used to assess the relationship between two dichotomous variables, the more common statistical procedure is a chi-square statistic, which is examined in Chapter 17.   ■

LEARNING CHECK

1. If the following scores are ranked, what rank is assigned to an individual with a score of $X = 7$?

    Scores: 1, 2, 2, 5, 6, 6, 6, 7, 10, 12

    a. 5
    b. 6
    c. 7
    d. 8

2. The Pearson and the Spearman correlations are both computed for the same set of data. If the Spearman correlation is $r_s = +1.00$, then what can you conclude about the Pearson correlation?

    a. It will be positive
    b. It will have a value of 1.00
    c. It will be positive and have a value of 1.00
    d. There is no predictable relationship between the Pearson and the Spearman correlations

3. The effect size for the data from an independent-measures t test can be measured by r^2, which is the percentage of variance accounted for. The value for r^2 can also be obtained by _____.

    a. squaring the Spearman correlation for the same data
    b. squaring the point-biserial correlation for the same data
    c. squaring the Pearson correlation for the same data
    d. None of the other options will produce r^2.

4. In what situations can the phi-coefficient be used?

    a. When an independent-measures t test would also be appropriate.
    b. When both X and Y are ranks.
    c. When a single-sample t test would also be appropriate.
    d. When both X and Y are dichotomous.

ANSWERS   **1.** D, **2.** A, **3.** B, **4.** D

SUMMARY

1. A correlation measures the relationship between two variables, X and Y. The relationship is described by three characteristics:
   a. *Direction.* A relationship can be either positive or negative. A positive relationship means that X and Y vary in the same direction. A negative relationship means that X and Y vary in opposite directions. The sign of the correlation ($+$ or $-$) specifies the direction.
   b. *Form.* The most common form for a relationship is a straight line. However, special correlations exist for measuring other forms. The form is specified by the type of correlation used. For example, the Pearson correlation measures linear form.
   c. *Strength or consistency.* The numerical value of the correlation measures the strength or consistency of the relationship. A correlation of 1.00 indicates a perfectly consistent relationship and 0.00 indicates no relationship at all. For the Pearson correlation, $r = 1.00$ (or -1.00) means that the data points fit perfectly on a straight line.

2. The most commonly used correlation is the Pearson correlation, which measures the degree of linear relationship. The Pearson correlation is identified by the letter r and is computed by

$$r = \frac{SP}{\sqrt{SS_X SS_Y}}$$

   In this formula, SP is the sum of products of deviations and can be calculated with either a definitional formula or a computational formula:

   definitional formula: $SP = \Sigma(X - M_X)(Y - M_Y)$

   computational formula: $SP = \Sigma XY - \dfrac{\Sigma X \Sigma Y}{n}$

3. A correlation between two variables should not be interpreted as implying a causal relationship. Simply because X and Y are related does not mean that X causes Y or that Y causes X.

4. To evaluate the strength of a relationship, you square the value of the correlation. The resulting value, r^2, is called the *coefficient of determination* because it measures the portion of the variability in one variable that can be determined using the relationship with the second variable.

5. A partial correlation measures the linear relationship between two variables by eliminating the influence of a third variable by holding it constant.

6. The Spearman correlation (r_S) measures the consistency of direction in the relationship between X and Y—that is, the degree to which the relationship is one-directional, or monotonic. The Spearman correlation is computed by a two-stage process:
   a. Rank the X scores and the Y scores separately.
   b. Compute the Pearson correlation using the ranks.

7. The point-biserial correlation is used to measure the strength of the relationship when one of the two variables is dichotomous. The dichotomous variable is coded using values of 0 and 1, and the regular Pearson formula is applied. Squaring the point-biserial correlation produces the same r^2 value that is obtained to measure effect size for the independent-measures t test. When both variables, X and $Y,$ are dichotomous, the phi-coefficient can be used to measure the strength of the relationship. Both variables are coded 0 and 1, and the Pearson formula is used to compute the correlation.

KEY TERMS

correlation (487)	Pearson correlation (490)	partial correlation (502)
positive correlation (488)	sum of products (*SP*) (490)	Spearman correlation (510)
negative correlation (488)	restricted range (498)	point-biserial correlation (516)
perfect correlation (489)	coefficient of determination (500)	phi-coefficient (518)

General instructions for using SPSS are presented in Appendix D. Following are detailed instructions for using SPSS to perform the **Pearson, Spearman, point-biserial, and partial correlations**. *Note:* We will focus on the Pearson correlation and then describe how slight modifications to this procedure can be made to compute the Spearman, point-biserial, and partial correlations. Separate instructions for the **phi-coefficient** are presented at the end of this section.

Data Entry

1. The data are entered into two columns in the data editor, one for the X values (VAR00001) and one for the Y values (VAR00002), with the two scores for each individual in the same row.

Data Analysis

1. Click **Analyze** on the tool bar, select **Correlate**, and click on **Bivariate**.
2. One by one, move the labels for the two data columns into the **Variables** box. (Highlight each label and click the arrow to move it into the box.)
3. The **Pearson** box should be checked but, at this point, you can switch to the Spearman correlation by clicking the appropriate box.
4. Click **OK**.

SPSS Output

We used SPSS to compute the correlation for the data in Example 15.3 and the output is shown in Figure 15.15. The program produces a correlation matrix showing all the possible correlations, including the correlation of X with X and the correlation of Y with Y (both are perfect correlations). You want the correlation of X and Y, which is contained in the upper right corner (or the lower left). The output includes the significance level (p value or alpha level) for the correlation.

Correlations

		VAR00001	VAR00002
VAR00001	Pearson Correlation	1	.875
	Sig. (2-tailed)		.052
	N	5	5
VAR00002	Pearson Correlation	.875	1
	Sig. (2-tailed)	.052	
	N	5	5

FIGURE 15.15
The SPSS output for the correlation in Example 15.3.

To compute a partial correlation, click **Analyze** on the tool bar, select **Correlate**, and click on **Partial**. Move the column labels for the two variables to be correlated into the **Variables** box and move the column label for the variable to be held constant into the **Controlling for** box and click **OK**.

To compute the **Spearman** correlation, enter either the X and Y ranks or the X and Y scores into the first two columns. Then follow the same Data Analysis instructions that were presented for the Pearson correlation. At step 3 in the instructions, click on the **Spearman** box before the final OK. (*Note:* If you enter X and Y scores into the data editor, SPSS converts the scores to ranks before computing the Spearman correlation.)

To compute the **point-biserial** correlation, enter the scores (X values) in the first column and enter the numerical values (usually 0 and 1) for the dichotomous variable in the second column. Then, follow the same Data Analysis instructions that were presented for the Pearson correlation.

The **phi-coefficient** can also be computed by entering the complete string of 0s and 1s into two columns of the SPSS data editor, then following the same Data Analysis instructions that were presented for the Pearson correlation. However, this can be tedious, especially with a large set of scores. The following is an alternative procedure for computing the phi-coefficient with large data sets.

Data Entry

1. Enter the values, 0, 0, 1, 1 (in order) into the first column of the SPSS data editor.

2. Enter the values 0, 1, 0, 1 (in order) into the second column.

3. Count the number of individuals in the sample who are classified with $X = 0$ and $Y = 0$. Enter this frequency in the top box in the third column of the data editor. Then, count how many have $X = 0$ and $Y = 1$ and enter the frequency in the second box of the third column. Continue with the number who have $X = 1$ and $Y = 0$, and finally the number who have $X = 1$ and $Y = 1$. You should end up with 4 values in column three.

4. Click **Data** on the Tool Bar at the top of the SPSS Data Editor page and select **Weight Cases** at the bottom of the list.

5. Click the circle labeled **weight cases by**, and then highlight the label for the column containing your frequencies (VAR00003) on the left and move it into the **Frequency Variable** box by clicking on the arrow.

6. Click **OK**.

7. Click **Analyze** on the tool bar, select **Correlate**, and click on **Bivariate**.

8. One by one move the labels for the two data columns containing the 0s and 1s (probably VAR00001 and VAR00002) into the **Variables** box. (Highlight each label and click the arrow to move it into the box.)

9. Verify that the **Pearson** box is checked.

10. Click **OK**.

SPSS Output

The program produces the same correlation matrix that was described for the Pearson correlation. Again, you want the correlation between X and Y which is in the upper right corner (or lower left). Remember, with the phi-coefficient, the sign of the correlation is meaningless.

FOCUS ON PROBLEM SOLVING

1. A correlation always has a value from $+1.00$ to -1.00. If you obtain a correlation outside this range, then you have made a computational error.

2. When interpreting a correlation, do not confuse the sign ($+$ or $-$) with its numerical value. The sign and the numerical value must be considered separately. Remember that the sign

indicates the direction of the relationship between X and Y. On the other hand, the numerical value reflects the strength of the relationship or how well the points approximate a linear (straight-line) relationship. Therefore, a correlation of -0.90 is as strong as a correlation of $+0.90$. The signs tell us that the first correlation is an inverse relationship.

3. Before you begin to calculate a correlation, sketch a scatter plot of the data and make an estimate of the correlation. (Is it positive or negative? Is it near 1 or near 0?) After computing the correlation, compare your final answer with your original estimate.

4. The definitional formula for the sum of products (SP) should be used only when you have a small set (n) of scores and the means for X and Y are both whole numbers. Otherwise, the computational formula produces quicker, easier, and more accurate results.

5. For computing a correlation, n is the number of individuals (and therefore the number of *pairs* of X and Y values).

DEMONSTRATION 15.1

CORRELATION

Calculate the Pearson correlation for the following data:

Person	X	Y	
A	0	4	$M_X = 4$ with $SS_X = 40$
B	2	1	$M_Y = 6$ with $SS_Y = 54$
C	8	10	$SP = 40$
D	6	9	
E	4	6	

STEP 1   **Sketch a scatter plot**   We have constructed a scatter plot for the data (Figure 15.16) and placed an envelope around the data points to make a preliminary estimate of the correlation. Note that the envelope is narrow and elongated. This indicates that the correlation is large—perhaps 0.80–0.90. Also, the correlation is positive because increases in X are generally accompanied by increases in Y.

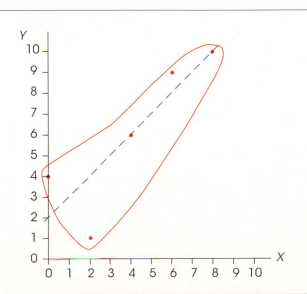

FIGURE 15.16

The scatter plot for the data in Demonstration 15.1. An envelope is drawn around the points and a line is drawn through the middle of the envelope.

STEP 2   **Compute the Pearson correlation**   For these data, the Pearson correlation is

$$r = \frac{SP}{\sqrt{SS_X SS_Y}} = \frac{40}{\sqrt{40(54)}} = \frac{40}{\sqrt{2160}} = \frac{40}{46.48} = 0.861$$

In Step 1, our preliminary estimate for the correlation was between $+0.80$ and $+0.90$. The calculated correlation is consistent with this estimate.

STEP 3   **Evaluate the significance of the correlation**   The null hypothesis states that, for the population, there is no linear relationship between X and Y, and that the value obtained for the sample correlation is simply the result of sampling error. Specifically, H_0 says that the population correlation is zero ($\rho = 0$). With $n = 5$ pairs of X and Y values the test has $df = 3$. Table B.6 lists a critical value of 0.878 for a two-tailed test with $\alpha = .05$. Because our correlation is smaller than this value, we fail to reject the null hypothesis and conclude that the correlation is not significant.

PROBLEMS

1. What information is provided by the sign ($+$ or $-$) of the Pearson correlation?

2. What information is provided by the numerical value of the Pearson correlation?

3. Calculate SP (the sum of products of deviations) for the following scores. *Note:* Both means are whole numbers, so the definitional formula works well:

X	Y
4	5
0	2
1	1
3	4

4. Calculate SP (the sum of products of deviations) for the following scores. *Note:* Both means are decimal values, so the computational formula works well:

X	Y
0	4
1	1
0	5
4	1
2	1
1	3

5. For the following scores,

X	Y
2	7
5	4
4	7
7	5
2	6
4	7

a. Sketch a scatter plot showing the six data points.
b. Just looking at the scatter plot, estimate the value of the Pearson correlation.
c. Compute the Pearson correlation.

6. For the following scores,

X	Y
0	4
2	9
1	6
1	9

a. Sketch a scatter plot and estimate the Pearson correlation.
b. Compute the Pearson correlation.

7. For the following scores,

X	Y
4	0
1	5
1	0
4	5

a. Sketch a scatter plot and estimate the Pearson correlation.
b. Compute the Pearson correlation.

8. For the following scores,

X	Y
3	6
5	5
6	0
6	2
5	2

a. Sketch a scatter plot and estimate the value of the Pearson correlation.
b. Compute the Pearson correlation.

9. With a small sample, a single point can have a large effect on the magnitude of the correlation. To create the following data, we started with the scores from problem 8 and changed the first X value from X = 3 to X = 8.

X	Y
8	6
5	5
6	0
6	2
5	2

a. Sketch a scatter plot and estimate the value of the Pearson correlation.
b. Compute the Pearson correlation.

10. For the following set of scores,

X	Y
4	5
6	5
3	2
9	4
6	5
2	3

a. Compute the Pearson correlation.
b. Add two points to each X value and compute the correlation for the modified scores. How does adding a constant to every score affect the value of the correlation?
c. Multiply each of the original X values by 2 and compute the correlation for the modified scores. How does multiplying each score by a constant affect the value of the correlation?

11. Correlation studies are often used to help determine whether certain characteristics are controlled more by genetic influences or by environmental influences. These studies often examine adopted children and compare their behaviors with the behaviors of their birth parents and their adoptive parents. One study examined how much time individuals spend watching TV (Plomin, Corley, DeFries, & Fulker, 1990). The following data are similar to the results obtained in the study.

Amount of Time Spent Watching TV		
Adopted Children	Birth Parents	Adoptive Parents
2	0	1
3	3	4
6	4	2
1	1	0
3	1	0
0	2	3
5	3	2
2	1	3
5	3	3

a. Compute the correlation between the children and their birth parents.
b. Compute the correlation between the children and their adoptive parents.
c. Based on the two correlations, does TV watching appear to be inherited from the birth parents or is it learned from the adoptive parents?

12. In the Chapter Preview we discussed a study by Judge and Cable (2010) demonstrating a negative relationship between weight and income for a group of women professionals. The following are data similar to those obtained in the study. To simplify the weight variable, the women are classified into five categories that measure actual weight relative to height, from 1 = thinnest to 5 = heaviest. Income figures are annual income (in thousands), rounded to the nearest $1,000.

a. Calculate the Pearson correlation for these data.
b. Is the correlation statistically significant? Use a two-tailed test with $\alpha = .05$.

Weight (X)	Income (Y)
1	115
1	78
4	53
3	63
5	37
2	84
5	41
3	51
1	94
5	44

13. The researchers cited in the previous problem also examined the weight/salary relationship for men and found a positive relationship, suggesting that we have very different standards for men than for women (Judge & Cable, 2010). The following are data similar to those obtained for a sample of male professionals. Again, weight relative to height is coded in five categories from 1 = thinnest to 5 = heaviest. Income is recorded as thousands earned annually.

a. Calculate the Pearson correlation for these data.
b. Is the correlation statistically significant? Use a two-tailed test with $\alpha = .05$.

Weight (X)	Income (Y)
4	151
5	88
3	52
2	73
1	49
3	92
1	56
5	143

14. Identifying individuals with a high risk of Alzheimer's disease usually involves a long series of cognitive tests. However, researchers have developed a 7-Minute Screen, which is a quick and easy way to accomplish the same goal. The question is whether the 7-Minute Screen is as effective as the complete series of tests. To address this question, Ijuin et al. (2008) administered both tests to a group of patients and compared the results. The following data represent results similar to those obtained in the study.

Patient	7-Minute Screen	Cognitive Series
A	3	11
B	8	19
C	10	22
D	8	20
E	4	14
F	7	13
G	4	9
H	5	20
I	14	25

a. Compute the Pearson correlation to measure the degree of relationship between the two test scores.
b. Is the correlation statistically significant? Use a two-tailed test with $\alpha = .01$.
c. What percentage of variance for the cognitive scores is predicted from the 7-Minute Screen scores? (Compute the value of r^2.)

15. For a two-tailed test with $\alpha = .05$, use Table B.6 to determine how large a Pearson correlation is necessary to be statistically significant for each of the following samples?
a. A sample of $n = 6$
b. A sample of $n = 12$
c. A sample of $n = 24$

16. As we have noted in previous chapters, even a very small effect can be significant if the sample is large enough. Suppose, for example, that a researcher obtains a correlation of $r = 0.60$ for a sample of $n = 10$ participants.
a. Is this sample sufficient to conclude that a significant correlation exists in the population? Use a two-tailed test with $\alpha = .05$.
b. If the sample had $n = 25$ participants, is the correlation significant? Again, use a two-tailed test with $\alpha = .05$.

17. A researcher measures three variables, X, Y, and Z for each individual in a sample of $n = 20$. The Pearson correlations for this sample are $r_{XY} = 0.6$, $r_{XZ} = 0.4$, and $r_{YZ} = 0.7$.
a. Find the partial correlation between X and Y, holding Z constant.
b. Find the partial correlation between X and Z, holding Y constant. (*Hint*: Simply switch the labels for the variables Y and Z to correspond with the labels in the equation.)

18. A researcher records the annual number of serious crimes and the amount spent on crime prevention for several small towns, medium cities, and large cities across the country. The resulting data show a strong positive correlation between the number of serious crimes and the amount spent on crime prevention. However, the researcher suspects that the positive correlation is actually caused by population; as population increases, both the amount spent on crime prevention and the number of crimes will also increase. If population is controlled, there probably should be a negative correlation between the amount spent on crime prevention and the number of serious crimes. The following data show the pattern of results obtained. Note that the municipalities are coded in three categories. Use a partial correlation holding population constant to measure the true relationship between crime rate and the amount spent on prevention.

Number of Crimes	Amount for Prevention	Population Size
3	6	1
4	7	1
6	3	1
7	4	1
8	11	2
9	12	2
11	8	2
12	9	2
13	16	3
14	17	3
16	13	3
17	14	3

19. A common concern for students (and teachers) is the assignment of grades for essays or term papers. Because there are no absolute right or wrong answers, these grades must be based on a judgment of quality. To demonstrate that these judgments actually are reliable, an English instructor asked a colleague to rank-order a set of term papers. The ranks and the instructor's grades for these papers are as follows:

Rank	Grade
1	A
2	B
3	A
4	B
5	B
6	C
7	D
8	C
9	C
10	D
11	F

a. Compute the Spearman correlation for these data. (*Note:* You must convert the letter grades to ranks, using tied ranks to represent tied grades.)

b. Is the Spearman correlation statistically significant? Use a two-tailed test with $\alpha = .05$.

20. It appears that there is a significant relationship between cognitive ability and social status, at least for birds. Boogert, Reader, and Laland (2006) measured social status and individual learning ability for a group of starlings. The following data represent results similar to those obtained in the study. Because social status is an ordinal variable consisting of five ordered categories, the Spearman correlation is appropriate for these data. Convert the social status categories and the learning scores to ranks, and compute the Spearman correlation.

Subject	Social Status	Learning Score
A	1	3
B	3	10
C	2	7
D	3	11
E	5	19
F	4	17
G	5	17
H	2	4
I	4	12
J	2	3

21. Problem 13 presented data showing a positive relationship between weight and income for a sample of professional men. However, weight was coded in five categories that could be viewed as an ordinal scale rather than an interval or ratio scale. If so, a Spearman correlation is more appropriate than a Pearson correlation.

a. Convert the weights and the incomes into ranks and compute the Spearman correlation for the scores in problem 13.

b. Is the Spearman correlation large enough to be significant?

22. Problem 13 in Chapter 10 presented data showing that varsity athletes in contact sports (football and hockey) who were regularly exposed to head impacts had significantly lower cognitive scores than varsity athletes from noncontact sports. The independent-measures t test produced $t = 2.25$ with $df = 14$ and a value of $r^2 = 0.265$ (26.5%).

a. Convert the data from this problem into a form suitable for the point-biserial correlation (use 1 for the noncontact athletes and 0 for the contact athletes), and then compute the correlation.

b. Square the value of the point-biserial correlation to verify that you obtain the same r^2 value that was computed in Chapter 10.

23. Problem 14 in Chapter 10 presented data demonstrating that handling money can reduce the perception of pain. In the study, one group counted money and another group counted blank pieces of paper. After the counting task, each participant dipped a hand into very hot water and rated how uncomfortable it was.

a. Convert the data from this problem into a form suitable for the point-biserial correlation (use 1 for the money and 0 for the plain paper), and then compute the correlation.

b. Square the value of the point-biserial correlation to obtain r^2.

c. The t test in chapter 10 produced $t = 3.57$ with $df = 16$. Use the equation on p. 518 to compute the value of r^2 directly from the t statistic and its df. Within rounding error, the value of r^2 from the equation should be equal to the value obtained from the point-biserial correlation.

24. Studies have shown people with high intelligence are generally more likely to volunteer as participants in research, but not for research that involves unusual experiences such as hypnosis. To examine this phenomenon, a researcher administers a questionnaire to a sample of college students. The survey asks for the student's grade point average (as a measure of intelligence) and whether the student would like to take part in a future study in which participants would be hypnotized. The results showed that 7 of the 10 lower intelligence people were willing to participant but only 2 of the 10 higher intelligence people were willing.

a. Convert the data to a form suitable for computing the phi-coefficient. (Code the two intelligence categories as 0 and 1 for the X variable, and code the willingness to participate as 0 and 1 for the Y variable.)

b. Compute the phi-coefficient for the data.

Introduction to Regression

Tools You Will Need

The following items are considered essential background material for this chapter. If you doubt your knowledge of any of these items, you should review the appropriate chapter or section before proceeding.

- Sum of squares (*SS*) (Chapter 4)
  - Computational formula
  - Definitional formula
- *z*-scores (Chapter 5)
- Analysis of variance (Chapter 12)
  - *MS* values and *F*-ratios
- Pearson correlation (Chapter 15)
  - Sum of products (*SP*)

© Deborah Batt

In Chapter 15, we noted that one common application of correlations is for purposes of prediction. Whenever there is a consistent relationship between two variables, it is possible to use the value of one variable to predict the value of another. Managers at the electric company, for example, can use the weather forecast to predict power demands for upcoming days. If exceptionally hot summer weather is forecast, they can anticipate an exceptionally high demand for electricity. In the field of psychology, a known relationship between certain environmental factors and specific behaviors can allow us to predict that individuals living with those environmental factors are more likely to display those behaviors. For example, Astudillo et al. (2014) examined the relationship between alcohol use and the concentration of alcohol outlets (bars, clubs, and other on-site drinking establishments) for different geographical regions of Switzerland. They obtained alcohol use scores from a previous study of 5519 Swiss men measuring both alcohol consumption and drinking consequences. The results showed a positive relationship between alcohol consumption and the density of outlets in an area, but no consistent association between outlets and drinking consequences. From these results, the researchers were able to predict the level of alcohol consumption within geographic regions based on the density of outlets in each region.

The correlations introduced in Chapter 15 allow researchers to measure and describe correlations, and the hypothesis tests allow researchers to evaluate the significance of correlations. However, we now want to go one step further and actually use correlations to make predictions.

In this chapter we introduce some of the statistical techniques that are used to make predictions based on correlations. Whenever there is a linear relationship (Pearson correlation) between two variables, it is possible to compute an equation that provides a precise, mathematical description of the relationship. With the equation, it is possible to plug in the known value for one variable (for example, the density of alcohol outlets), and then calculate a predicted value for the second variable (for example, the level of alcohol consumption). The general statistical process of finding and using a prediction equation is known as *regression*.

Beyond finding a prediction equation, however, it is reasonable to ask how good the predictions it makes are. For example, I can make predictions about the outcome of a coin toss by simply guessing. However, my predictions are correct only about 50% of the time. In statistical terms, my predictions are not significantly better than chance. In the same way, it is appropriate to challenge the significance of any prediction equation. In this chapter we introduce the techniques that are used to find prediction equations, as well as the techniques that are used to determine whether their predictions are statistically significant. Incidentally, although the results from the Astudillo et al. (2014) study may seem obvious and predictable, they are useful. Specifically, the results suggest that authorities should consider the density of alcohol outlets in an area when planning regional programs to prevent alcohol abuse.

16.1 | Introduction to Linear Equations and Regression

LEARNING OBJECTIVES

1. Define the equation that describes a linear relationship between two variables.

2. Compute the regression equation (slope and *Y*-intercept) for a set of *X* and *Y* scores.

In the previous chapter, we introduced the Pearson correlation as a technique for describing and measuring the linear relationship between two variables. Figure 16.1 presents hypothetical data showing the relationship between SAT scores and college grade point average (GPA). Note that the figure shows a good, but not perfect, positive relationship. Also note that we have drawn a line through the middle of the data points. This line serves several purposes.

1. The line makes the relationship between SAT and GPA easier to see.

2. The line identifies the center, or *central tendency,* of the relationship, just as the mean describes central tendency for a set of scores. Thus, the line provides a

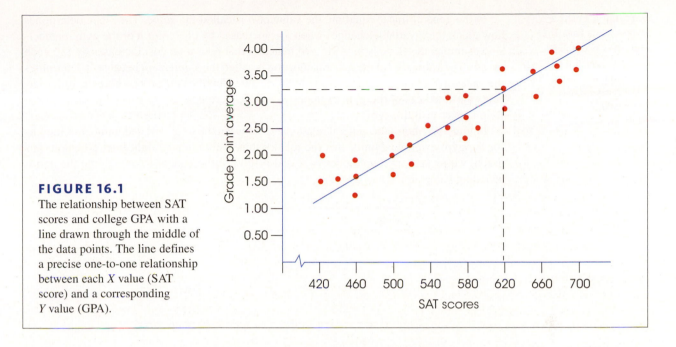

FIGURE 16.1

The relationship between SAT scores and college GPA with a line drawn through the middle of the data points. The line defines a precise one-to-one relationship between each *X* value (SAT score) and a corresponding *Y* value (GPA).

simplified description of the relationship. For example, if the data points were removed, the straight line would still give a general picture of the relationship between SAT and GPA.

3. Finally, the line can be used for prediction. The line establishes a precise, one-to-one relationship between each *X* value (SAT score) and a corresponding *Y* value (GPA). For example, an SAT score of 620 corresponds to a GPA of 3.25 (see Figure 16.1). Thus, the college admissions officers could use the straight-line relationship to predict that a student entering college with an SAT score of 620 should achieve a college GPA of approximately 3.25.

Our goal in this section is to develop a procedure that identifies and defines the straight line that provides the best fit for any specific set of data. This straight line does not have to be drawn on a graph; it can be presented in a simple equation. Thus, our goal is to find the equation for the line that best describes the relationship for a set of *X* and *Y* data.

■ Linear Equations

In general, a *linear relationship* between two variables *X* and *Y* can be expressed by the equation

$$Y = bX + a \qquad \text{(16.1)}$$

where *a* and *b* are fixed constants.

For example, a local gym charges a membership fee of $35 and a monthly fee of $15 for unlimited use of the facility. With this information, the total cost for the gym can be computed using a *linear equation* that describes the relationship between the total cost (*Y*) and the number months (*X*).

$$Y = 15X + 35$$

In the general linear equation, the value of b is called the *slope*. The slope determines how much the Y variable changes when X is increased by one point. For the gym membership example, the slope is $b = \$15$ and indicates that your total cost increases by $15 each month. The value of a in the general equation is called the Y-*intercept* because it determines the value of Y when $X = 0$. (On a graph, the a value identifies the point where the line intercepts the Y-axis.) For the gym example, $a = \$35$; there is a $35 membership charge even if you never visit the gym.

Figure 16.2 shows the general relationship between the total cost and number of months for the gym example. Notice that the relationship results in a straight line. To obtain this graph, we picked any two values of X and then used the equation to compute the corresponding values for Y. For example,

when $X = 3$:	when $X = 8$:
$Y = bX + a$	$Y = bX + a$
$= \$15(3) + \$35$	$= \$15(8) + \$35$
$= \$45 + \$35$	$= \$120 + \$35$
$= \$80$	$= \$155$

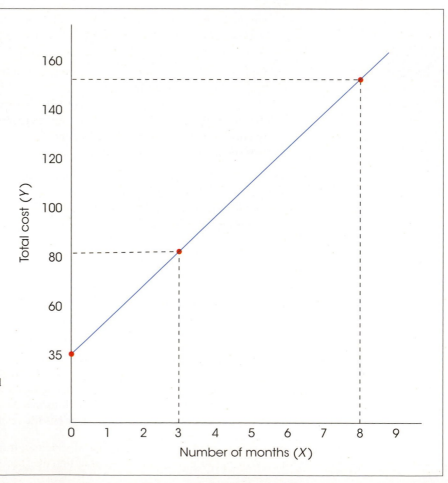

FIGURE 16.2

The relationship between total cost and number of months of gym membership. The gym charges a $35 membership fee and $15 per month. The relationship is described by a linear equation $Y = 15X + 35$ where Y is the total cost and X is the number of months.

When drawing a graph of a linear equation, it is wise to compute and plot at least three points to be certain you have not made a mistake.

Next, these two points are plotted on the graph: one point at $X = 3$ and $Y = 80$, the other point at $X = 8$ and $Y = 155$. Because two points completely determine a straight line, we simply drew the line so that it passed through these two points.

■ Regression

Because a straight line can be extremely useful for describing a relationship between two variables, a statistical technique has been developed that provides a standardized method for determining the best-fitting straight line for any set of data. The statistical procedure is *regression*, and the resulting straight line is called the *regression line*.

DEFINITION

> The statistical technique for finding the best-fitting straight line for a set of data is called **regression**, and the resulting straight line is called the **regression line**.

The goal for regression is to find the best-fitting straight line for a set of data. To accomplish this goal, however, it is first necessary to define precisely what is meant by "best fit." For any particular set of data, it is possible to draw lots of different straight lines that all appear to pass through the center of the data points. Each of these lines can be defined by a linear equation of the form $Y = bX + a$ where b and a are constants that determine the slope and Y-intercept of the line, respectively. Each individual line has its own unique values for b and a. The problem is to find the specific line that provides the best fit to the actual data points.

■ The Least-Squares Solution

To determine how well a line fits the data points, the first step is to define mathematically the distance between the line and each data point. For every X value in the data, the linear equation determines a Y value on the line. This value is the predicted Y and is called $\hat{Y}$ ("Y hat"). The distance between this predicted value and the actual Y value in the data is determined by

$$\text{distance} = Y - \hat{Y}$$

Note that we simply are measuring the vertical distance between the actual data point (Y) and the predicted point on the line. This distance measures the error between the line and the actual data (Figure 16.3).

Because some of these distances will be positive and some will be negative, the next step is to square each distance to obtain a uniformly positive measure of error. Finally, to determine the total error between the line and the data, we add the squared errors for all of the data points. The result is a measure of overall squared error between the line and the data:

$$\text{total squared error} = \Sigma(Y - \hat{Y})^2$$

Now we can define the *best-fitting* line as the one that has the smallest total squared error. For obvious reasons, the resulting line is commonly called the *least-squared-error* solution. In symbols, we are looking for a linear equation of the form

$$\hat{Y} = bX + a$$

For each value of X in the data, this equation determines the point on the line ($\hat{Y}$) *that gives the best prediction of Y*. The problem is to find the specific values for a and b that make this the best-fitting line.

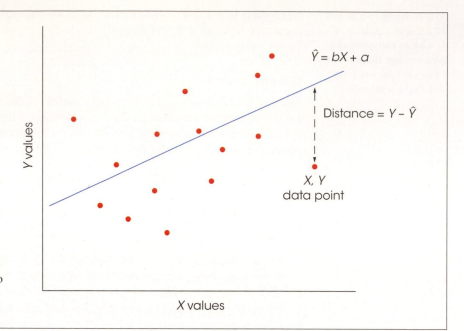

FIGURE 16.3

The distance between the actual data point (Y) and the predicted point on the line ($\hat{Y}$) is defined as $Y - \hat{Y}$. The goal of regression is to find the equation for the line that minimizes these distances.

The calculations that are needed to find this equation require calculus and some sophisticated algebra, so we will not present the details of the solution. The results, however, are relatively straightforward, and the solutions for b and a are as follows:

$$b = \frac{SP}{SS_X} \qquad (16.2)$$

where SP is the sum of products and SS_X is the sum of squares for the X scores.

A commonly used alternative formula for the slope is based on the standard deviations for X and Y. The alternative formula is

$$b = r\frac{s_Y}{s_X} \qquad (16.3)$$

where s_Y is the standard deviation for the Y scores, s_X is the standard deviation for the X scores, and r is the Pearson correlation for X and Y. After the value of b is computed, the value of the constant a in the equation is determined by

$$a = M_Y - bM_X \qquad (16.4)$$

Note that these formulas determine the linear equation that provides the best prediction of Y values. This equation is called the *regression equation for* Y.

The **regression equation for Y** is the linear equation

$$\hat{Y} = bX + a \qquad (16.5)$$

where the constant b is determined by Equation 16.2 or 16.3, and the constant a is determined by Equation 16.4. This equation results in the least squared error between the data points and the line.

EXAMPLE 16.1

The scores in the following table are used to demonstrate the calculation and use of the regression equation for predicting Y.

X	Y	$X - M_X$	$Y - M_Y$	$(X - M_X)^2$	$(Y - M_Y)^2$	$(X - M_X)(Y - M_Y)$
5	10	1	3	1	9	3
1	4	−3	−3	9	9	9
4	5	0	−2	0	4	0
7	11	3	4	9	16	12
6	15	2	8	4	64	16
4	6	0	−1	0	1	0
3	5	−1	−2	1	4	2
2	0	−2	−7	4	49	14
				$SS_X = 28$	$SS_Y = 156$	$SP = 56$

For these data, $\Sigma X = 32$, so $M_X = 4$. Also, $\Sigma Y = 56$, so $M_Y = 7$. These values have been used to compute the deviation scores for each X and Y value. The final three columns show the squared deviations for X and for Y, and the products of the deviation scores.

Our goal is to find the values for b and a in the regression equation. Using Equations 16.2 and 16.4, the solutions for b and a are

$$b = \frac{SP}{SS_X} = \frac{56}{28} = 2$$

$$a = M_Y - bM_X = 7 - 2(4) = -1$$

The resulting equation is

$$\hat{Y} = 2X - 1$$

The original data and the regression line are shown in Figure 16.4.   ■

The regression line shown in Figure 16.4 demonstrates some simple and very predictable facts about regression. First, the calculation of the Y-intercept (Equation 16.4) ensures that the regression line passes through the point defined by the mean for X and the mean for Y. That is, the point identified by the coordinates M_X, M_Y will always be on the line. We have included the two means in Figure 16.4 to show that the point they define is on the regression line. Second, the sign of the correlation (+ or −) is the same as the sign of the slope of the regression line. Specifically, if the correlation is positive, then the slope is also positive and the regression line slopes up to the right. On the other hand, if the correlation is negative, the slope is negative and the line slopes down to the right. A correlation of zero means that the slope is also zero and the regression equation produces a horizontal line that passes through the data at a level equal to the mean for the Y values. Note that the regression line in Figure 16.4 has a positive slope. One consequence of this fact is that that all of the points on the line that are above the mean for X are also above the mean for Y. Similarly, all of the points below the mean for X are also below the mean for Y. Thus, every individual with a positive deviation for X is predicted to have a positive deviation for Y, and everyone with a negative deviation for X is predicted to have a negative deviation for Y.

■ Using the Regression Equation for Prediction

As we noted at the beginning of this section, one common use of regression equations is for prediction. For any given value of X, we can use the equation to compute a predicted value

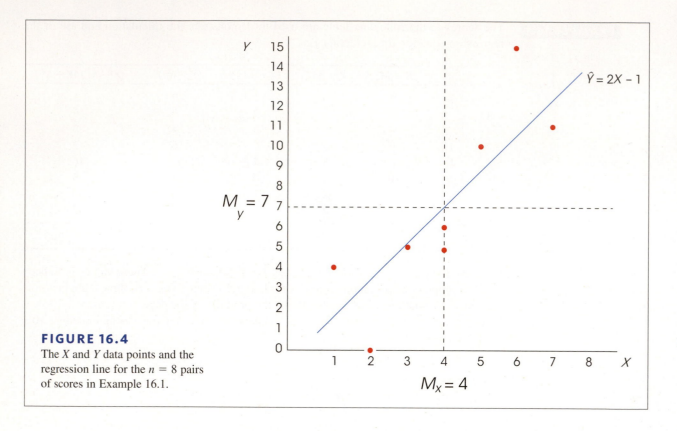

FIGURE 16.4

The X and Y data points and the regression line for the $n = 8$ pairs of scores in Example 16.1.

for Y. For the equation from Example 16.1, an individual with a score of $X = 3$ would be predicted to have a Y score of

$$\hat{Y} = 2X - 1 = 6 - 1 = 5$$

Although regression equations can be used for prediction, a few cautions should be considered whenever you are interpreting the predicted values.

1. The predicted value is not perfect (unless $r = +1.00$ or -1.00). If you examine Figure 16.4, it should be clear that the data points do not fit perfectly on the line. In general, there will be some error between the predicted Y values (on the line) and the actual data. Although the amount of error will vary from point to point, on average the errors will be directly related to the magnitude of the correlation. With a correlation near 1.00 (or -1.00), the data points will generally be clustered close to the line and the error will be small. As the correlation gets nearer to zero, the points will move away from the line and the magnitude of the error will increase.

2. The regression equation should not be used to make predictions for X values that fall outside the range of values covered by the original data. For Example 16.1, the X values ranged from $X = 1$ to $X = 7$, and the regression equation was calculated as the best-fitting line within this range. Because you have no information about the X-Y relationship outside this range, the equation should not be used to predict Y for any X value lower than 1 or greater than 7.

The following example is an opportunity to test your understand of the calculations needed to find a linear regression equation.

EXAMPLE 16.2

For the following data, find the regression equation for predicting Y from X.

X	Y
1	4
3	9
5	8

You should obtain $\hat{Y} = X + 4$.   ■

■ Standardized Form of the Regression Equation

So far we have presented the regression equation in terms of the original values, or raw scores, for X and Y. Occasionally, however, researchers standardize the scores by transforming the X and Y values into z-scores before finding the regression equation. The resulting equation is often called the standardized form of the regression equation and is greatly simplified compared to the raw-score version. The simplification comes from the fact that z-scores have standardized characteristics. Specifically, the mean for a set of z-scores is always zero and the standard deviation is always 1. As a result, the standardized form of the regression equation becomes

$$\hat{z}_Y = (\text{beta})z_X \tag{16.6}$$

First notice that we are now using the z-score for each X value (z_X) to predict the z-score for the corresponding Y value (z_Y). Also, note that the slope constant that was identified as b in the raw-score formula is now identified as beta. Because both sets of z-scores have a mean of zero, the constant a disappears from the regression equation. Finally, when one variable, X, is being used to predict a second variable, Y, the value of beta is equal to the Pearson correlation for X and Y. Thus, the standardized form of the regression equation can also be written as

$$\hat{z}_Y = rz_X \tag{16.7}$$

Because the process of transforming all of the original scores into z-scores can be tedious, researchers usually compute the raw-score version of the regression equation (Equation 16.4) instead of the standardized form. However, most computer programs report the value of beta as part of the output from linear regression, and you should understand what this value represents.

LEARNING CHECK

1. In the general linear equation, $Y = bX + a$, what is the value of b called?
   a. X intercept
   b. Y intercept
   c. correlation between X and Y
   d. slope

2. A set of $n = 10$ pairs of X and Y scores has $SS_X = 10$, $SS_Y = 60$, and $SP = 20$. What is the slope of the regression equation for these scores?
   a. 0.50
   b. 0.33
   c. 2.00
   d. 3.00

ANSWERS

1. D, 2. C

16.2 | The Standard Error of Estimate and Analysis of Regression: The Significance of the Regression Equation

LEARNING OBJECTIVES

3. Compute the standard error of estimate for a regression equation and explain what it measures.

4. Conduct an analysis of regression to evaluate the significance of a regression equation.

It is possible to determine a regression equation for any set of data by simply using the formulas already presented. The linear equation you obtain is then used to generate predicted Y values for any known value of X. However, it should be clear that the accuracy of this prediction depends on how well the points on the line correspond to the actual data points—that is, the amount of error between the predicted values, $\hat{Y}$, and the actual scores, Y values. Figure 16.5 shows two different sets of data that have exactly the same regression equation. In one case, there is a perfect correlation ($r = +1$) between X and Y, so the linear equation fits the data perfectly. For the second set of data, the predicted Y values on the line only approximate the real data points.

A regression equation, by itself, allows you to make predictions, but it does not provide any information about the accuracy of the predictions. To measure the precision of the regression, it is customary to compute a *standard error of estimate*.

DEFINITION

The **standard error of estimate** gives a measure of the standard distance between the predicted Y values on the regression line and the actual Y values in the data.

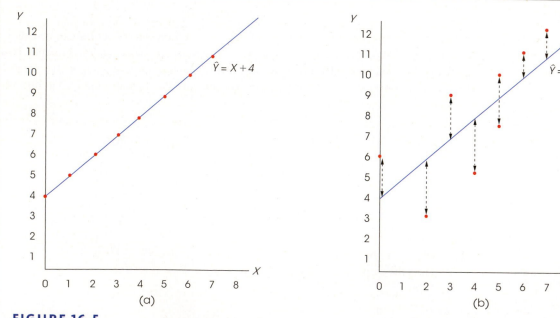

FIGURE 16.5

(a) A scatter plot showing data points that perfectly fit the regression line defined by $\hat{Y} = X + 4$. Note that the correlation is $r = +1.00$. (b) A scatter plot for another set of data with a regression equation of $\hat{Y} = X + 4$. Notice that there is error between the actual data points and the predicted Y values on the regression line.

Conceptually, the standard error of estimate is very much like a standard deviation: Both provide a measure of standard distance. Also, you will see that the calculation of the standard error of estimate is very similar to the calculation of standard deviation.

To calculate the standard error of estimate, we first find a sum of squared deviations (SS). Each deviation measures the distance between the actual Y value (from the data) and the predicted Y value (from the regression line). This sum of squares is commonly called $SS_{residual}$ because it is based on the remaining distance between the actual Y scores and the predicted values.

$$SS_{residual} = \Sigma(Y - \hat{Y})^2 \qquad (16.8)$$

The obtained SS value is then divided by its degrees of freedom to obtain a measure of variance. This procedure should be very familiar:

$$\text{Variance} = \frac{SS}{df}$$

The degrees of freedom for the standard error of estimate are $df = n - 2$. The reason for having $n - 2$ degrees of freedom, rather than the customary $n - 1$, is that we now are measuring deviations from a line rather than deviations from a mean. To find the equation for the regression line, you must know the means for both the X and the Y scores. Specifying these two means places two restrictions on the variability of the data, with the result that the scores have only $n - 2$ degrees of freedom. (*Note*: the $df = n - 2$ for $SS_{residual}$ is the same $df = n - 2$ that we encountered when testing the significance of the Pearson correlation on p. 508.)

Recall that variance measures the average squared distance.

The final step in the calculation of the standard error of estimate is to take the square root of the variance to obtain a measure of standard distance. The final equation is

$$\text{standard error of estimate} = \sqrt{\frac{SS_{residual}}{df}} = \sqrt{\frac{\Sigma(Y - \hat{Y})^2}{n - 2}} \qquad (16.9)$$

The following example demonstrates the calculation of this standard error.

EXAMPLE 16.3   The same data that were used in Example 16.1 are used here to demonstrate the calculation of the standard error of estimate. These data have the regression equation

$$\hat{Y} = 2X - 1$$

Using this regression equation, we have computed the predicted Y value, the residual, and the squared residual for each individual, using the data from Example 16.1.

Data		Predicted Y value	Residual	Squared Residual
X	Y	$\hat{Y} = 2X - 1$	$Y - \hat{Y}$	$(Y - \hat{Y})^2$
5	10	9	1	1
1	4	1	3	9
4	5	7	−2	4
7	11	13	−2	4
6	15	11	4	16
4	6	7	−1	1
3	5	5	0	0
2	0	3	−3	9
			0	$SS_{residual} = 44$

First note that the sum of the residuals is equal to zero. In other words, the sum of the distances above the line is equal to the sum of the distances below the line. This is true for any set of data and provides a way to check the accuracy of your calculations. The squared residuals are listed in the final column. For these data, the sum of the squared residuals is $SS_{residual} = 44$. With $n = 8$, the data have $df = n - 2 = 6$, so the standard error of estimate is

$$\text{standard error of estimate} = \sqrt{\frac{SS_{residual}}{df}} = \sqrt{\frac{44}{6}} = 2.708$$

Remember: The standard error of estimate provides a measure of how accurately the regression equation predicts the Y values. In this case, the standard distance between the actual data points and the regression line is measured by standard error of estimate = 2.708. ■

■ Relationship between the Standard Error and the Correlation

It should be clear from Example 16.3 that the standard error of estimate is directly related to the magnitude of the correlation between X and Y. If the correlation is near 1.00 (or −1.00), the data points are clustered close to the line, and the standard error of estimate is small. As the correlation gets nearer to zero, the data points become more widely scattered, the line provides less accurate predictions, and the standard error of estimate grows larger.

Earlier, (p. 500) we observed that squaring the correlation provides a measure of the accuracy of prediction. The squared correlation, r^2, is called the coefficient of determination because it determines what proportion of the variability in Y is predicted by the relationship with X. Because r^2 measures the predicted portion of the variability in the Y scores, we can use the expression $(1 - r^2)$ to measure the unpredicted portion. Thus,

$$\text{Predicted variability} = SS_{regression} = r^2 SS_Y \tag{16.10}$$

$$\text{Unpredicted variability} = SS_{residual} = (1 - r^2)SS_Y \tag{16.11}$$

For example, if $r = 0.80$, then $r^2 = 0.64$ (or 64%) of the variability for the Y scores is predicted by the relationship with X and the remaining 36% $(1 - r^2)$ is the unpredicted portion. Note that when $r = 1.00$, the prediction is perfect and there are no residuals. As the correlation approaches zero, the data points move farther off the line and the residuals grow larger. Using Equation 16.10 to compute $SS_{residual}$, the standard error of estimate can be computed as

$$\text{standard error of estimate} = \sqrt{\frac{SS_{residual}}{df}} = \sqrt{\frac{(1 - r^2)SS_Y}{n - 2}} \tag{16.12}$$

Because it is usually much easier to compute the Pearson correlation than to compute the individual $(Y - \hat{Y})^2$ values, Equation 16.11 is usually the easiest way to compute $SS_{residual}$, and Equation 16.12 is usually the easiest way to compute the standard error of estimate for a regression equation. The following example demonstrates this new formula.

EXAMPLE 16.4   We use the same data used in Examples 16.1 and 16.3, which produced $SS_X = 28$, $SS_Y = 156$, and $SP = 56$. For these data, the Pearson correlation is

$$r = \frac{56}{\sqrt{28(156)}} = \frac{56}{66.09} = 0.847$$

With $SS_Y = 156$ and a correlation of $r = 0.847$, the predicted variability from the regression equation is

$$SS_{regression} = r^2 SS_Y = (0.847^2)(156) = 0.718(156) = 112.01$$

Similarly, the unpredicted variability is

$$SS_{\text{residual}} = (1 - r^2)SS_Y = (1 - 0.847^2)(156) = 0.282(156) = 43.99$$

Notice that the new formula for SS_{residual} produces the same value, within rounding error, that we obtained by adding the squared residuals in Example 16.3. Also note that this new formula is generally much easier to use because it requires only the correlation value (r) and the SS for Y. The primary point of this example, however, is that SS_{residual} and the standard error of estimate are closely related to the value of the correlation. With a large correlation (near $+1.00$ or -1.00), the data points are close to the regression line, and the standard error of estimate is small. As a correlation gets smaller (near zero), the data points move away from the regression line, and the standard error of estimate gets larger. ■

Because it is possible to have the same regression line for sets of data that have different correlations, it is also important to examine r^2 and the standard error of estimate. The regression equation simply describes the best-fitting line and is used for making predictions. However, r^2 and the standard error of estimate indicate how accurate these predictions will be.

■ Analysis of Regression

As we noted in Chapter 15, a sample correlation is expected to be representative of its population correlation. For example, if the population correlation is zero, the sample correlation is expected to be near zero. Note that we do not expect the sample correlation to be exactly equal to zero. This is the general concept of *sampling error* that was introduced in Chapter 1 (p. 6). The principle of sampling error is that there is typically some discrepancy or error between the value obtained for a sample statistic and the corresponding population parameter. Thus, when there is no relationship whatsoever in the population, a correlation of $\rho = 0$, you are still likely to obtain a nonzero value for the sample correlation. In this situation, however, the sample correlation is meaningless and a hypothesis test usually demonstrates that the correlation is not significant.

Whenever you obtain a nonzero value for a sample correlation, you will also obtain real, numerical values for the regression equation. However, if there is no real relationship in the population, both the sample correlation and the regression equation are meaningless—they are simply the result of sampling error and should not be viewed as an indication of any relationship between X and Y. In the same way that we tested the significance of a Pearson correlation, we can test the significance of the regression equation. In fact, when a single variable X is being used to predict a single variable Y, the two tests are equivalent. In each case, the purpose for the test is to determine whether the sample correlation represents a real relationship or is simply the result of sampling error. For both tests, the null hypothesis states that there is no relationship between the two variables in the population. For a correlation,

$$H_0: \quad \text{the population correlation is } \rho = 0$$

For the regression equation,

$$H_0: \quad \text{the slope of the regression equation (b or beta) is zero}$$

The process of testing the significance of a regression equation is called *analysis of regression* and is very similar to the analysis of variance (ANOVA) presented in Chapter 12. As with ANOVA, the regression analysis uses an F-ratio to determine whether the variance predicted by the regression equation is significantly greater than would be expected if there were no relationship between X and Y. The F-ratio is a ratio of two variances, or mean square (MS) values, and each variance is obtained by dividing an SS value by its

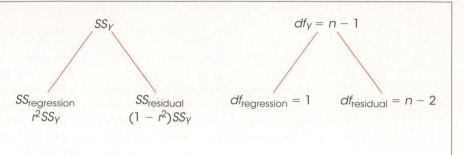

FIGURE 16.6

The partitioning of SS and df for analysis of regression. The variability for the original Y scores (both SS and df) is partitioned into two components: (1) the variability that is predicted by the regression equation and (2) the residual variability.

corresponding degrees of freedom. The numerator of the F-ratio is $MS_{\text{regression}}$, which is the variance in the Y scores that is predicted by the regression equation. This variance measures the systematic changes in Y that occur when the value of X increases or decreases. The denominator is MS_{residual}, which is the unpredicted variance in the Y scores. This variance measures the changes in Y that are independent of changes in X. The two MS value are defined as

$$MS_{\text{regression}} = \frac{SS_{\text{regression}}}{df_{\text{regression}}} \text{ with } df = 1 \quad \text{and} \quad MS_{\text{residuals}} = \frac{SS_{\text{residual}}}{df_{\text{residual}}} \text{ with } df = n - 2$$

The F-ratio is

$$F = \frac{MS_{\text{regression}}}{MS_{\text{residual}}} \quad \text{with } df = 1, n - 2 \tag{16.13}$$

The complete analysis of SS and degrees of freedom is diagrammed in Figure 16.6. The analysis of regression procedure is demonstrated in the following example, using the same data that we used in Examples 16.1, 16.3, and 16.4.

EXAMPLE 16.5   The data consist of $n = 8$ pairs of scores with a correlation of $r = 0.847$ and $SS_Y = 156$. The null hypothesis either states that there is no relationship between X and Y in the population or that the regression equation has $b = 0$ and does not account for a significant portion of the variance for the Y scores.

The F-ratio for the analysis of regression has $df = 1, n - 2$. For these data, $df = 1, 6$. With $\alpha = .05$, the critical value is 5.99.

As noted in the previous section, the SS for the Y scores can be separated into two components: the predicted portion corresponding to r^2 and the unpredicted, or residual, portion corresponding to $(1 - r^2)$. With $r = 0.847$, we obtain $r^2 = 0.718$ and

$$\text{predicted variability} = SS_{\text{regression}} = 0.718(156) = 112.01$$

$$\text{unpredicted variability} = SS_{\text{residual}} = (1 - 0.718)(156) = 0.282(156) = 43.99$$

Using these SS values and the corresponding df values, we calculate a variance or MS for each component. For these data the MS values are:

$$MS_{\text{regression}} = \frac{SS_{\text{regression}}}{df_{\text{regression}}} = \frac{112.01}{1} = 112.01$$

$$MS_{\text{residual}} = \frac{SS_{\text{residual}}}{df_{\text{residual}}} = \frac{43.99}{6} = 7.33$$

Finally, the F-ratio for evaluating the significance of the regression equation is

$$F = \frac{MS_{regression}}{MS_{residual}} = \frac{112.01}{7.33} = 15.28$$

The F-ratio is in the critical region, so we reject the null hypothesis and conclude that the regression equation does account for a significant portion of the variance for the Y scores. The complete analysis of regression is summarized in Table 16.1, which is a common format for computer printouts of regression analysis.

TABLE 16.1

A summary table showing the results of the analysis of regression in Example 16.5.

Source	SS	df	MS	F
Regression	112.01	1	112.01	15.28
Residual	43.99	6	7.33	
Total	156	7		

■ Significance of Regression and Significance of the Correlation

As noted earlier, in situation with a single X variable and a single Y variable, testing the significance of the regression equation is equivalent to testing the significance of the Pearson correlation. Therefore, whenever the correlation between two variables is significant, you can conclude that the regression equation is also significant. Similarly, if a correlation is not significant, the regression equation is also not significant. For the data in Example 16.5, we concluded that the regression equation is significant.

To demonstrate the equivalence of the two tests, we will show that the t statistic used to test the significance of a correlation (Chapter 15, p. 508) is equivalent to the F-ratio used to test the significance of the regression equation (Equation 16.12). We begin with the t statistic introduced in Chapter 15.

$$t = \frac{r - \rho}{\sqrt{\dfrac{(1 - r^2)}{(n - 2)}}}$$

First, we remove the population correlation, ρ, from the t equation. This value is always zero, as specified by the null hypothesis, and its removal does not affect the equation. Next, we square the t statistic to produce the corresponding F-ratio.

$$t^2 = F = \frac{r^2}{\dfrac{(1 - r^2)}{(n - 2)}}$$

Finally, multiply the numerator and the denominator by SS_Y to produce

$$t^2 = F = \frac{r^2(SS_Y)}{\dfrac{(1 - r^2)(SS_Y)}{(n - 2)}}$$

You should recognize the numerator as $SS_{regression}$, which is equivalent to $MS_{regression}$ because $df_{regression} = 1$. Also, the denominator is identical to $MS_{residual}$.

1. A researcher obtains a correlation of $r = 0.60$ for a set of $n = 18$ pairs of X and Y values. If the Y scores have $SS = 100$, then what is the standard error of estimate for the regression equation for predicting Y from X?

   a. 2 points

   b. 4 points

   c. 8 points

   d. 16 points

2. A researcher conducts an analysis of regression to evaluate the significance of the regression equation and obtains $F = 4.25$ with $df = 1, 24$. How many pairs of X and Y values were in the original data?

   a. 23

   b. 24

   c. 25

   d. 26

ANSWERS   **1.** A, **2.** D

16.3 | Introduction to Multiple Regression with Two Predictor Variables

Thus far, we have looked at regression in situations in which one variable is being used to predict a second variable. For example, IQ scores can be used to predict academic performance for a group of college students. However, a variable such as academic performance is usually related to a variety of other factors. For example, college GPA is probably related to motivation, self-esteem, SAT score, rank in high school graduating class, parents' highest level of education, and many other variables. In this case, it is possible to combine several predictor variables to obtain a more accurate prediction. For example, IQ predicts some of academic performance, but you can probably get a better prediction if you use IQ and SAT scores together. The process of using several predictor variables to help obtain more accurate predictions is called *multiple regression.*

Although it is possible to combine a large number of predictor variables in a single multiple-regression equation, we limit our discussion to the two-predictor case. There are two practical reasons for this limitation.

1. Multiple regression, even limited to two predictors, can be relatively complex. Although we present equations for the two-predictor case, the calculations are usually performed by a computer, so there is not much point in developing a set of complex equations when people are going to use a computer instead.

2. Usually, different predictor variables are related to each other, which means that they are often measuring and predicting the same thing. Because the variables may overlap with each other, adding another predictor variable to a regression equation does not always add to the accuracy of prediction. This situation is shown in Figure 16.7. In the figure, IQ overlaps with academic performance, which means

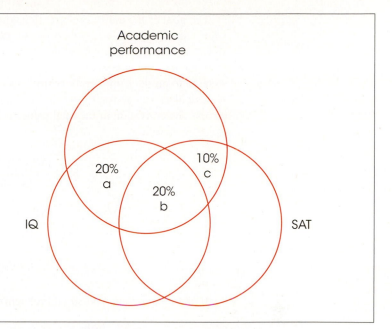

FIGURE 16.7

Predicting the variance in academic performance from IQ and SAT scores. Note that IQ predicts 40% of the variance in academic performance but adding SAT scores as a second predictor increases the predicted portion by only 10%.

that part of academic performance can be predicted by IQ. In this example, IQ overlaps (predicts) 40% of the variance in academic performance (combine sections a and b in the figure). The figure also shows that SAT scores overlap with academic performance, which means that part of academic performance can be predicted by knowing SAT scores. Specifically, SAT scores overlap or predict 30% of the variance (combine sections b and c). Thus, using both IQ and SAT to predict academic performance should produce better predictions than would be obtained from IQ alone. However, there is also a lot of overlap between SAT scores and IQ. In particular, much of the prediction from SAT scores overlaps with the prediction from IQ (section b). As a result, adding SAT scores as a second predictor only adds a small amount to the variance already predicted by IQ (section c). Because variables tend to overlap in this way, adding new variables beyond the first one or two predictors often does not add significantly to the quality of the prediction.

■ Regression Equations with Two Predictors

We identify the two predictor variables as X_1 and X_2. The variable we are trying to predict is identified as Y. Using this notation, the general form of the multiple regression equation with two predictors is

$$\hat{Y} = b_1 X_1 + b_2 X_2 + a \tag{16.14}$$

If all three variables, X_1, X_2, and Y, have been standardized by transformation into z-scores, then the standardized form of the multiple regression equation predicts the z-score for each Y value. The standardized form is

$$\hat{z}_Y = (\text{beta}_1)z_{X1} + (\text{beta}_2)z_{X2} \tag{16.15}$$

Researchers rarely transform raw X and Y scores into z-scores before finding a regression equation, however, the beta values are meaningful and are usually reported by computer programs conducting multiple regression. We return to the discussion of beta values later in this section.

The goal of the multiple-regression equation is to produce the most accurate estimated values for Y. As with the single predictor regression, this goal is accomplished with a least-squared solution. First, we define "error" as the difference between the predicted Y value from the regression equation and the actual Y value for each individual. Each error is then squared to produce uniformly positive values, and then we add the squared errors. Finally, we calculate values for b_1, b_2, and a that produce the smallest possible sum of squared errors. The derivation of the final values is beyond the scope of this text, but the final equations are as follows:

$$b_1 = \frac{(SP_{X1Y})(SS_{X2}) - (SP_{X1X2})(SP_{X2Y})}{(SS_{X1})(SS_{X2}) - (SP_{X1X2})^2} \qquad (16.16)$$

$$b_2 = \frac{(SP_{X2Y})(SS_{X1}) - (SP_{X1X2})(SP_{X1Y})}{(SS_{X1})(SS_{X2}) - (SP_{X1X2})^2} \qquad (16.17)$$

$$a = M_Y - b_1 M_{X1} - b_2 M_{X2} \qquad (16.18)$$

In these equations, you should recognize the following SS and SP values:

SS_{X1} is the sum of squared deviations for X_1

SS_{X2} is the sum of squared deviations for X_2

SP_{X1Y} is the sum of products of deviations for X_1 and Y

SP_{X2Y} is the sum of products of deviations for X_2 and Y

SP_{X1X2} is the sum of products of deviations for X_1 and X_2

Note: More detailed information about the calculation of SS is presented in Chapter 4 (pp. 112–113) and information concerning SP is in Chapter 15 (pp. 490–492). The following example demonstrates multiple regression with two predictor variables.

EXAMPLE 16.6

We use the data in Table 16.2 to demonstrate multiple regression. Note that each individual has a Y score and two X scores that are used as predictor variables. Also note that we have already computed the SS values for Y and for both of the X scores, as well as all of the SP values. These values are used to compute the coefficients, b_1 and b_2, and the constant, a, for the regression equation.

$$\hat{Y} = b_1 X_1 - b_2 X_2 + a$$

$$b_1 = \frac{(SP_{X1Y})(SS_{X2}) - (SP_{X1X2})(SP_{X2Y})}{(SS_{X1})(SS_{X2}) - (SP_{X1X2})^2} = \frac{54(64) - 42(47)}{62(64) - (42)^2} = 0.672$$

$$b_2 = \frac{(SP_{X2Y})(SS_{X1}) - (SP_{X1X2})(SP_{X1Y})}{(SS_{X1})(SS_{X2}) - (SP_{X1X2})^2} = \frac{47(62) - 42(54)}{62(64) - (42)^2} = 0.293$$

$$a = M_Y - b_1 M_{X1} - b_2 M_{X2} = 7 - 0.672(4) - 0.293(6) = 7 - 2.688 - 1.758 = 2.554$$

Thus, the final regression equation is

$$\hat{Y} = 0.672X_1 + 0.293X_2 + 2.554$$

TABLE 16.2

Sample data consisting of three scores for each person. Two of the scores, X_1 and X_2, are used to predict the Y score for each individual.

Person	Y	X_1	X_2	
A	11	4	10	$SP_{X1Y} = 54$
B	5	5	6	$SP_{X2Y} = 47$
C	7	3	7	$SP_{X1X2} = 42$
D	3	2	4	
E	4	1	3	
F	12	7	5	
G	10	8	8	
H	4	2	4	
I	8	7	10	
J	6	1	3	
	$M_Y = 7$	$M_{X1} = 4$	$M_{X2} = 6$	
	$SS_Y = 90$	$SS_{X1} = 62$	$SS_{X2} = 64$	

Example 16.6 also demonstrates that multiple regression can be a tedious process. As a result, multiple regression is usually conducted on a computer. To demonstrate this process, we used the SPSS computer program to perform a multiple regression on the data in Table 16.2 and the output from the program is shown in Figure 16.8. At this time, focus on the Coefficients Table at the bottom of the printout. The values in the first column of Unstandardized Coefficients include the constant, b_1 and b_2 for the regression equation. We will discuss other portions of the SPSS output later in this chapter.

■ R^2 and Residual Variance

In the same way that we computed an r^2 value to measure the percentage of variance accounted for with the single-predictor regression, it is possible to compute a corresponding percentage for multiple regression. For a multiple-regression equation, this percentage is identified by the symbol R^2. The value of R^2 describes the proportion of the total variability of the Y scores that is accounted for by the regression equation. In symbols,

$$R^2 = \frac{SS_{regression}}{SS_Y} \qquad \text{or} \qquad SS_{regression} = R^2 SS_Y$$

For a regression with two predictor variables, R^2 can be computed directly from the regression equation as follows:

$$R^2 = \frac{b_1 SP_{X1Y} + b_2 SP_{X2Y}}{SS_Y} \qquad \text{(16.19)}$$

For the data in Table 16.2, we obtain a value of

In the computer printout in Figure 16.8, the value of R^2 is reported in the Model Summary table as 0.557. Within rounding error, the two values are the same.

$$R^2 = \frac{0.672(54) + 0.293(47)}{90} = \frac{50.059}{90} = 0.5562 \text{ (or 55.62\%)}$$

Thus, 55.62% of the variance for the Y scores can be predicted by the regression equation. For the data in Table 16.2, $SS_Y = 90$, so the predicted portion of the variability is

$$SS_{regression} = R^2 SS_Y = 0.5562(90) = 50.06$$

Model Summary

Model	R	R Square	Adjusted R Square	Std. Error of the Estimate
1	.746[a]	.557	.430	2.38788

a. Predictors: (Constant), VAR00003, VAR00002

ANOVA[b]

Model		Sum of Squares	df	Mean Square	F	Sig.
1	Regression	50.086	2	25.043	4.392	.058[a]
	Residual	39.914	7	5.702		
	Total	90.000	9			

a. Predictors: (Constant), VAR00003, VAR00002
b. Dependent Variable: VAR00001

Coefficients[a]

Model		Unstandardized Coefficients		Standardized Coefficients	t	Sig.
		B	Std. Error	Beta		
1	(Constant)	2.552	1.944		1.313	.231
	VAR00002	.672	.407	.558	1.652	.142
	VAR00003	.293	.401	.247	.732	.488

a. Dependent Variable: VAR00001

FIGURE 16.8
The SPSS output for the multiple regression in Example 16.6.

The unpredicted, or residual, variance is determined by $1 - R^2$. For the data in Table 16.2, this is

$$SS_{residual} = (1 - R^2)SS_Y = 0.4438(90) = 39.94$$

■ The Standard Error of Estimate

On p. 538 we defined the standard error of estimate for a linear regression equation as the standard distance between the regression line and the actual data points. In more general terms, the standard error of estimate can be defined as the standard distance between the predicted Y values (from the regression equation) and the actual Y values (in the data). The more general definition applies equally well to both linear and multiple regression.

To find the standard error of estimate for either linear regression or multiple regression, we begin with $SS_{residual}$. For linear regression with one predictor, $SS_{residual} = (1 - r^2)SS_Y$ and has $df = n - 2$. For multiple regression with two predictors, $SS_{residual} = (1 - R^2)SS_Y$ and has $df = n - 3$. In each case, we can use the SS and df values to compute a variance or $MS_{residual}$.

$$MS_{residual} = \frac{SS_{residual}}{df}$$

The variance or MS value is a measure of the average squared distance between the actual Y values and the predicted Y values. By simply taking the square root, we obtain a measure of standard deviation or standard distance. This standard distance for the residuals is the standard error of estimate. Thus, for both linear regression and multiple regression

$$\text{the standard error of estimate} = \sqrt{MS_{residual}}$$

In the computer printout in Figure 16.8, the standard error of estimate is reported in the Model Summary table at the top.

For either linear or multiple regression, you do not expect the predictions from the regression equation to be perfect. In general, there will be some discrepancy between the predicted values of Y and the actual values. The standard error of estimate provides a measure of how much discrepancy, on average, there is between the $\hat{Y}$ values and the actual Y values.

■ Testing the Significance of the Multiple Regression Equation: Analysis of Regression

Just as we did with the single predictor equation, we can evaluate the significance of a multiple-regression equation by computing an F-ratio to determine whether the equation predicts a significant portion of the variance for the Y scores. The total variability of the Y scores is partitioned into two components, $SS_{regression}$ and $SS_{residual}$. With two predictor variables, $SS_{regression}$ has $df = 2$, and $SS_{residual}$ has $df = n - 3$. Therefore, the two MS values for the F-ratio are

$$MS_{regression} = \frac{SS_{regression}}{2} \tag{16.20}$$

and

$$MS_{residual} = \frac{SS_{residual}}{n - 3} \tag{16.21}$$

The data for the $n = 10$ people in Table 16.2 have produced $SS_{regression} = 50.06$ and $SS_{residual} = 39.94$.

Therefore, $\quad MS_{regression} = \dfrac{50.06}{2} = 25.03 \quad$ and $\quad MS_{residual} = \dfrac{39.94}{7} = 5.71$

and $\qquad\qquad F = \dfrac{MS_{regression}}{MS_{residual}} = \dfrac{25.03}{5.71} = 4.38$

With $df = 2, 7$, this F-ratio is not significant with $\alpha = .05$, so we cannot conclude that the regression equation accounts for a significant portion of the variance for the Y scores.

The analysis of regression is summarized in the following table, which is a common component of the output from most computer versions of multiple regression. In the computer printout in Figure 16.8, this summary table is reported in the ANOVA table in the center,

Source	SS	df	MS	F
Regression	50.06	2	25.03	4.38
Residual	39.94	7	5.71	
Total	90.00	9		

The following example is an opportunity to test your understanding of the standard error of estimate for a multiple-regression equation.

EXAMPLE 16.7

Data from a sample of $n = 15$ individuals are used to compute a multiple-regression equation with two predictor variables. The equation has $R^2 = 0.20$ and $SS_Y = 150$.

Find $SS_{residual}$ and compute the standard error of estimate for the regression equation. You should obtain $SS_{residual} = 120$ and a standard error of estimate equal to $\sqrt{10} = 3.16$. ■

■ Evaluating the Contribution of Each Predictor Variable

In addition to evaluating the overall significance of the multiple-regression equation, researchers are often interested in the relative contribution of each of the two predictor variables. Is one of the predictors responsible for more of the prediction than the other? Unfortunately, the b values in the regression equation are influenced by a variety of other factors and do not address this issue. If b_1 is larger than b_2, it does not necessarily mean that X_1 is a better predictor than X_2. However, in the standardized form of the regression equation, the relative size of the beta values is an indication of the relative contribution of the two variables. For the data in Table 16.2, the standardized regression equation is

$$\hat{z}_Y = (beta_1)z_{X1} + (beta_2)z_{X2}$$

$$= 0.558z_{X1} + 0.247z_{X2}$$

For the SPSS printout in Figure 16.8, the beta values are shown in the Coefficients table.

In this case, the larger beta value for the X_1 predictor indicates that X_1 predicts more of the variance than does X_2. The signs of the beta values are also meaningful. In this example, both betas are positive indicating the both X_1 and X_2 are positively related to Y.

Beyond judging the relative contribution for each of the predictor variables, it also is possible to evaluate the significance of each contribution. For example, does variable X_2 make a significant contribution beyond what is already predicted by variable X_1? The null hypothesis states that the multiple-regression equation (using X_2 in addition to X_1) is not any better than the simple regression equation using X_1 as a single predictor variable. An alternative view of the null hypothesis is that the b_2 (or beta$_2$) value in the equation is not significantly different from zero. To test this hypothesis, we first determine how much more variance is predicted using X_1 and X_2 together than is predicted using X_1 alone.

Earlier we found that the multiple regression equation with both X_1 and X_2 predicted $R^2 = 55.62\%$ of the variance for the Y scores. To determine how much is predicted by X_1 alone, we begin with the correlation between X_1 and Y, which is

$$r = \frac{SP_{X1Y}}{\sqrt{(SS_{X1})(SS_Y)}} = \frac{54}{\sqrt{(62)(90)}} = \frac{54}{74.70} = 0.7229$$

Squaring the correlation produces $r^2 = (0.7229)^2 = 0.5226$ or 52.26%. This means that the relationship with X_1 predicts 52.26% of the variance for the Y scores.

Therefore, the additional contribution made by adding X_2 to the regression equation can be computed as

$$(\% \text{ with both } X_1 \text{ and } X_2) - (\% \text{ with } X_1 \text{ alone})$$

$$= 55.62\% - 52.26\%$$

$$= 3.36\%$$

Because $SS_Y = 90$, the additional variability from adding X_2 as a predictor amounts to

$$SS_{additional} = 3.36\% \text{ of } 90 = 0.0336(90) = 3.024$$

This SS value has $df = 1$, and can be used to compute an F-ratio evaluating the significance of the contribution of X_2. First,

$$MS_{additional} = \frac{SS_{additional}}{1} = \frac{3.024}{1} = 3.024$$

This MS value is evaluated by computing an F-ratio with the $MS_{residual}$ value from the multiple regression as the denominator. (*Note:* This is the same denominator that was used in the F-ratio to evaluate the significance of the multiple-regression equation.) For these data, we obtain

$$F = \frac{MS_{additional}}{MS_{residual}} = \frac{3.024}{5.71} = 0.5296$$

With $df = 1, 7$, this F-ratio is not significant. Therefore, we conclude that adding X_2 to the regression equation does not significantly improve the prediction compared to using X_1 as a single predictor. The computer printout shown in Figure 16.8, reports a t statistic instead of an F-ratio to evaluate the contribution for each predictor variable. Each t value is simply the square root of the F-ratio and is reported in the right-hand side of the Coefficients table. Variable X_2, for example, is reported as VAR00003 in the table and has $t = 0.732$, which is within rounding error of the F-ratio we obtained; $\sqrt{F} = \sqrt{0.5296} = 0.728$.

■ Multiple Regression and Partial Correlations

In Chapter 15 we introduced *partial correlation* as a technique for measuring the relationship between two variables while eliminating the influence of a third variable. At that time, we noted that partial correlations serve two general purposes:

1. A partial correlation can demonstrate that an apparent relationship between two variables is actually caused by a third variable. Thus, there is no direct relationship between the original two variables.

2. Partial correlation can demonstrate that there is a relationship between two variables even after a third variable is controlled. Thus, there really is a relationship between the original two variables that is not being caused by a third variable.

Multiple regression provides an alternative procedure for accomplishing both of these goals. Specifically, the regression analysis evaluates the contribution of each predictor variable after the influence of the other predictor has been considered. Thus, you can determine whether each predictor variable contributes to the relationship by itself or simply duplicates the contribution already made by another variable.

For example, in Chapter 15 (p. 503) we presented data for a sample of 15 cities showing a logical relationship among the number of churches, the number of crimes, and the population size. Although the original data showed a strong, positive correlation between churches and crime, the partial correlation showed that the actual relationship was zero after the population size was controlled. Multiple regression produces the same conclusion. A multiple-regression analysis of these data using churches and population to predict crime, produced a beta value of $\beta = 0.961$ for the population variable and $\beta = 0$ for the church variable. Thus, there is a positive correlation between population and crime but the underlying correlation between churches and crime is zero (not positive and not negative).

LEARNING CHECK

1. A multiple regression equation with two predictor variables is calculated for a set of scores. If the constant values in the equation are $b_1 = 2$, $b_2 = -3$, and $a = 7$, then what Y value would be predicted for an individual with $X_1 = 2$ and $X_2 = 4$?
   - **a.** 1
   - **b.** −1
   - **c.** 17
   - **d.** 41

2. A multiple regression equation with two predictor variables produces $R^2 = 0.10$. What portion of the variability for the Y scores is predicted by the equation?
   - **a.** 1%
   - **b.** 10%
   - **c.** 90%
   - **d.** 99%

3. The multiple regression with two predictor variables is computed for a sample of $n = 25$ participants. What is the value for degrees of freedom for the predicted portion of the Y-score variance, $MS_{regression}$?
   - **a.** 2
   - **b.** 3
   - **c.** 22
   - **d.** 23

ANSWERS   **1.** B, **2.** B, **3.** A

SUMMARY

1. When there is a general linear relationship between two variables, X and Y, it is possible to construct a linear equation that allows you to predict the Y value corresponding to any known value of X:

$$\text{predicted } Y \text{ value} = \hat{Y} = bX + a$$

   The technique for determining this equation is called regression. By using a *least-squares* method to minimize the error between the predicted Y values and

the actual Y values, the best-fitting line is achieved when the linear equation has

$$b = \frac{SP}{SS_X} = r\frac{s_Y}{s_X} \quad \text{and} \quad a = M_Y - bM_X$$

2. The linear equation generated by regression (called the regression equation) can be used to compute a predicted Y value for any value of X. However, the prediction is not perfect, so for each Y value, there is

a predicted portion and an unpredicted, or residual, portion. Overall, the predicted portion of the Y score variability is measured by r^2, and the residual portion is measured by $1 - r^2$.

$$\text{Predicted variability} = SS_{\text{regression}} = r^2 SS_Y$$

$$\text{Unpredicted variability} = SS_{\text{residual}} = (1 - r^2)SS_Y$$

3. The residual variability can be used to compute the standard error of estimate, which provides a measure of the standard distance (or error) between the predicted Y values on the line and the actual data points. The standard error of estimate is computed by

$$\text{standard error of estimate} = \sqrt{\frac{SS_{\text{residual}}}{n - 2}} = \sqrt{MS_{\text{residual}}}$$

4. It is also possible to compute an F-ratio to evaluate the significance of the regression equation. The process is called analysis of regression and determines whether the equation predicts a significant portion of the variance for the Y scores. First a variance, or MS, value is computed for the predicted variability and the residual variability,

$$MS_{\text{regression}} = \frac{SS_{\text{regression}}}{df_{\text{regression}}} \qquad MS_{\text{residual}} = \frac{SS_{\text{residual}}}{df_{\text{residual}}}$$

where $df_{\text{regression}} = 1$ and $df_{\text{residual}} = n - 2$. Next, an F-ratio is computed to evaluate the significance of the regression equation.

$$F = \frac{MS_{\text{regression}}}{MS_{\text{residual}}} \qquad \text{with } df = 1, n - 2$$

5. Multiple regression involves finding a regression equation with more than one predictor variable. With two predictors (X_1 and X_2), the equation becomes

$$\hat{Y} = b_1 X_1 + b_2 X_2 + a$$

with the values for b_1, b_2, and a computed using Equations 16.16, 16.17, and 16.18.

6. For multiple regression, the value of R^2 describes the proportion of the total variability of the Y scores that is accounted for by the regression equation. With two predictor variables,

$$R^2 = \frac{b_1 SP_{X1Y} + b_2 SP_{X2Y}}{SS_Y}$$

The predicted variability $= SS_{\text{regression}} = R^2 SS_Y$. Unpredicted variability $= SS_{\text{residual}} = (1 - R^2)SS_Y$

7. The residual variability for the multiple-regression equation can be used to compute a standard error of estimate, which provides a measure of the standard distance (or error) between the predicted Y values from the equation and the actual data points. For multiple regression with two predictors, the standard error of estimate is computed by

$$\text{standard error of estimate} = \sqrt{\frac{SS_{\text{residual}}}{n - 3}} = \sqrt{MS_{\text{residual}}}$$

8. Evaluating the significance of the two-predictor multiple-regression equation involves computing an F-ratio that divides the $MS_{\text{regression}}$ (with $df = 2$) by the MS_{residual} (with $df = n - 3$). A significant F-ratio indicates that the regression equation accounts for a significant portion of the variance for the Y scores.

9. An F-ratio can also be used to determine whether a second predictor variable (X_2) significantly improves the prediction beyond what was already predicted by X_1. The numerator of the F-ratio measures the additional SS that is predicted by adding X_2 as a second predictor.

$$SS_{\text{additional}} = SS_{\text{regression with } X1 \text{ and } X2} - SS_{\text{regression with } X1 \text{ alone}}$$

This SS value has $df = 1$. The denominator of the F-ratio is the MS_{residual} from the two-predictor regression equation.

10. A partial correlation measures the relationship that exists between two variables after a third variable has been controlled or held constant.

KEY TERMS

SPSS®

General instructions for using SPSS are presented in Appendix D. Following are detailed instructions for using SPSS to perform the **Linear Regression** and **Multiple Regression** presented in this chapter.

LINEAR AND MULTIPLE REGRESSION

Data Entry

1. With one predictor variable (X), you enter the X values in one column and the Y values in a second column of the SPSS data editor. With two predictors (X_1 and X_2), enter the X_1 values in one column, X_2 in a second column, and Y in a third column.

Data Analysis

1. Click **Analyze** on the tool bar, select **Regression,** and click on **Linear.**
2. In the left-hand box, highlight the column label for the Y values, then click the arrow to move the column label into the **Dependent Variable** box.
3. For one predictor variable, highlight the column label for the X values and click the arrow to move it into the **Independent Variable(s)** box. For two predictor variables, highlight the X_1 and X_2 column labels, one at a time, and click the arrow to move them into the **Independent Variable(s)** box.
4. Click **OK.**

SPSS Output

We used SPSS to perform multiple regression for the data in Example 16.6 and the output is shown in Figure 16.8 (p. 548). The **Model Summary** table presents the values for R, R^2, and the standard error of estimate. (*Note:* For a single predictor, R is simply the Pearson correlation between X and Y.) The **ANOVA** table presents the analysis of regression evaluating the significance of the regression equation, including the F-ratio and the level of significance (the p value or alpha level for the test). The **Coefficients** table summarizes the unstandardized and the standardized coefficients for the regression equation. For one predictor, the table shows the values for the constant (a) and the coefficient (b). For two predictors, the table shows the constant (a) and the two coefficients (b_1 and b_2). The standardized coefficients are the beta values. For one predictor, beta is simply the Pearson correlation between X and Y. Finally, the table uses a t statistic to evaluate the significance of each predictor variable. For one predictor variable, this is identical to the significance of the regression equation and you should find that t is equal to the square root of the F-ratio from the analysis of regression. For two predictor variables, the t values measure the significance of the contribution of each variable beyond what is already predicted by the other variable.

FOCUS ON PROBLEM SOLVING

1. A basic understanding of the Pearson correlation, including the calculation of SP and SS values, is critical for understanding and computing regression equations.

2. You can calculate $SS_{residual}$ directly by finding the residual (the difference between the actual Y and the predicted Y for each individual), squaring the residuals, and adding the squared values. However, it usually is much easier to compute r^2 (or R^2) and then find $SS_{residual} = (1 - r^2)SS_Y$.

3. The F-ratio for analysis of regression is usually calculated using the actual $SS_{regression}$ and $SS_{residual}$. However, you can simply use r^2 (or R^2) in place of $SS_{regression}$ and you can use $1 - r^2$ or $(1 - R^2)$ in place of $SS_{residual}$. *Note:* You must still use the correct df value for the numerator and the denominator.

DEMONSTRATION 16.1

LINEAR REGRESSION

The following data will be used to demonstrate the process of linear regression. The scores and summary statistics are as follows.

Person	X	Y
A	0	4
B	2	1
C	8	10
D	6	9
E	4	6

$M_X = 4$ with $SS_X = 40$
$M_Y = 6$ with $SS_Y = 54$
$SP = 40$

These data produce a Pearson correlation of $r = 0.861$.

STEP 1   **Compute the values for the regression equation**   The general form of the regression equation is

$$\hat{Y} = bX + a \qquad \text{where } b = \frac{SP}{SS_X} \quad \text{and} \quad a = M_Y - bM_X$$

For these data, $b = \dfrac{40}{40} = 1.00$ and $a = 6 - 1(4) = +2.00$

Thus, the regression equation is $\hat{Y} = (1)X + 2.00$ or simply, $\hat{Y} = X + 2$.

STEP 2   **Evaluate the significance of the regression equation**   The null hypothesis states that the regression equation does not predict a significant portion of the variance for the Y scores. To conduct the test, the total variability for the Y scores, $SS_Y = 54$, is partitioned into the portion predicted by the regression equation and the residual portion.

$$SS_{\text{regression}} = r^2(SS_Y) = 0.741(54) = 40.01 \text{ with } df = 1$$

$$SS_{\text{residual}} = (1 - r^2)(SS_Y) = 0.259(54) = 13.99 \text{ with } df = n - 2 = 3$$

The two MS values (variances) for the F-ratio are

$$MS_{\text{regression}} = \frac{SS_{\text{regression}}}{df} = \frac{40.01}{1} = 40.01$$

$$MS_{\text{residual}} = \frac{SS_{\text{residual}}}{df} = \frac{13.99}{3} = 4.66$$

And the F-ratio is

$$F = \frac{MS_{\text{regression}}}{MS_{\text{residual}}} = \frac{40.01}{4.66} = 8.59$$

With $df = 1, 3$ and $\alpha = .05$, the critical value for the F-ratio is 10.13. Therefore, we fail to reject the null hypothesis and conclude that the regression equation does not predict a significant portion of the variance for the Y scores.

PROBLEMS

1. Sketch a graph showing the line for the equation $Y = 2X - 1$. On the same graph, show the line for $Y = -X + 8$.

2. The regression equation is intended to be the "best fitting" straight line for a set of data. What is the criterion for "best fitting"?

3. A set of $n = 18$ pairs of scores (X and Y values) has $SS_X = 20$, $SS_Y = 80$, and $SP = 10$. If the mean for the X values is $M_X = 8$ and the mean for the Y values is $M_Y = 10$.
   a. Calculate the Pearson correlation for the scores.
   b. Find the regression equation for predicting Y from the X values.

4. A set of $n = 15$ pairs of scores (X and Y values) produces a regression equation of $\hat{Y} = 2X + 6$. Find the predicted Y value for each of the following X scores: 0, 2, 3, and −4.

5. Briefly explain what is measured by the standard error of estimate.

6. In general, how is the magnitude of the standard error of estimate related to the value of the correlation?

7. For the following set of data, find the linear regression equation for predicting Y from X:

X	Y
2	1
7	10
5	8
3	0
3	4
4	13

8. For the following data:
   a. Find the regression equation for predicting Y from X.
   b. Calculate the Pearson correlation for these data. Use r^2 and SS_Y to compute $SS_{residual}$ and the standard error of estimate for the equation.

X	Y
3	3
6	9
5	8
4	3
7	10
5	9

9. Does the regression equation from problem 8 account for a significant portion of the variance in the Y scores? Use $\alpha = .05$ to evaluate the F-ratio.

10. For the following scores,

X	Y
3	8
5	8
2	6
2	3
4	6
1	4
4	7

   a. Find the regression equation for predicting Y from X.
   b. Calculate the predicted Y value for each X.

11. Problem 13 in Chapter 15 examined the relationship between weight and income for a sample of $n = 8$ men. Weights were classified in five categories and had a mean of $M = 3$ with $SS = 18$. Income, measured in thousands, had a mean score of $M = 88$ with $SS = 21,609$, and $SP = 330$.
   a. Find the regression equation for predicting income from weight. (Identify the weight scores as X values and the income scores as Y values.)
   b. What percentage of the variance in the income is accounted for by the regression equation? (Compute the correlation, r, then find r^2.)
   c. Does the regression equation account for a significant portion of the variance in income? Use $\alpha = .05$ to evaluate the F-ratio.

12. A professor obtains SAT scores and freshman grade point averages (GPAs) for a group of $n = 15$ college students. The SAT scores have a mean of $M = 580$ with $SS = 22,400$, and the GPAs have a mean of 3.10 with $SS = 1.26$, and $SP = 84$.
   a. Find the regression equation for predicting GPA from SAT scores.
   b. What percentage of the variance in GPAs is accounted for by the regression equation? (Compute the correlation, r, then find r^2.)
   c. Does the regression equation account for a significant portion of the variance in GPA? Use $\alpha = .05$ to evaluate the F-ratio.

13. Problem 14 in Chapter 15 described a study examining the effectiveness of a 7-Minute Screen test for Alzheimer's disease. The study evaluated the relationship between scores from the 7-Minute Screen and scores for the same patients from a set of cognitive exams that are typically used to test for

Alzheimer's disease. For a sample of $n = 9$ patients, the scores for the 7-Minute Screen averaged $M = 7$ with $SS = 92$. The cognitive test scores averaged $M = 17$ with $SS = 236$. For these data, $SP = 127$.
a. Find the regression equation for predicting the cognitive scores from the 7-Minute Screen score.
b. What percentage of variance in the cognitive scores is accounted for by the regression equation?
c. Does the regression equation account for a significant portion of the variance in the cognitive scores? Use $\alpha = .05$ to evaluate the F-ratio.

14. There appears to be some evidence suggesting that earlier retirement may lead to memory decline (Rohwedder & Willis, 2010). The researchers gave a memory test to men and women aged 60–64 in several countries that have different retirement ages. For each country, the researchers recorded the average memory score and the percentage of individuals in the 60–64 age range who were retired. Note that a higher percentage retired indicates a younger retirement age for that country. The following data are similar to the results from the study. Use the data to find the regression equation for predicting memory scores from the percentage of people aged 60–64 who are retired.

Country	% Retired (X)	Memory Score (Y)
Sweden	39	9.3
U.S.A.	48	10.9
England	59	10.7
Germany	70	9.1
Spain	74	6.4
Netherlands	78	9.1
Italy	81	7.2
France	87	7.9
Belgium	88	8.5
Austria	91	9.0

15. The regression equation is computed for a set of $n = 18$ pairs of X and Y values with a correlation of $r = +0.50$ and $SS_y = 48$.
a. Find the standard error of estimate for the regression equation.
b. How big would the standard error be if the sample size were $n = 66$?

16. a. One set of 10 pairs of scores, X and Y values, produces a correlation of $r = 0.60$. If $SS_y = 200$, find the standard error of estimate for the regression line.
b. A second set of 10 pairs of X and Y values produces of correlation of $r = 0.40$. If $SS_y = 200$, find the standard error of estimate for the regression line.

17. a. A researcher computes the linear regression equation for a sample of $n = 20$ pairs of scores, X and Y values. If an analysis of regression is used to test the significance of the equation, what are the df values for the F-ratio?
b. A researcher evaluating the significance of a regression equation obtains an F-ratio with $df = 1, 23$. How many pairs of scores, X and Y values, are in the sample?

18. For the following data:
a. Find the regression equation for predicting Y from X.
b. Use the regression equation to find a predicted Y for each X.
c. Find the difference between the actual Y value and the predicted Y value for each individual, square the differences, and add the squared values to obtain $SS_{residual}$.
d. Calculate the Pearson correlation for these data. Use r^2 and SS_y to compute $SS_{residual}$ with Equation 16.10. You should obtain the same value as in part c.

X	Y
3	2
5	6
4	9
2	3
5	6
5	4

19. A multiple-regression equation with two predictor variables produces $R^2 = .38$.
a. If $SS_y = 40$ for a sample of $n = 20$ individuals, does the equation predict a significant portion of the variance for the Y scores? Test with $\alpha = .05$.
b. If $SS_y = 40$ for a sample of $n = 10$ individuals, does the equation predict a significant portion of the variance for the Y scores? Test with $\alpha = .05$.

20. A researcher obtained the following multiple-regression equation using two predictor variables: $\hat{Y} = 1.1X_1 + 3X_2 + 4.6$. Given that $SS_y = 280$, the SP value for X_1Y is 48, and the SP value for X_2Y is 56, find R^2, the percentage of variance accounted for by the equation.

21. Problem 18 in Chapter 15 (p. 526) presented data showing the number of crimes, the amount spend on crime prevention, and the population for a set $n = 12$ cities. At that time, we used a partial correlation to evaluate the relationship between the amount spent on crime prevention and the number of crimes crime while controlling population. It is possible to use

multiple regression to accomplish essentially the same purpose. The data are as follows:

Number of Crimes	Amount for Prevention	Population Size
3	6	1
4	7	1
6	3	1
7	4	1
8	11	2
9	12	2
11	8	2
12	9	2
13	16	3
14	17	3
16	13	3
17	14	3

a. Find the multiple regression equation for predicting the number of crimes using the amount spent on prevention and population as the two predictor variables.
b. Find the value of R^2 for the regression equation.

22. For the data in problem 21, the correlation between the number of crimes and population is $r = 0.933$, which means that $r^2 = 0.87$ (87%) is the proportion of variance in the number of crimes that is predicted by population size. Does adding the amount spent on crime prevention as a second variable in the multiple regression equation add a significant amount to the prediction? Test with $\alpha = .05$.

23. For the following data, find the multiple-regression equation for predicting Y from X_1 and X_2.

X_1	X_2	Y
1	3	1
2	4	2
3	5	6
6	9	8
4	8	3
2	7	4
$M = 3$	$M = 6$	$M = 4$
$SS_{X1} = 16$	$SS_{X2} = 28$	$SS_Y = 34$

$$SP_{X1X2} = 18$$
$$SP_{X1Y} = 19$$
$$SP_{X2Y} = 21$$

24. A researcher evaluates the significance of a multiple-regression equation and obtains an F-ratio with $df = 2, 24$. How many participants were in the sample?

The Chi-Square Statistic: Tests for Goodness of Fit and Independence

© Deborah Batt

Tools You Will Need

The following items are considered essential background material for this chapter. If you doubt your knowledge of any of these items, you should review the appropriate chapter or section before proceeding.

- Proportions (Appendix A)
- Frequency distributions (Chapter 2)

Research suggests that romantic background music increases the likelihood that a woman will give her phone number to a man she has just met (Guéguen, Jacob, & Lamy, 2010). The participants in the study were female undergraduate students, 18–20 years old, who were recruited to take part in research on product evaluation. Each woman was taken to a waiting room with background music playing. For some of the women, the music was a popular love song and for the others, the music was a neutral song. After 3 minutes the participant was moved to another room in which a 20-year-old man was already waiting. The men were part of the study and were selected because they previously had been rated as average in attractiveness. The participant and the confederate were instructed to eat two cookies, one organic and one without organic ingredients, and then talk about the differences between the two for a few minutes. After 5 minutes, the experimenter returned to end the study and asked the pair to wait alone for a few minutes. During this time, the man used a scripted line to ask the woman for her phone number.

You should recognize this study as an example of an experiment. The researchers manipulated the type of music to create two treatment conditions and randomly assigned women to conditions to control extraneous variables. They recorded whether each woman gave her number to obtain a set of scores in each treatment and then compared the two sets of scores.

Table 17.1 shows the structure of this design represented by a matrix with the independent variable (different music) determining the rows of the matrix and the two categories for the dependent variable (yes/no) determining the columns. The number in each cell of the matrix is the frequency count showing how many participants are classified in that category. For example, of the 44 women who heard the romantic music, there

were 23 (52.3%) who gave their phone numbers. By comparison, only 12 of the 43 women (27.9%) who heard the neutral music gave their numbers. The researchers would like to use these data to support the argument that romantic background music increases the likelihood that a woman will give her number to a man she has just met.

Although the Guéguen et al. study involves an independent variable (the type of music) and a dependent variable (did she give her number), you should realize that this study is different from any experiment we have considered in the past. Specifically, the study does not produce a numerical score for each participant. Instead, each participant is simply classified into one of two categories (yes or no). The data consist of *frequencies* or *proportions* describing how many individuals are in each category. You should also note that Guéguen et al. want to use a hypothesis test to evaluate the data. The null hypothesis would state that the type of background music has no effect on whether a woman will give her phone number. The hypothesis test would determine whether the sample data provide enough evidence to reject this null hypothesis.

Because there are no numerical scores, it is impossible to compute a mean or a variance for the sample data. Therefore, it is impossible to use any of the familiar hypothesis tests (such as a *t* test or analysis of variance (ANOVA) to determine whether there is a significant difference between the treatment conditions. What is needed is a new hypothesis testing procedure that can be used with non-numerical data.

In this chapter we introduce two hypothesis tests based on the *chi-square* statistic. Unlike earlier tests that require numerical scores (*X* values), the chi-square tests use sample frequencies and proportions to test hypothesis about the corresponding population values.

TABLE 17.1

A frequency distribution table showing the number of participants who answered either yes or no when asked for their phone numbers. One group listened to romantic music while in a waiting room and second group listened to neutral music.

| | | Gave Phone Number? | |
		Yes	No
Type of Music	Romantic	23	21
	Neutral	12	31

17.1 | Introduction to Chi-Square: The Test for Goodness of Fit

■ Parametric and Nonparametric Statistical Tests

All the statistical tests we have examined thus far are designed to test hypotheses about specific population parameters. For example, we used t tests to assess hypotheses about a population mean (μ) or mean difference ($\mu_1 - \mu_2$). In addition, these tests typically make assumptions about other population parameters. Recall that, for analysis of variance (ANOVA), the population distributions are assumed to be normal and homogeneity of variance is required. Because these tests all concern parameters and require assumptions about parameters, they are called *parametric tests*.

Another general characteristic of parametric tests is that they require a numerical score for each individual in the sample. The scores then are added, squared, averaged, and otherwise manipulated using basic arithmetic. In terms of measurement scales, parametric tests require data from an interval or a ratio scale (see Chapter 1).

Often, researchers are confronted with experimental situations that do not conform to the requirements of parametric tests. In these situations, it may not be appropriate to use a parametric test. Remember that when the assumptions of a test are violated, the test may lead to an erroneous interpretation of the data. Fortunately, there are several hypothesis-testing techniques that provide alternatives to parametric tests. These alternatives are called *nonparametric tests*.

In this chapter, we introduce two commonly used examples of nonparametric tests. Both tests are based on a statistic known as chi-square and both tests use sample data to evaluate hypotheses about the proportions or relationships that exist within populations. Note that the two chi-square tests, like most nonparametric tests, do not state hypotheses in terms of a specific parameter and they make few (if any) assumptions about the population distribution. For the latter reason, nonparametric tests sometimes are called *distribution-free tests*.

One of the most obvious differences between parametric and nonparametric tests is the type of data they use. All of the parametric tests that we have examined so far require numerical scores. For nonparametric tests, on the other hand, the participants are usually just classified into categories such as Democrat and Republican, or High, Medium, and Low IQ. Note that these classifications involve measurement on nominal or ordinal scales, and they do not produce numerical values that can be used to calculate means and variances. Instead, the data for many nonparametric tests are simply frequencies—for example, the number of Democrats and the number of Republicans in a sample of $n = 100$ registered voters.

Occasionally, you have a choice between using a parametric and a nonparametric test. Changing to a nonparametric test usually involves transforming the data from numerical scores to nonnumerical categories. For example, you could start with numerical scores measuring self-esteem and create three categories consisting of high, medium, and low self-esteem. In most situations, the parametric test is preferred because it is more likely to detect a real difference or a real relationship. However, there are situations for which transforming scores into categories might be a better choice.

1. Occasionally, it is simpler to obtain category measurements. For example it is easier to classify students as high, medium, or low in leadership ability than to obtain a numerical score measuring each student's ability.

2. The original scores may violate some of the basic assumptions that underlie certain statistical procedures. For example, the *t* tests and ANOVA assume that the data come from normal distributions. Also, the independent-measures tests assume that the different populations all have the same variance (the homogeneity-of-variance assumption). If a researcher suspects that the data do not satisfy these assumptions, it may be safer to transform the scores into categories and use a nonparametric test to evaluate the data.

3. The original scores may have unusually high variance. Variance is a major component of the standard error in the denominator of *t* statistics and the error term in the denominator of *F*-ratios. Thus, large variance can greatly reduce the likelihood that these parametric tests will find significant differences. Converting the scores to categories essentially eliminates the variance. For example, all individuals fit into three categories (high, medium, and low) no matter how variable the original scores are.

4. Occasionally, an experiment produces an undetermined, or infinite, score. For example, a rat may show no sign of solving a particular maze after hundreds of trials. This animal has an infinite, or undetermined, score. Although there is no absolute number that can be assigned, you can say that this rat is in the highest category, and then classify the other scores according to their numerical values.

■ The Chi-Square Test for Goodness of Fit

Parameters such as the mean and the standard deviation are the most common way to describe a population, but there are situations in which a researcher has questions about the proportions or relative frequencies for a distribution. For example:

- How does the number of women lawyers compare with the number of men in the profession?
- Of the two leading brands of cola, which is preferred by most Americans?
- In the past 10 years, has there been a significant change in the proportion of college students who declare a business major?

The name of the test comes from the Greek letter χ (chi, pronounced "kye"), which is used to identify the test statistic.

Note that each of the preceding examples asks a question about proportions in the population. In particular, we are not measuring a numerical score for each individual. Instead, the individuals are simply classified into categories and we want to know what proportion of the population is in each category. The *chi-square test for goodness of fit* is specifically designed to answer this type of question. In general terms, this chi-square test uses the proportions obtained for sample data to test hypotheses about the corresponding proportions in the population.

DEFINITION

> The **chi-square test for goodness of fit** uses sample data to test hypotheses about the shape or proportions of a population distribution. The test determines how well the obtained sample proportions fit the population proportions specified by the null hypothesis.

Recall from Chapter 2 that a frequency distribution is defined as a tabulation of the number of individuals located in each category of the scale of measurement. In a frequency distribution graph, the categories that make up the scale of measurement are listed on the *X*-axis. In a frequency distribution table, the categories are listed in the first column. With chi-square tests, however, it is customary to present the scale of measurement as a series of boxes, with each box corresponding to a separate category on the scale. The frequency corresponding to each category is simply presented as a number written inside the box. Figure 17.1 shows how a distribution of eye colors for a set of $n = 40$ students can be presented as a graph, a table, or a series of boxes. The scale of measurement for this example consists of four categories of eye color (blue, brown, green, other).

■ The Null Hypothesis for the Goodness-of-Fit Test

For the chi-square test of goodness of fit, the null hypothesis specifies the proportion (or percentage) of the population in each category. For example, a hypothesis might state that 50% of all lawyers are men and 50% are women. The simplest way of presenting this hypothesis is to put the hypothesized proportions in the series of boxes representing the scale of measurement:

	Men	Women
H_0:	50%	50%

Although it is conceivable that a researcher could choose any proportions for the null hypothesis, there usually is some well-defined rationale for stating a null hypothesis. Generally, H_0 falls into one of the following categories:

1. **No Preference, Equal Proportions** The null hypothesis often states that there is no preference among the different categories. In this case, H_0 states that the population is divided equally among the categories. For example, a hypothesis stating that there is no preference among the three leading brands of soft drinks would specify a population distribution as follows:

	Brand X	Brand Y	Brand Z
H_0:	$\frac{1}{3}$	$\frac{1}{3}$	$\frac{1}{3}$

(Preferences in the population are equally divided among the three soft drinks.)

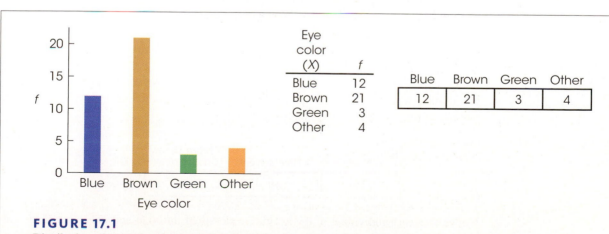

FIGURE 17.1

Distribution of eye colors for a sample of $n = 40$ individuals. The same frequency distribution is shown as a bar graph, as a table, and with the frequencies written in a series of boxes.

The no-preference hypothesis is used in situations in which a researcher wants to determine whether there are any preferences among the categories, or whether the proportions differ from one category to another.

Because the null hypothesis for the goodness-of-fit test specifies an exact distribution for the population, the alternative hypothesis (H_1) simply states that the population distribution has a different shape from that specified in H_0. If the null hypothesis states that the population is equally divided among three categories, the alternative hypothesis says that the population is not divided equally.

2. **No Difference from a Known Population**   The null hypothesis can state that the proportions for one population are not different from the proportions than are known to exist for another population. For example, suppose it is known that 28% of the licensed drivers in the state are younger than 30 years old and 72% are 30 or older. A researcher might wonder whether this same proportion holds for the distribution of speeding tickets. The null hypothesis would state that tickets are handed out equally across the population of drivers, so there is no difference between the age distribution for drivers and the age distribution for speeding tickets. Specifically, the null hypothesis would be

	Tickets Given to Drivers Younger Than 30	Tickets Given to Drivers 30 or Older
H_0:	28%	72%

(Proportions for the population of tickets are not different from proportions for drivers.)

The no-difference hypothesis is used when a specific population distribution is already known. For example, you may have a known distribution from an earlier time, and the question is whether there has been any change in the proportions. Or, you may have a known distribution for one population (drivers) and the question is whether a second population (speeding tickets) has the same proportions.

Again, the alternative hypothesis (H_1) simply states that the population proportions are not equal to the values specified by the null hypothesis. For this example, H_1 would state that the number of speeding tickets is disproportionately high for one age group and disproportionately low for the other.

■ The Data for the Goodness-of-Fit Test

The data for a chi-square test are remarkably simple. There is no need to calculate a sample mean or *SS;* you just select a sample of n individuals and count how many are in each category. The resulting values are called *observed frequencies*. The symbol for observed frequency is f_o. For example, the following data represent observed frequencies for a sample of 40 college students. The students were classified into three categories based on the number of times they reported exercising each week.

No Exercise	1 Time a Week	More Than Once a Week	
15	19	6	$n = 40$

Notice that each individual in the sample is classified into one and only one of the categories. Thus, the frequencies in this example represent three completely separate groups of students: 15 who do not exercise regularly, 19 who average once a week, and 6 who exercise more than once a week. Also note that the observed frequencies add up to the total sample size: $\Sigma f_o = n$. Finally, you should realize that we are not assigning individuals to categories. Instead, we are simply measuring individuals to determine the category in which they belong.

<table>
<tr><td>**DEFINITION**</td><td>The **observed frequency** is the number of individuals from the sample who are classified in a particular category. Each individual is counted in one and only one category.</td></tr>
</table>

■ Expected Frequencies

The general goal of the chi-square test for goodness of fit is to compare the data (the observed frequencies) with the null hypothesis. The problem is to determine how well the data fit the distribution specified in H_0—hence the name *goodness of fit*.

The first step in the chi-square test is to construct a hypothetical sample that represents how the sample distribution would look if it were in perfect agreement with the proportions stated in the null hypothesis. Suppose, for example, the null hypothesis states that the population is distributed in three categories with the following proportions:

	Category A	Category B	Category C
H_0:	25%	50%	25%

(The population is distributed across the three categories with 25% in category A, 50% in category B, and 25% in category C.)

If this hypothesis is correct, how would you expect a random sample of $n = 40$ individuals to be distributed among the three categories? It should be clear that your best strategy is to predict that 25% of the sample would be in category A, 50% would be in category B, and 25% would be in category C. To find the exact frequency expected for each category, multiply the sample size (n) by the proportion (or percentage) from the null hypothesis. For this example, you would expect

$$25\% \text{ of } 40 = 0.25(40) = 10 \text{ individuals in category A}$$

$$50\% \text{ of } 40 = 0.50(40) = 20 \text{ individuals in category B}$$

$$25\% \text{ of } 40 = 0.25(40) = 10 \text{ individuals in category C}$$

The frequency values predicted from the null hypothesis are called *expected frequencies*. The symbol for expected frequency is f_e, and the expected frequency for each category is computed by

$$\text{expected frequency} = f_e = pn \qquad (17.1)$$

where p is the proportion stated in the null hypothesis and n is the sample size.

<table>
<tr><td>**DEFINITION**</td><td>The **expected frequency** for each category is the frequency value that is predicted from the proportions in the null hypothesis and the sample size (n). The expected frequencies define an ideal, *hypothetical* sample distribution that would be obtained if the sample proportions were in perfect agreement with the proportions specified in the null hypothesis.</td></tr>
</table>

Note that the no-preference null hypothesis will always produce equal f_e values for all categories because the proportions (p) are the same for all categories. On the other hand, the no-difference null hypothesis typically will not produce equal values for the expected frequencies because the hypothesized proportions typically vary from one category to another. You also should note that the expected frequencies are calculated, hypothetical values and the numbers that you obtain may be decimals or fractions. The observed frequencies, on the other hand, always represent real individuals and always are whole numbers.

■ The Chi-Square Statistic

The general purpose of any hypothesis test is to determine whether the sample data support or refute a hypothesis about the population. In the chi-square test for goodness of fit, the sample is expressed as a set of observed frequencies (f_o values), and the null hypothesis is used to generate a set of expected frequencies (f_e values). The *chi-square statistic* simply measures how well the data (f_o) fit the hypothesis (f_e). The symbol for the chi-square statistic is χ^2. The formula for the chi-square statistic is

$$\text{chi-square} = \chi^2 = \Sigma \frac{(f_o - f_e)^2}{f_e} \tag{17.2}$$

As the formula indicates, the value of chi-square is computed by the following steps.

1. Find the difference between f_o (the data) and f_e (the hypothesis) for each category.

2. Square the difference. This ensures that all values are positive.

3. Next, divide the squared difference by f_e.

4. Finally, sum the values from all the categories.

The first two steps determine the numerator of the chi-square statistic and should be easy to understand. Specifically, the numerator measures how much difference there is between the data (the f_O values) and the hypothesis (represented by the f_e values). The final step is also reasonable: we add the values to obtain the total discrepancy between the data and the hypothesis. Thus, a large value for chi-square indicates that the data do not fit the hypothesis, and leads us to reject the null hypothesis.

However, the third step, which determines the denominator of the chi-square statistic, is not so obvious. Why must we divide by f_e before we add the category values? The answer to this question is that the obtained discrepancy between f_o and f_e is viewed as *relatively* large or *relatively* small depending on the size of the expected frequency. This point is demonstrated in the following analogy.

Suppose you were going to throw a party and you *expected* 1,000 people to show up. However, at the party you counted the number of guests and *observed* that 1,040 actually showed up. Forty more guests than expected are no major problem when all along you were planning for 1,000. There will still probably by enough beer and potato chips for everyone. On the other hand, suppose you had a party and you expected 10 people to attend but instead 50 actually showed up. Forty more guests in this case spell big trouble. How "significant" the discrepancy is depends in part on what you were originally expecting. With very large expected frequencies, allowances are made for more error between f_o and f_e. This is accomplished in the chi-square formula by dividing the squared discrepancy for each category, $(f_o - f_e)^2$, by its expected frequency.

LEARNING CHECK

1. Occasionally researchers will transform numerical scores into nonnumerical categories and use a nonparametric test instead of the standard parametric statistic. Which of the following is not a reason for making this transformation?
   a. The original scores have very high variance.
   b. The original scores form a very large sample.
   c. The original scores contain an undetermined or infinite value.
   d. The original scores violate an assumption of the parametric test.

2. The data for a chi-square test for goodness of fit are called _____.
   a. observed proportions
   b. observed frequencies
   c. expected proportions
   d. expected frequencies

3. Data from a recent census indicates that 54% of the adults in the city are women. A researcher selects a sample of $n = 200$ people who have been selected for jury duty during the past month to determine whether the gender distribution for jurors is significantly different from the distribution for the general population. For this test, what is the expected frequency for female jurors?
   a. 50
   b. 54
   c. 100
   d. 108

ANSWERS   1. B, 2. B, 3. D

17.2 | An Example of the Chi-Square Test for Goodness of Fit

LEARNING OBJECTIVES

4. Define the degrees of freedom for the chi-square test for goodness of fit and locate the critical value for a specific alpha level in the chi-square distribution.

5. Conduct a chi-square test for goodness of fit.

■ The Chi-Square Distribution and Degrees of Freedom

It should be clear from the chi-square formula that the numerical value of chi-square is a measure of the discrepancy between the observed frequencies (data) and the expected frequencies (H_0). As usual, the sample data are not expected to provide a perfectly accurate representation of the population. In this case, the proportions or observed frequencies in the sample are not expected to be exactly equal to the proportions in the population. Thus, if there are small discrepancies between the f_o and f_e values, we obtain a small value for chi-square and we conclude that there is a good fit between the data and the hypothesis (fail to reject H_0). However, when there are large discrepancies between f_o and f_e, we obtain a large value for chi-square and conclude that the data do not fit the hypothesis (reject H_0).

To decide whether a particular chi-square value is "large" or "small," we must refer to a *chi-square distribution*. This distribution is the set of chi-square values for all the possible random samples when H_0 is true. Much like other distributions we have examined (*t* distribution, *F* distribution), the chi-square distribution is a theoretical distribution with well-defined characteristics. Some of these characteristics are easy to infer from the chi-square formula.

1. The formula for chi-square involves adding squared values, so you can never obtain a negative value. Thus, all chi-square values are zero or larger.

2. When H_0 is true, you expect the data (f_o values) to be close to the hypothesis (f_e values). Thus, we expect chi-square values to be small when H_0 is true.

These two factors suggest that the typical chi-square distribution will be positively skewed (Figure 17.2). Note that small values, near zero, are expected when H_0 is true and large values (in the right-hand tail) are very unlikely. Thus, unusually large values of chi-square form the critical region for the hypothesis test.

Although the typical chi-square distribution is positively skewed, there is one other factor that plays a role in the exact shape of the chi-square distribution—the number of categories. Recall that the chi-square formula requires that you add values from every category. The more categories you have, the more likely it is that you will obtain a large sum for the chi-square value. On average, chi-square will be larger when you are adding values from 10 categories than when you are adding values from only 3 categories. As a result, there is a whole family of chi-square distributions, with the exact shape of each distribution determined by the number of categories used in the study. Technically, each specific chi-square distribution is identified by degrees of freedom (*df*) rather than the number of categories. For the goodness-of-fit test, the degrees of freedom are determined by

$$df = C - 1 \tag{17.3}$$

Caution: **The *df* for a chi-square test is *not* related to sample size (*n*), as it is in most other tests.**

where *C* is the number of categories. A brief discussion of this *df* formula is presented in Box 17.1. Figure 17.3 shows the general relationship between *df* and the shape of the chi-square distribution. Note that the chi-square values tend to get larger (shift to the right) as the number of categories and the degrees of freedom increase.

BOX 17.1 A Closer Look at Degrees of Freedom

Degrees of freedom for the chi-square test literally measure the number of free choices that exist when you are determining the null hypothesis or the expected frequencies. For example, when you are classifying individuals into three categories, you have exactly two free choices in stating the null hypothesis. You may select any two proportions for the first two categories, but then the third proportion is determined. If you hypothesize 25% in the first category and 50% in the second category, then the third category must be 25% to account for 100% of the population.

Category A	Category B	Category C
25%	50%	?

In general, you are free to select proportions for all but one of the categories, but then the final proportion is determined by the fact that the entire set must total 100%. Thus, you have $C - 1$ free choices, where *C* is the number of categories: Degrees of freedom, *df*, equal $C - 1$.

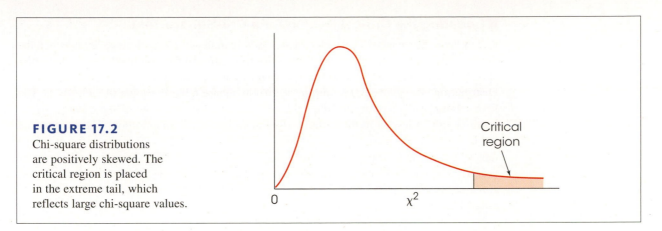

FIGURE 17.2
Chi-square distributions are positively skewed. The critical region is placed in the extreme tail, which reflects large chi-square values.

The following example is an opportunity to test your understanding of the expected frequencies and the *df* value for the chi-square test for goodness of fit.

EXAMPLE 17.1   A researcher has developed three different designs for a computer keyboard. A sample of $n = 60$ participants is obtained, and each individual tests all three keyboards and identifies his or her favorite. The frequency distribution of preferences is as follows:

Design A	Design B	Design C
23	12	25	$n = 60$

Assume that the null hypothesis states that there are no preferences among the three designs. Find the expected frequencies for the chi-square test and determine the *df* value for the chi-square statistic. You should find that $f_e = 20$ for all three keyboard designs and $df = 2$. ■

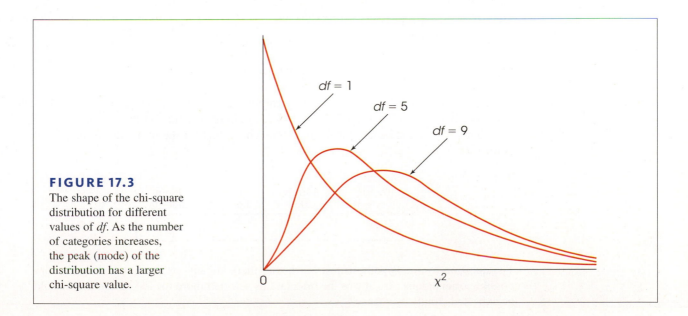

FIGURE 17.3
The shape of the chi-square distribution for different values of *df*. As the number of categories increases, the peak (mode) of the distribution has a larger chi-square value.

■ Locating the Critical Region for a Chi-Square Test

Recall that a large value for the chi-square statistic indicates a big discrepancy between the data and the hypothesis, and suggests that we reject H_0. To determine whether a particular chi-square value is significantly large, you must consult the table entitled The Chi-Square Distribution (Appendix B). A portion of the chi-square table is shown in Table 17.2. The first column lists df values for the chi-square test, and the top row of the table lists proportions (alpha levels) in the extreme right-hand tail of the distribution. The numbers in the body of the table are the critical values of chi-square. The table shows, for example, that when the null hypothesis is true and $df = 3$, only 5% (.05) of the chi-square values are greater than 7.81, and only 1% (.01) are greater than 11.34. Thus, with $df = 3$, any chi-square value greater than 7.81 has a probability of $p < .05$, and any value greater than 11.34 has a probability of $p < .01$.

TABLE 17.2

A portion of the table of critical values for the chi-square distribution.

df	Proportion in Critical Region				
	0.10	0.05	0.025	0.01	0.005
1	2.71	3.84	5.02	6.63	7.88
2	4.61	5.99	7.38	9.21	10.60
3	6.25	7.81	9.35	11.34	12.84
4	7.78	9.49	11.14	13.28	14.86
5	9.24	11.07	12.83	15.09	16.75
6	10.64	12.59	14.45	16.81	18.55
7	12.02	14.07	16.01	18.48	20.28
8	13.36	15.51	17.53	20.09	21.96
9	14.68	16.92	19.02	21.67	23.59

■ A Complete Chi-Square Test for Goodness of Fit

We use the same step-by-step process for testing hypotheses with chi-square as we used for other hypothesis tests. In general, the steps consist of stating the hypotheses, locating the critical region, computing the test statistic, and making a decision about H_0. The following example demonstrates the complete process of hypothesis testing with the goodness-of-fit test.

EXAMPLE 17.2

A psychologist examining art appreciation selected an abstract painting that had no obvious top or bottom. Hangers were placed on the painting so that it could be hung with any one of the four sides at the top. The painting was shown to a sample of $n = 50$ participants, and each was asked to hang the painting in the orientation that looked correct. The following data indicate how many people chose each of the four sides to be placed at the top:

Top Up (Correct)	Bottom Up	Left Side Up	Right Side Up
18	17	7	8

The question for the hypothesis test is whether there are any preferences among the four possible orientations. Are any of the orientations selected more (or less) often than would be expected simply by chance?

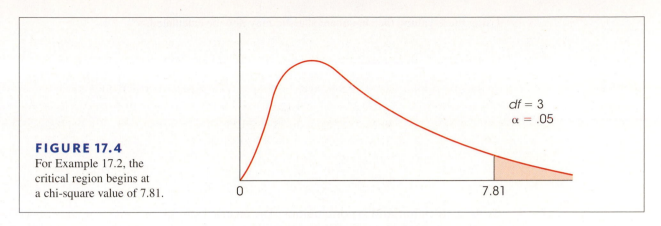

FIGURE 17.4
For Example 17.2, the critical region begins at a chi-square value of 7.81.

STEP 1 **State the hypotheses and select an alpha level** The hypotheses can be stated as follows:

H_0: In the general population, there is no preference for any specific orientation. Thus, the four possible orientations are selected equally often, and the population distribution has the following proportions:

Top Up (Correct)	Bottom Up	Left Side Up	Right Side Up
25%	25%	25%	25%

Expected frequencies are computed and may be decimal values. Observed frequencies are always whole numbers.

H_1: In the general population, one or more of the orientations is preferred over the others.

We will use $\alpha = .05$.

STEP 2 **Locate the critical region** For this example the value for degrees of freedom is

$$df = C - 1 = 4 - 1 = 3$$

For $df = 3$ and $\alpha = .05$, the table of critical values for chi-square indicates that the critical χ^2 has a value of 7.81. The critical region is sketched in Figure 17.4.

STEP 3 **Calculate the chi-square statistic** The calculation of chi-square is actually a two-stage process. First, you must compute the expected frequencies from H_0 and then calculate the value of the chi-square statistic. For this example, the null hypothesis specifies that one-quarter of the population ($p = 25\%$) will be in each of the four categories. According to this hypothesis, we should expect one-quarter of the sample to be in each category. With a sample of $n = 50$ individuals, the expected frequency for each category is

$$f_e = pn = \frac{1}{4}(50) = 12.5$$

The observed frequencies and the expected frequencies are presented in Table 17.3.

TABLE 17.3
The observed frequencies and the expected frequencies for the chi-square test in Example 17.2.

	Top Up (Correct)	Bottom Up	Left Side Up	Right Side Up
Observed Frequencies	18	17	7	8
Expected Frequencies	12.5	12.5	12.5	12.5

Using these values, the chi-square statistic may now be calculated.

$$\chi^2 = \Sigma \frac{(f_o - f_e)^2}{f_e}$$

$$= \frac{(18 - 12.5)^2}{12.5} + \frac{(17 - 12.5)^2}{12.5} + \frac{(7 - 12.5)^2}{12.5} + \frac{(8 - 12.5)^2}{12.5}$$

$$= \frac{30.25}{12.5} + \frac{20.25}{12.5} + \frac{30.25}{12.5} + \frac{20.25}{12.5}$$

$$= 2.42 + 1.62 + 2.42 + 1.62$$

$$= 8.08$$

STEP 4   State a decision and a conclusion   The obtained chi-square value is in the critical region. Therefore, H_0 is rejected, and the researcher may conclude that the four orientations are not equally likely to be preferred. Instead, there are significant differences among the four orientations, with some selected more often and others less often than would be expected by chance.   ◼

IN THE LITERATURE

Reporting the Results for Chi-Square

APA style specifies the format for reporting the chi-square statistic in scientific journals. For the results of Example 17.2, the report might state:

> The participants showed significant preferences among the four orientations for hanging the painting, $\chi^2(3, n = 50) = 8.08$, $p < .05$.

Note that the form of the report is similar to that of other statistical tests we have examined. Degrees of freedom are indicated in parentheses following the chi-square symbol. Also contained in the parentheses is the sample size (n). This additional information is important because the degrees of freedom value is based on the number of categories (C), not sample size. Next, the calculated value of chi-square is presented, followed by the probability that a Type I error has been committed. Because we obtained an extreme, very unlikely value for the chi-square statistic, the probability is reported as *less than* the alpha level. Additionally, the report may provide the observed frequencies (f_o) for each category. This information may be presented in a simple sentence or in a table.

◼ Goodness of Fit and the Single-Sample *t* Test

We began this chapter with a general discussion of the difference between parametric tests and nonparametric tests. In this context, the chi-square test for goodness of fit is an example of a nonparametric test; that is, it makes no assumptions about the parameters of the population distribution, and it does not require data from an interval or ratio scale. In contrast, the single-sample *t* test introduced in Chapter 9 is an example of a parametric test: It assumes a normal population, it tests hypotheses about the population mean (a parameter), and it requires numerical scores that can be added, squared, divided, and so on.

Although the chi-square test and the single-sample *t* are clearly distinct, they are also very similar. In particular, both tests are intended to use the data from a single sample to test hypotheses about a single population.

The primary factor that determines whether you should use the chi-square test or the *t* test is the type of measurement that is obtained for each participant. If the sample data consist of numerical scores (from an interval or ratio scale), it is appropriate to compute a sample mean and use a t test to evaluate a hypothesis about the population mean. For example, a researcher could measure the IQ for each individual in a sample of registered voters. A t test could then be used to evaluate a hypothesis about the mean IQ for the entire population of registered voters. On the other hand, if the individuals in the sample are classified into nonnumerical categories (on a nominal or ordinal scale), you would use a chi-square test to evaluate a hypothesis about the population proportions. For example, a researcher could classify people according to gender by simply counting the number of males and females in a sample of registered voters. A chi-square test would then be appropriate to evaluate a hypothesis about the population proportions.

LEARNING CHECK

1. A researcher uses a sample of 50 people to test whether they see any differences in picture quality for plasma, LED, and LCD televisions. If the data produce $\chi^2 = 5.75$, then what decision should the researcher make?
   a. Reject H_0 for $\alpha = .05$ but not for $\alpha = .01$.
   b. Reject H_0 for $\alpha = .01$ but not for $\alpha = .05$.
   c. Reject H_0 for either $\alpha = .05$ or $\alpha = .01$.
   d. Fail to reject H_0 for $\alpha = .05$ and $\alpha = .01$.

2. The chi-square distribution is _____.
   a. symmetrical with a mean of zero
   b. positively skewed with all values greater than or equal to zero
   c. negatively skewed with all values greater than or equal to zero
   d. symmetrical with a mean equal to $n - 1$

3. In a chi-square test for goodness of fit _____.
   a. $\Sigma f_e = n$
   b. $\Sigma f_e = \Sigma f_o$
   c. both $\Sigma f_e = n$ and $\Sigma f_e = \Sigma f_o$
   d. neither $\Sigma f_e = n$ nor $\Sigma f_e = \Sigma f_o$

ANSWERS   1. D, **2.** B, **3.** C

17.3 | The Chi-Square Test for Independence

LEARNING OBJECTIVES

6. Define the degrees of freedom for the chi-square test for independence and locate the critical value for a specific alpha level in the chi-square distribution.

7. Describe the hypotheses for a chi-square test for independence and explain how the expected frequencies are obtained.

8. Conduct a chi-square test for independence.

The chi-square statistic may also be used to test whether there is a relationship between two variables. In this situation, each individual in the sample is measured or classified on two separate variables. For example, a group of students could be classified in terms of personality (introvert, extrovert) and in terms of color preference (red, yellow, green, or blue). Usually, the data from this classification are presented in the form of a matrix, where the rows correspond to the categories of one variable and the columns correspond to the categories of the second variable. Table 17.4 presents hypothetical data for a sample of $n = 200$ students who have been classified by personality and color preference. The number in each box, or cell, of the matrix indicates the frequency, or number of individuals in that particular group. In Table 17.4, for example, there are 10 students who were classified as introverted and who selected red as their preferred color. To obtain these data, the researcher first selects a random sample of $n = 200$ students. Each student is then given a personality test and is asked to select a preferred color from among the four choices. Note that the classification is based on the measurements for each student; the researcher does not assign students to categories. Also, note that the data consist of frequencies, not scores, from a sample. The goal is to use the frequencies from the sample to test a hypothesis about the population frequency distribution. Specifically, are these data sufficient to conclude that there is a significant relationship between personality and color preference in the population of students.

TABLE 17.4

Color preferences according to personality types.

	Red	Yellow	Green	Blue	
Introvert	10	3	15	22	50
Extrovert	90	17	25	18	150
	100	20	40	40	$n = 200$

You should realize that the color preference study shown in Table 17.4 is an example of nonexperimental research (Chapter 1, page 13). The researcher did not manipulate any variable and the participants were not randomly assigned to groups or treatment conditions. However, similar data are often obtained from true experiments. A good example is the study described in the Preview, in which Guéguen, Jacob, and Lamy (2010) demonstrate that romantic background music increases the likelihood that a woman will give her phone number to a man she has just met. The researchers manipulated the type of background music and recorded the number of Yes and No responses for each type of music (see Table 17.1, page 560). As with the color preference data, the researchers would like to use the frequencies from the sample to test a hypothesis about the corresponding frequency distribution in the population. In this case, the researchers would like to know whether the sample data provide enough evidence to conclude that there is a significant relationship between the type of music and a woman's response to a request for her phone number.

The procedure for using sample frequencies to evaluate hypotheses concerning relationships between variables involves another test using the chi-square statistic. In this situation, however, the test is called the *chi-square test for independence*.

DEFINITION

The **chi-square test for independence** uses the frequency data from a sample to evaluate the relationship between two variables in the population. Each individual in the sample is classified on both of the two variables, creating a two-dimensional frequency distribution matrix. The frequency distribution for the sample is then used to test hypotheses about the corresponding frequency distribution in the population.

■ The Null Hypothesis for the Test for Independence

The null hypothesis for the chi-square test for independence states that the two variables being measured are independent; that is, for each individual, the value obtained for one variable is not related to (or influenced by) the value for the second variable. This general hypothesis can be expressed in two different conceptual forms, each viewing the data and the test from slightly different perspectives. The data in Table 17.4 describing color preference and personality are used to present both versions of the null hypothesis.

H_0 version 1:   For this version of H_0, the data are viewed as a single sample with each individual measured on two variables. The goal of the chi-square test is to evaluate the relationship between the two variables. For the example we are considering, the goal is to determine whether there is a consistent, predictable relationship between personality and color preference. That is, if I know your personality, will it help me to predict your color preference? The null hypothesis states that there is no relationship. The alternative hypothesis, H_1, states that there is a relationship between the two variables.

> H_0:   For the general population of students, there is no relationship between color preference and personality.

This version of H_0 demonstrates the similarity between the chi-square test for independence and a correlation. In each case, the data consist of two measurements (X and Y) for each individual, and the goal is to evaluate the relationship between the two variables. The correlation, however, requires numerical scores for X and Y. The chi-square test, on the other hand, simply uses frequencies for individuals classified into categories.

H_0 version 2:   For this version of H_0, the data are viewed as two (or more) separate samples representing two (or more) populations or treatment conditions. The goal of the chi-square test is to determine whether there are significant differences between the populations. For the example we are considering, the data in Table 17.4 would be viewed as a sample of $n = 50$ introverts (top row) and a separate sample of $n = 150$ extroverts (bottom row). The chi-square test will determine whether the distribution of color preferences for introverts is significantly different from the distribution of color preferences for extroverts. From this perspective, the null hypothesis is stated as follows:

> H_0:   In the population of students, the proportions in the distribution of color preferences for introverts are not different from the proportions in the distribution of color preferences for extroverts. The two distributions have the same shape (same proportions).

This version of H_0 demonstrates the similarity between the chi-square test and an independent-measures t test (or ANOVA). In each case, the data consist of two (or more) separate samples that are being used to test for differences between two (or more) populations. The t test (or ANOVA) requires numerical scores to compute means and mean differences. However, the chi-square test simply uses frequencies for individuals classified into categories. The null hypothesis for the chi-square test states that the populations have the same proportions (same shape). The alternative hypothesis, H_1, simply states that the populations have different proportions. For the example we are considering, H_1 states that the shape of the distribution of color preferences for introverts is different from the shape of the distribution of color preferences for extroverts.

Equivalence of H_0 version 1 and H_0 version 2:   Although we have presented two different statements of the null hypothesis, these two versions are equivalent. The first version of H_0 states that color preference is not related to personality. If this hypothesis is correct, then the distribution of color preferences should not depend on personality. In other words, the distribution of color preferences should have the same proportions for introverts and for extroverts, which is the second version of H_0.

For example, if we found that 60% of the introverts preferred red, then H_0 would predict that we also should find that 60% of the extroverts prefer red. In this case, knowing that an individual prefers red does not help you predict his/her personality. Note that finding the *same proportions* indicates *no relationship*.

On the other hand, if the proportions were different, it would suggest that there is a relationship. For example, if red is preferred by 60% of the extroverts but only 10% of the introverts, then there is a clear, predictable relationship between personality and color preference. (If I know your personality, I can predict your color preference.) Thus, finding *different proportions* means that there is *a relationship* between the two variables.

DEFINITION

> Two variables are **independent** when there is no consistent, predictable relationship between them. In this case, the frequency distribution for one variable is not related to (or dependent on) the categories of the second variable. As a result, when two variables are independent, the frequency distribution for one variable will have the same shape (same proportions) for all categories of the second variable.

Thus, stating that there is no relationship between two variables (version 1 of H_0) is equivalent to stating that the distributions have equal proportions (version 2 of H_0).

■ Observed and Expected Frequencies

The chi-square test for independence uses the same basic logic that was used for the goodness-of-fit test. First, a sample is selected, and each individual is classified or categorized. Because the test for independence considers two variables, every individual is classified on both variables, and the resulting frequency distribution is presented as a two-dimensional matrix (see Table 17.4). As before, the frequencies in the sample distribution are called *observed frequencies* and are identified by the symbol f_o.

The next step is to find the expected frequencies, or f_e values, for this chi-square test. As before, the *expected frequencies* define an ideal hypothetical distribution that is in perfect agreement with the null hypothesis. Once the expected frequencies are obtained, we compute a chi-square statistic to determine how well the data (observed frequencies) fit the null hypothesis (expected frequencies).

Although you can use either version of the null hypothesis to find the expected frequencies, the logic of the process is much easier when you use H_0 stated in terms of equal proportions. For the example we are considering, the null hypothesis states

H_0:   The frequency distribution of color preference has the same shape (same proportions) for both categories of personality.

To find the expected frequencies, we first determine the overall distribution of color preferences and then apply this distribution to both categories of personality. Table 17.5 shows an empty matrix corresponding to the data from Table 17.4. Notice that the empty matrix includes all of the row totals and column totals from the original sample data. The row totals and column totals are essential for computing the expected frequencies.

The column totals for the matrix describe the overall distribution of color preferences. For these data, 100 people selected red as their preferred color. Because the total sample

TABLE 17.5

An empty frequency distribution matrix showing only the row totals and column totals. (These numbers describe the basic characteristics of the sample from Table 17.4.)

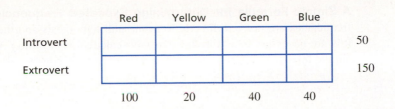

	Red	Yellow	Green	Blue	
Introvert					50
Extrovert					150
	100	20	40	40	

consists of 200 people, the proportion selecting red is 100 out of 200, or 50%. The complete set of color preference proportions is as follows:

100 out of 200 = 50% prefer red

20 out of 200 = 10% prefer yellow

40 out of 200 = 20% prefer green

40 out of 200 = 20% prefer blue

The row totals in the matrix define the two samples of personality types. For example, the matrix in Table 17.5 shows a total of 50 introverts (the top row) and a sample of 150 extroverts (the bottom row). According to the null hypothesis, both personality groups should have the same proportions for color preferences. To find the expected frequencies, we simply apply the overall distribution of color preferences to each sample. Beginning with the sample of 50 introverts in the top row, we obtain expected frequencies of

50% prefer red:	$f_e = 50\%$ of $50 = 0.50(50) = 25$
10% prefer yellow:	$f_e = 10\%$ of $50 = 0.10(50) = 5$
20% prefer green:	$f_e = 20\%$ of $50 = 0.20(50) = 10$
20% prefer blue:	$f_e = 20\%$ of $50 = 0.20(50) = 10$

Using exactly the same proportions for the sample of $n = 150$ extroverts in the bottom row, we obtain expected frequencies of

50% prefer red:	$f_e = 50\%$ of $150 = 0.50(50) = 75$
10% prefer yellow:	$f_e = 10\%$ of $150 = 0.10(50) = 15$
20% prefer green:	$f_e = 20\%$ of $150 = 0.20(50) = 30$
20% prefer blue:	$f_e = 20\%$ of $150 = 0.20(50) = 30$

The complete set of expected frequencies is shown in Table 17.6. Notice that the row totals and the column totals for the expected frequencies are the same as those for the original data (the observed frequencies) in Table 17.4.

TABLE 17.6

Expected frequencies corresponding to the data in Table 17.4. (This is the distribution predicted by the null hypothesis.)

	Red	Yellow	Green	Blue	
Introvert	25	5	10	10	50
Extrovert	75	15	30	30	150
	100	20	40	40	

A Simple Formula for Determining Expected Frequencies   Although expected frequencies are derived directly from the null hypothesis and the sample characteristics, it is not necessary to go through extensive calculations to find f_e values. In fact, there is a simple formula that determines f_e for any cell in the frequency distribution matrix:

$$f_e = \frac{f_c f_r}{n}$$

(17.4)

where f_c is the frequency total for the column (column total), f_r is the frequency total for the row (row total), and n is the number of individuals in the entire sample. To demonstrate this formula, we compute the expected frequency for introverts selecting yellow in Table 17.6. First, note that this cell is located in the top row and second column in the table. The column total is $f_c = 20$, the row total is $f_r = 50$, and the sample size is $n = 200$. Using these values in formula 17.4, we obtain

$$f_e = \frac{f_c f_r}{n} = \frac{20(50)}{200} = 5$$

This is identical to the expected frequency we obtained using percentages from the overall distribution.

■ The Chi-Square Statistic and Degrees of Freedom

The chi-square test of independence uses exactly the same chi-square formula as the test for goodness of fit:

$$\chi^2 = \Sigma \frac{(f_o - f_e)^2}{f_e}$$

As before, the formula measures the discrepancy between the data (f_o values) and the hypothesis (f_e values). A large discrepancy produces a large value for chi-square and indicates that H_0 should be rejected. To determine whether a particular chi-square statistic is significantly large, you must first determine degrees of freedom (df) for the statistic and then consult the chi-square distribution in the appendix. For the chi-square test of independence, degrees of freedom are based on the number of cells for which you can freely choose expected frequencies. Recall that the f_e values are partially determined by the sample size (n) and by the row totals and column totals from the original data. These various totals restrict your freedom in selecting expected frequencies. This point is illustrated in Table 17.7. Once three of the f_e values have been selected, all the other f_e values in the table are also determined. In general, the row totals and the column totals restrict the final choices in each row and column. As a result, we may freely choose all but one f_e in each row and all but one f_e in each column. If R is the number of rows and C is the number of columns, and you remove the last column and the bottom row from the matrix, you are left with a smaller matrix that has $C - 1$ columns and $R - 1$ rows. The number of cells in the smaller matrix determines the df value. Thus, the total number of f_e values that you can freely choose is $(R - 1)(C - 1)$, and the degrees of freedom for the chi-square test of independence are given by the formula

$$df = (R - 1)(C - 1)$$

(17.5)

Also note that once you calculate the expected frequencies to fill the smaller matrix, the rest of the f_e values can be found by subtraction.

TABLE 17.7

Degrees of freedom and expected frequencies. (Once three values have been selected, all the remaining expected frequencies are determined by the row totals and the column totals. This example has only three free choices, so $df = 3$.)

	Red	Yellow	Green	Blue	
	25	5	10	?	50
	?	?	?	?	150
	100	20	40	40	

The following example is an opportunity to test your understanding of the expected frequencies and the df value for the chi-square test for independence.

EXAMPLE 17.3

A researcher would like to know which factors are most important to people who are buying a new car. A sample of $n = 200$ customers between the ages of 20 and 29 are asked to identify the most important factor in the decision process: Performance, Reliability, or Style. The researcher would like to know whether there is a difference between the factors identified by women compared to those identified by men. The data are as follows:

Observed Frequencies of Most Important
Factor According to Gender

	Performance	Reliability	Style	Totals
Male	21	33	26	80
Female	19	67	34	120
Totals	40	100	60	

Compute the expected frequencies and determine the value for df for the chi-square test. You should find expected frequencies of 16, 40, and 24 for the males and 24, 60, and 36 for the females. $df = 2$. ∎

■ An Example of the Chi-Square Test for Independenedence

The following example demonstrates the complete hypothesis-testing procedure for the chi-square test for independence.

EXAMPLE 17.4

Many parents allow their underage children to drink alcohol in limited situations when an adult is present to supervise. The idea is that teens will learn responsible drinking habits if they first experience alcohol in a controlled environment. Other parents take a strict no-drinking approach with the idea that they are sending a clear message about what is right and what is wrong. Recent research, however, suggests that the more permissive approach may actually result in more negative consequences (McMorris et al., 2011). In the study, teens who were allowed to drink with their parents were significantly more likely to experienced alcohol-related problems than teens who were not allowed to drink. In an attempt to replicate this study, researchers surveyed a sample of 150 students each year from ages 14–17. The students were asked about their alcohol use and about alcohol-related problems such as binge drinking, fights, and blackouts. The results are shown in Table 17.8. Do the data show a significant relationship between the parents' rules about alcohol and subsequent alcohol-related problems?

TABLE 17.8

A frequency distribution showing experience with alcohol-related problems according to parents' rules concerning underage drinking.

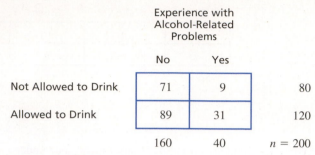

	Experience with Alcohol-Related Problems		
	No	Yes	
Not Allowed to Drink	71	9	80
Allowed to Drink	89	31	120
	160	40	$n = 200$

STEP 1 State the hypotheses, and select a level of significance. According to the null hypothesis, the two variables are independent. This general hypothesis can be stated in two different ways:

Version 1

H_0: In the general population, there is no relationship between parents' rules for alcohol use and the development of alcohol-related problems.

This version of H_0 emphasizes the similarity between the chi-square test and a correlation.

Version 2

H_0: In the general population, the distribution of alcohol-related problems has the same proportions for teenagers whose parents permit drinking and for those whose parents do not.

The second version of H_0 emphasizes the similarity between the chi-square test and the independent-measures t test.

   The alternative hypothesis states that there is a relationship between the two variables (version 1), or that the two distributions are different (version 2). Remember that the two versions for the hypotheses are equivalent. The choice between them is largely determined by how the researcher wants to describe the outcome. For example, a researcher may want to emphasize the *relationship* between variables or the *difference* between groups.

   For this test, we use $\alpha = .05$.

STEP 2 Determine the degrees of freedom and locate the critical region. For the chi-square test for independence,

$$df = (R - 1)(C - 1) = (2 - 1)(2 - 1) = 1$$

With $df = 1$ and $\alpha = .05$, the critical value for chi-square is 3.84 (see Table B.8, p. 659).

STEP 3 Determine the expected frequencies, and compute the chi-square statistic. The following table shows an empty matrix with the same row totals and column totals as the original data. The expected frequencies must maintain the same row totals and column totals, and create an ideal frequency distribution that perfectly represents the null hypothesis. Specifically, the proportions for the group of 80 teens for whom alcohol was not allowed must be the same as the proportions for the group of 120 who were permitted to drink.

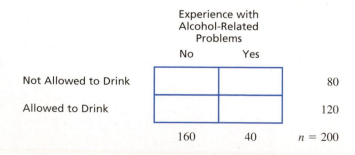

	Experience with Alcohol-Related Problems		
	No	Yes	
Not Allowed to Drink			80
Allowed to Drink			120
	160	40	$n = 200$

The column totals describe the overall distribution of alcohol-related problems. These totals indicate that 160 out of 200 participants reported that they did not experience any alcohol-related problems. This proportion corresponds to 160/200, or 80% of the total sample. Similarly, 40/200, or 20% reported that they did experience alcohol-related problems. The null hypothesis (version 2) states that these proportions are the same for both groups of participants. Therefore, we simply apply the proportions to each group to obtain the expected frequencies. For the group of 80 teens who were not allowed to drink (top row), we obtain

80% of 80 = 64 expected to experience problems

20% of 80 = 16 expected not to experience problems

For the group of 120 teens who were allowed to drink (bottom row), we expect

80% of 120 = 96 expected to experience problems

20% of 120 = 24 expected not to experience problems

These expected frequencies are summarized in Table 17.9. Note that the expected frequencies maintain the same row totals and column totals, and create an ideal frequency distribution that perfectly represents the null hypothesis. Specifically, the proportions for the group of 80 teens for whom alcohol was not allowed are the same as the proportions for the group of 120 who were permitted to drink.

TABLE 17.9

The expected frequencies (f_e values) if experience with alcohol-related problems is completely independent of parents' rules concerning underage drinking.

	Experience with Alcohol-Related Problems		
	No	Yes	
Not Allowed to Drink	64	16	80
Allowed to Drink	96	24	120
	160	40	$n = 200$

The chi-square statistic is now used to measure the discrepancy between the data (the observed frequencies in Table 17.8) and the null hypothesis that was used to generate the expected frequencies in Table 17.9.

$$\chi^2 = \frac{(71 - 64)^2}{64} + \frac{(9 - 16)^2}{16} + \frac{(89 - 96)^2}{96} + \frac{(31 - 24)^2}{24}$$

$$= 0.766 + 3.063 + 0.510 + 2.042$$

$$= 6.381$$

STEP 4   Make a decision regarding the null hypothesis and the outcome of the study. The obtained chi-square value exceeds the critical value (3.84). Therefore, the decision is to reject the null hypothesis. In the literature, this would be reported as a significant result with $\chi^2(1, n = 200) = 6.381$, $p < .05$. According to version 1 of H_0, this means that we have decided that there is a significant relationship between parents' rules about alcohol and subsequent problems. Expressed in terms of version 2 of H_0, the data show a significant difference in alcohol-related problems between teens whose parents allowed drinking and those whose parents did not. To describe the details of the significant result, you must compare the original data (Table 17.8) with the expected frequencies in Table 17.9. Looking at the two tables, it should be clear that teens whose parents allowed drinking experienced more alcohol-related problems than would be expected if the two variables were independent. ∎

LEARNING CHECK

1. If a chi-square test for goodness of fit has $df = 1$ then the individuals are classified into ___ categories and if a chi-square test for independence has $df = 1$ then the individuals are classified into ___ categories.

   a. 2, 2
   b. 2, 4
   c. 1, 4
   d. 1, 1

2. A sample of 100 people is classified by gender (male/female) and by whether or not they are registered voters. The sample consists of 60 females, of whom 50 are registered voters, and 40 males, of whom 25 are registered voters. If these data were used for a chi-square test for independence, the expected frequency for registered males would be _____.

   a. 15
   b. 25
   c. 30
   d. 45

3. A chi-square test for independence is conducted for data from a 2×2 data matrix with a total of $n = 40$ participants. If the data produce $\chi^2 = 4.00$, then what decision should be made?

   a. Reject H_0 with $\alpha = 0.5$ but not with $\alpha = .01$.
   b. Reject H_0 with $\alpha = 0.1$ but not with $\alpha = .05$.
   c. Reject H_0 with either $\alpha = 0.5$ or $\alpha = .01$.
   d. Fail to reject H_0 with $\alpha = 0.5$ or with $\alpha = .01$.

ANSWERS   1. B, 2. C, 3. A

17.4 | Effect Size and Assumptions for the Chi-Square Tests

LEARNING OBJECTIVES

9. Compute Cohen's w to measure effect size for both chi-square tests.

10. Compute the phi-coefficient or Cramér's V to measure effect size for the chi-square test for independence.

11. Identify the basic assumptions and restrictions for chi-square tests.

■ Cohen's w

Hypothesis tests, like the chi-square test for goodness of fit or for independence, evaluate the statistical significance of the results from a research study. Specifically, the intent of the test is to determine whether it is likely that the patterns or relationships observed in the sample data could have occurred without any corresponding patterns or relationships in the population. Tests of significance are influenced not only by the size or strength of the treatment effects but also by the size of the samples. As a result, even a small effect can be statistically significant if it is observed in a very large sample. Because a significant effect does not necessarily mean a large effect, it is generally recommended that the outcome of a hypothesis test be accompanied by a measure of the effect size. This general recommendation also applies to the chi-square tests presented in this chapter.

Cohen (1992) introduced a statistic called w that provides a measure of effect size for either of the chi-square tests. The formula for **Cohen's w** is very similar to the chi-square formula but uses proportions instead of frequencies.

$$w = \sqrt{\sum \frac{(P_o - P_e)^2}{P_e}} \qquad (17.6)$$

In the formula, the P_o values are the observed proportions in the data and are obtained by dividing each observed frequency by the total number of participants.

$$\text{observed proportion} = P_o = \frac{f_o}{n}$$

Similarly, the P_e values are the expected proportions that are specified in the null hypothesis. The formula instructs you to

1. Compute the difference between the observed proportion and the expected proportion for each cell (category).

2. For each cell, square the difference and divide by the expected proportion.

3. Add the values from step 2 and take the square root of the sum.

The following example demonstrates this process.

EXAMPLE 17.5

A researcher would like to determine whether students have any preferences among four pizza shops in town. A sample of $n = 40$ students is obtained and fresh pizza is ordered from each of the four shops. Each student tastes all four pizzas and then selects a favorite. The observed frequencies are as follows:

Shop A	Shop B	Shop C	Shop D	
6	12	8	14	40

The null hypothesis says that there are no preferences among the four shops so the expected proportion is $P = 0.25$ for each. The observed proportions are $6/40 = 0.15$ for shop A, $12/40 = 0.30$ for shop B, $8/40 = 0.20$ for shop C, and $14/40 = 0.35$ for shop D. The calculations for w are summarized in the table.

	P_o	P_e	$(P_o - P_e)$	$(P_o - P_e)^2$	$(P_o - P_e)^2/P_e$
Shop A	0.15	0.25	0.10	0.01	0.04
Shop B	0.30	0.25	−0.05	0.0025	0.01
Shop C	0.20	0.25	0.05	0.0025	0.01
Shop D	0.35	0.25	−0.10	0.01	0.04
					0.10

$$\sum \frac{(P_o - P_e)^2}{P_e} = 0.10 \text{ and } w = \sqrt{0.10} = 0.316$$

Cohen (1992) also suggested guidelines for interpreting the magnitude of w, with values near 0.10 indicating a small effect, 0.30 a medium effect, and 0.50 a large effect. By these standards, the value obtained in Example 17.5 is a medium effect.

The Role of Sample Size   You may have noticed that the formula for computing w does not contain any reference to the sample size. Instead, w is calculated using only the sample proportions and the proportions from the null hypothesis. As a result, the size of the sample has no influence on the magnitude of w. This is one of the basic characteristics of all measures of effect size. Specifically, the number of scores in the sample has little or no influence on effect size. On the other hand, sample size does have a large impact on the outcome of a hypothesis test. For example, the data in Example 17.5 produce $\chi^2 = 4.00$. With $df = 3$, the critical value for $\alpha = 0.5$ is 7.81 and we conclude that there are no significant preferences among the four pizza shops. However, if the number of individuals in each category is doubled, so that the observed frequencies become 12, 24, 16, and 28, then the new $\chi^2 = 8.00$. Now the statistic is in the critical region so we reject H_0 and conclude that there are significant preferences. Thus, increasing the size of the sample increases the likelihood of rejecting the null hypothesis. You should realize, however, that the proportions for the new sample are exactly the same as the proportions for the original sample, so the value of w does not change. For both sample sizes, $w = 0.316$.

Chi-square and w   Although the chi-square statistic and effect size as measured by w are intended for different purposes and are affected by different factors, they are algebraically related. In particular, the portion of the formula for w that is under the square root, can be obtained by dividing the formula for chi-square by n. Dividing by the sample size converts each of frequencies (observed and expected) into a proportion, which produces the formula for w. As a result, you can determine the value of w directly from the chi-square value by the following equation:

$$w = \sqrt{\frac{\chi^2}{n}} \qquad \text{(17.7)}$$

For the data in Example 15.3, we obtained $\chi^2 = 4.00$ and $w = 0.316$. Substituting in the formula, produces

$$w = \sqrt{\frac{\chi^2}{n}} = \sqrt{\frac{4.00}{40}} = \sqrt{0.10} = 0.316$$

Although Cohen's w statistic also can be used to measure effect size for the chi-square test for independence, two other measures have been developed specifically for this hypothesis test. These two measures, known as the phi-coefficient and Cramér's V, make allowances for the size of the data matrix and are considered to be superior to w, especially with very large data matrices.

■ The Phi-Coefficient and Cramér's V

In Chapter 15 (p. 518), we introduced the **phi-coefficient** as a measure of correlation for data consisting of two dichotomous variables (both variables have exactly two values). This same situation exists when the data for a chi-square test for independence form a 2×2 matrix (again, each variable has exactly two values). In this case, it is possible to compute the correlation phi (ϕ) in addition to the chi-square hypothesis test for the same set of data. Because phi is a correlation, it measures the strength of the relationship, rather than the significance, and thus provides a measure of effect size. The value for the phi-coefficient can be computed directly from chi-square by the following formula:

The value of χ^2 is already a squared value. Do not square it again.

$$\phi = \sqrt{\frac{\chi^2}{n}} \qquad \text{(17.8)}$$

Note that the phi-coefficient (Equation 17.8) and Cohen's w (Equation 17.7) are equivalent, and either can be used to measure effect size for a test for independence with a 2×2 matrix. The value obtained for the phi-coefficient (or Cohen's w) is determined entirely by the *proportions* in the 2×2 data matrix and is completely independent of the absolute size of the frequencies. The chi-square value, however, is influenced by the proportions and by the size of the frequencies. This distinction is demonstrated in the following example.

EXAMPLE 17.6

The following data show a frequency distribution evaluating the relationship between gender and preference between two candidates for student president.

	Candidate A	B
Male	5	10
Female	10	5

Note that the data show that males prefer candidate B by a 2-to-1 margin and females prefer candidate A by 2-to-1. Also note that the sample includes a total of 15 males and 15 females. We will not perform all the arithmetic here, but these data produce chi-square equal to 3.33 (which is not significant) and a phi-coefficient of 0.333.

Next we keep exactly the same proportions in the data, but double all of the frequencies. The resulting data are as follows:

	Candidate A	B
Male	10	20
Female	20	10

Once again, males prefer candidate B by 2-to-1 and females prefer candidate A by 2-to-1. However, the sample now contains 30 males and 30 females. For these new data, the value of chi-square is 6.66, twice as big as it was before (and now significant with $\alpha = .05$), but the value of the phi-coefficient is still 0.333.

Because the proportions are the same for the two samples, the value of the phi-coefficient is unchanged. However, the larger sample provides more convincing evidence than the smaller sample, so the larger sample is more likely to produce a significant result. ∎

The interpretation of ϕ follows the same standards used to evaluate a correlation (Table 9.3, p. 283 shows the standards for squared correlations): a correlation of 0.10 is a small effect, 0.30 is a medium effect, and 0.50 is a large effect. Occasionally, the value of ϕ is squared (ϕ^2) and is reported as a percentage of variance accounted for, exactly the same as r^2.

When the chi-square test involves a matrix larger than 2×2, a modification of the phi-coefficient, known as **Cramér's V**, can be used to measure effect size.

$$V = \sqrt{\frac{\chi^2}{n(df^*)}}$$ (17.9)

Note that the formula for Cramér's V (17.9) is identical to the formula for the phi-coefficient (17.8) except for the addition of df^* in the denominator. The df^* value is *not* the same as the degrees of freedom for the chi-square test, but it is related. Recall that the chi-square test for independence has $df = (R - 1)(C - 1)$, where R is the number of rows in the table and C is the number of columns. For Cramér's V, the value of df^* is the smaller of either $(R - 1)$ or $(C - 1)$.

Cohen (1988) also suggested standards for interpreting Cramér's V that are shown in Table 17.10. Note that when $df^* = 1$, as in a 2×2 matrix, the criteria for interpreting V are exactly the same as the criteria for interpreting a regular correlation or a phi-coefficient.

TABLE 17.10

Standards for interpreting Cramér's V as proposed by Cohen (1988).

	Small Effect	Medium Effect	Large Effect
For $df^* = 1$	0.10	0.30	0.50
For $df^* = 2$	0.07	0.21	0.35
For $df^* = 3$	0.06	0.17	0.29

In a research report, the measure of effect size appears immediately after the results of the hypothesis test. For the study in Example 17.4, for example, we obtained $\chi^2 = 6.381$ for a sample of $n = 200$ participants. Because the data form a 2×2 matrix, the phi-coefficient is the appropriate measure of effect size and the data produce

$$\phi = \sqrt{\frac{\chi^2}{n}} = \sqrt{\frac{6.381}{200}} = 0.179$$

For these data, the results from the hypothesis test and the measure of effect size would be reported as follows:

The results showed a significant relationship between parents' rules about alcohol and subsequent alcohol-related problems, $\chi^2(1, n = 200) = 6.381, p < .05, \phi = 0.179$. Specifically, teenagers whose parents allowed supervised drinking were more likely to experience problems.

■ Assumptions and Restrictions for Chi-Square Tests

To use a chi-square test for goodness of fit or a test of independence, several conditions must be satisfied. For any statistical test, violation of assumptions and restrictions casts doubt on the results. For example, the probability of committing a Type I error may be distorted when assumptions of statistical tests are not satisfied. Some important assumptions and restrictions for using chi-square tests are the following.

1. **Independence of Observations** This is *not* to be confused with the concept of independence between *variables,* as seen in the chi-square test for independence (Section 17.3). One consequence of independent observations is that each observed frequency is generated by a different individual. A chi-square test would be inappropriate if a person could produce responses that can be classified in more than one category or contribute more than one frequency count to a single category. (See p. 244 for more information on independence.)

2. **Size of Expected Frequencies** A chi-square test should not be performed when the expected frequency of any cell is less than 5. The chi-square statistic can be distorted when f_e is very small. Consider the chi-square computations for a single cell. Suppose that the cell has values of $f_e = 1$ and $f_o = 5$. Note that there is a 4-point difference between the observed and expected frequencies. However, the total contribution of this cell to the total chi-square value is

$$\text{cell} = \frac{(f_o - f_e)^2}{f_e} = \frac{(5 - 1)^2}{1} = \frac{4^2}{1} = 16$$

Now consider another instance, in which $f_e = 10$ and $f_o = 14$. The difference between the observed and the expected frequencies is still 4, but the contribution of this cell to the total chi-square value differs from that of the first case:

$$\text{cell} = \frac{(f_o - f_e)^2}{f_e} = \frac{(14 - 10)^2}{10} = \frac{4^2}{10} = 1.6$$

It should be clear that a small f_e value can have a great influence on the chi-square value. This problem becomes serious when f_e values are less than 5. When f_e is very small, what would otherwise be a minor discrepancy between f_o and f_e results in large chi-square values. The test is too sensitive when f_e values are extremely small. One way to avoid small expected frequencies is to use large samples.

LEARNING CHECK

1. Effect size, as measured Cohen's w, is determined by _____.
   a. the proportions in the sample data
   b. the proportions specified by the null hypothesis
   c. the proportions in the sample data and the proportions specified by the null hypothesis
   d. neither the proportions in the sample data nor the proportions specified by the null hypothesis

2. Under what circumstances is the phi-coefficient used instead of Cramér's V to measure effect size for the chi-square test for independence?
   a. Only when the data form a 2×2 matrix.
   b. Only when there are at least 3 rows or 3 columns in the data matrix.
   c. Only when there are at least 3 rows and at least 3 columns in the data matrix.
   d. Only when there are more than 3 rows and more than 3 columns in the data matrix.

3. Which of the following describes the assumptions and restrictions for a chi-square test for independence?
   a. Independent observations from a normal distribution.
   b. Independent observations and no expected frequency smaller than 5.
   c. Equal frequencies across each row of the data matrix.
   d. The observations within each row are from a normal distribution.

ANSWERS   **1.** C, **2.** A, **3.** B

17.5 | Special Applications of the Chi-Square Tests

LEARNING OBJECTIVES

12. Explain the similarities and the differences between the chi-square test for independence and the Pearson correlation.

13. Explain the similarities and the differences between the chi-square test for independence and the independent-measures t test or ANOVA.

At the beginning of this chapter, we introduced the chi-square tests as examples of non-parametric tests. Although nonparametric tests serve a function that is uniquely their own, they also can be viewed as alternatives to the common parametric techniques that were examined in earlier chapters. In general, nonparametric tests are used as substitutes for parametric techniques in situations in which one of the following occurs.

1. The data do not meet the assumptions needed for a standard parametric test.
2. The data consist of nominal or ordinal measurements, so that it is impossible to compute standard descriptive statistics such as the mean and standard deviation.

In this section, we examine some of the relationships between chi-square tests and the parametric procedures for which they may substitute.

■ Chi-Square and the Pearson Correlation

The chi-square test for independence and the Pearson correlation are both statistical techniques intended to evaluate the relationship between two variables. The type of data obtained in a research study determines which of these two statistical procedures is appropriate. Suppose, for example, that a researcher is interested in the relationship between self-esteem and academic performance for 10-year-old children. If the researcher obtained numerical scores for both variables, the resulting data would be similar to the values shown in Table 17.11(a) and the researcher could use a Pearson correlation to evaluate the relationship. On the other hand, if both variables are classified into non-numerical categories as in Table 17.11(b), then the data consist of frequencies and the relationship could be evaluated with a chi-square test for independence.

TABLE 17.11

Two possible data structures for research studies examining the relationship between self-esteem and academic performance. In part (a) there are numerical scores for both variable and the data are suitable for a correlation. In part (b) both variables are classified into categories and the data are frequencies suitable for a chi-square test.

(a)

Participant	Self-Esteem X	Academic Performance Y
A	13	73
B	19	88
C	10	71
D	22	96
E	20	90
F	15	82
.	.	.
.	.	.
.	.	.

(b)

		Level of Self-Esteem			
		High	Medium	Low	
Academic Performance	High	17	32	11	60
	Low	13	43	34	90
		30	75	45	$n = 150$

■ Chi-Square and the Independent-Measures t and ANOVA

Once again, consider a researcher investigating the relationship between self-esteem and academic performance for 10-year-old children. This time, suppose the researcher measured academic performance by simply classifying individuals into two categories, high and low, and then obtained a numerical score for each individual's self-esteem. The resulting data would be similar to the scores in Table 17.12(a), and an independent-measures *t* test would be used to evaluate the mean difference between the two groups of scores. Alternatively, the researcher could measure self-esteem by classifying individuals into three categories: high, medium, and low. If a numerical score is then obtained for each individual's academic performance, the resulting data would look like the scores in Table 17.12(b), and an ANOVA would be used to evaluate the mean differences among the three groups.

TABLE 17.12

Data appropriate for an independent-measures *t* test or an ANOVA. In part (a), self-esteem scores are obtained for two groups of students differing in level of academic performance. In part (b), academic performance scores are obtained for three groups of students differing in level of self-esteem.

(a) Self-esteem scores for two groups of students.

Academic Performance	
High	Low
17	13
21	15
16	14
24	20
18	17
15	14
19	12
20	19
18	16

(b) Academic performance scores for three groups of students.

Self-esteem		
High	Medium	Low
94	83	80
90	76	72
85	70	81
84	81	71
89	78	77
96	88	70
91	83	78
85	80	72
88	82	75

The point of these examples is that the chi-square test for independence, the Pearson correlation, and tests for mean differences can all be used to evaluate the relationship between two variables. One main distinction among the different statistical procedures is the form of the data. However, another distinction is the fundamental purpose of these different statistics. The chi-square test and the tests for mean differences (*t* and ANOVA) evaluate the *significance* of the relationship; that is, they determine whether the relationship observed in the sample provides enough evidence to conclude that there is a corresponding relationship in the population. You can also evaluate the significance of a Pearson correlation, however, the main purpose of a correlation is to measure the *strength* of the relationship. In particular, squaring the correlation, r^2, provides a measure of effect size, describing the proportion of variance in one variable that is accounted for by its relationship with the other variable.

■ The Median Test for Independent Samples

The *median test* provides a nonparametric alternative to the independent-measures *t* test (or ANOVA) to determine whether there are significant differences among two or more independent samples. The null hypothesis for the median test states that the different samples come from populations that share a common median (no differences). The alternative hypothesis states that the samples come from populations that are different and do not share a common median.

The median is the score that divides the population in half, with 50% scoring at or below the median.

The logic behind the median test is that whenever several different samples are selected from the same population distribution, roughly half of the scores in each sample should be above the population median and roughly half should be below. That is, all the separate samples should be distributed around the same median. On the other hand, if the samples come from populations with different medians, then the scores in some samples will be consistently higher and the scores in other samples will be consistently lower.

The first step in conducting the median test is to combine all the scores from the separate samples and then find the median for the combined group (see Chapter 3, page 81, for instructions for finding the median). Next, a matrix is constructed with a column for each of the separate samples and two rows: one for individuals above the median and one for individuals below the median. Finally, for each sample, you count how many individuals scored above the combined median and how many scored below. These values are the observed frequencies that are entered in the matrix.

The frequency distribution matrix is evaluated using a chi-square test for independence. The expected frequencies and a value for chi-square are computed exactly as described in Section 17.3. A significant value for chi-square indicates that the discrepancy between the individual sample distributions is greater than would be expected by chance.

The median test is demonstrated in the following example.

EXAMPLE 17.7

The following data represent self-esteem scores obtained from a sample of $n = 40$ children. The children are then separated into three groups based on their level of academic performance (high, medium, low). The median test will evaluate whether there is a significant relationship between self-esteem and level of academic performance.

Self-Esteem Scores for Children at Three Levels of Academic Performance						
High		Medium				Low	
22	14	22	13	24	20	11	19
19	18	18	22	10	16	13	15
12	21	19	15	14	19	20	16
20	18	11	18	11	10	10	18
23	20	12	19	15	12	15	11

The median for the combined group of $n = 40$ scores is $X = 17$ (exactly 20 scores are above this value and 20 are below). For the high performers, 8 out of 10 scores are above the median. For the medium performers, 9 out of 20 are above the median, and for the low performers, only 3 out of 10 are above the median. These observed frequencies are shown in the following matrix:

Academic Performance

	High	Medium	Low
Above Median	8	9	3
Below Median	2	11	7

The expected frequencies for this test are as follows:

Academic Performance

	High	Medium	Low
Above Median	5	10	5
Below Median	5	10	5

The chi-square statistic is

$$\chi^2 = \frac{9}{5} + \frac{9}{5} + \frac{1}{10} + \frac{1}{10} + \frac{4}{5} + \frac{4}{5} = 5.40$$

With $df = 2$ and $\alpha = .05$, the critical value for chi-square is 5.99. The obtained chi-square of 5.40 does not fall in the critical region, so we would fail to reject the null hypothesis. These data do not provide sufficient evidence to conclude that there are significant differences among the self-esteem distributions for these three groups of students. ■

A few words of caution are in order concerning the interpretation of the median test. First, the median test is *not* a test for mean differences. Remember: The mean for a distribution can be strongly affected by a few extreme scores. Therefore, the mean and the median for a distribution are not necessarily the same, and they may not even be related. The results from a median test *cannot* be interpreted as indicating that there is (or is not) a difference between means.

Second, you may have noted that the median test does not directly compare the median from one sample with the median from another. Thus, the median test is not a test for significant differences between medians. Instead, this test compares the distribution of scores for one sample versus the distribution for another sample. If the samples are distributed evenly around a common point (the group median), the test will conclude that there is no significant difference. On the other hand, finding a significant difference simply indicates that the samples are not distributed evenly around the common median. Thus, the best interpretation of a significant result is that there is a *difference in the distributions* of the samples.

LEARNING CHECK

1. Which of the following accurately describes the chi-square test for independence?
   a. It is similar to a single-sample *t* test because it uses one sample to test a hypothesis about one population.
   b. It is similar to a correlation because it uses one sample to evaluate the relationship between two variables.
   c. It is similar to an independent-measures *t* test because it uses separate samples to evaluate the difference between separate populations.
   d. It is similar to both a correlation and an independent-measures *t* test because it can be used to evaluate a relationship between variables or a difference between populations.

2. One version of the null hypothesis for the chi-square test for independence states that there is no relationship between the two variables. With this version of the null hypothesis, the chi-square test is most similar to what other statistical procedure?
   a. An independent-measures *t* test.
   b. A repeated-measures *t* test.
   c. A single-sample *t* test.
   d. A Pearson correlation.

ANSWERS   1. D, 2. D

SUMMARY

1. Chi-square tests are nonparametric techniques that test hypotheses about the form of the entire frequency distribution. Two types of chi-square tests are the test for goodness of fit and the test for independence. The data for these tests consist of the frequency or number of individuals who are located in each category.

2. The test for goodness of fit compares the frequency distribution for a sample to the population distribution

that is predicted by H_0. The test determines how well the observed frequencies (sample data) fit the expected frequencies (data predicted by H_0).

3. The expected frequencies for the goodness-of-fit test are determined by

$$\text{expected frequency} = f_e = pn$$

where p is the hypothesized proportion (according to H_0) of observations falling into a category and n is the size of the sample.

4. The chi-square statistic is computed by

$$\text{chi-square} = \chi^2 = \Sigma \frac{(f_o - f_e)^2}{f_e}$$

where f_o is the observed frequency for a particular category and f_e is the expected frequency for that category. Large values for χ^2 indicate that there is a large discrepancy between the observed (f_o) and the expected (f_e) frequencies and may warrant rejection of the null hypothesis.

5. Degrees of freedom for the test for goodness of fit are

$$df = C - 1$$

where C is the number of categories in the variable. Degrees of freedom measure the number of categories for which f_e values can be freely chosen. As can be seen from the formula, all but the last f_e value to be determined are free to vary.

6. The chi-square distribution is positively skewed and begins at the value of zero. Its exact shape is determined by degrees of freedom.

7. The test for independence is used to assess the relationship between two variables. The null hypothesis states that the two variables in question are independent of each other. That is, the frequency distribution for one variable does not depend on the categories of the second variable. On the other hand, if a relationship does exist, then the form of the distribution for one variable depends on the categories of the other variable.

8. For the test for independence, the expected frequencies for H_0 can be directly calculated from the marginal frequency totals,

$$f_e = \frac{f_c f_r}{n}$$

where f_c is the total column frequency and f_r is the total row frequency for the cell in question.

9. Degrees of freedom for the test for independence are computed by

$$df = (R - 1)(C - 1)$$

where R is the number of row categories and C is the number of column categories.

10. For the test of independence, a large chi-square value means there is a large discrepancy between the f_o and f_e values. Rejecting H_0 in this test provides support for a relationship between the two variables.

11. Both chi-square tests (for goodness of fit and independence) are based on the assumption that each observation is independent of the others. That is, each observed frequency reflects a different individual, and no individual can produce a response that would be classified in more than one category or more than one frequency in a single category.

12. The chi-square statistic is distorted when f_e values are small. Chi-square tests, therefore, should not be performed when the expected frequency of any cell is less than 5.

13. Cohen's w is a measure of effect size that can be used for both chi-square tests. For a chi-square test for independence, effect size is measured by computing a phi-coefficient, which is equivalent to Cohen's w, for data that form a 2×2 matrix or computing Cramér's V for a matrix that is larger than 2×2.

$$\text{phi} = \sqrt{\frac{\chi^2}{n}} \qquad \text{Cramér's } V = \sqrt{\frac{\chi^2}{n(df^*)}}$$

where df^* is the smaller of $(R - 1)$ and $(C - 1)$. Both phi and Cramér's V are evaluated using the criteria in Table 17.10.

KEY TERMS

parametric test (561)

nonparametric test 561)

chi-square test for goodness of fit (562)

observed frequencies (564)

expected frequencies (565)

chi-square statistic (566)

chi-square distribution (568)

test for independence (574)

Cohen's w (583)

phi-coefficient (584)

Cramér's V (585)

General instructions for using SPSS are presented in Appendix D. Following are detailed instructions for using SPSS to perform the **Chi-Square Tests for Goodness of Fit and for Independence** that are presented in this chapter.

THE CHI-SQUARE TEST FOR GOODNESS OF FIT

Data Entry

1. Enter the set of observed frequencies in the first column of the SPSS data editor. If there are four categories, for example, enter the four observed frequencies.
2. In the second column, enter the numbers 1, 2, 3, and so on, so there is a number beside each of the observed frequencies in the first column.

Data Analysis

1. Click **Data** on the tool bar at the top of the page and select **weight** cases at the bottom of the list.
2. Click the **weight cases by** circle, then highlight the label for the column containing the observed frequencies (VAR00001) on the left and move it into the **Frequency Variable** box by clicking on the arrow.
3. Click **OK**.
4. Click **Analyze** on the tool bar, select **Nonparametric Tests**, and click on **Chi-Square**.
5. Highlight the label for the column containing the digits 1, 2, 3, and move it into the Test Variables box by clicking on the arrow.
6. To specify the expected frequencies, you can either use the **all categories equal** option, which automatically computes expected frequencies, or you can enter your own values. To enter your own expected frequencies, click on the **values** option, and one by one enter the expected frequencies into the small box and click **Add** to add each new value to the bottom of the list.
7. Click **OK**.

SPSS Output

The program produces a table showing the complete set of observed and expected frequencies. A second table provides the value for the chi-square statistic, the degrees of freedom, and the level of significance (the p value or alpha level for the test).

THE CHI-SQUARE TEST FOR INDEPENDENCE

Data Entry

1. Enter the complete set of observed frequencies in one column of the SPSS data editor (VAR00001).
2. In a second column, enter a number (1, 2, 3, etc.) that identifies the row corresponding to each observed frequency. For example, enter a 1 beside each observed frequency that came from the first row.
3. In a third column, enter a number (1, 2, 3, etc.) that identifies the column corresponding to each observed frequency. Each value from the first column gets a 1, and so on.

Data Analysis

1. Click **Data** on the tool bar at the top of the page and select **weight cases** at the bottom of the list.
2. Click the **weight cases by** circle, then highlight the label for the column containing the observed frequencies (VAR00001) on the left and move it into the **Frequency Variable** box by clicking on the arrow.
3. Click **OK**.

4. Click **Analyze** on the tool bar at the top of the page, select **Descriptive Statistics**, and click on **Crosstabs**.
5. Highlight the label for the column containing the rows (VAR00002) and move it into the **Rows** box by clicking on the arrow.
6. Highlight the label for the column containing the columns (VAR00003) and move it into the **Columns** box by clicking on the arrow.
7. Click on **Statistics**, select **Chi-Square**, and click **Continue**.
8. Click **OK**.

SPSS Output

We used SPSS to conduct the chi-square test for independence for the data in Example 17.4, examining the relationship between parents' rules about alcohol use and subsequent alcohol-related problems for their children. The output is shown in Figure 17.5. The first table in the output simply lists the variables and is not shown in the figure. The **Crosstabulation** table simply shows the matrix of observed frequencies. The final table, labeled **Chi-Square Tests**, reports the results. Focus on the top row, the **Pearson Chi-Square**, which reports the calculated chi-square value, the degrees of freedom, and the level of significance (the *p* value or the alpha level for the test).

VAR00002 * VAR00003 Crosstabulation

Count

		VAR00003		Total
		1.00	2.00	
VAR00002	1.00	71	9	80
	2.00	89	31	120
Total		160	40	200

Chi-Square Tests

	Value	df	Asymp. Sig. (2-sided)	Exact Sig. (2-sided)	Exact Sig. (1-sided)
Pearson Chi-Square	6.380[a]	1	.012		
Continuity Correction[b]	5.501	1	.019		
Likelihood Ratio	6.774	1	.009		
Fisher's Exact Test				.012	.008
Linear-by-Linear Association	6.348	1	.012		
N of Valid Cases	200				

a. 0 cells (.0%) have expected count less than 5. The minimum expected count is 16.00.
b. Computed only for a 2x2 table

FIGURE 17.5

The SPSS output for the chi-square test for independence in Example 17.4.

FOCUS ON PROBLEM SOLVING

1. The expected frequencies that you calculate must satisfy the constraints of the sample. For the goodness-of-fit test, $\Sigma f_e = \Sigma f_o = n$. For the test of independence, the row totals and column totals for the expected frequencies should be identical to the corresponding totals for the observed frequencies.

2. It is entirely possible to have fractional (decimal) values for expected frequencies. Observed frequencies, however, are always whole numbers.

3. Whenever $df = 1$, the difference between observed and expected frequencies ($f_o = f_e$) will be identical (the same value) for all cells. This makes the calculation of chi-square easier.

4. Although you are advised to compute expected frequencies for all categories (or cells), you should realize that it is not essential to calculate all f_e values separately. Remember that df for chi-square identifies the number of f_e values that are free to vary. Once you have calculated that number of f_e values, the remaining f_e values are determined. You can get these remaining values by subtracting the calculated f_e values from their corresponding row or column totals.

5. Remember that, unlike previous statistical tests, the degrees of freedom (df) for a chi-square test are *not* determined by the sample size (n). Be careful!

DEMONSTRATION 17.1

TEST FOR INDEPENDENCE

A manufacturer of watches would like to examine preferences for digital versus analog watches. A sample of $n = 200$ people is selected, and these individuals are classified by age and preference. The manufacturer would like to know whether there is a relationship between age and watch preference. The observed frequencies (f_o) are as follows:

	Digital	Analog	Undecided	Totals
Younger than 30	90	40	10	140
30 or Older	10	40	10	60
Column totals	100	80	20	$n = 200$

STEP 1 **State the hypotheses, and select an alpha level**   The null hypothesis states that there is no relationship between the two variables.

> H_0:   Preference is independent of age. That is, the frequency distribution of preference has the same form for people younger than 30 as for people 30 or older.

The alternative hypothesis states that there is a relationship between the two variables.

> H_1:   Preference is related to age. That is, the type of watch preferred depends on a person's age.

We will set alpha to $\alpha = .05$.

STEP 2 **Locate the critical region**   Degrees of freedom for the chi-square test for independence are determined by

$$df = (C - 1)(R - 1)$$

For these data,

$$df = (3 - 1)(2 - 1) = 2(1) = 2$$

For $df = 2$ with $\alpha = .05$, the critical chi-square value is 5.99. Thus, our obtained chi-square must exceed 5.99 to be in the critical region and to reject H_0.

STEP 3 **Compute the test statistic** Computing chi-square requires two calculations: finding the expected frequencies and calculating the chi-square statistic.

Expected frequencies, f_e. For the test for independence, the expected frequencies can be found using the column totals (f_c), the row totals (f_r), and the following formula:

$$f_e = \frac{f_c f_r}{n}$$

For people younger than 30, we obtain the following expected frequencies:

$$f_e = \frac{100(140)}{200} = \frac{14{,}000}{200} = 70 \text{ for digital}$$

$$f_e = \frac{80(140)}{200} = \frac{11{,}200}{200} = 56 \text{ for analog}$$

$$f_e = \frac{20(140)}{200} = \frac{2800}{200} = 14 \text{ for undecided}$$

For individuals 30 or older, the expected frequencies are as follows:

$$f_e = \frac{100(60)}{200} = \frac{6000}{200} = 30 \text{ for digital}$$

$$f_e = \frac{80(60)}{200} = \frac{4800}{200} = 24 \text{ for analog}$$

$$f_e = \frac{20(60)}{200} = \frac{1200}{200} = 6 \text{ for undecided}$$

The following table summarizes the expected frequencies:

	Digital	Analog	Undecided
Younger than 30	70	56	14
30 or Older	30	24	6

The chi-square statistic. The chi-square statistic is computed from the formula

$$\chi^2 = \Sigma \frac{(f_o - f_e)^2}{f_e}$$

The following table summarizes the calculations:

Cell	f_o	f_e	$(f_o - f_e)$	$(f_o - f_e)^2$	$(f_o - f_e)^2/f_e$
Younger than 30—digital	90	70	20	400	5.71
Younger than 30—analog	40	56	−16	256	4.57
Younger than 30—undecided	10	14	−4	16	1.14
30 or Older—digital	10	30	−20	400	13.33
30 or Older—analog	40	24	16	256	10.67
30 or Older—undecided	10	6	4	16	2.67

Finally, we can add the last column to get the chi-square value.

$$\chi^2 = 5.71 + 4.57 + 1.14 + 13.33 + 10.67 + 2.67$$
$$= 38.09$$

STEP 4   Make a decision about H_0, and state the conclusion   The chi-square value is in the critical region. Therefore, we can reject the null hypothesis. There is a relationship between watch preference and age, $\chi^2(2, n = 200) = 38.09, p < .05$.

DEMONSTRATION 17.2

EFFECT SIZE WITH CRAMÉR'S *V*

Because the data matrix is larger than 2 × 2, we will compute Cramér's *V* to measure effect size.

$$\text{Cramér's } V = \sqrt{\frac{\chi^2}{n(df^*)}} = \sqrt{\frac{38.09}{200(1)}} = \sqrt{0.19} = 0.436$$

PROBLEMS

1. Parametric tests (such as *t* or ANOVA) differ from nonparametric tests (such as chi-square) primarily in terms of the assumptions they require and the data they use. Explain these differences.

2. The student population at the state college consists of 60% females and 40% males.
   a. The college theater department recently staged a production of a modern musical. A researcher recorded the gender of each student entering the theater and found a total of 37 females and 18 males. Is the gender distribution for theater goers significantly different from the distribution for the general college? Test at the .05 level of significance.
   b. The same researcher also recorded the gender of each student watching a men's basketball game in the college gym and found a total of 58 females and 82 males. Is the gender distribution for basketball fans significantly different from the distribution for the general college? Test at the .05 level of significance.

3. A developmental psychologist would like to determine whether infants display any color preferences. A stimulus consisting of four color patches (red, green, blue, and yellow) is projected onto the ceiling above a crib. Infants are placed in the crib, one at a time, and the psychologist records how much time each infant spends looking at each of the four colors. The color that receives the most attention during a 100-second test period is identified as the preferred color for that infant. The preferred colors for a sample of 80 infants are shown in the following table:

Red	Green	Blue	Yellow
25	18	23	14

   a. Do the data indicate any significant preferences among the four colors? Test at the .05 level of significance.
   b. Write a sentence demonstrating how the outcome of the hypothesis test would appear in a research report.

4. Data from the Motor Vehicle Department indicate that 80% of all licensed drivers are older than age 25.
   a. In a sample of $n = 50$ people who recently received speeding tickets, 33 were older than 25 years and the other 17 were age 25 or younger. Is the age distribution for this sample significantly different from the distribution for the population of licensed drivers? Use $\alpha = .05$.
   b. In a sample of $n = 50$ people who recently received parking tickets, 36 were older than 25 years and the other 14 were age 25 or younger. Is the age distribution for this sample significantly different from the distribution for the population of licensed drivers? Use $\alpha = .05$.

5. Research has demonstrated that people tend to be attracted to others who are similar to themselves. One study demonstrated that individuals are

disproportionately more likely to marry those with surnames that begin with the same last letter as their own (Jones, Pelham, Carvallo, & Mirenberg, 2004). The researchers began by looking at marriage records and recording the surname for each groom and the maiden name of each bride. From these records it is possible to calculate the probability of randomly matching a bride and a groom whose last names begin with the same letter. Suppose that this probability is only 6.5%. Next, a sample of $n = 200$ married couples is selected and the number who shared the same last initial at the time they were married is counted. The resulting observed frequencies are as follows:

Same Initial	Different Initials	
19	181	200

Do these date indicate that the number of couples with the same last initial is significantly different that would be expected if couples were matched randomly? Test with $\alpha = .05$.

6. Suppose that the researcher from the previous problem repeated the study of married couples' initials using twice as many participants and obtaining observed frequencies that exactly double the original values. The resulting data are as follows:

Same Initial	Different Initials	
38	362	400

a. Use a chi-square test to determine whether the number of couples with the same last initial is significantly different that would be expected if couples were matched randomly. Test with $\alpha = .05$.

b. You should find that the data lead to rejecting the null hypothesis. However, in problem 5 the decision was fail to reject. How do you explain the fact that the two samples have the same proportions but lead to different conclusions?

7. A professor in the psychology department would like to determine whether there has been a significant change in grading practices over the years. It is known that the overall grade distribution for the department in 1985 had 14% A's, 26% B's, 31% C's, 19% D's, and 10% F's. A sample of $n = 200$ psychology students from last semester produced the following grade distribution:

A	B	C	D	F
32	61	64	31	12

Do the data indicate a significant change in the grade distribution? Test at the .05 level of significance.

8. Automobile insurance is much more expensive for teenage drivers than for older drivers. To justify this cost difference, insurance companies claim that the younger drivers are much more likely to be involved in costly accidents. To test this claim, a researcher obtains information about registered drivers from the department of motor vehicles and selects a sample of $n = 300$ accident reports from the police department. The motor vehicle department reports the percentage of registered drivers in each age category as follows: 16% are younger than age 20; 28% are 20–29 years old; and 56% are age 30 or older. The number of accident reports for each age group is as follows:

Under Age 20	Age 20–29	Age 30 or Older
68	92	140

a. Do the data indicate that the distribution of accidents for the three age groups is significantly different from the distribution of drivers? Test with $\alpha = .05$.

b. Write a sentence demonstrating how the outcome of the hypothesis test would appear in a research report.

9. The color red is often associated with anger and male dominance. Based on this observation, Hill and Barton (2005) monitored the outcome of four combat sports (boxing, tae kwan do, Greco-Roman wrestling, and freestyle wrestling) during the 2004 Olympic games and found that participants wearing red outfits won significantly more often than those wearing blue.

a. In 50 wrestling matches involving red versus blue, suppose that the red outfit won 31 times and lost 19 times. Is this result sufficient to conclude that red wins significantly more than would be expected by chance? Test at the .05 level of significance.

b. In 100 matches, suppose red won 62 times and lost 38. Is this sufficient to conclude that red wins significantly more than would be expected by chance? Again, use $\alpha = .05$.

c. Note that the winning percentage for red uniforms in part a is identical to the percentage in part b (31 out of 50 is 62%, and 62 out of 100 is also 62%). Although the two samples have an identical winning percentage, one is significant and the other is not. Explain why the two samples lead to different conclusions.

10. A communications company has developed three new designs for a cell phone. To evaluate consumer

response, a sample of 120 college students is selected and each student is given all three phones to use for 1 week. At the end of the week, the students must identify which of the three designs they prefer. The distribution of preference is as follows:

Design 1	Design 2	Design 3
54	38	28

Do the results indicate any significant preferences among the three designs?

11. In Problem 10, a researcher asked college students to evaluate three new cell phone designs. However, the researcher suspects that college students may have criteria that are different from those used by older adults. To test this hypothesis, the researcher repeats the study using a sample of $n = 60$ older adults in addition to a sample of $n = 60$ students. The distribution of preference is as follows:

	Design 1	Design 2	Design 3	
Student	27	20	13	60
Older Adult	21	34	5	60
	48	54	18	

Do the data indicate that the distribution of preferences for older adults is significantly different from the distribution for college students? Test with $\alpha = .05$.

12. Many businesses use some type of customer loyalty program to encourage repeat customers. A common example is the buy-ten-get-one-free punch card. Drèze and Nunes (2006) examined a simple variation of this program that appears to give customers a head start on completing their cards. One group of customers at a car wash was given a buy-eight-get-one-free card and a second group was given a buy-ten-get-one-free card that had already been punched twice. Although both groups needed eight punches to earn a free wash, the group with the two free punches appeared to be closer to reaching their goal. A few months later, the researchers recorded the number of customers who had completed their cards and earned their free car wash. The following data are similar to the results obtained in the study. Do the data indicate a significant difference between the two card programs. Test with $\alpha = .05$.

	Completed	Not Completed	
Buy-Eight-Get-One-Free	10	40	50
Buy-Ten (with 2 Free Punches)	19	31	50
	29	71	

13. In a classic study, Loftus and Palmer (1974) investigated the relationship between memory for eyewitnesses and the questions they are asked. In the study, participants watched a film of an automobile accident and then were questioned about the accident. One group was asked how fast the cars were going when they "smashed into" each other. A second group was asked about the speed when the cars "hit" each other, and a third group was not asked any question about the speed of the cars. A week later, the participants returned to answer additional questions about the accident, including whether they recalled seeing any broken glass. Although there was no broken glass in the film, several students claimed to remember seeing it. The following table shows the frequency distribution of responses for each group.

Response to the Question
Did You See Any Broken Glass?

	Verb Used to Ask About the Speed	Response About Broken Glass	
		Yes	No
Verb Used to Ask About the Speed	Smashed into	16	34
	Hit	7	43
	Control (Not Asked)	6	44

a. Does the proportion of participants who claim to remember broken glass differ significantly from group to group? Test with $\alpha = .05$.

b. Compute Cramér's V to measure the size of the treatment effect.

c. Describe how the phrasing of the question influenced the participants' memories.

d. Write a sentence demonstrating how the outcome of the hypothesis test and the measure of effect size would be reported in a journal article.

14. In a study investigating freshman weight gain, the researchers also looked at gender differences in weight (Kasparek, Corwin, Valois, Sargent, & Morris, 2008). Using self-reported heights and weights, they computed the Body Mass Index (BMI) for each

student. Based on the BMI scores, the students were classified as either desirable weight or overweight. When the students were further classified by gender, the researchers found results similar to the frequencies in the following table.

	Desirable Weight	Overweight
Males	74	46
Females	62	18

a. Do the data indicate that the proportion of overweight men is significantly different from the proportion of overweight women? Test with $\alpha = .05$.
b. Compute the phi-coefficient to measure the strength of the relationship.
c. Write a sentence demonstrating how the outcome of the hypothesis test and the measure of effect size would be reported in a journal article.

15. Research results suggest that IQ scores for boys are more variable than IQ scores for girls (Arden & Plomin, 2006). A typical study looking at 10-year-old children classifies participants by gender and by low, average, or high IQ. Following are hypothetical data representing the research results. Do the data indicate a significant difference between the frequency distributions for males and females? Test at the .05 level of significance and describe the difference.

	IQ		
	Low	Average	High
Boys	18	42	20	80
Girls	12	54	14	80

$n = 160$

16. Gender differences in dream content are well documented (see Winget & Kramer, 1979). Suppose a researcher studies aggression content in the dreams of men and women. Each participant reports his or her most recent dream. Then each dream is judged by a panel of experts to have low, medium, or high aggression content. The observed frequencies are shown in the following matrix:

		Aggression Content		
		Low	Medium	High
Gender	Female	18	4	2
	Male	4	17	15

Is there a relationship between gender and the aggression content of dreams? Test with $\alpha = .01$.

17. In a study similar to one conducted by Fallon and Rozin (1985), a psychologist prepared a set of silhouettes showing different female body shapes ranging from somewhat thin to somewhat heavy and asked a group of women to indicate which body figure they thought men would consider the most attractive. Then a group of men were shown the same set of profiles and asked which image they considered the most attractive. The following hypothetical data show the number of individuals who selected each of the four body image profiles.
a. Do the data indicate a significant difference between the actual preferences for the men and the preferences predicted by the women? Test at the .05 level of significance.
b. Compute the phi-coefficient to measure the strength of the relationship.

	Body Image Profiles			
	Somewhat Thin	Slightly Thin	Slightly Heavy	Somewhat Heavy
Women	29	25	18	8	80
Men	11	15	22	12	60
	40	40	40	20

18. A recent study indicates that people tend to select video game avatars with characteristics similar to those of their creators (Bélisle & Bodur, 2010). Participants who had created avatars for a virtual community game completed a questionnaire about their personalities. An independent group of viewers examined the avatars and recorded their impressions of the avatars. One personality characteristic considered was introverted/extroverted. The following frequency distribution of personalities for participants and the avatars they created.

	Participant Personality	
	Introverted	Extroverted
Introverted Avatar	22	23	45
Extroverted Avatar	16	39	55
	38	62

a. Is there a significant relationship between the personalities of the participants and the personalities of their avatars? Test with $\alpha = .05$.
b. Compute the phi-coefficient to measure the size of the effect.

19. Research indicates that people who volunteer to participate in research studies tend to have higher intelligence than nonvolunteers. To test this phenomenon,

a researcher obtains a sample of 200 high school students. The students are given a description of a psychological research study and asked whether they would volunteer to participate. The researcher also obtains an IQ score for each student and classifies the students into high, medium, and low IQ groups. Do the following data indicate a significant relationship between IQ and volunteering? Test at the .05 level of significance.

IQ

	High	Medium	Low	
Volunteer	43	73	34	150
Not Volunteer	7	27	16	50
	50	100	50	

20. Cialdini, Reno, and Kallgren (1990) examined how people conform to norms concerning littering. The researchers wanted to determine whether a person's tendency to litter depended on the amount of litter already in the area. People were handed a handbill as they entered an amusement park. The entrance area had already been prepared with either no litter, a small amount of litter, or a lot of litter lying on the ground. The people were observed to determine whether they dropped their handbills. The frequency data are as follows:

Amount of Litter

	None	Small Amount	Large Amount
Littering	17	28	49
Not Littering	73	62	41

a. Do the data indicate that people's tendency to litter depends on the amount of litter already on the ground? That is, is there a significant relationship between littering and the amount of existing litter? Test at the .05 level of significance.
b. Compute Cramér's V to measure the size of the treatment effect.

21. Although the phenomenon is not well understood, it appears that people born during the winter months are slightly more likely to develop schizophrenia than people born at other times (Bradbury & Miller, 1985). The following hypothetical data represent a sample of 50 individuals diagnosed with schizophrenia and a sample of 100 people with no psychotic diagnosis. Each individual is also classified according to season in which he or she was born. Do the data indicate a significant relationship between schizophrenia and the season of birth? Test at the .05 level of significance.

Season of Birth

	Summer	Fall	Winter	Spring	
No Disorder	26	24	22	28	100
Schizophrenia	9	11	18	12	50
	35	35	40	40	

22. In the Preview for this chapter we presented a research study demonstrating that romantic background music increases the likelihood that a woman will give her phone number to a man she has just met (Guéguen, Jacob, & Lamy, 2010). The frequency distribution for the study is reproduced in the following table.
a. Use a chi-square test for independence with $\alpha = .05$ to determine whether the data show a significant relationship between the type of music and the behavior of the female participants.
b. Compute the phi-coefficient to measure the strength of the relationship.

Gave Phone Number?

		Yes	No
Type of Music	Romantic	23	21
	Neutral	12	31

23. Research has demonstrated strong gender differences in teenagers' approaches to dealing with mental health issues (Chandra & Minkovitz, 2006). In a typical study, eighth-grade students are asked to report their willingness to use mental health services in the event they were experiencing emotional or other mental health problems. Typical data for a sample of $n = 150$ students are shown in the following table.
a. Do the data show a significant relationship between gender and willingness to seek mental health assistance? Test with $\alpha = .05$.
b. Compute Cramér's V to measure the size of the treatment effect.

Willingness to Use Mental Health Services

	Probably No	Maybe	Probably Yes	
Males	17	32	11	60
Females	13	43	34	90
	30	75	45	$n = 150$

The Binomial Test

© Deborah Batt

Tools You Will Need

The following items are considered essential background material for this chapter. If you doubt your knowledge of any of these items, you should review the appropriate chapter or section before proceeding.

- Binomial distribution (Chapter 6)
- *z*-score hypothesis tests (Chapter 8)
- Chi-square test for goodness of fit (Chapter 17)

PREVIEW

In 1960, Gibson and Walk designed a classic piece of apparatus to test depth perception. Their device, called a *visual cliff,* consisted of a wide board with a deep drop (the cliff) to one side and a shallow drop on the other side. An infant was placed on the board and then observed to see whether he or she crawled off the shallow side or crawled off the cliff. Infants who moved to the deep side actually crawled onto a sheet of heavy glass, which prevented them from falling. Thus, the deep side only appeared to be a cliff—hence the name *visual cliff.*

Gibson and Walk reasoned that if infants are born with the ability to perceive depth, they would recognize the deep side and not crawl off the cliff. On the other hand, if depth perception is a skill that develops over time through learning and experience, then infants should not be able to perceive any difference between the shallow and the deep sides.

Out of 27 infants who moved off the board, only 3 ventured onto the deep side at any time during the experiment. The other 24 infants stayed exclusively on the shallow side. Gibson and Walk interpreted these data as convincing evidence that depth perception is innate. The infants showed a systematic preference for the shallow side.

You should notice that the data from this study are different from the data that we usually encounter. There are no scores. Gibson and Walk simply counted the number of infants who went off the deep side and the number who went to the shallow side. Still, we would like to use these data to make statistical decisions. Do these sample data provide sufficient evidence to make a confident conclusion about depth perception in the population? Suppose that 8 of the 27 infants had crawled to the deep side. Would you still be convinced that there is a significant preference for the shallow side? What about 12 out of 27?

We are asking a question about statistical significance and will need a hypothesis test to obtain an answer. The null hypothesis for the Gibson and Walk study would state that infants have no depth perception and cannot perceive a difference between the shallow and deep sides. In this case, their movement should be random with half going to either side. Notice that the data and the hypothesis both concern frequencies or proportions. This situation is perfect for the chi-square test introduced in Chapter 17, and a chi-square test can be used to evaluate the data. However, when individuals are classified into exactly two categories (for example, shallow and deep), a special statistical procedure exists. In this chapter, we introduce the *binomial test,* which is used to evaluate and interpret frequency data involving exactly two categories of classification.

18.1 | Introduction to the Binomial Test

LEARNING OBJECTIVES

1. Describe binomial data and identify situations for which a binomial test is appropriate.

2. Define the criteria that must be satisfied before the binomial distribution is normal and the z-score statistic can be used.

Data with exactly two categories are also known as dichotomous data.

In Chapter 6, we introduced the concept of *binomial data*. You should recall that binomial data exist whenever a measurement procedure classifies individuals into exactly two distinct categories. For example, the outcomes from tossing a coin can be classified as heads and tails; people can be classified as male or female; plastic products can be classified as recyclable or non-recyclable. In general, binomial data exist when:

1. the measurement scale consists of exactly two categories;

2. each individual observation in a sample is classified in only one of the two categories; and

3. the sample data consist of the frequency or number of individuals in each category.

The traditional notation system for binomial data identifies the two categories as A and B and identifies the probability (or proportion) associated with each category as p and q, respectively. For example, a coin toss results in either heads (A) or tails (B), with probabilities $p = \frac{1}{2}$ and $q = \frac{1}{2}$.

In this chapter, we examine the statistical process of using binomial data for testing hypotheses about the values of p and q for the population. This type of hypothesis test is called a *binomial test*.

DEFINITION

> A **binomial test** uses sample data to evaluate hypotheses about the values of p and q for a population consisting of binomial data.

Consider the following two situations.

1. In a sample of $n = 34$ color-blind students, 30 are male and only 4 are female. Does this sample indicate that color blindness is significantly more common for males in the general population?

2. In 2005, only 10% of American families had incomes below the poverty level. This year, in a sample of 100 families, 19 were below the poverty level. Does this sample indicate that there has been a significant change in the population proportions?

Notice that both of these examples have binomial data (exactly two categories). Although the data are relatively simple, we are asking the same statistical question about significance that is appropriate for a hypothesis test: Do the sample data provide sufficient evidence to make a conclusion about the population?

■ Hypotheses for the Binomial Test

In the binomial test, the null hypothesis specifies exact values for the population proportions p and q. Theoretically, you could choose any proportions for H_0, but usually there is a clear reason for the values that are selected. The null hypothesis typically falls into one of the following two categories.

1. **Just Chance**   Often the null hypothesis states that the two outcomes A and B occur in the population with the proportions that would be predicted simply by chance. If you were tossing a coin, for example, the null hypothesis might specify $p(\text{heads}) = \frac{1}{2}$ and $p(\text{tails}) = \frac{1}{2}$. Notice that this hypothesis states the usual, chance proportions for a balanced coin. Also notice that it is not necessary to specify both proportions. Once the value of p is identified, the value of q is determined by $1 - p$. For the coin toss example, the null hypothesis would simply state

$$H_0: \quad p = p(\text{heads}) = \frac{1}{2} \qquad \textbf{(The coin is balanced.)}$$

Similarly, if you were selecting cards from a deck and trying to predict the suit on each draw, the probability of predicting correctly would be $p = \frac{1}{4}$ for any given trial. (With four suits, you have a 1-out-of-4 chance of guessing correctly.) In this case, the null hypothesis would state

$$H_0: \quad p = p(\text{guessing correctly}) = \frac{1}{4} \qquad \textbf{(The outcome is simply due to chance.)}$$

In each case, the null hypothesis simply states that there is nothing unusual about the proportions in the population; that is, the outcomes are occurring by chance.

2. **No Change or No Difference**   Often you may know the proportions for one population and want to determine whether the same proportions apply to a different

population. In this case, the null hypothesis would simply specify that there is no difference between the two populations. Suppose that national statistics indicate that 1 out of 12 drivers will be involved in a traffic accident during the next year. Does this same proportion apply to 16-year-olds who are driving for the first time? According to the null hypothesis,

$$H_0: \quad \text{For 16-year-olds, } p = p(\text{accident}) = \frac{1}{12} \qquad \textbf{(Not different from the general population)}$$

Similarly, suppose that last year, 30% of the freshman class failed the college writing test. This year, the college is requiring all freshmen to take a writing course. Will the course have any effect on the number who fail the test? According to the null hypothesis,

$$H_0: \quad \text{For this year, } p = p(\text{fail}) = 30\% \qquad \textbf{(Not different from last year's class)}$$

■ The Data for the Binomial Test

For the binomial test, a sample of n individuals is obtained and you simply count how many are classified in category A and how many are classified in category B. We focus attention on category A and use the symbol X to stand for the number of individuals classified in category A. Recall from Chapter 6 that X can have any value from 0 to n and that each value of X has a specific probability. The distribution of probabilities for each value of X is called the *binomial distribution*. Figure 18.1 shows an example of a binomial distribution where X is the number of heads obtained in four tosses of a balanced coin.

■ The Test Statistic for the Binomial Test

As we noted in Chapter 6, when the values pn and qn are both equal to or greater than 10, the binomial distribution approximates a normal distribution. This fact is important because it allows us to compute z-scores and use the unit normal table to answer probability questions about binomial events. In particular, when pn and qn are both at least 10, the binomial distribution will have the following properties.

1. The shape of the distribution is approximately normal.
2. The mean of the distribution is $\mu = pn$.

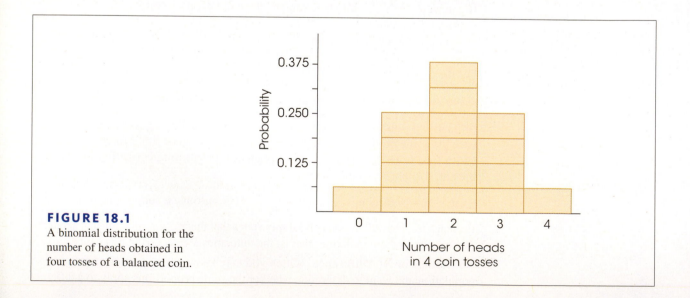

FIGURE 18.1

A binomial distribution for the number of heads obtained in four tosses of a balanced coin.

3. The standard deviation of the distribution is

$$\sigma = \sqrt{npq}$$

With these parameters in mind, it is possible to compute a z-score corresponding to each value of X in the binomial distribution.

$$z = \frac{X - \mu}{\sigma} = \frac{X - pn}{\sqrt{npq}} \qquad \textbf{(See Equation 6.3.)} \qquad \textbf{(18.1)}$$

This is the basic z-score formula that is used for the binomial test. However, the formula can be modified slightly to make it more compatible with the logic of the binomial hypothesis test. The modification consists of dividing both the numerator and the denominator of the z-score by n. (You should realize that dividing both the numerator and the denominator by the same value does not change the value of the z-score.) The resulting equation is

$$z = \frac{X/n - p}{\sqrt{pq/n}} \qquad \textbf{(18.2)}$$

For the binomial test, the values in this formula are defined as follows.

1. X/n is the proportion of individuals in the sample who are classified in category A.

2. p is the hypothesized value (from H_0) for the proportion of individuals in the population who are classified in category A.

3. $\sqrt{pq/n}$ is the standard error for the sampling distribution of X/n and provides a measure of the standard distance between the sample statistic (X/n) and the population parameter (p).

Thus, the structure of the binomial z-score (Equation 18.2) can be expressed as

$$z = \frac{X/n - p}{\sqrt{pq/n}} = \frac{\begin{array}{c}\text{sample} \\ \text{proportion} \\ \text{(data)}\end{array} - \begin{array}{c}\text{hypothesized} \\ \text{population} \\ \text{proportion}\end{array}}{\text{standard error}}$$

The logic underlying the binomial test is exactly the same as we encountered with the original z-score hypothesis test in Chapter 8. The hypothesis test involves comparing the sample data with the hypothesis. If the data are consistent with the hypothesis, we conclude that the hypothesis is reasonable. But if there is a big discrepancy between the data and the hypothesis, we reject the hypothesis. The value of the standard error provides a benchmark for determining whether the discrepancy between the data and the hypothesis is more than would be expected by chance. The alpha level for the test provides a criterion for deciding whether the discrepancy is significant. The hypothesis-testing procedure is demonstrated in the following section.

LEARNING CHECK

1. Which of the following is not an example of binomial data? Classifying college students according to

a. gender: male, female

b. age: 21 or younger, older than 21

c. college class: freshman, sophomore, junior, senior

d. registered voter: yes, no

2. If $pn = 40$ and $qn = 20$, then what is the shape of the binomial distribution?
   a. Normal
   b. Positively skewed
   c. Negatively skewed
   d. Cannot be determined without additional information

3. A binomial distribution has $p = 4/5$. How large a sample would be necessary to justify using the normal approximation to the binomial distribution?
   a. At least 10
   b. At least 13
   c. At least 40
   d. At least 50

ANSWERS   **1.** C, **2.** A, **3.** D

18.2 | An Example of the Binomial Test

LEARNING OBJECTIVES   **3.** Perform all of the necessary computations for the binomial test.

4. Explain how the decision in a binomial test can be influenced by the fact that each score actually corresponds to an interval in the distribution.

The binomial test follows the same four-step procedure presented earlier with other examples for hypothesis testing. The four steps are summarized as follows.

STEP 1   **State the hypotheses.**   In the binomial test, the null hypothesis specifies values for the population proportions p and q. Typically, H_0 specifies a value only for p, the proportion associated with category A. The value of q is directly determined from p by the relationship $q = 1 - p$. Finally, you should realize that the hypothesis, as always, addresses the probabilities or proportions for the *population*. Although we use a sample to test the hypothesis, the hypothesis itself always concerns a population.

STEP 2   **Locate the critical region.**   When both values for pn and qn are greater than or equal to 10, the z-scores defined by Equation 18.1 or 18.2 form an approximately normal distribution. Thus, the unit normal table can be used to find the boundaries for the critical region. With $\alpha = .05$, for example, you may recall that the critical region is defined as z-score values greater than $+1.96$ or less than -1.96.

STEP 3   **Compute the test statistic (z-score).**   At this time, you obtain a sample of n individuals (or events) and count the number of times category A occurs in the sample. The number of occurrences of A in the sample is the X value for Equation 18.1 or 18.2. Because the two z-score equations are equivalent, you may use either one for the hypothesis test. Usually Equation 18.1 is easier to use because it involves larger numbers (fewer decimals) and it is less likely to be affected by rounding error.

STEP 4   **Make a decision.**   If the z-score for the sample data is in the critical region, you reject H_0 and conclude that the discrepancy between the sample proportions and the hypothesized population proportions is significantly greater than chance. That is, the data are not consistent with the null hypothesis, so H_0 must be wrong. On the other hand, if the z-score is not in the critical region, you fail to reject H_0.

The following example demonstrates a complete binomial test.

EXAMPLE 18.1   In the Preview section, we described the *visual cliff* experiment designed to examine depth perception in infants. To summarize briefly, an infant is placed on a wide board that appears to have a deep drop on one side and a relatively shallow drop on the other. An infant who is able to perceive depth should avoid the deep side and move toward the shallow side. Without depth perception, the infant should show no preference between the two sides. Of the 27 infants in the experiment, 24 stayed exclusively on the shallow side and only 3 moved onto the deep side. The purpose of the hypothesis test is to determine whether these data demonstrate that infants have a significant preference for the shallow side.

This is a binomial hypothesis-testing situation. The two categories are

$$A = \text{move onto the deep side}$$

$$B = \text{move onto the shallow side}$$

STEP 1   The null hypothesis states that for the general population of infants, there is no preference between the deep and the shallow sides; the direction of movement is determined by chance. In symbols,

$$H_0: \quad p = p(\text{deep side}) = \frac{1}{2} \quad \left(\text{and } q = \frac{1}{2}\right)$$

$$H_1: \quad p \neq \frac{1}{2} \quad \textbf{(There is a preference.)}$$

We will use $\alpha = .05$.

STEP 2   With a sample of $n = 27$, $pn = 13.5$ and $qn = 13.5$. Both values are greater than 10, so the distribution of z-scores is approximately normal. With $\alpha = .05$, the critical region is determined by boundaries of $z = \pm 1.96$.

STEP 3   For this experiment, the data consist of $X = 3$ out of $n = 27$. Using Equation 18.1, these data produce a z-score value of

$$z = \frac{X - pn}{\sqrt{npq}} = \frac{3 - 13.5}{\sqrt{27\left(\frac{1}{2}\right)\left(\frac{1}{2}\right)}} = \frac{-10.5}{2.60} = -4.04$$

To use Equation 18.2, you first compute the sample proportion, $X/n = \frac{3}{27} = 0.111$. The z-score is then

$$z = \frac{X/n - p}{\sqrt{pq/n}} = \frac{0.111 - 0.5}{\sqrt{\frac{1}{2}\left(\frac{1}{2}\right)/27}} = \frac{-0.389}{0.096} = -4.05$$

Within rounding error, the two equations produce the same result.

STEP 4   Because the data are in the critical region, our decision is to reject H_0. These data do provide sufficient evidence to conclude that there is a significant preference for the shallow side. Gibson and Walk (1960) interpreted these data as convincing evidence that depth perception is innate. ∎

■ Score Boundaries and the Binomial Test

In Chapter 6, we noted that a binomial distribution forms a discrete histogram (see Figure 18.1), whereas the normal distribution is a continuous curve. The difference between the two distributions was illustrated in Figure 6.18, which is repeated here as Figure 18.2. In the binomial distribution (the histogram), each score is represented by a bar (Figure 18.2) and each bar has an upper and a lower boundary. For example, a score of $X = 6$ actually corresponds to a bar that reaches from $X = 5.5$ to $X = 6.5$. In the normal approximation to the binomial distribution, each of the bars corresponds to an interval with upper and lower real limits. For example, in the continuous distribution, a score of $X = 6$ corresponds to an interval that extends from a lower real limit of 5.5 to an upper real limit of 6.5.

When conducting a hypothesis test with the binomial distribution, the basic question is whether a specific score is located in the critical region. However, because each score actually corresponds to an interval in the normal distribution, it is possible that part of the score is in the critical region and part is not. Fortunately, this is usually not an issue. When pn and qn are both equal to or greater than 10 (the criteria for using the normal approximation), each interval in the binomial distribution is extremely small and it is very unlikely that the interval overlaps the critical boundary. For example, the experiment in Example 18.1 produced a score of $X = 3$, and we computed a z-score of $z = -4.04$. Because this value is in the critical region, beyond $z = -1.96$, we rejected H_0. If we had used the interval boundaries of $X = 2.5$ and $X = 3.5$, instead of $X = 3$, we would have obtained z-scores of

$$z = \frac{2.5 - 13.5}{2.60} \qquad \text{and} \qquad z = \frac{3.5 - 13.5}{2.60}$$
$$= -4.23 \qquad\qquad\qquad = -3.85$$

Note that $X = 3$ corresponds to $z = 4.04$, which is exactly in the middle of the interval.

Thus, a score of $X = 3$ actually corresponds to an interval of z-scores ranging from $z = -3.85$ to $z = -4.23$. However, this entire interval is in the critical region beyond $z = -1.96$, so the decision is still to reject H_0.

In most situations, if the whole number X value (in this case, $X = 3$) is in the critical region, then the entire interval will also be in the critical region and the correct decision is to reject H_0. The only exception to this general rule occurs when an X value produces a z-score that is in the critical region by a very small margin. In this situation, you should compute the z-scores corresponding to both ends of the interval to determine whether any part of the z-score interval is not located in the critical region. Suppose, for example, that

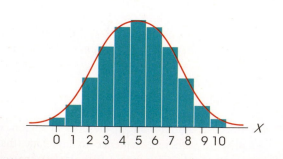

FIGURE 18.2

The relationship between the binomial distribution and the normal distribution. The binomial distribution is always a discrete histogram, and the normal distribution is a continuous, smooth curve. Each X value is represented by a bar in the histogram or a section of the normal distribution.

the researchers in Example 18.1 found that 8 out of 27 infants in the visual cliff experiment moved onto the deep side. A score of $X = 8$ corresponds to

$$z = \frac{8 - 13.5}{2.60}$$

$$z = \frac{-5.5}{2.60}$$

$$= -2.12$$

Because this value is beyond the -1.96 boundary, it appears that we should reject H_0. However, this z-score is only slightly beyond the critical boundary, so it would be wise to check both ends of the interval. For $X = 8$, the interval boundaries are 7.5 and 8.5, which correspond to z-scores of

$$z = \frac{7.5 - 13.5}{2.60} \qquad \text{and} \qquad z = \frac{8.5 - 13.5}{2.60}$$

$$= -2.31 \qquad\qquad\qquad\qquad = -1.92$$

Thus, a score of $X = 8$ corresponds to an interval extending from $z = -1.92$ to $z = -2.31$. However, the critical boundary is $z = -1.96$, which means that part of the interval (and part of the score) is not in the critical region for $\alpha = .05$. Because $X = 8$ is not completely beyond the critical boundary, the probability of obtaining $X = 8$ is greater than $\alpha = .05$. Therefore, the correct decision is to fail to reject H_0.

In general, it is safe to conduct a binomial test using the whole-number value for X. However, if you obtain a z-score that is only slightly beyond the critical boundary, you also should compute the z-scores for both ends of the interval. If any part of the z-score interval is not in the critical region, the correct decision is to fail to reject H_0.

The following example is an opportunity to test your understanding of the z-score statistic used in the binomial test.

EXAMPLE 18.2   If the results of the visual cliff study showed that 9 out of 36 infants crawled off the deep side, what z-score value would be obtained using Equation 18.1? You should find that the binomial distribution has $\mu = \frac{1}{2}(36) = 18$ and $\sigma = \sqrt{(\frac{1}{2})(\frac{1}{2})(36)} = \sqrt{9} = 3$. $X = 9$ corresponds $z = -\frac{9}{3} = -3.00$. ∎

IN THE LITERATURE

Reporting the Results of a Binomial Test

Reporting the results of the binomial test typically consists of describing the data and reporting the z-score value and the probability that the results are due to chance. It is also helpful to note that a binomial test was used because z-scores are used in other hypothesis-testing situations (see, for example, Chapter 8). For Example 18.1, the report might state:

> Three out of 27 infants moved to the deep side of the visual cliff. A binomial test revealed that there is a significant preference for the shallow side of the cliff, $z = -4.04$, $p < .05$.

Once again, p is *less than* .05. We have rejected the null hypothesis because it is very unlikely—probability less than 5%—that these results are simply due to chance.

■ Assumptions for the Binomial Test

The binomial test requires two very simple assumptions.

1. The sample must consist of *independent* observations (see Chapter 8, page 244).

2. The values for pn and qn must both be greater than or equal to 10 to justify using the unit normal table for determining the critical region.

LEARNING CHECK

1. What is the mean for the normal approximation to the binomial distribution with $n = 30$ and $p = .40$?
   a. .24
   b. 7.2
   c. 12
   d. 18

2. What is the standard deviation for the normal approximation to the binomial distribution with $n = 100$ and $p = .20$?
   a. 16
   b. 4
   c. 2
   d. 0.16

3. For a binomial test with $p = q = \frac{1}{2}$ and $n = 36$, is a score of $X = 24$ sufficient to reject the null hypothesis? Assume a two-tailed test with $\alpha = .05$.
   a. Yes, $X = 24$ is entirely in the critical region.
   b. No, none of $X = 24$ is in the critical region.
   c. No, $X = 24$ overlaps the critical boundary with the lower limit in the critical region but the upper limit outside the critical region.
   d. No, $X = 24$ overlaps the critical boundary with the upper limit in the critical region but the lower limit outside the critical region.

ANSWERS   1. C, 2. B, 3. D

18.3 | More about the Binomial Test: Relationship with Chi-Square and the Sign Test

LEARNING OBJECTIVES

5. Explain the relationship between the chi-square test for goodness of fit and the binomial test.

6. Recognize when a sign test is appropriate and be able to perform the necessary computations.

You may have noticed that the binomial test evaluates the same basic hypotheses as the chi-square test for goodness of fit; that is, both tests evaluate how well the sample proportions fit a hypothesis about the population proportions. When a research study produces binomial data, these two tests are equivalent, and either may be used. The relationship between the two tests can be expressed by the equation

$$\chi^2 = z^2$$

where χ^2 is the statistic from the chi-square test for goodness of fit and z is the z-score from the binomial test.

To demonstrate the relationship between the goodness-of-fit test and the binomial test, we reexamine the data from Example 18.1.

STEP 1   **Hypotheses.**   In the visual cliff experiment from Example 18.1, the null hypothesis states that there is no preference between the shallow side and the deep side. For the binomial test, the null hypothesis states

$$H_0: \quad p = \text{p(deep side)} = q = \text{p(shallow side)} = \frac{1}{2}$$

The chi-square test for goodness of fit would state the same hypothesis, specifying the population proportions as

	Shallow Side	Deep Side
H_0:	$\dfrac{1}{2}$	$\dfrac{1}{2}$

STEP 2   **Critical region.**   For the binomial test, the critical region is located by using the unit normal table. With $\alpha = .05$, the critical region consists of any z-score value beyond ± 1.96. The chi-square test would have $df = 1$, and with $\alpha = .05$, the critical region consists of chi-square values greater than 3.84. Notice that the basic relationship, $\chi^2 = z^2$, holds:

$$3.84 = (1.96)^2$$

STEP 3   **Test statistic.**   For the binomial test (Example 18.1), we obtained a z-score of $z = -4.04$. For the chi-square test, the expected frequencies are

	Shallow Side	Deep Side
f_0	13.5	13.5

With observed frequencies of 24 and 3, respectively, the chi-square statistic is

$$\chi^2 = \frac{(24 - 13.5)^2}{13.5} + \frac{(3 - 13.5)^2}{13.5}$$

$$= \frac{(10.5)^2}{13.5} + \frac{(-10.5)^2}{13.5}$$

$$= 8.167 + 8.167$$

$$= 16.33$$

With a little rounding error, the values obtained for the z-score and chi-square are related by the equation

Caution: The χ^2 value is already squared. Do not square it again.

$$\chi^2 = z^2$$
$$16.33 = (-4.04)^2$$

STEP 4   **Decision.**   Because the critical values for both tests are related by the equation $\chi^2 = z^2$ and the test statistics are related in the same way, these two tests *always* result in the same statistical conclusion.

The following example is an opportunity to test you understanding of the relationship between chi-square and the binomial test.

EXAMPLE 18.3

In Example 18.1 we demonstrated the binomial test for the visual cliff study presented in the Chapter Preview. If 10 out of 36 infants crawled off the deep side of the visual cliff, then what value would be obtained for the z-score in a binomial test and what value would be obtained for chi-square? You should obtain $z = 2.67$ and $\chi^2 = 7.11$. ■

■ The Sign Test

Although the binomial test can be used in many different situations, there is one specific application that merits special attention. For a repeated-measures study that compares two conditions, it is often possible to use a binomial test to evaluate the results. You should recall that a repeated-measures study involves measuring each individual in two different treatment conditions or at two different points in time. When the measurements produce numerical scores, the researcher can simply subtract to determine the difference between the two scores and then evaluate the data using a repeated-measures t test (see Chapter 11). Occasionally, however, a researcher may record only the *direction* of the difference between the two observations. For example, a clinician may observe patients before therapy and after therapy and simply note whether each patient got better or got worse. Note that there is no measurement of how much change occurred; the clinician is simply recording the direction of change. Also note that the direction of change is a binomial variable; that is, there are only two values. In this situation it is possible to use a binomial test to evaluate the data. Traditionally, the two possible directions are coded by signs, with a positive sign indicating an increase and a negative sign indicating a decrease. When the binomial test is applied to signed data, it is called a *sign test*.

An example of signed data is shown in Table 18.1. Notice that the data can be summarized by saying that seven out of eight patients showed a decrease in symptoms after therapy.

TABLE 18.1

Hypothetical data from a research study evaluating the effectiveness of a clinical therapy. For each patient, symptoms are assessed before and after treatment and the data record whether there is an increase or a decrease in symptoms following therapy.

Patient	Direction of Change After Treatment
A	− (decrease)
B	− (decrease)
C	− (decrease)
D	+ (increase)
E	− (decrease)
F	− (decrease)
G	− (decrease)
H	− (decrease)

The null hypothesis for the sign test states that there is no difference between the two treatment conditions being compared. Therefore, any change in a participant's score is due to chance. In terms of probabilities, this means that increases and decreases are equally likely, so

$$p = p(\text{increase}) = \frac{1}{2}$$

$$q = p(\text{decrease}) = \frac{1}{2}$$

A complete example of a sign test follows.

EXAMPLE 18.4   A researcher testing the effectiveness of acupuncture for treating the symptoms of arthritis obtains a sample of 36 people who have been diagnosed with arthritis. Each person's pain level is measured before treatment starts, and measured again after 4 months of acupuncture treatment. For this sample, 25 people experienced a reduction in pain, and 11 people had more pain after treatment. Do these data indicate a significant treatment effect?

STEP 1   **State the hypothesis.**   The null hypothesis states that acupuncture has no effect. Any change in the level of pain is due to chance, so increases and decreases are equally likely. Expressed as probabilities, the hypotheses are

$$H_0: \quad p = p(\text{increased pain}) = \frac{1}{2} \text{ and } q = p(\text{decreased pain}) = \frac{1}{2}$$

$$H_1: \quad p \neq q \quad \text{(Changes tend to be consistently in one direction.)}$$

Set $\alpha = .05$.

STEP 2   **Locate the critical region.**   With $n = 36$ people, both pn and qn are greater than 10, so the normal approximation to the binomial distribution is appropriate. With $\alpha = .05$, the critical region consists of z-scores greater than -1.96 at one extreme and z-scores less than -1.96 at the other.

STEP 3   **Compute the test statistic.**   For this sample we have $X = 25$ people with decreased pain. This score corresponds to a z-score of

$$z = \frac{X - pn}{\sqrt{npq}} = \frac{25 - 18}{\sqrt{36\left(\frac{1}{2}\right)\left(\frac{1}{2}\right)}} = \frac{7}{3} = 2.33$$

Because the z-score is only slightly beyond the 1.96 critical boundary, we will consider the interval boundaries for $X = 25$ to be certain that the entire interval is beyond the boundary. The interval for $X = 25$, extends from 24.5 to 25.5, which correspond to z-scores of

$$z = \frac{24.5 - 18}{3} \qquad \text{and} \qquad z = \frac{25.5 - 18}{3}$$
$$= 2.17 \qquad\qquad\qquad\qquad = 2.50$$

Thus, a score of $X = 25$ corresponds to an interval of z-scores ranging from $z = 2.17$ to $z = 2.50$. Note that this entire interval is beyond the 1.96 critical boundary.

STEP 4   **Make a decision.**   Because the data are in the critical region, we reject H_0 and conclude that acupuncture treatment has a significant effect on arthritis pain, $z = 2.33, p < .05$.   ■

The following example is an opportunity to test you understanding of the relationship between chi-square and the binomial test.

EXAMPLE 18.5   Suppose the researcher in Example 18.3 found that 24 people had reduced pain and 12 had increased pain. Is this result significant with $\alpha = .05$? Remember to look at both ends of the interval for $X = 24$. You should find that the interval boundaries correspond to $z = 1.83$ and $z = 2.17$. Because the interval overlaps the critical value (1.96), the result is not significant.   ■

■ Zero Differences in the Sign Test

You should notice that the null hypothesis in the sign test refers only to those individuals who show some difference between treatment 1 vs. treatment 2. The null hypothesis states that if there is any change in an individual's score, then the probability of an increase is equal to the probability of a decrease. Stated in this form, the null hypothesis does not consider individuals who show zero difference between the two treatments. As a result, the usual recommendation is that these individuals should be discarded from the data and the value of n be reduced accordingly. However, if the null hypothesis is interpreted more generally, it states that there is no difference between the two treatments. Phrased this way, it should be clear that individuals who show no difference actually are supporting the null hypothesis and should not be discarded. Therefore, an alternative approach to the sign test is to divide individuals who show zero differences equally between the positive and negative categories. (With an odd number of zero differences, discard one, and divide the rest evenly.) This alternative results in a more conservative test; that is, the test is more likely to fail to reject the null hypothesis.

EXAMPLE 18.6

It has been demonstrated that stress or exercise causes an increase in the concentration of certain chemicals in the brain called endorphins. Endorphins are similar to morphine and produce a generally relaxed feeling and a sense of wellbeing. The endorphins may explain the "high" experienced by long-distance runners. To demonstrate this phenomenon, a researcher tested pain tolerance for 40 athletes before and after they completed a mile run. Immediately after running, the ability to tolerate pain increased for 21 of the athletes, decreased for 12, and showed no change for the remaining 7.

Following the standard recommendation for handling zero differences, you would discard the 7 participants who showed no change and conduct a sign test with the remaining $n = 33$ athletes. With the more conservative approach, only 1 of the 7 who showed no difference would be discarded and the other 6 would be divided equally between the two categories. This would result in a total sample of $n = 39$ athletes with $21 + 3 = 24$ in the increased-tolerance category and $12 + 3 = 15$ in the decreased-tolerance category. ■

■ When to Use the Sign Test

In many cases, data from a repeated-measures experiment can be evaluated using either a sign test or a repeated-measures t test. In general, you should use the t test whenever possible. Because the t test uses the actual difference scores (not just the signs), it makes maximum use of the available information and results in a more powerful test. However, there are some cases in which a t test cannot or should not be used, and in these situations, the sign test can be valuable. Four specific cases in which a t test is inappropriate or inconvenient will now be described.

Before	After	Difference
20	23	+3
14	39	+25
27	Failed	+??
.	.	.
.	.	.
.	.	.

1. When you have infinite or undetermined scores, a t test is impossible, and the sign test is appropriate. Suppose, for example, that you are evaluating the effects of a sedative drug on problem-solving ability. A sample of rats is obtained, and each animal's performance is measured before and after receiving the drug. Hypothetical data are shown in the margin. Note that the third rat in this sample failed to solve the problem after receiving the drug. Because there is no score for this animal, it is impossible to compute a sample mean, an SS, or a t statistic. However, you could do a sign test because you know that the animal made more errors (an increase) after receiving the drug.

2. Often it is possible to describe the difference between two treatment conditions without precisely measuring a score in either condition. In a clinical setting, for example, a doctor can say whether a patient is improving, growing worse, or showing no change even though the patient's condition is not precisely measured by a score. In this situation, the data are sufficient for a sign test, but you could not compute a t statistic without individual scores.

3. Often a sign test is done as a preliminary check on an experiment before serious statistical analysis begins. For example, a researcher may predict that scores in treatment 2 should be consistently greater than scores in treatment 1. However, examination of the data after 1 week indicates that only 8 of 15 subjects showed the predicted increase. On the basis of these preliminary results, the researcher may choose to reevaluate the experiment before investing additional time.

4. Occasionally, the difference between treatments is not consistent across participants. This can create a very large variance for the difference scores. As we have noted in the past, large variance decreases the likelihood that a t test will produce a significant result. However, the sign test only considers the direction of each difference score and is not influenced by the variance of the scores.

LEARNING CHECK

1. In terms of the sample data, what other test does the binomial test most resemble?
   a. An analysis of variance.
   b. A test comparing two population means.
   c. The chi square test for goodness of fit.
   d. The Pearson correlation.

2. In the sign test, how are the difference scores coded?
   a. By their ranked size, with increases and decreases ranked separately.
   b. By their ranked size, independent of sign (increase or decrease).
   c. As increases ($+$) or decreases ($-$).
   d. As probabilities, p and q.

3. How should zero differences be interpreted in a sign test?
   a. They provide support for the null hypothesis.
   b. They provided evidence that the null hypothesis should be rejected.
   c. They are not relevant to the null hypothesis and should be discarded.
   d. They indicate that assumptions have been violated and the test should not be done.

ANSWERS   1. C, 2. C, 3. A

SUMMARY

1. The binomial test is used with dichotomous data—that is, when each individual in the sample is classified in one of two categories. The two categories are identified as A and B, with probabilities of

$$p(A) = p \text{ and } p(B) = q$$

2. The binomial distribution gives the probability for each value of X, where X equals the number of occurrences of category A in a sample of n events. For example, X equals the number of heads in $n = 10$ tosses of a coin.

3. When pn and qn are both at least 10, the binomial distribution is closely approximated by a normal distribution with

$$\mu = pn \quad \sigma = \sqrt{npq}$$

By using this normal approximation, each value of X has a corresponding z-score:

$$z = \frac{X - \mu}{\sigma} = \frac{X - pn}{\sqrt{npq}} \quad \text{or} \quad z = \frac{X/n - p}{\sqrt{pq/n}}$$

4. The binomial test uses sample data to test hypotheses about the binomial proportions, p and q, for a population. The null hypothesis specifies p and q, and the binomial distribution (or the normal approximation) is used to determine the critical region.

5. Usually the z-score in a binomial test is computed using the whole-number X value from the sample. However, if the z-score is only marginally in the critical region, you should compute the z-scores corresponding to both ends of the interval corresponding to the score. If either one of these z-score boundaries is not in the critical region, the correct decision is to fail to reject the null hypothesis.

6. One common use of the binomial distribution is for the sign test. This test evaluates the difference between two treatments using the data from a repeated measures design. The difference scores are coded as being either increases ($+$) or decreases ($-$). Without a consistent treatment effect, the increases and decreases should be mixed randomly, so the null hypothesis states that

$$p(\text{increase}) = \frac{1}{2} = p(\text{decrease})$$

With dichotomous data and hypothesized values for p and q, this is a binomial test.

KEY TERMS

binomial data (604)

binomial test (605)

binomial distribution (606)

sign test (614)

SPSS®

General instructions for using SPSS are presented in Appendix D. Following are detailed instructions for using SPSS to perform the **Binomial Test** that is presented in this chapter.

Data Entry

1. Enter the values 0 and 1 in the first column of the SPSS data editor.
2. In the second column, enter the frequencies that were obtained for the two binomial categories. For example, if 21 out of 25 people were classified in category A (and only 4 people in category B), you would enter the values 21 and 4 in the second column.

Data Analysis

1. Click **Data** on the tool bar at the top of the page and select **weight cases** at the bottom of the list.
2. Click the **weight cases by** circle, then highlight the label for the column containing the frequencies for the two categories and move it into the **Frequency Variable** box by clicking on the arrow.
3. Click OK.
4. Click Analyze on the tool bar, select **Nonparametric Tests,** and click on **Binomial**.
5. Highlight the label for the column containing the digits 0 and 1, and move it into the **Test Variables List** box by clicking on the arrow.
6. Specify the **Test Proportion**, which is the value of p from the null hypothesis. (The box is preset at .50 but you can change it to the value appropriate for your test.)
7. Click **OK**.

SPSS Output

The program will produce a table summarizing the results of the test. The table shows the number of individuals in each category, the proportion of individuals in each category, the test proportion from the null hypothesis, and a level of significance (the p value or alpha level for the test).

FOCUS ON PROBLEM SOLVING

1. For all binomial tests, the values of p and q must add up to 1.00 (or 100%).

2. Remember that both pn and qn must be at least 10 before you can use the normal distribution to determine critical values for a binomial test.

3. Although the binomial test usually specifies the critical region in terms of z-scores, it is possible to identify the X values that determine the critical region. With $\alpha = .05$, the critical region is determined by z-scores greater than 1.96 or less than -1.96. That is, to be significantly different from what is expected by chance, the individual score must be above (or below) the mean by at least 1.96 standard deviations. For example, in an ESP experiment in which an individual is trying to predict the suit of a playing card for a sequence of $n = 64$ trials, chance probabilities are

$$p = p(\text{right}) = \frac{1}{4} \quad q = p(\text{wrong}) = \frac{3}{4}$$

For this example, the binomial distribution has a mean of $pn = \left(\frac{1}{4}\right)(64) = 16$ right and a standard deviation of $\sqrt{npq} = \sqrt{64\left(\frac{1}{4}\right)\left(\frac{3}{4}\right)} = \sqrt{12} = 3.46$. To be significantly different from chance, a score must be above (or below) the mean by at least $1.96(3.46) = 6.78$. Thus, with a mean of 16, an individual would need to score above 22.78 $(16 + 6.78)$ or below 9.22 $(16 - 6.78)$ to be significantly different from chance.

DEMONSTRATION 18.1

THE BINOMIAL TEST

The population of students in the psychology department at State College consists of 60% females and 40% males. Last semester, the Psychology of Gender course had a total of 36 students, of whom 26 were female and only 10 were male. Are the proportions of females and males in this class significantly different from what would be expected by chance from a population with 60% females and 40% males? Test at the .05 level of significance.

STEP 1 **State the hypotheses, and specify alpha** The null hypothesis states that the male/female proportions for the class are not different from what is expected for a population with these proportions. In symbols,

$$H_0: \quad p = p(\text{female}) = 0.60 \text{ and } q = p(\text{male}) = 0.40$$

The alternative hypothesis is that the proportions for this class are different from what is expected for these population proportions.

$$H_1: \quad p \neq 0.60 \text{ (and } q \neq 0.40)$$

We will set alpha at $\alpha = .05$.

STEP 2 **Locate the critical region** Because pn and qn are both greater than 10, we can use the normal approximation to the binomial distribution. With $\alpha = .05$, the critical region is defined as any z-score value greater than -1.96 or less than -1.96.

STEP 3 **Calculate the test statistic** The sample has 26 females out of 36 students, so the sample proportion is

$$\frac{X}{n} = \frac{26}{36} = 0.72$$

The corresponding z-score (using Equation 18.2) is

$$z = \frac{X/n - p}{\sqrt{pq/n}} = \frac{0.72 - 0.60}{\sqrt{\dfrac{0.60(0.40)}{36}}} = \frac{0.12}{0.0816} = 1.47$$

STEP 4   **Make a decision about H_0, and state a conclusion**   The obtained z-score is not in the critical region. Therefore, we fail to reject the null hypothesis. On the basis of these data, you conclude that the male/female proportions in the gender class are not significantly different from the proportions in the psychology department as a whole.

PROBLEMS

1. Insurance companies charge young drivers more for automobile insurance because they tend to have more accidents than older drivers. To make this point, an insurance representative first determines that only 15% of licensed drivers are age 20 or younger. Because this age group makes up only 15% of the drivers, it is reasonable to predict that they should be involved in only 15% of the accidents. In a random sample of 80 accident reports, however, the representative finds 26 accidents that involved drivers who were 20 or younger. Is this sample sufficient to show that younger drivers have significantly more accidents than would be expected from the percentage of young drivers? Use a two-tailed test with $\alpha = .05$.

2. Güven, Elaimis, Binokay, and Tan (2003) studied the distribution of paw preferences in rats using a computerized food-reaching test. For a sample of $n = 144$ rats, they found 104 right-handed animals. Is this significantly more than would be expected if right- and left-handed rats are equally common in the population? Use a two-tailed test with $\alpha = .01$.

3. In problem 9 in Chapter 17, we described a study demonstrating that the color red is often associated with male dominance. Hill and Barton (2005) monitored the outcome of four combat sports (boxing, Tae Kwon Do, Greco-Roman wrestling, and freestyle wrestling) during the 2004 Olympic games and found that participants wearing red outfits won significantly more often than those wearing blue.
   a. If athletes wearing red won 31 out of 50 matches, is that sufficient to be significantly more than would be expected by chance? Use a two-tailed test with $\alpha = .05$.
   b. If 62 out of 100 wearing red won, is that enough to be significant using a two-tailed test with $\alpha = .05$.
   c. Note that the percentage of winning for red uniforms in part a is identical to the percentage in part b ($31/50 = 62/100 = 62\%$) however, one is significant and the other is not. Explain why the two samples lead to different conclusions.

4. Problems 5 and 6 in Chapter 17 cited a study showing that people tend to choose partners who are similar to themselves. Jones, Pelham, Carvallo, and Mirenberg, (2004) demonstrated that people have a tendency to select marriage partners with surnames that begin with the same last letter as their own. The probability of randomly matching two last names beginning with the same letter is only $p = 0.065$ (6.5%).
   a. If 19 out of 200 marriages involved brides and grooms with last names beginning with the same letter, is that enough to be significantly different than would be expected by chance? Use a two-tailed test with $\alpha = .05$.
   b. If 38 out of 400 marriages involved the same letter, is that enough to be significant using a two-tailed test with $\alpha = .05$?

5. A researcher would like to determine whether people really can tell the difference between bottled water and tap water. Participants are asked to taste two unlabeled glasses of water, one bottled and one tap, and identify the one they think tastes better. Out of 40 people in the sample, 28 picked the bottled water. Was the bottled water selected significantly more often than would be expected by chance? Use a two-tailed test with $\alpha = .05$.

6. A recent survey of practicing psychotherapists revealed that 25% of the individuals responding agreed with the statement, "Hypnosis can be used to recover accurate memories of past lives" (Yapko, 1994). A researcher would like to determine whether this same level of belief exists in the general population. A sample of 192 adults is surveyed and 65 believe that hypnosis can be used to recover accurate memories of past lives. Based on these data, can you conclude that beliefs held by the general population are significantly different from beliefs held by psychotherapists? Test with $\alpha = .05$.

7. In 2005, Fung et al. published a study reporting that patients prefer technical quality vs. interpersonal skills

when selecting a primary care physician. Participants were presented with report cards describing pairs of hypothetical physicians and were asked to select the one that they preferred. Suppose that this study is repeated with a sample of $n = 150$ participants, and the results show that physicians with greater technical skill are preferred by 92 participants and physicians with greater interpersonal skills are selected by 58. Are these results sufficient to conclude that there is a significant preference for technical skill? Test with $\alpha = .01$.

8. In problem 13 in Chapter 9, we presented a study examining the spotlight effect, which refers to overestimating the extent to which others notice your appearance or behavior, especially when you commit a social faux pas. Effectively, you feel as if you are suddenly standing in a spotlight with everyone looking. Gilovich, Medvec, and Savitsky (2000) asked college students to put on a Barry Manilow T-shirt that fellow students had previously judged to be embarrassing. The participants were then led into a room in which other students were already participating in an experiment. Later, each participant was asked to estimate how many people in the room had noticed the shirt and the individuals in the room were also asked whether they noticed the shirt. If 15 out of 20 participants overestimated the number who noticed the shirt, is this enough to be significantly more than chance? Use a two-tailed test with $\alpha = .05$.

9. Danner and Phillips (2008) report the results from a county-wide study showing that delaying high school start times by one hour significantly reduced the motor vehicle crash rate for teen drivers in the study. Suppose that the researchers monitored 500 student drivers for one year after the start time was delayed and found that 45 were involved in automobile accidents. Before delaying the start time, the accident rate was 12%.
   a. Use a binomial test to determine whether these results indicate a significant change in the accident rate following the change in school start time. Use a two-tailed test with $\alpha = .05$
   b. Because the outcome of the binomial test is a borderline z-score, use both ends of the interval corresponding to $X = 45$ to determine whether the entire interval is in the critical region.

10. For each of the following, assume that a two-tailed test using the normal approximation to the binomial distribution with $\alpha = .05$ is being used to evaluate the significance of the result.
   a. For a true-false test with 25 questions, how many would you have to get right to do significantly better than chance? That is, what X value is needed to produce a z-score greater than 1.96?

   b. How many would you need to get right on a 50-question true-false test?
   c. How many would you need to get right on a 75-question true-false test?

   Remember that each X value corresponds to an interval. Be sure that the entire interval is in the critical region.

11. On a multiple-choice exam with 80 questions and 4 possible answers for each question, you get a score of $X = 29$. Is your score significantly better than would be expected by chance (by simply guessing for each question)? Use a two-tailed test with $\alpha = .05$.

12. One of the original methods for testing ESP (extrasensory perception) used Zener cards, which were designed specifically for the testing process. Each card shows one of five symbols (square, circle, star, wavy lines, cross). The person being tested must predict the symbol before the card is turned over. Chance performance on this task would produce 1 out of 5 (20%) correct predictions. Use a binomial test to determine whether 27 correct predictions out of 100 attempts is significantly different from chance performance? Use a two-tailed test with $\alpha = .05$.

13. Reed, Vernon, and Johnson (2004) examined the relationship between brain nerve conduction velocity and intelligence in normal adults. Brain nerve conduction velocity was measured three separate ways and nine different measures were used for intelligence. The researchers then correlated each of the three nerve velocity measures with each of the nine intelligence measures for a total of 27 separate correlations. Unfortunately, none of the correlations were significant.
   a. For the 186 males in the study, however, 25 of the 27 correlations were positive. Is this significantly more than would be expected if positive and negative correlations were equally likely? Use a two-tailed test with $\alpha = .05$.
   b. For the 201 females in the study, 20 of the 27 correlations were positive. Is this significantly more than would be expected if positive and negative correlations were equally likely? Use a two-tailed test with $\alpha = .05$.

14. Most children and adults are able to learn the meaning of new words by listening to sentences in which the words appear. Shulman and Guberman (2007) tested the ability of children to learn word meaning from syntactical cues for three groups: children with autism, children with specific language impairment (SLI), and children with typical language development (TLD). Although the researchers used relatively small samples, their results indicate that the children with TLD and those with autism were able to learn novel words using the syntactical cues in sentences.

The children with SLI, on the other hand, experienced significantly more difficulty. Suppose a similar study is conducted in which each child listens to a set of sentences containing a novel word and then is given a choice of three definitions for the word.

a. If 25 out of 36 autistic children select the correct definition, is this significantly more than would be expected if they were simply guessing? Use a two-tailed test with $\alpha = .05$.

b. If only 16 out of 36 children with SLI select the correct definition, is this significantly more than would be expected if they were simply guessing? Use a two-tailed test with $\alpha = .05$.

15. In Example 11.2 (p. 343), we presented a repeated-measures research study demonstrating that swearing can help reduce ratings of pain (Stephens, Atkins, & Kingston, 2009). In the study, each participant was asked to plunge a hand into icy water and keep it there as long as the pain would allow. In one condition, the participants repeated their favorite curse words while their hands were in the water. In the other condition, the participants repeated a neutral word. Each participant rated the intensity of the pain on a 10-point scale. Of the $n = 25$ participants, 17 had a lower pain rating when swearing, 6 had a higher rating, and 2 showed no difference between the two conditions. Is this result significantly different from what would be expected by chance? Use a two-tailed test with $\alpha = .05$.

16. Stressful or traumatic experiences can often worsen other health-related problems such as asthma or rheumatoid arthritis. However, if patients are instructed to write about their stressful experiences, it can often lead to improvement in health (Smyth, Stone, Hurewitz, & Kaell, 1999). In a typical study, patients with asthma or arthritis are asked to write about the "most stressful event of your life." In a sample of $n = 112$ patients, suppose that 64 showed improvement in their symptoms, 12 showed no change, and 36 showed worsening symptoms.

a. If the 12 patients showing no change are discarded, are these results sufficient to conclude that the writing had a significant effect? Use a two-tailed test with $\alpha = .05$.

b. If the 12 patients who showed no change are split between the two groups, are the results sufficient to demonstrate a significant change? Use a two-tailed test with $\alpha = .05$.

17. Langewitz, Izakovic, and Wyler (2005) reported that self-hypnosis can significantly reduce hay-fever symptoms. Patients with moderate to severe allergic reactions were trained to focus their minds on specific locations where their allergies did not bother them, such as a beach or a ski resort. In a sample of 64 patients who received this training, suppose that 47 showed reduced allergic reactions and 17 showed an increase in allergic reactions. Are these results sufficient to conclude that the self-hypnosis has a significant effect? Use a two-tailed test with $\alpha = .05$.

18. In problem 11 of Chapter 11, we described a repeated-measures study showing that male college athletes significantly increased their attendance when they were required to text "in class" to their academic counselors when they arrived for class (Bicard, Lott, Mills, Bicard, & Baylot-Casey, 2012). If the study used a sample of $n = 24$ athletes with 19 showing increased attendance and only 5 showing poorer attendance, is the result sufficient to conclude that there is a significant effect? Use a two-tailed test with $\alpha = .05$.

19. Group-housed laying hens appear to prefer having more floor space than height in their cages. Albentosa and Cooper (2005) tested hens in groups of 10. The birds in each group were given free choice between a cage with a height of 38 cm (low) and a cage with a height of 45 cm (high). The results showed a tendency for the hens in each group to distribute themselves evenly between the two cages, showing no preference for either height. Suppose that a similar study tested a sample of $n = 80$ hens and found that 47 preferred the taller cage. Does this result indicate a significant preference? Use a two-tailed test with $\alpha = .05$.

20. In Problem 14 in Chapter 11, we described a study showing that students are likely to improve their test scores if they go back and change answers after reconsidering some of the questions on the exam (Johnston, 1975). In the study, one group of students was encouraged to reconsider each question and to change answers whenever they felt it was appropriate. The students were asked to record their original answers as well as the changes. For each student, the exam was graded based on the original answers and on the changed answers. For a group of $n = 40$ students, suppose that 26 had higher scores for the changed-answer version and only 14 had higher scores for the original-answer version. Is this result significantly different from chance? Use a two-tailed test with $\alpha = .01$.

21. In problem 13 in Chapter 17, we discussed a study by Loftus and Palmer (1974) examining how different phrasing of questions can influence eyewitness testimony. In the study, students watched a video of an automobile accident and then were questioned about what they had seen. One group of participants was asked to estimate the speed of the cars when "they smashed into each other." Another group of was asked to estimate the speed of the cars when "they hit each

other." Suppose that the actual speed of the cars was 22 miles per hour.

a. For the 50 people in the "smashed-into" group, assume that 32 overestimated the actual speed, 17 underestimated the speed, 1 was exactly right. Is this result significantly different from what would be expected by chance? Use a two-tailed test with $\alpha = .05$.

b. For the 50 people in the "hit" group, assume that 27 overestimated the actual speed, 22 underestimated the speed, 1 was exactly right. Again, use a two-tailed test with $\alpha = .05$ to determine whether this result significantly different from what would be expected by chance.

22. The habituation technique is one method that is used to examine memory for infants. The procedure involves presenting a stimulus to an infant (usually projected on the ceiling above the crib) for a fixed time period and recording how long the infant spends looking at the stimulus. After a brief delay, the stimulus is presented again. If the infant spends less time looking at the stimulus during the second presentation, it is interpreted as indicating that the stimulus is remembered and, therefore, is less novel and less interesting than it was on the first presentation. This procedure is used with a sample of $n = 30$ 2-week-old infants. For this sample, 22 infants spent less time looking at the stimulus during the second presentation than during the first. Do these data indicate a significant difference? Test at the .01 level of significance.

23. A researcher is testing the effectiveness of a skills-mastery imagery program for soccer players. A sample of $n = 25$ college varsity players is selected and each player is tested on a ball-handling obstacle course before beginning the imagery program and again after completing the 5-week program. Of the 25 players, 18 showed improved performance on the obstacle course after the imagery program and 7 were worse.

a. Is this result sufficient to conclude that there is a significant change in performance following the imagery program? Use a two-tailed test with $\alpha = .05$.

b. Because the outcome of the binomial test is a borderline z-score, use the interval boundaries for $X = 18$ and verify that the entire z-score interval is located in the critical region.

24. Last year the college counseling center offered a workshop for students who claimed to suffer from extreme exam anxiety. Of the 45 students who attended the workshop, 31 had higher grade point averages this semester than they did last year. Do these data indicate a significant difference from what would be expected by chance? Test at the .01 level of significance.

25. Biofeedback training is often used to help people who suffer migraine headaches. A recent study found that 29 out of 50 participants reported a decrease in the frequency and severity of their headaches after receiving biofeedback training. Of the remaining participants in this study, 10 reported that their headaches were worse, and 11 reported no change.

a. Discard the zero-difference participants, and use a sign test with $\alpha = .05$ to determine whether the biofeedback produced a significant difference.

b. Divide the zero-difference participants between the two groups, and use a sign test to evaluate the effect of biofeedback training.

Basic Mathematics Review

Preview

This appendix reviews some of the basic math skills that are necessary for the statistical calculations presented in this book. Many students already will know some or all of this material. Others will need to do extensive work and review. To help you assess your own skills, we include a skills assessment exam here. You should allow approximately 30 minutes to complete the test. When you finish, grade your test using the answer key on pages 644–645.

Notice that the test is divided into five sections. If you miss more than three questions in any section of the test, you probably need help in that area. Turn to the section of this appendix that corresponds to your problem area. In each section, you will find a general review, examples, and additional practice problems. After reviewing the appropriate section and doing the practice problems, turn to the end of the appendix. You will find another version of the skills assessment exam. If you still miss more than three questions in any section of the exam, continue studying. Get assistance from an instructor or a tutor if necessary. At the end of this appendix is a list of recommended books for individuals who need a more extensive review than can be provided here. We stress that mastering this material now will make the rest of the course much easier.

Skills Assessment Preview Exam

■ Section 1

(corresponding to Section A.1 of this appendix)

1. $3 + 2 \times 7 = ?$
2. $(3 + 2) \times 7 = ?$
3. $3 + 2^2 - 1 = ?$
4. $(3 + 2)^2 - 1 = ?$
5. $12/4 + 2 = ?$
6. $12/(4 + 2) = ?$
7. $12/(4 + 2)^2 = ?$
8. $2 \times (8 - 2^2) = ?$
9. $2 \times (8 - 2)^2 = ?$
10. $3 \times 2 + 8 - 1 \times 6 = ?$
11. $3 \times (2 + 8) - 1 \times 6 = ?$
12. $3 \times 2 + (8 - 1) \times 6 = ?$

■ Section 2

(corresponding to Section A.2 of this appendix)

1. The fraction $\frac{3}{4}$ corresponds to a percentage of
   _____.
2. Express 30% as a fraction.
3. Convert $\frac{12}{40}$ to a decimal.
4. $\frac{2}{13} + \frac{8}{13} = ?$
5. $1.375 + 0.25 = ?$
6. $\frac{2}{5} \times \frac{1}{4} = ?$
7. $\frac{1}{8} + \frac{2}{3} = ?$
8. $3.5 \times 0.4 = ?$
9. $\frac{1}{5} \div \frac{3}{4} = ?$
10. $3.75/0.5 = ?$
11. In a group of 80 students, 20% are psychology majors. How many psychology majors are in this group?
12. A company reports that two-fifths of its employees are women. If there are 90 employees, how many are women?

■ Section 3

(corresponding to Section A.3 of this appendix)

1. $3 + (-2) + (-1) + 4 = ?$
2. $6 - (-2) = ?$
3. $-2 - (-4) = ?$
4. $6 + (-1) - 3 - (-2) - (-5) = ?$
5. $4 \times (-3) = ?$

6. $-2 \times (-6) = ?$
7. $-3 \times 5 = ?$
8. $-2 \times (-4) \times (-3) = ?$
9. $12 \div (-3) = ?$
10. $-18 \div (-6) = ?$
11. $-16 \div 8 = ?$
12. $-100 \div (-4) = ?$

■ Section 4

(corresponding to Section A.4 of this appendix)
For each equation, find the value of X.

1. $X + 6 = 13$
2. $X - 14 = 15$
3. $5 = X - 4$
4. $3X = 12$
5. $72 = 3X$
6. $X/5 = 3$
7. $10 = X/8$
8. $3X + 5 = -4$
9. $24 = 2X + 2$
10. $(X + 3)/2 = 14$
11. $(X - 5)/3 = 2$
12. $17 = 4X - 11$

■ Section 5

(corresponding to Section A.5 of this appendix)

1. $4^3 = ?$
2. $\sqrt{25 - 9} = ?$
3. If $X = 2$ and $Y = 3$, then $XY^3 = ?$
4. If $X = 2$ and $Y = 3$, then $(X + Y)^2 = ?$
5. If $a = 3$ and $b = 2$, then $a^2 + b^2 = ?$
6. $(-3)^3 = ?$
7. $(-4)^4 = ?$
8. $\sqrt{4} \times 4 = ?$
9. $36/\sqrt{9} = ?$
10. $(9 + 2)^2 = ?$
11. $5^2 + 2^3 = ?$
12. If $a = 3$ and $b = -1$, then $a^2b^3 = ?$

   The answers to the skills assessment exam are at the end of the appendix (pages 697−698).

A.1 | Symbols and Notation

Table A.1 presents the basic mathematical symbols that you should know, along with examples of their use. Statistical symbols and notation are introduced and explained throughout this book as they are needed. Notation for exponents and square roots is covered separately at the end of this appendix.

Parentheses are a useful notation because they specify and control the order of computations. Everything inside the parentheses is calculated first. For example,

$$(5 + 3) \times 2 = 8 \times 2 = 16$$

Changing the placement of the parentheses also changes the order of calculations. For example,

$$5 + (3 \times 2) = 5 + 6 = 11$$

■ Order of Operations

Often a formula or a mathematical expression will involve several different arithmetic operations, such as adding, multiplying, squaring, and so on. When you encounter these situations, you must perform the different operations in the correct sequence. Following is a list of mathematical operations, showing the order in which they are to be performed.

1. Any calculation contained within parentheses is done first.
2. Squaring (or raising to other exponents) is done second.
3. Multiplying and/or dividing is done third. A series of multiplication and/or division operations should be done in order from left to right.
4. Adding and/or subtracting is done fourth.

The following examples demonstrate how this sequence of operations is applied in different situations.

To evaluate the expression

$$(3 + 1)^2 - 4 \times 7/2$$

first, perform the calculation within parentheses:

$$(4)^2 - 4 \times 7/2$$

Next, square the value as indicated:

$$16 - 4 \times 7/2$$

TABLE A.1

Symbol	Meaning	Example
$+$	Addition	$5 + 7 = 12$
$-$	Subtraction	$8 - 3 = 5$
$\times$, ( )	Multiplication	$3 \times 9 = 27$, $3(9) = 27$
$\div$, /	Division	$15 \div 3 = 5$, $15/3 = 5$, $\frac{15}{3} = 5$
$>$	Greater than	$20 > 10$
$<$	Less than	$7 < 11$
$\neq$	Not equal to	$5 \neq 6$

Then perform the multiplication and division:

$$16 - 14$$

Finally, do the subtraction:

$$16 - 14 = 2$$

A sequence of operations involving multiplication and division should be performed in order from left to right. For example, to compute $12/2 \times 3$, you divide 12 by 2 and then multiply the result by 3:

$$12/2 \times 3 = 6 \times 3 = 18$$

Notice that violating the left-to-right sequence can change the result. For this example, if you multiply before dividing, you will obtain

$$12/2 \times 3 = 12/6 = 2 \qquad \textbf{(This is wrong.)}$$

A sequence of operations involving only addition and subtraction can be performed in any order. For example, to compute $3 + 8 - 5$, you can add 3 and 8 and then subtract 5:

$$(3 + 8) - 5 = 11 - 5 = 6$$

or you can subtract 5 from 8 and then add the result to 3:

$$3 + (8 - 5) = 3 + 3 = 6$$

A mathematical expression or formula is simply a concise way to write a set of instructions. When you evaluate an expression by performing the calculation, you simply follow the instructions. For example, assume you are given these instructions:

1. First, add 3 and 8.
2. Next, square the result.
3. Next, multiply the resulting value by 6.
4. Finally, subtract 50 from the value you have obtained.

You can write these instructions as a mathematical expression.

1. The first step involves addition. Because addition is normally done last, use parentheses to give this operation priority in the sequence of calculations:

$$(3 + 8)$$

2. The instruction to square a value is noted by using the exponent 2 beside the value to be squared:

$$(3 + 8)^2$$

3. Because squaring has priority over multiplication, you can simply introduce the multiplication into the expression:

$$6 \times (3 + 8)^2$$

4. Addition and subtraction are done last, so simply write in the requested subtraction:

$$6 \times (3 + 8)^2 - 50$$

To calculate the value of the expression, you work through the sequence of operations in the proper order:

$$6 \times (3 + 8)^2 - 50 = 6 \times (11)^2 - 50$$
$$= 6 \times (121) - 50$$
$$= 726 - 50$$
$$= 676$$

As a final note, you should realize that the operation of squaring (or raising to any exponent) applies only to the value that immediately precedes the exponent. For example,

$$2 \times 3^2 = 2 \times 9 = 18 \qquad \textbf{(Only the 3 is squared.)}$$

If the instructions require multiplying values and then squaring the product, you must use parentheses to give the multiplication priority over squaring. For example, to multiply 2 times 3 and then square the product, you would write

$$(2 \times 3)^2 = (6)^2 = 36$$

LEARNING CHECK

1. Evaluate each of the following expressions:
   a. $4 \times 8/2^2$
   b. $4 \times (8/2)^2$
   c. $100 - 3 \times 12/(6 - 4)^2$
   d. $(4 + 6) \times (3 - 1)^2$
   e. $(8 - 2)/(9 - 8)^2$
   f. $6 + (4 - 1)^2 - 3 \times 4^2$
   g. $4 \times (8 - 3) + 8 - 3$

ANSWERS

1. **a.** 8, **b.** 64, **c.** 91, **d.** 40, **e.** 6, **f.** −33, **g.** 25

A.2 | Proportions: Fractions, Decimals, and Percentages

A proportion is a part of a whole and can be expressed as a fraction, a decimal, or a percentage. For example, in a class of 40 students, only 3 failed the final exam.

The proportion of the class that failed can be expressed as a fraction

$$\text{fraction} = \frac{3}{40}$$

or as a decimal value

$$\text{decimal} = 0.075$$

or as a percentage

$$\text{percentage} = 7.5\%$$

In a fraction, such as $\frac{3}{4}$, the bottom value (the denominator) indicates the number of equal pieces into which the whole is split. Here the "pie" is split into 4 equal pieces:

If the denominator has a larger value—say, 8—then each piece of the whole pie is smaller:

A larger denominator indicates a smaller fraction of the whole.

The value on top of the fraction (the numerator) indicates how many pieces of the whole are being considered. Thus, the fraction $\frac{3}{4}$ indicates that the whole is split evenly into 4 pieces and that 3 of them are being used:

A fraction is simply a concise way of stating a proportion: "Three out of four" is equivalent to $\frac{3}{4}$. To convert the fraction to a decimal, you divide the numerator by the denominator:

$$\frac{3}{4} = 3 \div 4 = 0.75$$

To convert the decimal to a percentage, simply multiply by 100, and place a percent sign (%) after the answer:

$$0.75 \times 100 = 75\%$$

The U.S. money system is a convenient way of illustrating the relationship between fractions and decimals. "One quarter," for example, is one-fourth $\left(\frac{1}{4}\right)$ of a dollar, and its decimal equivalent is 0.25. Other familiar equivalencies are as follows:

	Dime	Quarter	50 Cents	75 Cents
Fraction	$\frac{1}{10}$	$\frac{1}{10}$	$\frac{1}{2}$	$\frac{3}{4}$
Decimal	0.10	0.25	0.50	0.75
Percentage	10%	25%	50%	75%

■ Fractions

1. **Finding Equivalent Fractions**   The same proportional value can be expressed by many equivalent fractions. For example,

$$\frac{1}{2} = \frac{2}{4} = \frac{10}{20} = \frac{50}{100}$$

To create equivalent fractions, you can multiply the numerator and denominator by the same value. As long as both the numerator and the denominator of the fraction are multiplied by the same value, the new fraction will be equivalent to the original. For example,

$$\frac{3}{10} = \frac{9}{30}$$

because both the numerator and the denominator of the original fraction have been multiplied by 3. Dividing the numerator and denominator of a fraction by the same value will also result in an equivalent fraction. By using division, you can reduce a fraction to a simpler form. For example,

$$\frac{40}{100} = \frac{2}{5}$$

because both the numerator and the denominator of the original fraction have been divided by 20.

You can use these rules to find specific equivalent fractions. For example, find the fraction that has a denominator of 100 and is equivalent to $\frac{3}{4}$. That is,

$$\frac{3}{4} = \frac{?}{100}$$

Notice that the denominator of the original fraction must be multiplied by 25 to produce the denominator of the desired fraction. For the two fractions to be equal, both the numerator and the denominator must be multiplied by the same number. Therefore, we also multiply the top of the original fraction by 25 and obtain

$$\frac{3 \times 25}{4 \times 25} = \frac{75}{100}$$

2. **Multiplying Fractions** To multiply two fractions, you first multiply the numerators and then multiply the denominators. For example,

$$\frac{3}{4} \times \frac{5}{7} = \frac{3 \times 5}{4 \times 7} = \frac{15}{28}$$

3. **Dividing Fractions** To divide one fraction by another, you invert the second fraction and then multiply. For example,

$$\frac{1}{2} \div \frac{1}{4} = \frac{1}{2} \times \frac{4}{1} = \frac{1 \times 4}{2 \times 1} = \frac{4}{2} = \frac{2}{1} = 2$$

4. **Adding and Subtracting Fractions** Fractions must have the same denominator before you can add or subtract them. If the two fractions already have a common denominator, you simply add (or subtract as the case may be) *only* the values in the numerators. For example,

$$\frac{2}{5} + \frac{1}{5} = \frac{3}{5}$$

Suppose you divided a pie into five equal pieces (fifths). If you first ate two-fifths of the pie and then another one-fifth, the total amount eaten would be three-fifths of the pie:

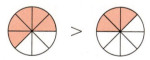

If the two fractions do not have the same denominator, you must first find equivalent fractions with a common denominator before you can add or subtract. The product of the two denominators will always work as a common denominator for equivalent fractions (although it may not be the lowest common denominator). For example,

$$\frac{2}{3} + \frac{1}{10} = ?$$

Because these two fractions have different denominators, it is necessary to convert each into an equivalent fraction and find a common denominator. We will use $3 \times 10 = 30$ as the common denominator. Thus, the equivalent fraction of each is

$$\frac{2}{3} = \frac{20}{30} \quad \text{and} \quad \frac{1}{10} = \frac{3}{30}$$

Now the two fractions can be added:

$$\frac{20}{30} + \frac{3}{30} = \frac{23}{30}$$

5. **Comparing the Size of Fractions**   When comparing the size of two fractions with the same denominator, the larger fraction will have the larger numerator. For example,

$$\frac{5}{8} > \frac{3}{8}$$

The denominators are the same, so the whole is partitioned into pieces of the same size. Five of these pieces are more than three of them:

When two fractions have different denominators, you must first convert them to fractions with a common denominator to determine which is larger. Consider the following fractions:

$$\frac{3}{8} \quad \text{and} \quad \frac{7}{16}$$

If the numerator and denominator of $\frac{3}{8}$ are multiplied by 2, the resulting equivalent fraction will have a denominator of 16:

$$\frac{3}{8} = \frac{3 \times 2}{8 \times 2} = \frac{6}{16}$$

Now a comparison can be made between the two fractions:

$$\frac{6}{16} < \frac{7}{16}$$

Therefore,

$$\frac{3}{8} < \frac{7}{16}$$

■ Decimals

1. **Converting Decimals to Fractions**   Like a fraction, a decimal represents part of the whole. The first decimal place to the right of the decimal point indicates how many tenths are used. For example,

$$0.1 = \frac{1}{10} \quad 0.7 = \frac{7}{10}$$

The next decimal place represents $\frac{1}{100}$, the next $\frac{1}{1000}$, the next $\frac{1}{10,000}$, and so on. To change a decimal to a fraction, just use the number without the decimal point for the numerator. Use the denominator that the last (on the right) decimal place represents. For example,

$$0.32 = \frac{32}{100} \quad 0.5333 = \frac{5333}{10,000} \quad 0.05 = \frac{5}{100} \quad 0.001 = \frac{1}{1000}$$

2. **Adding and Subtracting Decimals**   To add and subtract decimals, the only rule is that you must keep the decimal points in a straight vertical line. For example,

$$\begin{array}{r} 0.27 \\ +1.326 \\ \hline 1.596 \end{array} \qquad \begin{array}{r} 3.595 \\ -0.67 \\ \hline 2.925 \end{array}$$

3. **Multiplying Decimals**   To multiply two decimal values, you first multiply the two numbers, ignoring the decimal points. Then you position the decimal point in the answer so that the number of digits to the right of the decimal point is equal to the total number of decimal places in the two numbers being multiplied. For example,

$$\begin{array}{r} 1.73 \\ \times 0.251 \\ \hline 173 \\ 865 \\ 346 \\ \hline 0.43423 \end{array}$$ **(two decimal places)** **(three decimal places)** **(five decimal places)**

$$\begin{array}{r} 0.25 \\ \times 0.005 \\ \hline 125 \\ 00 \\ 00 \\ \hline 0.00125 \end{array}$$ **(two decimal places)** **(three decimal places)** **(five decimal places)**

4. **Dividing Decimals**   The simplest procedure for dividing decimals is based on the fact that dividing two numbers is identical to expressing them as a fraction:

$$0.25 \div 1.6 \text{ is identical to } \frac{0.25}{1.6}$$

You now can multiply both the numerator and the denominator of the fraction by 10, 100, 1000, or whatever number is necessary to remove the decimal places. Remember that multiplying both the numerator and the denominator of a fraction by the *same* value will create an equivalent fraction. Therefore,

$$\frac{0.25}{1.6} = \frac{0.25 \times 100}{1.6 \times 100} = \frac{25}{160} = \frac{5}{32}$$

The result is a division problem without any decimal places in the two numbers.

■ Percentages

1. **Converting a Percentage to a Fraction or a Decimal**   To convert a percentage to a fraction, remove the percent sign, place the number in the numerator, and use 100 for the denominator. For example,

$$52\% = \frac{52}{100} \qquad 5\% = \frac{5}{100}$$

To convert a percentage to a decimal, remove the percent sign and divide by 100, or simply move the decimal point two places to the left. For example,

$$83\% = 83. \;\; = 0.83$$
$$14.5\% = 14.5 = 0.145$$
$$5\% = \;\; 5. \;\; = 0.05$$

2. **Performing Arithmetic Operations with Percentages**   There are situations in which it is best to express percent values as decimals in order to perform certain arithmetic operations. For example, what is 45% of 60? This question may be stated as

$$45\% \times 60 = ?$$

The 45% should be converted to decimal form to find the solution to this question. Therefore,

$$0.45 \times 60 = 27$$

LEARNING CHECK

1. Convert $\frac{3}{25}$ to a decimal.

2. Convert $\frac{3}{8}$ to a percentage.

3. Next to each set of fractions, write "True" if they are equivalent and "False" if they are not:

   a. $\frac{3}{8} = \frac{9}{24}$ _____   **b.** $\frac{7}{9} = \frac{17}{19}$ _____

   c. $\frac{2}{7} = \frac{4}{14}$ _____

4. Compute the following:

   a. $\frac{1}{6} \times \frac{7}{10}$ **b.** $\frac{7}{8} - \frac{1}{2}$ **c.** $\frac{9}{10} \div \frac{2}{3}$ **d.** $\frac{7}{22} + \frac{2}{3}$

5. Identify the larger fraction of each pair:

   a. $\frac{7}{10}, \frac{21}{100}$ **b.** $\frac{3}{4}, \frac{7}{12}$ **c.** $\frac{22}{3}, \frac{19}{3}$

6. Convert the following decimals into fractions:

   a. 0.012   **b.** 0.77   **c.** 0.005

7. $2.59 \times 0.015 = ?$

8. $1.8 \div 0.02 = ?$

9. What is 28% of 45?

ANSWERS

1. 0.12, 2. 37.5%   3. **a.** True   **b.** False   **c.** True

4. **a.** $\frac{7}{60}$ **b.** $\frac{3}{8}$ **c.** $\frac{27}{20}$ **d.** $\frac{65}{66}$ 5. **a.** $\frac{7}{10}$ **b.** $\frac{3}{4}$ **c.** $\frac{22}{3}$

6. **a.** $\frac{12}{1000} = \frac{3}{250}$ **b.** $\frac{77}{100}$ **c.** $\frac{5}{1000} = \frac{1}{200}$ 7. 0.03885   8. 90   9. 12.6

A.3 | Negative Numbers

Negative numbers are used to represent values less than zero. Negative numbers may occur when you are measuring the difference between two scores. For example, a researcher may want to evaluate the effectiveness of a propaganda film by measuring people's attitudes with a test both before and after viewing the film:

	Before	After	Amount of Change
Person A	23	27	+4
Person B	18	15	−3
Person C	21	16	−5

Notice that the negative sign provides information about the direction of the difference: a plus sign indicates an increase in value, and a minus sign indicates a decrease.

Because negative numbers are frequently encountered, you should be comfortable working with these values. This section reviews basic arithmetic operations using negative numbers. You should also note that any number without a sign (+ or −) is assumed to be positive.

1. **Adding Negative Numbers**   When adding numbers that include negative values, simply interpret the negative sign as subtraction. For example,

$$3 + (-2) + 5 = 3 - 2 + 5 = 6$$

When adding a long string of numbers, it often is easier to add all the positive values to obtain the positive sum and then to add all of the negative values to obtain the negative sum. Finally, you subtract the negative sum from the positive sum. For example,

$$-1 + 3 + (-4) + 3 + (-6) + (-2)$$

positive sum = 6   negative sum = 13

Answer: $6 - 13 = -7$

2. **Subtracting Negative Numbers**   To subtract a negative number, change it to a positive number, and add. For example,

$$4 - (-3) = 4 + 3 = 7$$

This rule is easier to understand if you think of positive numbers as financial gains and negative numbers as financial losses. In this context, taking away a debt is equivalent to a financial gain. In mathematical terms, taking away a negative number is equivalent to adding a positive number. For example, suppose you are meeting a friend for lunch. You have $7, but you owe your friend $3. Thus, you really have only $4 to spend for lunch. But your friend forgives (takes away) the $3 debt. The result is that you now have $7 to spend. Expressed as an equation,

$$\$4 \text{ minus a } \$3 \text{ debt} = \$7$$

$$4 - (-3) = 4 + 3 = 7$$

3. Multiplying and Dividing Negative Numbers   When the two numbers being multiplied (or divided) have the same sign, the result is a positive number. When the two numbers have different signs, the result is negative. For example,

$$3 \times (-2) = -6$$

$$-4 \times (-2) = +8$$

The first example is easy to explain by thinking of multiplication as repeated addition. In this case,

$$3 \times (-2) = (-2) + (-2) + (-2) = -6$$

You add three negative 2s, which results in a total of negative 6. In the second example, we are multiplying by a negative number. This amounts to repeated subtraction. That is,

$$-4 \times (-2) = -(-2) - (-2) - (-2) - (-2)$$

$$= 2 + 2 + 2 + 2 = 8$$

By using the same rule for both multiplication and division, we ensure that these two operations are compatible. For example,

$$-6 \div 3 = -2$$

which is compatible with

$$3 \times (-2) = -6$$

Also,

$$8 \div (-4) = -2$$

which is compatible with

$$-4 \times (-2) = +8$$

LEARNING CHECK

1. Complete the following calculations:
   a. $3 + (-8) + 5 + 7 + (-1) + (-3)$
   b. $5 - (-9) + 2 - (-3) - (-1)$
   c. $3 - 7 - (-21) + (-5) - (-9)$
   d. $4 - (-6) - 3 + 11 - 14$
   e. $9 + 8 - 2 - 1 - (-6)$
   f. $9 \times (-3)$
   g. $-7 \times (-4)$
   h. $-6 \times (-2) \times (-3)$
   i. $-12 \div (-3)$
   j. $18 \div (-6)$

ANSWERS

1. a. 3   b. 20   c. 21   d. 4   e. 20
   f. −27   g. 28   h. −36   i. 4   j. −3

A.4 | Basic Algebra: Solving Equations

An equation is a mathematical statement that indicates two quantities are identical. For example,

$$12 = 8 + 4$$

Often an equation will contain an unknown (or variable) quantity that is identified with a letter or symbol, rather than a number. For example,

$$12 = 8 + X$$

In this event, your task is to find the value of X that makes the equation "true," or balanced. For this example, an X value of 4 will make a true equation. Finding the value of X is usually called *solving the equation*.

To solve an equation, there are two points to keep in mind

1. Your goal is to have the unknown value (X) isolated on one side of the equation. This means that you need to remove all of the other numbers and symbols that appear on the same side of the equation as the X.

2. The equation remains balanced, provided you treat both sides exactly the same. For example, you could add 10 points to *both* sides, and the solution (the X value) for the equation would be unchanged.

■ Finding the Solution for an Equation

We will consider four basic types of equations and the operations needed to solve them.

1. **When X Has a Value Added to It**   An example of this type of equation is

$$X + 3 = 7$$

Your goal is to isolate X on one side of the equation. Thus, you must remove the $+3$ on the left-hand side. The solution is obtained by subtracting 3 from *both* sides of the equation:

$$X + 3 - 3 = 7 - 3$$
$$X = 4$$

The solution is $X = 4$. You should always check your solution by returning to the original equation and replacing X with the value you obtained for the solution. For this example,

$$X + 3 = 7$$
$$4 + 3 = 7$$
$$7 = 7$$

2. **When X Has a Value Subtracted From It**   An example of this type of equation is

$$X - 8 = 12$$

In this example, you must remove the -8 from the left-hand side. Thus, the solution is obtained by adding 8 to *both* sides of the equation:

$$X - 8 + 8 = 12 + 8$$
$$X = 20$$

Check the solution:

$$X - 8 = 12$$

$$20 - 8 = 12$$

$$12 = 12$$

3. When X Is Multiplied by a Value   An example of this type of equation is

$$4X = 24$$

In this instance, it is necessary to remove the 4 that is multiplied by X. This may be accomplished by dividing both sides of the equation by 4:

$$\frac{4X}{4} = \frac{24}{4}$$

$$X = 6$$

Check the solution:

$$4X = 24$$

$$4(6) = 24$$

$$24 = 24$$

4. When X Is Divided by a Value   An example of this type of equation is

$$\frac{X}{3} = 9$$

Now the X is divided by 3, so the solution is obtained by multiplying by 3. Multiplying both sides yields

$$3\left(\frac{X}{3}\right) = 9(3)$$

$$X = 27$$

For the check,

$$\frac{X}{3} = 9$$

$$\frac{27}{3} = 9$$

$$9 = 9$$

■ Solutions for More Complex Equations

More complex equations can be solved by using a combination of the preceding simple operations. Remember that at each stage you are trying to isolate X on one side of the equation. For example,

$$3X + 7 = 22$$

$$3X + 7 - 7 = 22 - 7 \qquad \textbf{(Remove +7 by subtracting 7 from both sides.)}$$

$$3X = 15$$

$$\frac{3X}{3} = \frac{15}{3} \qquad \textbf{(Remove 3 by dividing both sides by 3.)}$$

$$X = 5$$

To check this solution, return to the original equation, and substitute 5 in place of X:

$$3X + 7 = 22$$
$$3(5) + 7 = 22$$
$$15 + 7 = 22$$
$$22 = 22$$

Following is another type of complex equation frequently encountered in statistics:

$$\frac{X + 3}{4} = 2$$

First, remove the 4 by multiplying both sides by 4:

$$4\left(\frac{X + 3}{4}\right) = 2(4)$$
$$X + 3 = 8$$

Now remove the +3 by subtracting 3 from both sides:

$$X + 3 - 3 = 8 - 3$$
$$X = 5$$

To check this solution, return to the original equation, and substitute 5 in place of X:

$$\frac{X + 3}{4} = 2$$
$$\frac{5 + 3}{4} = 2$$
$$\frac{8}{4} = 2$$
$$2 = 2$$

LEARNING CHECK

1. Solve for X, and check the solutions:

a. $3X = 18$ **b.** $X + 7 = 9$ **c.** $X - 4 = 18$ **d.** $5X - 8 = 12$

e. $\frac{X}{9} = 5$ **f.** $\frac{X + 1}{6} = 4$ **g.** $X + 2 = -5$ **h.** $\frac{X}{5} = -5$

i. $\frac{2X}{3} = 12$ **j.** $\frac{X}{3} + 1 = 3$

ANSWERS

1. a. $X = 6$ **b.** $X = 2$ **c.** $X = 22$ **d.** $X = 4$ **e.** $X = 45$
f. $X = 23$ **g.** $X = -7$ **h.** $X = -25$ **i.** $X = 18$ **j.** $X = 6$

A.5 | Exponents and Square Roots

■ Exponential Notation

A simplified notation is used whenever a number is being multiplied by itself. The notation consists of placing a value, called an *exponent*, on the right-hand side of and raised above another number, called a *base*. For example,

$$7^3 \leftarrow \text{exponent}$$
$$\uparrow$$
$$\text{base}$$

The exponent indicates how many times the base is used as a factor in multiplication. Following are some examples:

$7^3 = 7(7)(7)$	**(Read "7 cubed" or "7 raised to the third power")**
$5^2 = 5(5)$	**(Read "5 squared")**
$2^5 = 2(2)(2)(2)(2)$	**(Read "2 raised to the fifth power")**

There are a few basic rules about exponents that you will need to know for this course. They are outlined here.

1. **Numbers Raised to One or Zero**   Any number raised to the first power equals itself. For example,

$$6^1 = 6$$

Any number (except zero) raised to the zero power equals 1. For example,

$$9^0 = 1$$

2. **Exponents for Multiple Terms**   The exponent applies only to the base that is just in front of it. For example,

$$XY^2 = XYY$$

$$a^2b^3 = aabbb$$

3. **Negative Bases Raised to an Exponent**   If a negative number is raised to a power, then the result will be positive for exponents that are even and negative for exponents that are odd. For example,

$$(-4)^3 = -4(-4)(-4)$$
$$= 16(-4)$$
$$= -64$$

and

$$(-3)^4 = -3(-3)(-3)(-3)$$
$$= 9(-3)(-3)$$
$$= 9(9)$$
$$= 81$$

Note: The parentheses are used to ensure that the exponent applies to the entire negative number, including the sign. Without the parentheses there is some ambiguity as to how the exponent should be applied. For example, the expression -3^2 could have two interpretations:

$$-3^2 = (-3)(-3) = 9 \quad \text{or} \quad -3^2 = -(3)(3) = -9$$

4. **Exponents and Parentheses** If an exponent is present outside of parentheses, then the computations within the parentheses are done first, and the exponential computation is done last:

$$(3 + 5)^2 = 8^2 = 64$$

Notice that the meaning of the expression is changed when each term in the parentheses is raised to the exponent individually:

$$3^2 + 5^2 = 9 + 25 = 34$$

Therefore,

$$X^2 + Y^2 \neq (X + Y)^2$$

5. **Fractions Raised to a Power** If the numerator and denominator of a fraction are each raised to the same exponent, then the entire fraction can be raised to that exponent. That is,

$$\frac{a^2}{b^2} = \left(\frac{a}{b}\right)^2$$

For example,

$$\frac{3^2}{4^2} = \left(\frac{3}{4}\right)^2$$

$$\frac{9}{16} = \frac{3}{4}\left(\frac{3}{4}\right)$$

$$\frac{9}{16} = \frac{9}{16}$$

■ Square Roots

The square root of a value equals a number that when multiplied by itself yields the original value. For example, the square root of 16 equals 4 because 4 times 4 equals 16. The symbol for the square root is called a *radical*, $\sqrt{}$. The square root is taken for the number under the radical. For example,

$$\sqrt{16} = 4$$

Finding the square root is the inverse of raising a number to the second power (squaring). Thus,

$$\sqrt{a^2} = a$$

For example,

$$\sqrt{3^2} = \sqrt{9} = 3$$

Also,

$$(\sqrt{b})^2 = b$$

For example,

$$(\sqrt{64})^2 = 8^2 = 64$$

Computations under the same radical are performed *before* the square root is taken. For example,

$$\sqrt{9 + 16} = \sqrt{25} = 5$$

Note that with addition (or subtraction), separate radicals yield a different result:

$$\sqrt{9} + \sqrt{16} = 3 + 4 = 7$$

Therefore,

$$\sqrt{X} + \sqrt{Y} \neq \sqrt{X + Y}$$
$$\sqrt{X} - \sqrt{Y} \neq \sqrt{X - Y}$$

If the numerator and denominator of a fraction each have a radical, then the entire fraction can be placed under a single radical:

$$\frac{\sqrt{16}}{\sqrt{4}} = \sqrt{\frac{16}{4}}$$

$$\frac{4}{2} = \sqrt{4}$$

$$2 = 2$$

Therefore,

$$\frac{\sqrt{X}}{\sqrt{Y}} = \sqrt{\frac{X}{Y}}$$

Also, if the square root of one number is multiplied by the square root of another number, then the same result would be obtained by taking the square root of the product of both numbers. For example,

$$\sqrt{9} \times \sqrt{16} = \sqrt{9 \times 16}$$

$$3 \times 4 = \sqrt{144}$$

$$12 = 12$$

Therefore,

$$\sqrt{a} \times \sqrt{b} = \sqrt{ab}$$

LEARNING CHECK

1. Perform the following computations:

   a. $(-6)^3$

   b. $(3 + 7)^2$

   c. a^3b^2 when $a = 2$ and $b = -5$

   d. a^4b^3 when $a = 2$ and $b = 3$

   e. $(XY)^2$ when $X = 3$ and $Y = 5$

   f. $X^2 + Y^2$ when $X = 3$ and $Y = 5$

   g. $(X + Y)^2$ when $X = 3$ and $Y = 5$

   h. $\sqrt{5 + 4}$

   i. $(\sqrt{9})^2$

   j. $\dfrac{\sqrt{16}}{\sqrt{4}}$

ANSWERS

1. **a.** -216 **b.** 100   **c.** 200   **d.** 432   **e.** 225
   f. 34   **g.** 64   **h.** 3   **i.** 9   **j.** 2

Problems for Appendix A   Basic Mathematics Review

1. $50/(10 - 8) = ?$

2. $(2 + 3)^2 = ?$

3. $20/10 \times 3 = ?$

4. $12 - 4 \times 2 + 6/3 = ?$

5. $24/(12 - 4) + 2 \times (6 + 3) = ?$

6. Convert $\frac{7}{20}$ to a decimal.

7. Express $\frac{9}{25}$ as a percentage.

8. Convert 0.91 to a fraction.

9. Express 0.0031 as a fraction.

10. Next to each set of fractions, write "True" if they are equivalent and "False" if they are not:

    a. $\dfrac{4}{1000} = \dfrac{2}{100}$ _____

    b. $\dfrac{5}{6} = \dfrac{52}{62}$ _____

    c. $\dfrac{1}{8} = \dfrac{7}{56}$ _____

11. Perform the following calculations:

    a. $\dfrac{4}{5} \times \dfrac{2}{3} = ?$ **b.** $\dfrac{7}{9} \div \dfrac{2}{3} = ?$

    c. $\dfrac{3}{8} + \dfrac{1}{5} = ?$ **d.** $\dfrac{5}{18} - \dfrac{1}{6} = ?$

12. $2.51 \times 0.017 = ?$

13. $3.88 \times 0.0002 = ?$

14. $3.17 + 17.0132 = ?$

15. $5.55 + 10.7 + 0.711 + 3.33 + 0.031 = ?$

16. $2.04 \div 0.2 = ?$

17. $0.36 \div 0.4 = ?$

18. $5 + 3 - 6 - 4 + 3 = ?$

19. $9 - (-1) - 17 + 3 - (-4) + 5 = ?$

20. $5 + 3 - (-8) - (-1) + (-3) - 4 + 10 = ?$

21. $8 \times (-3) = ?$

22. $-22 \div (-2) = ?$

23. $-2(-4) - (-3) = ?$

24. $84 \div (-4) = ?$

Solve the equations in problems $25 - 32$ for X.

25. $X - 7 = -2$

26. $9 = X + 3$

27. $\dfrac{X}{4} = 11$

28. $-3 = \dfrac{X}{3}$

29. $\dfrac{X + 3}{5} = 2$

30. $\dfrac{X + 1}{3} = -8$

31. $6X - 1 = 11$

32. $2X + 3 = -11$

33. $(-5)^2 = ?$

34. $(-5)^3 = ?$

35. If $a = 4$ and $b = 3$, then $a^2 + b^4 = ?$

36. If $a = -1$ and $b = 4$, then $(a + b)^2 = ?$

37. If $a = -1$ and $b = 5$, then $ab^2 = ?$

38. $\dfrac{18}{\sqrt{4}} = ?$

39. $\sqrt{\dfrac{20}{5}} = ?$

Skills Assessment Final Exam

■ Section 1

1. $4 + 8/4 = ?$
2. $(4 + 8)/4 = ?$
3. $4 \times 3^2 = ?$
4. $(4 \times 3)^2 = ?$
5. $10/5 \times 2 = ?$
6. $10/(5 \times 2) = ?$
7. $40 - 10 \times 4/2 = ?$
8. $(5 - 1)^2/2 = ?$
9. $3 \times 6 - 3^2 = ?$
10. $2 \times (6 - 3)^2 = ?$
11. $4 \times 3 - 1 + 8 \times 2 = ?$
12. $4 \times (3 - 1 + 8) \times 2 = ?$

■ Section 2

1. Express $\frac{14}{80}$ as a decimal.
2. Convert $\frac{6}{25}$ to a percentage.
3. Convert 18% to a fraction.
4. $\frac{3}{5} \times \frac{2}{3} = ?$
5. $\frac{5}{24} + \frac{5}{6} = ?$
6. $\frac{7}{12} \div \frac{5}{6} = ?$
7. $\frac{5}{9} - \frac{1}{3} = ?$
8. $6.11 \times 0.22 = ?$
9. $0.18 \div 0.9 = ?$
10. $8.742 + 0.76 = ?$
11. In a statistics class of 72 students, three-eighths of the students received a B on the first test. How many Bs were earned?
12. What is 15% of 64?

■ Section 3

1. $3 - 1 - 3 + 5 - 2 + 6 = ?$
2. $-8 - (-6) = ?$
3. $2 - (-7) - 3 + (-11) - 20 = ?$
4. $-8 - 3 - (-1) - 2 - 1 = ?$
5. $8(-2) = ?$
6. $-7(-7) = ?$
7. $-3(-2)(-5) = ?$
8. $-3(5)(-3) = ?$
9. $-24 \div (-4) = ?$
10. $36 \div (-6) = ?$
11. $-56/7 = ?$
12. $-7/(-1) = ?$

■ Section 4

Solve for X.

1. $X + 5 = 12$
2. $X - 11 = 3$
3. $10 = X + 4$
4. $4X = 20$
5. $\dfrac{X}{2} = 15$
6. $18 = 9X$
7. $\dfrac{X}{5} = 35$
8. $2X + 8 = 4$
9. $\dfrac{X + 1}{3} = 6$
10. $4X + 3 = -13$
11. $\dfrac{X + 3}{3} = -7$
12. $23 = 2X - 5$

■ Section 5

1. $5^3 = ?$
2. $(-4)^3 = ?$
3. $(-2)^5 = ?$
4. $(-2)^6 = ?$
5. If $a = 4$ and $b = 2$, then $ab^2 = ?$
6. If $a = 4$ and $b = 2$, then $(a + b)^3 = ?$
7. If $a = 4$ and $b = 2$, then $a^2 + b^2 = ?$
8. $(11 + 4)^2 = ?$
9. $\sqrt{7^2} = ?$
10. If $a = 36$ and $b = 64$, then $\sqrt{a + b} = ?$
11. $\dfrac{25}{\sqrt{25}} = ? = ?$
12. If $a = -1$ and $b = 2$, then $a^3 b^4 = ?$

Answer Key Skills Assessment Exams

PREVIEW EXAM

■ Section 1

1. 17
2. 35
3. 6
4. 24
5. 5
6. 2
7. $\dfrac{1}{3}$
8. 8
9. 72
10. 8
11. 24
12. 48

FINAL EXAM

■ Section 1

1. 6
2. 3
3. 36
4. 144
5. 4
6. 1
7. 20
8. 8
9. 9
10. 18
11. 27
12. 80

PREVIEW EXAM

■ Section 2

1. 75%   **2.** $\dfrac{30}{100}$, or $\dfrac{3}{10}$ **3.** 0.3

4. $\dfrac{10}{13}$ **5.** 1.625   **6.** $\dfrac{2}{20}$, or $\dfrac{1}{10}$

7. $\dfrac{19}{24}$ **8.** 1.4   **9.** $\dfrac{4}{15}$

10. 7.5   **11.** 16   **12.** 36

■ Section 3

1. 4   **2.** 8   **3.** 2

4. 9   **5.** -12 **6.** 12

7. -15 **8.** -24 **9.** -4

10. 3   **11.** -2 **12.** 25

■ Section 4

1. $X = 7$ **2.** $X = 29$ **3.** $X = 9$

4. $X = 4$ **5.** $X = 24$ **6.** $X = 15$

7. $X = 80$ **8.** $X = -3$ **9.** $X = 11$

10. $X = 25$ **11.** $X = 11$ **12.** $X = 7$

■ Section 5

1. 64   **2.** 4   **3.** 54

4. 25   **5.** 13   **6.** -27

7. 256   **8.** 8   **9.** 12

10. 121   **11.** 33   **12.** -9

FINAL EXAM

■ Section 2

1. 0.175   **2.** 24%   **3.** $\dfrac{18}{100}$, or $\dfrac{9}{50}$

4. $\dfrac{6}{15}$, or $\dfrac{2}{5}$ **5.** $\dfrac{25}{24}$ **6.** $\dfrac{42}{60}$, or $\dfrac{7}{10}$

7. $\dfrac{2}{9}$ **8.** 1.3442   **9.** 0.2

10. 9.502   **11.** 27   **12.** 9.6

■ Section 3

1. 8   **2.** -2 **3.** -25

4. -13 **5.** -16 **6.** 49

7. -30 **8.** 45   **9.** 6

10. -6 **11.** -8 **12.** 7

■ Section 4

1. $X = 7$ **2.** $X = 14$ **3.** $X = 6$

4. $X = 5$ **5.** $X = 30$ **6.** $X = 2$

7. $X = 175$ **8.** $X = -2$ **9.** $X = 17$

10. $X = -4$ **11.** $X = -24$ **12.** $X = 14$

■ Section 5

1. 125   **2.** -64 **3.** -32

4. 64   **5.** 16   **6.** 216

7. 20   **8.** 225   **9.** 7

10. 10   **11.** 5   **12.** -16

Solutions to Selected Problems for Appendix A   Basic Mathematics Review

1. 25   **3.** 6   **17.** 0.9   **19.** 5

5. 21   **6.** 0.35   **21.** -24 **22.** 11

7. 36%   **9.** $\dfrac{31}{10,000}$ **25.** $X = 5$ **28.** $X = -9$

10. b. False   **30.** $X = -25$ **31.** $X = 2$

11. a. $\dfrac{8}{15}$ **b.** $\dfrac{21}{18}$ **c.** $\dfrac{23}{40}$ **34.** -125 **36.** 9

12. 0.04267   **14.** 20.1832   **37.** -25 **39.** 2

Suggested Review Books

There are many basic mathematics books available if you need a more extensive review than this appendix can provide. Several are probably available in your library. The following books are but a few of the many that you may find helpful:

Karr, R., Massey, M., & Gustafson, R. D. (2013). *Beginning Algebra: A Guided Approach* (10th ed.). Belmont, CA: Brooks/Cole.

Lial, M. L., Salzman, S.A., & Hestwood, D.L. (2010). *Basic College Mathematics* (8th ed). Reading MA: Addison-Wesley.

McKeague, C. P. (2013). *Basic Mathematics: A Text/ Workbook*. (8th ed.). Belmont, CA: Brooks/Cole.

Statistical Tables

TABLE B.1 The Unit Normal Table*

*Column A lists z-score values. A vertical line drawn through a normal distribution at a z-score location divides the distribution into two sections.
Column B identifies the proportion in the larger section, called the *body*.
Column C identifies the proportion in the smaller section, called the *tail*.
Column D identifies the proportion between the mean and the z-score.
Note: Because the normal distribution is symmetrical, the proportions for negative z-scores are the same as those for positive z-scores.

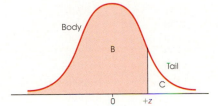
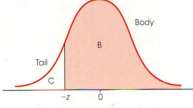
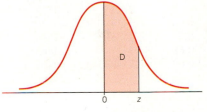

(A) z	(B) Proportion in Body	(C) Proportion in Tail	(D) Proportion Between Mean and z	(A) z	(B) Proportion in Body	(C) Proportion in Tail	(D) Proportion Between Mean and z
0.00	.5000	.5000	.0000	0.25	.5987	.4013	.0987
0.01	.5040	.4960	.0040	−0.26	.6026	.3974	.1026
0.02	.5080	.4920	.0080	0.27	.6064	.3936	.1064
0.03	.5120	.4880	.0120	0.28	.6103	.3897	.1103
0.04	.5160	.4840	.0160	0.29	.6141	.3859	.1141
0.05	.5199	.4801	.0199	0.30	.6179	.3821	.1179
0.06	.5239	.4761	.0239	0.31	.6217	.3783	.1217
0.07	.5279	.4721	.0279	0.32	.6255	.3745	.1255
0.08	.5319	.4681	.0319	0.33	.6293	.3707	.1293
0.09	.5359	.4641	.0359	0.34	.6331	.3669	.1331
0.10	.5398	.4602	.0398	0.35	.6368	.3632	.1368
0.11	.5438	.4562	.0438	0.36	.6406	.3594	.1406
0.12	.5478	.4522	.0478	0.37	.6443	.3557	.1443
0.13	.5517	.4483	.0517	0.38	.6480	.3520	.1480
0.14	.5557	.4443	.0557	0.39	.6517	.3483	.1517
0.15	.5596	.4404	.0596	0.40	.6554	.3446	.1554
0.16	.5636	.4364	.0636	0.41	.6591	.3409	.1591
0.17	.5675	.4325	.0675	0.42	.6628	.3372	.1628
0.18	.5714	.4286	.0714	0.43	.6664	.3336	.1664
0.19	.5753	.4247	.0753	0.44	.6700	.3300	.1700
0.20	.5793	.4207	.0793	0.45	.6736	.3264	.1736
0.21	.5832	.4168	.0832	0.46	.6772	.3228	.1772
0.22	.5871	.4129	.0871	0.47	.6808	.3192	.1808
0.23	.5910	.4090	.0910	0.48	.6844	.3156	.1844
0.24	.5948	.4052	.0948	0.49	.6879	.3121	.1879

(A) z	(B) Proportion in Body	(C) Proportion in Tail	(D) Proportion Between Mean and z	(A) z	(B) Proportion in Body	(C) Proportion in Tail	(D) Proportion Between Mean and z
0.50	.6915	.3085	.1915	1.00	.8413	.1587	.3413
0.51	.6950	.3050	.1950	1.01	.8438	.1562	.3438
0.52	.6985	.3015	.1985	1.02	.8461	.1539	.3461
0.53	.7019	.2981	.2019	1.03	.8485	.1515	.3485
0.54	.7054	.2946	.2054	1.04	.8508	.1492	.3508
0.55	.7088	.2912	.2088	1.05	.8531	.1469	.3531
0.56	.7123	.2877	.2123	1.06	.8554	.1446	.3554
0.57	.7157	.2843	.2157	1.07	.8577	.1423	.3577
0.58	.7190	.2810	.2190	1.08	.8599	.1401	.3599
0.59	.7224	.2776	.2224	1.09	.8621	.1379	.3621
0.60	.7257	.2743	.2257	1.10	.8643	.1357	.3643
0.61	.7291	.2709	.2291	1.11	.8665	.1335	.3665
0.62	.7324	.2676	.2324	1.12	.8686	.1314	.3686
0.63	.7357	.2643	.2357	1.13	.8708	.1292	.3708
0.64	.7389	.2611	.2389	1.14	.8729	.1271	.3729
0.65	.7422	.2578	.2422	1.15	.8749	.1251	.3749
0.66	.7454	.2546	.2454	1.16	.8770	.1230	.3770
0.67	.7486	.2514	.2486	1.17	.8790	.1210	.3790
0.68	.7517	.2483	.2517	1.18	.8810	.1190	.3810
0.69	.7549	.2451	.2549	1.19	.8830	.1170	.3830
0.70	.7580	.2420	.2580	1.20	.8849	.1151	.3849
0.71	.7611	.2389	.2611	1.21	.8869	.1131	.3869
0.72	.7642	.2358	.2642	1.22	.8888	.1112	.3888
0.73	.7673	.2327	.2673	1.23	.8907	.1093	.3907
0.74	.7704	.2296	.2704	1.24	.8925	.1075	.3925
0.75	.7734	.2266	.2734	1.25	.8944	.1056	.3944
0.76	.7764	.2236	.2764	1.26	.8962	.1038	.3962
0.77	.7794	.2206	.2794	1.27	.8980	.1020	.3980
0.78	.7823	.2177	.2823	1.28	.8997	.1003	.3997
0.79	.7852	.2148	.2852	1.29	.9015	.0985	.4015
0.80	.7881	.2119	.2881	1.30	.9032	.0968	.4032
0.81	.7910	.2090	.2910	1.31	.9049	.0951	.4049
0.82	.7939	.2061	.2939	1.32	.9066	.0934	.4066
0.83	.7967	.2033	.2967	1.33	.9082	.0918	.4082
0.84	.7995	.2005	.2995	1.34	.9099	.0901	.4099
0.85	.8023	.1977	.3023	1.35	.9115	.0885	.4115
0.86	.8051	.1949	.3051	1.36	.9131	.0869	.4131
0.87	.8078	.1922	.3078	1.37	.9147	.0853	.4147
0.88	.8106	.1894	.3106	1.38	.9162	.0838	.4162
0.89	.8133	.1867	.3133	1.39	.9177	.0823	.4177
0.90	.8159	.1841	.3159	1.40	.9192	.0808	.4192
0.91	.8186	.1814	.3186	1.41	.9207	.0793	.4207
0.92	.8212	.1788	.3212	1.42	.9222	.0778	.4222
0.93	.8238	.1762	.3238	1.43	.9236	.0764	.4236
0.94	.8264	.1736	.3264	1.44	.9251	.0749	.4251
0.95	.8289	.1711	.3289	1.45	.9265	.0735	.4265
0.96	.8315	.1685	.3315	1.46	.9279	.0721	.4279
0.97	.8340	.1660	.3340	1.47	.9292	.0708	.4292
0.98	.8365	.1635	.3365	1.48	.9306	.0694	.4306
0.99	.8389	.1611	.3389	1.49	.9319	.0681	.4319

(A) z	(B) Proportion in Body	(C) Proportion in Tail	(D) Proportion Between Mean and z	(A) z	(B) Proportion in Body	(C) Proportion in Tail	(D) Proportion Between Mean and z
1.50	.9332	.0668	.4332	2.00	.9772	.0228	.4772
1.51	.9345	.0655	.4345	2.01	.9778	.0222	.4778
1.52	.9357	.0643	.4357	2.02	.9783	.0217	.4783
1.53	.9370	.0630	.4370	2.03	.9788	.0212	.4788
1.54	.9382	.0618	.4382	2.04	.9793	.0207	.4793
1.55	.9394	.0606	.4394	2.05	.9798	.0202	.4798
1.56	.9406	.0594	.4406	2.06	.9803	.0197	.4803
1.57	.9418	.0582	.4418	2.07	.9808	.0192	.4808
1.58	.9429	.0571	.4429	2.08	.9812	.0188	.4812
1.59	.9441	.0559	.4441	2.09	.9817	.0183	.4817
1.60	.9452	.0548	.4452	2.10	.9821	.0179	.4821
1.61	.9463	.0537	.4463	2.11	.9826	.0174	.4826
1.62	.9474	.0526	.4474	2.12	.9830	.0170	.4830
1.63	.9484	.0516	.4484	2.13	.9834	.0166	.4834
1.64	.9495	.0505	.4495	2.14	.9838	.0162	.4838
1.65	.9505	.0495	.4505	2.15	.9842	.0158	.4842
1.66	.9515	.0485	.4515	2.16	.9846	.0154	.4846
1.67	.9525	.0475	.4525	2.17	.9850	.0150	.4850
1.68	.9535	.0465	.4535	2.18	.9854	.0146	.4854
1.69	.9545	.0455	.4545	2.19	.9857	.0143	.4857
1.70	.9554	.0446	.4554	2.20	.9861	.0139	.4861
1.71	.9564	.0436	.4564	2.21	.9864	.0136	.4864
1.72	.9573	.0427	.4573	2.22	.9868	.0132	.4868
1.73	.9582	.0418	.4582	2.23	.9871	.0129	.4871
1.74	.9591	.0409	.4591	2.24	.9875	.0125	.4875
1.75	.9599	.0401	.4599	2.25	.9878	.0122	.4878
1.76	.9608	.0392	.4608	2.26	.9881	.0119	.4881
1.77	.9616	.0384	.4616	2.27	.9884	.0116	.4884
1.78	.9625	.0375	.4625	2.28	.9887	.0113	.4887
1.79	.9633	.0367	.4633	2.29	.9890	.0110	.4890
1.80	.9641	.0359	.4641	2.30	.9893	.0107	.4893
1.81	.9649	.0351	.4649	2.31	.9896	.0104	.4896
1.82	.9656	.0344	.4656	2.32	.9898	.0102	.4898
1.83	.9664	.0336	.4664	2.33	.9901	.0099	.4901
1.84	.9671	.0329	.4671	2.34	.9904	.0096	.4904
1.85	.9678	.0322	.4678	2.35	.9906	.0094	.4906
1.86	.9686	.0314	.4686	2.36	.9909	.0091	.4909
1.87	.9693	.0307	.4693	2.37	.9911	.0089	.4911
1.88	.9699	.0301	.4699	2.38	.9913	.0087	.4913
1.89	.9706	.0294	.4706	2.39	.9916	.0084	.4916
1.90	.9713	.0287	.4713	2.40	.9918	.0082	.4918
1.91	.9719	.0281	.4719	2.41	.9920	.0080	.4920
1.92	.9726	.0274	.4726	2.42	.9922	.0078	.4922
1.93	.9732	.0268	.4732	2.43	.9925	.0075	.4925
1.94	.9738	.0262	.4738	2.44	.9927	.0073	.4927
1.95	.9744	.0256	.4744	2.45	.9929	.0071	.4929
1.96	.9750	.0250	.4750	2.46	.9931	.0069	.4931
1.97	.9756	.0244	.4756	2.47	.9932	.0068	.4932
1.98	.9761	.0239	.4761	2.48	.9934	.0066	.4934
1.99	.9767	.0233	.4767	2.49	.9936	.0064	.4936

(A) z	(B) Proportion in Body	(C) Proportion in Tail	(D) Proportion Between Mean and z	(A) z	(B) Proportion in Body	(C) Proportion in Tail	(D) Proportion Between Mean and z
2.50	.9938	.0062	.4938	2.95	.9984	.0016	.4984
2.51	.9940	.0060	.4940	2.96	.9985	.0015	.4985
2.52	.9941	.0059	.4941	2.97	.9985	.0015	.4985
2.53	.9943	.0057	.4943	2.98	.9986	.0014	.4986
2.54	.9945	.0055	.4945	2.99	.9986	.0014	.4986
2.55	.9946	.0054	.4946	3.00	.9987	.0013	.4987
2.56	.9948	.0052	.4948	3.01	.9987	.0013	.4987
2.57	.9949	.0051	.4949	3.02	.9987	.0013	.4987
2.58	.9951	.0049	.4951	3.03	.9988	.0012	.4988
2.59	.9952	.0048	.4952	3.04	.9988	.0012	.4988
2.60	.9953	.0047	.4953	3.05	.9989	.0011	.4989
2.61	.9955	.0045	.4955	3.06	.9989	.0011	.4989
2.62	.9956	.0044	.4956	3.07	.9989	.0011	.4989
2.63	.9957	.0043	.4957	3.08	.9990	.0010	.4990
2.64	.9959	.0041	.4959	3.09	.9990	.0010	.4990
2.65	.9960	.0040	.4960	3.10	.9990	.0010	.4990
2.66	.9961	.0039	.4961	3.11	.9991	.0009	.4991
2.67	.9962	.0038	.4962	3.12	.9991	.0009	.4991
2.68	.9963	.0037	.4963	3.13	.9991	.0009	.4991
2.69	.9964	.0036	.4964	3.14	.9992	.0008	.4992
2.70	.9965	.0035	.4965	3.15	.9992	.0008	.4992
2.71	.9966	.0034	.4966	3.16	.9992	.0008	.4992
2.72	.9967	.0033	.4967	3.17	.9992	.0008	.4992
2.73	.9968	.0032	.4968	3.18	.9993	.0007	.4993
2.74	.9969	.0031	.4969	3.19	.9993	.0007	.4993
2.75	.9970	.0030	.4970	3.20	.9993	.0007	.4993
2.76	.9971	.0029	.4971	3.21	.9993	.0007	.4993
2.77	.9972	.0028	.4972	3.22	.9994	.0006	.4994
2.78	.9973	.0027	.4973	3.23	.9994	.0006	.4994
2.79	.9974	.0026	.4974	3.24	.9994	.0006	.4994
2.80	.9974	.0026	.4974	3.30	.9995	.0005	.4995
2.81	.9975	.0025	.4975	3.40	.9997	.0003	.4997
2.82	.9976	.0024	.4976	3.50	.9998	.0002	.4998
2.83	.9977	.0023	.4977	3.60	.9998	.0002	.4998
2.84	.9977	.0023	.4977	3.70	.9999	.0001	.4999
2.85	.9978	.0022	.4978	3.80	.99993	.00007	.49993
2.86	.9979	.0021	.4979	3.90	.99995	.00005	.49995
2.87	.9979	.0021	.4979	4.00	.99997	.00003	.49997
2.88	.9980	.0020	.4980				
2.89	.9981	.0019	.4981				
2.90	.9981	.0019	.4981				
2.91	.9982	.0018	.4982				
2.92	.9982	.0018	.4982				
2.93	.9983	.0017	.4983				
2.94	.9984	.0016	.4984				

TABLE B.2   The *t* Distribution

Table entries are values of *t* corresponding to proportions in one tail or in two tails combined.

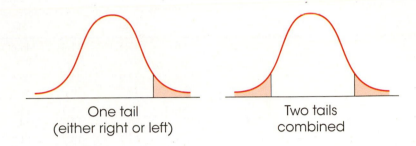

One tail
(either right or left)

Two tails
combined

	Proportion in One Tail					
	0.25	0.10	0.05	0.025	0.01	0.005
			Proportion in Two Tails Combined			
df	0.50	0.20	0.10	0.05	0.02	0.01
1	1.000	3.078	6.314	12.706	31.821	63.657
2	0.816	1.886	2.920	4.303	6.965	9.925
3	0.765	1.638	2.353	3.182	4.541	5.841
4	0.741	1.533	2.132	2.776	3.747	4.604
5	0.727	1.476	2.015	2.571	3.365	4.032
6	0.718	1.440	1.943	2.447	3.143	3.707
7	0.711	1.415	1.895	2.365	2.998	3.499
8	0.706	1.397	1.860	2.306	2.896	3.355
9	0.703	1.383	1.833	2.262	2.821	3.250
10	0.700	1.372	1.812	2.228	2.764	3.169
11	0.697	1.363	1.796	2.201	2.718	3.106
12	0.695	1.356	1.782	2.179	2.681	3.055
13	0.694	1.350	1.771	2.160	2.650	3.012
14	0.692	1.345	1.761	2.145	2.624	2.977
15	0.691	1.341	1.753	2.131	2.602	2.947
16	0.690	1.337	1.746	2.120	2.583	2.921
17	0.689	1.333	1.740	2.110	2.567	2.898
18	0.688	1.330	1.734	2.101	2.552	2.878
19	0.688	1.328	1.729	2.093	2.539	2.861
20	0.687	1.325	1.725	2.086	2.528	2.845
21	0.686	1.323	1.721	2.080	2.518	2.831
22	0.686	1.321	1.717	2.074	2.508	2.819
23	0.685	1.319	1.714	2.069	2.500	2.807
24	0.685	1.318	1.711	2.064	2.492	2.797
25	0.684	1.316	1.708	2.060	2.485	2.787
26	0.684	1.315	1.706	2.056	2.479	2.779
27	0.684	1.314	1.703	2.052	2.473	2.771
28	0.683	1.313	1.701	2.048	2.467	2.763
29	0.683	1.311	1.699	2.045	2.462	2.756
30	0.683	1.310	1.697	2.042	2.457	2.750
40	0.681	1.303	1.684	2.021	2.423	2.704
60	0.679	1.296	1.671	2.000	2.390	2.660
120	0.677	1.289	1.658	1.980	2.358	2.617
∞	0.674	1.282	1.645	1.960	2.326	2.576

Table III of R. A. Fisher and F. Yates, *Statistical Tables for Biological, Agricultural and Medical Research*, 6th ed. London: Longman Group Ltd., 1974 (previously published by Oliver and Boyd Ltd., Edinburgh). Copyright ©1963 R. A. Fisher and F. Yates. Adapted and reprinted with permission of Pearson Education Limited.

TABLE B.3   Critical Values for the *F*-Max Statistic*

*The critical values for α = .05 are in lightface type, and for α = .01, they are in boldface type.

n − 1	\multicolumn{11}{c}{*k* = Number of Samples}
	2	3	4	5	6	7	8	9	10	11	12
4	9.60	15.5	20.6	25.2	29.5	33.6	37.5	41.4	44.6	48.0	51.4
	23.2	**37.**	**49.**	**59.**	**69.**	**79.**	**89.**	**97.**	**106.**	**113.**	**120.**
5	7.15	10.8	13.7	16.3	18.7	20.8	22.9	24.7	26.5	28.2	29.9
	14.9	**22.**	**28.**	**33.**	**38.**	**42.**	**46.**	**50.**	**54.**	**57.**	**60.**
6	5.82	8.38	10.4	12.1	13.7	15.0	16.3	17.5	18.6	19.7	20.7
	11.1	**15.5**	**19.1**	**22.**	**25.**	**27.**	**30.**	**32.**	**34.**	**36.**	**37.**
7	4.99	6.94	8.44	9.70	10.8	11.8	12.7	13.5	14.3	15.1	15.8
	8.89	**12.1**	**14.5**	**16.5**	**18.4**	**20.**	**22.**	**23.**	**24.**	**26.**	**27.**
8	4.43	6.00	7.18	8.12	9.03	9.78	10.5	11.1	11.7	12.2	12.7
	7.50	**9.9**	**11.7**	**13.2**	**14.5**	**15.8**	**16.9**	**17.9**	**18.9**	**19.8**	**21.**
9	4.03	5.34	6.31	7.11	7.80	8.41	8.95	9.45	9.91	10.3	10.7
	6.54	**8.5**	**9.9**	**11.1**	**12.1**	**13.1**	**13.9**	**14.7**	**15.3**	**16.0**	**16.6**
10	3.72	4.85	5.67	6.34	6.92	7.42	7.87	8.28	8.66	9.01	9.34
	5.85	**7.4**	**8.6**	**9.6**	**10.4**	**11.1**	**11.8**	**12.4**	**12.9**	**13.4**	**13.9**
12	3.28	4.16	4.79	5.30	5.72	6.09	6.42	6.72	7.00	7.25	7.48
	4.91	**6.1**	**6.9**	**7.6**	**8.2**	**8.7**	**9.1**	**9.5**	**9.9**	**10.2**	**10.6**
15	2.86	3.54	4.01	4.37	4.68	4.95	5.19	5.40	5.59	5.77	5.93
	4.07	**4.9**	**5.5**	**6.0**	**6.4**	**6.7**	**7.1**	**7.3**	**7.5**	**7.8**	**8.0**
20	2.46	2.95	3.29	3.54	3.76	3.94	4.10	4.24	4.37	4.49	4.59
	3.32	**3.8**	**4.3**	**4.6**	**4.9**	**5.1**	**5.3**	**5.5**	**5.6**	**5.8**	**5.9**
30	2.07	2.40	2.61	2.78	2.91	3.02	3.12	3.21	3.29	3.36	3.39
	2.63	**3.0**	**3.3**	**3.5**	**3.6**	**3.7**	**3.8**	**3.9**	**4.0**	**4.1**	**4.2**
60	1.67	1.85	1.96	2.04	2.11	2.17	2.22	2.26	2.30	2.33	2.36
	1.96	**2.2**	**2.3**	**2.4**	**2.4**	**2.5**	**2.5**	**2.6**	**2.6**	**2.7**	**2.7**

TABLE B.4   The *F* Distribution*

*Table entries in lightface type are critical values for the .05 level of significance.
Boldface type values are for the .01 level of significance.

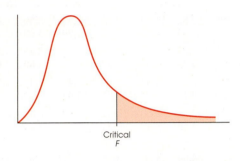

Degrees of Freedom: Denominator	Degrees of Freedom: Numerator														
	1	2	3	4	5	6	7	8	9	10	11	12	14	16	20
1	161	200	216	225	230	234	237	239	241	242	243	244	245	246	248
	4052	**4999**	**5403**	**5625**	**5764**	**5859**	**5928**	**5981**	**6022**	**6056**	**6082**	**6106**	**6142**	**6169**	**6208**
2	18.51	19.00	19.16	19.25	19.30	19.33	19.36	19.37	19.38	19.39	19.40	19.41	19.42	19.43	19.44
	98.49	**99.00**	**99.17**	**99.25**	**99.30**	**99.33**	**99.34**	**99.36**	**99.38**	**99.40**	**99.41**	**99.42**	**99.43**	**99.44**	**99.45**
3	10.13	9.55	9.28	9.12	9.01	8.94	8.88	8.84	8.81	8.78	8.76	8.74	8.71	8.69	8.66
	34.12	**30.92**	**29.46**	**28.71**	**28.24**	**27.91**	**27.67**	**27.49**	**27.34**	**27.23**	**27.13**	**27.05**	**26.92**	**26.83**	**26.69**
4	7.71	6.94	6.59	6.39	6.26	6.16	6.09	6.04	6.00	5.96	5.93	5.91	5.87	5.84	5.80
	21.20	**18.00**	**16.69**	**15.98**	**15.52**	**15.21**	**14.98**	**14.80**	**14.66**	**14.54**	**14.45**	**14.37**	**14.24**	**14.15**	**14.02**
5	6.61	5.79	5.41	5.19	5.05	4.95	4.88	4.82	4.78	4.74	4.70	4.68	4.64	4.60	4.56
	16.26	**13.27**	**12.06**	**11.39**	**10.97**	**10.67**	**10.45**	**10.27**	**10.15**	**10.05**	**9.96**	**9.89**	**9.77**	**9.68**	**9.55**
6	5.99	5.14	4.76	4.53	4.39	4.28	4.21	4.15	4.10	4.06	4.03	4.00	3.96	3.92	3.87
	13.74	**10.92**	**9.78**	**9.15**	**8.75**	**8.47**	**8.26**	**8.10**	**7.98**	**7.87**	**7.79**	**7.72**	**7.60**	**7.52**	**7.39**
7	5.59	4.74	4.35	4.12	3.97	3.87	3.79	3.73	3.68	3.63	3.60	3.57	3.52	3.49	3.44
	12.25	**9.55**	**8.45**	**7.85**	**7.46**	**7.19**	**7.00**	**6.84**	**6.71**	**6.62**	**6.54**	**6.47**	**6.35**	**6.27**	**6.15**
8	5.32	4.46	4.07	3.84	3.69	3.58	3.50	3.44	3.39	3.34	3.31	3.28	3.23	3.20	3.15
	11.26	**8.65**	**7.59**	**7.01**	**6.63**	**6.37**	**6.19**	**6.03**	**5.91**	**5.82**	**5.74**	**5.67**	**5.56**	**5.48**	**5.36**
9	5.12	4.26	3.86	3.63	3.48	3.37	3.29	3.23	3.18	3.13	3.10	3.07	3.02	2.98	2.93
	10.56	**8.02**	**6.99**	**6.42**	**6.06**	**5.80**	**5.62**	**5.47**	**5.35**	**5.26**	**5.18**	**5.11**	**5.00**	**4.92**	**4.80**
10	4.96	4.10	3.71	3.48	3.33	3.22	3.14	3.07	3.02	2.97	2.94	2.91	2.86	2.82	2.77
	10.04	**7.56**	**6.55**	**5.99**	**5.64**	**5.39**	**5.21**	**5.06**	**4.95**	**4.85**	**4.78**	**4.71**	**4.60**	**4.52**	**4.41**
11	4.84	3.98	3.59	3.36	3.20	3.09	3.01	2.95	2.90	2.86	2.82	2.79	2.74	2.70	2.65
	9.65	**7.20**	**6.22**	**5.67**	**5.32**	**5.07**	**4.88**	**4.74**	**4.63**	**4.54**	**4.46**	**4.40**	**4.29**	**4.21**	**4.10**
12	4.75	3.88	3.49	3.26	3.11	3.00	2.92	2.85	2.80	2.76	2.72	2.69	2.64	2.60	2.54
	9.33	**6.93**	**5.95**	**5.41**	**5.06**	**4.82**	**4.65**	**4.50**	**4.39**	**4.30**	**4.22**	**4.16**	**4.05**	**3.98**	**3.86**
13	4.67	3.80	3.41	3.18	3.02	2.92	2.84	2.77	2.72	2.67	2.63	2.60	2.55	2.51	2.46
	9.07	**6.70**	**5.74**	**5.20**	**4.86**	**4.62**	**4.44**	**4.30**	**4.19**	**4.10**	**4.02**	**3.96**	**3.85**	**3.78**	**3.67**
14	4.60	3.74	3.34	3.11	2.96	2.85	2.77	2.70	2.65	2.60	2.56	2.53	2.48	2.44	2.39
	8.86	**6.51**	**5.56**	**5.03**	**4.69**	**4.46**	**4.28**	**4.14**	**4.03**	**3.94**	**3.86**	**3.80**	**3.70**	**3.62**	**3.51**
15	4.54	3.68	3.29	3.06	2.90	2.79	2.70	2.64	2.59	2.55	2.51	2.48	2.43	2.39	2.33
	8.68	**6.36**	**5.42**	**4.89**	**4.56**	**4.32**	**4.14**	**4.00**	**3.89**	**3.80**	**3.73**	**3.67**	**3.56**	**3.48**	**3.36**
16	4.49	3.63	3.24	3.01	2.85	2.74	2.66	2.59	2.54	2.49	2.45	2.42	2.37	2.33	2.28
	8.53	**6.23**	**5.29**	**4.77**	**4.44**	**4.20**	**4.03**	**3.89**	**3.78**	**3.69**	**3.61**	**3.55**	**3.45**	**3.37**	**3.25**

TABLE B.4   **The *F* Distribution**   *(continued)*

Degrees of Freedom: Denominator	Degrees of Freedom: Numerator														
	1	2	3	4	5	6	7	8	9	10	11	12	14	16	20
17	4.45	3.59	3.20	2.96	2.81	2.70	2.62	2.55	2.50	2.45	2.41	2.38	2.33	2.29	2.23
	8.40	**6.11**	**5.18**	**4.67**	**4.34**	**4.10**	**3.93**	**3.79**	**3.68**	**3.59**	**3.52**	**3.45**	**3.35**	**3.27**	**3.16**
18	4.41	3.55	3.16	2.93	2.77	2.66	2.58	2.51	2.46	2.41	2.37	2.34	2.29	2.25	2.19
	8.28	**6.01**	**5.09**	**4.58**	**4.25**	**4.01**	**3.85**	**3.71**	**3.60**	**3.51**	**3.44**	**3.37**	**3.27**	**3.19**	**3.07**
19	4.38	3.52	3.13	2.90	2.74	2.63	2.55	2.48	2.43	2.38	2.34	2.31	2.26	2.21	2.15
	8.18	**5.93**	**5.01**	**4.50**	**4.17**	**3.94**	**3.77**	**3.63**	**3.52**	**3.43**	**3.36**	**3.30**	**3.19**	**3.12**	**3.00**
20	4.35	3.49	3.10	2.87	2.71	2.60	2.52	2.45	2.40	2.35	2.31	2.28	2.23	2.18	2.12
	8.10	**5.85**	**4.94**	**4.43**	**4.10**	**3.87**	**3.71**	**3.56**	**3.45**	**3.37**	**3.30**	**3.23**	**3.13**	**3.05**	**2.94**
21	4.32	3.47	3.07	2.84	2.68	2.57	2.49	2.42	2.37	2.32	2.28	2.25	2.20	2.15	2.09
	8.02	**5.78**	**4.87**	**4.37**	**4.04**	**3.81**	**3.65**	**3.51**	**3.40**	**3.31**	**3.24**	**3.17**	**3.07**	**2.99**	**2.88**
22	4.30	3.44	3.05	2.82	2.66	2.55	2.47	2.40	2.35	2.30	2.26	2.23	2.18	2.13	2.07
	7.94	**5.72**	**4.82**	**4.31**	**3.99**	**3.76**	**3.59**	**3.45**	**3.35**	**3.26**	**3.18**	**3.12**	**3.02**	**2.94**	**2.83**
23	4.28	3.42	3.03	2.80	2.64	2.53	2.45	2.38	2.32	2.28	2.24	2.20	2.14	2.10	2.04
	7.88	**5.66**	**4.76**	**4.26**	**3.94**	**3.71**	**3.54**	**3.41**	**3.30**	**3.21**	**3.14**	**3.07**	**2.97**	**2.89**	**2.78**
24	4.26	3.40	3.01	2.78	2.62	2.51	2.43	2.36	2.30	2.26	2.22	2.18	2.13	2.09	2.02
	7.82	**5.61**	**4.72**	**4.22**	**3.90**	**3.67**	**3.50**	**3.36**	**3.25**	**3.17**	**3.09**	**3.03**	**2.93**	**2.85**	**2.74**
25	4.24	3.38	2.99	2.76	2.60	2.49	2.41	2.34	2.28	2.24	2.20	2.16	2.11	2.06	2.00
	7.77	**5.57**	**4.68**	**4.18**	**3.86**	**3.63**	**3.46**	**3.32**	**3.21**	**3.13**	**3.05**	**2.99**	**2.89**	**2.81**	**2.70**
26	4.22	3.37	2.98	2.74	2.59	2.47	2.39	2.32	2.27	2.22	2.18	2.15	2.10	2.05	1.99
	7.72	**5.53**	**4.64**	**4.14**	**3.82**	**3.59**	**3.42**	**3.29**	**3.17**	**3.09**	**3.02**	**2.96**	**2.86**	**2.77**	**2.66**
27	4.21	3.35	2.96	2.73	2.57	2.46	2.37	2.30	2.25	2.20	2.16	2.13	2.08	2.03	1.97
	7.68	**5.49**	**4.60**	**4.11**	**3.79**	**3.56**	**3.39**	**3.26**	**3.14**	**3.06**	**2.98**	**2.93**	**2.83**	**2.74**	**2.63**
28	4.20	3.34	2.95	2.71	2.56	2.44	2.36	2.29	2.24	2.19	2.15	2.12	2.06	2.02	1.96
	7.64	**5.45**	**4.57**	**4.07**	**3.76**	**3.53**	**3.36**	**3.23**	**3.11**	**3.03**	**2.95**	**2.90**	**2.80**	**2.71**	**2.60**
29	4.18	3.33	2.93	2.70	2.54	2.43	2.35	2.28	2.22	2.18	2.14	2.10	2.05	2.00	1.94
	7.60	**5.42**	**4.54**	**4.04**	**3.73**	**3.50**	**3.33**	**3.20**	**3.08**	**3.00**	**2.92**	**2.87**	**2.77**	**2.68**	**2.57**
30	4.17	3.32	2.92	2.69	2.53	2.42	2.34	2.27	2.21	2.16	2.12	2.09	2.04	1.99	1.93
	7.56	**5.39**	**4.51**	**4.02**	**3.70**	**3.47**	**3.30**	**3.17**	**3.06**	**2.98**	**2.90**	**2.84**	**2.74**	**2.66**	**2.55**
32	4.15	3.30	2.90	2.67	2.51	2.40	2.32	2.25	2.19	2.14	2.10	2.07	2.02	1.97	1.91
	7.50	**5.34**	**4.46**	**3.97**	**3.66**	**3.42**	**3.25**	**3.12**	**3.01**	**2.94**	**2.86**	**2.80**	**2.70**	**2.62**	**2.51**
34	4.13	3.28	2.88	2.65	2.49	2.38	2.30	2.23	2.17	2.12	2.08	2.05	2.00	1.95	1.89
	7.44	**5.29**	**4.42**	**3.93**	**3.61**	**3.38**	**3.21**	**3.08**	**2.97**	**2.89**	**2.82**	**2.76**	**2.66**	**2.58**	**2.47**
36	4.11	3.26	2.86	2.63	2.48	2.36	2.28	2.21	2.15	2.10	2.06	2.03	1.98	1.93	1.87
	7.39	**5.25**	**4.38**	**3.89**	**3.58**	**3.35**	**3.18**	**3.04**	**2.94**	**2.86**	**2.78**	**2.72**	**2.62**	**2.54**	**2.43**
38	4.10	3.25	2.85	2.62	2.46	2.35	2.26	2.19	2.14	2.09	2.05	2.02	1.96	1.92	1.85
	7.35	**5.21**	**4.34**	**3.86**	**3.54**	**3.32**	**3.15**	**3.02**	**2.91**	**2.82**	**2.75**	**2.69**	**2.59**	**2.51**	**2.40**
40	4.08	3.23	2.84	2.61	2.45	2.34	2.25	2.18	2.12	2.07	2.04	2.00	1.95	1.90	1.84
	7.31	**5.18**	**4.31**	**3.83**	**3.51**	**3.29**	**3.12**	**2.99**	**2.88**	**2.80**	**2.73**	**2.66**	**2.56**	**2.49**	**2.37**

TABLE B.4   **The F Distribution**   *(continued)*

Degrees of Freedom: Denominator	Degrees of Freedom: Numerator														
	1	2	3	4	5	6	7	8	9	10	11	12	14	16	20
42	4.07	3.22	2.83	2.59	2.44	2.32	2.24	2.17	2.11	2.06	2.02	1.99	1.94	1.89	1.82
	7.27	5.15	4.29	3.80	3.49	3.26	3.10	2.96	2.86	2.77	2.70	2.64	2.54	2.46	2.35
44	4.06	3.21	2.82	2.58	2.43	2.31	2.23	2.16	2.10	2.05	2.01	1.98	1.92	1.88	1.81
	7.24	5.12	4.26	3.78	3.46	3.24	3.07	2.94	2.84	2.75	2.68	2.62	2.52	2.44	2.32
46	4.05	3.20	2.81	2.57	2.42	2.30	2.22	2.14	2.09	2.04	2.00	1.97	1.91	1.87	1.80
	7.21	5.10	4.24	3.76	3.44	3.22	3.05	2.92	2.82	2.73	2.66	2.60	2.50	2.42	2.30
48	4.04	3.19	2.80	2.56	2.41	2.30	2.21	2.14	2.08	2.03	1.99	1.96	1.90	1.86	1.79
	7.19	5.08	4.22	3.74	3.42	3.20	3.04	2.90	2.80	2.71	2.64	2.58	2.48	2.40	2.28
50	4.03	3.18	2.79	2.56	2.40	2.29	2.20	2.13	2.07	2.02	1.98	1.95	1.90	1.85	1.78
	7.17	5.06	4.20	3.72	3.41	3.18	3.02	2.88	2.78	2.70	2.62	2.56	2.46	2.39	2.26
55	4.02	3.17	2.78	2.54	2.38	2.27	2.18	2.11	2.05	2.00	1.97	1.93	1.88	1.83	1.76
	7.12	5.01	4.16	3.68	3.37	3.15	2.98	2.85	2.75	2.66	2.59	2.53	2.43	2.35	2.23
60	4.00	3.15	2.76	2.52	2.37	2.25	2.17	2.10	2.04	1.99	1.95	1.92	1.86	1.81	1.75
	7.08	4.98	4.13	3.65	3.34	3.12	2.95	2.82	2.72	2.63	2.56	2.50	2.40	2.32	2.20
65	3.99	3.14	2.75	2.51	2.36	2.24	2.15	2.08	2.02	1.98	1.94	1.90	1.85	1.80	1.73
	7.04	4.95	4.10	3.62	3.31	3.09	2.93	2.79	2.70	2.61	2.54	2.47	2.37	2.30	2.18
70	3.98	3.13	2.74	2.50	2.35	2.23	2.14	2.07	2.01	1.97	1.93	1.89	1.84	1.79	1.72
	7.01	4.92	4.08	3.60	3.29	3.07	2.91	2.77	2.67	2.59	2.51	2.45	2.35	2.28	2.15
80	3.96	3.11	2.72	2.48	2.33	2.21	2.12	2.05	1.99	1.95	1.91	1.88	1.82	1.77	1.70
	6.96	4.88	4.04	3.56	3.25	3.04	2.87	2.74	2.64	2.55	2.48	2.41	2.32	2.24	2.11
100	3.94	3.09	2.70	2.46	2.30	2.19	2.10	2.03	1.97	1.92	1.88	1.85	1.79	1.75	1.68
	6.90	4.82	3.98	3.51	3.20	2.99	2.82	2.69	2.59	2.51	2.43	2.36	2.26	2.19	2.06
125	3.92	3.07	2.68	2.44	2.29	2.17	2.08	2.01	1.95	1.90	1.86	1.83	1.77	1.72	1.65
	6.84	4.78	3.94	3.47	3.17	2.95	2.79	2.65	2.56	2.47	2.40	2.33	2.23	2.15	2.03
150	3.91	3.06	2.67	2.43	2.27	2.16	2.07	2.00	1.94	1.89	1.85	1.82	1.76	1.71	1.64
	6.81	4.75	3.91	3.44	3.14	2.92	2.76	2.62	2.53	2.44	2.37	2.30	2.20	2.12	2.00
200	3.89	3.04	2.65	2.41	2.26	2.14	2.05	1.98	1.92	1.87	1.83	1.80	1.74	1.69	1.62
	6.76	4.71	3.88	3.41	3.11	2.90	2.73	2.60	2.50	2.41	2.34	2.28	2.17	2.09	1.97
400	3.86	3.02	2.62	2.39	2.23	2.12	2.03	1.96	1.90	1.85	1.81	1.78	1.72	1.67	1.60
	6.70	4.66	3.83	3.36	3.06	2.85	2.69	2.55	2.46	2.37	2.29	2.23	2.12	2.04	1.92
1000	3.85	3.00	2.61	2.38	2.22	2.10	2.02	1.95	1.89	1.84	1.80	1.76	1.70	1.65	1.58
	6.66	4.62	3.80	3.34	3.04	2.82	2.66	2.53	2.43	2.34	2.26	2.20	2.09	2.01	1.89
∞	3.84	2.99	2.60	2.37	2.21	2.09	2.01	1.94	1.88	1.83	1.79	1.75	1.69	1.64	1.57
	6.64	4.60	3.78	3.32	3.02	2.80	2.64	2.51	2.41	2.32	2.24	2.18	2.07	1.99	1.87

Table A14 of *Statistical Methods*, 7th ed. by George W. Snedecor and William G. Cochran. Copyright © 1980 by the Iowa State University Press, 2121 South State Avenue, Ames, Iowa 50010. Reprinted with permission of the Iowa State University Press.

TABLE B.5   The Studentized Range Statistic (q)*

*The critical values for q corresponding to α = .05 (lightface type) and α = .01 (boldface type).

df for Error Term	K		k = Number of Treatments								
	2	3	4	5	6	7	8	9	10	11	12
5	3.64	4.60	5.22	5.67	6.03	6.33	6.58	6.80	6.99	7.17	7.32
	5.70	**6.98**	**7.80**	**8.42**	**8.91**	**9.32**	**9.67**	**9.97**	**10.24**	**10.48**	**10.70**
6	3.46	4.34	4.90	5.30	5.63	5.90	6.12	6.32	6.49	6.65	6.79
	5.24	**6.33**	**7.03**	**7.56**	**7.97**	**8.32**	**8.61**	**8.87**	**9.10**	**9.30**	**9.48**
7	3.34	4.16	4.68	5.06	5.36	5.61	5.82	6.00	6.16	6.30	6.43
	4.95	**5.92**	**6.54**	**7.01**	**7.37**	**7.68**	**7.94**	**8.17**	**8.37**	**8.55**	**8.71**
8	3.26	4.04	4.53	4.89	5.17	5.40	5.60	5.77	5.92	6.05	6.18
	4.75	**5.64**	**6.20**	**6.62**	**6.96**	**7.24**	**7.47**	**7.68**	**7.86**	**8.03**	**8.18**
9	3.20	3.95	4.41	4.76	5.02	5.24	5.43	5.59	5.74	5.87	5.98
	4.60	**5.43**	**5.96**	**6.35**	**6.66**	**6.91**	**7.13**	**7.33**	**7.49**	**7.65**	**7.78**
10	3.15	3.88	4.33	4.65	4.91	5.12	5.30	5.46	5.60	5.72	5.83
	4.48	**5.27**	**5.77**	**6.14**	**6.43**	**6.67**	**6.87**	**7.05**	**7.21**	**7.36**	**7.49**
11	3.11	3.82	4.26	4.57	4.82	5.03	5.20	5.35	5.49	5.61	5.71
	4.39	**5.15**	**5.62**	**5.97**	**6.25**	**6.48**	**6.67**	**6.84**	**6.99**	**7.13**	**7.25**
12	3.08	3.77	4.20	4.51	4.75	4.95	5.12	5.27	5.39	5.51	5.61
	4.32	**5.05**	**5.50**	**5.84**	**6.10**	**6.32**	**6.51**	**6.67**	**6.81**	**6.94**	**7.06**
13	3.06	3.73	4.15	4.45	4.69	4.88	5.05	5.19	5.32	5.43	5.53
	4.26	**4.96**	**5.40**	**5.73**	**5.98**	**6.19**	**6.37**	**6.53**	**6.67**	**6.79**	**6.90**
14	3.03	3.70	4.11	4.41	4.64	4.83	4.99	5.13	5.25	5.36	5.46
	4.21	**4.89**	**5.32**	**5.63**	**5.88**	**6.08**	**6.26**	**6.41**	**6.54**	**6.66**	**6.77**
15	3.01	3.67	4.08	4.37	4.59	4.78	4.94	5.08	5.20	5.31	5.40
	4.17	**4.84**	**5.25**	**5.56**	**5.80**	**5.99**	**6.16**	**6.31**	**6.44**	**6.55**	**6.66**
16	3.00	3.65	4.05	4.33	4.56	4.74	4.90	5.03	5.15	5.26	5.35
	4.13	**4.79**	**5.19**	**5.49**	**5.72**	**5.92**	**6.08**	**6.22**	**6.35**	**6.46**	**6.56**
17	2.98	3.63	4.02	4.30	4.52	4.70	4.86	4.99	5.11	5.21	5.31
	4.10	**4.74**	**5.14**	**5.43**	**5.66**	**5.85**	**6.01**	**6.15**	**6.27**	**6.38**	**6.48**
18	2.97	3.61	4.00	4.28	4.49	4.67	4.82	4.96	5.07	5.17	5.27
	4.07	**4.70**	**5.09**	**5.38**	**5.60**	**5.79**	**5.94**	**6.08**	**6.20**	**6.31**	**6.41**
19	2.96	3.59	3.98	4.25	4.47	4.65	4.79	4.92	5.04	5.14	5.23
	4.05	**4.67**	**5.05**	**5.33**	**5.55**	**5.73**	**5.89**	**6.02**	**6.14**	**6.25**	**6.34**
20	2.95	3.58	3.96	4.23	4.45	4.62	4.77	4.90	5.01	5.11	5.20
	4.02	**4.64**	**5.02**	**5.29**	**5.51**	**5.69**	**5.84**	**5.97**	**6.09**	**6.19**	**6.28**
24	2.92	3.53	3.90	4.17	4.37	4.54	4.68	4.81	4.92	5.01	5.10
	3.96	**4.55**	**4.91**	**5.17**	**5.37**	**5.54**	**5.69**	**5.81**	**5.92**	**6.02**	**6.11**
30	2.89	3.49	3.85	4.10	4.30	4.46	4.60	4.72	4.82	4.92	5.00
	3.89	**4.45**	**4.80**	**5.05**	**5.24**	**5.40**	**5.54**	**5.65**	**5.76**	**5.85**	**5.93**
40	2.86	3.44	3.79	4.04	4.23	4.39	4.52	4.63	4.73	4.82	4.90
	3.82	**4.37**	**4.70**	**4.93**	**5.11**	**5.26**	**5.39**	**5.50**	**5.60**	**5.69**	**5.76**
60	2.83	3.40	3.74	3.98	4.16	4.31	4.44	4.55	4.65	4.73	4.81
	3.76	**4.28**	**4.59**	**4.82**	**4.99**	**5.13**	**5.25**	**5.36**	**5.45**	**5.53**	**5.60**
120	2.80	3.36	3.68	3.92	4.10	4.24	4.36	4.47	4.56	4.64	4.71
	3.70	**4.20**	**4.50**	**4.71**	**4.87**	**5.01**	**5.12**	**5.21**	**5.30**	**5.37**	**5.44**
∞	2.77	3.31	3.63	3.86	4.03	4.17	4.28	4.39	4.47	4.55	4.62
	3.64	**4.12**	**4.40**	**4.60**	**4.76**	**4.88**	**4.99**	**5.08**	**5.16**	**5.23**	**5.29**

TABLE B.6   Critical Values for the Pearson Correlation*

*To be significant, the sample correlation, r, must be greater than or equal to the critical value in the table.

	Level of Significance for One-Tailed Test			
	.05	.025	.01	.005
	Level of Significance for Two-Tailed Test			
$df = n - 2$	.10	.05	.02	.01
1	.988	.997	.9995	.9999
2	.900	.950	.980	.990
3	.805	.878	.934	.959
4	.729	.811	.882	.917
5	.669	.754	.833	.874
6	.622	.707	.789	.834
7	.582	.666	.750	.798
8	.549	.632	.716	.765
9	.521	.602	.685	.735
10	.497	.576	.658	.708
11	.476	.553	.634	.684
12	.458	.532	.612	.661
13	.441	.514	.592	.641
14	.426	.497	.574	.623
15	.412	.482	.558	.606
16	.400	.468	.542	.590
17	.389	.456	.528	.575
18	.378	.444	.516	.561
19	.369	.433	.503	.549
20	.360	.423	.492	.537
21	.352	.413	.482	.526
22	.344	.404	.472	.515
23	.337	.396	.462	.505
24	.330	.388	.453	.496
25	.323	.381	.445	.487
26	.317	.374	.437	.479
27	.311	.367	.430	.471
28	.306	.361	.423	.463
29	.301	.355	.416	.456
30	.296	.349	.409	.449
35	.275	.325	.381	.418
40	.257	.304	.358	.393
45	.243	.288	.338	.372
50	.231	.273	.322	.354
60	.211	.250	.295	.325
70	.195	.232	.274	.302
80	.183	.217	.256	.283
90	.173	.205	.242	.267
100	.164	.195	.230	.254

Table VI of R. A. Fisher and F. Yates, *Statistical Tables for Biological, Agricultural and Medical Research*, 6th ed. London: Longman Group Ltd., 1974 (previously published by Oliver and Boyd Ltd., Edinburgh). Copyright ©1963 R. A. Fisher and F. Yates. Adapted and reprinted with permission of Pearson Education Limited.

TABLE B.7   Critical Values for the Spearman Correlation*

*To be significant, the sample correlation, r_s, must be greater than or equal to the critical value in the table.

	Level of Significance for One-Tailed Test			
	.05	.025	.01	.005
	Level of Significance for Two-Tailed Test			
n	.10	.05	.02	.01
4	1.000			
5	0.900	1.000	1.000	
6	0.829	0.886	0.943	1.000
7	0.714	0.786	0.893	0.929
8	0.643	0.738	0.833	0.881
9	0.600	0.700	0.783	0.833
10	0.564	0.648	0.745	0.794
11	0.536	0.618	0.709	0.755
12	0.503	0.587	0.671	0.727
13	0.484	0.560	0.648	0.703
14	0.464	0.538	0.622	0.675
15	0.443	0.521	0.604	0.654
16	0.429	0.503	0.582	0.635
17	0.414	0.485	0.566	0.615
18	0.401	0.472	0.550	0.600
19	0.391	0.460	0.535	0.584
20	0.380	0.447	0.520	0.570
21	0.370	0.435	0.508	0.556
22	0.361	0.425	0.496	0.544
23	0.353	0.415	0.486	0.532
24	0.344	0.406	0.476	0.521
25	0.337	0.398	0.466	0.511
26	0.331	0.390	0.457	0.501
27	0.324	0.382	0.448	0.491
28	0.317	0.375	0.440	0.483
29	0.312	0.368	0.433	0.475
30	0.306	0.362	0.425	0.467
35	0.283	0.335	0.394	0.433
40	0.264	0.313	0.368	0.405
45	0.248	0.294	0.347	0.382
50	0.235	0.279	0.329	0.363
60	0.214	0.255	0.300	0.331
70	0.190	0.235	0.278	0.307
80	0.185	0.220	0.260	0.287
90	0.174	0.207	0.245	0.271
100	0.165	0.197	0.233	0.257

Zar, J. H. (1972). Significance testing of the Spearman rank correlation coefficient. *Journal of the American Statistical Association*, 67, 578–580. Reprinted with permission from the *Journal of the American Statistical Association*. Copyright © 1972 by the American Statistical Association. All rights reserved.

TABLE B.8   The Chi-Square Distribution*

*The table entries are critical values of χ^2.

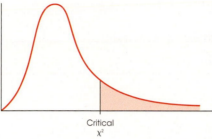

Critical
χ^2

df	Proportion in Critical Region				
	0.10	0.05	0.025	0.01	0.005
1	2.71	3.84	5.02	6.63	7.88
2	4.61	5.99	7.38	9.21	10.60
3	6.25	7.81	9.35	11.34	12.84
4	7.78	9.49	11.14	13.28	14.86
5	9.24	11.07	12.83	15.09	16.75
6	10.64	12.59	14.45	16.81	18.55
7	12.02	14.07	16.01	18.48	20.28
8	13.36	15.51	17.53	20.09	21.96
9	14.68	16.92	19.02	21.67	23.59
10	15.99	18.31	20.48	23.21	25.19
11	17.28	19.68	21.92	24.72	26.76
12	18.55	21.03	23.34	26.22	28.30
13	19.81	22.36	24.74	27.69	29.82
14	21.06	23.68	26.12	29.14	31.32
15	22.31	25.00	27.49	30.58	32.80
16	23.54	26.30	28.85	32.00	34.27
17	24.77	27.59	30.19	33.41	35.72
18	25.99	28.87	31.53	34.81	37.16
19	27.20	30.14	32.85	36.19	38.58
20	28.41	31.41	34.17	37.57	40.00
21	29.62	32.67	35.48	38.93	41.40
22	30.81	33.92	36.78	40.29	42.80
23	32.01	35.17	38.08	41.64	44.18
24	33.20	36.42	39.36	42.98	45.56
25	34.38	37.65	40.65	44.31	46.93
26	35.56	38.89	41.92	45.64	48.29
27	36.74	40.11	43.19	46.96	49.64
28	37.92	41.34	44.46	48.28	50.99
29	39.09	42.56	45.72	49.59	52.34
30	40.26	43.77	46.98	50.89	53.67
40	51.81	55.76	59.34	63.69	66.77
50	63.17	67.50	71.42	76.15	79.49
60	74.40	79.08	83.30	88.38	91.95
70	85.53	90.53	95.02	100.42	104.22
80	96.58	101.88	106.63	112.33	116.32
90	107.56	113.14	118.14	124.12	128.30
100	118.50	124.34	129.56	135.81	140.17

TABLE B.9A   Critical Values of the Mann-Whitney U for $\alpha = .05$*

*Critical values are provided for a *one-tailed* test at $\alpha = .05$ (lightface type) and for a *two-tailed* test at $\alpha = .05$ (boldface type). To be significant for any given n_A and n_B, the obtained U must be *equal to* or *less than* the critical value in the table. Dashes (—) in the body of the table indicate that no decision is possible at the stated level of significance and values of n_A and n_B.

n_B \ n_A	1	2	3	4	5	6	7	8	9	10	11	12	13	14	15	16	17	18	19	20
1	—	—	—	—	—	—	—	—	—	—	—	—	—	—	—	—	—	—	0	0
	—	—	—	—	—	—	—	—	—	—	—	—	—	—	—	—	—	—	—	—
2	—	—	—	—	0	0	0	1	1	1	1	2	2	2	3	3	3	4	4	4
	—	—	—	—	—	—	—	**0**	**0**	**0**	**0**	**1**	**1**	**1**	**1**	**1**	**2**	**2**	**2**	**2**
3	—	—	0	0	1	2	2	3	3	4	5	5	6	7	7	8	9	9	10	11
	—	—	—	—	**0**	**1**	**1**	**2**	**2**	**3**	**3**	**4**	**4**	**5**	**5**	**6**	**6**	**7**	**7**	**8**
4	—	—	0	1	2	3	4	5	6	7	8	9	10	11	12	14	15	16	17	18
	—	—	—	**0**	**1**	**2**	**3**	**4**	**4**	**5**	**6**	**7**	**8**	**9**	**10**	**11**	**11**	**12**	**13**	**13**
5	—	0	1	2	4	5	6	8	9	11	12	13	15	16	18	19	20	22	23	25
	—	—	**0**	**1**	**2**	**3**	**5**	**6**	**7**	**8**	**9**	**11**	**12**	**13**	**14**	**15**	**17**	**18**	**19**	**20**
6	—	0	2	3	5	7	8	10	12	14	16	17	19	21	23	25	26	28	30	32
	—	—	**1**	**2**	**3**	**5**	**6**	**8**	**10**	**11**	**13**	**14**	**16**	**17**	**19**	**21**	**22**	**24**	**25**	**27**
7	—	0	2	4	6	8	11	13	15	17	19	21	24	26	28	30	33	35	37	39
	—	—	**1**	**3**	**5**	**6**	**8**	**10**	**12**	**14**	**16**	**18**	**20**	**22**	**24**	**26**	**28**	**30**	**32**	**34**
8	—	1	3	5	8	10	13	15	18	20	23	26	28	31	33	36	39	41	44	47
	—	**0**	**2**	**4**	**6**	**8**	**10**	**13**	**15**	**17**	**19**	**22**	**24**	**26**	**29**	**31**	**34**	**36**	**38**	**41**
9	—	1	3	6	9	12	15	18	21	24	27	30	33	36	39	42	45	48	51	54
	—	**0**	**2**	**4**	**7**	**10**	**12**	**15**	**17**	**20**	**23**	**26**	**28**	**31**	**34**	**37**	**39**	**42**	**45**	**48**
10	—	1	4	7	11	14	17	20	24	27	31	34	37	41	44	48	51	55	58	62
	—	**0**	**3**	**5**	**8**	**11**	**14**	**17**	**20**	**23**	**26**	**29**	**33**	**36**	**39**	**42**	**45**	**48**	**52**	**55**
11	—	1	5	8	12	16	19	23	27	31	34	38	42	46	50	54	57	61	65	69
	—	**0**	**3**	**6**	**9**	**13**	**16**	**19**	**23**	**26**	**30**	**33**	**37**	**40**	**44**	**47**	**51**	**55**	**58**	**62**
12	—	2	5	9	13	17	21	26	30	34	38	42	47	51	55	60	64	68	72	77
	—	**1**	**4**	**7**	**11**	**14**	**18**	**22**	**26**	**29**	**33**	**37**	**41**	**45**	**49**	**53**	**57**	**61**	**65**	**69**
13	—	2	6	10	15	19	24	28	33	37	42	47	51	56	61	65	79	75	80	84
	—	**1**	**4**	**8**	**12**	**16**	**20**	**24**	**28**	**33**	**37**	**41**	**45**	**50**	**54**	**59**	**63**	**67**	**72**	**76**
14	—	2	7	11	16	21	26	31	36	41	46	51	56	61	66	71	77	82	87	92
	—	**1**	**5**	**9**	**13**	**17**	**22**	**26**	**31**	**36**	**40**	**45**	**50**	**55**	**59**	**64**	**67**	**74**	**78**	**83**
15	—	3	7	12	18	23	28	33	39	44	50	55	61	66	72	77	83	88	94	100
	—	**1**	**5**	**10**	**14**	**19**	**24**	**29**	**34**	**39**	**44**	**49**	**54**	**59**	**64**	**70**	**75**	**80**	**85**	**90**
16	—	3	8	14	19	25	30	36	42	48	54	60	65	71	77	83	89	95	101	107
	—	**1**	**6**	**11**	**15**	**21**	**26**	**31**	**37**	**42**	**47**	**53**	**59**	**64**	**70**	**75**	**81**	**86**	**92**	**98**
17	—	3	9	15	20	26	33	39	45	51	57	64	70	77	83	89	96	102	109	115
	—	**2**	**6**	**11**	**17**	**22**	**28**	**34**	**39**	**45**	**51**	**57**	**63**	**67**	**75**	**81**	**87**	**93**	**99**	**105**
18	—	4	9	16	22	28	35	41	48	55	61	68	75	82	88	95	102	109	116	123
	—	**2**	**7**	**12**	**18**	**24**	**30**	**36**	**42**	**48**	**55**	**61**	**67**	**74**	**80**	**86**	**93**	**99**	**106**	**112**
19	0	4	10	17	23	30	37	44	51	58	65	72	80	87	94	101	109	116	123	130
	—	**2**	**7**	**13**	**19**	**25**	**32**	**38**	**45**	**52**	**58**	**65**	**72**	**78**	**85**	**92**	**99**	**106**	**113**	**119**
20	0	4	11	18	25	32	39	47	54	62	69	77	84	92	100	107	115	123	130	138
	—	**2**	**8**	**13**	**20**	**27**	**34**	**41**	**48**	**55**	**62**	**69**	**76**	**83**	**90**	**98**	**105**	**112**	**119**	**127**

TABLE B.9B   **Critical Values of the Mann-Whitney *U* for α = .01***

*Critical values are provided for a *one-tailed* test at α = .01 (lightface type) and for a *two-tailed* test at α = .01 (boldface type). To be significant for any given n_A and n_B, the obtained *U* must be *equal to* or *less than* the critical value in the table. Dashes (—) in the body of the table indicate that no decision is possible at the stated level of significance and values of n_A and n_B.

n_B \ n_A	1	2	3	4	5	6	7	8	9	10	11	12	13	14	15	16	17	18	19	20
1	—	—	—	—	—	—	—	—	—	—	—	—	—	—	—	—	—	—	—	—
2	—	—	—	—	—	—	—	—	—	—	—	—	0	0	0	0	0	0	1	1
	—	—	—	—	—	—	—	—	—	—	—	—	—	—	—	—	—	—	**0**	**0**
3	—	—	—	—	—	—	0	0	1	1	1	2	2	2	3	3	4	4	4	5
	—	—	—	—	—	—	—	—	**0**	**0**	**0**	**1**	**1**	**1**	**2**	**2**	**2**	**2**	**3**	**3**
4	—	—	—	—	0	1	1	2	3	3	4	5	5	6	7	7	8	9	9	10
	—	—	—	—	**0**	**0**	**0**	**1**	**1**	**2**	**2**	**3**	**3**	**4**	**5**	**5**	**6**	**6**	**7**	**8**
5	—	—	—	0	1	2	3	4	5	6	7	8	9	10	11	12	13	14	15	16
	—	—	—	**0**	**1**	**1**	**2**	**3**	**4**	**5**	**6**	**7**	**7**	**8**	**9**	**10**	**11**	**12**	**13**
6	—	—	—	1	2	3	4	6	7	8	9	11	12	13	15	16	18	19	20	22
	—	—	—	**0**	**1**	**2**	**3**	**4**	**5**	**6**	**7**	**9**	**10**	**11**	**12**	**13**	**15**	**16**	**17**	**18**
7	—	—	0	1	3	4	6	7	9	11	12	14	16	17	19	21	23	24	26	28
	—	—	—	**0**	**1**	**3**	**4**	**6**	**7**	**9**	**10**	**12**	**13**	**15**	**16**	**18**	**19**	**21**	**22**	**24**
8	—	—	0	2	4	6	7	9	11	13	15	17	20	22	24	26	28	30	32	34
	—	—	—	**1**	**2**	**4**	**6**	**7**	**9**	**11**	**13**	**15**	**17**	**18**	**20**	**22**	**24**	**26**	**28**	**30**
9	—	—	1	3	5	7	9	11	14	16	18	21	23	26	28	31	33	36	38	40
	—	—	**0**	**1**	**3**	**5**	**7**	**9**	**11**	**13**	**16**	**18**	**20**	**22**	**24**	**27**	**29**	**31**	**33**	**36**
10	—	—	1	3	6	8	11	13	16	19	22	24	27	30	33	36	38	41	44	47
	—	—	**0**	**2**	**4**	**6**	**9**	**11**	**13**	**16**	**18**	**21**	**24**	**26**	**29**	**31**	**34**	**37**	**39**	**42**
11	—	—	1	4	7	9	12	15	18	22	25	28	31	34	37	41	44	47	50	53
	—	—	**0**	**2**	**5**	**7**	**10**	**13**	**16**	**18**	**21**	**24**	**27**	**30**	**33**	**36**	**39**	**42**	**45**	**48**
12	—	—	2	5	8	11	14	17	21	24	28	31	35	38	42	46	49	53	56	60
	—	—	**1**	**3**	**6**	**9**	**12**	**15**	**18**	**21**	**24**	**27**	**31**	**34**	**37**	**41**	**44**	**47**	**51**	**54**
13	—	0	2	5	9	12	16	2	23	27	31	35	39	43	47	51	55	59	63	67
	—	—	**1**	**3**	**7**	**10**	**13**	**17**	**20**	**24**	**27**	**31**	**34**	**38**	**42**	**45**	**49**	**53**	**56**	**60**
14	—	0	2	6	10	13	17	22	26	30	34	38	43	47	51	56	60	65	69	73
	—	—	**1**	**4**	**7**	**11**	**15**	**18**	**22**	**26**	**30**	**34**	**38**	**42**	**46**	**50**	**54**	**58**	**63**	**67**
15	—	0	3	7	11	15	19	24	28	33	37	42	47	51	56	61	66	70	75	80
	—	—	**2**	**5**	**8**	**12**	**16**	**20**	**24**	**29**	**33**	**37**	**42**	**46**	**51**	**55**	**60**	**64**	**69**	**73**
16	—	0	3	7	12	16	21	26	31	36	41	46	51	56	61	66	71	76	82	87
	—	—	**2**	**5**	**9**	**13**	**18**	**22**	**27**	**31**	**36**	**41**	**45**	**50**	**55**	**60**	**65**	**70**	**74**	**79**
17	—	0	4	8	13	18	23	28	33	38	44	49	55	60	66	71	77	82	88	93
	—	—	**2**	**6**	**10**	**15**	**19**	**24**	**29**	**34**	**39**	**44**	**49**	**54**	**60**	**65**	**70**	**75**	**81**	**86**
18	—	0	4	9	14	19	24	30	36	41	47	53	59	65	70	76	82	88	94	100
	—	—	**2**	**6**	**11**	**16**	**21**	**26**	**31**	**37**	**42**	**47**	**53**	**58**	**64**	**70**	**75**	**81**	**87**	**92**
19	—	1	4	9	15	20	26	32	38	44	50	56	63	69	75	82	88	94	101	107
	—	**0**	**3**	**7**	**12**	**17**	**22**	**28**	**33**	**39**	**45**	**51**	**56**	**63**	**69**	**74**	**81**	**87**	**93**	**99**
20	—	1	5	10	16	22	28	34	40	47	53	60	67	73	80	87	93	100	107	114
	—	**0**	**3**	**8**	**13**	**18**	**24**	**30**	**36**	**42**	**48**	**54**	**60**	**67**	**73**	**79**	**86**	**92**	**99**	**105**

TABLE B.10   Critical Values of *T* for the Wilcoxon Signed-Ranks Test*

*To be significant, the obtained *T* must be *equal to* or *less than* the critical value. Dashes (—) in the columns indicate that no decision is possible for the stated α and *n*.

	Level of Significance for One-Tailed Test					Level of Significance for One-Tailed Test			
	.05	.025	.01	.005		.05	.025	.01	.005
	Level of Significance for Two-Tailed Test					Level of Significance for Two-Tailed Test			
n	.10	.05	.02	.01	*n*	.10	.05	.02	.01
5	0	—	—	—	30	151	137	120	109
6	2	0	—	—	31	163	147	130	118
7	3	2	0	—	32	175	159	140	128
8	5	3	1	0	33	187	170	151	138
9	8	5	3	1	34	200	182	162	148
10	10	8	5	3	35	213	195	173	159
11	13	10	7	5	36	227	208	185	171
12	17	13	9	7	37	241	221	198	182
13	21	17	12	9	38	256	235	211	194
14	25	21	15	12	39	271	249	224	207
15	30	25	19	15	40	286	264	238	220
16	35	29	23	19	41	302	279	252	233
17	41	34	27	23	42	319	294	266	247
18	47	40	32	27	43	336	310	281	261
19	53	46	37	32	44	353	327	296	276
20	60	52	43	37	45	371	343	312	291
21	67	58	49	42	46	389	361	328	307
22	75	65	55	48	47	407	378	345	322
23	83	73	62	54	48	426	396	362	339
24	91	81	69	61	49	446	415	379	355
25	100	89	76	68	50	466	434	397	373
26	110	98	84	75					
27	119	107	92	83					
28	130	116	101	91					
29	140	126	110	100					

Adapted from F. Wilcoxon, S. K. Katti, and R. A. Wilcox, *Critical Values and Probability Levels of the Wilcoxon Rank-Sum Test and the Wilcoxon Signed-Ranks Test*. Wayne, NJ: American Cyanamid Company, 1963. Adapted and reprinted with permission of the American Cyanamid Company.

Solutions for Odd-Numbered Problems in the Text

Note: Many of the problems in the text require several stages of computation. At each stage there is an opportunity for rounding answers. Depending on the exact sequence of operations used to solve a problem, different individuals will round their answers at different times and in different ways. As a result, you may obtain answers that are slightly different from those presented here. To help minimize this problem, we have tried to include the numerical values obtained at different stages of complex problems rather than presenting a single final answer.

CHAPTER 1 | Introduction to Statistics

1. **a.** The population is the entire set of high school students in the United States.

   b. The sample is the group of 100 students who were tested in the study.

3. Descriptive statistics are used to simplify and summarize data. Inferential statistics use sample data to make general conclusions about populations.

5. A correlational study has only one group of individuals and measures two (or more) different variables for each individual. Other research methods evaluating relationships between variables compare two (or more) different groups of scores.

7. The independent variable is the type of word that the participants shouted. The dependent variable is the amount of pain they tolerated.

9. This is a nonexperimental study. The researcher is simply observing two preexisting groups, not manipulating variables.

11. This is not an experiment because there is no manipulation. Instead, the study is comparing two preexisting groups (state university and religious college students).

13. Annual income is continuous. Number of dependents and social security number are both discrete.

15. The independent variable is the amount of control over office design. The dependent variables are productivity and well-being.

17. **a.** The independent variable is whether or not the motivational signs were posted, and the dependent variable is amount of use of the stairs.

   b. Posting vs. not posting is measured on a nominal scale.

19. **a.** $\Sigma X^2 = 48$

   b. $(\Sigma X)^2 = 14^2 = 196$

   c. $\Sigma(X - 1) = 9$

   d. $\Sigma(X - 1)^2 = 25$

21. **a.** $\Sigma X = 4$

   b. $\Sigma Y = 18$

   c. $\Sigma XY = 11$

23. **a.** $\Sigma X^2 = 50$

   b. $(\Sigma X)^2 = 12^2 = 144$

   c. $\Sigma(X - 3) = 0$

   d. $\Sigma(X - 3)^2 = 14$

CHAPTER 2 | Frequency Distributions

1.

X	f
7	2
6	3
5	1
4	1
3	4
2	6
1	3

3. a. $n = 15$
   b. $\Sigma X = 41$
   c. $\Sigma X^2 = 131$

5. a.

X	f
50–54	2
45–49	2
40–44	2
35–39	1
30–34	3
25–29	5
20–24	3
15–19	5
10–14	1

b.

X	f
50–50	2
40–49	4
30–39	4
20–29	8
10–19	6

7. a. 5 points wide and around 7 intervals
   b. 2 points wide and around 9 intervals
   c. 10 points wide and around 8 intervals

9. A bar graph leaves a space between adjacent bars and is used with data from nominal or ordinal scales. In a histogram, adjacent bars touch at the real limits. Histograms are used to display data from interval or ratio scales.

11. a. histogram or polygon (ratio scale)
   b. bar graph (ordinal scale)
   c. bar graph (nominal scale)
   d. bar graph (nominal scale)

13. a. $n = 17$
   b. $\Sigma X = 55$
   c. $\Sigma X^2 = 197$

15. a.

X	f
10	2
9	4
8	5
7	4
6	3
5	2
4	2
3	1
2	1

b. Figure C1

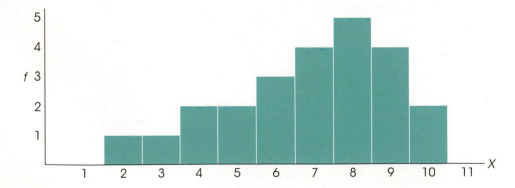

c. 1) The distribution is negatively skewed.
   2) The scores are centered around $X = 7$.
   3) The scores are clustered at the high end of the scale.

17.

X	f	cf	c%
5	2	20	100
4	5	18	90
3	6	13	65
2	4	7	35
1	3	3	15

   a. The percentile rank for $X = 2.5$ is 35%

   b. The percentile rank for $X = 4.5$ is 90%

   c. The 15th percentile is $X = 1.5$.

   d. The 65th percentile is $X = 3.5$.

19. a. 26%

   b. 48%

   c. $X = 5.7$

   d. $X = 8.0$

21. a. 97%

   b. 70.8%

   c. $X = 5.75$

   d. $X = 12.83$

23.

1	978368
2	35654475
3	2046
4	4793
5	12

25.

1	4916
2	58223
3	692437
4	77294
5	2875
6	5

CHAPTER 3 | Central Tendency

1. The mean is $\frac{32}{8} = 4$, the median is 4.5, and the mode is 5.

3. For this sample, the mean is $\frac{39}{14} = 2.79$, the median is 2.5, and the mode is 1.

5. a. Median = 5

   b. Median = 4.83

7. $\Sigma X = 96$.

9. The first sample has $\Sigma X = 84$ and the second has $\Sigma X = 96$. The combined sample has $n = 20$ and $\Sigma X = 180$. The combined sample mean is $M = 9$.

11. a. The combined sample mean is $M = 7.5$.

   b. The combined sample mean is $\frac{(20 + 60)}{10} = 8$.

   c. The combined sample mean is $\frac{(30 + 40)}{10} = 7$.

13. The original sample has $n = 7$ and $\Sigma X = 112$. The new sample has $n = 7$ and $\Sigma X = 126$. The new mean is $M = 18$.

15. The original sample has $n = 7$ and $\Sigma X = 63$. The new sample has $n = 8$ and $\Sigma X = 64$. The new mean is $M = 8$.

17. The original sample has $n = 15$ and $\Sigma X = 90$. The new sample has $n = 16$ and $\Sigma X = 112$. The new mean is $M = 7$.

19. The original sample has $n = 7$ and $\Sigma X = 35$. The new sample has $n = 8$ and $\Sigma X = 48$. The total increased by 13 points, so $X = 13$.

21. The original sample has $n = 9$ and $\Sigma X = 117$. The new sample has $n = 10$ and $\Sigma X = 120$. The score that was added must be $X = 3$.

23. a. For treatment 1 the mean is $M = 9$ and for treatment 2 the mean is $M = 7$. Based on the means, the scores are higher in treatment 1.

   b. For treatment 1 the median is 6.5 and for treatment 2 the median is 7.5. Based on the medians, the scores are higher in treatment 2.

   c. Both treatments have a mode of 6. Based on the mode, there is no difference between the two treatments.

25. The participants were able to withstand the pain for $M = \frac{694}{10} = 69.4$ seconds when they were swearing comparing to $M = \frac{554}{10} = 55.4$ seconds when yelling a neutral word. Swearing does appear to help improve pain tolerance.

CHAPTER 4 | Variability

1. a. *SS* is the sum of squared deviation scores.

   b. Variance is the mean squared deviation.

   c. Standard deviation is the square root of the variance. It provides a measure of the standard distance from the mean.

3. A standard deviation of zero indicates there is no variability. In this case, all of the scores in the sample have exactly the same value.

5. Sample A has $SS = 10$. The definitional formula works well. Sample B has $SS = 13$. The computational formula is better when the mean is not a whole number.

7. $SS = 28$, variance is 4, and the standard deviation is 2.

9. **a.** The mean is 4, $SS = 20$, $\sigma^2 = 4$, and $\sigma = 2$.

   b. $SS = 20$, $s^2 = 5$, and $s = \sqrt{5} = 2.24$.

11. **a.** Figure C2

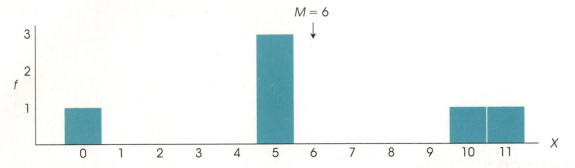

b. The mean is $\frac{36}{6} = 6$. The three scores of $X = 5$ are one point from the mean. The score $X = 0$ is farthest from the mean (6 points). The standard deviation should be between 1 and 6 points, probably around 3.5.

c. $SS = 80$, $s^2 = 16$, $s = 4$, which agrees with the estimate.

13. $SS = 36$, $s^2 = 9$, and $s = 3$.

15. **a.** The original mean is $M = 64$ and the standard deviation is $s = 13$.

   b. The original mean is $M = 16$ and the standard deviation is $s = 6$.

17. **a.** The simplified scores had $M = 1.5$, $SS = 28$, and $s = 2$.

   b. The original scores had $M = 0.75$ and $s = 1$.

19. **a.** The range is either 14 or 15, and the standard deviation is $s = 5$.

   b. After spreading out the two scores in the middle, the range is still 14 or 15 but the standard deviation is now $s = 7$.

c. The two distributions are the same according to the range. The range is completely determined by the two extreme scores and is insensitive to the variability of the rest of the scores. The second distribution has more variability according to the standard deviation, which measures variability for the complete set.

21. **a.** For the older adults, the mean is $M = 5.47$, $SS = 65.73$, $s^2 = 4.70$ and $s = 2.17$. For the younger adults, the mean is $M = 7.27$, $SS = 18.93$, $s^2 = 1.35$, and $s = 1.16$.

   b. The younger adults had a higher average and were much less variable.

23. **a.** With $\sigma = 20$, $X = 70$ is not an extreme score. It is located above the mean by a distance of only one standard deviation.

   b. With $\sigma = 5$, $X = 70$ is an extreme score. It is located above the mean by a distance equal to four times the standard deviation.

CHAPTER 5 | z-Scores

1. The sign of the z-score tells whether the location is above (+) or below (−) the mean, and the magnitude tells the distance from the mean in terms of the number of standard deviations.

3. **a.** above the mean by 40 points

   b. above the mean by 10 points

   c. below the mean by 20 points

   d. below the mean by 5 points

5.

X	z		X	z		X	z
45	0.45		52	1.09		41	0.09
30	−0.91		25	−1.36		38	−0.18

7. a.

X	z		X	z		X	z
69	0.75		84	2.00		63	0.25
54	−0.50		48	−1.00		45	−1.25

b.

X	z		X	z		X	z
66	0.50		78	1.50		30	−2.50
57	−0.25		54	−0.50		75	1.25

9. **a.** $X = 41$

b. $X = 42$

c. $X = 44$

d. $X = 48$

11. $\sigma = 6$

13. $\sigma = 4$

15. **a.** central ($z = 0.67$)

b. extreme ($z = 2.50$)

c. central ($z = -0.25$)

d. extreme ($z = -2.67$)

17. **a.** $X = 70$ corresponds to $z = -1.50$, and $X = 60$ corresponds to $z = -1.00$ (better grade).

b. $X = 58$ corresponds to $z = 1.50$, and $X = 85$ corresponds to $z = 1.50$. The two scores have the same relative position and should receive the same grade.

c. $X = 32$ corresponds to $z = 2.00$, and $X = 26$ corresponds to $z = 3.00$ (better grade).

19. **a.** $X = 75$ ($z = -1.25$)

b. $X = 90$ ($z = -0.50$)

c. $X = 115$ ($z = 0.75$)

d. $X = 130$ ($z = 1.50$)

21.

X	z		X	z		X	z
32	0.25		34	0.50		36	0.75
28	−0.25		20	−1.25		18	−1.50

23. $M = 53$

25. $\mu = 50$ and $\sigma = 6$. The distance between the two scores is 9 points which is equal to 1.5 standard deviations.

CHAPTER 6 | Probability

1. The two requirements for a random sample are: (1) each individual has an equal chance of being selected, and (2) if more than one individual is selected, the probabilities must stay constant for all selections.

3. **a.** $p = \frac{1}{3} = 0.33$

b. $p = \frac{2}{3} = 0.67$

c. $p = \frac{2}{3} = 0.67$

5. **a.** tail to the right, $p = 0.1587$

b. tail to the right, $p = 0.3085$

c. tail to the left, $p = 0.1056$

d. tail to the left, $p = 0.3446$

7. **a.** $p(z > 1.25) = 0.1056$

b. $p(z > -0.60) = 0.7257$

c. $p(z < 0.70) = 0.7580$

d. $p(z < -1.30) = 0.0968$

9. **a.** $p = 0.5762$

b. $p = 0.5328$

c. $p = 0.3539$

d. $p = 0.0968$

11. **a.** $z = \pm 0.39$

b. $z = \pm 0.52$

c. $z = \pm 0.67$

d. $z = \pm 0.84$

13. **a.** body to the left, $p = 0.5987$

b. body to the right, $p = 0.8413$

c. body to the right, $p = 0.6915$

d. body to the left, $p = 0.7734$

15. **a.** $z = 1.38, p = 0.0838$

b. $z = -1.13, p = 0.1292$

17. **a.** $p(z < -1.00) = 0.1587$

b. $p(z > 1.50) = 0.0668$

c. $z = 0.84, X = 584$

19. **a.** $p(z > 1.80) = 0.0359$

b. $p(-1.00 < z < 1.00) = 0.6826$

21. a. $p = \frac{1}{4}$

   b. $\mu = 8$

   c. $\sigma = \sqrt{6} = 2.45$ and for $X = 12.5$, $z = 1.84$, and $p = 0.0329$

23. a. $p = q = \frac{1}{2}$, and with $n = 64$ the normal approximation has $\mu = 32$ and $\sigma = 4$.
   $p(X > 39.5) = p(z > 1.88) = 0.0301$.

   b. $p(X > 40.5) = p(z > 2.13) = 0.0166$.

c. $p(X = 40)$ is equal to the difference between the probabilities in parts a and b. $p(X = 40) = 0.0135$.

25. a. $p = 0.30$.

   b. $\mu = pn = 25.2$ and $\sigma = 4.2$. For $X = 30.5$, $z = 1.26$ and $p = 0.1038$.

   c. For $X = 20.5$, $z = -1.12$ and $p = 0.1314$.

CHAPTER 7 | The Distribution of Sample Means

1. a. The distribution of sample means consists of the sample means for all the possible random samples of a specific size (n) from a specific population.

   b. The expected value of M is the mean of the distribution of sample means (μ).

   c. The standard error of M is the standard deviation of the distribution of sample means $\left(\sigma_M = \dfrac{\sigma}{\sqrt{n}}\right)$.

3. The distribution will be normal because $n > 30$, with an expected value of $\mu = 90$ and a standard error of $\frac{32}{\sqrt{64}} = 4$ points.

5. a. Standard error $= \frac{24}{\sqrt{4}} = 12$ points

   b. Standard error $= \frac{24}{\sqrt{9}} = 8$ points

   c. Standard error $= \frac{24}{\sqrt{16}} = 6$ points

7. a. $n > 9$

   b. $n > 16$

   c. $n > 36$

9. a. $\sigma = 32$

   b. $\sigma = 16$

   c. $\sigma = 4$

11. a. $\sigma_M = 2$ points and $z = 4.00$

   b. $\sigma_M = 4$ points and $z = 2.00$

   c. $\sigma_M = 8$ points and $z = 1.00$

13. a. With a standard error of 5, $M = 65$ corresponds to $z = 1.00$, $p(M > 65) = 0.1587$.

   b. With a standard error of 4, $M = 65$ corresponds to $z = 1.25$, $p(M > 65) = 0.1056$.

   c. With a standard error of 2, $M = 65$ corresponds to $z = 2.50$, $p(M > 65) = 0.0062$.

15. a. $z = -0.50$ and $p = 0.3085$

   b. $\sigma_M = 3$, $z = -1.00$ and $p = 0.1587$

   c. $\sigma_M = 1$, $z = -3.00$ and $p = 0.0013$

17. a. $\sigma_M = 4$, $z = \pm 1.96$ and the range is 42.16 to 57.84

   b. $\sigma_M = 4$, $z = \pm 2.58$ and the range is 39.68 to 60.32

19. $\sigma_M = 0.20$, $z = 2.50$ and $p = 0.0062$

21. a. With a standard error of 9, $M = 67$ corresponds to $z = 0.78$, which is not extreme.

   b. With a standard error of 3, $M = 67$ corresponds to $z = 2.33$, which is extreme.

23. a. With a standard error of 10, $M = 74$ corresponds to $z = 0.90$, which is not extreme.

   b. With a standard error of 4, $M = 74$ corresponds to $z = 2.25$, which is extreme.

CHAPTER 8 | Introduction to Hypothesis Testing

1. The four steps are: (1) State the hypotheses and select an alpha level, (2) Locate the critical region, (3) Compute the test statistic, and (4) Make a decision.

3. A Type I error is rejecting a true null hypothesis (deciding that there is a treatment effect when there is not). A Type II error is failing to reject a false null hypothesis (failing to detect a real treatment effect).

5. a. A larger difference will produce a larger value in the numerator which will produce a larger z-score.

   b. A larger standard deviation will produce larger standard error in the denominator which will produce a smaller z-score.

   c. A larger sample will produce a smaller standard error in the denominator which will produce a larger z-score.

7. H_0: $\mu = 14$ (there has been no change). H_1: $\mu \neq 14$ (the mean has changed). The critical region consists of z-scores beyond ± 1.96. For these data, the standard error is 0.6 and $z = -\frac{1.5}{0.6} = -2.50$. Fail to reject the null hypothesis. There has been a significant change in study hours.

9. a. H_0: $\mu = 71$. With $\sigma = 12$, the sample mean corresponds to $z = \frac{5}{2} = 2.50$. This is sufficient to reject the null hypothesis. Conclude that the online course has a significant effect.

   b. H_0: $\mu = 71$. With $\sigma = 18$, the sample mean corresponds to $z = \frac{5}{3} = 1.67$. This is not sufficient to reject the null hypothesis. Conclude that the online course does not have a significant effect.

   c. There is a 5 point difference between the sample mean and the hypothesis. In part a, the standard error is 2 points and the 5-point difference is significant. However, in part b, the standard error is 3 points and the 5-point difference is not significantly more than is expected by chance. In general, a larger standard deviation produces a larger standard error, which reduces the likelihood of rejecting the null hypothesis.

11. a. With $\sigma = 5$, the standard error is 1, and $z = \frac{4}{1} = 4.00$. Reject H_0.

   b. With $\sigma = 15$, the standard error is 3, and $z = \frac{4}{3} = 1.33$. Fail to reject H_0.

   c. Larger variability reduces the likelihood of rejecting H_0.

13. a. With a 4-point treatment effect, for the z-score to be greater than 1.96, the standard error must be smaller than 2.03. The sample size must be greater than 96.12; a sample of $n = 97$ or larger is needed.

   b. With a 2-point treatment effect, for the z-score to be greater than 1.96, the standard error must be smaller than 1.02. The sample size must be greater than 384.47; a sample of $n = 385$ or larger is needed.

15. a. The null hypothesis states that there is no effect on reaction time, $\mu = 400$. The critical region consists of z-scores beyond $z = \pm 1.96$. For these data, the standard error is 6.67 and $z = -\frac{8}{6.67} = -1.20$. Fail to reject H_0. There is no significant change in reaction time.

   b. Cohen's $d = 8/40 = 0.20$.

   c. The results indicate that caffeine had no significant effect on response time, $z = 1.20$, $p > .05$, $d = 0.20$.

17. a. The null hypothesis states that the new course has no effect on SAT scores, $\mu = 500$. The critical region consists of z-scores beyond $z = +2.33$. For these data, the standard error is 22.37 and $z = \frac{62}{22.37} = 2.77$. Reject H_0. There is a significant change in SAT scores.

   b. Cohen's $d = \frac{62}{100} = 0.62$.

   c. The new course had a significant effect on SAT scores, $z = 2.77$, $p < .01$, $d = 0.62$.

19. a. For a sample of $n = 9$ the standard error is 4 points, and the critical boundary for $z = 1.96$ corresponds to a sample mean of $M = 47.84$. With a 6-point effect, the distribution of sample means would be centered at $\mu = 46$. In this distribution, the critical boundary of $M = 47.84$ corresponds to $z = 0.46$. The power for the test is $p(z > 0.46) = 0.3228$ or 32.28%.

   b. For a sample of $n = 16$ the standard error would be 3 points, and the critical boundary for $z = 1.96$ corresponds to a sample mean of $M = 45.88$. With a 6-point effect, the distribution of sample means would be centered at $\mu = 46$. In this distribution, the critical boundary of $M = 45.88$ corresponds to $z = -0.12$. The power for the test is $p(z > -0.12) = 0.5478$ or 54.78%.

21. a. With no treatment effect the distribution of sample means is centered at $\mu = 100$ with a standard error of 3 points. The critical boundary of $z = 1.96$ corresponds to a sample mean of $M = 105.88$. With a 7-point treatment effect, the distribution of sample means is centered at $\mu = 107$. In this distribution a mean of $M = 105.88$ corresponds to $z = -0.37$. The power for the test is the probability of obtaining a z-score greater than -0.37, which is $p = 0.6443$.

   b. With a one-tailed test, a critical boundary of $z = 1.65$ corresponds to a sample mean of $M = 104.95$. With a 7-point treatment effect, the distribution of sample means is centered at $\mu = 107$. In this distribution a mean of $M = 104.95$ corresponds to $z = -0.68$. The power for the test is the probability of obtaining a z-score greater than -0.68, which is $p = 0.7517$.

23. a. Increasing alpha increases power.

   b. Changing from one- to two-tailed decreases power (if the effect is in the predicted direction).

CHAPTER 9 | Introduction to the t Statistic

1. A z-score is used when the population standard deviation (or variance) is known. The t statistic is used when the population variance or standard deviation is unknown. The t statistic uses the sample variance or standard deviation in place of the unknown population values.

3. **a.** The sample variance is 144 and the estimated standard error is 4.

   b. The sample variance is 36 and the estimated standard error is 1.5.

   c. The sample variance is 25 and the estimated standard error is 1.

5. **a.** $t = \pm 3.182$

   b. $t = \pm 2.145$

   c. $t = \pm 2.069$

7. **a.** $M = 3$ and $s = \sqrt{4} = 2$.

   b. $s_M = 1$.

9. **a.** 4.5 points

   b. The sample variance is 27 and the estimated standard error is $s_M = 1.5$.

   c. For these data, $t = 3.00$. With $df = 11$ the critical value is $t = \pm 2.201$. Reject H_0 and conclude that there is a significant effect.

11. **a.** With $n = 16$, $s_M = 0.75$ and $t = \frac{1.3}{0.75} = 1.73$. This is not greater than the critical value of 2.131, so there is no significant effect.

    b. With $n = 36$, $s_M = 0.50$ and $t = \frac{1.3}{0.50} = 2.60$. This value is greater than the critical value of 2.042 (using $df = 30$), so we reject the null hypothesis and conclude that there is a significant treatment effect.

    c. As the sample size increases, the likelihood of rejecting the null hypothesis also increases.

13. **a.** With a two-tailed test, the critical boundaries are ± 2.306 and the obtained value of $t = \frac{3.3}{1.5} = 2.20$ is not sufficient to reject the null hypothesis.

    b. For the one-tailed test the critical value is 1.860, so we reject the null hypothesis and conclude that participants significantly overestimated the number who noticed.

15. **a.** With $df = 15$, the critical values are ± 2.947. For these data, the sample variance is 16, the estimated standard error is 1, and $t = \frac{8.2}{1} = 8.20$. Reject the null hypothesis and conclude that there has been a significant change in the level of anxiety.

    b. With $df = 15$, the t values for 90% confidence are ± 1.753, and the interval extends from 21.547 to 25.053.

    c. The data indicate a significant change in the level of anxiety, $t(15) = 8.20$, $p < .01$, 90% CI [21.547, 25.053].

17. **a.** The estimated standard error is 1.9, and $t = -4.5/1.9 = -2.37$. For a two-tailed test, the critical value is 2.306. Reject the null hypothesis, scores for students with e-books are significantly different.

    b. For 90% confidence, use $t = \pm 1.860$. The interval is $77.2 \pm (1.860)1.9$ and extends from 73.666 to 80.734.

    c. The results show that exam scores were significantly different for students using e-books than for other students, $t(8) = 2.37$, $p < .05$, 90%CI[73.666, 80.734]. 0.637.

19. **a.** With $n = 9$ the estimated standard error is 4 and $t = \frac{4}{4} = 1$. $r^2 = \frac{1}{9} = 0.111$. Cohen's $d = \frac{4}{12} = 0.333$.

    b. With $n = 16$ the estimated standard error is 3 and $t = \frac{4}{3} = 1.33$. $r^2 = \frac{1.77}{16.77} = 0.106$. Cohen's $d = \frac{4}{12} = 0.333$.

    c. The sample size does not have any influence on Cohen's d and has only a minor effect on r^2.

21. **a.** $H_0: \mu \leq 4$ (not greater than neutral). The estimated standard error is 0.26 and $t = 2.04$. With a critical value of 1.753, reject H_0 and conclude that the males with a great sense of humor were rated significantly higher than neutral.

    b. $H_0: \mu \geq 4$ (not lower than neutral). The estimated standard error is 0.295 and $t = -2.37$. With a critical value of -1.753, reject H_0 and conclude that the males with a no sense of humor were rated significantly lower than neutral.

23. **a.** $H_0: \mu = 50$. With $df = 9$ the critical values are $t = \pm 3.250$. For these data, $M = 55.5$, $SS = 162.5$, $s^2 = 18.06$, the standard error is 1.34, and $t = 4.10$. Reject H_0 and conclude that mathematical achievement scores for children with a history of daycare are significantly different from scores for other children

    b. Cohen's $d = \frac{5.5}{4.25} = 1.29$.

    c. The results indicate that mathematics test scores for children with a history of daycare are significantly different from scores for children without daycare experience, $t(9) = 4.10$, $p < .01$, $d = 1.29$.

CHAPTER 10 | The *t* Test for Two Independent Samples

There are several possible solutions to the matchstick puzzle in Problem 15 but all involve destroying two of the existing squares. One square is destroyed by removing two matchsticks from one of the corners and a second square is destroyed by removing one matchstick. The three removed matchsticks are then used to build a new square using a line that already exists in the figure as the fourth side. One solution is shown in the following figure.

Figure C3

Original pattern with 5 squares (arrows note matchsticks to remove)

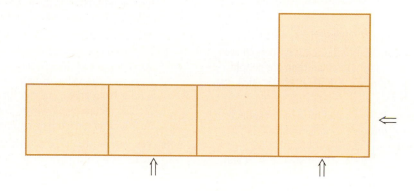

New pattern with 4 squares (arrows note new locations for matchsticks)

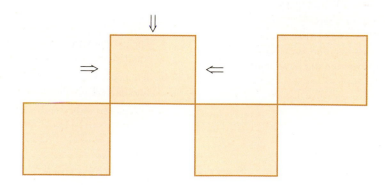

1. An independent-measures study uses a separate sample for each of the treatments or populations being compared.

3. **a.** The first sample has $s^2 = 12$ and the second has $s^2 = 6$. The pooled variance is $\frac{54}{6} = 9$ (halfway between).

   b. The first sample has $s^2 = 12$ and the second has $s^2 = 3$. The pooled variance is $\frac{54}{9} = 6$ (closer to the variance for the larger sample).

5. **a.** The pooled variance is 120.

   b. The estimated standard error is 4.00.

   c. A mean difference of 8 would produce $t = \frac{8}{4} = 2.00$. With $df = 28$ the critical values are ± 2.048. Fail to reject H_0.

7. The null hypothesis says that there is no mean difference between the two populations of students. The pooled variance is 20, the estimated standard error is 2, and $t = \frac{8}{2} = 4.00$. With $df = 18$, the critical boundaries are ± 2.878. Reject the null hypothesis and conclude that there is a significant difference in the high school performance of students who watched Sesame Street as children and those who did not.

9. a. The pooled variance is 7753, the estimated standard error is 12.45, and $t = \frac{14}{12.45} = 1.12$. With $df = 198$ the critical value is 1.98 (using $df = 120$). Fail to reject the null hypothesis and conclude that there was no significant change in calorie consumption after the mandatory posting.

 b. $r^2 = \frac{1.25}{199.25} = 0.0063$ or 0.63%

11. a. The pooled variance is 30.375, the estimated standard error is 1.5, and $t(52) = 1.75$. For a one-tailed test with $df = 52$ the critical value is 1.684 (using $df = 40$). Reject the null hypothesis. There is a significant difference between the two creativity groups.

 b. For these data, the estimated $d = \frac{2.63}{5.51} = 0.477$.

 c. The results indicate that high creativity people are significantly more likely to cheat than are people with low creativity, $t(52) = 1.75, p < .05$, one tailed, $d = 0.477$.

13. a. The null hypothesis states that the type of sport does not affect neurological performance. For a one-tailed test, the critical boundary is $t = 1.761$. For the noncontact athletes, $M = 9$ and $SS = 44$. For the contact athletes, $M = 6$ and $SS = 56$. The pooled variance is 7.14 and $t(14) = 2.25$. Reject H_0. The data show that the contact athletes have significantly lower scores.

 b. For these data, $r^2 = 0.265$ (26.5%).

15. a. The null hypothesis states that the method of instruction does not affect learning. The pooled variance is 16, the standard error is 2, and $t(14) = 2.17$. With $df = 14$ the critical values are ± 2.145. Reject the null hypothesis and conclude that the method of instruction did have an effect on new learning.

b. For 90% confidence, the t values are ± 1.761 and the interval extends from 0.818 to 7.862 points higher for the participants who found the solution on their own.

17. a. With $df = 22$, the critical region consists of t values beyond ± 2.074. The pooled variance is 0.96, the estimated standard error is 0.4, and $t(22) = \frac{0.9}{0.4} = 2.25$. The t statistic is in the critical region. Reject H_0 and conclude that there is a significant difference.

 b. For 80% confidence, the t values are ± 1.321 and the interval extends from 0.372 to 1.428 points higher for the lower economic class participants.

19. a. The size of the two samples influences the magnitude of the estimated standard error in the denominator of the t statistic. As sample size increases, the value of t also increases (moves farther from zero), and the likelihood of rejecting H_0 also increases, however sample size has little or no effect on measures of effect size.

 b. The variability of the scores influences the estimated standard error in the denominator. As the variability of the scores increases, the value of t decreases (becomes closer to zero), the likelihood of rejecting H_0 decreases, and measures of effect size decrease.

21. a. The estimated standard error is 5.

 b. The estimated standard error is 3.

 c. As sample size increases, the magnitude of the standard error decreases.

23. a. The pooled variance is 24 and the estimated standard error is 3.

 b. The pooled variance is 96 and the estimated standard error is 6.

 c. Larger variability produces a larger standard error.

CHAPTER 11 | The *t* Test for Two Related Samples

1. a. Independent-measures: The researcher is comparing 2 separate groups.

 b. Repeated-measures: There are two scores (humorous and not humorous) for each individual.

 c. Repeated-measures: There are two scores (before and after) for each individual.

3. For a repeated-measures design the same subjects are used in both treatment conditions. In a matched-subjects design, two different sets of subjects are used. However, in a matched-subjects design, each subject in one condition is matched with respect to a specific variable with a subject in the second condition so that the two separate samples are equivalent with respect to the matching variable.

5. a. The standard deviation is 4 points and measures the average distance between an individual score and the sample mean.

 b. The estimated standard error is 1.33 points and measures the average distance between a sample mean and the population mean.

7. a. The difference scores are 1, 7, 2, 2, 1, and 5. $M_D = 3$.

b. $SS = 30$, sample variance is 6, and the estimated standard error is 1.

c. With $df = 5$ and $\alpha = .05$, the critical values are $t = \pm 2.571$. For these data, $t = 3.00$. Reject H_0. There is a significant treatment effect.

9. a. The null hypothesis states that the ratings are for masculine-themed ads are not different from ratings for neutral ads. For these data the variance is 6.25, the estimated standard error is 0.50 and $t = 2.64$. With $df = 24$, the critical value is 2.064. Reject the null hypothesis.

b. $r^2 = \frac{2.64^2}{(2.64^2 + 24)} = 0.225$.

c. Participants rated the masculine-themed ads significantly less appealing than the neutral ads, $t(24) = 2.64$, $p < .05$, $r^2 = 0.225$.

11. a. The null hypothesis states that texting has no effect on lateness for class. The sample variance is 196, the estimated standard error is 3.5, and $t = 6.00$. With $df = 15$, the critical boundaries are ± 2.947. Reject the null hypothesis and conclude that there is a significant effect.

b. For 95% confidence, the t values are ± 2.131 and the interval boundaries are $21 \pm 2.131(3.5)$. The interval extends from 13.541 to 28.459.

13. The null hypothesis states that there is no difference in the perceived intelligence between attractive and unattractive photos. For these data, the estimated standard error is 0.4 and $t = \frac{2.7}{0.4} = 6.75$. With $df = 24$, the critical value is 2.064. Reject the null hypothesis.

15. The null hypothesis states that the background color has no effect on judged attractiveness. For these data,

$M_D = 3$, $SS = 18$, the sample variance is 2.25, the estimated standard error is 0.50, and $t(8) = 6.00$. With $df = 8$ and $\alpha = .01$, the critical values are $t = \pm 3.355$. Reject the null hypothesis, the background color does have a significant effect.

17. a. The estimated standard error is 1 point and $t(15) = 3.00$. With a critical boundary of ± 2.131, reject the null hypothesis.

b. The estimated standard error is 3 points and $t(15) = 1.00$. With a critical boundary of ± 2.131, fail to reject the null hypothesis.

c. A larger standard deviation (or variance) reduces the likelihood of finding a significant difference.

19. a. With $n = 4$, the estimated standard error is 2 and $t(3) = \frac{3}{2} = 1.50$. With critical boundaries of ± 3.182, fail to reject H_0.

b. With $n = 16$, the estimated standard error is 1 and $t(15) = \frac{3}{1} = 3.00$. With critical boundaries of ± 2.131, reject H_0.

c. If other factors are held constant, a larger sample increases the likelihood of finding a significant mean difference.

21. a. The sample data produce $M_1 = 7$ with $SS = 68$ and $M_2 = 9$ with $SS = 100$. The pooled variance is 12, the estimated standard error is 1.73 and $t = 1.16$. With $df = 14$ there is no significant difference.

b. The mean difference is $M_D = 2$ with $SS = 26$ for the difference scores. The variance is 3.71, the estimated standard error is 0.68, and $t = 2.94$. With $df = 7$ and $\alpha = .05$, this mean difference is significant.

CHAPTER 12 | Introduction to Analysis of Variance

1. When there is no treatment effect, the numerator and the denominator of the F-ratio are both measuring the same sources of variability (random, unsystematic differences from sampling error). In this case, the F-ratio is balanced and should have a value near 1.00.

3. With 3 or more treatment conditions, you need three or more t tests to evaluate all the mean differences. Each test involves a risk of a Type I error. The more tests you do, the more risk there is of a Type I error. The ANOVA performs all of the tests simultaneously with a single, fixed alpha level.

5. $df_{total} = 23$; $df_{between} = 2$; $df_{within} = 21$

7. a. $k = 3$ treatment conditions.

b. The study used a total of $N = 30$ participants.

9. a. The three sample variances are 7.00, 7.33, and 9.67

b. $SS_{within} = 216$, $df_{within} = 27$, and $MS_{within} = 8$. The average of the three sample variances is also equal to 8.

11.

Source	SS	df	MS	
Between Treatments	30	3	10	$F = 2.5$
Within Treatments	144	36	4	
Total	174	39		

13.

Source	SS	df	MS	
Between Treatments	16	2	8	$F(2, 21) = 2.80$
Within Treatments	60	21	2.86	
Total	76	23		

With $df = 2, 21$, the critical value for $\alpha = .05$ is 3.47. Fail to reject the null hypothesis.

15. With $G = 60$ and $N = 15$, $SS_{between} = 30$.

17. **a.** If the mean for treatment III were changed to $M = 25$, it would reduce the size of the mean differences (the three means would be closer together). This would reduce the size of $MS_{between}$ and would reduce the size of the F-ratio.

   b. If the SS in treatment I were increased to $SS = 1400$, it would increase the size of the variability within treatments. This would increase MS_{within} and would reduce the size of the F-ratio.

19. **a.** The sample variances are 2.67, 3.33, and 6.00. These values are much larger than the variances in problem 18.

   b. The larger variances should result in a smaller F-ratio that is less likely to reject the null hypothesis.

   c.

Source	SS	df	MS	
Between Treatments	32	2	16	$F(2, 9) = 4.00$
Within Treatments	36	9	4	
Total	68	11		

With $\alpha = .05$, the critical value is $F = 4.26$. Fail to reject the null hypothesis and conclude that there are no significant differences among the three treatments. The F-ratio is much smaller as predicted.

21. **a.** Larger samples should increase the F-ratio.

Source	SS	df	MS	
Between Treatments	50	2	25	$F(2, 72) = 12.50$
Within Treatments	144	72	2	
Total	194	74		

The F-ratio is much larger than it was in problem 20.

   b. Increasing the sample size should have little or no effect on η^2. $\eta^2 = \frac{50}{144} = 0.258$, which is about the same as the value obtained in problem 20.

23. If the F-ratio has $df = 1, 34$, then the experiment compared only two treatments, and you could use a t statistic to evaluate the data. The t statistic would have $df = 34$.

CHAPTER 13 | Repeated-Measures Anova

1. For an independent measures design, the variability within treatments is the appropriate error term. For repeated measures, however, you must subtract out variability due to individual differences from the variability within treatments to obtain a measure of error.

3. **a.** A total of 48 participants is needed; four separate samples, each with $n = 12$. The F-ratio has $df = 3, 44$.

   b. One sample of $n = 12$ is needed. The F-ratio has $df = 3, 33$.

5.

Source	SS	df	MS	
Between Treatments	70	2	35	$F(2, 8) = 28$
Within Treatments	40	12		
Between Subjects	30	4		
Error	10	8	1.25	
Total	110	14		

With $df = 2, 8$, the critical value is 4.46. Reject H_0. There are significant differences among the three treatments.

7.

Source	SS	df	MS	
Between Treatments	64	1	64	$F(1, 7) = 112.28$
Within Treatments	36	14		
Between Subjects	32	7		
Error	4	7	0.57	
Total	100	15		

With $df = 1, 7$, the critical value is 5.59. Reject H_0. There is a significant difference between the two treatments.

9. a. $SS_{between} = 70$ and $SS_{total} = 96$

b. The new data are as follows.

Person	Treatments I	II	III	Person Totals	
A	3	2	1	$P = 6$	
B	5	5	5	$P = 15$	$N = 15$
C	2	3	4	$P = 9$	$G = 45$
D	2	1	3	$P = 6$	$\Sigma X^2 = 161$
E	3	4	2	$P = 9$	
	$M = 3$	$M = 3$	$M = 3$		
	$T = 15$	$T = 15$	$T = 15$		
	$SS = 6$	$SS = 10$	$SS = 10$		

c. $SS_{total} = 26$ for the new scores.

d. The SS is reduced by 70 points which is exactly equal to the original $SS_{between}$.

11. The means and standard deviations for the four measurement days are as follows:

7 days	5 days	3 days	1 day
$M = 39$	$M = 40$	$M = 46$	$M = 55$
$s = 3.61$	$s = 4.73$	$s = 3.67$	$s = 3.32$

Source	SS	df	MS	
Between Treatments	810	3	270	$F(3, 12) = 28$
Within Treatments	240	16		
Between Subjects	208	4		
Error	32	12	2.67	
Total	1050	19		

With $\alpha = .01$, the critical value is 5.95. There are significant differences.

13.

Source	SS	df	MS	
Between Treatments	26	2	13	$F(2, 14) = 6.50$
Within Treatments	70	21		
Between Subjects	42	7		
Error	28	14	2	
Total	96	23		

15 a. 4 treatments

b. 9 participants

17. a.

Source	SS	df	MS	
Between Treatments	56	2	28	$F(2, 6) = 16.77$
Within Treatments	22	9		
Between Subjects	12	3		
Error	10	6	1.67	
Total	78	11		

$$\eta^2 = \frac{56}{66} = 0.848$$

b.

Source	SS	df	MS	
Between Treatments	112	2	56	$F(2, 14) = 39.16$
Within Treatments	44	21		
Between Subjects	24	7		
Error	20	14	1.43	
Total	156	23		

$$\eta^2 = \frac{112}{132} = 0.848$$

c. Doubling the sample size greatly increased the F-ratio but had little or no effect on η^2.

19. a. The null hypothesis states that there are no differences among the three treatments, $H_0: \mu_1 = \mu_2 = \mu_3$. With $df = 2, 6$, the critical value is 5.14.

Source	SS	df	MS	
Between Treatments	8	2	4	$F(2, 6) = 0.27$
Within Treatments	94	9		
Between Subjects	4	3		
Error	90	6	15	
Total	102	11		

Fail to reject H_0. There are no significant differences among the three treatments.

b. In problem 18 the individual differences were relatively large and consistent. When the individual differences were subtracted out, the error was greatly reduced.

21. The null hypothesis states that there are no differences among the three weeks. With $df = 2, 15$, the critical value is 3.68.

Source	SS	df	MS	
Between Treatments	12	2	6	$F(2, 15) = 0.90$
Within Treatments	100	15	6.67	
Total	112	17		

Fail to reject H_0. There are no significant differences among the three weeks. For these data, $\eta^2 = \frac{12}{112} = 0.107$.

23. a. The null hypothesis states that there are no differences between treatments, $H_0: \mu_1 = \mu_2 = \mu_3$. For an independent-measures design, the critical value is 3.88.

Source	SS	df	MS	
Between Treatments	40	2	20.00	$F(2, 12) = 3.24$
Within Treatments	74	12	6.17	
Total	114	14		

b. The null hypothesis states that there are no differences between treatments, $H_0: \mu_1 = \mu_2 = \mu_3$. For a repeated-measures design, the critical value is 4.46.

Source	SS	df	MS	
Between Treatments	40	2	20.0	$F(2, 8) = 8.00$
Within Treatments	74	12		
Between Subjects	54	4		
Error	20	8	2.5	
Total	114	14		

c. The independent measures design includes all the individual differences in the error term (MS_{within}).

As a result the F-ratio, $F(2,12) = 3.24$ is not significant. With a repeated measures design, the individual differences are removed and the result is a significant F-ratio, $F(2,8) = 8.00$, $p < .05$.

25. a.

Source	SS	df	MS	
Between Treatments	40.5	1	40.5	$F(1, 8) = 20.25$
Within Treatments	38	16		
Between Subjects	22	8		
Error	16	8	2	
Total	78.5	17		

The F-ratio has $df = 1, 8$ and the critical value is 5.32. Reject H_0. There is a significant difference between the two tattoo conditions.

b. For the t test, the mean difference is $M_D = 3$, SS for the difference scores is 32, the variance is 4, and the standard error is 0.667. $t = \frac{3}{0.667} = 4.50$. With $df = 8$, the critical value is 2.306. Reject H_0. There is a significant difference between the two tattoo conditions. Note that $F = t^2$.

27. a. The null hypothesis states that there is no difference between the two treatments, $H_0: \mu_D = 0$. The critical region consists of t values beyond ± 3.182. The mean difference is $M_D = 4.5$. The variance for the difference scores is 9, the estimated standard error is 1.50, and $t(3) = 3.00$. Fail to reject H_0.

b. The null hypothesis states that there is no difference between treatments, $H_0: \mu_1 = \mu_2$. The critical value is $F = 10.13$.

Source	SS	df	MS	
Between Treatments	40.5	1	40.5	$F(1, 3) = 9.00$
Within Treatments	31	6		
Between Subjects	17.5	3		
Error	13.5	3	4.5	
Total	71.5	7		

Fail to reject H_0. Note that $F = t^2$.

CHAPTER 14 | Two-Factor ANOVA

1. a. In analysis of variance, an independent variable (or a quasi-independent variable) is called a *factor*.

b. The values of a factor that are used to create the different groups or treatment conditions are called the *levels* of the factor.

c. A research study with two independent (or quasi-independent) variables is called a *two-factor study*.

3. During the second stage of the two-factor ANOVA the mean differences between treatments are analyzed into differences from each of the two main effects and differences from the interaction.

5. a. $M = 5$

b. $M = 1$

c. $M = 9$

7. a. The scores in treatment 1 are consistently higher than the scores in treatment 2. There is a main effect for treatment.

b. The overall mean is around $M = 15$ for all three age groups. There is no main effect for age.

c. The two lines are not parallel. Instead, the difference between the treatments increases as the participants get older. Yes, there is an interaction.

9. a. $df = 1, 54$

b. $df = 2, 54$

c. $df = 2, 54$

11. a. 3

b. 4

c. 6, 108

13. a.

Source	SS	df	MS	
Between Treatments	400	5		
A	120	1	120	$F(1, 24) = 15.00$
B	260	2	130	$F(2, 24) = 16.25$
A × B	20	2	10	$F(2, 24) = 1.25$
Within Treatments	192	24	8	
Total	592	29		

The F-ratio for factor A has $df = 1, 24$ and the critical value is $F = 4.26$. Factor B and the interaction have $df = 2, 24$ and the critical value is 3.40. The main effect is significant for both factors but not for the interaction.

b. For level B1 $F = \frac{10}{8} = 1.25$, for B2 $F = \frac{40}{8} = 5.00$, and for B3 $F = \frac{90}{8} = 11.25$. For all tests, $df = 1, 24$ and the critical value is 4.26. There are significant differences at levels B2 and B3 but not at B1.

15. a.

Source	SS	df	MS	
Between Treatments	40	3		
Type of Program	20	1	20	$F(1, 16) = 8.89$
Amount of Time	0	1	0	$F(1, 16) = 0$
A × B	20	1	20	$F(1, 16) = 8.89$
Within Treatments	36	16	2.25	
Total	76	19		

With $df = 1, 16$, the critical value for all three F-ratios is 8.53. The main effect for the type of program and the interaction are significant. The main effect for amount of TV watching is not significant.

b. For the type program factor and the interaction, eta squared is $\frac{20}{56} = 0.357$. For the time factor, eta squared is 0.

c. For educational TV, increased watching time results in significantly higher performance, but for noneducational TV, increased watching results in lower performance.

17.

Source	SS	df	MS	
Between Treatments	74	7		
A	20	1	20	$F(1, 32) = 10$
B	36	3	12	$F(3, 32) = 6$
A × B	18	3	6	$F(3, 32) = 3$
Within Treatments	65	32	2	
Total	138	39		

19.

Source	SS	df	MS	
Between Treatments	52	3		
A	16	1	16	$F(1, 56) = 8.00$
B	24	1	24	$F(3, 56) = 12.00$
A × B	12	1	12	$F(3, 56) = 6.00$
Within Treatments	112	56	2	
Total	164	59		

21. a.

Source	SS	df	MS	
Between Treatments	360	5		
Gender	72	1	72	$F(1, 12) = 9.00$
Treatments	252	2	126	$F(2, 12) = 15.75$
Gender × Treat.	36	2	18	$F(2, 12) = 2.25$
Within Treatments	96	12	8	
Total	456	17		

With $df = 1, 12$, the critical value for the gender main effect is 4.75. The main effect for gender is significant. With $df = 2, 12$, the critical value for the treatment main effect and the interaction is 3.88. The main effect for treatments is significant but the interaction is not.

b. For treatment I, $F = 0$; for treatment II, $F = \frac{54}{8} = 6.75$; and for treatment III, $F = \frac{54}{8} = 6.75$. With $df = 1, 12$, the critical value for all three tests is 4.75. The results indicate a significant difference between males and females in treatments II and III, but not in treatment I.

23. a. The means for the six groups are as follows:

	Middle School	High School	College
Non-User	4.00	4.00	4.00
User	3.00	2.00	1.00

Source	SS	df	MS	
Between Treatments	32	5		
Use	24	1	24	$F(1, 18) = 14.4$
School level	4	2	2	$F(2, 18) = 1.2$
Interaction	4	2	2	$F(2, 48) = 1.2$
Within Treatments	30	18	1.67	
Total	62	23		

For $df = 1, 18$ the critical value is 4.41 and for $df = 2, 18$ it is 3.55. The main effect for Facebook use is significant. For $df = 2, 18$ the critical value is 3.55. The other main effect and the interaction are not significant.

b. Grades are significantly lower for Facebook users. A difference exists for all three grade levels but appears to increase as the students get older although there is no significant interaction.

25. a.

Source	SS	df	MS	
Between Treatments	270	3		
Self-Esteem	150	1	150	$F(1, 20) = 34.88$
Audience	96	1	96	$F(1, 20) = 22.33$
Interaction	24	1	24	$F(1, 20) = 5.58$
Within Treatments	86	20	4.3	
Total	356	23		

With $df = 1, 20$ the critical value is 4.35 for all three tests. Both main effects and the interaction are significant. Overall, there are fewer errors for the high self-esteem participants and for those working alone. The audience condition has very little effect on the high self-esteem participants but a very large effect on those with low self-esteem.

b. For the self-esteem main effect, $\eta^2 = \frac{150}{236} = 0.636$, for the audience main effect, $\eta^2 = \frac{96}{182} = 0.527$, and for the interaction, $\eta^2 = \frac{24}{110} = 0.218$.

CHAPTER 15 | Correlation

1. A positive correlation indicates that X and Y change in the same direction: As X increases, Y also increases. A negative correlation indicates that X and Y tend to change in opposite directions: As X increases, Y decreases.

3. $SP = 9$

5. a. The scatter plot shows points widely scattered around a line sloping down to the right.

   b. The correlation is small and negative; around -0.4 to -0.6.

   c. For these scores, $SS_X = 18$, $SS_Y = 8$, and $SP = -7$. The correlation is $r = -\frac{7}{12} = -0.583$.

7. a. The scatter plot shows four points forming a square. There is no linear trend so the correlation is near zero.

   b. $SS_X = 9$, $SS_Y = 25$, and $SP = 0$. The correlation is $r = \frac{0}{15} = 0$.

9. a. The scatter plot shows points scattered around a line sloping up to the right. The correlation should be 0.4 to 0.6.

   b. $SS_X = 6$, $SS_Y = 24$, and $SP = 5$. The correlation is $r = \frac{5}{12} = 0.42$.

11. a. For the children, $SS = 32$ and for the birth parents, $SS = 14$. $SP = 15$. The correlation is $r = 0.709$.

   b. For the children, $SS = 32$ and for the adoptive parents $SS = 16$. $SP = 3$. The correlation is $r = 0.133$.

c. The children's behavior is strongly related to their birth parents and only weakly related to their adoptive parents. The data suggest that the behavior is inherited rather than learned.

13. a. For the men's weights, $SS = 18$ and for their incomes, $SS = 11,076$. $SP = 330$. The correlation is $r = 0.739$.

   b. With $n = 8$, $df = 6$ and the critical value is 0.707. The correlation is significant.

15. a. $r = 0.811$

   b. $r = 0.576$

   c. $r = 0.404$

17. a. $r_{XY-Z} = \frac{0.32}{0.655} = 0.489$

   b. $r_{XZ-Y} = -\frac{0.02}{0.571} = -0.035$

19. a. $r_s = +0.907$

   b. With $n = 11$, the critical value is 0.618. The correlation is significant.

21. a. $r_s = 0.732$ using the special formula or $r_s = 0.727$ using the Pearson formula on the ranks.

   b. For $n = 8$, the critical value is 0.738 for $\alpha = .05$. The correlation is not significant.

23. a. Using the pain ratings as the X variable and coding money as 1 and plain paper as 0 for the Y variable produces $SS_X = 144.44$, $SS_Y = 4.5$, and $SP = -17$. The point-biserial correlation is $r = -0.667$.

   b. $r^2 = 0.445$

   c. The formula produces $r^2 = 0.443$.

CHAPTER 16 | Introduction To Regression

1. Figure C4

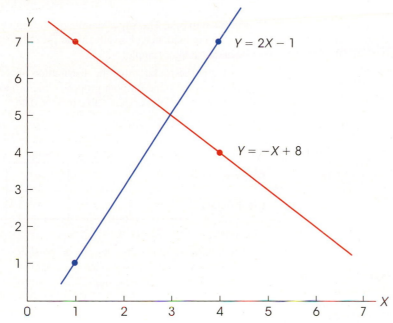

3. **a.** $r = 0.25$

   b. $\hat{Y} = 0.5X + 6$

5. The standard error of estimate is a measure of the average distance between the predicted Y points from the regression equation and the actual Y points in the data.

7. $SS_X = 16$, $SS_Y = 134$, $SP = 32$. With $M_X = 4$ and $M_Y = 6$, the regression equation is $\hat{Y} = 2X - 2$.

9. $SS_{\text{regression}} = r^2 SS_Y = 39.95$ with $df = 1$. $MS_{\text{residual}} = \frac{10.05}{4} = 2.51$. $F = \frac{39.95}{2.51} = 15.92$. With $df = 1, 4$, the F-ratio is significant with $\hat{Y} = .05$.

11. **a.** $SS_{\text{weight}} = 18$, $SS_{\text{income}} = 21{,}609$, $SP = 330$.
   $\hat{Y} = 18.33X + 33$

   b. $r = -0.529$ and $r^2 = 0.280$

   c. $F = 2.33$ with $df = 1, 6$. The regression equation is not significant.

13. **a.** $\hat{Y} = 1.38X + 7.34$

   b. $r^2 = 0.743$ or 74.3%

   c. $F = 20.23$ with $df = 1, 7$. The equation accounts for a significant portion of the variance.

15. **a.** The standard error of estimate is $\sqrt{\frac{36}{16}} = 1.50$.

   b. The standard error of estimate is $\sqrt{\frac{36}{64}} = 0.75$.

17. **a.** $df = 1, 18$

   b. $n = 25$ pairs of scores

19. **a.** $F = \frac{7.6}{1.46} = 5.21$. With $df = 2, 17$, the critical value is 3.59. The equation does account for a significant portion of the variance.

   b. $F = \frac{7.6}{3.54} = 2.15$. With $df = 2, 7$, the equation is not significant.

21. **a.** $SS_{\text{crime}} = 230$, $SS_{\text{prevention}} = 230$, $SS_{\text{population}} = 8$, and. SP for crime and prevention is 40, SP for crime and population is 40, and SP for prevention and population is 176, and. The regression equation is $\hat{Y} = -0.8X_1 + 9X_2$, where X_1 is the amount spent on prevention and X_2 is the population.

   b. $R^2 = 0.953$ or 93.5%

23. $\hat{Y} = 1.242X_1 - 0.048X_2 + 0.565$

CHAPTER 17 | Chi-Square Tests

1. Nonparametric tests make few if any assumptions about the populations from which the data are obtained. For example, the populations do not need to form normal distributions, nor is it required that different populations in the same study have equal variances (homogeneity of variance assumption). Parametric tests require data measured on an interval or ratio scale. For nonparametric tests, any scale of measurement is acceptable.

3. **a.** The null hypothesis states that there is no preference among the four colors; $p = \frac{1}{4}$ for all categories. The expected frequencies are $f_e = 20$ for all categories, and chi-square = 3.70. With $df = 3$, the critical value is 7.81. Fail to reject H_0 and conclude that there are no significant preferences.

   b. The results indicate that there are no significant preferences among the four colors, $\chi^2(3, N = 80) = 3.70, p > .05$.

5. The null hypothesis states that couples with the same initial do not occur more often than would be expected by chance. For a sample of 200, the expected frequencies are 13 with the same initial and 187 with different initials. With $df = 1$ the critical value is 3.84, and the data produce a chi-square of 2.96. Fail to reject the null hypothesis.

7. The null hypothesis states that the grade distribution for last semester has the same proportions as it did in 1985. For a sample of $n = 200$, the expected frequencies are 28, 52, 62, 38, and 20 for grades of A, B, C, D, and F, respectively. With $df = 4$, the critical value for chi-square is 9.49. For these data, the chi-square statistic is 6.68. Fail to reject H_0 and conclude that there is no evidence that the distribution has changed.

9. **a.** The null hypothesis states that there is no advantage (no preference) for red or blue. With $df = 1$, the critical value is 3.84. The expected frequency is 25 wins for each color, and chi-square = 2.88. Fail to reject H_0 and conclude that there is no significant advantage for one color over the other.

   b. The null hypothesis states that there is no advantage (no preference) for red or blue. With $df = 1$, the critical value is 3.84. The expected frequency is 50 wins for each color, and chi-square = 5.76. Reject H_0 and conclude that there is a significant advantage for the color red.

   c. Although the proportions are identical for the two samples, the sample in part b is twice as big as the sample in part a. The larger sample provides more convincing evidence of an advantage for red than does the smaller sample.

11. The null hypothesis states that the distribution of preferences is the same for both groups (same proportions). With $df = 2$, the critical value is 5.99. The expected frequencies are:

	Design 1	Design 2	Design 3
Students	24	27	9
Older Adults	24	27	9

Chi-square = 7.94. Reject H_0.

13. **a.** The null hypothesis states that the proportion who falsely recall seeing broken glass should be the same for all three groups. The expected frequency of saying yes is 9.67 for all groups, and the expected frequency for saying no is 40.33 for all groups. With $df = 2$, the critical value is 5.99. For these data, chi-square = 7.78. Reject the null hypothesis and conclude that the likelihood of recalling broken glass is depends on the question that the participants were asked.

   b. Cramér's V = 0.228.

   c. Participants who were asked about the speed with the cars "smashed into" each other, were more than two times more likely to falsely recall seeing broken glass.

   d. The results of the chi-square test indicate that the phrasing of the question had a significant effect on the participants' recall of the accident, $\chi^2(2, N = 150) = 7.78, p < .05, V = 0.228$.

15. The null hypothesis states that IQ and gender are independent. The distribution of IQ scores for boys should be the same as the distribution for girls. With $df = 2$ and and $\alpha = .05$, the critical value is 5.99. The expected frequencies are 15 low IQ, 48 medium, and 17 high for both boys and girls. For these data, chi-square is 3.76. Fail to reject the null hypothesis. These data do not provide evidence for a significant relationship between IQ and gender.

17. The null hypothesis states that there is no difference between the distribution of preferences predicted by women and the actual distribution for men. With $df = 3$ and $\alpha = .05$, the critical value is 7.81. The expected frequencies are:

	Somewhat thin	Slightly thin	Slightly heavy	Somewhat heavy
Women	22.9	22.9	22.9	11.4
Men	17.1	17.1	17.1	8.6

Chi-square $= 9.13$. Reject H_0 and conclude that there is a significant difference in the preferences predicted by women and the actual preferences expressed by men.

19. a. The null hypothesis states that there is no relationship between IQ and volunteering. With $df = 2$ and $\alpha = .05$, the critical value is 5.99. The expected frequencies are:

	IQ		
	High	Medium	Low
Volunteer	37.5	75	37.5
Not volunteer	12.5	25	12.5

The chi-square statistic is 4.75. Fail to reject H_0 with $\alpha = .05$ and $df = 2$.

21. The null hypothesis states that there is no relationship between the season of birth and schizophrenia. With $df = 3$ and $\alpha = .05$, the critical value is 7.81. The expected frequencies

	Summer	Fall	Winter	Spring
No Disorder	23.33	23.33	26.67	26.67
Schizophrenia	11.67	11.67	13.33	13.33

Chi-square $= 3.62$. Fail to reject H_0 and conclude that these data do not provide enough evidence to conclude that there is a significant relationship between the season of birth and schizophrenia.

23. a. The null hypothesis states that there is no relationship between gender and willingness to use mental health serviced. With $df = 2$, the critical value is 5.99. The expected frequencies are:

	Willingness to Use Mental Health Services			
	Probably No	Maybe	Probably Yes	
Males	12	30	18	60
Females	18	45	27	90
	30	75	45	

Chi-square $= 8.23$. Reject the null hypothesis.

b. Cramér's $V = 0.234$.

CHAPTER 18 | The Binomial Test

1. H_0: p(accident involves a driver aged 20 or younger) $= 0.15$ (15%): The distribution of ages for accidents is not different from the distribution of ages for licensed drivers. The critical boundaries are $z = \pm 1.96$. With $X = 26$, $\mu = 12$, and $\sigma = 3.19$, we obtain $z = 4.39$. Reject H_0 and conclude that there is a significant difference.

3. a. H_0: $p = q = \frac{1}{2}$ (no advantage for one color over the other). The critical boundaries are $z = \pm 1.96$. With $X = 31$, $\mu = 25$, and $\sigma = 3.54$, we obtain $z = 1.69$. Because the z-score is not in the critical region, the data do not show a significant difference in winning percentage between the two colors.

b. With $X = 62$, $\mu = 50$, and $\sigma = 5$, we obtain $z = 2.40$. Because the z-score is in the critical region, the data show a significant difference in winning percentage between the two colors.

c. With a small sample (part a) a winning percentage of 62% is not significantly different from 50%, but with a larger sample (part b) the difference is enough to be significant.

5. H_0: $p = q = \frac{1}{2}$ (no preference). The critical boundaries are $z = \pm 1.96$. With $X = 28$, $\mu = 20$, and $\sigma = 3.16$, we obtain $z = 2.53$. Reject H_0. There is evidence of a significant preference between bottled water and tap water.

7. H_0: $p = q = \frac{1}{2}$ (no preference). The critical boundaries are $z = \pm 2.58$. With $X = 92$, $\mu = 75$, and $\sigma = 6.12$, we obtain $z = 2.77$. Reject H_0, there is a significant preference for technical skill.

9. a. H_0: p(accident) $= 0.12$ (no change). The critical boundaries are $z = \pm 1.96$. With $X = 45$, $\mu = 60$, and $\sigma = 7.27$, we obtain $z = -2.06$. Reject H_0, there has been a significant change in the accident rate.

b. The boundaries for $X = 45$ are 44.5 and 45.5, which correspond to $z = -2.13$ and $z = -1.99$. The entire interval is in the critical region and the null hypothesis is rejected.

11. $H_0: p = \frac{1}{4} = p$(guessing correctly). The critical boundaries are $z = \pm 1.96$. With $X = 29$, $\mu = 20$, and $\sigma = 3.87$, we obtain $z = 2.33$. Reject H_0 and conclude that this level of performance is significantly different from chance.

13. a. $H_0: p = q = \frac{1}{2}$ (positive and negative correlations are equally likely). The critical boundaries are $z = \pm 1.96$. With $X = 25$, $\mu = 13.5$, and $\sigma = 2.60$, we obtain $z = 4.42$. Reject H_0, positive and negative correlations are not equally likely.

b. The critical boundaries are $z = \pm 1.96$. With $X = 20$, $\mu = 13.5$, and $\sigma = 2.60$, we obtain $z = 2.50$. Reject H_0, positive and negative correlations are not equally likely.

15. $H_0: p = 0.50$ (no difference between the two conditions). The critical boundaries are $z = \pm 1.96$. Dividing the zero differences between the two groups produces $X = 18$, $\mu = 12.5$, and $\sigma = 2.5$, we obtain $z = 2.20$. Reject H_0 and conclude that the difference between conditions is significant with $\alpha = .05$.

17. $H_0: p = \frac{1}{2} = p$(reduced reactions). The critical boundaries are $z = \pm 1.96$. The binomial distribution has $\mu = 32$ and $\sigma = 4$. With $X = 47$ we obtain $z = 3.75$. Reject H_0 and conclude that there is evidence of significantly reduced allergic reactions.

19. $H_0: p = \frac{1}{2} = p$(no preference). The critical boundaries are $z = \pm 1.96$. The binomial distribution has $\mu = 40$ and $\sigma = 4.47$. With $X = 47$ we obtain $z = 1.57$. Fail to reject H_0 and conclude that there is no preference for either of the two cages.

21. a. $H_0: p$(overestimate) $= 0.50$ (just chance). The critical boundaries are $z = \pm 1.96$. With

$n = 49$ because one participant is removed, $X = 32$, $\mu = 24.5$, and $\sigma = 3.5$, we obtain $z = 2.14$. Because the z-score is borderline, test the real limits for $X = 32$. $X = 31.5$ produces $z = 2.00$ and $X = 32.5$ produces $z = 2.29$. The entire interval is in the critical region so we reject H_0.

b. $H_0: p$(overestimate) $= 0.50$ (just chance). The critical boundaries are $z = \pm 1.96$. With $n = 49$ because one participant is removed, $X = 27$, $\mu = 24.5$, and $\sigma = 3.5$, we obtain $z = 0.71$. Fail to reject H_0 and conclude that the speed estimates are not significantly different from chance.

23. a. $H_0: p = q = \frac{1}{2}$ (increases and decreases are equally likely). The critical boundaries are $z = \pm 1.96$. The binomial distribution has $\mu = 12.5$ and $\sigma = 2.5$. With $X = 18$ we obtain $z = 2.20$. Reject H_0 and conclude that there is significant improvement following the imagery program.

b. The real limits for $X = 18$ are 17.5 and 18.5 which correspond to z-scores of $z = 2.00$ and $z = 2.40$. Because this entire interval is in the critical region (beyond 1.96) the correct decision is to reject the null hypothesis.

25. a. $H_0: p = q = \frac{1}{2}$ (the training has no effect). The critical boundaries are $z = \pm 1.96$. Discarding the 11 people who showed no change, the binomial distribution has $\mu = 19.5$ and $\sigma = 3.12$. With $X = 29$ we obtain $z = 3.05$. Reject H_0 and conclude that biofeedback training has a significant effect.

b. Discarding only 1 subject and dividing the others equally, the binomial distribution has $\mu = 24.5$ and $\sigma = 3.50$. With $X = 34$ we obtain $z = 2.71$. Reject H_0.

General Instructions for Using SPSS

The Statistical Package for the Social Sciences, commonly known as SPSS, is a computer program that performs statistical calculations, and is widely available on college campuses. Detailed instructions for using SPSS for specific statistical calculations (such as computing sample variance or performing an independent-measures t test) are presented at the end of the appropriate chapter in the text. Look for the SPSS logo in the Resources section at the end of each chapter. In this appendix, we provide a general overview of the SPSS program.

SPSS consists of two basic components: A data editor and a set of statistical commands. The **data editor** is a huge matrix of numbered rows and columns. To begin any analysis, you must type your data into the data editor. Typically, the scores are entered into columns of the editor. Before scores are entered, each of the columns is labeled "var." After scores are entered, the first column becomes VAR00001, the second column becomes VAR00002, and so on. To enter data into the editor, the **Data View** tab must be set at the bottom left of the screen. If you want to name a column (instead of using VAR00001), click on the **Variable View** tab at the bottom of the data editor. You will get a description of each variable in the editor, including a box for the name. You may type in a new name using up to 8 lowercase characters (no spaces, no hyphens). Click the **Data View** tab to go back to the data editor.

The **statistical commands** are listed in menus that are made available by clicking on **Analyze** in the tool bar at the top of the screen. When you select a statistical command, SPSS typically asks you to identify exactly where the scores are located and exactly what other options you want to use. This is accomplished by identifying the column(s) in the data editor that contain the needed information. Typically, you are presented with a display similar to the following figure. On the left is a box that lists all of the columns in the data editor that contain information. In this example, we have typed values into columns 1, 2, 3, and 4. On the right is an empty box that is waiting for you to identify the correct column. For example, suppose that you wanted to do a statistical calculation using the scores in column 3. You should highlight VAR00003 by clicking on it in the left-hand box, then click the arrow to move the column label into the right hand box. (If you make a mistake, you can highlight the variable in the right-hand box, which will reverse the arrow so that you can move the variable back to the left-hand box.)

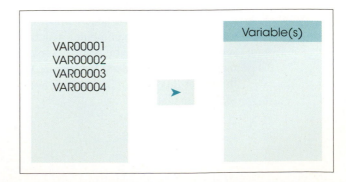

SPSS Data Formats

The SPSS program uses two basic formats for entering scores into the data matrix. Each is described and demonstrated as follows:

1. The first format is used when the data consist of several scores (more than one) for each individual. This includes data from a repeated-measures study, in which each person is measured in all of the different treatment conditions, and data from a correlational study where there are two scores, X and Y, for each individual. Table D1 illustrates this kind of data and shows how the scores would appear in the SPSS data matrix. Note that the scores in the data matrix have exactly the same structure as the scores in the original data. Specifically, each row of the data matrix contains the scores for an individual participant, and each column contains the scores for one treatment condition.

TABLE D1

Data for a repeated-measures or correlational study with several scores for each individual. The left half of the table (a) shows the original data, with three scores for each person; and the right half (b) shows the scores as they would be entered into the SPSS data matrix. Note: SPSS automatically adds the two decimal points for each score. For example, you type in 10 and it appears as 10.00 in the matrix.

(a) Original data

Person	I	II	III
A	10	14	19
B	9	11	15
C	12	15	22
D	7	10	18
E	13	18	20

(b) Data as entered into the SPSS data matrix

	VAR0001	VAR0002	VAR0003	var
1	10.00	14.00	19.00	
2	9.00	11.00	15.00	
3	12.00	15.00	22.00	
4	7.00	10.00	18.00	
5	13.00	18.00	20.00	

2. The second format is used for data from an independent-measures study using a separate group of participants for each treatment condition. This kind of data is entered into the data matrix in a *stacked* format. Instead of having the scores from different treatments in different columns, all of the scores from all of the treatment conditions are entered into a single column so that the scores from one treatment condition are literally stacked on top of the scores from another treatment condition. A code number is then entered into a second column beside each score to tell the computer which treatment condition corresponds to each score. For example, you could enter a value of 1 beside each score from treatment #1, enter a 2 beside each score from treatment #2, and so on. Table D2 illustrates this kind of data and shows how the scores would be entered into the SPSS data matrix.

TABLE D2

Data for an independent-measures study with a different group of participants in each treatment condition. The left half of the table shows the original data, with three separate groups, each with five participants, and the right half shows the scores as they would be entered into the SPSS data matrix. Note that the data matrix lists all 15 scores in the same column, then uses code numbers in a second column to indicate the treatment condition corresponding to each score.

(a) Original data

Treatments		
I	II	III
10	14	19
9	11	15
12	15	22
7	10	18
13	18	20

(b) Data as entered into the SPSS data matrix

	VAR0001	VAR0002	var
1	10.00	1.00	
2	9.00	1.00	
3	12.00	1.00	
4	7.00	1.00	
5	13.00	1.00	
6	14.00	2.00	
7	11.00	2.00	
8	15.00	2.00	
9	10.00	2.00	
10	18.00	2.00	
11	19.00	3.00	
12	15.00	3.00	
13	22.00	3.00	
14	18.00	3.00	
15	20.00	3.00	

Hypothesis Tests for Ordinal Data: Mann-Whitney, Wilcoxon, Kruskal-Wallis, and Friedman Tests

E.1 | Data from an Ordinal Scale

Occasionally, a research study generates data that consist of measurements on an ordinal scale. Recall from Chapter 1 that an ordinal scale simply produces a *rank ordering* for the individuals being measured. For example, a kindergarten teacher may rank children in terms of their maturity, or a business manager may classify job applicants as outstanding, good, and average.

■ Ranking Numerical Scores

In addition to obtaining measurements from an ordinal scale, a researcher may begin with a set of numerical measurements and convert these scores into ranks. For example, if you had a listing of the actual heights for a group of individuals, you could arrange the numbers in order from greatest to least. This process converts data from an interval or a ratio scale into ordinal measurements. In Chapter 17 (page 562), we identify several reasons for converting numerical scores into nominal categories. These same reasons also provide justification for transforming scores into ranks. The following list should give you an idea of why there can be an advantage to using ranks instead of scores.

1. Ranks are simpler. If someone asks you how tall your sister is, you could reply with a specific numerical value, such as 5 feet $7\frac{3}{4}$ inches tall. Or you could answer, "She is a little taller than I am." For many situations, the relative answer would be better.

2. The original scores may violate some of the basic assumptions that underlie certain statistical procedures. For example, the *t* tests and ANOVA assume that the data come from normal distributions. Also, the independent-measures tests assume that the different populations all have the same variance (the homogeneity-of-variance assumption). If a researcher suspects that the data do not satisfy these assumptions, it may be safer to convert the scores to ranks and use a statistical technique designed for ranks.

3. The original scores may have unusually high variance. Variance is a major component of the standard error in the denominator of t statistics and the error term in the denominator of F-ratios. Thus, large variance can greatly reduce the likelihood that these parametric tests will find significant differences. Converting the scores to ranks essentially eliminates the variance. For example, 10 scores have ranks from 1 to 10 no matter how variable the original scores are.

4. Occasionally, an experiment produces an undetermined, or infinite, score. For example, a rat may show no sign of solving a particular maze after hundreds of trials. This animal has an infinite, or undetermined, score. Although there is no absolute score that can be assigned, you can say that this rat has the highest score for the sample and then rank the rest of the scores by their numerical values.

■ Ranking Tied Scores

Whenever you are transforming numerical scores into ranks, you may find two or more scores that have exactly the same value. Because the scores are tied, the transformation process should produce ranks that are also tied. The procedure for converting tied scores into tied ranks was presented in Chapter 15 (page 513) when we introduced the Spearman correlation, and is repeated briefly here. First, you list the scores in order, including tied values. Second, you assign each position in the list a rank (1st, 2nd, and so on). Finally, for any scores that are tied, you compute the mean of the tied ranks, and use the mean value as the final rank. The following set of scores demonstrates this process for a set of $n = 8$ scores.

Original scores:	3	4	4	7	9	9	9	12
Position ranks:	1	2	3	4	5	6	7	8
Final ranks:	1	2.5	2.5	4	6	6	6	8

■ Statistics for Ordinal Data

You should recall from Chapter 1 that ordinal values tell you only the direction from one score to another, but provide no information about the distance between scores. Thus, you know that first place is better than second or third, but you do not know how much better. Because the concept of *distance* is not well defined with ordinal data, it generally is considered unwise to use traditional statistics such as t tests and analysis of variance with scores consisting of ranks or ordered categories. Therefore, statisticians have developed special techniques that are designed specifically for use with ordinal data.

In this chapter we introduce four hypothesis-testing procedures that are used with ordinal data. Each of the new tests can be viewed as an alternative for a commonly used parametric test. The four tests and the situations in which they are used are as follows:

1. The Mann-Whitney test uses data from two separate samples to evaluate the difference between two treatment conditions or two populations. The Mann-Whitney test can be viewed as an alternative to the independent-measures t hypothesis test introduced in Chapter 10.

2. The Wilcoxon test uses data from a repeated-measures design to evaluate the difference between two treatment conditions. This test is an alternative to the repeated-measures t test from Chapter 11.

3. The Kruskal-Wallis test uses data from three or more separate samples to evaluate the differences between three or more treatment conditions (or populations). The Kruskal-Wallis test is an alternative to the single-factor, independent-measures ANOVA introduced in Chapter 12.

4. The Friedman test uses data from a repeated-measures design to compare the differences between three or more treatment conditions. This test is an alternative to the repeated-measures ANOVA from Chapter 13.

In each case, you should realize that the new, ordinal-data tests are back-up procedures that are available in situations in which the standard, parametric tests cannot be used. In general, if the data are appropriate for an ANOVA or one of the t tests, then the standard test is preferred to its ordinal-data alternative.

E.2 | The Mann-Whitney *U*-Test: An Alternative to the Independent-Measures *t* Test

Recall that a study using two separate samples is called an *independent-measures* study or a *between-subjects* study. The Mann-Whitney test is designed to use the data from two separate samples to evaluate the difference between two treatments (or two populations). The calculations for this test require that the individual scores in the two samples be rank-ordered. The mathematics of the Mann-Whitney test are based on the following simple observation:

> A real difference between the two treatments should cause the scores in one sample to be generally larger than the scores in the other sample. If the two samples are combined and all the scores are ranked, then the larger ranks should be concentrated in one sample and the smaller ranks should be concentrated in the other sample.

■ The Null Hypothesis for the Mann-Whitney Test

Because the Mann-Whitney test compares two distributions (rather than two means), the hypotheses tend to be somewhat vague. We state the hypotheses in terms of a consistent, systematic difference between the two treatments being compared.

H_0:   There is no difference between the two treatments. Therefore, there is no tendency for the ranks in one treatment condition to be systematically higher (or lower) than the ranks in the other treatment condition.

H_1:   There is a difference between the two treatments. Therefore, the ranks in one treatment condition are systematically higher (or lower) than the ranks in the other treatment condition.

■ Calculating the Mann-Whitney *U*

Again, we begin by combining all the individuals from the two samples and then rank ordering the entire set. If you have a numerical score for each individual, combine the two sets of scores and rank order them. The Mann-Whitney U is then computed as if the two samples were two teams of athletes competing in a sports event. Each individual in sample A (the A team) gets one point whenever he or she is ranked ahead of an individual from sample B. The total number of points accumulated for sample A is called U_A. In the same way, a U value, or team total, is computed for sample B. The final Mann-Whitney U is the smaller of these two values. This process is demonstrated in the following example.

EXAMPLE E.1   We begin with two separate samples with $n = 6$ scores in each.

Sample A (treatment 1):	27	2	9	48	6	15
Sample B (treatment 2):	71	63	18	68	94	8

Next, the two samples are combined, and all 12 scores are ranked.

Ranks for Sample A	7	1	4	8	2	5
Ranks for Sample B	11	9	6	10	12	3

Each individual in sample A is assigned 1 point for every score in sample B that has a higher rank. $U_A = 4 + 6 + 5 + 4 + 6 + 5 = 30$. Similarly, $U_B = 0 + 0 + 2 + 0 + 0 + 4 = 6$.

Thus, the Mann-Whitney U is 6. As a simple check on your arithmetic, note that $U_A + U_B = n_A n_B$. For these data, $30 + 6 = 6(6)$. ∎

■ Computing *U* for Large Samples

Because the process of counting points to determine the Mann-Whitney U can be tedious, especially with large samples, there is a formula that will generate the U value for each sample. To use this formula, you combine the samples and rank-order all the individuals as before. Then you must find ΣR_A, which is the sum of the ranks for individuals in sample A, and the corresponding ΣR_B for sample B. The U value for each sample is then computed as follows: For sample A,

$$U_A = n_A n_B + \frac{n_A(n_A + 1)}{2} - \Sigma R_A$$

and for sample B,

$$U_B = n_A n_B + \frac{n_B(n_B + 1)}{2} - \Sigma R_B$$

These formulas are demonstrated using the data from Example E.1. For sample A, the sum of the ranks is

$$\Sigma R_A = 1 + 2 + 4 + 5 + 7 + 8 = 27$$

For sample B, the sum of the ranks is

$$\Sigma R_B = 3 + 6 + 9 + 10 + 11 + 12 = 51$$

By using the special formula, for sample A,

$$U_A = n_A n_B + \frac{n_A(n_A + 1)}{2} - \Sigma R_A$$

$$= 6(6) + \frac{6(7)}{2} - 27$$

$$= 36 + 21 - 27$$

$$= 30$$

For sample B,

$$U_B = n_A n_B + \frac{n_B(n_B + 1)}{2} - \Sigma R_B$$

$$= 6(6) + \frac{6(7)}{2} - 51$$

$$= 36 + 21 - 51$$

$$= 6$$

Notice that these are the same U values we obtained in Example E.1 using the counting method. The Mann-Whitney U value is the smaller of these two, $U = 6$.

■ Evaluating the Significance of the Mann-Whitney U

Table B9 in Appendix B lists critical value of U for $\alpha = .05$ and $\alpha = .01$. The null hypothesis is rejected when the sample data produce a U that is *less than or equal to* the table value.

In Example E.1 both samples have $n = 6$ scores, and the table shows a critical value of $U = 5$ for a two-tailed test with $\alpha = .05$. This means that a value of $U = 5$ or smaller is very unlikely to occur (probability less than .05) if the null hypothesis is true. The data actually produced $U = 6$, which is not in the critical region. Therefore, we fail to reject H_0 because the data do not provide enough evidence to conclude that there is a significant difference between the two treatments.

There are no strict rules for reporting the outcome of a Mann-Whitney U-test. However, APA guidelines suggest that the report include a summary of the data (including such information as the sample size and the sum of the ranks) and the obtained statistic and p value. For the study presented in Example E.1, the results could be reported as follows:

> The original scores were ranked ordered and a Mann-Whitney U-test was used to compare the ranks for the $n = 6$ participants in treatment A and the $n = 6$ participants in treatment B. The results indicate no significant difference between treatments, $U = 6$, $p > .05$, with the sum of the ranks equal to 27 for treatment A and 51 for treatment B.

■ Normal Approximation for the Mann-Whitney U

When the two samples are both large (about $n = 20$) and the null hypothesis is true, the distribution of the Mann-Whitney U statistic tends to approximate a normal shape. In this case, the Mann-Whitney hypotheses can be evaluated using a z-score statistic and the unit normal distribution. The procedure for this normal approximation is as follows.

1. Find the U values for sample A and sample B as before. The Mann-Whitney U is the smaller of these two values.

2. When both samples are relatively large (around $n = 20$ or more), the distribution of the Mann-Whitney U statistic tends to form a normal distribution with

$$\mu = \frac{n_A n_B}{2} \quad \text{and} \quad \sigma = \sqrt{\frac{n_A n_B (n_A + n_B + 1)}{12}}$$

The Mann-Whitney U obtained from the sample data can be located in this distribution using a z-score:

$$z = \frac{X - \mu}{\sigma} = \frac{U - \dfrac{n_A n_B}{2}}{\sqrt{\dfrac{n_A n_B (n_A + n_B + 1)}{12}}}$$

3. Use the unit normal table to establish the critical region for this z-score. For example, with $\alpha = .05$, the critical values would be ± 1.96.

Usually the normal approximation is used with samples of $n = 20$ or larger; however, we will demonstrate the formulas with the data that were used in Example E.1. This study compared two treatments, A and B, using a separate sample of $n = 6$ for each treatment. The data produced a value of $U = 6$. The z-score corresponding to $U = 6$ is

$$z = \frac{U - \dfrac{n_A n_B}{2}}{\sqrt{\dfrac{n_A n_B (n_A + n_B + 1)}{12}}}$$

$$= \frac{6 - \dfrac{6(6)}{2}}{\sqrt{\dfrac{6(6)(6 + 6 + 1)}{12}}}$$

$$= \frac{-12}{\sqrt{\dfrac{468}{12}}}$$

$$= -1.92$$

With $\alpha = .05$, the critical value is $z = \pm 1.96$. Our computed z-score, $z = -1.92$, is not in the critical region, so the decision is to fail to reject H_0. Note that we reached the same conclusion for the original test using the critical values in the Mann-Whitney U table.

E.3 | The Wilcoxon Signed-Ranks Test: An Alternative to the Repeated-Measures *t* Test

The Wilcoxon test is designed to evaluate the difference between two treatments, using the data from a repeated-measures experiment. Recall that a repeated-measures study involves only one sample, with each individual in the sample being measured twice. The difference between the two measurements for each individual is recorded as the score for that individual. The Wilcoxon test requires that the differences be rank-ordered from smallest to largest in terms of their absolute magnitude, without regard for sign or direction. For example, Table E.1 shows differences scores and ranks for a sample of $n = 8$ participants.

TABLE E.1

Ranking difference scores. Note that the differences are ranked by magnitude, independent of direction.

Participant	Difference from Treatment 1 to Treatment 2	Rank
A	+4	1
B	−14	5
C	+5	2
D	−20	7
E	−6	3
F	−16	6
G	−8	4
H	−24	8

■ Zero Differences and Tied Scores

For the Wilcoxon test, there are two possibilities for tied scores.

1. A participant may have the same score in treatment 1 and in treatment 2, resulting in a difference score of zero.

2. Two (or more) participants may have identical difference scores (ignoring the sign of the difference).

When the data include individuals with difference scores of zero, one strategy is to discard these individuals from the analysis and reduce the sample size (n). However, this procedure ignores the fact that a difference score of zero is evidence for retaining the null hypothesis. A better procedure is to divide the zero differences evenly between the positives and negatives. (With an odd number of zero differences, discard one and divide the rest evenly.) When there are ties among the difference scores, each of the tied scores should be assigned the average of the tied ranks. This procedure was presented in detail in an earlier section of this appendix (see page 690).

■ Hypotheses for the Wilcoxon Test

The null hypothesis for the Wilcoxon test simply states that there is no consistent, systematic difference between the two treatments.

If the null hypothesis is true, any differences that exist in the sample data must be due to chance. Therefore, we would expect positive and negative differences to be intermixed evenly. On the other hand, a consistent difference between the two treatments should cause the scores in one treatment to be consistently larger than the scores in the other. This should produce difference scores that tend to be consistently positive or consistently negative. The Wilcoxon test uses the signs and the ranks of the difference scores to decide whether there is a significant difference between the two treatments.

H_0: There is no difference between the two treatments. Therefore, in the general population there is no tendency for the difference scores to be either systematically positive or systematically negative.

H_1: There is a difference between the two treatments. Therefore, in the general population the difference scores are systematically positive or systematically negative.

■ Calculation and Interpretation of the Wilcoxon T

After ranking the absolute values of the difference scores, the ranks are separated into two groups: those associated with positive differences (increases) and those associated with negative differences (decreases). Next, the sum of the ranks is computed for each group. The smaller of the two sums is the test statistic for the Wilcoxon test and is identified by the letter T. For the difference scores in Table E.1, the positive differences have ranks of 1 and 2, which add to $\Sigma R = 3$ and the negative difference scores have ranks of 5, 7, 3, 6, 4, and 8, which add up to $\Sigma R = 33$. For these scores, $T = 3$.

Table B10 in Appendix B lists critical values of T for $\alpha = .05$ and $\alpha = .01$. The null hypothesis is rejected when the sample data produce a T that is *less than or equal to* the table value. With $n = 8$ and $\alpha = .05$ for a two-tailed test, the table lists a critical value of 3. For the data in Table E.1, we obtained $T = 3$ so we would reject the null hypothesis and conclude that there is a significant difference between the two treatments.

As with the Mann-Whitney U-test, there is no specified format for reporting the results of a Wilcoxon T-test. It is suggested, however, that the report include a summary of the data and the value obtained for the test statistic as well as the p value. If there are zero-difference scores in the data, it is also recommended that the report describe how they were treated. For the data in Table E.1, the report could be as follows:

> The eight participants were rank ordered by the magnitude of their difference scores and a Wilcoxon T was used to evaluate the significance of the difference between treatments. The results showed a significant difference, $T = 3$, $p < .05$, with the positive ranks totaling 3 and the negative ranks totaling 33.

■ Normal Approximation for the Wilcoxon T-Test

When a sample is relatively large, the values for the Wilcoxon T statistic tend to form a normal distribution. In this situation, it is possible to perform the test using a z-score statistic and the normal distribution rather than looking up a T value in the Wilcoxon table. When the sample size is greater than 20, the normal approximation is very accurate and can be used. For samples larger than $n = 50$, the Wilcoxon table typically does not provide any critical values, so the normal approximation must be used. The procedure for the *normal approximation for the Wilcoxon* T is as follows:

1. Find the total for the positive ranks and the total for the negative ranks as before. The Wilcoxon T is the smaller of the two values.

2. With n greater than 20, the Wilcoxon T values form a normal distribution with a mean of

$$\mu = \frac{n(n + 1)}{4}$$

and a standard deviation of

$$\sigma = \sqrt{\frac{n(n + 1)(2n + 1)}{24}}$$

The Wilcoxon T from the sample data corresponds to a z-score in this distribution defined by

$$z = \frac{X - \mu}{\sigma} = \frac{T - \dfrac{n(n + 1)}{4}}{\sqrt{\dfrac{n(n + 1)(2n + 1)}{24}}}$$

3. The unit normal table is used to determine the critical region for the z-score. For example, the critical values are ± 1.96 for $\alpha = .05$.

Although the normal approximation is intended for samples with at least $n = 20$ individuals, we will demonstrate the calculations with the data in Table E.1. These data have $n = 8$ and produced $T = 3$. Using the normal approximation, these values produce

$$\mu = \frac{n(n + 1)}{4} = \frac{8(9)}{4} = 18$$

$$\sigma = \sqrt{\frac{n(n + 1)(2n + 1)}{24}} = \sqrt{\frac{8(9)(17)}{24}} = \sqrt{51} = 7.14$$

With these values, the obtained $T = 3$ corresponds to a z-score of

$$z = \frac{T - \mu}{\sigma} = \frac{3 - 18}{7.14} = \frac{-15}{7.14} = -2.10$$

With critical boundaries of ± 1.96, the obtained z-score is close to the boundary, but it is enough to be significant at the .05 level. Note that this is exactly the same conclusion we reached using the Wilcoxon T table.

E.4 | The Kruskal-Wallis Test: An Alternative to the Independent-Measures ANOVA

The *Kruskal-Wallis test* is used to evaluate differences among three or more treatment conditions (or populations) using ordinal data from an independent-measures design. You should recognize that this test is an alternative to the single-factor ANOVA introduced in Chapter 12. However, the ANOVA requires numerical scores that can be used to calculate means and variances. The Kruskal-Wallis test, on the other hand, simply requires that you are able to rank-order the individuals for the variable being measured. You also should recognize that the Kruskal-Wallis test is similar to the Mann-Whitney test introduced earlier in this chapter. However, the Mann-Whitney test is limited to comparing only two treatments, whereas the Kruskal-Wallis test is used to compare three or more treatments.

■ The Data for a Kruskal-Wallis Test

The Kruskal-Wallis test requires three or more separate samples. The samples can represent different treatment conditions or they can represent different preexisting populations. For example, a researcher may want to examine how social input affects creativity. Children are asked to draw pictures under three different conditions: (1) working alone without supervision, (2) working in groups where the children are encouraged to examine and criticize each other's work, and (3) working alone but with frequent supervision and comments from a teacher. Three separate samples are used to represent the three treatment conditions with $n = 6$ children in each condition. At the end of the study, the researcher collects the drawings from all 18 children and rank-orders the complete set of drawings in terms of creativity. The purpose for the study is to determine whether one treatment condition produces drawings that are ranked consistently higher (or lower) than another condition. Notice that the researcher does not need to determine an absolute creativity score for each painting. Instead, the data consist of relative measures; that is, the researcher must decide which painting shows the most creativity, which shows the second most creativity, and so on.

The creativity study that was just described is an example of a research study comparing different treatment conditions. It also is possible that the three groups could be defined by a subject variable so that the three samples represent different populations. For example, a researcher could obtain drawings from a sample of 5-year-old children, a sample of 6-year-old children, and a third sample of 7-year-old children. Again, the Kruskal-Wallis test would begin by rank-ordering all of the drawings to determine whether one age group showed significantly more (or less) creativity than another.

Finally, the Kruskal-Wallis test can be used if the original, numerical data are converted into ordinal values. The following example demonstrates how a set of numerical scores is transformed into ranks to be used in a Kruskal-Wallis analysis.

EXAMPLE E.2

Table E.2(a) shows the original data from an independent-measures study comparing three treatment conditions. To prepare the data for a Kruskal-Wallis test, the complete set of original scores is rank-ordered using the standard procedure for ranking tied scores. Each of the original scores is then replaced by its rank to create the transformed data in Table E.2(b) that are used for the Kruskal-Wallis test. ■

TABLE E.2

Preparing a set of data for analysis using the Kruskal-Wallis test. The original data consisting of numerical scores are shown in table (a). The original scores are combined into one group and rank ordered using the standard procedure for ranking tied scores. The ranks are then substituted for the original scores to create the set of ordinal data shown in table (b).

(a) Original Numerical Scores

I	II	III	
14	2	26	$N = 15$
3	14	8	
21	9	14	
5	12	19	
16	5	20	
$n_1 = 5$	$n_2 = 5$	$n_3 = 5$	

(b) Ordinal Data (Ranks)

I	II	III	
9	1	15	$N = 15$
2	9	5	
14	6	9	
3.5	7	12	
11	3.5	13	
$T_1 = 39.5$	$T_2 = 26.5$	$T_3 = 54$	
$n_1 = 5$	$n_2 = 5$	$n_3 = 5$	

■ The Null Hypothesis for the Kruskal-Wallis Test

As with the other tests for ordinal data, the null hypothesis for the Kruskal-Wallis test tends to be somewhat vague. In general, the null hypothesis states that there are no differences among the treatments being compared. Somewhat more specifically, H_0 states that there is no tendency for the ranks in one treatment condition to be systematically higher (or lower) than the ranks in any other condition. Generally, we use the concept of "systematic differences" to phrase the statement of H_0 and H_1. Thus, the hypotheses for the Kruskal-Wallis test are phrased as follows:

H_0: There is no tendency for the ranks in any treatment condition to be systematically higher or lower than the ranks in any other treatment condition. There are no differences between treatments.

H_1: The ranks in at least one treatment condition are systematically higher (or lower) than the ranks in another treatment condition. There are differences between treatments.

Table E.2(b) presents the notation that is used in the Kruskal-Wallis formula along with the ranks. The notation is relatively simple and involves the following values.

1. The ranks in each treatment are added to obtain a total or T value for that treatment condition. The T values are used in the Kruskal-Wallis formula.

2. The number of subjects in each treatment condition is identified by a lowercase n.

3. The total number of subjects in the entire study is identified by an uppercase N.

The Kruskal-Wallis formula produces a statistic that is usually identified with the letter H and has approximately the same distribution as chi-square, with degrees of freedom defined by the number of treatment conditions minus one. For the data in Table E.2(b), there are 3 treatment conditions, so the formula produces a chi-square value with $df = 2$. The formula for the Kruskal-Wallis statistic is

$$H = \frac{12}{N(N + 1)} \left(\Sigma \frac{T^2}{n} \right) - 3(N + 1)$$

Using the data in Table E.2(b), the Kruskal-Wallis formula produces a chi-square value of

$$H = \frac{12}{15(16)}\left(\frac{39.5^2}{5} + \frac{26.5^2}{5} + \frac{54^2}{5}\right) - 3(16)$$

$$= 0.05(312.05 + 140.45 + 583.2) - 48$$

$$= 0.05(1035.7) - 48$$

$$= 51.785 - 48$$

$$= 3.785$$

With $df = 2$, the chi-square table lists a critical value of 5.99 for $\alpha = .05$. Because the obtained chi-square value (3.785) is not greater than the critical value, our statistical decision is to fail to reject H_0. The data do not provide sufficient evidence to conclude that there are significant differences among the three treatments.

As with the Mann-Whitney and the Wilcoxon tests, there is no standard format for reporting the outcome of a Kruskal-Wallis test. However, the report should provide a summary of the data, the value obtained for the chi-square statistic, as well as the value of df, N, and the p value. For the Kruskal-Wallis test that we just completed, the results could be reported as follows:

After ranking the individual scores, a Kruskal-Wallis test was used to evaluate differences among the three treatments. The outcome of the test indicated no significant differences among the treatment conditions, $H = 3.785$ (2, $N = 15$), $p > .05$.

There is one assumption for the Kruskal-Wallis test that is necessary to justify using the chi-square distribution to identify critical values for H. Specifically, each of the treatment conditions must contain at least five scores.

E.5 | The Friedman Test: An Alternative to the Repeated-Measures ANOVA

The *Friedman test* is used to evaluate the differences between three or more treatment conditions using data from a repeated-measures design. This test is an alternative to the repeated-measures ANOVA that was introduced in Chapter 13. However, the ANOVA requires numerical scores that can be used to compute means and variances. The Friedman test simply requires ordinal data. The Friedman test is also similar to the Wilcoxon test that was introduced earlier in this chapter. However, the Wilcoxon test is limited to comparing only two treatments, whereas the Friedman test is used to compare three or more treatments.

■ The Data for a Friedman Test

The Friedman test requires only one sample, with each individual participating in all of the different treatment conditions. The treatment conditions must be rank-ordered for each individual participant. For example, a researcher could observe a group of children diagnosed with ADHD in three different environments: at home, at school, and during unstructured play time. For each child, the researcher observes the degree to which the disorder interferes with normal activity in each environment, and then ranks the three environments from most disruptive to least disruptive. In this case, the ranks are obtained by comparing the individual's behavior across the three conditions. It is also possible for each individual to produce his or her own rankings. For example, each individual could be asked to evaluate three different designs for a new smart phone. Each person practices with each phone and then ranks them, 1st, 2nd, and 3rd in terms of ease of use.

Finally, the Friedman test can be used if the original data consist of numerical scores. However, the scores must be converted to ranks before the Friedman test is used. The following example demonstrates how a set of numerical scores is transformed into ranks for the Friedman test.

EXAMPLE E.3

To demonstrate the Friedman test, we will use the results from a repeated-measures study in which each of $n = 5$ participants watched a television program from four different viewing distances. The data consist of the participants' rating scores for each of the four viewing distances. To convert the data for the Friedman test, the four scores for each participant are replaced with ranks 1, 2, 3, and 4, corresponding to the size of the original scores. As usual, tied scores are assigned the mean of the tied ranks. The complete set of ranks is shown in part (b) of the table.   ■

■ The Hypotheses for the Friedman Test

In general terms, the null hypothesis for the Friedman test states that there are no differences between the treatment conditions being compared so the ranks in one treatment should not be systematically higher (or lower) than the ranks in any other treatment condition. Thus, the hypotheses for the Friedman test can be phrased as follows:

H_0:   There is no difference between treatments. Thus, the ranks in one treatment condition are not systematically higher or lower than the ranks in any other treatment condition.

H_1:   There are differences between treatments. Thus, the ranks in at least one treatment condition are systematically higher or lower than the ranks in another treatment condition.

■ Notation and Calculation for the Friedman Test

The first step in the Friedman test is to compute the sum of the ranks for each treatment condition. The ΣR values are shown in Table E.3(b). In addition to the ΣR values, the

TABLE E.3

Results from a repeated-measures study comparing four television-viewing distances. Part (a) shows the original rating score for each of the distances. In part (b), the four distances are rank-ordered according to the preferences for each participant.

(a) The original rating scores

Person	9 Feet	12 Feet	15 Feet	18 Feet
A	3	4	7	6
B	0	3	6	3
C	2	1	5	4
D	0	1	4	3
E	0	1	3	4

(b) The ranks of the treatment conditions for each participant

Person	9 Feet	12 Feet	15 Feet	18 Feet
A	1	2	4	3
B	1	2.5	4	2.5
C	2	1	4	3
D	1	2	4	3
E	1	2	3	4
	$\Sigma R_1 = 6$	$\Sigma R_2 = 9.5$	$\Sigma R_3 = 19$	$\Sigma R_4 = 15.5$

calculations for the Friedman test require the number of individuals in the sample (n) and the number of treatment conditions (k). For the data in Table E.3(b), $n = 5$ and $k = 4$. The Friedman test evaluates the differences between treatments by computing the following test statistic:

$$\chi_r^2 = \frac{12}{nk(k + 1)} (\Sigma R)^2 - 3n(k + 1)$$

Note that the statistic is identified as chi-square (χ^2) with a subscript r, and corresponds to a chi-square statistic for ranks. This chi-square statistic has degrees of freedom determined by $df = k - 1$, and is evaluated using the critical values in the chi-square distribution shown in Table B8 in Appendix B.

For the data in Table E.3(b), the statistic is

$$\chi_r^2 = \frac{12}{5(4)(5)} (6^2 + 9.5^2 + 15.5^2 + 19^2) - 3(5)(5)$$

$$= \frac{12}{100} (36 + 90.25 + 240.25 + 361) - 75$$

$$= 0.12(727.5) - 75$$

$$= 12.3$$

With $df = k - 1 = 3$, the critical value of chi-square is 7.81. Therefore, the decision is to reject the null hypothesis and conclude that there are significant differences among the four treatment conditions.

As with most of the tests for ordinal data, there is no standard format for reporting the results from a Friedman test. However, the report should provide the value obtained for the chi-square statistic as well as the values for df, n, and p. For the data that were used to demonstrate the Friedman test, the results would be reported as follows:

After ranking the original scores, a Friedman test was used to evaluate the differences among the four treatment conditions. The outcome indicated that there are significant differences, $\chi_r^2 = 12.3$ (3, $n = 5$), $p < .05$.

Statistics Organizer: Finding the Right Statistics for Your Data

Overview: Three Basic Data Structures

After students have completed a statistics course, they occasionally are confronted with situations in which they have to apply the statistics they have learned. For example, in the context of a research methods course, or while working as a research assistant, students are presented with the results from a study and asked to do the appropriate statistical analysis. The problem is that many of these students have no idea where to begin. Although they have learned the individual statistics, they cannot match the statistical procedures to a specific set of data. The Statistics Organizer attempts to help you find the right statistics by providing an organized overview for most of the statistical procedures presented in this book.

We assume that you know (or can anticipate) what your data look like. Therefore, we begin by presenting some basic categories of data so you can find that one that matches your own data. For each data category, we then present the potential statistical procedures and identify the factors that determine which are appropriate for you based on the specific characteristics of your data. Most research data can be classified in one of three basic categories.

Category 1: A single group of participants with one score per participant.

Category 2: A single group of participants with two variables measured for each participant.

Category 3: Two (or more) groups of scores with each score a measurement of the same variable.

In this section we present examples of each structure. Once you match your own data to one of the examples, you can proceed to the section of the chapter in which we describe the statistical procedures that apply to that example.

■ Scales of Measurement

Before we begin discussion of the three categories of data, there is one other factor that differentiates data within each category and helps to determine which statistics are appropriate. In Chapter 1 we introduced four scales of measurement and noted that different measurement scales allow different kinds of mathematical manipulation, which result in different statistics. For most statistical applications, however, ratio and interval scales are equivalent so we group them together for the following review.

Ratio scales and **interval scales** produce numerical scores that are compatible with the full range of mathematical manipulation. Examples include measurements of height in inches, weight in pounds, the number of errors on a task, and IQ scores.

Ordinal scales consist of ranks or ordered categories. Examples include classifying cups of coffee as small, medium, and large or ranking job applicants as 1st, 2nd, and 3rd.

Nominal scales consist of named categories. Examples include gender (male/female), academic major, or occupation.

Within each category of data, we present examples representing these three measurement scales and discuss the statistics that apply to each.

■ Category 1: A Single Group of Participants with One Score per Participant

This type of data often exists in research studies that are conducted simply to describe individual variables as they exist naturally. For example, a recent news report stated that half of American teenagers, ages 12–17, send 50 or more text messages a day. To get this number, the researchers had to measure the number of text messages for each individual in a large sample of teenagers. The resulting data consist of one score per participant for a single group.

It is also possible that the data are a portion of the results from a larger study examining several variables. For example, a college administrator may conduct a survey to obtain information describing the eating, sleeping, and study habits of the college's students. Although several variables are being measured, the intent is to look at them one at a time. For example, the administrator will look at the number of hours each week that each student spends studying. These data consist of one score for each individual in a single group. The administrator will then shift attention to the number of hours per day that each student spends sleeping. Again, the data consist of one score for each person in a single group. The identifying feature for this type of research (and this type of data) is that there is no attempt to examine relationships between different variables. Instead, the goal is to describe individual variables, one at a time.

Table 1 presents three examples of data in this category. Note that the three data sets differ in terms of the scale of measurement used to obtain the scores. The first set (a) shows numerical scores measured on an interval or ratio scale. The second set (b) consists of ordinal, or rank-ordered categories, and the third set shows nominal measurements. The statistics used for data in this category are discussed in Section I.

TABLE 1

Three examples of data with one score per participant for one group of participants.

(a) Number of Text Messages Sent in Past 24 Hours	(b) Rank in Class for High School Graduation	(c) Got a Flu Shot Last Season
X	X	X
6	23rd	No
13	18th	No
28	5th	Yes
11	38th	No
9	17th	Yes
31	42nd	No
18	32nd	No

■ Category 2: A Single Group of Participants with Two Variables Measured for Each Participant

These research studies are specifically intended to examine relationships between variables. Note that different variables are being measured, so each participant has two or more scores, each representing a different variable. Typically, there is no attempt to manipulate or control the variables; they are simply observed and recorded as they exist naturally.

Although several variables may be measured, researchers usually select pairs of variables to evaluate specific relationships. Therefore, we present examples showing pairs of variables and focus on statistics that evaluate relationships between two variables. Table 2 presents four examples of data in this category. Once again, the four data sets differ in terms of the scales of measurement that are used. The first set of data (a) shows numerical scores for each set of measurements. For the second set (b) we have ranked the scores from the first set and show the resulting ranks. The third data set (c) shows numerical scores for one variable and nominal scores for the second variable. In the fourth set (d), both scores are measured on a nominal scale of measurement. The appropriate statistical analyses for these data are discussed in Section II.

TABLE 2

Examples of data with two scores for each participant for one group of participants.

(a) SAT Score (X) and College Freshman GPA (Y)

X	Y
620	3.90
540	3.12
590	3.45
480	2.75
510	3.20
660	3.85
570	3.50
560	3.24

(b) Ranks for the Scores in Set (a)

X	Y
7	8
3	2
6	5
1	1
2	3
8	7
5	6
4	4

(c) Age (X) and Wrist Watch Preference (Y)

X	Y
27	digital
43	analogue
19	digital
34	digital
37	digital
49	analogue
22	digital
65	analogue
46	digital

(d) Gender (X) and Academic Major (Y)

X	Y
M	Sciences
M	Humanities
F	Arts
M	Professions
F	Professions
F	Humanities
F	Arts
M	Sciences
F	Humanities

■ Category 3: Two or More Groups of Scores with Each Score a Measurement of the Same Variable

A second method for examining relationships between variables is to use the categories of one variable to define different groups and then measure a second variable to obtain a set of scores within each group. The first variable, defining the groups, usually falls into one of the following general categories:

a. **Participant Characteristic:** For example, gender or age.

b. **Time:** For example, before versus after treatment.

c. **Treatment Conditions:** For example, with caffeine vs. without caffeine.

If the scores in one group are consistently different from the scores in another group, then the data indicate a relationship between variables. For example, if the performance scores for a group of females are consistently higher than the scores for a group of males, then there is a relationship between performance and gender.

Another factor that differentiates data sets in this category is the distinction between independent-measures and repeated-measures designs. Independent-measures designs were introduced in Chapters 10 and 12, and repeated-measures designs were presented in Chapters 11 and 13. You should recall that an *independent-measures design*, also known as a *between-subjects design*, requires a separate group of participants for each group of scores. For example, a study comparing scores for males with scores for females would require two groups of participants. On the other hand, a *repeated-measures design*, also known as a *within-subjects design*, obtains several groups of scores from the same group of participants. A common example of a repeated-measures design is a before/after study in which one group of individuals is measured before a treatment and then measured again after the treatment.

Examples of data sets in this category are presented in Table 3. The table includes a sampling of independent-measures and repeated-measures designs as well as examples representing measurements from several different scales of measurement. The appropriate statistical analyses for data in this category are discussed in Section III.

TABLE 3

Examples of data comparing two or more groups of scores with all scores measuring the same variable.

(a) Attractiveness Ratings for a Woman in a Photograph Shown on a Red or a White Background

White	Red
5	7
4	5
4	4
3	5
4	6
3	4
4	5

(b) Performance Scores Before and After 24 Hours of Sleep Deprivation

Participant	Before	After
A	9	7
B	7	6
C	7	5
D	8	8
E	5	4
F	9	8
G	8	5

(c) Success or Failure on a Task for Participants Working Alone or in a Group

Alone	Group
Fail	Succeed
Succeed	Succeed
Succeed	Succeed
Succeed	Succeed
Fail	Fail
Fail	Succeed
Succeed	Succeed
Fail	Succeed

(d) Amount of Time Spent on Facebook (Small, Medium, Large) for Students from Each High School Class

Freshman	Sophomore	Junior	Senior
med	small	med	large
small	large	large	med
small	med	large	med
med	med	large	large
small	med	med	large
large	large	med	large
med	large	small	med
small	med	large	large

Section I: Statistical Procedures for Data from a Single Group of Participants with One Score Per Participant

One feature of this data category is that the researcher typically does not want to examine a relationship between variables but rather simply intends to describe individual variables as they exist naturally. Therefore, the most commonly used statistical procedures for these data are descriptive statistics that are used to summarize and describe the group of scores.

■ Scores from Ratio or Interval Scales: Numerical Scores

When the data consist of numerical values from interval or ratio scales, there are several options for descriptive and inferential statistics. We consider the most likely statistics and mention some alternatives.

Descriptive Statistics   The most often used descriptive statistics for numerical scores are the mean (Chapter 3) and the standard deviation (Chapter 4). If there are a few extreme scores or the distribution is strongly skewed, the median (Chapter 3) may be better than the mean as a measure of central tendency.

Inferential Statistics   If there is a basis for a null hypothesis concerning the mean of the population from which the scores were obtained, a single-sample t test (Chapter 9) can be used to evaluate the hypothesis. Some potential sources for a null hypothesis are as follows.

1. If the scores are from a measurement scale with a well-defined neutral point, then the t test can be used to determine whether the sample mean is significantly different from (higher than or lower than) the neutral point. On a 7-point rating scale, for example, a score of $X = 4$ is often identified as neutral. The null hypothesis would state that the population mean is equal to (greater than or less than) $\mu = 4$.

2. If the mean is known for a comparison population, then the t test can be used to determine whether the sample mean is significantly different from (higher than or lower than) the known value. For example, it may be known that the average score on a standardized reading achievement test for children finishing first grade is $\mu = 20$. If a researcher uses a sample of second-grade children to determine whether there is a significant difference between the two grade levels, then the null hypothesis would state that the mean for the population of second-grade children is also equal to 20. The known mean could also be from an earlier time, for example 10 years ago. The hypothesis test would then determine whether a sample from today's population indicates a significant change in the mean during the past 10 years.

The single-sample t test evaluates the statistical significance of the results. A significant result means that the data are very unlikely ($p < \alpha$) to have been produced by random, chance factors. However, the test does not measure the size or strength of the effect. Therefore, a t test should be accompanied by a measure of effect size such as Cohen's d or the percentage of variance accounted for, r^2.

■ Scores from Ordinal Scales: Ranks or Ordered Categories

Descriptive Statistics   Occasionally, the original scores are measurements on an ordinal scale. It is also possible that the original numerical scores have been transformed into ranks or ordinal categories (for example, small, medium, and large). In either case, the median is appropriate for describing central tendency for ordinal measurements and proportions can be used to describe the distribution of individuals across categories. For example, a researcher might report that 60% of the students were in the high self-esteem category, 30% in the moderate self-esteem category, and only 10% in the low self-esteem category.

Inferential Statistics   If there is a basis for a null hypothesis specifying the proportions in each ordinal category for the population from which the scores were obtained, then a chi-square test for goodness of fit (Chapter 17) can be used to evaluate the hypothesis. With only two categories, the binomial test (Chapter 18) also can be used. For example, it may be reasonable to hypothesize that the categories occur equally often (equal proportions) in the population and the test would determine whether the sample proportions are significantly different. If the original data were converted from numerical values into ordered categories using z-score values to define the category boundaries, then the null hypothesis could state that the population distribution is normal, using proportions obtained from the unit normal table. A chi-square test for goodness of fit would determine whether the shape of the sample distribution is significantly different from a normal distribution. For example, the null hypothesis would state that the distribution has the following proportions, which describe a normal distribution according to the unit normal table:

$z < -1.5$	$-1.5 < z < -0.5$	$-0.5 < z < 0.5$	$0.5 < z < 1.5$	$z > 1.50$
6.68%	24.17%	38.30%	24.17%	6.68%

■ Scores from a Nominal Scale

For these data, the scores simply indicate the nominal category for each individual. For example, individuals could be classified as male/female or grouped into different occupational categories.

Descriptive Statistics   The only descriptive statistics available for these data are the mode (Chapter 3) for describing central tendency or using proportions (or percentages) to describe the distribution across categories.

Inferential Statistics   If there is a basis for a null hypothesis specifying the proportions in each category for the population from which the scores were obtained, then a chi-square test for goodness of fit (Chapter 17) can be used to evaluate the hypothesis. With only two categories, the binomial test (Chapter 18) also can be used. For example, it may be reasonable to hypothesize that the categories occur equally often (equal proportions) in the population. If proportions are known for a comparison population or for a previous time, the null hypothesis could specify that the proportions are the same for the population from which the scores were obtained. For example, if it is known that 35% of the adults in the United States get a flu shot each season, then a researcher could select a sample of college students and count how many got a shot and how many did not (see the data in Table 1(c)). The null hypothesis for the chi-square test would state that the distribution for college students is not different from the distribution for the general population. For the chi-square test

	Flu Shot	No Flu Shot
H_0:	35%	65%

For the binomial test, H_0: $p = p(\text{shot}) = 0.35$ and $q = p(\text{no shot}) = 0.65$
    Figure 1 summarizes the statistical procedures used for data in category 1.

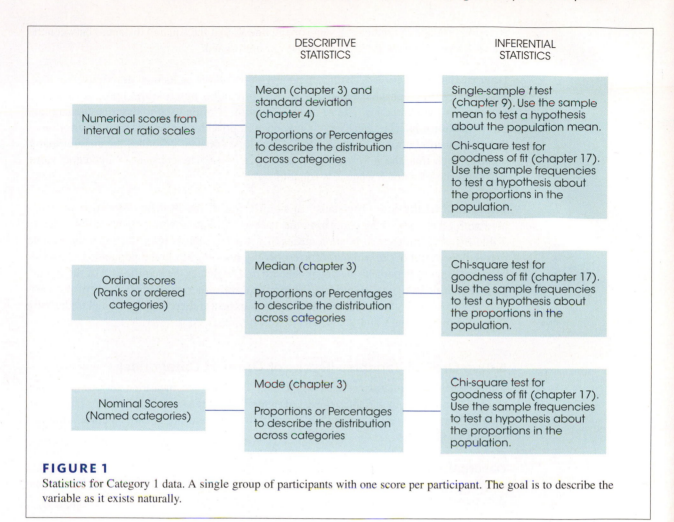

FIGURE 1

Statistics for Category 1 data. A single group of participants with one score per participant. The goal is to describe the variable as it exists naturally.

Section II: Statistical Procedures for Data from a Single Group of Participants with Two Variables Measured for Each Participant

The goal of the statistical analysis for data in this category is to describe and evaluate the relationships between variables, typically focusing on two variables at a time. With only two variables, the appropriate statistics are correlations and regression (Chapters 15 and 16), and the chi-square test for independence (Chapter 17). With three variables, another alternative is a partial correlation (Chapter 15), and multiple regression (Chapter 16).

■ Two Numerical Variables from Interval or Ratio Scales

The Pearson correlation measures the degree and direction of linear relationship between the two variables (see Example 15.3 on p. 494). Linear regression determines the equation for the straight line that gives the best fit to the data points. For each X value in the data, the equation produces a predicted Y value on the line so that the squared distances between the actual Y values and the predicted Y values are minimized.

Descriptive Statistics   The Pearson correlation serves as its own descriptive statistic. Specifically, the sign and magnitude of the correlation describe the linear relationship between the two variables. The squared correlation is often used to describe the strength of the relationship. The linear regression equation provides a mathematical description of the relationship between X values and Y. The slope constant describes the amount that Y changes each time the X value is increased by 1 point. The constant (Y intercept) value describes the value of Y when X is equal to zero.

Inferential Statistics   The statistical significance of the Pearson correlation is evaluated with a t statistic or by comparing the sample correlation with critical values listed in Table B6. A significant correlation means that it is very unlikely ($p < \alpha$) that the sample correlation would occur without a corresponding relationship in the population. Analysis of regression is a hypothesis-testing procedure that evaluates the significance of the regression equation. Statistical significance means that the equation predicts more of the variance in the Y scores than would be reasonable to expect if there were not a real underlying relationship between X and Y.

■ Two Ordinal Variables (Ranks or Ordered Categories)

The Spearman correlation is used when both variables are measured on ordinal scales (ranks). If one or both variables consist of numerical scores from an interval or ratio scale, then the numerical values can be transformed to ranks and the Spearman correlation can be computed.

Descriptive Statistics   The Spearman correlation describes the degree and direction of monotonic relationship; that is the degree to which the relationship is consistently one directional.

Inferential Statistics   The statistical significance of the Spearman correlation is evaluated by comparing the sample correlation with critical values listed in Table B7. A significant correlation means that it is very unlikely ($p < \alpha$) that the sample correlation would occur without a corresponding relationship in the population.

■ One Numerical Variable and One Dichotomous Variable (A Variable with Exactly 2 Values)

The point-biserial correlation measures the relationship between a numerical variable and a dichotomous variable. The two categories of the dichotomous variable are coded as numerical values, typically 0 and 1, to calculate the correlation.

Descriptive Statistics   Because the point-biserial correlation uses arbitrary numerical codes, the direction of relationship is meaningless. However, the size of the correlation, or the squared correlation, describes the degree of relationship.

Inferential Statistics   The data for a point-biserial correlation can be regrouped into a format suitable for an independent-measures t hypothesis test, or the t value can be computed directly from the point-biserial correlation (see the example on pp. 516-518). The t value from the hypothesis test determines the significance of the relationship.

■ Two Dichotomous Variables

The phi-coefficient is used when both variables are dichotomous. For each variable, the two categories are numerically coded, typically as 0 and 1, to calculate the correlation.

Descriptive Statistics   Because the phi-coefficient uses arbitrary numerical codes, the direction of relationship is meaningless. However, the size of the correlation, or the squared correlation, describes the degree of relationship.

Inferential Statistics   The data from a phi-coefficient can be regrouped into a format suitable for a 2 × 2 chi-square test for independence, or the chi-square value can be computed directly from the phi-coefficient (see Chapter 17, p. 586). The chi-square value determines the significance of the relationship.

■ Two Variables from any Measurement Scales

The chi-square test for independence (Chapter 17) provides an alternative to correlations for evaluating the relationship between two variables. For the chi-square test, each of the two variables can be measured on any scale, provided that the number of categories is reasonably small. For numerical scores covering a wide range of value, the scores can be grouped into a smaller number of ordinal intervals. For example, IQ scores ranging from 93–137 could be grouped into three categories described as high, medium, and low IQ.

For the chi-square test, the two variables are used to create a matrix showing the frequency distribution for the data. The categories for one variable define the rows of the matrix and the categories of the second variable define the columns. Each cell of the matrix contains the frequency or number of individuals whose scores correspond to the row and column of the cell. For example, the gender and academic major scores in Table 2(d) could be reorganized in a matrix as follows.

	Arts	Humanities	Sciences	Professions
Female				
Male				

The value in each cell is the number of students with the gender and major identified by the cell's row and column. The null hypothesis for the chi-square test would state that there is no relationship between gender and academic major.

Descriptive Statistics   The chi-square test is an inferential procedure that does not include the calculation of descriptive statistics. However, it is customary to describe the data by listing or showing the complete matrix of observed frequencies. Occasionally researchers describe the results by pointing out cells that have exceptionally large discrepancies. For example, in Chapter 17 we described a study investigating the effect of background music on the likelihood that a woman will give her phone number to a man she has just met. Female participants spent time in a waiting room with either romantic or neutral background music before beginning the study. At the end of the study, each participant was left alone in a room with a male confederate who used a scripted line to ask for her phone number. The description of the results focused on the "Yes" responses. Specifically, women who had heard romantic music were almost twice a likely to give their numbers.

Inferential Statistics   The chi-square test evaluates the significance of the relationship between the two variables. A significant result means that the distribution of frequencies in the data is very unlikely to occur ($p < \alpha$) if there is no underlying relationship between variables in the population. As with most hypothesis tests, a significant result does not provide information about the size or strength of the relationship. Therefore, either a phi-coefficient or Cramér's V is used to measure effect size.

■ Three Numerical Variables from Interval or Ratio Scales

To evaluate the relationship among three variables the appropriate statistics are partial correlation (Chapter 15) and multiple regression (Chapter 16). A partial correlation measures the relationship between two variables while controlling the third variable. Multiple regression determines the linear equation that gives the best fit to the data points. For each pair of X values in the data, the equation produces a predicted Y value so that the squared distances between the actual Y values and the predicted Y values are minimized.

Descriptive Statistics   A partial correlation describes the direction and degree of linear relationship between two variables while the influence of a third variable is controlled. This technique determines the degree to which the third variable is responsible for what appears to be a relationship between the first two. The multiple regression equation provides a mathematical description of the relationship between the two X values and Y. Each of the two slope constants describes the amount that Y changes each time the corresponding X value is increased by 1 point. The constant value describes the value of Y when both X values are equal to zero.

Inferential Statistics   The statistical significance of a partial correlation is evaluated by comparing the sample correlation with critical values listed in Table B6 using $df = n - 3$ instead of the $n - 2$ value that is used for a routine Pearson correlation. A significant correlation means that it is very unlikely ($p < \alpha$) that the sample correlation would occur without a corresponding relationship in the population. Analysis of regression evaluates the significance of the multiple regression equation. Statistical significance means that the equation predicts more of the variance in the Y scores than would be reasonable to expect if there was not a real underlying relationship between the two Xs and Y.

■ Three Variables Including Numerical Values and Dichotomous Variables

Partial correlation (Chapter 15) and multiple regression (Chapter 16) can also be used to evaluate the relationship among three variables, including one or more dichotomous variables. For each dichotomous variable, the two categories are numerically coded, typically as 0 and 1, before the partial correlation or multiple regression is done. The *descriptive statistics* and the *inferential statistics* for the two statistical procedures are identical to those for numerical scores except that direction of relationship (or sign of the slope constant) is meaningless for the dichotomous variables.

Figure 2 summarizes the statistical procedures used for data in category 2.

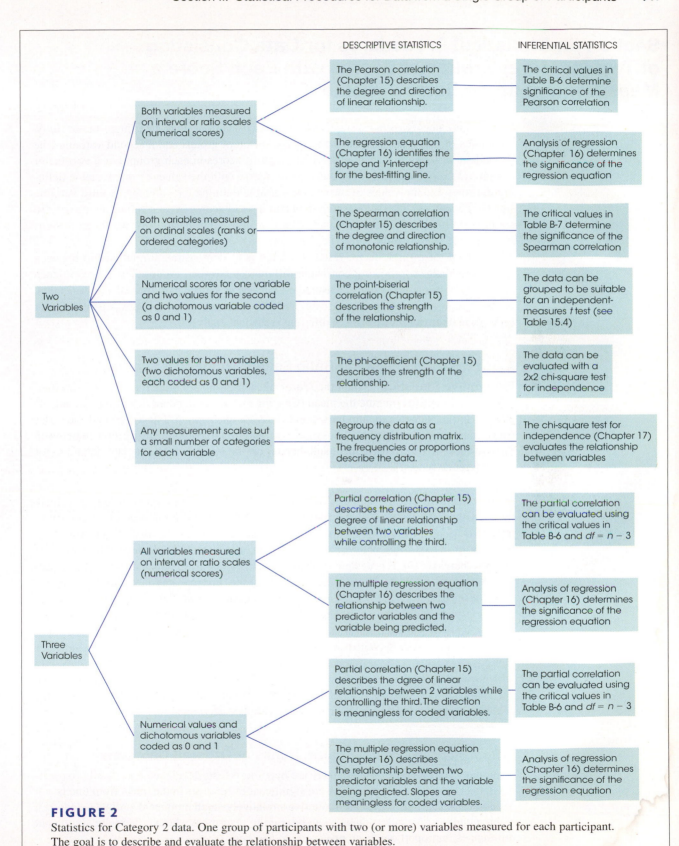

FIGURE 2

Statistics for Category 2 data. One group of participants with two (or more) variables measured for each participant. The goal is to describe and evaluate the relationship between variables.

Section III: Statistical Procedures for Data Consisting of Two (or More) Groups of Scores with Each Score a Measurement of the Same Variable

Data in this category includes single-factor and two-factor designs. In a single-factor study, the values of one variable are used to define different groups and a second variable (the dependent variable) is measured to obtain a set of scores in each group. For a two-factor design, two variables are used to construct a matrix with the values of one variable defining the rows and the values of the second variable defining the columns. A third variable (the dependent variable) is measured to obtain a set of scores in each cell of the matrix. To simplify discussion, we focus on single-factor designs now and address two-factor designs in a separate section at the end of this section.

The goal for a single-factor research design is to demonstrate a relationship between the two variables by showing consistent differences between groups. The scores in each group can be numerical values measured on interval or ratio scales, ordinal values (ranks), or simply categories on a nominal scale. The different measurement scales permit different types of mathematics and result in different statistical analyses.

■ Scores from Interval or Ratio Scales: Numerical Scores

Descriptive Statistics   When the scores in each group are numerical values, the standard procedure is to compute the mean (Chapter 3) and the standard deviation (Chapter 4) as descriptive statistics to summarize and describe each group. For a repeated-measures study comparing exactly two groups, it also is common to compute the difference between the two scores for each participant and then report the mean and the standard deviation for the difference scores.

Inferential Statistics   Analysis of variance (ANOVA) and t tests are used to evaluate the statistical significance of the mean differences between the groups of scores. With only two groups, the two tests are equivalent and either may be used. With more than two groups, mean differences are evaluated with an ANOVA. For independent-measures designs (between-subjects designs), the independent-measures t (Chapter 10) and independent-measures ANOVA (Chapter 12) are appropriate. For repeated-measures designs, the repeated-measures t (Chapter 11) and repeated-measures ANOVA (Chapter 13) are used. For all tests, a significant result indicates that the sample mean differences in the data are very unlikely ($p < \alpha$) to occur if there are not corresponding mean differences in the population. For an ANOVA comparing more than two means, a significant F-ratio indicates that post tests such as Scheffé or Tukey (Chapter 12) are necessary to determine exactly which sample means are significantly different. Significant results from a t test should be accompanied by a measure of effect size such as Cohen's d or r^2. For ANOVA, effect size is measured by computing the percentage of variance accounted for, η^2.

■ Scores from Ordinal Scales: Ranks or Ordered Categories

If the scores can be rank ordered, there are hypothesis tests developed specifically for ordinal data to determine whether there are significant differences in the ranks from one group to another. Also, if the scores are limited to a relatively small number of ordinal categories, a chi-square test for independence can be used to determine whether there are significant differences in proportions from one group to another.

Descriptive Statistics   Ordinal scores can be described by the set of ranks or ordinal categories within each group. For example, the ranks in one group may be consistently larger (or smaller) than ranks in another group. Or, the "large" ratings may be concentrated in one group and the "small" ratings in another.

Inferential Statistics   Appendix E presents a set of hypothesis tests developed for evaluating differences between groups of ordinal data.

1. The Mann-Whitney U test evaluates the difference between two groups of scores from an independent-measures design. The scores are the ranks obtained by combining the two groups and rank ordering the entire set of participants from smallest to largest.

2. The Wilcoxon signed ranks test evaluates the difference between two groups of scores from a repeated-measures design. The scores are the ranks obtained by rank ordering the magnitude of the differences, independent of sign ($+$ or $-$).

3. The Kruskal-Wallis test evaluates differences between three or more groups from an independent-measures design. The scores are the ranks obtained by combining all the groups and rank ordering the entire set of participants from smallest to largest.

4. The Friedman test evaluates differences among three or more groups from a repeated-measures design. The scores are the ranks obtained by rank ordering the scores for each participant. With three conditions, for example, each participant is measured three times and would receive ranks of 1, 2, and 3.

For all tests, a significant result indicates that the differences between groups are very unlikely ($p < \alpha$) to have occurred unless there are consistent differences in the population.

A chi-square test for independence (Chapter 17) can be used to evaluate differences between groups for an independent-measures design with a relatively small number of ordinal categories for the dependent variable. In this case, the data can be displayed as a frequency distribution matrix with the groups defining the rows and the ordinal categories defining the columns. For example, a researcher could group high school students by class (Freshman, Sophomore, Junior, Senior) and measure the amount of time each student spends on Facebook by classifying students into three ordinal categories (small, medium, large). An example of the resulting data is shown in Table 3(d). However, the same data could be regrouped into a frequency distribution matrix as follows:

Amount of time on Facebook		
Small	Medium	Large
Freshman		
Sophomore		
Junior		
Senior		

The value in each cell is the number of students with the high school class and amount of Facebook time identified by the cell's row and column. A chi-square test for independence would evaluate the differences between groups. A significant result indicates that the frequencies (proportions) in the sample data would be very unlikely ($p < \alpha$) unless the proportions for the population distributions are different from one group to another.

■ Scores from a Nominal Scale

Descriptive Statistics   As with ordinal data, data from nominal scales are usually described by the distribution of individuals across categories. For example, the scores in one group may be clustered in one category or set of categories and the scores in another group are clustered in different categories.

Inferential Statistics   With a relatively small number of nominal categories, the data can be displayed as a frequency distribution matrix with the groups defining the rows and the nominal categories defining the columns. The number in each cell is the frequency, or number of individuals in the group identified by the cell's row, with scores corresponding to the cell's column. For example, the data in Table 3(c) show success or failure on a task for participants who are working alone or working in a group. These data could be regrouped as follows:

	Success	Failure
Work alone		
Work in a group		

A chi-square test for independence (Chapter 17) can be used to evaluate differences between groups. A significant result indicates that the two sample distributions would be very unlikely ($p < \alpha$) to occur if the two population distributions have the same proportions (same shape).

■ Two-Factor Designs with Scores from Interval or Ratio Scales

Research designs with two independent (or quasi-independent) variables are known as two-factor designs. These designs can be presented as a matrix with the levels of one factor defining the rows and the levels of the second factor defining the columns. A third variable (the dependent variable) is measured to obtain a group of scores in each cell of the matrix (see Example 14.2 on page 461).

Descriptive Statistics   When the scores in each group are numerical values, the standard procedure is to compute the mean (Chapter 3) and the standard deviation (Chapter 4) as descriptive statistics to summarize and describe each group.

Inferential Statistics   A two-factor ANOVA is used to evaluate the significance of the mean differences between cells. The ANOVA separates the mean differences into three categories and conducts three separate hypothesis tests.

1. The main effect for factor *A* evaluates the overall mean differences for the first factor; that is, the mean differences between rows in the data matrix.

2. The main effect for factor *B* evaluates the overall mean differences for the second factor; that is, the mean differences between columns in the data matrix.

3. The interaction between factors evaluates the mean differences between cells that are not accounted for by the main effects.

For each test, a significant result indicates that the sample mean differences in the data are very unlikely ($p < \alpha$) to occur if there are not corresponding mean differences in the population. For each of the three tests, effect size is measured by computing the percentage of variance accounted for, η^2.

Figure 3 summarizes the statistical procedures used for data in category 3.

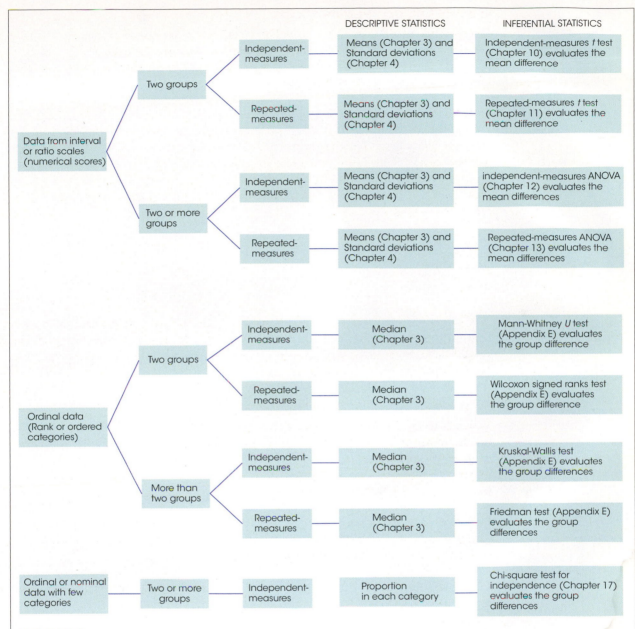

FIGURE 3

Statistics for Category 3 data. Two or more groups of scores, which are all measurements of the same variable. The goal is to describe and evaluate the differences between groups of scores.

References

Ackerman, R. (2011). Metacognitive regulation of text learning: On screen versus on \ paper. *Journal of Experimental Psychology: Applied, 17,* 18–32. DOI: 10.1037/a0022086

Ackerman, P. L., & Beier, M. E. (2007). Further explorations of perceptual speed abilities in the context of assessment methods, cognitive abilities, and individual differences during skill acquisition. *Journal of Experimental Psychology: Applied, 13,* 249–272.

Ackerman, R., & Goldsmith, M. (2011). Metacognitive regulation of text learning: On screen versus on paper. *Journal of Experimental Psychology: Applied, 17,* 18–32. DOI:10.1037/a0022086

Albentosa, M. J., & Cooper, J. J. (2005). Testing resource value in group-housed animals: An investigation of cage height preference in laying hens. *Behavioural Processes, 70,* 113–121.

American Psychological Association (APA). (2010). *Publication Manual of the American Psychological Association* (6th ed.) Washington, DC: Author.

Anderson, D. R., Huston, A. C., Wright, J. C., & Collins, P. A. (1998). Initial findings on the long term impact of Sesame Street and educational television for children: The recontact study. In R. Noll & M. Price (Eds.), *A Communication Cornucopia: Markle Foundation Essays on Information Policy* (pp. 279–296). Washington, DC: Brookings Institution.

Arden, R., & Plomin, R. (2006). Sex differences in variance of intelligence across childhood. *Personality and Individual Differences, 41,* 39–48.

Astudillo, M., Kuendig, H., Centeno-Gil, A., Wicki, M., & Gmel, G. (2014). Regional abundance of on-premise outlets and drinking pattern among Swiss young men: District level analysis and geographic adjustments. *Drug and Alcohol Review, 33,* 526–533.

Athos, E. A., Levinson, B., Kistler, A., Zemansky, J., Bostrom, A., Freimer, N., & Gitschier, J. (2007). Dichotomy and perceptual distortions in absolute pitch ability. *Proceedings of the National Academy of Science of the United States of America, 104,* 14795–14800.

Babcock, P., & Marks, M. (2010). Leisure college, USA: The decline in student study time. *Education Outlook,* No. 7. Retrieved from http:// www.aei.org/article /education/higher-education/leisure-college-usa/

Bartholow, B. D., & Anderson, C. A. (2002). Effects of violent video games on aggressive behavior: Potential sex differences. *Journal of Experimental Social Psychology, 38,* 283–290.

Bartus, R. T. (1990). Drugs to treat age-related neuro-degenerative problems: The final frontier of medical science? *Journal of the American Geriatrics Society, 38,* 680–695.

Bélisle, J., & Bodur, H. O. (2010). Avatars as information: Perception of consumers based on the avatars in virtual worlds. *Psychology & Marketing, 27,* 741–765.

Bem, D. J. (2011). Feeling the future: Experimental evidence for anomalous retroactive influences on cognition and affect. *Journal of Personality and Social Psychology, 100,* 407–425. DOI:10.1037/a0021524

Bicard, D. F., Lott, V., Mills, J., Bicard, S., & Baylot-Casey, L. (2012). Effects of text messaged self-monitoring on class attendance and punctuality of at-risk college athletes. *Journal of Applied Behavior Analysis,* 45, 205–210. DOI:10.1901/jaba.2012.45-205

Blum, J. (1978). *Pseudoscience and Mental Ability: The Origins and Fallacies of the IQ Controversy.* New York: Monthly Review Press.

Boogert, N. J., Reader, S. M., & Laland, K. N. (2006). The relation between social rank, neophobia and individual learning in starlings. *Behavioural Biology, 72,* 1229–1239.

Bradbury, T. N., & Miller, G. A. (1985). Season of birth in schizophrenia: A review of evidence, methodology, and etiology. *Psychological Bulletin, 98,* 569–594.

Broberg, A. G., Wessels, H., Lamb, M. E., & Hwang, C. P. (1997). Effects of day care on the development of cognitive abilities in 8-year-olds: A longitudinal study. *Development Psychology, 33,* 62–69.

Brunt, A., Rhee, Y., & Zhong, L. (2008). Differences in dietary patterns among college students according to body mass index. *Journal of American College Health, 56,* 629–634.

Callahan, L. F. (2009). Physical activity programs for chronic arthritis. *Current Opinion in Rheumatology, 21,* 177–182.

Candappa, R. (2000). *The Little Book of Wrong Shui.* Kansas City: Andrews McMeel Publishing.

Chandra, A., & Minkovitz, C. S. (2006). Stigma starts early: Gender differences in teen willingness to use mental health services. *Journal of Adolescent Health, 38,* 754.el–754.e8.

Cialdini, R. B., Reno, R. R., & Kallgren, C. A. (1990). A focus theory of normative conduct: Recycling the concept of norms to reduce littering in public places. *Journal of Personality and Social Psychology, 58,* 1015–1026.

Cohen, J. (1988). *Statistical power analysis for the behavioral sciences.* Hillsdale, NJ: Lawrence Erlbaum Associates.

Cohen, J. (1992). A power primer. *Psychological Bulletin, 112,* 155–159.

Cowles, M., & Davis, C. (1982). On the origins of the .05 level of statistical significance. *American Psychologist, 37,* 553–558.

Danner F., & Phillips B. (2008). Adolescent sleep, school start times, and teen motor vehicle crashes. *Journal of Clinical Sleep Medicine, 4,* 533–535.

Drèze, X., and Nunes, J. (2006). The endowed progress effect: How artificial advancement increases effort. *Journal of Consumer Research, 32,* 504–512.

Eagly, A. H., Ashmore, R. D., Makhijani, M. G., & Longo, L. C. (1991). What is beautiful is good but . . . : meta-analytic review of research on the physical attractiveness stereotype. *Psychological Bulletin, 110,* 109–128.

Elbel, B., Gyamfi, J., & Kersh, R. (2011). Child and adolescent fast-food choice and the influence of calorie labeling: a natural experiment. *International Journal of Obesity, 35,* 493–500. DOI: 10.1038/ijo.2011.4.

Elliot, A. J., Niesta, K., Greitemeyer, T., Lichtenfeld, S., Gramzow, R., Maier, M. A., & Liu, H. (2010). Red, rank, and romance in women viewing men. *Journal of Experimental Psychology: General, 139,* 399–417.

Elliot, A. J., & Niesta, D. (2008). Romantic red: Red enhances men's attraction to women. *Journal of Personality and Social Psychology, 95,* 1150–1164.

Evans, S. W., Pelham, W. E., Smith, B. H., Bukstein, O., Gnagy, E. M., Greiner, A. R., Atenderfer, L. B., & Baron-Myak, C. (2001). Dose-response effects of methylphenidate on ecologically valid measures of academic performance and classroom behavior in adolescents with ADHD. *Experimental and Clinical Psychopharmacology, 9,* 163–175.

Fallon, A. E., & Rozin, P. (1985). Sex differences in perceptions of desirable body shape. *Journal of Abnormal Psychology, 94,* 102–105.

Flynn, J. R. (1984). The mean IQ of Americans: Massive gains 1932 to 1978. *Psychological Bulletin. 95,* 29–51.

Flynn, J. R. (1999). Searching for justice: The discovery of IQ gains over time. *American Psychologist, 54,* 5–20.

Ford, A. M., & Torok, D. (2008). Motivational signage increases physical activity on a college campus. *Journal of American College Health, 57,* 242–244.

Fung, C. H., Elliott, M. N., Hays, R. D., Kahn, K. L., Kanouse, D. E., McGlynn, E. A., Spranca, M. D., & Shekelle, P. G. (2005). Patient's preferences for technical versus interpersonal quality when selecting a primary care physician. *Health Services Research, 40,* 957–977.

Gaucher, D., Friesen, J., and Kay, A. C. (2011). Evidence that gendered wording in job advertisements exists and sustains gender inequality. *Journal of Personality and Social Psychology, 101,* 109–128.

Gentile, D. A., Lynch, P. J., Linder, J. R., & Walsh, D. A. (2004). The effects of video game habits on adolescent hostility, aggressive behaviors, and school performance. *Journal of Adolescence, 27,* 5–22. doi:10.1016/j.adolescence.2003.10.002

Gibson, E. J., & Walk, R. D. (1960). The "visual cliff." *Scientific American, 202,* 64–71.

Gillen-O'Neel, C., Huynh, V. W., and Fuligni, A. J. (2013). To study or to sleep? The academic costs of extra studying at the expense of sleep. *Child Development, 84,* 133–142. DOI:10/1111/j.1467-8624.2012.01834.x

Gilovich, T., Medvec, V. H., & Savitsky, K. (2000). The spotlight effect in social judgment: An egocentric bias in estimates of the salience of one's own actions and appearance. *Journal of Personality and Social Psychology, 78,* 211–222.

Gino, F., & Ariely, D. (2012). The dark side of creativity: original thinkers can be more dishonest. *Journal of Personality and Social Psychology, 102,* 445–459. DOI: 10.1037/a0026406.

Gintzler, A. R. (1980). Endorphin-mediated increases in pain threshold during pregnancy. *Science, 210,* 193–195.

Green, L., Fry, A. F., & Myerson, J. (1994). Discounting of delayed rewards: A lifespan comparison. *Psychological Science, 5,* 33–36.

Guéguen, N., Jacob, C., & Lamy, L. (2010). 'Love is in the air': Effects of songs with romantic lyrics on compliance with a courtship request. *Psychology of Music, 38,* 303–307.

Guéguen, N., & Jacob, C. (2012, April). Clothing color and tipping: Gentlemen patrons give more tips to waitresses with red clothes. *Journal of Hospitality & Tourism Research.* DOI:10.1177/1096348012442546 .

Güven, M., Elaimis, D. D., Binokay, S., & Tan, O. (2003). Population-level right-paw preference in rats assessed by a new computerized food-reaching test. *International Journal of Neuroscience, 113,* 1675–1689.

Hill, R. A., & Barton, R. A. (2005). Red enhances human performance in contests. *Nature, 435,* 293.

Hunter, J. E. (1997). Needed: A ban on the significance test. *Psychological Science, 8,* 3–7.

Ijuin, M., Homma, A., Mimura, M., Kitamura, S., Kawai, Y., Imai, Y., & Gondo, Y. (2008). Validation of the 7-minute screen for the detection of early-stage

Alzheimer's disease. *Dementia and Geriatric Cognitive Disorders, 25,* 248–255.

Jackson, E. M., & Howton, A. (2008). Increasing walking in college students using a pedometer intervention: Differences according to body mass index. *Journal of American College Health, 57,* 159–164.

Johnston, J. J. (1975). Sticking with first responses on multiple-choice exams: For better or worse? *Teaching of Psychology, 2,* 178–179.

Jones, B. T, Jones, B. C. Thomas, A. P., & Piper, J. (2003). Alcohol consumption increases attractiveness ratings of opposite sex faces: a possible third route to risky sex. *Addiction, 98,* 1069–1075. DOI: 10.1046/j.1360-0443.2003.00426.x

Jones, J. T, Pelham, B. W., Carvallo, M., & Mirenberg, M. C. (2004). How do I love thee, let me count the Js: Implicit egotism and interpersonal attraction. *Journal of Personality and Social Behavior, 87,* 665–683.

Joseph. J. A., Shukitt-Hale. B., Denisova. N. A., Bielinuski, D., Martin, A., McEwen. J. J., & Bickford, P. C. (1999). Reversals of age-related declines in neuronal signal transduction, cognitive, and motor behavioral deficits with blueberry, spinach, or strawberry dietary supplementation. *Journal of Neuroscience, 19,* 8114–8121.

Judge, T. A., & Cable, D. M. (2010). When it comes to pay, do the thin win? The effect of weight on pay for men and women. *Journal of Applied Psychology, 96,* 95–112. DOI: 10.1037/a0020860

Kasparek, D. G., Corwin, S. J., Valois, R. F., Sargent, R. G., & Morris, R. L. (2008). Selected health behaviors that influence college freshman weight change. *Journal of American College Health, 56,* 437–444.

Katona, G. (1940). *Organizing and memorizing.* New York: Columbia University Press.

Keppel, G. (1973). *Design and Analysis: A Researcher's Handbook.* Englewood Cliffs, NJ: Prentice-Hall.

Killeen. P. R. (2005). An alternative to null-hypothesis significance tests. *Psychological Science, 16,* 345–353.

Kirschner, P. A., & Karpinski, A. C. (2010). Facebook and academic performance. *Computers in Human Behavior, 26,* 1237–1245. DOI: 10.1016/j.chb.2010.03.024

Knight, C., Haslam, S. A. (2010). The relative merits of lean, enriched, and empowered offices: An experimental examination of the impact of workspace management strategies on well-being and productivity. *Journal of Experimental Psychology: Applied, 16,* 158–172.

Kosfeld, M., Heinrichs, M., Zak, P. J., Fischblacher, U., & Fehr, R. (2005). Oxytocin increases trust in humans. *Nature, 435,* 673–676.

Kuo, M., Adlaf, E. M., Lee, H., Gliksman, L., Demers, A., & Wechsler, H. (2002). More Canadian students drink but American students drink more: Comparing college alcohol use in two countries. *Addiction, 97,* 1583–1592.

Langewitz, W., Izakovic, J., & Wyler, J. (2005). Effect of self-hypnosis on hay fever symptoms—a randomized controlled intervention. *Psychotherapy and Psychosomatics, 74,* 165–172.

Levine, S. C., Suriyakham, L. W., Rowe, M. L., Huttenlocher, J., & Gunderson, E. A. (2010). What counts in the development of young children's number knowledge? *Developmental Psychology, 46,* 1309–1319.

Liger-Belair, G., Bourget, M., Villaume, S., Jeandet, P., Pron, H., & and Polidori, G. (2010). On the losses of dissolved CO_2 during Champagne serving. *Journal of Agricultural and Food Chemistry, 58,* 8768–8775. DOI: 10.1021/jf101239w

Liguori, A., & Robinson, J. H. (2001). Caffeine antagonism of alcohol-induced driving impairment. *Drug and Alcohol Dependence, 63,* 123–129. DOI:10.1016/S0376-8716(00)00196-4

Loftus, E. F., & Palmer, J. C. (1974). Reconstruction of automobile destruction: An example of the interaction between language and memory. *Journal of Verbal Learning & Verbal Behavior, 13,* 585–589.

Loftus, G. R. (1996). Psychology will be a much better science when we change the way we analyze data. *Current Directions in Psychological Science, 5,* 161–171.

Madera, J. M., and Hebl, M. R. (2012). Discrimination against facially stigmatized applicants in interviews: An eye-tracking and face-to-face investigation. *Journal of Applied Psychology, 97,* 317–330. DOI: 10.1037/a0025799

McAllister, T. W., Flashman, L. A., Maerlender, A., Greenwald, R. M., Beckwith, J. G., Tosteson, T. D., Crisco, J. J.; Brolinson, P. G., Duma, S. M., Duhaime, A. C., Grove, M. R., & Turco, J. H. (2012). Cognitive effects of one season of head impacts in a cohort of collegiate contact sport athletes. *Neurology, 78,* 1777–1784. DOI: 10.1212/WNL.0b013e3182582fe7

McGee, E., & Shevliln, M. (2009). Effect of humor on interpersonal attraction and mate selection. *Journal of Psychology, 143,* 67–77.

McGee, R., Williams, S., Howden-Chapman, P., Martin, J., & Kawachi. I. (2006). Participation in clubs and groups from childhood to adolescence and its effects on attachment and self-esteem. *Journal of Adolescence, 29,* 1–17.

McMorris, B. J., Catalano, R. F., Kim, M. J. Toumbourou, J. W., & Hemphill, S. A. (2011). Influence of family factors and supervised alcohol use on adolescent alcohol use and harms: similarities between youth in different alcohol policy contexts. *Journal of Studies on Alcohol and Drugs, 72,* 418–428.

Morse, C. K. (1993). Does variability increase with age? An archival study of cognitive measures. *Psychology and Aging, 8,* 156–164.

Oishi, S., & Schimmack, U. (2010). Residential mobility, well-being, and mortality. *Journal of Personality and Social Psychology, 98,* 980–994.

Persson, J., Bringlov, E., Nilsson, L., & Nyberg. L. (2004). The memory-enhancing effects of Ginseng and Ginkgo biloba in health volunteers. *Psychopharmacology, 172,* 430–434.

Piff, P. K., Kraus, M. W., Cote, S., Cheng, B. H., & Keitner, D. (2010). Having less, giving more: the influence of social class on prosocial behavior. *Journal of Personality and Social Psychology, 99,* 771–784.

Plomin, R., Corley, R., DeFries, J. C., & Fulker, D. W. (1990). Individual differences in television viewing in early childhood: Nature as well as nurture. *Psychological Science, 1,* 371–377.

Polman, H., de Castro, B. O., & van Aken, M. A.G. (2008). Experimental study of the differential effects of playing versus watching violent video games on children's aggressive behavior. *Aggressive Behavior, 34,* 256–264. doi:10.1002/ab.20245

Reed, E. T., Vernon, P. A., & Johnson, A. M. (2004). Confirmation of correlation between brain nerve conduction velocity and intelligence level in normal adults. *Intelligence, 32,* 563–572.

Resenhoeft, A., Villa, J., & Wiseman, D. (2008). Tattoos can harm perceptions: A study and suggestions. *Journal of American College Health, 56,* 593–596.

Rohwedder, S., & Willis, R. J. (2010). Mental retirement. *Journal of Economic Perspectives, 24,* 119–138. DOI: 10.1257/jep.24.1.119

Scaife, M. (1976). The response to eye-like shapes by birds. I. The effect of context: A predator and a strange bird. *Animal Behaviour, 24,* 195–199.

Schachter. S. (1968). Obesity and eating. *Science, 161,* 751–756.

Scharf, R., Demmer, R., & DeBoer, M. M. (2013). Longitudinal evaluation of milk type consumed and weight status in preschoolers. *Archives of Disease in Childhood, 98,* 335–340. Doi 10.1136/archdischild-2012-302941

Schmidt, S. R. (1994). Effects of humor on sentence memory. *Journal of Experimental Psychology: Learning, Memory, & Cognition, 20,* 953–967.

Seery, M. D., Holman, E. A., & Silver, R. C. (2010). Whatever does not kill us: Cumultive lifetime adversity, vulnerability, and resilience. *Journal of Personality and Social Psychology, 99,* 1025–1041. DOI: 10.1037/a0021344

Shrauger, J. S. (1972). Self-esteem and reactions to being observed by others. *Journal of Personality and Social Psychology, 23,* 192–200.

Shulman, C., & Guberman, A. (2007). Acquisition of verb meaning through syntactic cues: A comparison of children with autism, children with specific language impairment (SLI) and children with typical language development (TLD). *Journal of Child Language, 34,* 411–423.

Slater, A., Von der Schulenburg, C., Brown, E., Badenoch, M., Butterworth, G., Parsons, S., & Samuels, C. (1998). Newborn infants prefer attractive faces. *Infant Behavior and Development, 21,* 345–354.

Smyth, J. M., Stone, A. A., Hurewitz, A., & Kaell, A. (1999). Effects of writing about stressful experiences on symptom reduction in patients with asthma or rheumatoid arthritis: A randomized trial. *Journal of the American Medical Association, 281,* 1304–1309.

Stephens, R., Atkins, J., & Kingston, A. (2009). Swearing as a response to pain. *NeuroReport: For Rapid Communication of Neuroscience Research, 20,* 1056–1060. DOI: 10.1097/WNR.0b013e32832e64b1

Strack, F., Martin, L. L., & Stepper, S. (1988). Inhibiting and facilitating conditions of the human smile: A non-obtrusive test of the facial feedback hypothesis. *Journal of Personality & Social Psychology, 54,* 768–777.

Trockel, M. T., Barnes, M. D., & Egget, D. L. (2000). Health-related variables and academic performance among first-year college students: Implications for sleep and other behaviors. *Journal of American College Health, 49,* 125–131.

Trombetti, A, Hars, M. Herrmann, F.R., Kressig, R. W., Ferrari, S., & Rizzoli, R. (2011). Effect of music-based multitask training on gait, balance, and fall risk in elderly people: a randomized controlled trial. *Archives of Internal Medicine, 171,* 525–533.

Tukey, J. W. (1977). *Exploratory Data Analysis.* Reading, MA: Addison-Wesley.

Twenge, J. M. (2000). The age of anxiety? Birth cohort change in anxiety and neuroticism, 1952–1993. *Journal of Personality and Social Psychology, 79,* 1007–1021.

U.S. Census Bureau. (2005). *Americans spend more than 100 hours commuting to work each year, Census Bureau reports.* Retrieved January 14, 2009, from www.census.gov/Press-Release/www/releases/archives/american_community_survey_acs/004489.html

von Hippel, P. T. (2005). Mean, median, and skew: Correcting a textbook rule. *Journal of Statistics Education, 13.*

Wegesin, D. J., & Stern, Y. (2004). Inter- and intra-individual variability in recognition memory: Effects of aging and estrogen use. *Neuropsychology, 18,* 646–657.

Weinberg, G. H., Schumaker, J. A., & Oltman, D. (1981). *Statistics: An Intuitive Approach* (p. 14). Belmont, CA: Wadsworth.

Weinstein, Y., McDermott, K. B., & Roediger, H. L. III (2010). A comparison of study strategies for passages: Rereading, answering questions, and generating questions. *Journal of Experimental Psychology: Applied, 16,* 308–316.

Wells, G. M. (2010). The effect of religiosity and campus alcohol culture on collegiate alcohol consumption, *Journal of American College Health. 58,* 295–304.

Wilkinson, L., & the Task Force on Statistical Inference. (1999). Statistical methods in psychology journals. *American Psychologist, 54,* 594–604.

Winget, C., & Kramer, M. (1979). *Dimensions of Dreams.* Gainesville, FL: University Press of Florida.

Xu, F., & Garcia, V. (2008). Intuitive statistics by 8-month-old infants. *Proceedings of the National Academy of Sciences of the United States of America, 105,* 5012–5015.

Yan, D., & Sengupta, J. (2011). Effects of construal level on the price-quality relationship. *Journal of Consumer Research, 38,* 376–389. DOI: 10.1086/659755

Yapko, M. D. (1994). Suggestibility and repressed memories of abuse: A survey of psychotherapists' beliefs. *American Journal of Clinical Hypnosis, 36,* 163–171.

Yong, E. (2012). Replication studies: Bad copy. *Nature, 485,* 298–300. DOI:10.1038/485298a

Zhong, C., Bohns, V. K., & Gino, F. (2010). Good lamps are the best police: Darkness increases dishonesty and self-interested behavior. *Psychological Science, 21,* 311–314.

Zhou, X., Vohs, K. D., & Baumeister, R. F. (2009). The symbolic power of money: Reminders of money after social distress and physical pain. *Psychological Science, 20,* 700–706.

Name Index

Subject Index